Culinary Arts

PRINCIPLES AND APPLICATIONS

SECOND EDITION

AMERICAN TECHNICAL PUBLISHERS
ORLAND PARK, ILLINOIS 60467-5756

Michael J. McGreal

Cover Photo: Irinox USA

American Technical Publishers, Inc., Editorial Staff

Editor in Chief:
        Jonathan F. Gosse
Vice President—Production:
        Peter A. Zurlis
Director of Product Development:
        Cathy A. Scruggs
Art Manager:
        James M. Clarke
Multimedia Manager:
        Carl R. Hansen
Technical Editors:
        Cathy A. Scruggs
        Larry E. Pierce
        Sara M. Marconi
Copy Editor:
        Jeana M. Platz

Cover Design:
        Jennifer M. Hines
Illustration/Layout:
        Jennifer M. Hines
        Mark S. Maxwell
        Samuel T. Tucker
        Melanie G. Doornbos
        Nicholas W. Basham
        Thomas E. Zabinski
Photo Acquisition:
        Kimberly Sienko
        Amy B. Weissenburger
CD-ROM Development:
        Daniel Kundrat
        Robert E. Stickley
        Nicole S. Polak

 This book is printed on recycled paper.

About the Author

Michael J. McGreal, CEC, CCE, CHE, FMP, CHA, MCFE, has worked in the foodservice industry for 30 years, holding chef positions at some of Chicago's premier restaurants and hotels. He earned his chef training degree from Washburne Culinary Institute in Chicago and his bachelor's degree in hospitality organizational management from the University of St. Francis. He is currently pursuing a master's degree in management at Robert Morris University. Chef McGreal joined the prestigious Culinary Arts and Hospitality Management program at Joliet Junior College as an instructor in 1996 and currently serves as the department chair.

Throughout his career Chef McGreal has earned many industry certifications, including Certified Executive Chef, Certified Culinary Educator, Certified Hospitality Educator, Foodservice Management Professional, Certified Hotel Administrator, and Master Certified Food Executive. He is an active member of the American Culinary Federation (ACF), the International Association of Culinary Professionals (IACP), the Council on Hotel, Restaurant, and Institutional Education (CHRIE), and the International Food Service Executives Association (IFSEA). He also is a chef partner in the Chefs Move to Schools initiative and serves as the Central Region representative on the ACF Chefs Move to Schools task force. Chef McGreal currently serves on the national ACF Education Committee.

Chef McGreal has received many awards including a 2011 ACF Presidential Medallion, the 2009 FENI Postsecondary Educator of the Year, an American Culinary Federation local chapter Culinary Educator of the Year and Chef of the Year, an Illinois Pork Producer's Taste of Elegance competition winner, and a ProStart Mentor of the Year by the National Restaurant Association. Chef McGreal has also received the Illinois Federation of Teachers Everyday Hero award and the Professional Achievement Award from the University of St. Francis. Chef McGreal serves on numerous advisory committees and frequently gives presentations and demonstrations on healthy cooking, nutrition, and using food to fight illness. He has also been featured as a guest chef on cruise lines, offering cooking and wine-tasting demonstrations for passengers at sea. Chef McGreal's true passion for cooking and teaching is reflected in this foundational textbook.

Acknowledgments

The author and publisher are grateful for the technical information and assistance provided by the following individuals.

Timothy Bucci, CEC, CCE, CHE, CCJ
Culinary Arts Instructor
Joliet Junior College, Joliet, IL

Robert Bifulco
Culinary Program Chair
Remington College, Garland, TX

Kristy Begley, MNM
Director of Education and Professional Development
American Culinary Federation, St. Augustine, FL

J. Desmond Keefe, III, CEC, CCE
Culinary Program Chair
Southern New Hampshire University, Manchester, NH

Christopher C. Misiak, CEC, CCE
Executive Chef
Schoolcraft College, Livonia, MI

Jeffrey A. Bricker, CEC, CCE, AAC
Hospitality Program Chair
Ivy Tech Community College, Indianapolis, IN

Keith Vonhoff, CEPC, CCE, CHE, FMP, CHA, CCP
Culinary and Hospitality Instructor
Joliet Junior College, Joliet, IL

Kyle D. Richardson, CEC, CCE, AAC, CHE
Culinary Arts Instructor
Joliet Junior College, Joliet, IL

Scott R. Smith, PhD, CEC, CCE
Assistant Professor, The Hospitality College
Johnson & Wales University, Denver, CO

Linda J. Trakselis
Assistant Professor, Culinary Arts
Illinois Institute of Art, Chicago, IL

Peter Sproul, CEC
Culinary Arts Department Director
Utah Valley University, Orem, UT

Michael A. Sodaro
Personal Chef
Downers Grove, IL

John Johnson, CEC, CCE
Culinary Arts Instructor
Madison Area Technical College, Madison, WI

Mark Muszynski, CEPC, CHE
Baking and Pastry Instructor
Joliet Junior College, Joliet, IL

The author and publisher would also like to thank the following companies, organizations, and individuals for providing images.

2008 ACF Culinary Team USA
Advance Tabco
Agricultural Research Service, USDA
Alinea/Photo by Lara Kastner
All-Clad Metalcrafters
Alpha Baking Co., Inc.
Amana Commercial Products

American Egg Board
American Lamb Board
American Metalcraft, Inc.
Barilla America, Inc.
Barker Company
Basic American Foods
Beef Checkoff

Blodgett Oven Company
Bridgeford Foods Corporation
Browne-Halco (NJ)
Bunn-O-Matic Corporation
Cabot Creamery Cooperative
California Fresh Apricot Council
California Strawberry Commission

Canadian Beef, Beef Information Centre
Carlisle FoodService Products
Charlie Trotter's
Chef Daniel Pliska
Chef Gui Alinat
Chef's Choice® by EdgeCraft Corporation
Cooper-Atkins Corporation
Cres Cor
CROPP Cooperative
CSI Hospitality Systems, Inc.
Czimer's Game & Seafoods, Inc.
Daniel NYC
D'Artagnan, Photography by Doug Adams Studio
Daymark® Safety Systems
Detecto, A Division of Cardinal Scale Manufacturing Co.
Dexter-Russell, Inc.
Earthstone Ovens, Inc.
Edlund Co.
Edward Don & Company
Eloma Combi Ovens
Emu Today and Tomorrow
Entourage
Florida Department of Agriculture and Consumer Services
Florida Department of Agriculture and Consumer Services, Bureau of Seafood and Aquaculture Marketing
Florida Department of Citrus
Florida Tomato Committee
Fluke Corporation
Fortune Fish Company
Frieda's Specialty Produce
Harbor Seafood, Inc.
Henny Penny Corporation
HerbThyme Farms
Hobart
House Foods
Idaho Potato Commission

Indian Harvest Specialtifoods, Inc./ Rob Yuretich
In-Sink-Erator
InterMetro Industries Corporation
Irinox USA
Kolpak
Kyocera Advanced Ceramics
Lauren Frisch
Lincoln Foodservice Products, Inc.
L. Isaacson and Stein Fish Company
Lodge Manufacturing
MacArthur Place Hotel, Sonoma
MacFarlane Pheasants, Inc.
Matfer Bourgeat USA
McCain Foods USA
Melissa's Produce
Mercer Cutlery
Messermeister
Mushroom Council
National Cancer Institute
National Cancer Institute, Daniel Sone (photographer)
National Cancer Institute, Renee Comet (photographer)
National Cattlemen's Beef Association
National Cherry Growers and Industries Foundation
National Chicken Council
National Honey Board
National Oceanic and Atmospheric Administration
National Onion Association
National Pasta Association
The National Pork Board
National Pork Producers Council
National Turkey Federation
National Watermelon Promotion Board
New Zealand Greenshell™ Mussels
NSF International
Oregon Raspberry & Blackberry Commission

Paderno World Cuisine
Pear Bureau Northwest
Pepper-Passion, Inc.
Perdue Foodservice, Perdue Farms Incorporated
Planet Hollywood International, Inc.
Plitt Seafood
PolyScience
Rishi Tea
Robot Coupe USA
San Jamar
Service Ideas, Inc.
Shenandoah Growers
Southern Pride
The Spice House
Strauss Free Raised
Sullivan University
Tanimura & Antle®
Trails End Chestnuts
Tru, Chicago
True FoodService Equipment, Inc.
Tyco/ANSUL
United States Department of Agriculture
United States Potato Board
U.S. Apple Association
USA Rice Federation
USDA National Nutrient Database
U.S. Fish and Wildlife Service
U.S. Geological Survey
U.S. Highbush Blueberry Council
U.S. Range
U.S. Wellness Meats
Venison World
Viking Commercial
Vita-Mix® Corporation
The Vollrath Company, LLC
Vulcan-Hart, a Division of the ITW Food Equipment Group, LLC
Wisconsin Milk Marketing Board, Inc.

Contents

Charlie Trotter's

Carlisle FoodService Products

Edlund Co.

Carlisle FoodService Products

Edlund Co.

Courtesy of The National Pork Board

Contents

National Turkey Federation

National Turkey Federation

Frieda's Specialty Produce

Contents

Photo Courtesy of Perdue Foodservice,
Perdue Farms Incorporated

Daniel NYC

Contents

All-Clad Metalcrafters

Interactive DVD Contents

- Quick Quizzes®
- Illustrated Glossary
- Flash Cards
- Chapter Reviews
- Application Scoresheets
- Recipes
- Culinary Math Tutorials
- Media Clips
- ATPeResources.com

Procedures Index

Procedures Index

Recipe Index

Chapter 10
Sandwiches

Chapter 11
Eggs and Breakfast

Chapter 12
Fruits

Recipe Index

Chapter 16
Poultry, Ratites, and Related Game

Recipe Index

Book Features

Chapter Introductions provide a general overview of the chapter content.

Chapter Objectives provide a list of learning goals for the chapter.

Key Terms are listed at the beginning of each chapter.

Nutrition Notes provide nutiritional information and tips for planning healthy menus.

Nutrition Note

Oxalic acid is a naturally occurring chemical found in vegetables such as spinach, sorrel, rhubarb, and beets. It combines with calcium and magnesium to form insoluble salts, which prevent absorption of these essential minerals. High concentrations of oxalic acid are toxic.

Chef's Tips highlight additional information to enhance understanding of concepts and techniques.

Chef's Tip

When clarifying a consommé, the acid added to the clearmeat depends on the main ingredient of the broth. Tomato purée is used as the acid ingredient for a dark consommé such as beef, veal, or lamb. Lemon juice is used as the acid ingredient for a light-colored consommé such as poultry or fish.

Cuisine Notes provide insight into how food is prepared in various cultures.

Cuisine Note

Phrases such as "multigrain," "assorted grains," and "100% wheat" do not guarantee that a product is a whole grain. For example, 100% wheat can mean bleached and refined white flour produced from 100% wheat.

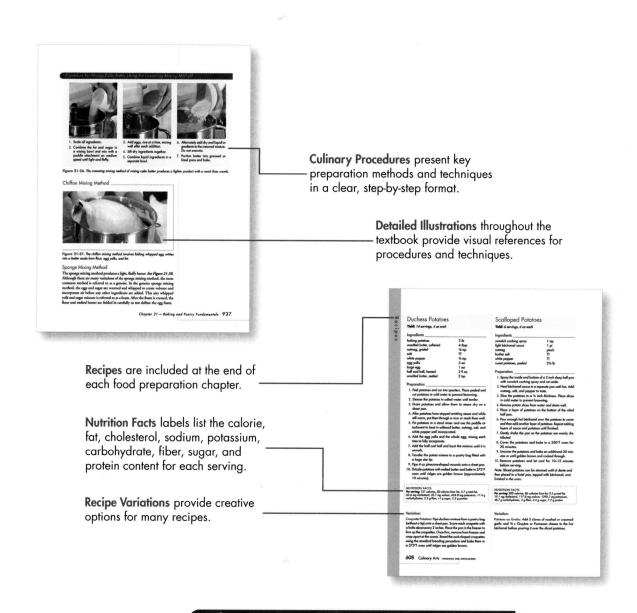

Culinary Procedures present key preparation methods and techniques in a clear, step-by-step format.

Detailed Illustrations throughout the textbook provide visual references for procedures and techniques.

Recipes are included at the end of each food preparation chapter.

Nutrition Facts labels list the calorie, fat, cholesterol, sodium, potassium, carbohydrate, fiber, sugar, and protein content for each serving.

Recipe Variations provide creative options for many recipes.

Tips provide suggestions to ensure successful execution of culinary procedures.

Tips for Preparing Meringues

- Egg whites must be beaten in a clean bowl that is free of fat, especially egg yolks. Fat prohibits egg whites from increasing in volume.
- Allow egg whites to stand at room temperature for at least 30 minutes prior to whipping so the whites will whip faster.
- Add cream of tartar at the beginning of the whipping process to help stabilize and add volume to the egg whites.

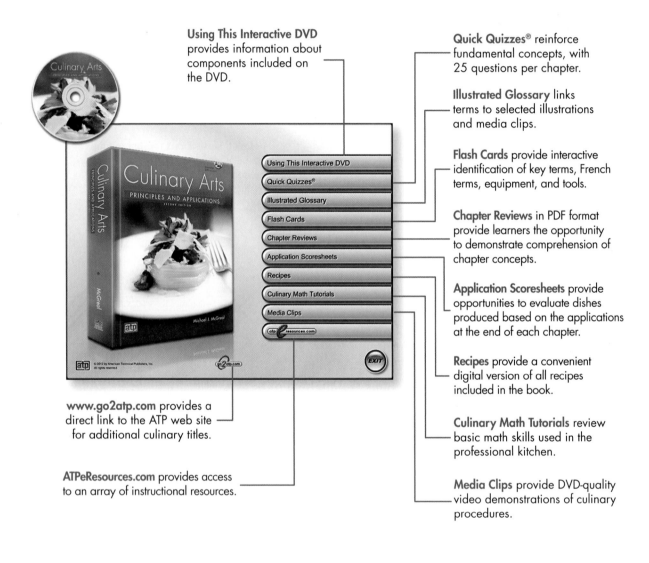

Using This Interactive DVD provides information about components included on the DVD.

Quick Quizzes® reinforce fundamental concepts, with 25 questions per chapter.

Illustrated Glossary links terms to selected illustrations and media clips.

Flash Cards provide interactive identification of key terms, French terms, equipment, and tools.

Chapter Reviews in PDF format provide learners the opportunity to demonstrate comprehension of chapter concepts.

Application Scoresheets provide opportunities to evaluate dishes produced based on the applications at the end of each chapter.

Recipes provide a convenient digital version of all recipes included in the book.

Culinary Math Tutorials review basic math skills used in the professional kitchen.

Media Clips provide DVD-quality video demonstrations of culinary procedures.

www.go2atp.com provides a direct link to the ATP web site for additional culinary titles.

ATPeResources.com provides access to an array of instructional resources.

Introduction

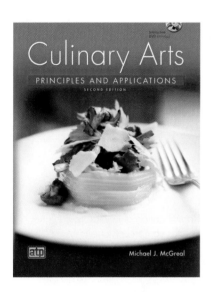

Culinary Arts Principles and Applications presents the core content and skills required to be successful in the culinary arts industry with a highly visual and learner-friendly format. The companion DVD is packed with interactive learning tools that reinforce and enhance content knowledge and skills. This new edition includes:

- Detailed product identification images and skill-based procedures
- Expanded coverage of equipment and cooking techniques
- Updated HACCP, Food Code, and nutrition information
- Relevant application assignments for each chapter
- Recipes that reinforce foundational cooking techniques
- New media clips featuring knife skills and cooking methods

Large illustrations provide visual references for fundamental culinary concepts and procedures. Also included are special book features that enhance the learning experience in the context of the topic covered.

- *Culinary Procedures* present key preparation methods and techniques in a clear, step-by-step format.
- *Tips* provide suggestions to ensure successful execution of culinary procedures.
- *Chef's Tips* highlight additional information to enhance understanding of principles and techniques.
- *Nutrition Notes* provide nutritional information and tips for planning healthy menus.
- *Cuisine Notes* provide insight into how food is prepared in various countries.

Culinary Arts Principles and Applications incorporates an instructional design in each chapter that helps learners understand and apply key culinary concepts in a professional kitchen. The objectives at the beginning of each chapter establish the learning goals that are supported by the written and visual content presented. The summary provides a synopsis of the core content. The chapter review correlates directly back to the objectives and the applications at the end of each chapter offer opportunities to practice the skills learned. Application scoresheets provided on the Interactive DVD in the back of the textbook allow both students and instructors to assess the proficiency level of preparation skills and prepared dishes.

The Interactive DVD included in the back of this textbook provides an array of learning tools that reinforce and enhance the knowledge and skills detailed within the book. Information about using the *Culinary Arts Principles and Applications DVD* is included on the last page. To obtain information about related products from American Technical Publishers, visit www.go2atp.com.

The Publisher

Culinary Arts
PRINCIPLES AND APPLICATIONS

Charlie Trotter's

Foodservice Professionals

Although food knowledge is essential in the foodservice industry, those who succeed are driven by the desire to provide outstanding service to guests. Meeting guest expectations requires dedication and teamwork. Choosing a career in foodservice offers a diverse range of opportunities from entry-level positions to restaurant ownership. There are a variety of education and training options available to help individuals find the career path that meets their aspirations. Preparing for employment in the foodservice industry requires networking, research, and attention to detail.

Chapter Objectives

1. Contrast external and internal customers.
2. Describe six common types of dining environments.
3. Describe nine meal service styles.
4. Summarize six types of cuisine that have evolved over time.
5. Describe front-of-house (FOH) foodservice career opportunities.
6. Describe back-of-house (BOH) foodservice career opportunities.
7. Summarize management and specialized foodservice careers.
8. Describe four types of education and training options.
9. Identify five essential employability skills.
10. Identify five essential FOH skills.
11. Explain how FOH staff can accommodate guests.
12. Describe FOH workflow.
13. Explain the importance of the FOH interaction with the BOH.
14. Describe math applications used by FOH staff.
15. Identify five essential BOH skills.
16. Explain the importance of BOH workflow.
17. Describe math applications used by BOH staff.
18. Explain how BOH staff can accommodate guests.
19. Describe the components of an effective résumé and portfolio.
20. Identify three types of job search tools.
21. Explain the multiple purposes of job interviews.

Refer to DVD
for **Flash Cards**

Key Terms

- **external customer**
- **internal customer**
- **meal service style**
- **sidework**
- **grande cuisine**
- **classical cuisine**
- **nouvelle cuisine**
- **new American cuisine**
- **fusion cuisine**
- **avant-garde cuisine**
- **front-of-house (FOH)**
- **back-of-house (BOH)**
- **brigade system**
- **apprentice**
- **accommodation**
- **mise en place**
- **expediting**
- **POS system**
- **portfolio**
- **résumé**

THE FOOD "SERVICE" INDUSTRY

Creating an atmosphere where exceptional service is the prime objective is essential to succeeding in the foodservice industry. Whether greeting guests at the door or preparing food and beverages, providing high-quality service is imperative. Guest satisfaction is achieved when guests receive quality meals from polite and efficient foodservice professionals. Regardless of the dining environment, meal service style, or type of cuisine, service is the key to meeting and exceeding guest expectations.

Delivering quality service is vital for foodservice operations. If the service is poor, guests may choose not to return to a foodservice operation, even if the food is memorable. Likewise, if the food is superior, guests are less likely to return for a second meal if the service is poor. The key to providing quality service is for staff to work together to provide guests with a positive dining experience. *See Figure 1-1.* Service involves both external and internal customers.

Customer Service

Figure 1-1. *The key to providing quality service is for staff to work together to provide guests with a positive dining experience.*

An *external customer* is an individual who uses the products or services of a business. In foodservice operations, the external customer is commonly referred to as the guest. Guests expect to be able to relax and enjoy their dining experience. When that experience is positive, guests are inclined to come back, and they may even tell their friends and family about the experience. Creating a positive dining experience for external customers requires teamwork among internal customers.

An *internal customer* is an individual who is the recipient of other products or services within the same organization. For example, dining room staff receive food from the kitchen staff within the foodservice operation. Foodservice operations also build relationships with vendors, suppliers, and distributors to acquire quality goods and services that can be passed on to guests. It is essential that internal customers communicate effectively and work as a team to ensure that external customers have a positive dining experience.

Dining Environments

To serve both external and internal customers requires the talents of many individuals working together in different dining environments. Common dining environments include quick service operations, quick-casual operations, casual dining operations, fine dining operations, catering and banquet operations, and institutional operations.

Quick Service Operations. In a quick service operation guests typically order or select items and pay for them prior to eating. Food and beverages can be eaten on the premises, taken out, or sometimes delivered. In quick service operations, guests often place and receive an order at a counter or a drive-thru window. *See Figure 1-2.* Quick service operations are often part of a restaurant chain featuring a limited menu that remains fairly constant.

Quick-Casual Operations. In a quick-casual operation, guests are served food to order. The orders of guests are placed at a counter and then a runner brings the food to the table. Some quick-casual operations also offer carry-out.

Casual Dining Operations. A casual dining operation offers moderately priced food in an informal environment. *See Figure 1-3.* Guests are seated at tables, and their orders are taken by service staff. Many casual dining operations are developed around a particular theme, such as sports, or a particular type of cuisine, such as Mexican. The theme is reflected in the décor, music, menu, and the staff uniforms.

Quick Service Drive-Thrus

Figure 1-2. *In quick service operations, guests often place and receive orders at a counter or a drive-thru window.*

Casual Dining

Daniel NYC

Figure 1-3. *A casual dining operation offers moderately priced food in an informal environment.*

Fine Dining Operations. A fine dining operation focuses on providing seated guests with personalized service and refined food presentations. Attention to detail is often demonstrated through elegant décor and beautifully plated dishes. *See Figure 1-4.* The cost for this level of service and type of food is reflected in higher menu prices.

Fine Dining

Figure 1-4. *In a fine dining operation, attention to detail is often demonstrated through elegant décor and beautifully plated dishes.*

Cafeteria-Style Dining

Figure 1-5. *Institutional foodservice operations often feature a cafeteria-style dining environment.*

Catering and Banquet Operations. A catering service provides food to a group of people at a remote site or special event. Food can be prepared at the site of the event or prepared and delivered. Meals are catered to many different settings. Organizations and individuals hosting a special event may hire a catering service to prepare meals and serve the dining needs of guests. Catering services can range from donuts and coffee to formal dinners. Banquet halls are facilities that serve a large number of guests for special events such as wedding receptions.

Institutional Operations. Foodservice operations, such as those in hospitals and correctional centers, often feature a cafeteria-style dining environment. *See Figure 1-5.* This type of foodservice operation generally serves a large number of people throughout the course of a day, requiring volumes of food to be prepared in advance.

Meal Service Styles

The type of foodservice operation often dictates the manner in which food is prepared and served. A *meal service style* is the way in which guests are served food and beverages. For example, some foodservice operations feature foods that are prepared in front of guests. Other operations feature foods prepared in the kitchen. Regardless of the meal service style, all staff members must work together to create a positive dining experience for guests. For example, servers depend on the kitchen staff to prepare high-quality dishes. Likewise, the kitchen staff depends on servers to present items in a manner that reflects the quality of the food.

Common meal service styles include booth, banquette, buffet, plated, English, Russian, French, butler, and banquet service. Although some service styles have names affiliated with specific countries, these styles are commonly used around the world.

Booth Service. *Booth service* is a style of meal service in which food is served to guests seated at a table positioned against a wall with benches on either side. *See Figure 1-6.* Because the server cannot walk to each guest, all food is served from the end of the table. Guests seated closest to the wall are served first. The guests on the right side of the booth should be served with the left hand, while the guests on the left should be served with the right hand.

Booth Service

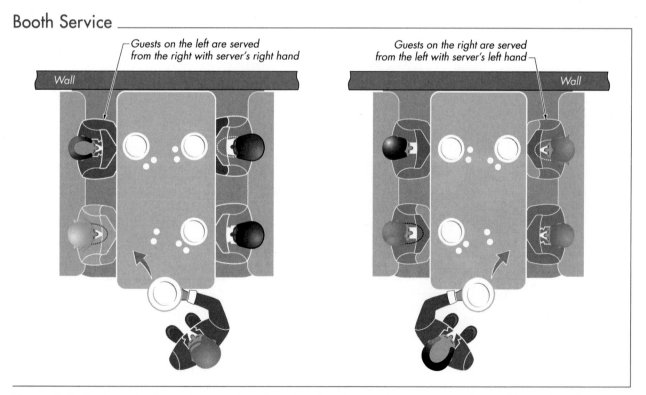

Figure 1-6. Booth service is a style of meal service where food is served to guests seated at a table positioned against a wall with benches on either side.

Banquette Service. *Banquette service* is a style of meal service in which food is served to guests seated at a table with a bench on one side and chairs on the opposite side. *See Figure 1-7.* To serve food, both ends of the table are utilized. Guests seated along the bench are served first. With banquette service, all food and beverages are served with the right hand and cleared with the left.

Banquette Service

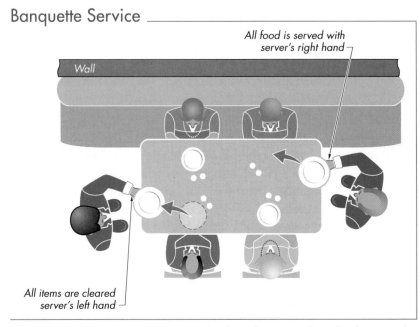

Figure 1-7. *Banquette service is a style of meal service where food is served to guests seated at a table with a bench on one side and chairs on the opposite side.*

Buffet Service

Figure 1-8. *Buffet service is a style of meal service where a variety of food is displayed on a table that guests approach for self-service.*

Buffet Service. *Buffet service* is a style of meal service in which food is displayed on a table that guests approach for self-service. *See Figure 1-8.* It is common for guests to return to the buffet for repeat visits. Because buffets promote self-service, the service staff can attend to a large number of guests. Although buffets are usually self-service, some have stations where staff prepare or carve and serve items. Staff must keep a buffet clean and well-stocked. Guests should be provided with a clean plate each time they return to the buffet.

Plated Service. *Plated service,* also known as American service, is a style of meal service in which individual portions of food are plated before being brought to guests. *See Figure 1-9.* Beverages and soups are presented from the right side of the guest with the right hand in a clockwise direction. Other foods are served from the left side of the guest using the left hand in a counterclockwise direction. Tables are cleared from the right side of the guest going clockwise.

English Service. *English service* is a style of meal service in which food is carved by a server and placed on a preset plate in front of the guest, yet side dishes are passed around the table for self-service. For example, a portion of roast beef may be carved by a server and placed on a preset plate for each guest. Side dishes, such as mashed potatoes and asparagus, are placed in serving dishes with utensils and set on the table for guests to serve themselves. English service is similar to family-style service. *Family-style service* is a style of meal service where all food is placed on the table for guests to pass for self-service.

Russian Service. *Russian service* is a style of meal service in which a server holds a tray of food and serves guests food from the tray. *See Figure 1-10.* In Russian service, all food is prepared, portioned, and garnished in the kitchen. Food is commonly presented on silver trays, and servers place a portion of each food item on preset plates. One server is responsible for serving the entrée and another server brings the side dishes. Servers hold the platter or tray along the left forearm and serve using the right hand. Service starts from the left side of the guest and moves counterclockwise around the table. Servers clear plates from the right side of the guest in a clockwise direction.

Plated Service

Daniel NYC

Figure 1-9. *Plated service is a style of meal service where individual portions of food are plated before being brought to guests.*

Russian Service

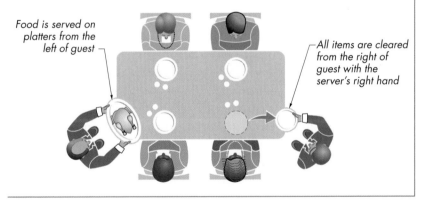

Food is served on platters from the left of guest

All items are cleared from the right of guest with the server's right hand

Figure 1-10. *Russian service is a style of meal service where a server holds a tray of food and serves guests food from the tray.*

French Service. *French service* is a style of meal service in which food is fully prepared or finished in front of guests. For example, a salad may be assembled tableside, a fish deboned in front of the guest, or a leg of lamb carved as the guests watch. French service can also include cooking foods tableside as guests observe. Foods are commonly brought to the table on a gueridon. A *gueridon* is a cart equipped with the items necessary to prepare foods tableside. Because dishes are prepared in front of guests, service staff are skilled in cooking techniques and have strong communication skills.

Banquet Service

Figure 1-11. *In banquet service, servers present food to guests attending a special function.*

Early Restaurants

Figure 1-12. *Pub, inn, and tavern owners prepared meals for sale, beginning the restaurant industry.*

Carême

Figure 1-13. *Carême was one of the first chefs to standardize the recipes, cooking methods, and techniques used in the professional kitchen.*

Butler Service. *Butler service* is a style of meal service in which servers present prepared food on a tray to standing or seated guests for self-service. Butler service is commonly used at receptions or cocktail parties where bite-sized portions of food are offered. When butler service is used for a meal, each course is displayed on a platter. The platter is presented to the left side of the guest for self-service.

Banquet Service. *Banquet service* is a style of meal service in which servers present food to guests attending a special function. *See Figure 1-11.* Servers are assigned to specific tables but must work as a team to ensure the event flows in an organized and timely manner. Prior to service, the banquet captain holds a meeting with all of the servers to review the reason for the function and to relay any special instructions. Although a variety of dining venues offer banquet service, hotels and reception halls often specialize in this service style.

The Evolution of Cuisine

In 1765, a French pub owner by the name of Boulanger put a sign on the front of his pub advertising a meal of sheep feet in white wine sauce. He called the dish a "restaurant" based on the French word *restaurer,* meaning "to restore." At that time, many people considered hearty foods, such as soups and stews, to be "restoratives" that boosted health, strength, and vitality.

Pub, inn, and tavern owners who witnessed the success of Boulanger decided to also prepare meals for sale. Thus, the restaurant industry began. *See Figure 1-12.* However, the quality of food served in some of the first restaurants was not very good because many of the owners did not have cooking experience.

There were specific guilds for everything, from meat roasters to bakers and caterers. A *guild* is an organization of craftsmen that has exclusive control of a particular craft and the production and distribution of its products. Guilds were commissioned by French royalty to prepare foods for royal celebrations. However, when the French Revolution put an end to the guilds, many guild members were left without an income. This resulted in former guild members opening restaurants and training employees.

Throughout the rest of the 18th century, restaurants opened across Europe. As the demand in the restaurant industry grew, chefs began experimenting with different cooking methods and presentations. Grande cuisine, classical cuisine, nouvelle cuisine, new American cuisine, fusion cuisine, and avant-garde cuisine evolved as chefs prepared foods in new and interesting ways by blending ingredients and flavors.

Grande Cuisine. *Grande cuisine,* also known as haute cuisine, is a style of cuisine in which intricate food preparation methods and large, elaborate presentations are used. Food items are decorated, sculpted, carved, sauced, and beautifully arranged. At the start of the 19th century, Marie-Antoine (Antonin) Carême was one of the first chefs to standardize the recipes, cooking methods, and techniques used in the professional kitchen. *See Figure 1-13.*

Classical Cuisine. *Classical cuisine* is a style of cuisine in which foods are prepared using a formalized system of cooking techniques and are presented in courses. Classical cuisine stresses the use of the finest ingredients and the most appropriate preparation methods to produce the best results, but does not focus on presentation. French chef Auguste Escoffier is credited with developing classical cuisine during the latter half of the 19th century. Escoffier is often referred to as the father of culinary education. He emphasized that all who enter the culinary profession should learn classical cooking methods. *See Figure 1-14.*

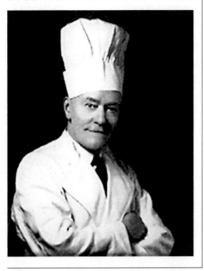

Escoffier

Figure 1-14. *Escoffier is often referred to as the father of culinary education.*

A significant contribution Escoffier made to classical cuisine was the simplification of the many varieties of sauces presented by Carême. According to Escoffier, all varieties of sauces stem from five mother sauces: béchamel, velouté, espagnole, tomato, and hollandaise.

Nouvelle Cuisine. *Nouvelle cuisine* is a style of cuisine in which foods are cooked quickly, seasoned lightly, and artistically presented in smaller portions. French chefs Paul Bocuse and Michel Guérard lead this movement at the turn of the 20th century. It gained momentum in the United States in the early 1970s.

Rejecting a reliance on cream sauces and butter, chefs that developed nouvelle cuisine established a trend of serving lower-fat foods and letting the true flavors of food stand out. Chefs also began experimenting with new methods, such as sous vide (slow, vacuum-sealed cooking), and tools, such as food processors, to create dishes using fresh ingredients.

New American Cuisine. *New American cuisine* is a style of cuisine that emphasizes the use of foods that are grown in America. *See Figure 1-15.* New American cuisine emerged in the United States as culinary education increased in popularity during the 1980s. With a focus on classical cooking techniques, New American cuisine is known for reinventing classic dishes from around the world using seasonal ingredients.

Fusion Cuisine. *Fusion cuisine* is a style of cuisine that uses a variety of cooking techniques to combine the flavors of two or more cultural regions. This style blends food, spices, and cooking methods from around the world. By fusing together flavors from different cultures, chefs are able to create unique dishes that represent aspects of diverse cultural traditions.

New American Cuisine

Figure 1-15. *New American cuisine emphasizes the use of foods that are grown in America.*

Avant-Garde Cuisine. *Avant-garde cuisine,* also known as modernist cuisine, is a style of cuisine, based on food chemistry, that involves the manipulation of the textures and temperatures of familiar dishes to reinvent and present them in new and creative ways. The presentation of food is often abstract, yet the flavors are usually familiar. *See Figure 1-16.* Avant-garde cuisine evolved in great part due to the talents of chef Ferran Adrià of Barcelona, Spain, who used his kitchen laboratory to create items such as liquid ravioli, Parmesan snow, and caviar made from olive oil. Although no one knows what style of cuisine will seize the attention of the world next, emphasis is being placed on whole, natural foods.

Avant-Garde Cuisine _____

Alinea/Photo by Lara Kastner

Figure 1-16. *In the avant-garde style of cuisine, the presentation of food is often abstract, yet the flavors are usually familiar.*

FOODSERVICE CAREERS

From entry-level positions to ownership, there are a variety of careers in customer service, food production, management, sales, research, and education. Foodservice employment opportunities can be found in restaurants, banquet halls, catering operations, hotels, resorts, spas, health care facilities, the military, sports stadiums, casinos, correctional facilities, educational facilities, corporate facilities, convention centers, recreational parks, shopping centers, and retail stores.

Foodservice Positions

The success of a foodservice operation is dependent upon the efforts of many different individuals employed in a variety of positions in the front-of-house and the back-of-house. The *front-of-house (FOH)* is the portion of a foodservice operation that is open to guests. This includes the entry area, dining room, bar area, and public restrooms. FOH staff have direct contact with guests. The *back-of-house (BOH)* is the portion of a foodservice operation that is typically not open to guests. This includes the delivery area, storerooms, kitchen, and employee-only areas. BOH staff work primarily in the kitchen and have infrequent, if any, interaction with guests.

FOH Positions. FOH staff are responsible for executing duties that provide direct service to guests. Common FOH positions include, but are not limited to, maître d', sommelier, bartender, host, server, and busser.

- A *maître d'*, also known as a dining room supervisor, is the person responsible for overseeing and coordinating all FOH activities. ***See Figure 1-17.*** This includes creating a welcoming atmosphere for guests, managing reservations, managing the dining room staff, interacting with guests to ensure a positive dining experience, and following up with unsatisfied guests.

- A *sommelier,* also known as a wine steward, is the person responsible for all aspects of wine service as well as wine and food pairings. ***See Figure 1-18.***

- A *bartender* is the person responsible for serving alcoholic beverages from behind a bar. Guests may make direct requests or servers may bring requests to the bartender.

- A *host* is the person responsible for greeting and seating guests. The host is often the first person a guest encounters and sets the tone for the dining experience. A host is also responsible for keeping wait-times to a minimum.

- A *server* is the person responsible for taking the orders of guests and bringing food and beverages to those guests. ***See Figure 1-19.*** Servers have the most contact with guests and greatly influence the dining experience.

- A *busser* is the person responsible for setting tables and removing dirty dishes from the dining area. Attentive bussers can make the dining room run more smoothly.

Maître d's

Charlie Trotter's

Figure 1-17. *A maître d' oversees and coordinates all FOH activities.*

Sommeliers

Daniel NYC

Figure 1-18. *A sommelier is responsible for all aspects of wine service as well as wine and food pairings.*

Servers

Bunn-O-Matic Corporation

Figure 1-19. *A server is responsible for taking the orders of guests and bringing food and beverages to those guests.*

BOH Positions. BOH staff are responsible for preparing the food for guests and providing clean dinnerware for service. In many foodservice operations, the BOH operates within a hierarchy designed to organize tasks based on the brigade system developed by Escoffier. The *brigade system* is a structured chain of command in which specific duties are aligned with the stations to which staff are assigned. ***See Figure 1-20.*** Common BOH positions include, but are not limited to, chef, sous chef, station chef, pastry chef, line cook, expediter, porter, and dishwasher.

Kitchen Brigade System	
French Terms	**English Meaning**
Boucher (boo-SHAY)	Butcher of meats and poultry
Boulanger (boo-lawn-ZHAY)	Bread baker
Chef de partie (chef-duh-par-TEE)	Station chef
Commis (co-MEE)	Apprentice cook
Confiseur (cone-fiss-UHR)	Petits fours and specialty candy maker
Decorateur (duh-kur-AHTUR)	Showplace and specialty cake maker
Entremetier (ehn-tra-meh-tee-YAY)	Vegetable and soup station chef
Friturier (free-too-ree-YAY)	Fry station chef
Garde manger (gahrd-mahn-ZHAY)	Pantry chef (responsible for cold foods)
Glacier (GLAH-see-yay)	Chilled- and frozen-dessert chef
Grillardin (gree-yar-DAHN)	Grill station chef
Légumier (lay-GOO-mee-yay)	Vegetable station chef
Pâtisser (pah-tees-ee-YAY)	Pastry chef
Poissonier (pwah-sawng-ee-YAY)	Fish station chef
Potager (pah-tah-ZHAY)	Soup station chef
Rotisseur (roh-tees-UHR)	Roast station chef; also responsible for related sauces
Saucier (SAW-see-yay)	Responsible for all sautéed items and most sauces
Sous chef (soo-chef)	Second chef; second in command after the chef
Tournant (toor-NAHN)	Roundsman or swing cook; works wherever needed in the kitchen

Figure 1-20. *The brigade system is a structured chain of command in which specific duties are aligned with the stations to which staff are assigned.*

- A *chef* is the person responsible for all kitchen operations, including menu management, purchasing, scheduling, and food production.
- A *sous chef* is the person responsible for carrying out objectives, as determined by the chef, regarding all aspects of kitchen operations. When the chef is absent, the sous chef is in charge of the kitchen.
- A *station chef* is the person responsible for overseeing a specific production area of the kitchen. *See Figure 1-21.* Production areas include stations such as grill stations and cold foods stations.
- A *pastry chef* is the person responsible for making and decorating the sweets and desserts offered by the foodservice operation.
- A *line cook* is the person responsible for preparing foods that are assigned to a particular station within the hot production line. *See Figure 1-22.* A line cook may also "prep" foods used in other stations.
- An *expediter* is the person responsible for ensuring each dish is acceptable before it leaves the kitchen. The duties of an expediter commonly include relaying dining room orders to various kitchen stations and reviewing each plate for accuracy and presentation before it leaves the kitchen.

Station Chefs

Vulcan-Hart, a division of the ITW Food Equipment Group LLC

Figure 1-21. *A station chef oversees a specific production area of the kitchen.*

Line Cooks

Charlie Trotter's

Figure 1-22. *A line cook prepares foods that are assigned to a particular station within the hot production line.*

- A *porter* is the person who ensures that the kitchen area is clean and in order, including the dish area, garbage area, and floors. A porter commonly supervises dishwashers to ensure all items are clean and put away properly.

- A *dishwasher* is the person who operates the warewashing machine and cleans all of the pots, pans, dinnerware, glassware, and flatware. *See Figure 1-23.*

Management Positions. Managers are responsible for daily operations, financial activities, staff supervision, and the overall guest experience. Successful foodservice managers put in countless hours and typically have prior work experience in the foodservice industry. A restaurateur and general manager are examples of management positions commonly found throughout the foodservice industry. Individuals that pursue management positions are responsible for organizing and overseeing all of the activities of the foodservice operation.

Dishwashers

Advance Tabco

Figure 1-23. *A dishwasher operates the warewashing machine and cleans all of the pots, pans, dinnerware, glassware, and flatware.*

Restaurateurs

Daniel NYC

Figure 1-24. *A restaurateur, also known as an owner, holds the legal title of a foodservice operation.*

A *restaurateur,* also known as an owner, is the person who holds the legal title of a foodservice operation. *See Figure 1-24.* The owner may have an active role in running the operation or may rely on a general manager to handle the daily responsibilities.

A *general manager* is a person who directs an operation and oversees food production, sales, and service. A general manager also is responsible for scheduling personnel and for determining food and labor costs.

Other Positions. There are also specialized foodservice careers within the foodservice industry. For example, some individuals may choose to pursue a career in sales and distribution in which they sell food, supplies, or equipment to foodservice operations. Others may choose to work in food research and development to create and test new products or flavors. Food stylists express their creativity by arranging foods for photo shoots. Foodservice educators impart culinary knowledge and skills to students. *See Figure 1-25.*

Foodservice Educators

Figure 1-25. *Foodservice educators impart culinary knowledge and skills to students.*

Foodservice Education and Training

Advances in technology, food products, preparation techniques, equipment, and management methods make it a necessity to attain the education and training needed to stay current in the field. There are many foodservice and hospitality education and training options available including apprenticeships, certificate programs, degree programs, and on-the-job training.

Apprenticeship Programs. Apprenticeship programs consist of both classroom instruction and hands-on application instruction with a chef-mentor. An *apprentice* is an individual enrolled in a formal training program who learns by practical experience under the supervision of a skilled professional. *See Figure 1-26.* Apprentices gain valuable work experience while honing their knowledge and skills.

Apprenticeships

Figure 1-26. *An apprentice learns by practical experience under the supervision of a skilled professional.*

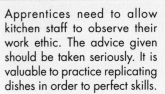
Certificate Programs. Certificate programs offer individuals an opportunity to receive specialized training in a particular area of the industry. After successful completion of a certificate program, students may be equipped with the necessary skills to earn an entry-level position. For example, a certificate is earned by completing a food safety and sanitation program. Some individuals enroll in a certificate program simply to broaden their knowledge or improve upon an existing skill.

Degree Programs. Many foodservice operations now require a college degree for various positions. Postsecondary culinary arts programs range in length from two to four years and offer education and training in areas such as culinary arts, food production, nutrition, pastry, cost control, hospitality, and management. *See Figure 1-27.* Some college programs also offer scholarships to students. Before enrolling in a degree program, it is important to research the reputation of the institution, its accreditation, and the background of its faculty.

On-the-Job Training. Some individuals enter the foodservice industry in entry-level positions and learn through on-the-job training. While this method may not lead to a desired position as quickly, some individuals have turned this opportunity into very successful careers.

Figure 1-27. *Postsecondary programs offer education and training in areas such as culinary arts, food production, nutrition, pastry, cost control, hospitality, and management.*

EMPLOYABILITY SKILLS

In addition to skills and knowledge, foodservice employers look for individuals who show respect for self and others by being professionals. Wearing appropriate attire, being dependable, and having a positive attitude are key employability skills. Foodservice professionals take pride in personal hygiene and health, wear professional attire, have a good attitude, demonstrate professional ethics, and exhibit reliability. They are also good communicators who interact appropriately, work as a team, and show initiative on the job.

Personal Hygiene and Health

Guests expect safe food in a clean environment offered by staff members who are well-groomed and healthy. Employees with poor personal hygiene are not only unattractive to guests, but place others at risk by spreading germs. Personal hygiene practices such as bathing daily and following proper handwashing techniques are essential for all foodservice staff. It is also important to refrain from working when seriously ill to prevent jeopardizing the health of others. Maintaining personal health minimizes the frequency of illness and requires adequate rest, nutrition, and exercise.

Professional Attire and Attitude

Wearing appropriate work attire represents a commitment to professionalism by the individual and the foodservice operation. Work attire should always be clean and neatly pressed as well as fit properly. *See Figure 1-28.* Staff who look professional make a positive first impression that reflects the quality of service guests can expect during their dining experience.

Professional Attire

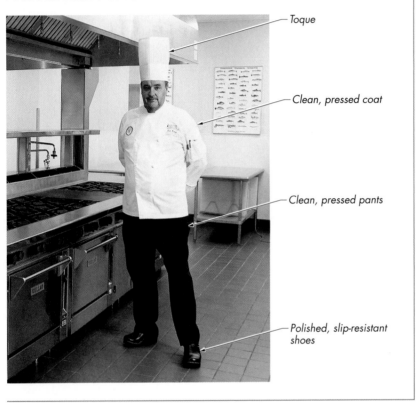

Toque

Clean, pressed coat

Clean, pressed pants

Polished, slip-resistant shoes

Vulcan-Hart, a division of the ITW Food Equipment Group LLC

Figure 1-28. *Work attire should always be clean and neatly pressed as well as fit properly.*

Having a positive attitude in the workplace makes a staff member stand out. Instead of seeing obstacles, individuals with a positive attitude see challenges with solutions. A positive attitude leads to increased productivity, influences the attitude of others, and results in the best products and services for guests.

Professional Ethics and Reliability

An ethical person demonstrates honesty, trustworthiness, character, and responsibility. Exhibiting ethics in the workplace helps build a foundation of respect that can promote success and advancement. In addition to being ethical, professional staff need to be reliable, punctual, and prepared. Fellow staff and managers count on every staff member to be present and on task in order to run a successful foodservice operation.

Viking Commercial

Communication and Interaction

Foodservice operations are totally dependent upon communication and interactions among staff and guests. Effective communication skills are important for giving and understanding instructions, sharing information, and providing feedback. *See Figure 1-29.* In order for communication to be successful, each message must be clear. Effective communication requires speaking, listening, writing, and reading.

Communication

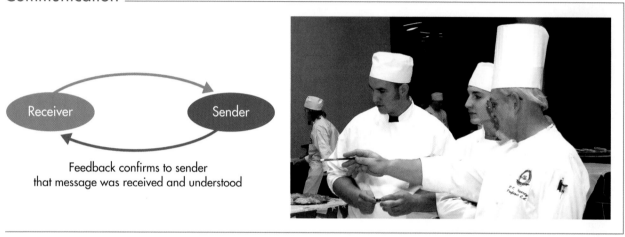

Feedback confirms to sender
that message was received and understood

Figure 1-29. *Effective communication skills are used to give and understand instructions, share information, and provide feedback.*

Speaking. Staff should always speak in a clear and pleasant tone that is loud enough to be heard, yet not loud enough to disturb others. To develop speaking skills, individuals can practice by speaking slowly and properly enunciating each word so that they are easily understood. A speaking tone should always be professional, respectful, and sound welcoming.

Listening. An effective listener shows respect by allowing not only guests, but fellow staff to speak without being interrupted. Making eye contact with the speaker shows the person that the message being conveyed is important. Effective listeners also summarize what they hear in order to reinforce the message and allow for any required clarification.

Another aspect of listening includes the ability to read the nonverbal cues of guests and coworkers. For example, guests who raise a glass or push a dish to the side are indicating that attention is needed. Likewise, a coworker may signal that a table needs to be cleared or that an order is ready. Correctly interpreting nonverbal cues leads to exceptional service.

Writing. Purchase orders, memos, résumés, and job applications are forms of written communication used in foodservice operations. Handwritten guest orders must accurately convey requests so that dishes are prepared according to guest specifications. When communicating in writing, it is important to write legibly so that the message is not misunderstood.

Reading. Cookbooks and professional journals contain valuable information that must be read. Reading skills are also used to interpret menus, follow recipes, and check the accuracy of guest checks.

Teamwork and Initiative

Each job within a foodservice operation cannot exist alone. Teamwork among staff creates an environment that fosters mutual respect and collaboration. *See Figure 1-30.* If one member of the team does not perform, it reflects on the rest of the team. Working in a cooperative manner creates an atmosphere in which staff show respect for one another and guests.

Teamwork

Figure 1-30. *Teamwork among staff creates an environment that fosters mutual respect and collaboration.*

In addition to teamwork, taking the initiative to perform tasks without being asked is a hallmark of professionalism. Recognizing and fulfilling a need such as refilling water glasses or straightening a work station before being asked shows initiative. Managers recognize staff who exhibit teamwork and initiative when promoting within the organization.

FRONT-OF-HOUSE SKILLS

When FOH staff have the necessary skills to execute their jobs with confidence, guest satisfaction improves. In order to fulfill guest expectations, FOH staff need to have menu knowledge, accommodate guests, maintain dining room workflow, interact with kitchen staff, and use appropriate math applications.

Menu Knowledge

Guests often ask for recommendations, inquire about special dishes, and request nutritional information. Having menu knowledge enables staff to respond accurately and with confidence about which dishes are available and how they are prepared. FOH staff who taste menu items are even better prepared to interact with guests. When staff members taste a dish, the likelihood of selling that menu item increases. For example, instead of asking guests if they would like dessert, servers could offer a specific suggestion and increase the likelihood guests will order that item.

Guest Accommodations

Foodservice operations that provide accommodations leave a favorable impression on guests. An *accommodation* is the act of modifying something in response to a need or request. For example, a guest may dislike a particular ingredient and ask for that item to be removed from a dish. When feasible, some operations prepare in advance for accommodations. This allows the staff to handle these situations with confidence. Meeting the needs of guests often means making accommodations for persons with disabilities and for children, as well as handling complaints.

Persons with Disabilities. According to the Americans with Disabilities Act (ADA), businesses serving the public must make reasonable accommodations for guests with disabilities. For example, removing a barrier such as a chair may be required for a guest with a wheelchair. FOH staff who are sensitive to the needs of persons with disabilities can help put guests at ease, make them feel welcome, and enhance overall guest satisfaction.

Children. Booster seats and child-friendly menu items are accommodations that can make dining with children an enjoyable event. *See Figure 1-31.* Disruptive children can negatively impact the dining experience of all guests. Providing other accommodations, such as crayons and paper, can also help promote a pleasurable dining experience for all.

Complaints. Regardless of the well-trained service staff and high-quality food, there will be occasions when things go wrong. Failure to accommodate a guest with a concern may lead to lost business. Asking open-ended questions and observing nonverbal cues are two ways of identifying an unhappy guest. For example, a server could ask an open-ended question such as, "How was the seafood special?" to encourage a detailed answer that could expose any concerns.

If a guest is dissatisfied, the following guidelines may be used to address the complaint:

- Listen carefully to fully understand any concerns the guest may have.

- Never ignore a dissatisfied guest. Address the concern immediately.

- Remain calm so that the guest will also remain calm.

- Do not become defensive, and never blame a guest for a problem. Regardless of who is at fault, make the situation right.

- Once a solution has been implemented, follow up to ensure the guest is satisfied, and thank the guest for being understanding and patient.

Accommodating Children

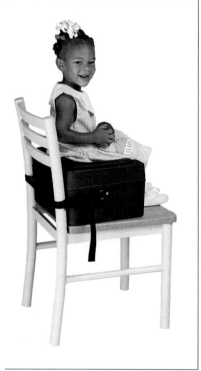

Carlisle FoodService Products

Figure 1-31. *Booster seats and child-friendly menu items are accommodations made for children.*

Responsible Alcohol Service

There are many risks and liability concerns associated with serving alcohol, which can result in fines, loss of a liquor license, imprisonment, or the closing of a business. When serving alcohol, taking appropriate precautions is necessary to protect guests, employees, and the foodservice operation. *See Figure 1-32.*

Responsible Alcohol Service

Daniel NYC

Figure 1-32. *When serving alcohol, taking appropriate precautions is necessary to protect guests, employees, and the foodservice operation.*

Many states have mandatory or voluntary regulations such as training through an alcohol-service training program, which addresses topics such as alcohol laws, preventing intoxication, and checking identification. Serving alcohol responsibly provides a level of service that accommodates guests while promoting a pleasurable experience for all.

FOH Workflow

FOH staff are responsible for preparing the dining room before guests arrive. Preparing for service often includes tasks such as setting tables, filling salt and pepper shakers, and making sure menus are clean. The dining room layout should provide an area for servers to stock essential service items. Workflow is most efficient when FOH staff implement mise en place and complete sidework frequently.

Setting the Stage. The FOH staff is responsible for organizing and assembling items necessary for dining room service. This includes setting tables, stocking side areas, and rearranging furniture to accommodate large parties. Tables should always make a positive impression by being clean and set appropriately. They should also provide adequate space for guests and servers. *See Figure 1-33.* Side areas contain tableware and condiments to minimize the number of trips to the kitchen.

Setting Tables

Daniel NYC

Figure 1-33. *Tables should always make a positive impression by being clean, being set appropriately, and providing adequate space for guests and servers.*

Sidework. FOH staff often have gaps between peak service hours to perform sidework. *Sidework* is the process of cleaning, restocking, and preparing the items needed to keep service running smoothly. For example, sidework may consist of brewing a fresh batch of coffee or restocking glassware. *See Figure 1-34.* Assigning specific tasks to each FOH staff member enables sidework to be executed quickly.

Sidework

Bunn-O-Matic Corporation

Figure 1-34. *Sidework may consist of brewing a fresh batch of coffee or restocking glassware.*

Kitchen Interactions

FOH staff frequently interact with back-of-house (BOH) staff for a variety of reasons, such as checking on the status of an order and relaying specific requests from guests. When all staff members work together in a professional manner and keep one another informed, the efforts of the entire foodservice operation are supported. Two factors influencing successful interaction are clear communication and the expediting of orders to ensure guests receive both quality meals and service.

Communication. The goal of pleasing guests can be accomplished when all staff members communicate effectively. An open line of communication helps ensure that orders are ready in a timely manner and prepared as requested. Strong communication starts with mutual respect and teamwork. A breakdown in communication typically results in dissatisfied guests.

Expediting. *Expediting* is the process of speeding up the ordering, preparation, and delivery of food to guests. A server or dining room manager may expedite plates to ensure guests receive efficient service. *See Figure 1-35.* This process often involves reading food orders out loud to the cooks, checking plated dishes for accuracy, and organizing completed plates on a tray.

FOH Math Applications

Math skills are essential for all individuals working in the foodservice industry. For example, FOH staff must place the accurate number of orders per table, collect payment for service, make change, and often balance the register at the end of a shift. Addition, subtraction, multiplication, and division are often used by FOH staff to confirm orders and verify transactions made on a point-of-sale system as well as when making change.

Point-of-Sale Systems. A *POS (point-of-sale) system* is a computerized network that compiles data on all sales incurred by a business. *See Figure 1-36.* A POS system creates a guest check that documents what is sold to the guest.

A guest check lists the items ordered by a customer, the prices to be charged for those items, and the total amount of money due. In addition to the amount due for food and beverages, the guest check may also include additional charges for items such as sales tax and service charges. It may also reflect discounts that reduce the total amount owed based on coupons or other promotions offered by the foodservice operation. The processing of a guest check ends once payment has been received by the foodservice operation and the proper amount of change and a receipt for payment have been returned to the guest.

Making Change. When a guest pays with cash, it is often necessary to return change. The amount of change given to the guest should be equal to the difference between the amount presented for payment and the total amount of the guest check. For example, if a guest uses a $20 bill to pay for a guest check with a total of $13.75, the guest is given $6.25 in change ($20.00 − $13.75 = $6.25).

Instead of simply handing the guest the change all at once, it is appropriate to return it to the guest in an orderly process. *See Figure 1-37.* The first step is to return any coins. In this example, the cashier would hand the guest a quarter ($0.25) and say, "Twenty-five cents makes fourteen dollars." The next step would be to hand the guest a $1.00 bill and say, "fifteen," and then hand the guest a $5.00 bill and say, "and five makes twenty." The process always ends with the total amount of money originally presented by the guest for payment.

In addition to cash, most foodservice operations offer guests the option of paying with a credit or debit card. Some restaurants even have their own gift cards that can be used for payment. The method for processing card payments depends on the type of system used by the foodservice operation. Sometimes cards are processed through a separate, stand-alone terminal while others are processed through a POS system.

Front-of-House (FOH) Expediting

Figure 1-35. *A server or dining room manager may expedite plates to ensure guests receive efficient service.*

Point-of-Sale (POS) Systems

POS device

CSI Hospitality Systems, Inc.

Figure 1-36. *A POS system is a computerized network that compiles data on all sales incurred by a business.*

Making Change

GUEST CHECK

Date	Table	Guests	Server	
1/5	10	1	#123	47221

1	Sandwich	7	99
1	Cup of Soup	3	75
1	Iced Tea		99
	Subtotal	12	73
	Discount		
	Sales Tax	1	02
	Total	13	75

A 15% gratuity will be added to groups of 8 or more.

Cashier says to the guest,
"Your total is thirteen seventy-five."
($13.75)

Guest pays for the $13.75 check
with a $20.00 bill.

Cashier returns a quarter and says,
"Twenty-five cents makes fourteen."
($13.75 + $0.25 = $14.00)

Cashier hands guest a dollar and says,
"Fifteen."
($14.00 + $1.00 = $15.00)

Cashier hands guest a $5.00 bill and says,
"And five makes twenty."
($15.00 + $5.00 = $20.00)

Figure 1-37. *It is appropriate to return change to the guest in an orderly process.*

Food Knowledge

Kolpak

Figure 1-38. *BOH staff need a thorough knowledge of ingredients, cooking methods, and nutritional values.*

BACK-OF-HOUSE SKILLS

Both FOH and BOH skill sets are equally important to the success of a food-service operation. When these skills are combined effectively, collaboration and teamwork result in a positive dining experience for guests.

Essential skills for BOH staff include food knowledge, the ability to create and maintain an effective workflow, math applications, interfacing effectively with FOH staff, and accommodating guest requests.

Food Knowledge

In order to prepare meals that meet and exceed guest expectations, BOH staff members need a thorough knowledge of ingredients, cooking methods, and nutritional values. *See Figure 1-38.* Food knowledge also includes knowing which cooking methods and ingredients should be used to effectively prepare and enhance the flavor of each dish. In addition, good food nourishes the body. It is also important to understand the nutritional values of foods in order to create flavorful and healthy dishes. New discoveries about food make it essential to continuously increase food knowledge.

BOH Workflow

A productive foodservice operation depends on high-quality food that is prepared efficiently. When food is prepared in a timely manner, the quality of the dish is preserved, and guests are more likely to be satisfied. To help ensure the appropriate timing of each dish, BOH staff must maintain a productive workflow. This can be achieved through effective mise en place and the use of appropriate food preparation skills.

Mise en Place. *Mise en place* is a French term meaning "put in place." Mise en place makes efficient use of time and keeps preparation areas organized. *See Figure 1-39.* To help ensure food preparation is carried out in an orderly and efficient manner, it is essential for BOH staff to complete the necessary mise en place prior to receiving the orders of guests. For example, prior to cooking an omelet to order, the appropriate pans should be chosen, the cheese shredded, and ingredients, such as mushrooms, onions, and peppers diced.

Preparation Skills. BOH staff must know how to execute cooking methods and how to properly portion and plate food. It is also imperative to understand how to store and prepare foods in a safe and sanitary manner. Preparation skills such as the ability to read recipes, measure ingredients accurately, and use knives to properly cut ingredients ranging from fruits and vegetables to breads and meats are essential BOH skills.

BOH Math Applications

In food production, math is used daily to measure ingredients, and recipes are often adjusted frequently to reflect the number of portions needed. Working with fractions, decimals, and percentages as well as adding, subtracting, multiplying, and dividing are the math skills used by BOH staff.

Measuring Ingredients. Properly measuring ingredients is vital to successful cooking. *See Figure 1-40.* Knowing how and when to use different types of measuring spoons, measuring cups, scales, and portioning equipment is necessary in order to produce quality dishes. Knowing the units of measurement, such as grams, ounces, and pounds, and the difference between weight and volume are vital to accurately measuring ingredients.

Adjusting Recipes. Adjusting a recipe requires the calculation of new amounts for each ingredient. For example, a recipe that normally serves 12 may be adjusted to serve 50. Converting measurements is often required when adjusting recipes. In addition to converting measurements, BOH staff need to understand how those adjustments can affect the equipment needs, cooking times, and cooking temperatures.

Mise en Place

Charlie Trotter's

Figure 1-39. *Mise en place makes efficient use of time and keeps preparation areas organized.*

Measuring Ingredients

Figure 1-40. *Properly measuring ingredients is vital to successful cooking.*

Back-of-House (BOH) Expediting

Figure 1-41. *BOH staff expedite dishes by checking to make sure the food is correctly portioned, plated, and the proper temperature.*

Dining Room Interactions

To keep food production and service efficient, BOH staff must effectively interact with FOH staff. This interaction is necessary to keep servers informed as to the status of an order. Guests receive higher quality meals and service when each foodservice staff member communicates effectively and expedites orders.

Communications. When communication is effective, the BOH staff member knows what is expected and dishes can be prepared efficiently and to exact specifications. Effective communication also includes informing the FOH staff when dishes are nearing completion and ready to be presented to guests.

Expediting. BOH staff expedite dishes by checking to make sure food is correctly portioned, plated correctly, and the proper temperature. *See Figure 1-41.* To assist FOH staff, BOH staff may carry a tray of food to the appropriate table.

Guest Accommodations

Guests often have special dietary considerations that require them to limit or restrict the consumption of certain foods. Accommodating guests who have specific requests may result in leaving out ingredients or changing cooking methods. Accommodations are often made by BOH staff for guests with food allergies and for individuals following special diets.

Food Allergies. Accommodating guests with food allergies is vital because an allergic reaction can be life-threatening. Knowledge of menu items and ingredients is a must. Providing guests with a list of ingredients used to prepare dishes, being able to substitute or eliminate specific ingredients, and preparing food in a special area of the kitchen are ways BOH staff accommodate guests with food allergies.

Special Diets. Weight management, cultural traditions, and the choice to consume plant-based meals cause some guests to request food preparation accommodations from BOH staff. Fulfilling special diet requests is a crucial aspect of providing a positive dining experience. *See Figure 1-42.*

Special Diet Requests

Daymark® Safety Systems

Figure 1-42. *Fulfilling special diet requests is a crucial aspect of providing a positive dining experience.*

EMPLOYMENT PREPARATION

Finding jobs in the foodservice industry requires organization and attention to detail. Before the application process can begin, the first step is to create a résumé and compile a professional portfolio. These items are the best way to make a good impression with potential employers. Once the résumé and portfolio have been created, a variety of resources are available to help potential employees identify job openings that match their skills and abilities. Once the job application has been completed, arriving at the interview prepared and confident is very important. It is also essential to handle the acceptance or rejection of employment in a professional manner.

Résumés and Portfolios

A *résumé* is a document listing the education, professional experience, and interests of a job applicant. A résumé should be short, factual, and accurate. It should outline all work experience, accomplishments, and special skills that relate to the position being sought. Misspellings and improper grammar send a negative message to the employer. A professional résumé can make a positive impression and show the employer the applicant is prepared and organized. *See Figure 1-43.* An effective résumé includes a career objective and lists the education and training, work experience, and accomplishments of the applicant.

A *career objective* is a direct statement expressing the type of employment goal being sought. For example, someone who has no experience in a restaurant, yet is looking for a first job may have the following career objective: To acquire a position in a quality foodservice operation in order to gain real-life work experience.

A résumé should also list the level of education, date of graduation, school or college name, and diploma or certificate title. A résumé should describe the work experience related to the job being sought. Starting with the most recent employment, the title of each position, start and end dates of employment, name of workplace, and the primary duties of the position should be listed. It is also important to include a list of professional accomplishments, including competitions and awards.

A *portfolio* is a collection of items that depict the knowledge, skills, and accomplishments of an individual. Portfolio items give potential employers an overview of the skills the applicant will bring to the job. A portfolio may include letters of recommendation, certificates earned, and images. For example, if an applicant is looking for a pastry job, that portfolio would likely include images of pastries the applicant has made. Portfolio materials need to be well organized, neatly presented, up-to-date, and complete.

Job Searches

Searching for a job can be a time consuming task. It can be challenging to sort through job postings to find one that matches the interests and abilities of the applicant. *See Figure 1-44.* However, there are a variety of resources available that can help streamline the search and make the process more manageable. Job-search tools such as networking, online resources, and printed publications all present opportunities that could lead to a potential career in foodservice.

Résumés

Alison Altmann
a.altmann@inet.com

Career Objective
To be a line cook

Education
- Little City High School, 2010 Honors Graduate

Work Experience
March 2011–present, Prep cook, The Pancake Shop
- Prepares all pancake batters
- Cleans and prepares all produce for service
- Assists on hot line as needed

March 2010–2011, Order taker, The Pizza Palace
- Answered phones and took pizza orders
- Ran the cash register
- Organized pizza delivery orders

Accomplishments
- Holds a valid state sanitation license
- Member of the Little City Culinary Team
- Won 2nd place in the state chili cook-off

Figure 1-43. *A professional résumé can show an employer the applicant is prepared and organized.*

Job Postings

RESTAURANT / CATERING MANAGER

Northwest suburban catering company seeks Restaurant/Catering Mgr. Duties include direct and supervise employee activities, control and monitor quality of environment, facilitate corporate events. Selling banquet functions and budgeting. Requires minimum 3 years experience in fast-paced kitchen operation. Salary commensurate with experience plus bonus, 401K, medical.

www.chefsville.net

RESTAURANT HIRING

South Beach area country club now hiring Asst Bar Manager. Position requires minimum 3 years experience bartending with short order cook experience, strong front-house skills, procuring music entertainment and ability to work weekend hours.

REF # 78455783
Employer-paid ad

RESTAURANT OPENINGS

Angelo's Family Dining

Now Hiring: Servers • Bartenders • Host Staff
Espresso Bar • Bakers • Cashiers
Line & Prep Cooks • Asst Kitchen Manager

Full & Part Time, Top $

APPLY IN PERSON
Mon-Sat 9am to 6pm at our temporary interviewing site:

PARAGON INN & SUITES

RESTAURANT COOK TRAINER

Specialty cook trainer, Italian cuisine. 40-hours per week, M-F, 6am-2pm. HS grad, 2 yr experience in position or as chef of Italian cuisine; overseeing & implementing training program and preparing & cooking in traditional Italian cuisine according to original recipes. Educating employees in fundamentals of preparing and cooking fine cuisine and in testing ingredients and final products. Applicants must show proof of legal authorization to work in U.S.

REF # 22568413
Equal Opportunity Employer

Figure 1-44. *It can be challenging to sort through job postings to find one that matches the interests and abilities of the applicant.*

Networking. *Networking* is a means of using personal connections, such as contacts through professional organizations, friends, teachers, and acquaintances, to locate possible employment opportunities. Networking is the most common means of finding employment. For example, a culinary student may speak to chefs to determine if they are aware of any job openings.

Online Resources. The Internet is an accessible and quick way to search for jobs. Many websites allow individuals to search by city, state, and position. They also allow individuals to post résumés for potential employers to review. A job seeker can also view online job postings and search online trade publications.

Printed Resources. Foodservice publications are written for members of the foodservice industry and provide valuable information about many different areas of the industry. For example, foodservice trade journals often list job postings. Newspapers also list employment opportunities.

Employment Applications

An employment application requires applicants to list basic information such as name, address, relevant work experience, educational background, position desired, and availability. An employment application is used to quickly review applicant qualifications and should be completed neatly and accurately. ***See Figure 1-45.*** Background checks reflect the accuracy of a completed application. Some interview questions are related to the employment application.

Completing Employment Applications

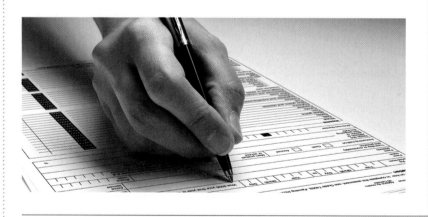

Canadian Beef, Beef Information Centre

Figure 1-45. *An employment application is used to quickly review applicant qualifications and should be completed neatly and accurately.*

Interviews

A job interview is an opportunity to discover if the skills and background of the applicant match the requirements of the position. ***See Figure 1-46.*** The interview also reveals the communication skills, confidence, and integrity of the applicant. A job interview allows a manager to find out more about the applicant, to provide more information about the job, and to answer any questions the applicant may have. A simple way to make a good impression at an interview is to arrive early and dress like a professional.

Interviewing

Figure 1-46. *A job interview is an opportunity to discover if the skills and background of the applicant match the requirements of the position.*

The following guidelines can help job seekers interview with confidence:

- Maintain appropriate eye contact with the interviewer at all times.
- Think before speaking and do not answer questions inaccurately.
- Ask questions about the foodservice operation.
- Thank the interviewer when the interview ends.
- After the interview, send a thank-you letter to the interviewer.

Career Decisions

In the event that the job is offered, an applicant may accept the position, decline the position, or ask for time to consider the offer. If the position is declined, the applicant should tactfully thank the employer for the offer. If time is needed to consider a job offer, the applicant must still decide in a timely manner. Employers deserve a respectful and prompt response. It is important to react professionally, regardless of whether a job is offered.

SUMMARY

The foodservice industry is centered around customer service. Quality service is the result of positive interactions between external and internal customers. When dining, there are a variety of environments available, and choosing one often depends on the meal service style and type of cuisine desired. From quick service operations to fine dining and institutional operations, the evolution of cuisines has presented many combinations of ingredients and unique presentations.

There are a multitude of career opportunities within the foodservice industry and a variety of available education and training programs. Achieving a successful career requires a commitment to personal hygiene, a professional appearance, a strong work ethic, effective communication, and teamwork.

Daniel NYC

Both FOH and BOH staff need menu and food knowledge and the necessary skills to maintain an efficient workflow, accommodate guests, and apply appropriate math functions in order to provide guests with a positive dining experience.

To prepare for employment, a portfolio and résumé need to be developed that highlight the education and training, work experience, and accomplishments of the applicant. Personal networks and online and printed resources can lead to job interviews and rewarding careers in the foodservice industry. It is important to be professional and accurate when seeking employment and to thank employers for opportunities offered.

Refer to DVD for
Quick Quiz® questions

Review

Refer to DVD for
Review Questions

1. Explain why customer service is more than waiting on guests.
2. Contrast external and internal customers.
3. Describe six common types of dining environments.
4. Describe nine meal service styles.
5. Explain the difference between a booth and a banquette.
6. Explain the proper way to serve guests using plated service.
7. Contrast English, Russian, and French service.
8. Explain the role guilds played in the evolution of cuisine.
9. Summarize six types of cuisine that have evolved over time.
10. Explain the contributions of Escoffier to classical cuisine.
11. Name two chefs noted for nouvelle cuisine.
12. Name the chef credited for avant-garde cuisine.
13. Describe six front-of-house (FOH) foodservice positions.
14. Explain the purpose of the brigade system developed by Escoffier.
15. Describe eight back-of-house (BOH) foodservice positions.
16. Describe two foodservice management positions.
17. Name four foodservice career areas that are not part of the restaurant environment.
18. Describe four types of education and training options.
19. Explain what is meant by an apprenticeship program.
20. Contrast a certificate program and a degree program.
21. Identify five essential employability skills.
22. Identify four types of communication required to effectively interact with others.
23. Identify five FOH skills required to fulfill guest expectations.
24. Explain why it is important for FOH staff to have menu knowledge.
25. Identify four ways FOH staff can accommodate guests.
26. List five guidelines to follow when guests are dissatisfied.
27. Explain why it is important to provide responsible alcohol service.
28. Describe FOH workflow.
29. Define sidework.
30. Explain the importance of the FOH interaction with the BOH.

31. Describe two primary math applications used by FOH staff.

32. Identify five essential BOH skills.

33. Explain why it is important for BOH staff to have food knowledge.

34. Describe the importance of BOH workflow.

35. Define mise en place.

36. Describe two primary math applications used by BOH staff.

37. Explain the importance of the BOH interaction with the FOH.

38. Explain how BOH staff can accommodate guests.

39. Describe the components of an effective résumé.

40. Describe the value of a professional portfolio.

41. Identify three types of job search tools.

42. Explain why it is important to complete a job application neatly and accurately.

43. Explain the various purposes of job interviews.

44. List five guidelines to follow during job interviews.

45. Explain why it is important to respond to a job offer in a timely manner.

Applications

1. Categorize the foodservice operations found in the local area according to the six dining environments.

2. Demonstrate each of the nine meal service styles.

3. Research a chef who made a significant contribution to the evolution of cuisine.

Refer to DVD for
Application Scoresheets

4. Download online postings for three front-of-house (FOH) positions of interest. Identify the types of education and training required for each position.

5. Download online postings for three back-of-house (BOH) positions of interest. Identify the types of education and training required for each position.

6. Download an online posting for a foodservice management position. Identify the type of education and training required for this position.

7. Research available apprenticeship, certificate, and degree programs in the local area.

8. Research a well-known, living chef who developed knowledge and skills through on-the-job training.

9. Write an essay explaining the value of employability skills in the foodservice industry.

10. Research the certificate program for food safety and sanitation.

11. Research the certificate program for responsible alcohol service.

12. Set a table for four guests and identify the type of sidework that would be required.

13. Use a POS system to process a guest order and a credit or debit payment.

14. Make change for two $20 bills for a guest check of $27.36.

15. Demonstrate how to expedite an order from the kitchen to the dining room.

16. Prepare a résumé.

17. Prepare a portfolio.

18. Complete a job application.

19. Participate in a mock interview for a foodservice job posting.

20. Write a thank-you letter to the person who conducted the mock interview.

Culinary Arts
PRINCIPLES AND APPLICATIONS

Carlisle FoodService Products

Food Safety and Sanitation

Chapter 2

*F*ood safety and sanitation are paramount to successful foodservice operations. Customers expect to be served safe, flavorful foods and beverages in a safe and clean environment. It is the responsibility of every foodservice operation to establish and follow food safety and sanitation standards. It is also the duty of every foodservice employee to follow those safety and sanitation standards in both the kitchen and dining room.

Refer to DVD
for **Flash Cards**

Chapter Objectives

1. Identify seven agencies and organizations responsible for regulating food safety standards.
2. Explain how pathogens cause foodborne illnesses.
3. Explain why some people are more susceptible to foodborne illness.
4. Describe three biological contaminants that can be found in food.
5. Describe six factors that contribute to bacterial growth.
6. Define the temperature range known as the temperature danger zone.
7. Identify three types of physical contaminants that can be found in food.
8. Identify two types of chemical contaminants that can be found in food.
9. List the accepted personal hygiene practices for foodservice staff.
10. Demonstrate proper handwashing.
11. List circumstances under which gloves should be worn.
12. Describe common sanitizers in foodservice operations.
13. Demonstrate warewashing in a compartment sink and using a dish machine.
14. Describe how to keep food safe as it moves through the flow of food.
15. Explain the importance of properly preparing food.
16. Describe the Hazard Analysis Critical Control Points (HACCP) principles.
17. Identify five components of fire safety.
18. Explain the purpose of a safety data sheet (SDS).
19. List common injuries that occur in the professional kitchen.

Key Terms

- **contamination**
- **foodborne illness**
- **biological contaminant**
- **pathogen**
- **parasite**
- **virus**
- **bacteria**
- **temperature danger zone**
- **food spoilage indicator**
- **physical contaminant**
- **chemical contaminant**
- **personal hygiene**
- **sanitizing**
- **warewashing**
- **flow of food**
- **FIFO**
- **HACCP**
- **critical control point**
- **safety data sheet**

FOOD SAFETY

The reputation of a foodservice operation is earned based on the quality of the food and service provided to its customers. A vital step in earning that reputation is carefully inspecting, properly storing, and safely handling food during the entire journey from the farm to the table. Food safety standards are established and regulated by agencies, such as the Food and Drug Administration (FDA), and organizations, such as NSF International. *See Figure 2-1.*

Regulation of Food Safety Standards	
Agency/Organization	**Mission**
U.S. Food and Drug Administration (FDA)	Government agency that ensures the safety of all food except meat, poultry, and egg products; publishes the *Food Code*, a document that recommends licensing, inspection, and enforcement regulations for the foodservice industry
USDA Food Safety and Inspection Service (FSIS)	Government agency that ensures the safety of meat, poultry, and egg products
USDA Animal and Plant Health Inspection Service (APHIS)	Government agency that ensures protection of animals and plants from diseases or pests
United States Environmental Protection Agency (EPA)	Government agency that establishes the levels of pesticide residue that can be tolerated by humans
Centers for Disease Control (CDC)	Government agency that investigates foodborne illnesses and plays a key role in supporting state and local health departments
NSF International	Not-for-profit organization that sets safety and sanitation standards for tools and equipment used in the foodservice industry; publishes safety standards for food, water, and consumer goods
Underwriters Laboratories (UL)	Independent safety certification organization that develops standards for various tools and equipment used by the foodservice industry

Figure 2-1. *Food safety standards are established and regulated by agencies, such as the Food and Drug Administration (FDA), and organizations, such as NSF International.*

Food safety standards are enforced by local health departments and other authorities having jurisdiction. Entities that enforce food safety codes have the authority to shut down foodservice operations that fail to meet local safety codes. Most states also require foodservice managers to complete ServSafe® manager certification training and pass the certification exam. ServSafe® certification is valid for five years. However, some states and employers have even more stringent guidelines regarding certification renewal.

FOOD CONTAMINATION

Contamination is a process of corrupting or infecting by direct contact or association with an intermediate carrier. For example, direct contamination can occur if lettuce is placed directly next to raw poultry. Cross-contamination occurs when food, water, or equipment comes in contact with contaminated elements through an intermediate carrier. It most commonly occurs when ready-to-eat foods become contaminated by coming in contact with a contaminated surface, such as a cutting board, previously contaminated by raw food. Unsafe food handling and poor sanitation practices, such as staff not washing their hands, can also cause cross-contamination.

Foodborne Illnesses

A *foodborne illness* is an illness that is carried or transmitted to people through contact with or consumption of contaminated food. Foodborne illnesses can come on rapidly or over the course of many days. Symptoms may include vomiting, cramping, headache, sweats, chills, diarrhea, and fever.

Foodborne illnesses are most often caused by pathogens. A *pathogen* is a microorganism that can cause disease. There are many types of pathogens that can cause foodborne illnesses. *See Figure 2-2.* Pathogens can be transferred through a variety of sources including raw foods, contact surfaces, garbage, pests, and people. Pathogens can cause foodborne illness shortly after contaminated food is eaten or up to several days later. In some cases, a foodborne illness can be life threatening.

Some people are more susceptible to foodborne illness than others. Infants, young children, pregnant women, the elderly, people with compromised immune systems, and those who have life-threatening allergies are considered highly susceptible to foodborne illness. Menu items that contain raw or undercooked eggs, fish, or meat must be marked clearly on menus so that these items can be avoided by guests who are highly susceptible to foodborne illness. Pathogens that cause foodborne illness are often categorized as biological, chemical, or physical contaminants.

Biological Contaminants

A *biological contaminant* is a living microorganism that can be hazardous if it is inhaled, swallowed, or otherwise absorbed. Biological contaminants that can be found in food include parasites, viruses, and bacteria.

- **Parasites.** A *parasite* is a pathogen that relies on a host for survival in a way that benefits the organism and harms the host. Parasites inhabit the stomachs and intestinal tracts of animals and are passed on to people when contaminated food is consumed. Anisakis simplex is an example of a parasite caused by a roundworm that is found in contaminated fish and squid.

- **Viruses.** A *virus* is a pathogen that grows inside the cells of a host. Human beings are often hosts to viruses. Sneezing or coughing on food is one way that viruses can spread. Viruses can also survive on the surface of food or production equipment.

  Hepatitis A is a virus most commonly spread by contaminated shellfish and feces-contaminated water. Noroviruses stem from foods that have been in contact with feces and are often spread by inadequate handwashing by food handlers. A person who consumes contaminated shellfish will immediately become a carrier of the virus but may not become sick for many months. A person infected with a virus can infect others.

  Viruses can withstand freezing temperatures. However, heat can destroy viruses. Ready-to-use foods, such as deli meats, do not require heating, making them more susceptible to viruses.

Common Pathogens that Cause Foodborne Illnesses		
Pathogen	**Methods of Transmission**	**Common Symptoms**
Amisakis simplex	Herring; cod; halibut; mackerel; Pacific salmon; certain shellfish	*Noninvasive type:* throat tingle and coughing up worms *Invasive type:* stomach pain, diarrhea, nausea, and vomiting
Bacillus cereus	Contaminated meats; cooked corn; cooked vegetables; baked potatoes; cooked rice	*Diarrhea type:* watery diarrhea and abdominal cramps *Vomiting type:* nausea and vomiting
Campylobacter jejuni	Contaminated water; poultry	Fever, headache, and abdominal cramps followed by diarrhea
Clostridium botulinum	Improperly canned foods; untreated garlic and oil mixtures; reduced-oxygen-packaged (ROP) food; baked potatoes	Vomiting and nausea; double vision, trouble speaking and swallowing, and weakness
Clostridium perfringens	Food left for long periods in the temperature danger zone; poultry; meats; stews; gravies	Diarrhea and abdominal pains
Cryptosporidium parvum	Food contaminated by food handlers or herd animals; contaminated water; contaminated produce	Severe watery diarrhea, nausea, and weight loss
Cyclospora cayetanensis	Contaminated water; contaminated produce; food contaminated by infected food handlers	Low fever, nausea, abdominal cramping, and diarrhea that alternates with constipation
E. coli O157:H7	Raw or rare ground beef; contaminated fruits and vegetables	Diarrhea or bloody diarrhea, abdominal cramps; can cause hemolytic uremic syndrome (HUS)
Giardia duodenalis	Contaminated water; food prepared by infected food handlers	Fever, diarrhea, abdominal cramps, and nausea
Hepatitis A virus	Contaminated water; food contaminated by food handlers; fruits, vegetables, shellfish, and salads; ready-to-eat food	Fever, nausea, abdominal pain, and weakness; jaundice
Listeria monocytogenes	Ready-to-eat food; raw meats; soft cheeses; unpasteurized milk	Spontaneous abortion in third trimester of pregnancy; sepsis, pneumonia, and meningitis in newborns
Noroviruses (Norwalk and Norwalk-like viruses)	Contaminated water; contaminated shellfish; food contaminated by infected food handlers, ready-to-eat food	Nausea, vomiting, diarrhea, and abdominal cramps
Salmonella spp.	Raw or undercooked eggs, poultry, and meat; unpasteurized milk and dairy products; food contaminated by infected food handlers	Abdominal pain, diarrhea, vomiting, and fever
Shigella spp.	Fecal contamination of food and water; food contacted by food handlers with poor personal hygiene; salads that contain potentially hazardous foods such as eggs	Diarrhea containing blood, fever, and abdominal cramps
Staphylococcus aureus	Improperly handled food; salads that contain potentially hazardous foods; deli meat	Severe nausea, abdominal cramps, vomiting
Vibrio parahaemolyticus	Raw or partially cooked oysters	Low fever, chills, nausea, vomiting, diarrhea, abdominal cramps
Vibrio vulnificus	Raw or partially cooked oysters	*Septicemia:* diarrhea, nausea, vomiting, skin lesions, fever, and sudden chills *Gastroenteritis:* diarrhea and abdominal cramps

Figure 2-2. *There are many types of pathogens that can cause foodborne illnesses.*

- **Bacteria.** *Bacteria* are pathogens that live in soil, water, or organic matter. Organic matter includes the bodies of plants and animals. Bacteria divide every 12–20 minutes and can multiply into millions in a matter of hours. *See Figure 2-3.* Bacterial contamination is the leading cause of foodborne illness and may result in outbreaks of pathogens such as E. coli and Salmonella. However, not all the bacteria found in food will cause illness. Some beneficial bacteria are added to prepared foods, such as probiotics that are added to yogurt.

Six factors contribute to bacterial growth. FAT TOM is the acronym that identifies these factors. FAT TOM stands for food, acidity, temperature, time, oxygen, and moisture.

E. Coli Bacteria

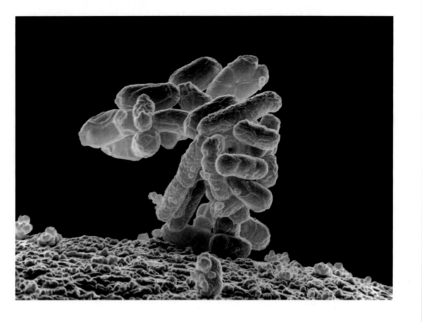

Agricultural Research Service, USDA

Figure 2-3. *Bacteria, such as E. coli, divide every 12–20 minutes and can multiply into millions in a matter of hours.*

- **Food.** High-protein foods such as poultry, meats, seafood, and dairy products easily support bacterial growth. Bacteria thrive on protein-rich foods for the energy they need to reproduce.

- **Acidity.** Every food falls between 0 and 14 on the pH scale, where 0 is the highest in acidity and 14 is the highest in alkalinity. *See Figure 2-4.* Most bacteria growth is inhibited by very acidic conditions, below a 4.6 pH. Acidic ingredients, such as lemon juice, do not provide a favorable climate for bacteria. The middle of the pH scale is neither acidic nor alkaline and is considered neutral. For example, distilled water has a pH of 7.

pH Scale

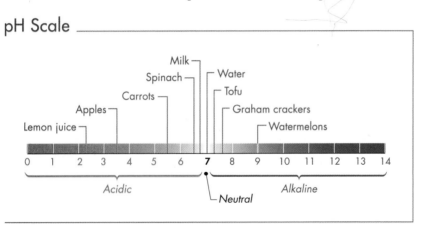

Figure 2-4. *Every food falls between 0 and 14 on the pH scale, where 0 is the highest in acidity and 14 is the highest in alkalinity.*

- **Temperature.** Bacteria reproduce best in a temperature range known as the temperature danger zone. The *temperature danger zone* is a range of temperature, between 41°F and 135°F, in which bacteria thrive. *See Figure 2-5.* Most bacteria levels are safe at temperatures of 165°F or above. Bacteria multiply very slowly at temperatures below 41°F and stop completely at temperatures of 0°F and below. However, the bacteria do not die. When frozen bacteria are returned to temperatures between 41°F and 135°F they begin to grow again.

Temperature Danger Zone

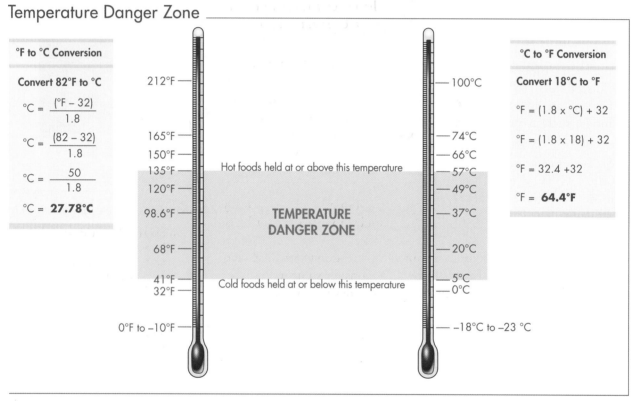

°F to °C Conversion
Convert 82°F to °C
$°C = \dfrac{(°F - 32)}{1.8}$
$°C = \dfrac{(82 - 32)}{1.8}$
$°C = \dfrac{50}{1.8}$
$°C = \mathbf{27.78°C}$

°C to °F Conversion
Convert 18°C to °F
$°F = (1.8 \times °C) + 32$
$°F = (1.8 \times 18) + 32$
$°F = 32.4 + 32$
$°F = \mathbf{64.4°F}$

212°F — — 100°C

165°F — — 74°C
150°F — — 66°C
Hot foods held at or above this temperature
135°F — — 57°C
120°F — — 49°C

TEMPERATURE DANGER ZONE

98.6°F — — 37°C

68°F — — 20°C

41°F — — 5°C
Cold foods held at or below this temperature
32°F — — 0°C

0°F to −10°F — — −18°C to −23 °C

Figure 2-5. *The temperature danger zone is a range of temperature, between 41°F and 135°F, in which bacteria thrive.*

Bacterial Growth Phases

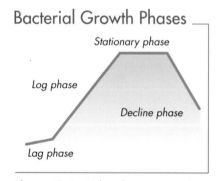

Stationary phase

Log phase

Decline phase

Lag phase

Figure 2-6. *When bacteria are introduced to a new surface or environment, they go through several stages of growth.*

- **Time.** The less time that foods are left in the temperature danger zone, the less opportunity there is for bacteria to multiply and reach dangerous levels. When bacteria are introduced to a new surface or environment, they go through several phases. *See Figure 2-6.*

  The lag phase is the period of time (1–4 hours) where bacteria are introduced to a new environment and reproduce slowly. During the log phase, bacteria begin to reproduce rapidly. Then, bacteria reproduce to such an extent that they compete for food, moisture, and space during the stationary phase. During the decline phase, bacteria die and leave behind high levels of oxygen. The less time that low-acid and high-protein foods are in the temperature danger zone, the less opportunity there is for bacteria to multiply and reach dangerous levels.

- **Oxygen.** Almost all foodborne pathogens require oxygen to grow. For example, campylobacter and E. coli require oxygen to grow. Most bacteria found in foods, such as Salmonella and Bacillus cereus, are facultative bacteria. *Facultative bacteria* are bacteria that can grow either with or without oxygen. Clostridium botulinum, the pathogen that causes botulism, does not require oxygen to grow.

- **Moisture.** Foods such as poultry, seafood, meats, soft cheeses, and produce contain high amounts of water and are therefore more susceptible to fast-growing bacteria. If moisture is extracted from food, the food can be stored with less chance of bacterial growth. However, when dried or dehydrated foods are rehydrated, bacteria can multiply.

Some biological contaminants can cause detectable changes in food as food begins to spoil. For example, bacteria decompose food and cause changes that can be seen, smelled, or tasted. A *food spoilage indicator* is a condition that signifies that food is deteriorating and is no longer safe for consumption. Food spoilage indicators include changes in color, odor, and texture. If any of these indicators are present, food should be thrown away to prevent foodborne illnesses. Foods eaten that have these indicators can cause gastrointestinal disturbances.

Although mold can be a natural component of some foods, it is usually a food spoilage indicator. *See Figure 2-7.* Molds produce green or black spores that are visible to the human eye. Molds multiply on acidic foods that have low water content, such as breads. In addition to food spoilage, molds can cause minor to severe allergic reactions for some individuals. Freezing can reduce mold growth, but does not destroy mold.

Chef's Tip

Moldy foods should be discarded, with the exception of items such as hard cheeses, hard salami, dry-cured country hams, and firm vegetables, such as bell peppers. These foods can be used if 1 inch of the food is removed from the perimeter of the moldy area.

Molds

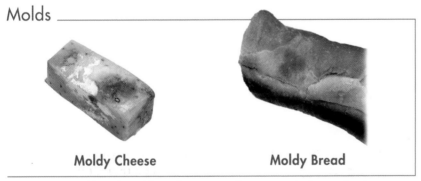

Moldy Cheese   Moldy Bread

Figure 2-7. *Although mold can be a natural component of some foods, it is usually a food spoilage indicator.*

Physical Contaminants

A *physical contaminant* is any object that can be hazardous if it is inhaled, swallowed, or otherwise absorbed into the body. Examples of physical contaminants found in food include bone fragments, bits of metal or glass, and tiny stones. Most physical contaminants in food are accidental. Physical contamination can be minimized by carefully inspecting food items and storing them in safe, well-maintained locations. *See Figure 2-8.*

Inspecting Food for Physical Contaminants

Figure 2-8. *Physical contamination can be minimized by inspecting food items for foreign objects such as bone fragments.*

Chemical Contaminants

A *chemical contaminant* is any chemical substance that can be hazardous if it is inhaled, swallowed, or otherwise absorbed into the body. Chemical contaminants include cleaning supplies, pesticides, and other chemical-based compounds used in a foodservice operation. *See Figure 2-9.* These items should be stored separately from food products and in properly labeled containers. Containers used for chemicals cannot be recycled for any food preparation task or used for future storage.

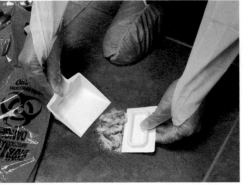

Daymark® Safety Systems

Chemical Contaminants

Hobart

Figure 2-9. *Chemical contaminants include cleaning supplies, pesticides, and other chemical-based compounds used in a foodservice operation.*

Most food contamination is accidental. However, foodservice operations can encounter intentional acts of food contamination. *Food bioterrorism* is the purposeful act of releasing toxins into foods with the intent to cause harm. In 2002 the Public Health Security and Bioterrorism Preparedness and Response Act was passed to help safeguard food from bioterrorist acts.

SANITATION PRACTICES

The foodservice staff is the most important part of any food safety and sanitation program. Sanitary work conditions reduce the threat of contaminants and foodborne illness. Following good personal hygiene standards, adhering to handwashing procedures, properly cleaning and sanitizing the work environment, and managing pests help promote sanitary work conditions.

Charlie Trotter's

Personal Hygiene

Personal hygiene is the physical care maintained by an individual. Foodservice staff must always practice good personal hygiene to help eliminate the spread of pathogens. *See Figure 2-10.* The following are universally accepted personal hygiene practices for staff in foodservice operations:

- Shower or bathe daily.
- Keep hair washed, combed, and covered as required.
- Keep facial hair clean, trimmed, and covered as required.
- Wear the proper clean clothing for the job.
- Do not wear nail polish or artificial fingernails.
- Except for a plain wedding band, do not wear jewelry.
- Keep hands and fingernails clean.
- Handle food only as required and always with clean or gloved hands.
- Do not allow dirty utensils or equipment to touch food.
- Eat and smoke only in designated areas.
- Never work around food with open cuts or sores. Bandage cuts and sores, and wear gloves over injured hands.
- Do not cough, spit, or sneeze near food or in food preparation areas. Always cover your mouth with your arm when coughing, and sneeze into a handkerchief. Wash hands immediately after using a handkerchief or coughing.
- Notify a supervisor and stay home when sick or if experiencing vomiting, diarrhea, jaundice, or sore throat and fever.

Personal Hygiene

Head covering

Clean and pressed jacket

Gloved hands

Apron

Sullivan University

Figure 2-10. *Foodservice staff must always practice good personal hygiene to help eliminate the spread of pathogens.*

Handwashing

Every member of the foodservice staff must follow FDA-approved handwashing procedures. ***See Figure 2-11.*** The FDA *Food Code* § 2-301.12, *Personal Cleanliness—Hands and Arms,* specifies the following handwashing procedure for all foodservice employees.

Foodservice staff must wash their hands at a handwashing station when working in the professional kitchen. Handwashing stations must be located in preparation, service, and dishwashing areas. Handwashing stations must be equipped with soap and an approved means for drying hands. Hand sanitizers must be FDA-approved for contact with food and food preparation utensils. Washrooms used by foodservice employees must display signs notifying employees that they must wash their hands before returning to work.

1. Wet hands and arms with hot water (at least 100°F).

2. Lather fingers, fingertips, areas between the fingers, hands, and lower arms with soap.

3. Scrub vigorously for at least 20 seconds.

4. Clean under the fingernails and between fingers.

5. Rinse hands and arms thoroughly with warm water.

6. Dry hands with disposable paper towels, a heated-air hand drying device, or a drying device that delivers high-velocity pressurized air.

Figure 2-11. *Every member of the foodservice staff must follow FDA-approved handwashing procedures.*

Foodservice staff must wash their hands immediately before beginning any food preparation task. Handwashing should also occur after the following:

- handling raw poultry, seafood, or meat
- handling money
- using the restroom
- touching the hair, face, or body
- coughing, sneezing, or using a tissue
- eating, drinking, chewing gum, and using tobacco products
- handling chemicals
- disposing of garbage

Refer to DVD for
Handwashing
Media Clip

Gloves

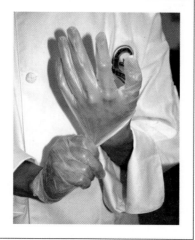

Figure 2-12. *Gloves need to be changed when an employee leaves a station, after handling raw food, and when gloves are torn.*

Gloves are often worn in the professional kitchen, but they are not a substitute for handwashing. Gloves are worn when serving ready-to-eat foods such as salads and sandwiches. Gloves need to be changed when an employee leaves a station, after handling raw foods, and when gloves are torn. *See Figure 2-12.* Gloves should never be worn for more than 4 hours. Each time gloves are removed, hands should be washed again. Before applying gloves, hands should be washed and dried using the proper handwashing procedure.

Cleaning and Sanitizing

All food contact surfaces must be cleaned and sanitized after each use, before working with any food, or a minimum of every 4 hours. *See Figure 2-13.* *Cleaning* is the process of removing food and residue from a surface. There are different types of cleaning agents depending on the item to be cleaned. For example, detergents can remove residue from food contact surfaces, equipment, and floors. Solvent cleaners cut through grease. Abrasive cleaners are often used to remove baked-on foods from pots and pans.

Cleaning Surfaces

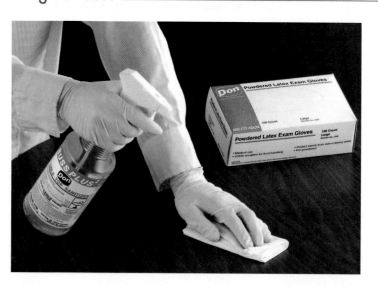

Edward Don & Company

Figure 2-13. *All food contact surfaces must be cleaned and sanitized after each use, before working with any food, or a minimum of every 4 hours.*

An item must be cleaned before it can be sanitized. *Sanitizing* is the process of destroying or reducing harmful microorganisms to a safe level. An object can be sanitized using heat or chemicals.

Heat sanitizing is the process of using very hot water to reduce harmful pathogens to a safe level. An example of heat sanitizing is immersing an object in water that is at least 171°F for a minimum of 30 seconds. *Chemical sanitizing* is the process of using a chemical solution to reduce harmful pathogens to a safe level. Quaternary ammonium compounds (quats), chlorine, and iodine are types of chemical sanitizers used in the professional kitchen. It is important to know which sanitizers work best with local water sources.

Edward Don & Company

Warewashing

Warewashing is the process of cleaning and sanitizing all items used to prepare and serve food. Warewashing can be accomplished using a three- or four-compartment sink or a dish machine. Each compartment of a warewashing sink and the surrounding workstation should be properly cleaned and sanitized before washing items. A compartment sink is often used to wash and sanitize large items and items that are difficult to clean due to foods that have adhered to the items. *See Figure 2-14.* The following procedure can be used to warewash items in a compartment sink:

Procedure for Warewashing in Three-Compartment Sinks

1. Scrape and rinse off each item.
2. In the first compartment, wash each item in a detergent and water solution that is at least 110°F.
3. In the second compartment, rinse the items by immersing them under water.
4. In the third compartment, immerse each item for 30 seconds in a chemical sanitizing solution mixed, per manufacturer recommendations, with 75°F to 120°F water.
5. Allow items to air-dry.

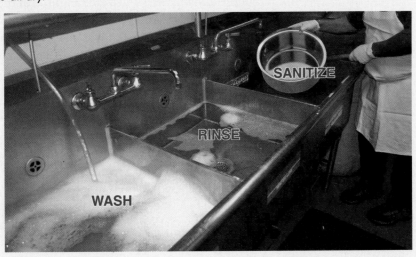

Figure 2-14. *Each compartment of a warewashing sink and the surrounding workstation should be properly cleaned and sanitized before washing items.*

A dish machine is an efficient way to wash and sanitize items that do not require special attention. Each dish is scraped, rinsed, and then placed in a rack so that all sides of the dish will be exposed to the spray action of the dish machine. *See Figure 2-15.*

Dish Machines

Figure 2-15. *Each dish is scraped, rinsed, and then placed in a rack so that all sides of the dish will be exposed to the spray action of the dish machine.*

After washing, items are checked to ensure that they are clean. Items that come out of the dish machine soiled are run through the machine a second time. The water temperature in a high-temperature dish machine should reach 180°F to properly sanitize items. In a low-temperature dish machine, the chemical sanitizer needs to reach at least 120°F and the instructions for the chemical being used need to be followed carefully.

Pest Management

Food contamination can be caused by insects and rodents, which are carriers of disease and bacteria. Most of these pests live in colonies. If one is spotted, it is likely there are more on the premises.

To eliminate these pests and prevent infestations, a regular maintenance schedule should be followed. Incoming supplies should also be checked for any indication of pests. Garbage and recycling areas should be in designated areas and not allowed to accumulate with excess. A routine maintenance program should include checking that all openings in doors, windows, air vent screens, and other openings are sealed. Foodservice operations contract pest control operators to control pests proactively.

THE FLOW OF FOOD

Preventing foodborne illness starts long before a meal is served. Contaminants can enter at any stage in the flow of food. The *flow of food* is the path food takes in a foodservice operation as it moves from purchasing to service. The flow of food includes purchasing, receiving, storing, preparing, thawing, cooking, cooling, holding, reheating, and serving. *See Figure 2-16.*

Hobart

The Flow of Food

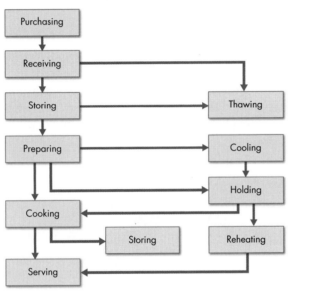

Figure 2-16. *The flow of food includes purchasing, receiving, storing, preparing, thawing, cooking, cooling, holding, reheating, and serving.*

Purchasing Food

When choosing a supplier, it is critical to purchase goods and products from reputable suppliers that follow food safety guidelines. This includes confirming that suppliers are getting products from approved sources, have staff trained in food safety practices, and follow a good manufacturing practice (GMP) program that addresses the minimum sanitation and processing requirements set forth by the FDA to help ensure the safe production of food.

Receiving Food

Once food is received, inspections are necessary to uphold quality and safety standards. Items should be rejected if they are damaged, past their expiration date, or do not meet temperature requirements. Frozen foods that show signs of thawing or refreezing, such as ice crystals or freezer burn, should be rejected. Unless otherwise specified by the manufacturer, ready-to-eat foods should be received at 41°F or less with packaging intact. This pertains to foods such as precut meats, salads containing potentially hazardous foods such as eggs or cold cuts, and refrigerated entrees that require heating before service.

Potentially Hazardous Foods

Figure 2-17. *Potentially hazardous foods, such as raw chicken, require temperature control in order to keep it safe for consumption.*

Refer to DVD for **Stock Rotation** Media Clip

Special attention needs to be given to potentially hazardous foods. A *potentially hazardous food* is a food that requires temperature control in order to keep it safe for consumption. *See Figure 2-17.* Potentially hazardous foods have a neutral or slightly acidic pH level and contain moisture and protein. Potentially hazardous foods must be maintained at a temperature at or below 41°F. An exception to this is shell eggs, which are allowed to maintain a minimum temperature of 45°F. Other exceptions include milk and shellfish, which may have alternate minimum temperatures specified by state laws.

Storing Food

Once food has been properly inspected, it needs to be stored in a safe and sanitary manner. Stock records should be kept to uphold safety and quality. All food should be properly covered, labeled, and dated. To promote quality and avoid hazards, stock rotation is important when storing deliveries. *First-in, first-out (FIFO)* is the process of dating new items as they are placed into inventory and placing them behind or below older items to ensure that older items are used first. *See Figure 2-18.*

Storing Food

Sullivan University

Figure 2-18. *The FIFO method can be accomplished by storing new items behind or below older items.*

Goods used in foodservice operations commonly fall into three storage categories. These categories include refrigerated storage, frozen storage, and dry storage.

Refrigerated Storage. Refrigerator temperature settings must be cold enough to keep the internal temperature of food at 41°F or below. This often requires setting the refrigeration unit a few degrees lower in order to compensate for the doors being opened frequently. *See Figure 2-19.*

Refrigerated Storage

Kolpak

Figure 2-19. *Refrigerators are set a few degrees below 41° F to compensate for the doors being opened frequently.*

In top-to-bottom refrigerated storage, cooked and ready-to-use items are stored on the top shelf of the refrigerator above raw seafood, meat, and poultry. This is done to prevent the juices of raw food from potentially dripping on cooked and ready-to-use items. Food that requires a higher cooking temperature to destroy harmful bacteria should be stored below food that requires a lower cooking temperature to destroy bacteria.

Frozen Storage. Freezers should be set at −10°F to keep all products frozen. To keep freezer temperatures in an acceptable range, items should be placed below the freezer load line, and the freezer doors should only be opened when necessary. Freezer temperature should be checked regularly. In addition to temperature, it is essential to monitor dates on labels of frozen food to maintain safety and quality.

Dry Storage. Dry storage facilities should be kept dry, clean, and between 50°F and 70°F. All dry storage items must be kept at a minimum of 6 inches above the floor. *See Figure 2-20.* The condition of dry storage items and the dates on labels need to be checked regularly. Items should be discarded if they are damaged. The condition of the dry storage area should also be checked frequently to prevent pest infestations and promote general upkeep.

Dry Storage

Sullivan University

Figure 2-20. *All dry storage items must be kept at a minimum of 6 inches above the floor.*

Preparing Food

While preparing food, it is important to keep food out of the temperature danger zone (41°F to 135°F) as much as possible. Preparing food efficiently and using proper time and temperature controls can limit the amount of time food spends in the temperature danger zone. *See Figure 2-21.* Time and temperature controls are vital during the processes of thawing, cooking, cooling, reheating, and holding food.

Preparing Food _____

Figure 2-21. *Preparing food efficiently and using proper time and temperature controls can limit the amount of time food spends in the temperature danger zone.*

Thawing. Raw food that has been frozen requires special care when thawing. Proper thawing helps prevent contamination and the growth of bacteria. The following are ways to thaw foods properly:

- Food can be thawed in the refrigerator as long as the food maintains a temperature of 41°F or below once it thaws.

- Food can be submerged under clean, running water that is 70°F or below. No portion of the food should reach 41°F or above. An overflow drain must be used to catch any particles from the food, and the work area must be cleaned and sanitized immediately after thawing.

- Food can thaw during the cooking process as long as the minimum internal temperature for the cooked food is met and held.

Cooking. The presence of harmful pathogens can be reduced to a safe level by thoroughly cooking foods. The look, smell, or taste of a food is not always enough to determine if it is thoroughly cooked. A thermometer inserted into the thickest part of the food is the best way to determine its temperature. Food needs to reach a minimum internal cooking temperature and hold that temperature for a specific amount of time. Different foods have different internal temperature and time requirements. *See Figure 2-22.*

Minimum Internal Cooked Temperatures and Times		
Food	**Temperature**	**Time**
Eggs	145°F	15 seconds
Poultry	165°F	15 seconds
Fish and shellfish	145°F	15 seconds

Beef, lamb, and veal:
   Steaks and chops
   Roasts
   Ground
	
	145°F
	145°F
	155°F
		
		15 seconds
		4 minutes
		15 seconds

Pork:
   Chops and cutlets
   Roasts
   Ground
	
	145°F
	145°F
	160°F
		
		15 seconds
		4 minutes
		15 seconds

Stuffed poultry, fish, shellfish, meat, and pasta	165°F	15 seconds
Reheated foods	165°F	15 seconds

Figure 2-22. *Food needs to reach a minimum internal cooking temperature and hold that temperature for a specific amount of time.*

Cooling. When cooling a hot food, it must be cooled to 70°F within 2 hours. Over the next 4 hours, the food must be cooled to a minimum of 41°F. The following methods and procedures are recommended for cooling foods:

- Use a cold paddle to cool foods such as soups, sauces, and stews. *See Figure 2-23.* These types of foods will cool quicker if placed in a flat container rather than left in a stockpot. When the food has cooled, cover the container and place it in the refrigerator.

- Place food in a quick chill unit such as a blast chiller. When the food has cooled, place the covered container in the refrigerator.

- Place food in a shallow stainless steel or aluminum pan. If appropriate, cut the food into smaller pieces before placing it in the pan. Loosely cover the pan and place it on an upper shelf in the refrigerator. When the food has cooled, tightly cover the pan.

- Place hot food in a shallow stainless steel or aluminum pan. Create an ice bath by filling a larger pan or a sink with ice. Place the pan of food in the ice bath. Stir the food every 10–15 minutes, replacing the ice as it melts. When the food has cooled, cover the pan and place it in the refrigerator.

Holding. Whether a food is served hot or cold, its temperature needs to stay out of the temperature danger zone (41°F to 135°F) while it is being held for service. A thermometer should to be used to check the internal temperature every 2–4 hours. *See Figure 2-24.* If the temperature of the food falls within the temperature danger zone, the item needs to be discarded.

San Jamar

Figure 2-23. *A cold paddle can be used to cool foods such as soups.*

Holding Food

Fluke Corporation

Figure 2-24. *Foods need to stay out of the temperature danger zone (41°F to 135°F) while being held for service.*

Reheating. The internal temperature of a cooked and cooled food needs to reach 165°F for 15 seconds within 2 hours before it can be served. Reheated food that does not reach this temperature within the allotted time frame must be discarded.

Serving Food

After food has been prepared safely, it must be served safely. Special consideration needs to be given to food that is prepared for self-service areas and to food that will be transported for off-site service.

Self-Service. Self-service areas, such as salad bars or food bars, need to be properly maintained to reduce the threat of contamination. Barriers, such as sneeze guards, help protect food from contaminants. Monitoring the temperature of foods on the food bar helps prevent foods from entering the temperature danger zone. *See Figure 2-25.* Self-service areas should also be monitored to ensure that guests receive a fresh plate for each return visit, food is replenished, and the area is kept clean and sanitary.

Self-Service Areas

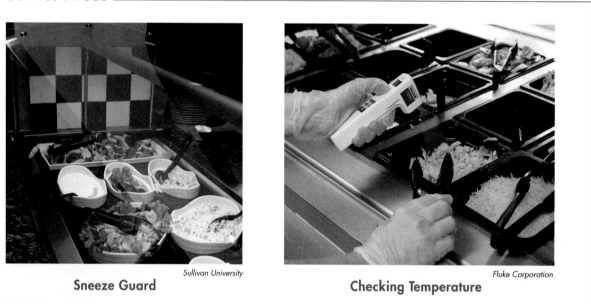

Sullivan University

Sneeze Guard

Fluke Corporation

Checking Temperature

Figure 2-25. *Using sneeze guards and monitoring the temperature of foods on a self-service food bar help reduce the threat of contamination.*

Transporting Food. Foodservice operations often prepare food in one location and transport the food to another location for service. When transporting food, steps must be taken to ensure the food is not contaminated while in transit. The following precautions should be followed when transporting foods:

• A vehicle used for transporting food must be clean and designed to maintain safe food temperatures.

- Transport containers used must be clean, tightly sealed, and designed for efficient cleaning.

- Transport containers must be well-insulated so that cold foods are held at 41°F or below and hot foods are held at 135°F or above. *See Figure 2-26.*

- The shortest route possible to the area where the food will be served must be taken.

- The loading and unloading time must be minimized as much as possible.

- Upon arrival, cold foods are kept cold using ice or a refrigeration unit and hot foods are kept hot using steam tables or chafing dishes.

HAZARD ANALYSIS CRITICAL CONTROL POINTS SYSTEM PRINCIPLES

The *Hazard Analysis Critical Control Points (HACCP) system* is a food safety management system that aims to identify, evaluate, and control contamination hazards throughout the flow of food. In order to implement an effective HACCP system, a prerequisite program should be in place that identifies the basic operational procedures and sanitation conditions within that operation. This program may include vendor certification, staff training, allergen management, buyer specifications, recipe and process instructions, and supply rotation procedures that protect products from physical, chemical, and biological contaminants.

A successful HACCP system includes a HACCP plan. A *HACCP plan* is a written document detailing what policies and procedures will be followed to help ensure the safety of food. Some local authorities require use of a HACCP plan for foodservice operations that use high-risk processes, such as smoking foods, as a method of preservation. The FDA *Food Code* establishes the implementation of a HACCP plan as a voluntary effort for foodservice operations.

Although a HACCP plan is different for each operation, there are seven basic principles that must be followed in sequential order. The seven principles include conducting a hazard analysis, determining critical control points, establishing critical limits, setting monitoring procedures, identifying corrective actions, verifying the system, and maintaining documentation. The first two principles are designed to help identify hazards. The next three principles help control hazards, and the last two principles help evaluate the effectiveness of an operation's HACCP plan.

Hazard Analysis

A *hazard analysis* is the process of assessing potential risks in the flow of food in order to establish what must be addressed in the HACCP plan. All food safety hazards that exist throughout the flow of food are identified. This includes purchasing, receiving, storing, preparing, and serving foods in a way that minimizes physical, chemical, and biological hazards.

Transporting Food

Carlisle FoodService Products

Figure 2-26. *Transport containers must be well-insulated so that cold foods are held at 41°F or below and hot foods are held at 135°F or above.*

A hazard analysis includes process-specific lists identifying all menu items and food products that use a particular process. *See Figure 2-27.* Processes may include preparing food with ready-to-use items or cooking food for immediate service. Some food, such as soup, may be cooked, held, cooled, and reheated before it is served.

Examples of Process-Specific Lists		
Process #1 **Food Preparation without Cooking**	**Process #2** **Food Preparation with Cooking**	**Process #3** **Complex Food Preparation**
Caeser salad	Broiled fish	Soups
Fresh vegetables	Steamed mussels	Large roasts
Oysters on the half shell	Grilled steak	Chili
Tuna salad	Fried chicken	Egg rolls
Caesar salad dressing	Soup du jour	House-made chicken salad
Cole slaw	Grilled vegetables	Beef stew
Sliced sandwich meats	Cooked eggs	Lasagna
Fruit salad	Baked potato	Dim sum
Cheese plate	Risotto	Roasted turkey
Vegetable dip	Pancakes and waffles	Stuffed squab

Figure 2-27. *A hazard analysis includes process-specific lists identifying all menu items and food products that use a particular process.*

Critical Control Points

Carlisle FoodService Products

Figure 2-28. *Some foodservice operations use a CCP logo to denote critical control points.*

Critical Control Points

A *critical control point (CCP)* is the point where a hazard can be prevented, eliminated, or reduced. Some foodservice operations use a CCP logo to denote critical control points. *See Figure 2-28.* After identifying how a food is processed, it must be determined where hazards are likely to occur. To determine CCPs, risk factors are considered. Common risk factors include purchasing food from unsafe sources, inadequately cooking food, holding foods at improper temperatures, using contaminated equipment, and poor personal hygiene. For example, a menu item might feature a grilled-chicken pasta dish that is cooked for immediate service. In order to eliminate or reduce levels of harmful bacteria, cooking the chicken to the proper temperature would be a CCP.

Critical Limits

A *critical limit* is the point in a HACCP plan where a minimum or maximum value is established for a CCP in order to prevent, eliminate, or reduce a hazard to a safe level. A critical limit needs to be established for each CCP. These limits must be met to eliminate that hazard or reduce it to a safe limit. An example of a critical limit would be meeting minimum internal temperature and holding times when cooking food. *See Figure 2-29.*

Hazard Analysis and Critical Control Points (HACCP) Control Measures		
Critical Control Point	**Preventive Measure**	**Critical Limit**
Receiving	Receive foods at proper temperatures	41°F or lower for potentially hazardous foods
	Put perishable foods in cold storage quickly	4 hour maximum time in temperature danger zone
	Obtain supplies from approved sources	Suppliers must be regulated and inspected by proper authorities
Storage	Maintain temperature control	41°F for perishable foods 0°F for frozen foods
	Prevent cross-contamination	Ready-to-eat foods do not contact raw meat, seafood, or eggs
Preparation	Minimize bacterial growth	4 hour maximum time in the temperature danger zone
	Minimize contamination from staff and equipment	Ready-to-eat foods do not contact raw meat, seafood, or eggs Employees wash hands when changing tasks and at least once every 4 hours
Cooking	Cook foods to their proper temperatures	145°F for eggs, fish, meat 165°F for poultry 135°F for fruits and vegetables
Cooling	Use rapid chill refrigeration equipment	Cool to 70°F in 2 hours following cooking
	Stir hot foods in an ice water bath	Cool to 41°F in 6 hours following cooking
Holding	Keep hot foods at temperatures above the temperature danger zone	Hold at temperature of 135°F or above
	Keep cold foods at temperatures below the temperature danger zone	Hold at temperature of 41°F or below

Figure 2-29. *An example of a critical limit would be meeting minimum internal temperatures and holding times when cooking food.*

Monitoring Procedures

A HACCP plan must identify what must be monitored, how those items are monitored, the times and frequency at which those items should be monitored, and the individual responsible for the monitoring. Monitoring procedures should address each CCP. Monitoring procedures might also require determining the individual responsible for checking cooking temperatures and times and how they are checked.

Corrective Actions

A *corrective action* is the point in an HACCP plan that identifies the steps that must be taken when food does not meet a critical limit. These corrective actions must be identified in advance and implemented if critical limits are not met. For example, a corrective action could be disposing of food if the minimum cooking temperature is not met. Corrective actions must be clearly defined and easy to implement. Corrective actions are documented to allow modifications that can result in the elimination of the problem.

Carlisle FoodService Products

System Verification

A HACCP plan needs to be verified to determine if it is preventing, reducing, or eliminating the identified hazards. Verification is conducted by someone other than the person directly responsible for performing the activities specified in the food safety management system. The person conducting the verification checks receiving, cooling, handwashing, and cooking logbooks to see that the required entries were made by each shift.

Documentation

Documents should be kept and maintained in order for a HACCP plan to be the most effective. Logs, graphs, charts, receipts, and notes are examples of these documents. These documents can pertain to suppliers, equipment, CCPs, monitoring activities, and any additional component essential to the operation's HACCP plan.

A relatively new documentation source for foodservice operations comes in the form of radio frequency identification (RFID) technology. *Radio frequency identification (RFID)* is a form of technology that uses electronic tags to store data that can be monitored from remote distances. This means that foods can be tagged and monitored for temperature requirements during transportation as well as during storage. *See Figure 2-30.*

RFID tags promote compliance with the Food Safety Modernization Act (FSMA) of 2011, which focuses on tracing high-risk foods, such as raw fruits and vegetables. RFID tags also can alert foodservice operators if a food has been recalled and can be used to manage inventory from remote locations.

RFID Tags

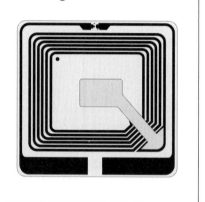

Figure 2-30. *Foods with an RFID tag can be monitored during transportation as well as during storage.*

FACILITY SAFETY

It is important for foodservice staff to be informed of all safety procedures upon employment. Employers are responsible for safety training and ensuring that employees follow Occupational Safety and Health Administration (OSHA) standards. OSHA is responsible for setting and enforcing workplace safety standards in the United States. Foodservice staff members must report safety hazards, accidents, and injuries to a supervisor. Understanding fire safety, chemical safety, and what to do in the event of an accident promotes a safe work environment.

Fire Safety

Foodservice staff should be trained in how to prevent fires and what to do in the event of a fire. OSHA standards require employers to provide a fire protection plan, fire exits, fire extinguishers, a fire-suppression system, and an emergency action plan.

Fire Protection Plans. A fire protection plan typically includes housekeeping procedures. Housekeeping procedures include proper storage of flammable materials and the proper maintenance of heat producing equipment such as ovens, steamers, and fryers.

Fire Exits. Each foodservice operation facility must have at least two properly marked fire exits that are clear of obstructions. To keep the fire exits clear, all items should be stored in designated areas.

Fire Extinguishers. Employees must know where fire extinguishers are located and how to use them properly. Class A, Class C, and Class K fire extinguishers are commonly used in foodservice operations. *See Figure 2-31.* Class A extinguishers are for wood, paper, and trash fires, and Class C extinguishers are used for electrical fires. Class K extinguishers are used for grease fires.

Fire Extinguishers

Class A: Ordinary Combustibles

Class C: Electrical Equipment

Class K: Combustible Oils and Fats

Figure 2-31. *Class A, Class C, and Class K fire extinguishers are commonly used in foodservice operations.*

Fire-Suppression Systems. A *fire-suppression system* is an automatic fire extinguishing system that is activated by the intense heat generated by a fire. A fire suppression system is designed to provide fire protection in areas with open flames, such as flames from ranges, grills, and broilers. Foodservice fire suppression systems extinguish fires by spraying special chemicals from overhead nozzles when the system is activated.

Emergency Action Plans. An *emergency action plan* is a written plan intended to organize employees during an emergency situation. In the event of a fire, an emergency action plan includes procedures for reporting fires, evacuation and escape routes, how to account for all employees after an evacuation, medical duties to be performed, and contact information.

Storing Chemicals

Figure 2-32. *Hazardous materials must be stored in an area that will not contaminate food.*

Minor Cuts

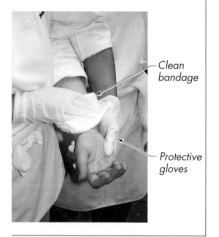

Clean bandage

Protective gloves

Figure 2-33. *When minor cuts occur, they should be treated immediately to prevent infection.*

Chemical Safety

A *hazardous material* is a chemical present in the workplace that is capable of causing harm. Improper handling of chemicals can have harmful effects such as respiratory tract irritation, burns, and eye damage. Examples of hazardous materials include cleaning supplies, sanitizers, and pesticides. Hazardous materials must be stored in an area that will not contaminate food. *See Figure 2-32.* If a chemical is placed in a separate container, it must be labeled with the name and address of the manufacturer, the name of the chemical in the container, and the hazardous warnings for that chemical.

Each hazardous material is required to have a safety data sheet that is accessible for review. A *safety data sheet (SDS)* is a document that provides detailed information describing a chemical, instructions for its safe use, the potential hazards, and appropriate first-aid measures.

Accidents

Accidents that occur in the professional kitchen can lead to injuries such as burns, cuts, strains, sprains, and falls. When an accident occurs, a supervisor should be notified immediately. Following an injury, proper medical attention should be immediately sought and precautions taken to avoid further injury. Precautions also must be taken to protect food and kitchen equipment from contamination due to accidents.

Burns. Burns may be caused by splashed grease, escaping steam, and gas that is ignited incorrectly. Burns are classified as first-degree, second-degree, or third-degree burns. First-degree burns affect only the top layer of skin and appear red and swollen. Second-degree burns occur when the burned area extends beyond the upper layer of skin into the second layer, known as the dermis. Blisters, intense pain, and swelling result from second-degree burns. Third-degree burns involve damage to body tissue well beyond the first and second layers of skin. Emergency medical personnel should be contacted immediately to address third-degree burns.

Cuts. Cuts can easily occur in the professional kitchen. The frequency of cuts can be reduced by using proper knife skills. When minor cuts occur, they should be treated immediately to prevent infection. *See Figure 2-33.* For serious cuts, emergency medical help should be contacted immediately. Before returning to work, cuts on the hands and fingers must be covered with a waterproof covering such as a finger cot or a disposable glove. A *finger cot* is a protective sleeve placed over the finger to prevent contamination of a cut.

Strains and Sprains. Strains and sprains can cause pain to muscles and tissue. These injuries can be caused by lifting heavy or oversized objects, carrying large numbers of objects at once, or repetitive reaching across tables or counters. Strains and sprains can be prevented by using carts to transport objects, using proper lifting techniques, and wearing slip-resistant shoes.

Proper lifting prevents back strain by using the legs to power the lift. *See Figure 2-34.* Before lifting heavy or large objects, it is important to ensure that the path to be used is clear of obstacles.

1. Stand close to the item to gain a safe lifting position. Keep a wide stance with the feet turned out and heels down.
2. Squat by bending at the hips and the knees so that the shoulders and hips nearly form a straight, vertical line.
3. Pull the load close to the body and grasp the object firmly.
4. Rise from the squatting position, using the legs to power the lift. *Note:* Do not bend over at the neck, shoulders, or waist while lifting.
5. To unload, face the chosen location and lower the item slowly using the legs and not the back.
6. Bend at the knees to lower the load at the same time as the body lowers. *Note:* Keep the back comfortably straight.

Figure 2-34. *Proper lifting prevents back strain by using the legs to power the lift.*

Falls. Falls can cause serious accidents in the professional kitchen. Falls may be caused by incorrect ladder use, wet floors, spilled food, grease, torn mats, and damaged floors. Care should be taken when using a ladder to retrieve an item from storage shelves. *See Figure 2-35.* Spills should be mopped up immediately and wet floor signs placed as a warning. Torn mats and damaged floors should be reported to a supervisor immediately. All steps should be indicated with hazard tape.

SUMMARY

Established food safety and sanitation standards must be enforced in foodservice operations in order to provide a quality product. Foodborne illnesses can occur because of improper food handling. The elderly and young children are especially at risk for foodborne illness. Food contamination may result from biological, physical, or chemical hazards. Six factors foster bacterial growth. These factors are food, acidity, temperature, time, oxygen, and moisture.

Sanitation practices include personal hygiene, proper handwashing, good cleaning and sanitizing practices, and pest management. The flow of food must be monitored from purchasing through service in order to keep food out of the temperature danger zone as much as possible. HACCP is a food safety management system that aims to identify, evaluate, and control contamination hazards throughout the flow of food. Foodservice operations are responsible for maintaining a clean and safe environment that promotes fire and chemical safety and prevents accidents and injuries.

Ladder Usage

Figure 2-35. *Care should be taken when using a ladder to retrieve items.*

Refer to DVD for
Quick Quiz® questions

1. Identify seven agencies and organizations responsible for regulating food safety standards.
2. Explain how pathogens cause foodborne illnesses.
3. Explain why some people are more susceptible to foodborne illness than others.
4. Describe three types of biological contaminants that can be found in food.
5. Name the acronym that describes the six factors that promote bacterial growth.
6. Describe how food contributes to bacterial growth.
7. Describe how acidity contributes to bacterial growth.
8. Describe how temperature contributes to bacterial growth.
9. Define the temperature range known as the temperature danger zone.
10. Describe how food time contributes to bacterial growth.
11. Describe how oxygen contributes to bacterial growth.
12. Describe how moisture contributes to bacterial growth.
13. Define food spoilage indicator.
14. Identify three types of physical contaminants that can be found in food.
15. Identify two types of chemical contaminants that can be found in food.
16. List the accepted personal hygiene practices for foodservice staff.
17. Describe the six steps involved in proper handwashing.
18. List ten situations when handwashing should occur.
19. List circumstances in which gloves should be worn.
20. Explain how heat is used to sanitize items in the professional kitchen.
21. Identify three types of chemical sanitizers used in foodservice operations.
22. Explain how warewashing is done in a compartment sink.
23. Explain how warewashing is done in a dish machine.
24. Identify ways to manage pests in a foodservice operation.
25. Describe the steps included in the flow of food.
26. Define potentially hazardous food.
27. Define first-in, first-out (FIFO).
28. Identify three methods of storing food and the temperature settings at which those storage areas must be maintained.
29. Describe three ways to thaw food.
30. Explain why a minimum internal cooking temperature is important to food safety.
31. Describe four ways to cool food.
32. List the seven Hazard Analysis Critical Control Point (HACCP) principles.
33. Explain the purpose of a HACCP plan.
34. Define a critical control point (CCP) and a critical limit.
35. Describe a corrective action.
36. Identify at least five types of documentation that can be used in a HACCP plan.
37. Explain how foodservice operations can use radio frequency identification (RFID).
38. Identify five components of fire safety.

Refer to DVD for
Review Questions

39. Name the three types of fire extinguishers commonly used in foodservice operations.
40. Define hazardous material.
41. Explain the purpose of a safety data sheet (SDS).
42. Identify five injuries that can result from accidents that occur in a foodservice facility.
43. Describe the proper procedure for lifting.
44. List five ways that falls can occur in a foodservice facility.

Applications

1. Using the "Regulation of Food Safety Standards" chart, access the websites of each of the food safety agencies and organizations. Locate at least one piece of new information about each group and share the information with a supervising instructor.

Refer to DVD for
Application Scoresheets

2. Using the "Common Pathogens that Cause Foodborne Illnesses" chart, research three pathogens that can cause foodborne illnesses. Share the findings with a supervising instructor.
3. Create a food experiment demonstrating how each of the six factors in FAT TOM promote bacterial growth.
4. Demonstrate universally accepted personal hygiene practices.
5. Wash hands using the proper handwashing procedure.
6. After washing hands properly, rub a quarter-sized amount of Glo Germ™ gel over both hands like putting on a lotion. Then, place hands under an ultraviolet (UV) lamp to view the bacteria that remain.
7. Dust the surface of a stainless steel table with Glo Germ™ powder. Then, clean the surface of the table until all of the powder disappears.
8. Wash, sanitize, and dry tools used to prepare food. Dust each tool with Glo Germ™ powder. Then, clean and sanitize the tool again before storing it properly.
9. Warewash food preparation and service items in a compartment sink.
10. Warewash food preparation and service items in a dish machine.
11. Use pHydrion™ or Krowne Metal Corp. test strips to verify that a warewashing station is within allowable range as specified by local health codes.
12. Prepare a cold dish and a hot dish and outline the flow of food used to prepare each. Compare the flow charts.
13. Cool a hot food using an ice bath.
14. Monitor the temperature of a hot food that is being held for 3 hours prior to service.
15. Test the temperature of items on a salad bar to determine if they are in the temperature danger zone.
16. Pack a hot food and a cold food for transportation. Then, test the temperature of each food after 30 minutes.
17. Create a HACCP plan for a high-risk menu item. Identify critical control points (CCPs) for each food process. Chart the critical limits for each CCP and identify a potential corrective action as appropriate. Provide example documentation to support the HACCP plan. Ask a supervising instructor to verify the plan.
18. Fill out a safety data sheet (SDS) for a hazardous material used in the professional kitchen.
19. Treat a minor cut.
20. Demonstrate the proper lifting procedure.

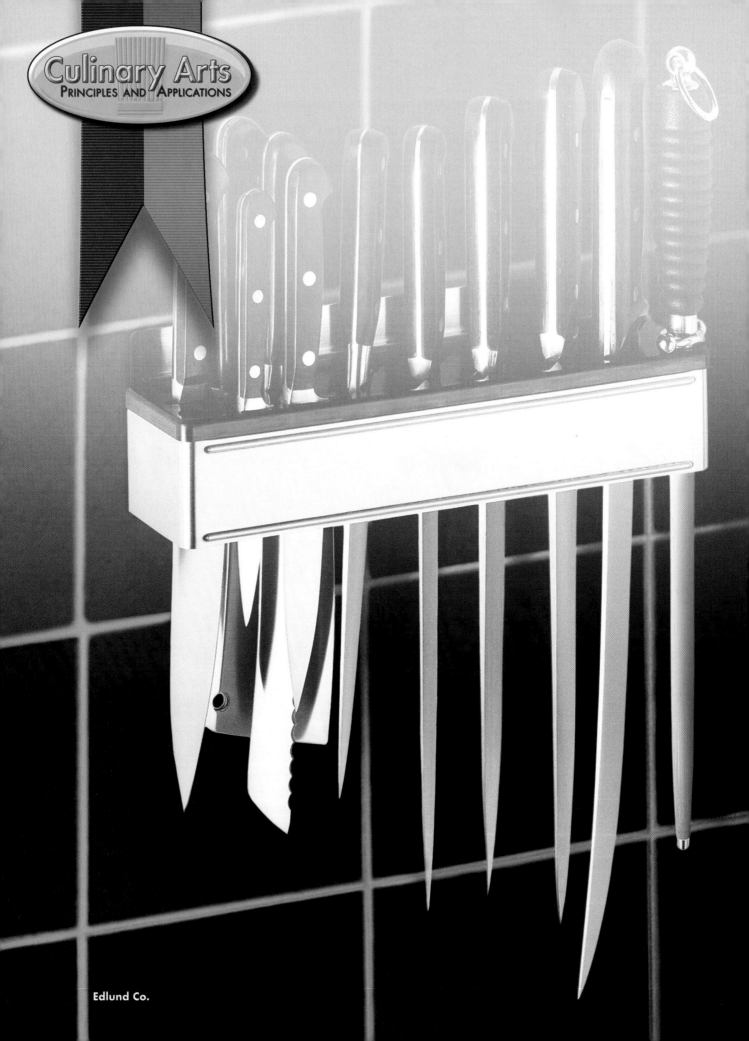

Culinary Arts
PRINCIPLES AND APPLICATIONS

Edlund Co.

Knife Skills

*K*nife skills are among the most important skills a foodservice professional needs to develop. Knives are used to perform a variety of tasks, from dicing vegetables to cutting up poultry. Each type of knife is used to perform specific tasks, and it is important to know which knife to use for each task. Handling knives safely and using the proper cutting technique are extremely important. Uniform knife cuts aid even cooking as well as add visual interest to the plate. Producing basic knife cuts with precision and speed requires a lot of practice.

Refer to DVD for **Flash Cards**

Chapter Objectives

1. Describe the parts of a knife and the function of each.
2. Differentiate among the four types of blade edges.
3. Describe the distinguishing features of large knives.
4. Describe the distinguishing features of small knives.
5. Describe the distinguishing features of specialty cutting tools.
6. Demonstrate the safe handling of knives.
7. Grip and position a chef's knife properly.
8. Use a rocking motion to cut food using a chef's knife.
9. Sharpen a chef's knife.
10. Hone a chef's knife.
11. Demonstrate rondelle, diagonal, oblique, and chiffonade cuts.
12. Demonstrate batonnet, julienne, and fine julienne cuts.
13. Demonstrate large dice, medium dice, small dice, brunoise, fine brunoise, and paysanne cuts.
14. Demonstrate mincing and chopping.
15. Demonstrate fluted cuts and tourné cuts.

Key Terms

- **blade**
- **tang**
- **bolster**
- **whetstone**
- **honing**
- **steel**
- **rondelle cut**
- **diagonal cut**
- **oblique cut**
- **chiffonade cut**
- **batonnet cut**
- **julienne cut**
- **dice cut**
- **paysanne cut**
- **brunoise cut**
- **mincing**
- **fluted cut**
- **tourné cut**

KNIFE CONSTRUCTION

Knives are the most fundamental tool used in the professional kitchen. The use of a sharp knife in skilled hands can accomplish a wide variety of cutting tasks with great efficiency. Well-constructed knives are comfortable and balanced in the hand. Each part of a knife, including the blade, tang, handle, bolster, and rivets, has a specific function. *See Figure 3-1.*

Parts of a Knife

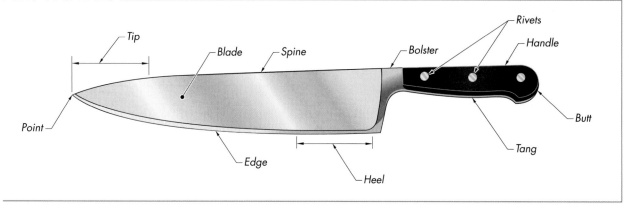

Figure 3-1. *Each part of a knife, including all parts of the blade, as well as the tang, handle, bolster, and rivets, has a specific function.*

Edlund Co.

Blades

A knife blade has five parts: the heel, tip, point, spine, and edge. The *heel* is the rear portion of the blade and is most often used to cut thick items where more force is required. The *tip* is the front quarter of the knife blade. Most cutting is accomplished with the section of the blade between the tip and the heel. The point of the blade is used as a piercing tool. The *spine* is the unsharpened top part of the knife blade that is opposite the edge. The *edge* is the sharpened part of the knife blade that extends from the heel to the tip.

A knife with a sharp edge is safer than a knife with a dull edge because it requires less pressure to use. There are four basic types of blade edges: straight, serrated, granton, and hollow ground. *See Figure 3-2.*

- Straight edge blades are the most common type of knife blade.

- Serrated edge blades have scallop-shaped teeth that easily penetrate tough outer crusts or skins of food products such as breads and fruits.

- Granton edge blades have hollowed out grooves running along both sides that reduce the amount of friction as the edge of the blade cuts the food, allowing maximum contact. Granton edge blades are often used to cut meats and poultry.

- Hollow ground edge blades have been ground just below the midpoint of the blade to form a very thin cutting edge that is easily dulled. Hollow ground edge blades are ideal for skinning fish, peeling fruits, and preparing sushi.

Knife Blade Edges

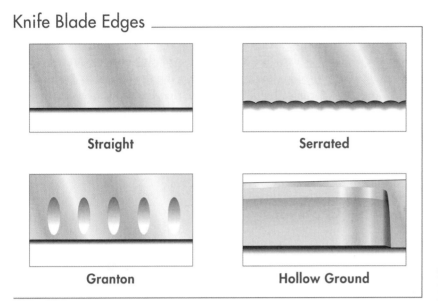

Straight

Serrated

Granton

Hollow Ground

Figure 3-2. *Each of the four basic types of blade edges (straight, serrated, granton, and hollow ground) offers an advantage when cutting specific foods.*

Knife blades are made from a variety of materials. In the past, carbon steel was widely used because the soft metal makes knives easy to sharpen. However, soft metal also makes it hard to keep a sharp edge for long periods of use. Carbon steel knives also discolor over time if they come in contact with highly acidic foods such as tomatoes or lemons. This blade discoloration can cause some foods to oxidize or turn brown when cut and also can leave a metallic taste on these foods because carbon steel reacts with acid. In contrast, knife blades constructed from stainless steel do not discolor or react with acidic foods. The hardness of stainless steel makes the blade more difficult to sharpen, yet it keeps a sharper edge much longer than a carbon steel blade.

Most knives currently used in the professional kitchen are made of high-carbon stainless steel or ceramic material. High-carbon stainless steel combines the best qualities of carbon steel and stainless steel. High-carbon stainless steel produces a blade that is easy to keep sharp, does not change color, and does not transfer a metallic taste to foods. Likewise, ceramic blades provide a sharper edge for a longer period than any other material, do not react with acidic foods, and are very easy to keep clean. Ceramic blades are made from zirconium oxide, a less flexible material than stainless steel. Ceramic blades may chip or break if they strike hard surfaces, such as large bones or are dropped on tile floors. The sharpest of all blades, ceramic knives must be professionally sharpened on diamond wheels. *See Figure 3-3.*

All types of knife blades are either stamped or forged. Stamped blades are thinner, lighter blades cut from a flat sheet of metal and then ground to form a sharp edge. Forged blades are thicker, heavier blades formed from red-hot steel that is hammered into shape and then ground to create a sharp edge. Forged blade knives have a bolster between the heel and the handle and are also better balanced than stamped blade knives. Forged knives are also more expensive.

Ceramic Knives

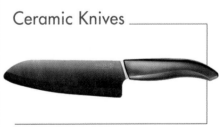

Kyocera Advanced Ceramics

Figure 3-3. *The sharpest of all blades, ceramic knives must be professionally sharpened.*

Tangs

The *tang* is the unsharpened tail of a knife blade that extends into the handle. The highest-quality knives have a tang that extends all the way to the end of the handle. The tang contains holes for securing the handle to the blade. A *partial tang* is a shorter tail of a knife blade that has fewer rivets than a full tang. Partial tang knives are less durable than full-tang knives, but may be acceptable for infrequent or light use. A *rat-tail tang* is a narrow rod of metal that runs the length of the knife handle but is not as wide as the handle. Rat-tail tangs are fully enclosed in the handle and are less durable than full or partial tangs. *See Figure 3-4.*

Knife Tangs

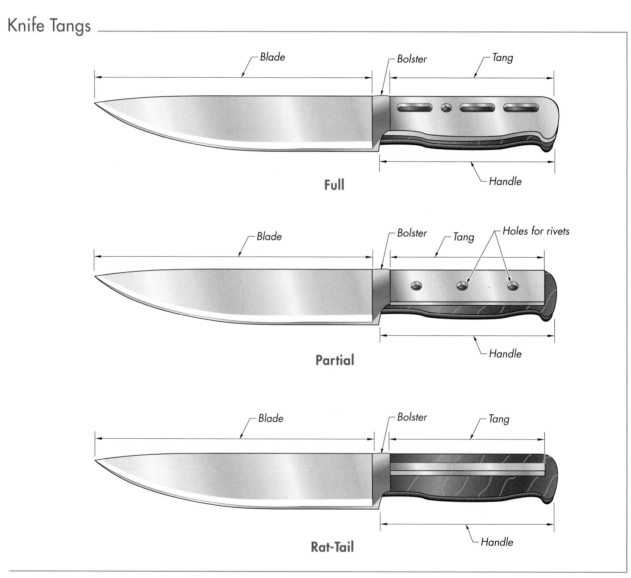

Full

Partial

Rat-Tail

Figure 3-4. *The tang is the unsharpened tail of a knife blade that extends into the handle. Full and partial tangs are more durable than rat-tail tangs.*

Handles

The handle of a knife can be made from wood, stainless steel, or synthetic materials. Wood handles are becoming less common due to their lack of durability and how easily they trap bacteria. Stainless steel handles are virtually maintenance free, however, they become slippery when wet—making them a less than optimal choice. Synthetic handles made from plastic, nylon, styrene, resin, or polypropylene are popular because they are easy to clean, last longer than wood, and are easier to grip when wet than stainless steel. However, synthetic materials crack over time and when they are exposed to extreme temperature changes.

It is important to keep knife handles clean to ensure a good grip. For safety and sanitation reasons, knives should always be washed by hand. They should never be left in standing water or placed in a commercial dishwasher because this can cause the handles to crack or warp. The end of a knife handle is referred to as the butt of the knife.

Dexter-Russell, Inc.

Bolsters and Rivets

A *bolster* is a thick band of metal located where the blade joins the handle. The purpose of the bolster is to provide strength to the blade and prevent food from entering the seam between the blade and the handle. *A rivet* is a metal fastener used to attach the tang of a knife to the handle.

Some knives do not have bolsters and rivets. High-quality knives have a bolster and several rivets that are flush with the surface of the handle.

KNIFE TYPES

A chef uses many different types of knives and special cutting tools in the professional kitchen. Knowing which knife or special cutting tool to use in a given application makes working with knives safer and more efficient. Professional knives can be grouped into large knives and small knives.

Large Knives

Large knives used constantly in the professional kitchen include chef's knives, utility knives, cleavers, santoku knives, butcher's knives, scimitars, boning knives, slicers, and bread knives. *See Figure 3-5.*

Chef's Knives. A *chef's knife,* also known as a French knife, is a large and very versatile knife with a tapering blade used for slicing, dicing, and mincing. The heel of the blade is wide and tapers to a point. The most popular blade lengths are 8, 10, and 12 inches. The weight of a chef's knife should be evenly balanced between the blade and the handle to prevent hand and wrist fatigue.

Utility Knives. A *utility knife* is a multipurpose knife with a stiff 6–10 inch blade that is similar in shape to a chef's knife but much narrower at the heel. The blade edge may be straight or serrated. This knife is a cross between a chef's knife and a paring knife.

Mercer Cutlery

Large Knives

Figure 3-5. *Large knives, such as chef's knives, utility knives, cleavers, santoku knives, butcher's knives, scimitars, boning knives, slicers, and bread knives, are constantly used in the professional kitchen.*

Cleavers. A *cleaver* is a heavy, rectangular-bladed knife that is used to cut through bones and thick meat.

Santoku Knives. A *santoku knife* is a knife with a razor-sharp edge and a heel that is perpendicular to the spine. A santoku knife resembles a small cleaver with a pointed tip. It usually has a granton edge blade 5–8 inches in length, which prevents food from sticking to the blade.

Butcher's Knives. A *butcher's knife* is a heavy knife with a curved tip and a blade that is 7–14 inches in length. The tip of the blade curves upward about 25°. A butcher's knife is used to cut, section, and portion raw meats.

Scimitars. A *scimitar* is a long knife with an upward curved tip that is used to cut steaks and primal cuts of meat. The shape of the scimitar blade resembles a boning knife, yet is much larger.

Boning Knives. A *boning knife* is a thin knife with a pointed 6–8 inch blade used to separate meat from bones with minimal waste. The blade may be either stiff (curved) or flexible (straight). Boning knives with stiff blades are used on larger cuts of meat. Those with flexible blades are used for filleting fish.

Slicers. A *slicer,* also known as a carving knife, is a knife with a narrow blade 10–14 inches long that is used to slice roasted meats. Slicers are available with a straight, serrated, or granton blade edge. The blade may be stiff or flexible. Slicers with stiff blades often have a rounded, blunt tip and are used to slice hot meats such as roasts. Flexible-blade slicers are better suited for cutting cold meats such as ham.

Bread Knives. A *bread knife* is a knife with a serrated blade 8–12 inches long and is used to cut through the crusts of breads without crushing the soft interior. A sawing motion is required to use a bread knife without smashing the bread as it is sliced. Serrated knives are hard to sharpen and chefs may opt to replace bread knives rather than have them sharpened. For this reason, the bread knives used may be of lesser quality than the other knives used in the professional kitchen.

Small Knives

Small knives offer the user the ability to make precise cuts in small areas or to open food items such as shellfish. Small knives commonly used in the professional kitchen include paring knives, tourné knives, clam knives, and oyster knives. *See Figure 3-6.*

Small Knives

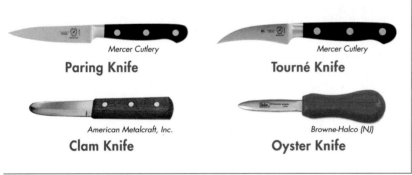

Mercer Cutlery
Paring Knife

Mercer Cutlery
Tourné Knife

American Metalcraft, Inc.
Clam Knife

Browne-Halco (NJ)
Oyster Knife

Figure 3-6. *Small knives commonly used in the professional kitchen include paring knives, tourné knives, clam knives, and oyster knives.*

Paring Knives. A *paring knife* is a short knife with a stiff 2–4 inch blade used to trim and peel fruits and vegetables. A paring knife is often used in conjunction with a chef's knife to remove stems from items.

Tourné Knives. A *tourné knife,* also known as a bird's beak knife, is a short knife with a curved blade that is primarily used to carve vegetables into a specific shape called a tourné, which is a seven-sided football shape with flat ends.

Clam Knives. A *clam knife* is a small knife with a short, flat, round-tipped sharp blade that is used to open clams. The proper use of a clam knife makes the task of opening clams an efficient and safe process.

Oyster Knives. An *oyster knife* is a small knife with a short, dull-edged blade with a tapered point that is used to open oysters. The proper use of an oyster knife makes the task of opening oysters an efficient and safe process.

Special Cutting Tools

In addition to knives, special cutting tools are used to cut food items for specific applications. Although there are many special cutting tools, those commonly used in the professional kitchen include channel knives, zesters, peelers, parisienne scoops, and mandolines. *See Figure 3-7.*

Special Cutting Tools

Dexter-Russell, Inc. **Channel Knife** Browne-Halco (NJ) **Zester** Carlisle Foodservice Products **Peeler** Paderno World Cuisine **Parisienne Scoop** Paderno World Cuisine **Mandoline**

Figure 3-7. *Special cutting tools such as channel knives, zesters, peelers, parisienne scoops, and mandolines are used to cut food items for specific applications.*

Channel Knives. Although not an actual knife, a *channel knife* is a special cutting tool with a thin metal blade within a raised channel that is used to remove a large string from the surface of a food item. A channel knife leaves a decorative pattern on the surface of an item, such as a cucumber.

Zesters. A *zester* is a special cutting tool with tiny blades inside of five or six sharpened holes that are attached to a handle. To use a zester, the cutting holes are drawn across the peel of a citrus fruit such as a lemon to yield small strings or "zest" that can be added to foods as a natural flavoring.

Peelers. A *peeler* is a special cutting tool with a swiveling, double-edged blade that is attached to a handle and is used to remove the skin or peel from fruits and vegetables. The double-edged blade contours to the shape of the fruit or vegetable, such as a carrot or a potato.

Parisienne Scoops. A *parisienne scoop* is a special cutting tool that has a half-ball cup with a blade edge attached to a handle and is used to cut fruits and vegetables into uniform spheres.

Mandolines. A *mandoline* is a special cutting tool with adjustable steel blades used to cut food into consistently thin slices. A mandoline can cut foods paper thin and also produce julienne cuts and waffle cuts. A hand guard needs to be in place when using a mandoline.

KNIFE SAFETY

Because of their sharp edges, knives are dangerous and improper use can lead to injury. Always adhere to the following safety precautions when holding, using, carrying, washing, and storing knives:

- Always grip a knife properly to ensure safety and control. When using a knife, the more pressure that is applied the higher the risk of the knife slipping and of personal injury occurring.

- Always position the guiding hand properly when using, sharpening, and honing a knife.

- Always cut food items on a nonporous cutting board because the nonporous surface greatly reduces the risk of cross-contamination. Color-coded cutting boards may be used for specific types of foods. *See Figure 3-8.*

Edlund Co.

Cutting Boards

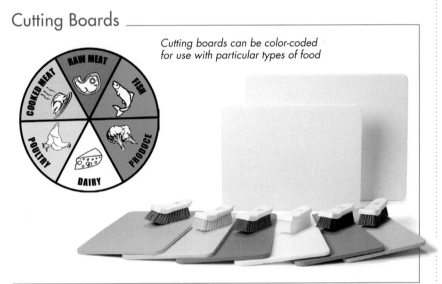

Cutting boards can be color-coded for use with particular types of food

Carlisle FoodService Products

Figure 3-8. *Color-coded, nonporous cutting boards may be used for specific types of foods to reduce the risk of cross-contamination.*

- Always wipe a knife blade with the edge facing away from the hand.

- Always pass a knife to a person by laying it on a table and sliding it forward.

- When walking with a knife, always keep the knife pointing down and hold it along the side of the body.

- Use only clean, sanitized knives on a whetstone or sharpening steel to avoid cross-contamination.

- Always wipe a blade after using a whetstone or sharpening steel to remove any metal residue.

- Always keep knives sharp. Injury is more likely to occur with a dull knife than a sharp one. Hone knives after each use to maintain a smooth, sharp edge.

- Always clean and sanitize knives before storing them.

Refer to DVD for
Knife Safety
Media Clip

- Store knives in sleeves, guards, or knife holders to avoid injury.
- Never leave knives in a sink as someone could reach in and be injured.
- Never wash knives in a commercial dish machine because the heat and chemicals can ruin the handles.
- Never use a knife to pry a lid off of any type of container.
- Never attempt to catch a falling knife.

Knife Grip and Positioning

There are different acceptable methods for gripping a knife, but there is a common method used by culinary professionals that provides control and stability. To begin, the knife is held by the handle while resting the side of the index finger against one side of the blade and placing the thumb on the other side of the blade. The hand not holding the knife is referred to as the guiding hand. The guiding hand is responsible for guiding the item to be cut into the knife. To correctly position the fingers of the guiding hand, imitate the shape of a spider on the table. The fingertips should all be slightly tucked, yet touch the surface of the table. This guiding hand position is used to safely hold the food next to the blade of the knife.

Using the proper knife grip, the tip of the knife is placed on the cutting board. The guiding hand is placed next to the knife blade in the proper position, with fingertips slightly tucked under near the back half of the blade. The side of the blade should rest against the knuckle of the middle finger of the guiding hand. This position reduces the chances of cutting fingers. *See Figure 3-9.*

Knife Grip and Positioning

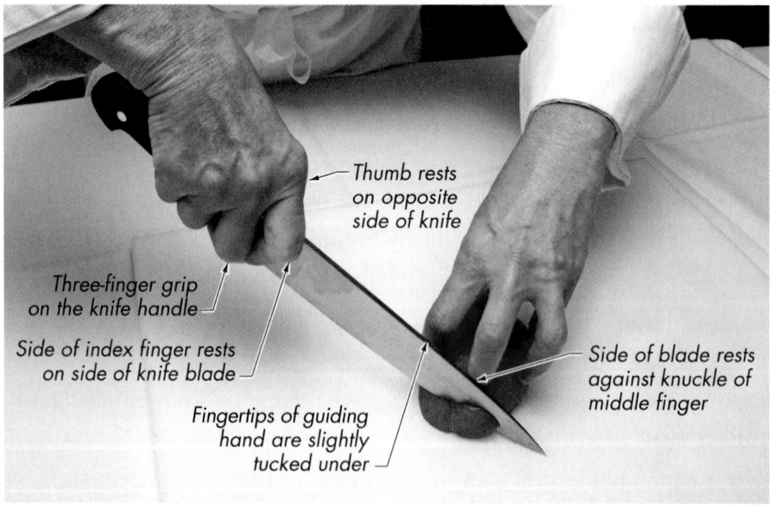

Figure 3-9. *Using the proper knife grip with the knife hand, and with fingertips slightly tucked under with the guiding hand, the side of the blade should rest against the knuckle of the middle finger of the guiding hand.*

With the proper knife grip and hand position, a rocking motion is used to cut with a chef's knife. Using the wrist as a fulcrum, the handle is brought down as the tip of the knife slides forward. Likewise, the handle is raised up as the tip of the knife slides backward. This rocking movement, coupled with the correct position of the guiding hand, creates a controlled motion that can be used to efficiently cut through food. *See Figure 3-10.*

Refer to DVD for
Proper Cutting Techniques
Media Clip

Procedure for Using Proper Cutting Techniques

Note: Be sure to keep the knife in contact with the cutting board at all times.

Note: When cutting using a rocking motion, continually rest the blade of the knife against the knuckle of the middle finger of the guiding hand.

1. Using the proper cutting grip, place the tip of the knife on the cutting board and press down on the knife handle.

2. Continue pressing down on the handle and slide the blade forward, following the curve of the blade.

3. With the heel of the blade on the cutting surface, slide the blade backward and then raise the handle slightly to position the knife for the next slice.

Figure 3-10. *Using the wrist as a fulcrum, a rocking motion is used to cut with a chef's knife. As the handle is brought down the tip of the knife slides forward. Then, the handle is raised up as the tip of the knife slides backward.*

Sharpening Knives

Always check the edge of a knife to make sure it is sharp and properly maintained before using it. A sharp knife is much safer than a dull knife. Having a sharp knife helps to prevent injury because less pressure is required to use a sharp knife as compared to a dull knife.

Although hand-held or electric sharpeners can be used, a whetstone is typically used to sharpen professional knives. A *whetstone* is a stone used to grind the edge of a blade to the proper angle for sharpness. A three-sided whetstone has a coarse-grit, a medium-grit, and a fine-grit side. Two-sided whetstones have a medium-grit side and a fine-grit side.

To sharpen a knife, the blade of the knife is held at a specific angle to the stone. To achieve this angle, the knife blade is held at a 90° angle straight above the whetstone as if it were cutting the stone in half. Then, the knife is tilted halfway toward the stone, at a 45° angle, and then halfway again to find the perfect sharpening angle between 20° and 25°. *See Figure 3-11.*

Sharpening Angle

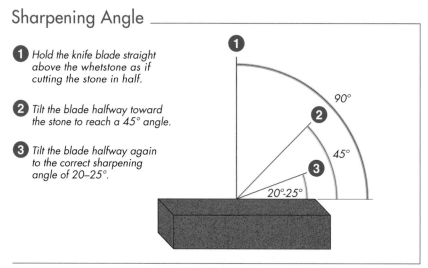

1 Hold the knife blade straight above the whetstone as if cutting the stone in half.

2 Tilt the blade halfway toward the stone to reach a 45° angle.

3 Tilt the blade halfway again to the correct sharpening angle of 20–25°.

Figure 3-11. *To sharpen a knife, the blade is held at a 20–25° angle against the whetstone.*

After the proper angle is achieved, the knife blade is then slowly dragged across the stone from tip to heel while applying light pressure. *See Figure 3-12.*

Sharpening Knives

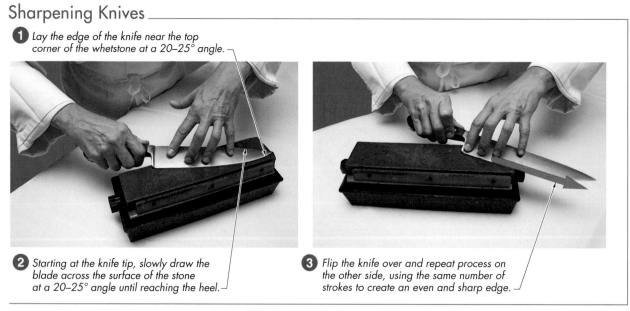

1 Lay the edge of the knife near the top corner of the whetstone at a 20–25° angle.

2 Starting at the knife tip, slowly draw the blade across the surface of the stone at a 20–25° angle until reaching the heel.

3 Flip the knife over and repeat process on the other side, using the same number of strokes to create an even and sharp edge.

Figure 3-12. *To sharpen a knife, the blade is held at a 20–25° angle to the whetstone and then slowly dragged across the stone from tip to heel while applying light pressure.*

Honing Knives

After using a whetstone, it is important to "hone" or align the edge of a knife blade. *Honing,* also known as truing, is the process of aligning a blade's edge and removing any burrs or rough spots on the blade. A steel is used to hone professional knives. A *steel,* also known as a butcher's steel, is a steel rod approximately 18 inches long attached to a handle and is used to align the edge of knife blades.

The 20–25° angle used to sharpen knives is also used to hone knives. To achieve this angle, hold the steel perpendicular or pointed toward the floor with the guiding hand and hold the knife blade at a 20–25° angle in relation to the steel. This can be done by first holding the blade at a 90° angle to the steel, then adjusting it to about half that angle (45°), and finally adjusting it about half of that angle again to a 20–25° angle. *See Figure 3-13.*

Refer to DVD for
Honing Knives
Media Clip

Positioning Knife for Honing

1 *Hold the blade at a 90° angle to the steel.*

2 *Adjust the blade to half that angle (45°).*

3 *Adjust the blade about halfway again to reach the correct 20–25° angle.*

Figure 3-13. *To hone a knife, the blade is held at a 20–25° angle against the steel.*

A steel is usually made of hardened steel, but may also have a ceramic or diamond-impregnated surface. The tip of a steel is magnetic and catches the metal fragments as they are removed from the blade. A steel should always be used to hone the blade of a knife after sharpening as well as between sharpenings to maintain a sharp, smooth edge. *See Figure 3-14.*

1. Place the heel of the knife at the top of the steel while maintaining a 20–25° angle. *Note:* This process can be reversed by starting the knife at the bottom of the steel.

2. With gentle pressure, slide the knife down the steel, moving the blade in an arc along the steel. Finish the stroke with the tip of the knife at the bottom of the steel.

3. Starting at the top of the steel again, place the heel of the knife behind the steel at a 20–25° angle.

4. Slide the knife down the steel, moving the blade in an arc along the steel. Finish the stroke with the tip of the knife at the bottom of the steel.

5. Repeat the process 3–5 times, using the same number of strokes on each side of the blade, until the knife is finely honed.

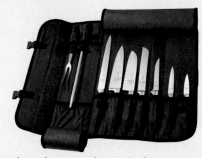

6. With the blade facing away from the body, use a folded towel to wipe metal residue from the knife blade.

7. Wash and sanitize knives before using or storing them.

Figure 3-14. *A steel should always be used to hone a knife after it has been sharpened as well as between sharpenings to maintain a sharp, smooth edge.*

BASIC KNIFE CUTS

Every foodservice worker must know the dimensions of the basic knife cuts and be able to execute them accurately and efficiently. Basic knife cuts are designated with standard measurements that are accepted throughout the foodservice industry. These uniform cuts ensure that items cook evenly and look appealing in the finished product. Common cuts used in the professional kitchen can be grouped into slicing cuts, stick cuts, dice cuts, mincing and chopping, fluted cuts, and tourné cuts.

Slicing Cuts

Slicing involves passing the blade of the knife slowly through an item to make long, thin pieces. In slicing, the knife is pulled backward or slid forward through the item. Slicing cuts include the rondelle, diagonal, oblique, and chiffonade.

Chef's Tip

Appropriately shaped vegetables need to be selected for the desired cut. For example, a poorly shaped bell pepper is not the best choice for julienne cuts.

Rondelle Cuts. A *rondelle cut,* also known as a round cut, is a slicing cut that produces disks. Rondelle cuts are produced from slicing cylindrical vegetables such as cucumbers and carrots straight through. To make a rondelle cut, the cylindrical vegetable is placed perpendicular to blade and then sliced to create disks. ***See Figure 3-15.***

Chef's Tip

Remember to keep part of the knife blade on the cutting board at all times.

Procedure for Making Rondelle Cuts

1. Position the knife blade perpendicular to a washed and peeled cylindrical vegetable.
2. With fingers of the guiding hand tucked, use a rocking motion to slice the item into ¼ inch, ⅛ inch, or ¹⁄₁₆ inch disks.

Finished carrot rondelles can be ¼ inch, ⅛ inch, or ¹⁄₁₆ inch.

Refer to DVD for **Rondelle Cuts** Media Clip

Figure 3-15. *A rondelle cut, also known as a round cut, is a slicing cut that produces disks.*

Diagonal Cuts. A *diagonal cut* is a slicing cut that produces flat-sided, oval slices. Diagonal cuts are made from cylindrical vegetables that are cut on the bias. To make a diagonal cut, place the item at a 45° angle to the knife blade. Then, guide the item toward the blade as each cut is made. ***See Figure 3-16.***

Procedure for Making Diagonal Cuts

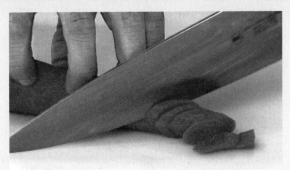

1. Position the knife blade at a 45° angle to a washed and peeled cylindrical vegetable.
2. With fingers of the guiding hand tucked, use a rocking motion to slice the item into ¼ inch, ⅛ inch, or ¹⁄₁₆ inch diagonals.

Finished carrot diagonals can be ¼ inch, ⅛ inch, or ¹⁄₁₆ inch.

Refer to DVD for **Diagonal Cuts** Media Clip

Figure 3-16. *A diagonal cut is a slicing cut that produces flat-sided, oval slices.*

Refer to DVD for
Oblique Cuts
Media Clip

Oblique Cuts. An *oblique cut,* also known as a rolled cut, is a slicing cut that produces wedge-shaped pieces with two angled sides. The sides are neither parallel nor perpendicular. The oblique cut is similar to the diagonal cut in that 45° angle slices are also made on a cylindrical item. However, the oblique cut produces larger pieces that are more wedge-shaped. ***See Figure 3-17.***

Procedure for Making Oblique Cuts

1. With fingers of the guiding hand tucked, guide the cylindrical vegetable toward the blade at a 45° angle and slice. Reserve this first cut for later use.

2. Roll the vegetable 180° and slice again, keeping the blade at a 45° angle. *Note:* This slice produces the first oblique cut.

3. Roll the item back 180° to the original position and slice again.

4. Continue rolling the item 180° between cuts to make wedge-shaped oblique cuts.

Figure 3-17. *An oblique cut, also known as a rolled cut, is a slicing cut that produces small, rounded pieces with two angled sides.*

Chiffonade Cuts. A *chiffonade cut* is a slicing cut that produces thin shreds of leafy greens or herbs. Chiffonade-cut items can be used as ingredients or as a base under displayed foods. To make a chiffonade cut, first wash the leafy items, stack the leaves on top of one another, and roll the stack lengthwise like a cigar. Place the cigar-shaped roll on the cutting board perpendicular to the knife blade. Use a rocking motion to thinly slice the roll as it is fed with the guiding hand into the knife blade. The result is finely shredded leaves or herbs. ***See Figure 3-18.***

1. Place washed, dry leaves in a neat stack.

2. Roll the stack into a tight cylinder and place the cylinder perpendicular to the knife blade.

3. With the fingers of the guiding hand tucked, use a rocking motion to thinly slice the leaves.

Finished basil chiffonades are very finely cut.

Figure 3-18. *A chiffonade cut is a slicing cut that produces thinly sliced leafy greens or herbs.*

Refer to DVD for
Chiffonade Cuts
Media Clip

Stick Cuts

Stick cuts are used for a wide variety of food preparations in the professional kitchen. Many fruits and vegetables are cut into sticklike shapes to create a uniform appearance and to ensure even cooking. The terms "batonnet," "julienne," and "fine julienne" refer to three different stick cuts. All stick cuts begin by squaring off the item to be cut. There are industry-accepted dimensions for each stick cut. However, the length of stick cuts may vary depending on the desired result for a specific dish.

Refer to DVD for
Batonnet Cuts
Media Clip

Batonnet Cuts. A *batonnet cut* is a stick cut that produces a stick-shaped item ¼ × ¼ × 2 inches long. ***See Figure 3-19.***

Procedure for Making Batonnet Cuts

1. Cut washed and peeled vegetables, such as carrots, into pieces 2 inches in length.

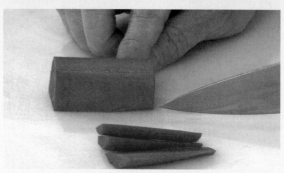

2. Carefully square off three sides, leaving the fourth side rounded. Save scraps for later use.

3. With the rounded side farthest away from the knife, cut even, ¼ inch thick planks. Save scraps.

4. Stack a few ¼ inch planks.

5. Carefully slice again into ¼ inch sticks.

Finished batonnets measure ¼ × ¼ × 2 inches.

Figure 3-19. *A batonnet cut is a stick cut that produces a stick-shaped item ¼ × ¼ × 2 inches long.*

Julienne Cuts. A *julienne cut* is a stick cut that produces a stick-shaped item ⅛ × ⅛ × 2 inches long. ***See Figure 3-20.***

Fine Julienne Cuts. A *fine julienne cut* is a stick cut that produces a stick-shaped item 1/16 × 1/16 × 2 inches long. ***See Figure 3-21.***

Procedure for Making Julienne Cuts

1. Cut washed and peeled vegetables, such as carrots, into pieces 2 inches in length.
2. Carefully square off three sides, leaving the fourth side rounded. Save scraps for later use.

3. With the rounded side farthest away from the blade, cut an even ⅛ inch thick plank.
4. Continue cutting ⅛ inch thick planks toward the rounded side of the piece. Save scraps.
5. Stack a few of the ⅛ inch thick planks and then carefully slice again into ⅛ inch sticks. The resulting sticks should measure ⅛ × ⅛ × 2 inches.

Finished juliennes measure ⅛ × ⅛ × 2 inches.

Figure 3-20. *A julienne cut is a stick cut that produces a stick-shaped item ⅛ × ⅛ × 2 inches long.*

Procedure for Making Fine Julienne Cuts

1. Cut washed and peeled vegetables, such as carrots, into pieces 2 inches in length.
2. Carefully square off three sides, leaving the fourth side rounded. Save scraps for later use.

3. With the rounded side farthest away from the blade, cut an even 1/16 inch thick plank.
4. Continue cutting 1/16 inch thick planks toward the rounded side of the piece. Save scraps.
5. Stack a few of the 1/16 inch thick planks and then carefully slice again into 1/16 inch sticks. The resulting sticks should measure 1/16 × 1/16 × 2 inches.

Finished fine juliennes measure 1/16 × 1/16 × 2 inches.

Figure 3-21. *A fine julienne cut is a stick cut that produces a stick-shaped item 1/16 × 1/16 × 2 inches long.*

Refer to DVD for
Julienne Cuts and **Fine Julienne Cuts**
Media Clips

Refer to DVD for
Dice Cuts
Media Clip

Dice Cuts

Dice cuts are precise cubes cut from uniform stick cuts. Common dice cuts include large dice, medium dice, small dice, brunoise, and fine brunoise. To produce a dice cut, a stick cut of the appropriate dimension is cut into cubes with six equal sides.

Tips for Using Stick Cuts to Make Dice Cuts

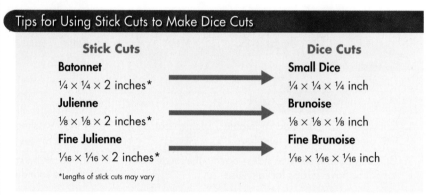

Stick Cuts	Dice Cuts
Batonnet	**Small Dice**
¼ × ¼ × 2 inches*	¼ × ¼ × ¼ inch
Julienne	**Brunoise**
⅛ × ⅛ × 2 inches*	⅛ × ⅛ × ⅛ inch
Fine Julienne	**Fine Brunoise**
¹⁄₁₆ × ¹⁄₁₆ × 2 inches*	¹⁄₁₆ × ¹⁄₁₆ × ¹⁄₁₆ inch

*Lengths of stick cuts may vary

Large, Medium, and Small Dice. A large dice is ¾ × ¾ × ¾ inch cubes cut from ¾ × ¾ × 2 inch sticks. A medium dice is ½ × ½ × ½ inch cubes cut from ½ × ½ × 2 inch sticks. A small dice is ¼ × ¼ × ¼ inch cubes cut from ¼ × ¼ × 2 inch sticks, or batonnets. *See Figure 3-22.*

Procedure for Making Dice Cuts

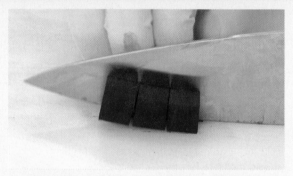

1. Choose a stick cut of the appropriate dimensions, such as ¾ × ¾ × 2 inches to create a large dice.

2. Align several sticks into a uniform bundle against the side of the knife blade.

3. To make a large dice, cut through the bundle to produce ¾ inch cubes with six equal sides.

Large Dice   **Medium Dice**   **Small Dice**

Finished dice cuts can be large, medium, or small dice.

Figure 3-22. *Large, medium, and small dice cuts are precise cubes cut from uniform stick cuts.*

Some items, such as onions, consist of many layers, preventing it from being diced in the same manner as solid items such as carrots. For this reason, a modified procedure is used to dice onions to any desired size. *See Figure 3-23.*

Refer to DVD for
Dicing Onions
Media Clip

Procedure for Dicing Onions

1. Using a chef's knife, cut off the stem end and lightly trim the root end of an onion. *Note:* Do not cut the root end off completely as it holds the layers of the onion in place, preventing it from falling apart.

2. Cut the onion in half from the stem end to root end.

3. Make a thin slice from root end to stem end through the outer peel only.
4. Use the tip of the paring knife to pull off the top layer of the peel.

5. Position onion half on the cutting board with the flat side down. Use the chef's knife to make two or three horizontal cuts through the onion, leaving the root end intact.

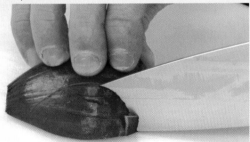

6. Make vertical slices through the onion from stem end to root end, again leaving the root end intact. *Note:* The closer together the slices, the smaller the finished dice.

7. Turn the onion a quarter turn and make cuts the thickness of the desired dice, slicing all the way through from stem end to root end. Repeat the dicing process on the other half of the onion.

Figure 3-23. *Onions consist of many layers, requiring a modified dicing procedure.*

Refer to DVD for
Brunoise Cuts
Media Clip

Brunoise Cuts. A *brunoise cut* is a dice cut that produces a cube-shaped item with six equal sides measuring ⅛ inch each. A brunoise is cut from a julienne stick. ***See Figure 3-24.***

Fine Brunoise Cuts. A *fine brunoise cut* is a dice cut that produces a cube-shaped item with six equal sides measuring ¹⁄₁₆ inch. A fine brunoise is cut from a fine julienne stick. ***See Figure 3-25.***

Procedure for Making Brunoise Cuts

1. Place a small bundle of julienne sticks perpendicular to the knife blade.

2. With a slicing motion, cut through the bundle at equally spaced intervals to produce six-sided cubes that are ⅛ × ⅛ × ⅛ inch.

Figure 3-24. *A brunoise cut is a dice cut that uses a julienne stick to produce a cube-shaped item with six equal sides measuring ⅛ inch each.*

Procedure for Making Fine Brunoise Cuts

1. Place a small bundle of fine julienne sticks perpendicular to the knife blade.

2. With a slicing motion, cut through the bundle at equally spaced intervals to produce six-sided cubes that are ¹⁄₁₆ × ¹⁄₁₆ × ¹⁄₁₆ inch.

Figure 3-25. *A fine brunoise cut is a dice cut that uses a fine julienne stick to produce a cube-shaped item with six equal sides measuring ¹⁄₁₆ inch each.*

Refer to DVD for
Fine Brunoise Cuts
Media Clip

Paysanne Cuts

Figure 3-26. *A paysanne cut is a dice cut ½ × ½ × ⅛ inch thick.*

Paysanne Cuts. A *paysanne cut* is a dice cut that produces a flat square, round, or triangular cut ½ × ½ × ⅛ inch thick. Paysanne cuts are generally used to give prepared dishes a rustic appeal. Cutting a tile-shaped paysanne is done in the same manner as a medium dice except that each batonnet stick is cut into ⅛ inch slices. ***See Figure 3-26.***

Mincing and Chopping

Mincing and chopping have fewer applications in the professional kitchen than the other knife cuts. *Mincing* is finely chopping an item to yield a very small cut, yet not entirely uniform, product. Shallots, garlic, and fresh herbs are commonly minced. *See Figure 3-27.*

Refer to DVD for
Mincing
Media Clip

Procedure for Mincing

1. Using a paring knife, cut off the stem end and lightly trim the root end of a vegetable, such as a shallot. *Note:* Do not cut the root end off completely as it holds the layers of the shallot in place, preventing it from falling apart.

2. Make a thin slice from root end to stem end through the outer peel only.
3. Use the tip of the paring knife to pull off the top layer of the peel.

4. Cut the shallot in half lengthwise from stem end to root end.
5. Lay half of the shallot on the cutting board with the flat side down. Use the tip of the knife to make two or three horizontal cuts through the shallot, leaving the root end intact.

6. Make vertical slices through the shallot from stem end to root end, again leaving the root end intact. *Note:* The closer together the slices, the smaller the finished dice.

7. Using a chef's knife, turn the shallot and make cuts all the way through from stem end to root end until only the root remains.

8. Place the guiding hand, opened and flat, on the top of the blade to help pivot the knife back and forth.

Figure 3-27. *Mincing is finely chopping an item to yield a very small cut, yet not entirely uniform, product.*

Chopping is rough-cutting an item so that there are relatively small pieces throughout, although there is no uniformity in shape or size. Parsley, hard-cooked eggs, and a rough-cut mix of vegetables called mirepoix are often chopped because a uniform shape is not required. *See Figure 3-28.*

Procedure for Chopping

1. Gather washed greens, such as parsley, into a tight bundle with the guiding hand.

2. Draw the knife across the greens in a rocking motion, shaving off thin strips.

3. With the blade pointing away, gently remove the greens from the knife blade.

4. Gather the greens into a pile.

5. Place the guiding hand, opened and flat, on the top of the knife blade to help pivot the knife back and forth while chopping. Gather the greens into a pile again and repeat the chopping process until the shavings are very fine. *Note:* Keep the blade in constant contact with the cutting board while chopping.

6. Place the finely chopped greens in a clean towel or double-layered cheesecloth and ring the cloth to remove excess water from the parsley. *Note:* This step does not apply to chopping hard-cooked eggs or mirepoix.

Finished chopped parsley is dry and airy.

Figure 3-28. *Parsley, hard-cooked eggs, and mirepoix are often chopped because a uniform shape or size is not required.*

Fluted Cuts

A *fluted cut* is a specialty cut that leaves a spiral pattern on the surface of an item by removing only a sliver with each cut. Button mushrooms are often fluted. Making fluted cuts requires good hand-eye coordination as each sliver is cut away and then the mushroom is turned slightly clockwise or counter clockwise before making the next cut from the same central point at the top of the mushroom. *See Figure 3-29.*

Refer to DVD for
Flutes
Media Clip

Procedure for Cutting Flutes

1. Wipe mushroom with a damp towel. *Note:* Always start with a cold button mushroom.
2. Hold the mushroom with the index finger and thumb.
3. While holding a paring knife, place the thumb on the top of the blade, the index finger on the bottom of the blade, and the middle finger underneath the knife to hold it steady.

4. Hold the mushroom level and place the front third of knife blade on the center point of the top of the mushroom facing 12 o'clock.

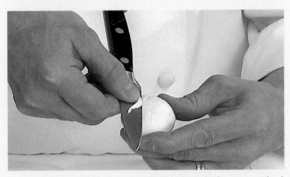

5. Turn the knife to 7 o'clock (right-handed) or 5 o'clock (left-handed), keeping the knife in contact with the mushroom at all times. *Note:* A thin strip of mushroom will be removed with each cut, leaving a fluted edge.

6. Rotate the mushroom clockwise (right-handed) or counter clockwise (left-handed) and continue cutting flutes until they intersect at the top center of the mushroom.

Figure 3-29. *A fluted cut is a specialty cut that leaves a spiral pattern on the surface of an item by removing only a sliver with each cut.*

Tourné Cuts

Tourné is a French word from the verb "to turn." A *tourné cut* is a carved, football-shaped cut with seven sides and flat ends. It is a difficult specialty cut to master without lots of practice. *See Figure 3-30.*

1. Cut a washed and peeled root vegetable, such as a beet or a carrot, into 2 inch long pieces. *Note:* Wider vegetables, such as potatoes, may need to be cut into 4–6 sections before being cut into 2 inch long pieces.

2. Cut each 2 inch piece into 1 inch widths.

3. Holding the item in the guiding hand, place the index finger on one end and the thumb on the opposite end.

4. Using a tourné knife, slowly carve the item from one end to the other in a smooth, continuous stroke that creates a slightly rounded surface.

5. Turn the vegetable slightly and continue to carve the item in smooth, curved strokes until a seven-sided football shape is achieved. *Note:* As each slice is carved, the ends will become narrower than the middle.

A finished tourné has seven sides and is flat on each end.

Figure 3-30. *A tourné cut is a carved, football-shaped cut with seven sides and flat ends.*

Refer to DVD for
Tournés
Media Clip

SUMMARY

It is important to understand the construction and purpose of each knife used in the professional kitchen. Safe knife handling and proper sanitation are as important as having the ability to produce accurate and consistent knife cuts. Using the correct grip, hand position, and rocking motion ensures safe and efficient use of the most important tool in the professional kitchen. Knives must be kept sharp and stored safely so they are ready when needed. A sharp knife is safer than a dull knife. Batonnet, julienne, brunoise, large dice, medium dice, small dice, paysanne, fine julienne, and fine brunoise cuts require specific dimensions for consistency. Fluted cuts and tourné cuts require a lot of hand-eye coordination. Mastering knife skills requires practice and shows pride in one's work.

Refer to DVD for
Quick Quiz® questions

1. Identify the main parts of a knife and explain the function of each.
2. Explain why a knife with a sharp edge is safer to use than a knife with a dull edge.
3. Explain the advantage of using a knife with a granton edge blade.
4. Contrast high-carbon stainless steel blades and ceramic blades.
5. Differentiate between a partial tang and a rat-tail tang.
6. Describe eight types of large knives used in the professional kitchen.
7. Describe four types of small knives used in the professional kitchen.
8. Describe four special cutting tools used in the professional kitchen.
9. Explain why nonporous cutting boards are used in the professional kitchen.
10. Describe how to safely carry a knife when walking.
11. Describe how to safely pass a knife to another person.
12. Describe how knives should be stored.
13. Provide four examples of what should never be done with a knife.
14. Describe the commonly accepted way of gripping and positioning a knife.
15. Identify the term used to describe the hand that is not used to hold the knife.
16. Identify the angle at which a knife blade is held against a whetstone when being sharpened.
17. Explain the process of honing a knife.
18. Explain why uniform knife cuts are used in the professional kitchen.
19. Identify the four slicing cuts.
20. Describe the difference between a diagonal cut and an oblique cut.
21. Identify the three stick cuts and each of their dimensions.
22. Identify the dimensions of large dice, medium dice, and small dice cuts.
23. Identify the dimensions of a paysanne dice cut.
24. Identify the dimensions of brunoise and fine brunoise dice cuts.
25. Describe how to make the cut typically used on shallots, garlic, and fresh herbs.
26. Describe how to flute mushrooms.
27. Describe how to tourné vegetables.

Refer to DVD for
Review Questions

Applications

1. Sharpen and hone a chef's knife.
2. Slice a root vegetable into rondelles, diagonals, and obliques.
3. Chiffonade a bunch of fresh greens.
4. Cut a root vegetable into batonnet, julienne, and fine julienne stick cuts.
5. Cut a whole fruit into large dice, medium dice, and small dice.
6. Dice an onion.
7. Cut julienne sticks into brunoise cubes and fine julienne sticks into fine brunoise cubes.
8. Mince an onion or a shallot.
9. Chop a fresh herb such as parsley or cilantro.
10. Flute a button mushroom.
11. Tourné a potato, a turnip, a beet, or a carrot.

Refer to DVD for
Application Scoresheets

Culinary Arts
PRINCIPLES AND APPLICATIONS

Carlisle FoodService Products

Tools and Equipment

A professional kitchen is organized, well-equipped, and laid out for efficient food preparation. Each area of the kitchen contains the appropriate tools and equipment needed to perform assigned tasks. Specific work areas and work stations enable efficient execution of assigned tasks. It is important to be able to identify and use the hand tools and pieces of equipment used to safely receive, store, sanitize, and prepare foods and beverages in the professional kitchen.

Refer to DVD for **Flash Cards**

Chapter Objectives

1. Identify specialized cutting and shaping tools.
2. Describe volume measuring tools.
3. Describe strainers, sieves, and skimmers.
4. Describe mixing and blending tools.
5. Describe turning and lifting tools.
6. Describe cookware and ovenware.
7. Explain the meaning of NSF-certified tools and equipment.
8. List the safety guidelines for operating and maintaining equipment.
9. Identify the major areas of the professional kitchen.
10. Describe safety equipment used in the professional kitchen.
11. Describe receiving equipment used in the professional kitchen.
12. Describe storage equipment used in the professional kitchen.
13. Describe sanitation equipment used in the professional kitchen.
14. Identify common work sections and stations in the professional kitchen.
15. Describe preparation equipment used in the professional kitchen.
16. Describe cooking equipment used in the professional kitchen.
17. Describe baking equipment used in the professional kitchen.

Key Terms

- **hand tool**
- **ventilation system**
- **fire-suppression system**
- **receiving area**
- **work section**
- **work station**
- **storage area**
- **sanitation area**
- **preparation area**

PROFESSIONAL HAND TOOLS

Foodservice professionals use hundreds of different hand tools in the professional kitchen. A *hand tool* is any of a variety of manual tools used to cut, shape, measure, strain, sift, mix, blend, turn, or lift food items. In addition to knives, common hand tools used in the professional kitchen include specialized cutting and shaping tools; volume measuring tools; strainers, sieves, and skimmers; mixing and blending tools; and turning and holding tools.

Specialized Cutting and Shaping Tools

In addition to the knives and specialty tools described in Chapter 3, mandolines, grinders, graters, and kitchen shears are used to cut, grind, or grate foods. *See Figure 4-1.* Meat mallets, rings, and molds are used to shape foods.

Specialized Cutting Tools

Mandoline — *Browne-Halco (NJ)*

Grinder — *Bunn-O-Matic Corporation*

Rasp Grater — *Browne-Halco (NJ)*

Box Grater — *Browne-Halco (NJ)*

Rotary Graters — *American Metalcraft, Inc.*

Kitchen Shears — *Mercer Cutlery*

Figure 4-1. Mandolines, grinders, graters, and kitchen shears can be used to cut foods.

Mandolines. A mandoline is a manual slicing tool with adjustable steel blades used to cut food in consistently thin slices. The mandoline can slice items nearly paper thin and can also produce julienne and waffle cuts. A mandoline must always be used with a hand guard in place or while wearing a cut-resistant glove.

Grinders. Coffee grinders are used to grind coffee beans. They are also used to grind spices in the professional kitchen.

Graters. Graters come in a wide variety of styles and sizes. A *rasp grater* is a nearly flat, razor-sharp, handheld grater that shaves food into fine or very fine pieces. A *box grater* is a stainless steel box with grids of various sizes on each side that are used to cut food into small pieces. A *rotary grater* is a sharp, perforated stainless steel cylinder attached to a handle that is used to shave hard cheeses, such as Parmesan, tableside.

Kitchen Shears. *Kitchen shears* are heavy-gauge scissors used to trim foods during the preparation process. They can be used for various tasks such as cutting butcher's twine, trimming artichoke leaves, and cutting through soft bones.

Meat Mallets. A *meat mallet* is a hand tool used to tenderize meats prior to cooking. Pounding meat with a meat mallet also changes the shape of the meat.

Rings and Molds. Rings and molds are used to give foods specific shapes during cooking or plating. For example, metal rings and molds are often used to shape foods, such as eggs, as they cook. Rings and molds can also be used to shape rice, polenta, or other items as they are being plated. Other types of molds are used to form items such as petit fours, aspic, and brioche. *See Figure 4-2.*

Shaping Tools

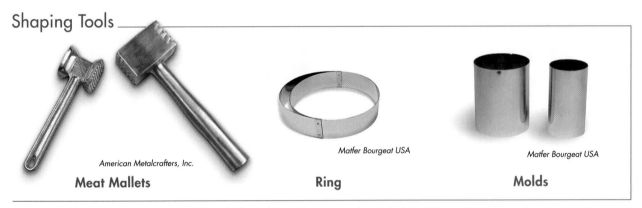

American Metalcrafters, Inc.

Meat Mallets

Matfer Bourgeat USA

Ring

Matfer Bourgeat USA

Molds

Figure 4-2. *Meat mallets, rings, and molds can be used to shape foods before, during, or after cooking.*

Volume Measuring Tools

In the professional kitchen, volume is measured with various tools sized to contain a specific volume or to divide an ingredient into smaller volumes. Volume measuring tools include measuring spoons, three types of measuring cups, portion control scoops, ladles, and spoodles. Although not a volume measuring tool, funnels are used to transfer substances from one container to another, which is useful for measuring.

Measuring Spoons. A *measuring spoon* is a stainless steel spoon used to measure a small volume of an ingredient. Sets of measuring spoons often include a ¼ tsp, ½ tsp, 1 tsp, and 1 tbsp. *See Figure 4-3.* Sets may also be stamped with the metric equivalents of these units. The top edge of a measuring spoon is the actual measurement so it must be filled to the brim and then leveled off for the contents to equal the full measure.

Measuring Spoons

Carlisle FoodService Products

Figure 4-3. *Measuring spoons are used to measure dry and liquid ingredients in increments from ¼ tsp to 1 tbsp.*

Measuring Cups. The types of measuring cups used in the professional kitchen include dry, liquid, and volume measures. *See Figure 4-4.*

- A *dry measuring cup* is a metal cup with a straight handle that is used to measure dry ingredients. A common set of dry measuring cups often consists of ¼ cup, ⅓ cup, ½ cup, ¾ cup, and 1 cup measures. A dry measuring cup does not have a pour lip, as the top edge of the cup is the actual measurement. The cup must be filled to the brim and then leveled off for the contents to equal the full measure.

- A *liquid measuring cup* is a transparent cup with a pouring lip and a loop handle and is used to measure liquid ingredients. Liquid measuring cups are available in many sizes. The cups are often graduated in ounce increments as well as in milliliters.

- A *volume measure* is a large, graduated aluminum container with a pouring lip and a loop handle and is used to measure larger volumes of ingredients. Volume measures are available in 1 pt, 1 qt, ½ gal., and 1 gal. capacities as well as metric equivalents.

Measuring Cups

The Vollrath Company, LLC

Dry Measuring Cups

Carlisle FoodService Products

Liquid Measuring Cups

The Vollrath Company, LLC

Volume Measures

Figure 4-4. *Different sizes of dry, liquid, and volume measures are used in the professional kitchen.*

Portion Control Scoops. A *portion control scoop,* also known as a disher, is a stainless steel scoop of a specific size attached to a handle with a thumb-operated release lever. Portion control scoops are used to serve food in equal amounts. Scoops are sized by numbers typically ranging from 6 to 40. The number on the scoop indicates the number of level scoopfuls that equal 1 qt. As the number of the scoop increases, scoop capacity decreases. *See Figure 4-5.* Each scoop size has an approximate capacity in ounces as well as an equivalent volume in cups or tablespoons. Scoops are often used to portion batters, mashed potatoes, bound salads, and ice cream.

Ladles. A *ladle* is a stainless steel, cuplike bowl attached to a long handle that is often used to serve soups, sauces, and salad dressings. Ladles range in size from ½–32 fl oz. The capacity is usually stamped on the handle in ounces or milliliters for easy reference. *See Figure 4-6.*

Portion Control Scoops

Scoop Capacity		
Handle Color	Scoop No.	Volume
White	6	4¾ fl oz
Gray	8	3¾ fl oz
Ivory	10	3¼ fl oz
Green	12	2¾ fl oz
Blue	16	2 fl oz
Yellow	20	1¾ fl oz
Red	24	1½ fl oz
Black	30	1 fl oz
Purple	40	¾ fl oz

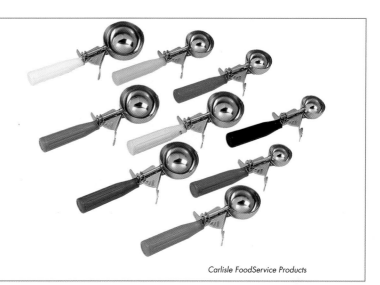

Carlisle FoodService Products

Figure 4-5. *Scoops are sized by numbers from 6–40, indicating the number of level scoopfuls equal to 1 qt. As the number of the scoop increases, the scoop capacity decreases.*

Ladles

Ladle Sizes	
Ladle Marking	Equivalent Volume
½ fl oz	1 tbsp
1 fl oz	⅛ cup
2 fl oz	¼ cup
3 fl oz	⅜ cup
4 fl oz	½ cup
6 fl oz	¾ cup
8 fl oz	1 cup
12 fl oz	1½ cup
24 fl oz	3 cups
32 fl oz	4 cups

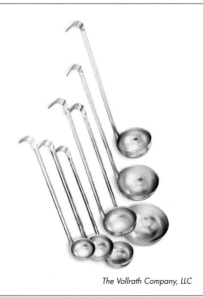

The Vollrath Company, LLC

Figure 4-6. *Ladles range in size from ½–32 fl oz, and the capacity is usually stamped on the handle in ounces or milliliters for easy reference.*

Spoodles. A *spoodle* is a solid or perforated flat-bottomed ladle. *See Figure 4-7.* Spoodles range in size from 1–8 oz. Spoodles are often color-coded.

Funnels. A *funnel* is a tapered bowl attached to a short tube that is used to transfer substances from one container to another container without spilling. *See Figure 4-8.* For example, a funnel can be used to transfer 1 gal. of olive oil in to 4 quart-sized containers.

Spoodles

Carlisle FoodService Products

Figure 4-7. *A spoodle is a solid or perforated flat-bottomed ladle.*

Funnels

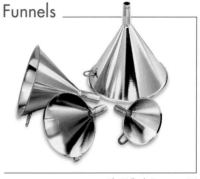

The Vollrath Company, LLC

Figure 4-8. *Funnels are used to transfer substances from one container to another container.*

Paderno World Cuisine

Handheld Strainer

Carlisle FoodService Products

Colander

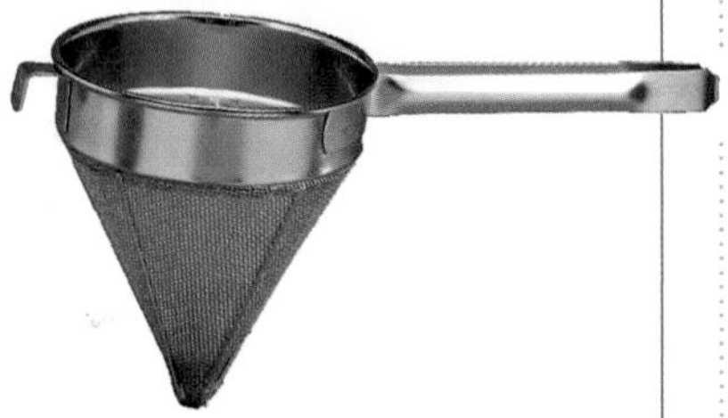

American Metalcraft, Inc.

China Cap

Lincoln Foodservice Products, Inc.

Chinois

Figure 4-9. *Strainers include colanders, china caps, and chinois.*

Strainers, Sieves, and Skimmers

Strainers, sieves, and skimmers are hand tools that are used to separate items during the preparation or cooking process. Strainers are typically used to separate solids from liquids. Sieves are used to remove lumps from dry ingredients and to purée soft foods. Skimmers are used to remove floating items from liquids.

Strainers. A *strainer* is a bowl-shaped woven mesh screen, often with a handle, that is used to strain or drain foods. For example, a strainer may be used to hold grapes that are being rinsed under a running faucet. Types of strainers include colanders, china caps, and chinois. ***See Figure 4-9.***

- A *colander* is a bowl-shaped perforated strainer. Colanders are commonly used for draining and rinsing foods such as cooked pasta, rice, legumes, fruits, and vegetables.

- A *china cap* is a perforated cone-shaped metal strainer used to strain gravies, soups, stocks, sauces, and other liquids. A china cap has a clip on the outside rim so that it can be clipped onto a pot or bain-marie while straining. A china cap may also be used to purée soft foods by forcing food through the strainer with a pestle. Cheesecloth may be used to line a china cap in order to use it in place of a chinois.

- A *chinois* is a china cap that strains liquids through a very fine-mesh screen. A chinois is often called a bouillon strainer because the mesh screen makes it ideal for straining smooth sauces and consommés. Care should be taken with a chinois, as the fine-mesh screen is very fragile.

Sieves. A *sieve* is a fine-mesh sifter used to sift, aerate, and remove lumps or impurities from dry ingredients. There are several types of sieves, including a tamis, sifter, food mill, and ricer. ***See Figure 4-10.***

- A *tamis,* also known as a drum sieve, is a flat, round sieve with a wood or aluminum frame and a mesh screen bottom. A tamis is used to sift lumps from dry ingredients. It may also be used to purée soft foods by using a spatula or bowl scraper to force the food through the wire mesh bottom.

- A *sifter* is a cylindrical metal sieve that is hand-cranked and used to aerate and remove lumps from dry ingredients such as flour. This hand tool has a woven screen bottom.

- A *food mill* is a hand-cranked sieve with a bowl-shaped body that is used to purée soft or cooked foods. Food is placed in the body of the food mill. A hand-operated crank controls a blade that presses food against the top of a perforated disk. Turning the crank rotates the blade, which in turn forces food through the holes in the disk. Various disks produce coarse to very fine purées, depending on the size of the holes on the disk.

- A *ricer* is a sieve with an attached plunger that is used to purée food by pushing it through a perforated metal plate. For example, a soft food, such as cooked potatoes, is placed in the well of a ricer between the plunger and the perforated metal plate. The handles of the ricer are then squeezed together, forcing the food through the perforated plate.

Sieves

Tamis
Browne-Halco (NJ)

Sifter
Browne-Halco (NJ)

Food Mill
All-Clad Metalcrafters

Ricer
Browne-Halco (NJ)

Figure 4-10. *A tamis, a sifter, a food mill, and a ricer are each a type of sieve.*

Skimmers. A *skimmer* is a flat, stainless steel perforated disk connected to a long handle and is used to skim impurities from soups, stocks, and sauces. A *spider* is a skimmer with an open-wire design that makes it perfect for removing hot foods from a fryer. *See Figure 4-11.*

Skimmers

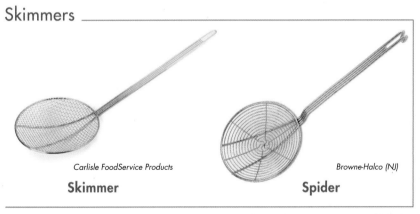

Skimmer
Carlisle FoodService Products

Spider
Browne-Halco (NJ)

Figure 4-11. *A skimmer and a spider are both used to remove items from hot liquids.*

Mixing and Blending Tools

Ingredients most often need to be mixed or blended. Tools used for mixing and blending include mixing bowls, whisks, kitchen spoons, and mixing paddles. *See Figure 4-12.*

Mixing Bowls. A *mixing bowl* is a stainless steel or aluminum bowl used for mixing ingredients. Mixing bowls are available in various sizes from ¾–20 qt.

Whisks. A *whisk* is a mixing tool made of stainless steel or silicone wires bent into loops and attached to a stainless steel handle. Common wire whisks include the balloon whisk and the rigid wire whisk. A balloon whisk has very flexible wires, allowing the user to whip a great amount of air into items such as egg whites. A rigid wire whisk (French whisk) has heavier gauge wire loops and is longer than a balloon whisk. It is often used to stir thick substances such as heavy batters.

Kitchen Spoons. A *kitchen spoon* is a large stainless steel or silicone spoon that is used to stir or serve foods. Kitchen spoons may be solid, perforated, or slotted. Slotted and perforated spoons drain liquid from foods as they are lifted from a container.

Mixing Paddles. A *mixing paddle* is a long-handled paddle used to stir foods in deep pots or steam kettles. Mixing paddles are typically made of stainless steel or polyurethane. The long handles enable them to reach to the bottom of deep pots or kettles.

Mixing and Blending Tools

American Metalcraft, Inc.

Mixing Bowls

Carlisle FoodService Products

Whisks

The Vollrath Company, LLC

Kitchen Spoons

Carlisle FoodService Products

Mixing Paddles

Figure 4-12. *Mixing bowls, whisks, kitchen spoons, and mixing paddles are mixing and blending tools.*

Baking and Pastry Tools

Special hand tools are required when baking and making pastries. Common baking and pastry tools include bench brushes, dough cutters, dough dockers, markers, palette knives, pastry bags and tips, pastry brushes, pastry wheels, rolling pins, and silicone mats. ***See Figure 4-13.***

Baking and Pastry Tools

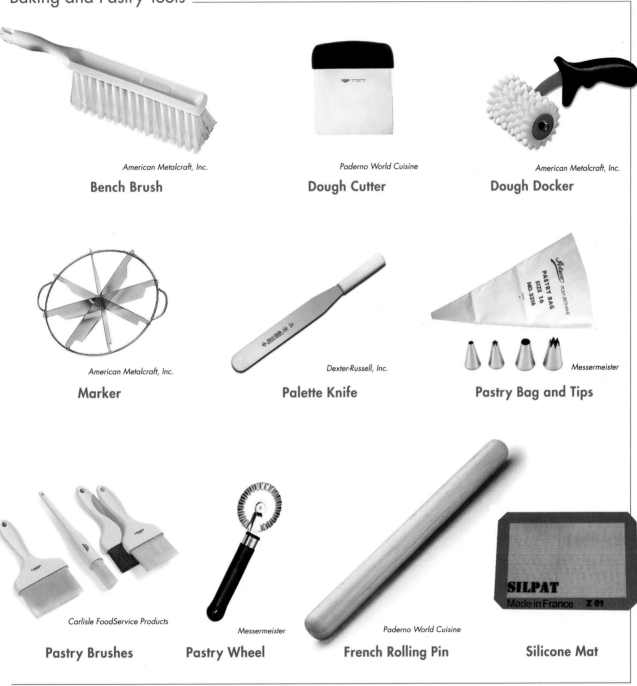

American Metalcraft, Inc.

Bench Brush

Paderno World Cuisine

Dough Cutter

American Metalcraft, Inc.

Dough Docker

American Metalcraft, Inc.

Marker

Dexter-Russell, Inc.

Palette Knife

Messermeister

Pastry Bag and Tips

Carlisle FoodService Products

Pastry Brushes

Messermeister

Pastry Wheel

Paderno World Cuisine

French Rolling Pin

Silicone Mat

Figure 4-13. *Common baking and pastry tools include bench brushes, dough cutters, dough dockers, markers, palette knives, pastry bags and tips, pastry brushes, pastry wheels, rolling pins, and silicone mats.*

Bench Brushes. A *bench brush* is a brush with long bristles set in vulcanized rubber attached to a handle. Bench brushes are used to brush excess flour from the baker's bench.

Dough Cutters. A *dough cutter*, also known as a bench knife, is a flat, stainless steel blade attached to a sturdy handle. Dough cutters are used to cut dough into portions and to scrape dough off the surface of the baker's bench.

Dough Dockers. A *dough docker* is a roller with pins that is used to perforate dough so that it will bake evenly without blistering in the oven heat.

Markers. A *marker* is a round tool that has wire guides that leave marks indicating where to cut pies, round cakes, or pizzas into equal portions. Markers are available in various diameters and portion sizes.

Palette Knives. A *palette knife*, also known as a cake spatula, is a flat, narrow knife with a rounded, 3½–12 inch blade that varies in flexibility. Palette knives are most often used to ice cakes.

Pastry Bags and Tips. A *pastry bag* is a cone-shaped paper, canvas, or plastic bag that is fitted with a pastry tip. Pastry bags are used to pipe icings or soft foods such as whipped potatoes. A *pastry tip* is a cone-shaped tip that is fitted into the narrow end of a pastry bag. Pastry tips are used to create decorative shapes and patterns with icings and soft foods.

Pastry Brushes. A *pastry brush* is a small, narrow brush that is used to apply liquids, such as egg wash or butter, onto baked products. Pastry brushes are available in natural, nylon, or silicone bristles. Nylon brushes cannot be used with hot items. Silicone bristles can withstand temperatures up to 650°F.

Pastry Wheels. A *pastry wheel* is a dough-cutting tool with a rotating disk attached to a handle. Pastry wheels are used to cut dough into desired shapes.

Rolling Pins. A *rolling pin* is a slim cylinder that is used to flatten pastry dough, bread crumbs, or other foods. Rolling pins are available in wood, marble, ceramic, and metal and may have a handle on each end. French rolling pins have tapered ends instead of handles.

Silicone Mats. A *silicone mat* is a woven, nonstick mat that may be used in the refrigerator, freezer, or oven and can withstand temperatures between −40°F and 580°F.

Scraping Tools

Scraping tools are used frequently in the professional kitchen to scrape batter and food from containers, mixing bowls, pots, and pans. Scraping tools are commonly made from flexible rubber, plastic, or thin metal blades. Common scraping tools include flat spatulas and bowl scrapers. *See Figure 4-14.*

Spatulas. A *spatula* is a scraping tool consisting of a rubber or silicone blade attached to a long handle that is used to mix foods and to scrape food from bowls, pots, and pans. Rubber spatulas are not used with hot foods because they can melt. Silicone spatulas, also known as high-temperature spatulas, can withstand temperatures up to 650°F.

Scraping Tools

Spatula

Bowl Scraper

American Metalcraft, Inc.

Figure 4-14. *Scraping tools are used to remove batter and food from containers, mixing bowls, pots, and pans.*

Bowl Scrapers. A *bowl scraper* is a curved, flexible scraping tool that is used to scrape batter or dough out of curved containers. Its flexible structure allows it to curve with the shape of the container being scraped.

Turning and Lifting Tools

When preparing food in the professional kitchen, hand tools are often needed to turn or lift food items during preparation or plating. Turning and lifting tools include tongs, kitchen forks, offset spatulas, and peels. *See Figure 4-15.*

Tongs. *Tongs* are a spring-type, long metal tool used to pick up foods while retaining their shape. Tongs come in various lengths and some can be locked in place.

Kitchen Forks. A *kitchen fork,* also known as a chef's fork, is a large fork with two long prongs and is used to hold meats steady while they are being carved.

Offset Spatulas. An *offset spatula,* also known as an offset turner, is a tool with a wide metal blade that bends up and back toward a handle. It is used to turn foods such as pancakes or hamburgers over to cook on the other side. There are solid and slotted offset spatulas. Slotted offset spatulas, also known as fish spatulas, allow fat to drain off foods before serving and are thin and flexible. They are well-suited for turning delicate items such as fish.

Peels. A *peel* is a long, flat, narrow piece of wood or metal shaped like a wide, thin paddle that is used to lift items and place them into and remove them from ovens. Peels are often used when cooking pizzas.

Turning and Lifting Tools

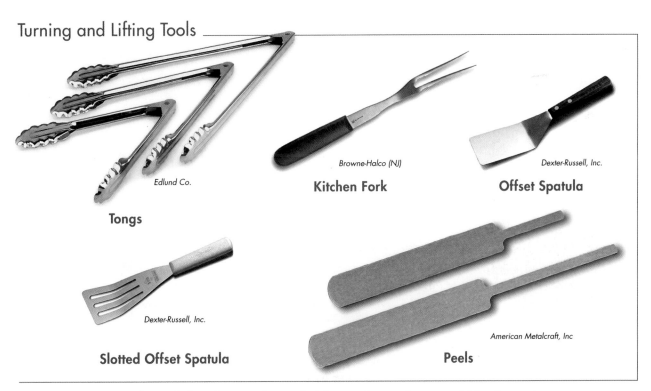

Edlund Co.

Tongs

Browne-Halco (NJ)

Kitchen Fork

Dexter-Russell, Inc.

Offset Spatula

Dexter-Russell, Inc.

Slotted Offset Spatula

American Metalcraft, Inc

Peels

Figure 4-15. *Tongs, kitchen forks, offset spatulas, and peels are often used to turn or lift food items during preparation or plating.*

PROFESSIONAL COOKWARE

Professional cookware is constructed for heavy use and intense heat and is sized to accommodate different quantities of food. Cookware used in the professional kitchen includes a variety of skillets, pots, steamware, and ovenware.

Pans

Many types of pans are used in the professional kitchen. Some pans can be used in the oven. *See Figure 4-16.* Common pans include sautè pans, saucepans, crêpe pans, cast-iron skillets, and woks.

Sauté Pans. A *sauté pan,* also known as a skillet, is a round, shallow-walled pan with a long handle that is used to sauté foods. Common types of sauté pans include sautoirs and sauteuses. A *sautoir* is a sauté pan with straight sides. A *sauteuse* is a sauté pan with sloped sides. The sloped walls of the sauteuse enable foods to be flipped in the pan without using an offset spatula. A small sauteuse is often used to cook omelets.

Saucepans. A *saucepan* is a small, slightly shallow skillet with straight or slightly sloped sides. Saucepans are used to cook small amounts of food in a liquid. Saucepans are commonly used for preparing small amounts of a sauce and for shallow poaching. The shallow depth and wide surface area of a saucepan help reduce the risk of scorching liquids and also make it easy to retrieve poached items without breaking or damaging them.

The Vollrath Company, LLC

Pans

Carlisle FoodService Products

Sautoir

Sauté Pans

Carlisle FoodService Products

Sauteuses

Carlisle FoodService Products

Saucepans

The Vollrath Company, LLC

Crêpe Pan

Lodge Manufacturing

Cast-Iron Skillet

Paderno World Cuisine

Wok

Figure 4-16. *Many types of pans are used in the professional kitchen, including sauté pans, saucepans, crêpe pans, cast-iron skillets, and woks.*

Crêpe Pans. A *crêpe pan* is a small skillet with very short, sloped sides that is used to prepare crêpes. Crêpe pans are usually made from rolled (blue) steel, which is thinner than that used in other commercial skillets. This type of steel heats very quickly, enabling crêpes to cook without sticking to the pan.

Cast-Iron Skillets. A *cast-iron skillet* is a shallow-walled pan made of cast iron that can be used for pan-broiling, pan-frying, or baking. Cast-iron skillets can withstand extreme heat and are available in diameters ranging from 6–15 inches.

Woks. A *wok* is a round-bottom pan that is used to stir-fry, steam, braise, stew, or deep fry foods. Woks can be made of rolled (blue) steel, stainless steel, or aluminum and come in various sizes and weights. They may have one or two handles and are usually accompanied by a ring-shaped stand that allows the wok to sit on the cooktop. Woks require less oil than other pans.

Pots

Pots are cookware used to cook foods such as stocks, sauces, and various meats and fish. Common pots used in the professional kitchen include stockpots, saucepots, rondeaus, and fish poachers. *See Figure 4-17.*

Pots _____

Carlisle FoodService Products
Stockpots

Paderno World Cuisine
Spigot Stockpot

Carlisle FoodService Products
Saucepots

Carlisle FoodService Products
Rondeaus

Paderno World Cuisine
Fish Poacher

Figure 4-17. *Common pots used in the professional kitchen include stockpots, saucepots, rondeaus, and fish poachers.*

Stockpots. A *stockpot* is a large, round, high-walled pot that is taller than it is wide. It has loop-style handles on each side for easy lifting. A stockpot is used for simmering items such as soups and stocks. The tall, narrow shape of the stockpot helps reduce evaporation by leaving a smaller surface area exposed. Some stockpots are fitted with a spigot-style drain at the base to drain off liquids. Stockpots are made of aluminum or stainless steel and range in size from 6–100 qt.

Saucepots. A *saucepot* is a small stockpot. Like the stockpot, it also has loop handles for easy lifting and is used to prepare soups and sauces when smaller a quantity is needed. A saucepot is often used for thicker items such as cream soups or chili. Its shallower depth makes it easier to stir all the way to the bottom, reducing the risk of scorching.

Rondeaus. A *rondeau,* also known as a braiser, is a wide, shallow-walled, round pot that is used for braising, stewing, and searing meats. It has a heavy metal base, allowing for longer cooking times. Rondeaus have side loop handles similar to a stockpot.

Fish Poachers. A *fish poacher* is a thin, oblong pot with loop handles that is used to poach fish. A raised mesh screen is placed in the bottom of the pan to lift poached fish from the pan without breaking it.

Warming and Steaming Cookware

Warming and steaming cookware is cookware that is used to keep foods warm or to steam foods. Common cookware used to warm or steam foods include bain-marie, inserts, double boilers, and steamer inserts. *See Figure 4-18.*

Warming and Steaming Cookware

Top pot is a tiered insert for holding food items

Top pot of a steamer holds food items

Perforations allow steam to surround food items

Bottom pot holds hot water

Bottom pot holds hot water

Bain-Marie Inserts

Double Boiler

Steamer Inserts

Carlisle FoodService Products

Figure 4-18. *Common cookware used to warm or steam foods include bain-marie inserts, double boilers, and steamer inserts.*

Bain-Marie Inserts. A *bain-marie insert* is a round stainless steel food storage container with high walls used for holding sauces or soups in a water bath or steam table. The term bain-marie refers to the container used to hold or store the food as well as to the hot water bath used to keep food warm. Bain-marie inserts have many applications in the professional kitchen such as holding soup in a soup-warming unit and holding hot fudge in a sundae bar.

Double Boilers. A *double boiler* is a round, stainless steel pot that sits inside another slightly larger pot. Water is added to the bottom pot and heated to either cook or heat the food that is placed in the top pot. A double boiler is often used to heat food without scorching or drying it out. It will only heat foods to the temperature of the water and steam beneath the upper vessel, no hotter than 212°F. A double boiler is commonly used to melt chocolate or to make a hollandaise sauce.

Steamers Inserts. A *steamer insert* is a round stainless steel vessel with a perforated liner. A tiered steamer is similar to a double boiler except that the steamer insert is perforated. The perforations allow steam from the simmering or boiling water below to rise into the insert and cook the food inside.

All-Clad Metalcrafters

Ovenware

Special cookware known as ovenware can withstand extremely high oven temperatures. Common types of ovenware used in the professional kitchen include sheet pans, roasting pans, cake pans, loaf pans, muffin pans, pie pans, springform pans, tart pans, and hotel pans. *See Figure 4-19.*

Sheet Pans. A *sheet pan* is a flat pan with very low sides. Sheet pans come in either full (17¾ × 25 × ¾ inches) or half pan (17¾ × 12 × ¾ inches) sizes. They are used for cooking meats such as bacon, sausage links, or chicken pieces in an oven. Sheet pans are also used for baking cookies, sheet cakes, and rolls.

Roasting Pans. A *roasting pan* is a rectangular pan with 4–5 inch sides. A roasting pan is similar to a sheet pan in length and width. Strapped roasting pans that come with reinforced straps are used to roast larger pieces of poultry, fish, or meats. Roasting pans may or may not be covered.

Cake Pans. A *cake pan* is a round, square, or specially shaped pan with short or tall sides and is used to bake cakes. Several cake pans may be required to make multiple layers of a cake.

Loaf Pans. A *loaf pan* is a deep rectangular pan that is used to bake loaves of bread.

Muffin Pans. A *muffin pan* is a rectangular pan with cuplike wells and is used to bake teacakes, muffins, or cupcakes. The diameter of the cuplike wells varies from miniature to jumbo in size.

Pie Pans. A *pie pan* is a round, shallow pan with sloped sides and is used for baking pies.

Springform Pans. A *springform pan* is a round pan with a metal clamp on the side that allows the bottom of the pan to be separated from the sides. Springform pans are typically used to bake cheesecakes.

Tart Pans. A *tart pan* is a round, shallow baking pan with sloped sides that are smooth or fluted and may have a removable bottom.

Ovenware

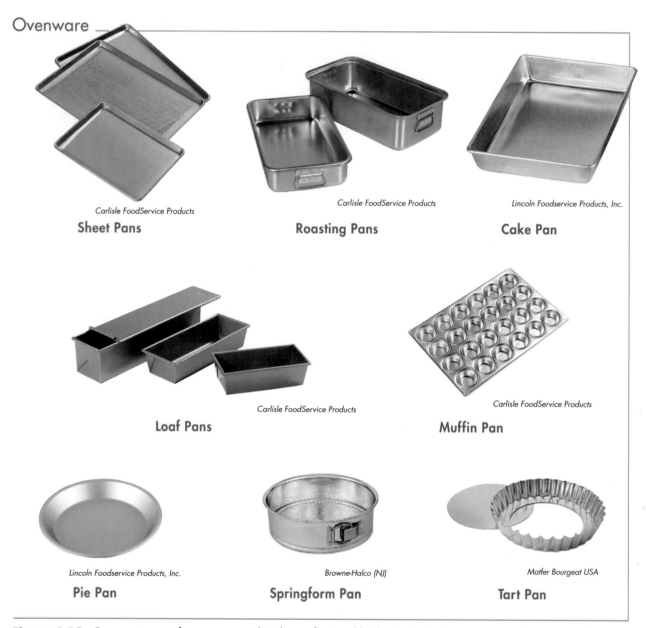

Carlisle FoodService Products
Sheet Pans

Carlisle FoodService Products
Roasting Pans

Lincoln Foodservice Products, Inc.
Cake Pan

Carlisle FoodService Products
Loaf Pans

Carlisle FoodService Products
Muffin Pan

Lincoln Foodservice Products, Inc.
Pie Pan

Browne-Halco (NJ)
Springform Pan

Matfer Bourgeat USA
Tart Pan

Figure 4-19. *Common types of ovenware used in the professional kitchen include sheet pans, roasting pans, cake pans, loaf pans, muffin pans, pie pans, springform pans, tart pans, and hotel pans.*

Hotel Pans. A *hotel pan* is a stainless steel pan that is used to cook, hold, or serve food. Hotel pans have a lip around the outer edge to hold them above hot water in a steam table or a chafing dish. Hotel pans come in a variety of shapes and sizes such as half pans, quarter pans, and third pans.

See Figure 4-20. Hotel pans are available in solid and perforated forms and can be stored in refrigerators and freezers. It is important to note that hotel pans should only be used to cook foods in an oven or a steamer. They should never be placed directly on an open-burner.

Hotel Pans

Hotel Pan Capacity		
Pan Size	**Depth***	**Capacity†**
Full	2½	8
	4	13
	6	20
⅔	2½	5½
	4	6½
	6	10
½	2½	3½
	4	5½
	6	8
½ long	2½	3½
	4	5½
	6	8
⅓	2½	2½
	4	4
	6	6
¼	2½	2
	4	3
	6	4½
⅙	2½	1
	4	2
	6	2½
⅑	2½	⅝
	4	1⅛

* in inches
† in quarts

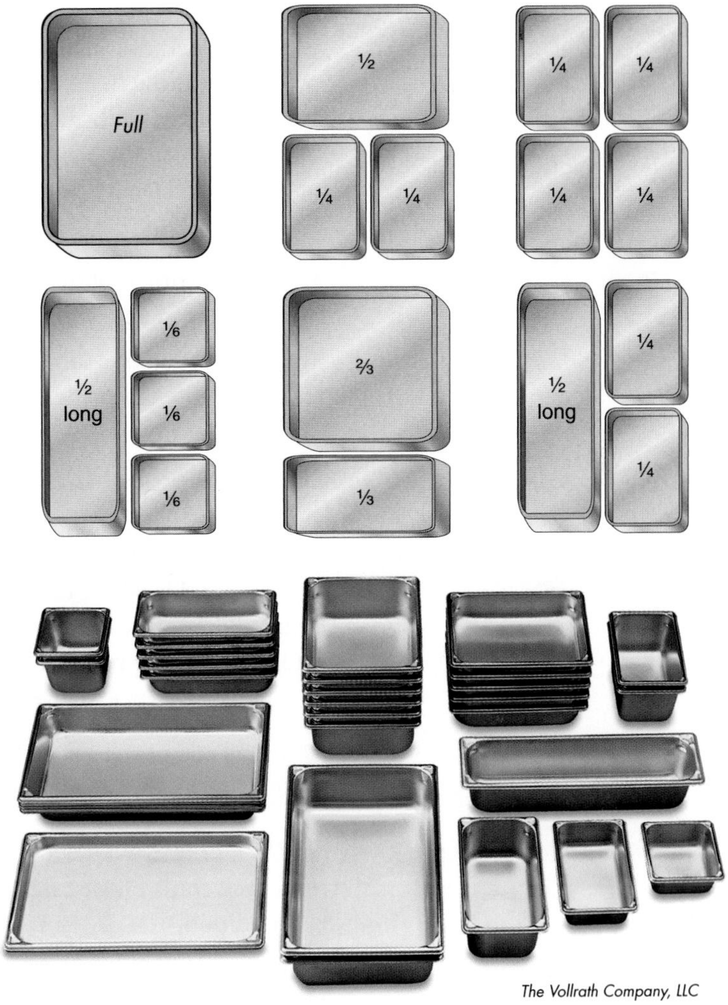

The Vollrath Company, LLC

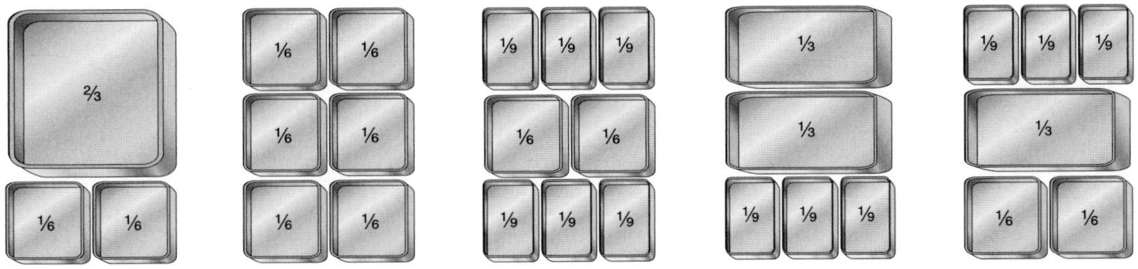

Figure 4-20. *Hotel pans are used to cook, hold, and serve food and come in a variety of shapes and sizes such as half pans, quarter pans, and third pans.*

NSF Mark

NSF-certified products bear the NSF mark

Edlund Co.

Figure 4-21. *NSF-certified products bear the NSF logo, indicating the product is able to withstand the daily wear and tear of a professional kitchen.*

PROFESSIONAL EQUIPMENT

The equipment used in a professional kitchen must be commercial grade to withstand the wear and tear and must be able to be easily cleaned. Some pieces of equipment are designed to reduce preparation and cooking times and increase consistency, while other items simply provide a safe and durable work surface. Professional equipment is designed according to NSF sanitation standards.

NSF International, formerly known as the National Sanitation Foundation, is an organization focused on standards development, product certification, education, and risk management for public health and safety. NSF-certified products bear the NSF mark, indicating the product has passed rigorous inspection, can be easily maintained, and can withstand the daily wear and tear of a professional kitchen. *See Figure 4-21.* The NSF mark indicates that the product meets the following specifications:

- The item has a smooth, nonporous, nontoxic, corrosion-resistant surface.
- Internal corners of the item are sealed and smooth, and the external edges are rounded and smooth.
- The item can be easily cleaned and easily taken apart for routine cleaning and maintenance.

Safe Equipment Operation

Prior to operating equipment in the professional kitchen, safety guidelines must be understood to protect against personal injury and damage to equipment. Unsafe or careless operation of equipment can lead to serious injury. The following safety guidelines are recommended for operating and maintaining professional kitchen equipment:

- Prior to use, read manufacturer instructions for safe equipment operation.
- Install and use all available safety features, such as guards, on equipment.
- Securely station or anchor equipment to prevent slipping or falling.
- Turn off and unplug all equipment before cleaning or disassembling.
- Check cords on electrical appliances to verify they are in safe operating condition.
- Never use extension cords to operate commercial equipment.
- Sanitize equipment after cleaning to prevent foodborne illness.
- Turn off and unplug equipment before reassembling after cleaning or service.
- Notify a supervisor immediately if equipment malfunctions or damage is detected.
- Post a caution notice on malfunctioning or damaged equipment.

Proper cleaning and sanitation of equipment is essential for safe food handling. The presence of bacteria on equipment in the professional kitchen can create cross-contamination and spread foodborne illnesses. Each piece of equipment should be regularly disassembled and thoroughly washed, rinsed, and sanitized.

Hobart

Areas of the Professional Kitchen

Each area within the professional kitchen requires equipment that meets the demands of a fast-paced work environment. The professional kitchen is typically divided into several major areas including safety, receiving, storage, sanitation, and preparation. Each area is equipped to serve a particular function. *See Figure 4-22.*

\multicolumn{2}{c}{**Areas of the Professional Kitchen**}	
Area	**Function**
Safety	Areas where ventilation systems, fire suppression systems, fire extinguishers, safety data sheets (SDSs), and first aid kits are located
Receiving	Areas where items ordered by the foodservice operation enter the facility
Storage	Areas where dry items, cold foods, and hot foods are stored
Sanitation	Areas where used items are cleaned and sanitized
Preparation	Areas where food is prepared, cooked, and plated

Figure 4-22. *The professional kitchen is typically divided into safety, receiving, storage, sanitation, and preparation areas.*

SAFETY AREAS

Safety areas include areas such as first aid stations, fire extinguisher stations, and safety data sheet (SDS) information areas. These areas are checked frequently to ensure that all safety equipment and materials are in working order and easily accessible. Having specific safety areas in the operation ensures that safety equipment can be easily located in emergency situations.

Safety Equipment

Standards are established by local health departments to help foodservice operations serve food safely and to provide a safe workplace for employees. Common safety equipment in the professional kitchen includes thermometers, timers, ventilation systems, fire-suppression systems, fire extinguishers, and first aid kits.

Thermometers. For food safety, it is important that food cooked in the professional kitchen is stored at and cooked to required temperatures. Many types of thermometers are used to measure the temperatures of cooked and stored foods. Common types of thermometers used in the professional kitchen are instant-read thermometers, candy/deep-fat thermometers, electronic probe thermometers, and infrared thermometers. *See Figure 4-23.* All thermometers should be calibrated on a regular basis.

- An *instant-read thermometer* is a stem-like thermometer attached to either a digital or mechanical display. The instant-read thermometer is small enough to carry in a pocket and has a pocket clip. The stem of the instant-read thermometer is briefly inserted into foods during cooking to determine the internal temperature. However, this thermometer cannot be left in food as it cooks because doing so will damage the thermometer. The stem should be sanitized after each use to prevent cross-contamination.

- A *candy/deep-fat thermometer* is a thermometer with a long, stainless steel stem and a large display. Candy/deep-fat thermometers are used to measure the temperature of very hot substances as they are being cooked. This thermometer has a clip that allows it to be clipped to the side of a pot during cooking. When removing a candy/deep-fat thermometer from a hot substance, care must be taken to avoid placing it into something cold, as the sudden temperature change can damage or break the thermometer. The stem should be sanitized after each use to prevent cross-contamination.

Thermometers

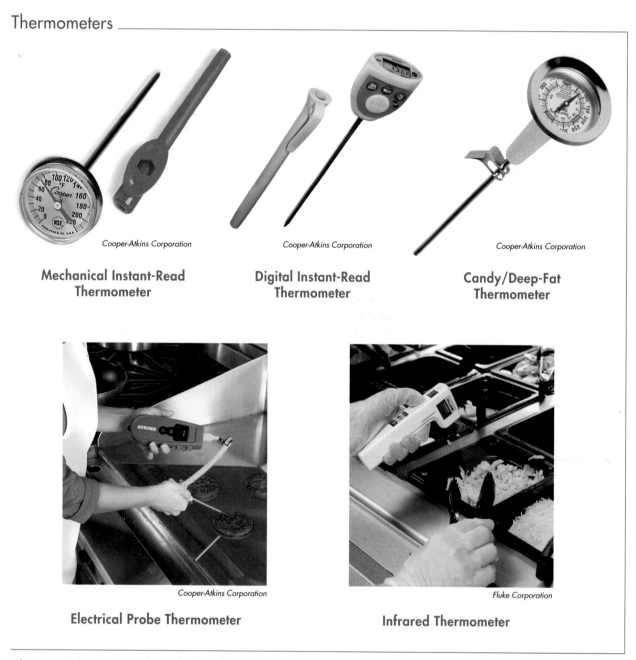

Cooper-Atkins Corporation

Mechanical Instant-Read Thermometer

Cooper-Atkins Corporation

Digital Instant-Read Thermometer

Cooper-Atkins Corporation

Candy/Deep-Fat Thermometer

Cooper-Atkins Corporation

Electrical Probe Thermometer

Fluke Corporation

Infrared Thermometer

Figure 4-23. *Instant-read, candy/deep-fat, electronic probe, and infrared thermometers are used to measure the temperatures of cooked and stored foods.*

- An *electronic probe thermometer* is a thermocouple thermometer with a thin, stainless steel stem that is attached by wires to a battery-operated readout device. The stem is placed into a food item, and the internal temperature of the item is displayed on the handheld readout. The stem of an electronic probe thermometer is much thinner than a traditional stem thermometer and provides immediate temperature readings in both Fahrenheit and Celsius. The stem should be sanitized after each use to prevent cross-contamination.

- An *infrared thermometer* is a thermometer that measures the surface temperature of an item through the use of infrared laser technology. Infrared thermometers are noncontact thermometers, meaning that the infrared laser is pointed at a food item and the external temperature is taken immediately and shown on the digital display. Infrared thermometers are often used to check the temperature of items in the receiving area as well as on salad bars and in steam tables.

Timers. A *timer* is a measuring tool that indicates the amount of time that has passed or sounds an alarm when a specified time period has ended. Timers come in a variety of styles including hanging timers, countertop timers, and digital timers. *See Figure 4-24.*

Ventilation Systems. A *ventilation system* is a large exhaust system that draws heat, smoke, and fumes out of the kitchen and into the outside air. A properly working ventilation system is essential in any foodservice operation to remove excess heat, smoke, and fumes that can build up in the kitchen. A large fan located at the end of the ventilation system ductwork creates a vacuum that draws exhaust air out from the building. As hot, greasy air is drawn up into the hood, grease tends to build up on the interior surface and the filters. This grease must be cleaned off of the hood and the fire-suppression components regularly to prevent potential fires. *See Figure 4-25.*

Timers

Hanging Timer

Countertop Timer

Digital Timer

Figure 4-24. *Timers come in a variety of styles including hanging units, countertop timers, and digital timers.*

Ventilation Systems

Photo Courtesy of Tyco/ANSUL

Figure 4-25. *A ventilation system removes excess heat, smoke, and fumes that can build up in the professional kitchen.*

Fire-Suppression Systems. A fire-suppression system is required over any open-flame cooking surface or combustible surface such as a gas cooktop or fryer. A *fire-suppression system* is an automatic fire-extinguishing system that is activated by the intense heat generated by a fire. Fire-suppression systems include discharge nozzles located above each piece of cooking equipment. These spray heads are connected to a highly pressurized tank containing fire-extinguishing chemicals. If a fire occurs under the suppression system, intense heat triggers the system to shut off the gas and releases chemicals, thus extinguishing the fire. *See Figure 4-26.* Fire codes vary by locale, and some municipalities have stricter codes for fire-suppression systems.

Fire-Supression Systems

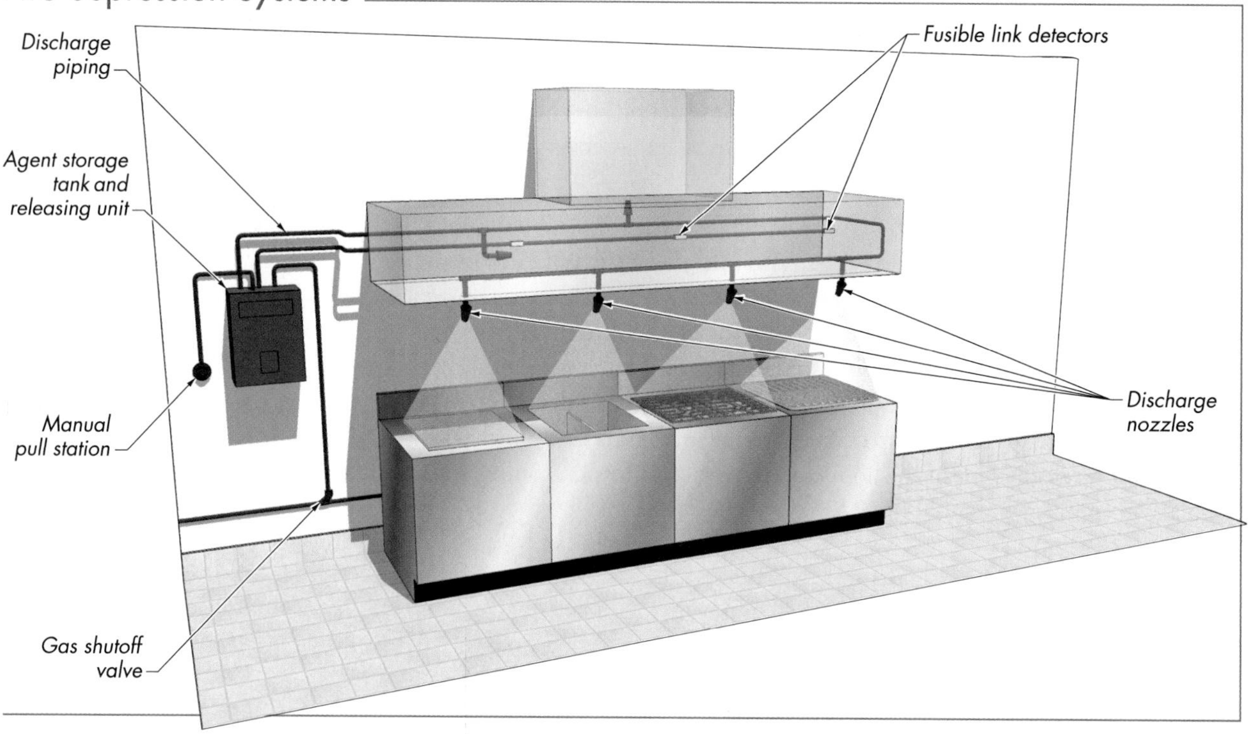

Discharge piping

Agent storage tank and releasing unit

Manual pull station

Gas shutoff valve

Fusible link detectors

Discharge nozzles

Figure 4-26. *A fire-suppression system is an automatic fire-extinguishing system with discharge nozzles located above each piece of cooking equipment.*

Fire Extinguishers. Fire extinguishers must be stationed at various areas throughout the kitchen and the entire foodservice operation. These metal canisters are filled with pressurized dry chemicals, foam, or water. Fire extinguishers are operated manually, meaning that a person must squeeze the trigger to dispense the pressurized material inside the extinguisher. The extinguisher should always be directed at the base of a fire to extinguish it.

The three major classes of fire extinguishers used in the professional kitchen are Class A, Class C, and Class K. *See Figure 4-27.* Each class is designed to eliminate a particular type of fire. Class A fire extinguishers are for fires involving common combustible materials such as trash, wood, or paper. Class C fire extinguishers are used to extinguish electrical fires. Class K fire extinguishers are used specifically for grease fires. It is important to check fire extinguishers frequently to make sure they are adequately charged in case of an emergency. Checking them at the end of each month during inventory is a good way to include a safety check in a foodservice operation.

Fire Extinguishers

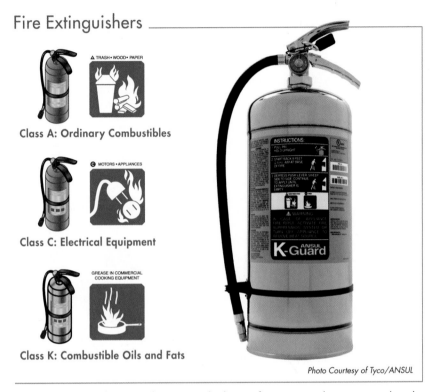

Class A: Ordinary Combustibles

Class C: Electrical Equipment

Class K: Combustible Oils and Fats

Photo Courtesy of Tyco/ANSUL

Figure 4-27. *Class A, Class C, and Class K fire extinguishers are used in the professional kitchen.*

First Aid Kits. Even when safe tool and equipment use are practiced in the professional kitchen, it is inevitable that some injuries will occur. First aid kits should be stored in plain view and never locked. In the event that an injury cannot be adequately treated by available first aid equipment, the paramedics should be contacted immediately. *See Figure 4-28.*

RECEIVING AREAS

The first steps for safe food handling in a professional kitchen occur the moment food enters the building through the receiving area. The *receiving area* is the area of the professional kitchen where all delivered items are checked for freshness, proper amount, correct temperature, and accurate price. If a problem is detected with an item at the time it is received, the item is immediately returned to the supplier.

First Aid Kits

Service Ideas, Inc.

Figure 4-28. *First aid kits should be stored in plain view and never locked.*

Receiving Equipment

A receiving area should contain moving equipment, measurement equipment, inspection tables, and labeling equipment.

Moving Equipment. Dollies, pallet jacks, and utility carts are all used to move received items. *See Figure 4-29.* Dollies are used in receiving areas to move heavy items or large boxes from one area to another. Pallet jacks are used to move large amounts of stock from the receiving area to storage areas. Utility carts make transporting food, equipment, and other small items from one area to another easier and faster. Foodservice utility carts are constructed of stainless steel or polyurethane and are equipped with heavy-duty wheels. They have two or three shelves and may also have side bins that can be used to transport different types of items at the same time.

Moving Equipment

Cres Cor

Dolly

Carlisle FoodService Products

Pallet Jack

InterMetro Industries Corporation

Utility Cart

Figure 4-29. *Dollies, pallet jacks, and utility carts are all used to move received items.*

Infrared Thermometers

Fluke Corporation

Figure 4-30. *An infrared, or noncontact, thermometer instantly measures the surface temperature of an item.*

Measurement Equipment. Thermometers and scales are used to measure items as they are received. The most common thermometer used in the receiving area is an infrared, or noncontact, thermometer. *See Figure 4-30.* Frozen foods should be delivered at a temperature of 0°F or below. Fresh perishable foods should be delivered at a temperature of 41°F or below. Foods that are delivered at temperatures above the maximum allowed temperature are rejected and returned to the supplier.

Many delivered items such as meat and seafood are sold and priced by weight and these items need to be weighed upon arrival. Weight, or the heaviness of a substance, is measured using hanging, platform, bench, and portion scales. *See Figure 4-31.* Scales can be calibrated in pounds, ounces, or grams. Platform and bench scales are used to weigh large or heavy boxes and bags. Mechanical and digital portion scales are used to weigh smaller items such as portion-controlled cuts of meat. A balance scale, also known as a baker's scale, is used in the bakeshop.

Scales

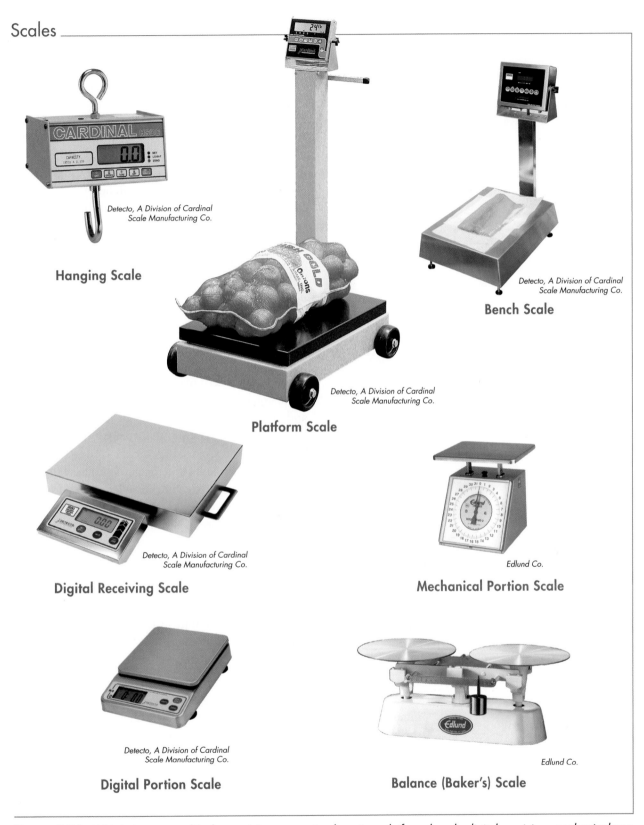

Hanging Scale

Detecto, A Division of Cardinal Scale Manufacturing Co.

Platform Scale

Detecto, A Division of Cardinal Scale Manufacturing Co.

Bench Scale

Detecto, A Division of Cardinal Scale Manufacturing Co.

Digital Receiving Scale

Detecto, A Division of Cardinal Scale Manufacturing Co.

Mechanical Portion Scale

Edlund Co.

Digital Portion Scale

Detecto, A Division of Cardinal Scale Manufacturing Co.

Balance (Baker's) Scale

Edlund Co.

Figure 4-31. *Weight is measured in the receiving area using hanging, platform, bench, digital receiving, mechanical portion, digital portion, and balance scales.*

Inspection Tables

Figure 4-32. *Stainless steel inspection tables allow boxes to be opened for inspection without having to bend down.*

Labeling Equipment

Figure 4-33. *Markers and adhesive labels are used to label and date items.*

When using any type of scale, the scale is always tared, or set to zero, before weighing any items. When using a container to hold food that is being weighed, the empty container is placed on the scale and the scale is set to zero again. Food is then added to the empty container until the desired amount registers on the scale.

Inspection Tables. Stainless steel inspection tables in the receiving area allow the receiving clerk to open boxes for inspection without having to bend down. *See Figure 4-32.*

Labeling Equipment. All items must be labeled and dated upon arrival. Markers are used for labeling boxes. Adhesive labels are used as needed to identify containers. *See Figure 4-33.* After an item is received, it is marked with the delivery date prior to being moved into an appropriate storage area. Dating each received item is essential to ensure older items are used before newly delivered items. New items should always be placed at the back of the storage area, and older items are rotated to the front for accessibility using the first-in first-out (FIFO) stock rotation method.

STORAGE AREAS

All food, beverage, and supply items need to be stored for security, safety, and inventory control. Separate storage areas are designated within the professional kitchen for dry, cold, and hot items. Dry storage areas are used to hold dry or packaged items that do not require refrigeration. Dry storage rooms are kept at temperatures between 50°F and 70°F. Cold storage areas include refrigeration and freezer units to hold perishable items at safe temperatures. These areas may be as small as an under-counter refrigerated cabinet or as large as a refrigerated warehouse. Hot storage areas are used to hold hot foods at safe temperatures while serving or until needed for service. These areas may be stationary or mobile units that can be moved closer to the where the food will be served.

Dry Storage Equipment

The equipment used in dry storage areas makes it easier to keep food items clean and free of pests. Dry storage equipment consists of shelving units, storage containers and bins, and security cages.

Shelving Units. Shelving units are available in a variety of styles, sizes, and shapes. Shelving units must be a minimum of 6 inches above the floor to allow access for cleaning underneath the unit. Styles vary depending on the intended use and include can rack shelving, wire shelving, dunnage rack shelving, overhead shelving, and overhead racks. *See Figure 4-34.*

- *Can rack shelving* is shelving with rails in which cans of product can be loaded from the top. These rails allow cans to roll toward the bottom for easy removal.

- *Wire shelving* is shelving made of wire and is primarily used to store boxed items. Wire shelving units are available in many different sizes and may be used in almost any size room or area.

- *Dunnage rack shelving* is shelving consisting of reinforced platforms that keep heavy items at least 12 inches above the floor. Dunnage racks are often stacked like wire shelving.

- *Overhead shelving* is shelving mounted on the wall or above a work surface. Overhead shelving is commonly located in the preparation area of the professional kitchen and is designed to hold items such as kitchen tools or other items needed at a given work station.

- An *overhead rack* is an overhead, suspended rack with hooks that allow utensils, pots, and pans to be easily accessible.

Shelving Units

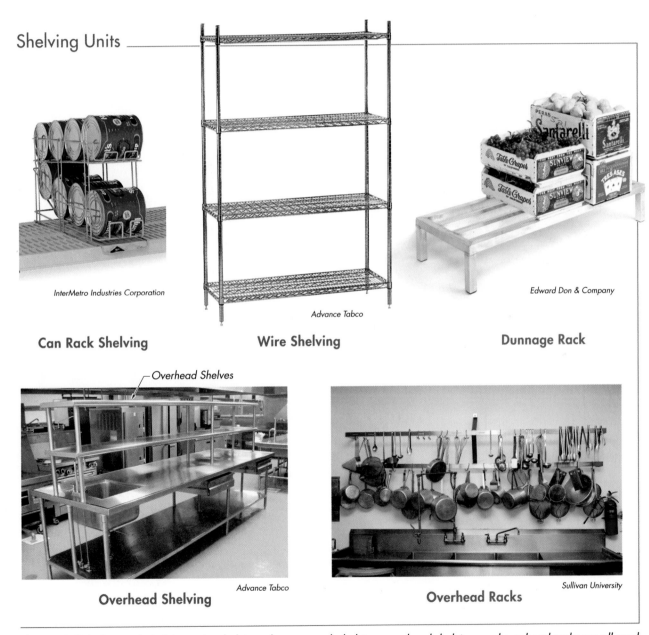

InterMetro Industries Corporation

Can Rack Shelving

Advance Tabco

Wire Shelving

Edward Don & Company

Dunnage Rack

Overhead Shelves

Overhead Shelving

Advance Tabco

Overhead Racks

Sullivan University

Figure 4-34. *Can rack shelving, wire shelving, dunnage rack shelving, overhead shelving, and overhead racks are all used to store items in the professional kitchen.*

Speed Racks

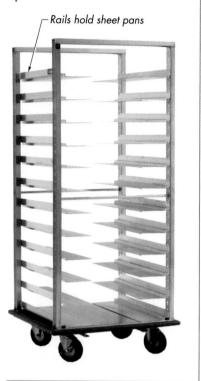

Rails hold sheet pans

Cres Cor

Figure 4-35. *Speed racks hold entire sheet pans of food and can be easily moved as needed.*

Speed Racks. A *speed rack,* also known as a tallboy, is a tall cart on wheels with rails that hold entire sheet pans of food. These storage units can be moved to various areas of the kitchen as needed. *See Figure 4-35.*

Storage Containers and Bins. Storage containers and bins come in many shapes and sizes and are used throughout the professional kitchen. Most storage containers usually have flat, tight-fitting, snap-on lids to prevent the absorption of odors from other foods. Storage containers often have graduated measurements printed on the sides. Bus boxes are heavy-duty polyurethane containers that are often used for many purposes in the kitchen as well as for clearing tables in the dining area.

Large polyurethane storage bins on wheels are used to store bulk items such as flour, sugar, and grains. They can easily be moved to where they are needed. *See Figure 4-36.* All storage containers and bins should be labeled with the contents and the time and date that the item was placed in the container.

Storage Containers and Bins

Storage Containers

Storage Bins

Bus Boxes

Carlisle FoodService Products

Figure 4-36. *Storage containers and bins come in many shapes and sizes.*

Security Cages. A *security cage* is a lockable, wire-cage storage unit on wheels used to hold expensive items such as fine china or restricted items such as alcohol. Security cages on wheels can be moved to various locations as needed. *See Figure 4-37.*

Figure 4-37. *A security cage is a lockable storage unit.*

Cold Storage Equipment

Having enough cold storage is essential for the safe handling of foods. Four major types of refrigeration and freezer units used in the professional kitchen are walk-in units, roll-in units, reach-in units, and blast chillers. *See Figure 4-38.* Regardless of the style, all refrigeration units must be kept at 41°F or below and all freezers must be kept at 0°F or below. The interior walls, shelves, and racks of refrigeration and freezer units should be cleaned regularly with a solution of baking soda and water.

Walk-in Units. A *walk-in unit* is a room-size insulated storage unit used to store bulk quantities of food. Walk-in units are often outfitted with adjustable shelving. The shelving units used in refrigerators and freezers are either stainless steel or metal that has been dipped into a plastic-like resin to prevent rust or corrosion. Carts or speed racks can be wheeled directly into these units. Walk-in units can be built in any size or shape that fits the needs of a foodservice operation.

Roll-in Units. A *roll-in unit* is an individual refrigeration unit that allows speed racks to be rolled in and out of the unit through a door opening that is just above floor height. Roll-in units are similar to reach-in units in size and dimensions, but have a small ramp at the door opening to allow speed racks to be easily rolled in or out.

Reach-in Units. A *reach-in unit* is a temperature-controlled cabinet for storing cold or frozen food items. Reach-in units can be individual units or may have two or three doors, each with shelves the size of a standard sheet pan. Reach-in units are usually located throughout the kitchen, creating convenient refrigerated or frozen storage near where items are needed.

A *lowboy* is a reach-in refrigerated unit located beneath a work surface. This allows the foodservice worker to prepare foods on the work surface and have adequate refrigerated storage beneath.

Blast Chillers. A *blast chiller* is a specialized cooling unit that rapidly reduces the temperature of foods, rendering them safe for immediate storage. A blast chiller can reduce the temperature of a cooked food to below 41°F very quickly. Items can also be quickly frozen in a blast chiller, resulting in a high-quality frozen product.

Cold Storage Equipment

Kolpak

Walk-in Unit

True FoodService Equipment, Inc.

Reach-in Unit

True FoodService Equipment, Inc.

Roll-in Unit

True FoodService Equipment, Inc.

Lowboy

Irinox USA

Blast Chiller

Figure 4-38. *Walk-in units, reach-in units, roll-in units, lowboys, and blast chillers are used for cold storage.*

Hot Storage Equipment

Storing hot foods requires equipment that provides a safe environment for hot foods during cooking, holding, transporting, and service. Hot foods must be held at temperatures above 140°F to prevent bacteria growth.

Bain-Maries. A *bain-marie* is a hot water bath used to keep foods such as sauces and soups hot. A bain-marie uses the same principle as a double boiler. A bain-marie consists of two vessels nested inside one another with the source of heat being hot water beneath the vessel. *See Figure 4-39.*

Steam Tables. A *steam table* is an open-top table with heated wells that are filled with water. Foods are placed in hotel pans and the pans are placed into the wells of the steam table. *See Figure 4-40.* The heated wells are controlled by a thermostat so temperature can be adjusted as needed. Foods are kept hot by the heat from the hot water in each well of the steam table. Foods in a steam table must be kept covered to prevent heat loss.

Bain-Maries

Carlisle FoodService Products

Figure 4-39. *A kettle-style warmer is one type of bain-marie.*

Steam Tables

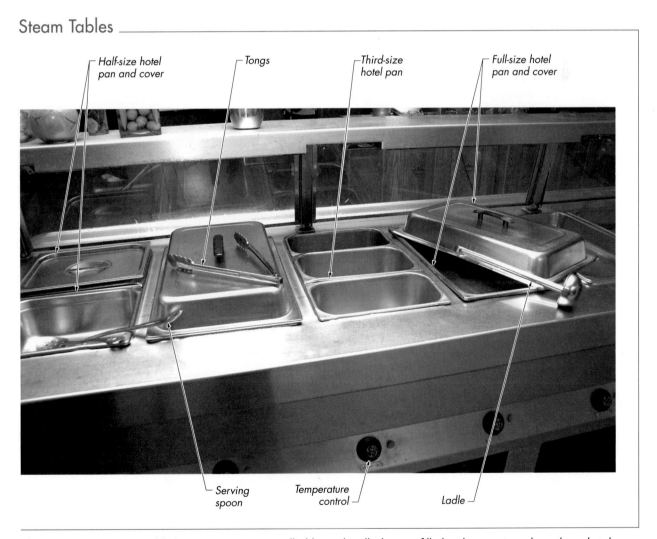

Half-size hotel pan and cover — Tongs — Third-size hotel pan — Full-size hotel pan and cover

Serving spoon — Temperature control — Ladle

Figure 4-40. *A steam table has temperature-controlled heated wells that are filled with water in order to keep hotel pans of food warm.*

Proofing Cabinets

Cres Cor

Figure 4-41. *A proofing cabinet is used for proofing dough and for holding hot food without drying it out.*

Holding and Proofing Cabinets. A *holding cabinet,* also known as a hot box, is a tall and narrow stainless steel box on wheels that accommodates standard sheet pans and contains temperature controls. A *proofing cabinet,* also known as a proofer, is a holding cabinet that contains both temperature and humidity controls. A proofing cabinet is used for proofing dough and for holding hot food without drying it out. *See Figure 4-41. Proofing* is the process of letting yeast dough rise in a warm (85°F), and moist (80% humidity) environment until the dough doubles in size.

Overhead Food Warmers. An *overhead warmer* is a heat source located above a prepared food that needs to be kept hot for service. These warmers contain electric or infrared rod-style elements or bulbs that radiate intense heat downward, keeping the food beneath it hot. *See Figure 4-42.*

Overhead Warmers

Carlisle FoodService Products
French Fry Station

Cres Cor
Carving Station

Figure 4-42. *An overhead warmer keeps prepared food hot for service.*

Chafing Dishes. A *chafing dish* is a hotel pan inside of a stand with a water reservoir and a portable heat source, such as canned fuel, underneath. *Canned fuel* is a flammable gel that provides several hours of heat once it is lit. *See Figure 4-43.*

Chafing Dishes

American Metalcraft, Inc.

Figure 4-43. *A chafing dish is a hotel pan inside of a stand with a water reservoir and a portable heat source, such as canned fuel, underneath.*

Insulated Carriers. An *insulated carrier* is an insulated container made of heavy polyurethane that is designed to hold hotel pans of hot or cold foods during transport. Insulated carrier lids and doors have insulated gaskets that seal and lock securely in place. *See Figure 4-44.*

Insulated Carriers

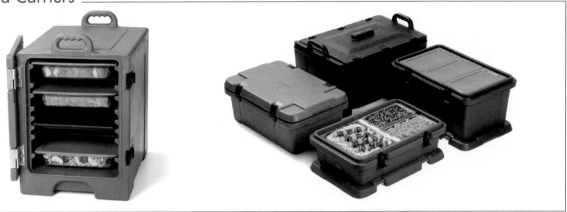

Carlisle FoodService Products

Figure 4-44. *An insulated carrier is designed to hold hotel pans of hot or cold foods during transport.*

SANITATION AREAS

A *sanitation area* is a location in the professional kitchen where sanitation equipment is kept. Sanitation areas include locations in the kitchen and dining room where items are cleaned and sanitized, such as a warewashing station and a custodial closet where cleaning supplies and equipment are stored. Foodservice operations have designated areas where chemicals, cleaning supplies, and sanitation tools are stored.

Sanitation pails and brooms are two of the most frequently used sanitation tools. *See Figure 4-45.* Staff should be well trained in proper sanitation procedures to ensure the operation is kept clean and sanitary. Most municipalities require sanitation supplies and equipment to be stored behind closed doors.

Sanitation Tools

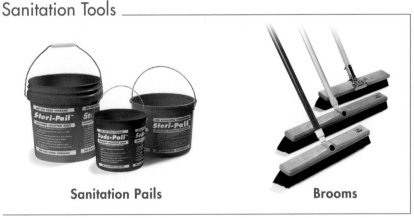

Sanitation Pails

Brooms

Carlisle FoodService Products

Figure 4-45. *Sanitation pails and brooms are two of the most frequently used sanitation tools.*

Handwashing Stations

Figure 4-46. *Handwashing stations are equipped with approved soaps and hand-drying materials or equipment.*

Sanitation Equipment

Customers visiting a foodservice operation expect clean conditions, clean washrooms, and no evidence of pests. Common sanitation equipment in the professional kitchen includes handwashing sinks, three-compartment sinks, commercial dishwashers, and food waste disposers.

Handwashing Sinks. Small sinks designated for handwashing are located in the preparation, service, and dishwashing areas. Handwashing stations are equipped with approved soaps and hand-drying materials or equipment. *See Figure 4-46.*

Compartment Sinks. Commercial sinks are constructed of stainless steel and have rounded corners, making them easier to clean. Typical warewashing sinks have at least three compartments. *See Figure 4-47.* The first compartment is for hot soapy wash water and the second is for hot rinse water. The last compartment is for a mixture of sanitizing solution and warm water. In a four-compartment sink, the first compartment is used to rinse off large debris before following the warewashing sequence.

Three-Compartment Sinks

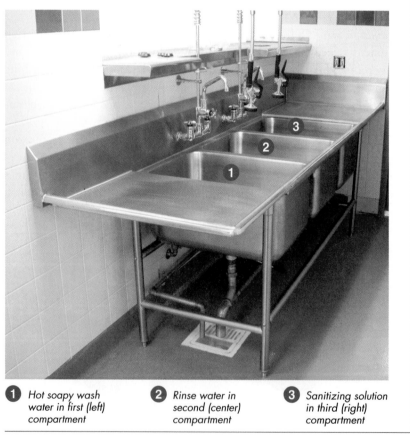

1 Hot soapy wash water in first (left) compartment

2 Rinse water in second (center) compartment

3 Sanitizing solution in third (right) compartment

Advance Tabco

Figure 4-47. *Warewashing sinks typically have three compartments to accommodate washing, rinsing, and sanitizing.*

Compartment sinks should always be cleaned and sanitized prior to being used for food preparation tasks. The process of cleaning and sanitizing pots, pans, and hand tools requires very hot water. While very hot water does destroy bacteria, it is also necessary to continually refill each of the sinks when washing, rinsing, and sanitizing food preparation items in order to keep the water fresh. Refilling the sinks every few hours helps remove bacteria as well as remove the soils and fats that build up in the water from the food debris removed from the pots, pans, and hand tools.

Compartment sinks should not be used for handwashing. Separate handwashing sinks are located elsewhere in the professional kitchen.

Food Waste Disposers. A *food waste disposer* is a food grinder mounted beneath warewashing sinks to eliminate solid food material. *See Figure 4-48.* Solid food material is rinsed from plates into sinks prior to being loaded into a commercial dishwasher. The majority of solid foods should be scraped into a garbage can and not rinsed into the disposer. The disposer is only designed to eliminate smaller bits of food that would otherwise be caught in the sink drain.

Commercial Dish Machines. Commercial dish machines make cleaning soiled dishes fast and efficient. There are many different styles and sizes of commercial dish machines. Both low-temperature and high-temperature dish machines are available. A single-tank dish machine has a door that is raised up to load racks of scraped dirty dishes and glassware. After the rack is loaded, the door is closed and the washing cycle automatically begins. A carousel or multitank dish machine takes racks by conveyor through the prewash, wash, rinse, sanitize, and dry cycle. *See Figure 4-49.* The racks emerge from the other end of the dish machine where they can be removed and stored.

Food Waste Disposers

In-Sink-Erator

Figure 4-48. *A food waste disposer is mounted beneath warewashing sinks to eliminate solid food material.*

Commercial Dish Machines

Hobart

Figure 4-49. *A carousel or multitank dish machine takes racks by conveyor through the prewash, wash, rinse, sanitize, and dry cycle.*

PREPARATION AREAS

A *preparation area* is an area in a professional kitchen where food items are prepared and cooked. The preparation area is typically divided into work sections and stations where specific tasks are performed. Large operations may have duplicate work sections and stations. No matter the size of the operation, the equipment in each work section is placed in a way that enables efficient production.

Work Sections and Work Stations

The preparation area is divided into work sections. A *work section* is an area where members of the kitchen staff are all working toward the same goal at the same time. For example, a baking work section is dedicated solely to the production of baked goods. This area is not used for other purposes such as meat fabrication.

Work sections are further divided into work stations. A *work station* is an area within a work section where specific tasks are performed by specific people. For example, several line cooks are often assigned to each work section. Individual line cooks are then assigned to specific work stations, such as a broiler station, within that work section. *See Figure 4-50.* Work stations are relatively self-sufficient and provide the necessary tools, equipment, and work space needed to complete specific tasks. For example, a broiler station would be equipped with a broiler, related hand tools such as tongs and offset spatulas, cold storage to hold uncooked food items, seasonings, and plates to hold the prepared food. Thus, the broiler cook should not need to leave the broiler station while preparing broiled foods.

Broiler Stations

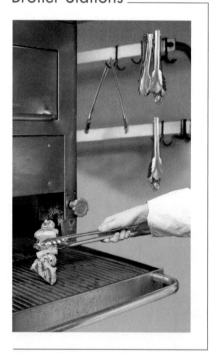

Figure 4-50. *A broiler station would be equipped with the equipment and hand tools needed to prepare broiled foods without leaving that station.*

Prep Sections. The prep section of the kitchen is where food items that are needed by one or more work stations are prepped. For example, a prep station cook may cut vegetables and fruits for use at a salad station. *See Figure 4-51.* A prep cook may also make stocks that will be given to someone else who will use them to make soups and sauces. A prep section typically contains sinks, work tables, slicers, mixers, and a range.

Prep Sections

Figure 4-51. *A prep section is where items such as vegetables are cut for use at other stations.*

Hot Foods Sections. The hot foods section of the kitchen is where foods are actually cooked. Typical stations within a hot foods section include a broiler station, grill station, griddle station, sauté/sauce station, fry station, and vegetable station. Each work station in this section contains the appropriate cooking equipment, refrigeration units, and storage. The greatest challenge for a hot foods section is delivering all of the orders prepared at different work stations to a given table at the same time.

Short Order Sections. The short order section of the professional kitchen is where foods such as hot sandwiches and breakfast items are prepared. Typical stations in a short order section include griddle, grill, broiler, and fry stations. A prep station is often connected to the short order section. *See Figure 4-52.*

Cold Foods Sections. The cold foods section, also known as a garde manger section, is where salads and salad dressings, cheese trays, cold appetizers and sandwiches, buffet centerpieces, and charcuterie are prepared. Work tables, slicers, mixers, blenders, and multiple refrigeration units are located in this section. Typical work stations within a cold foods section include salads, sandwiches, and plated desserts.

Griddle Stations

Figure 4-52. *A griddle station, where breakfast items are prepared, is part of the short order section.*

Baking and Pastry Sections. The baking and pastry section is where baked goods and pastries are produced. *See Figure 4-53.* Scales, mixers, proofing cabinets, sheeters, ovens, refrigeration units, and special work tables are located in this section. Typical work stations within a baking and pastry section include dough preparation, baking, dessert preparation, and plated desserts.

Baking Sections

Figure 4-53. *The baking section is where baked goods such as breads and rolls are produced.*

Banquet Sections. The banquet section is a special work section where the food and beverages for private dining functions are prepared and staged. A banquet section often contains all of the equipment found among the various work sections in the main kitchen, because it often functions as a self-sufficient, independent operation. Additional work stations beyond those often found in the main kitchen include meat carving and holding stations.

Beverage Sections. The beverage section is where beverages are prepared for service. Equipment in a beverage section includes ice bins, soft drink dispensers, glass racks, drink blenders, and coffee and tea brewers. *See Figure 4-54.* Typical stations within a beverage section include hot beverage, cold beverage, and alcoholic beverage stations.

Sanitation Sections. The sanitation section contains warewashing equipment, dish machines, dish racks, and nearby access to all of the cleaning supplies, cleaning equipment, and wet floor signs. Typical stations within a sanitation section include the warewashing station, dish machine station, and maintenance station.

Hot Beverage Equipment

Automatic Coffee Urns　　　**Automatic Coffee Brewers**　　　**Thermal Server System**

Figure 4-54. *Hot beverage stations require equipment such as coffee urns, brewers, and thermal server systems.*

Preparation Equipment

Most food preparation requires the use of equipment such as work tables, slicers, food processors, food choppers, vertical cutter/mixers, blenders, juicers, and mixers.

Work Tables. Work tables provide a surface with the perfect height for all preparation tasks and safe and efficient operation of preparation and processing tools. Stainless steel tables are used throughout the kitchen because they have very durable surfaces and are easy to clean. *See Figure 4-55.* In addition to stainless steel tables, marble- or quartz-topped and wood-topped work tables are used in the bakeshop.

Work Tables

Figure 4-55. *Stainless steel tables are used throughout the professional kitchen because they have very durable surfaces and are easy to clean.*

Can Openers. Heavy-duty can openers used in the professional kitchen can be table-mounted or free-standing counter models. Can opener blades must be kept sharp to prevent metal shavings from falling into cans as they are opened. Cleaning and sanitizing can opener blades also helps prevent cross-contamination. Canned goods are available in standard can sizes. ***See Figure 4-56.***

Can Openers and Can Sizes

Edlund Co.

Table-Mounted

Edlund Co.

Counter Model

Standard Can Capacity		
Can Size	**Average Weight**	**Average Volume**
#1 picnic	10½ oz	1¼ cup
#211	12 oz	1½ cup
#300	15 oz	1¾ cup
#1 tall (or 303)	1 lb	2 cups
#2	1 lb 4 oz	2½ cups
#2½	1 lb 12 oz	3½ cups
#3	2 lb 2 oz	4¼ cups
#3 cylinder	2 lb 14 oz	5¾ cups
#5	3 lb 8 oz	7 cups
#10	6 lb 8 oz	13 cups

Figure 4-56. *Table-mounted and counter model can openers are used to open standard canned goods.*

Slicers. A *slicer* is an appliance that is used to uniformly slice foods such as meat and cheese. It has a regulator providing a wide range of slice thicknesses and a feed grip that firmly holds the top of the food item or serves as a pusher plate for slicing small end pieces. ***See Figure 4-57.*** A slicer has a circular blade that rotates at a high speed, slicing items as they move across it. Electric slicers can be manually operated or set on automatic. All slicers have safety guards that help protect the user from the revolving blade. It is extremely important to ensure that these guards remain in place when using a slicer.

Slicers

Chef's Choice® by EdgeCraft Corporation

Figure 4-57. *A slicer is used to uniformly slice foods such as meat and cheese.*

Mixers. A *mixer* is a versatile electric appliance with U-shaped arms that secure one of several stainless steel mixing bowls of various sizes under a rotating head that accommodates various attachments. Mixers can be operated at different speeds for light or aggressive mixing and come in both bench and floor models. Bench mixers range in size from 4½–20 qt capacities. Floor mixers range in size from 30–140 qt capacities. *See Figure 4-58.*

Mixers

Bench Mixer

Floor Mixer

Paddle Attachment

Hook Attachment

Whip Attachment

Hobart

Figure 4-58. *Bench mixers range in size from 4½–20 qt capacities. Floor mixers range in size from 30–140 qt capacities.*

A paddle attachment is used for mixing and creaming. A whip attachment is used for whipping volume into products. A dough hook attachment is used for kneading dough. Mixers also have additional attachments such as grinders and shredders. A grinder attachment is used to grind meat or breadcrumbs, and a shredder attachment is used to shred items such as cabbage for coleslaw.

Vegetable Dryers

Paderno World Cuisine

Figure 4-59. *A vegetable dryer is a manual piece of equipment used to spin moisture out of vegetables such as leafy greens.*

Food Processors

Hobart

Figure 4-61. *A food processor comes with additional blades for shredding, grating, slicing, and julienning foods.*

Vegetable Dryers. A vegetable dryer is a manual piece of equipment used to spin moisture out of vegetables such as leafy greens. *See Figure 4-59.* Vegetable dryers are useful when making salads.

Blenders and Juicers. Different types of blenders and juicers are used to process a variety of foods quickly and evenly. *See Figure 4-60.* A *blender* is a tall appliance with a slender canister that is used to chop, blend, purée, or liquefy food. It is designed for puréeing soups, soft foods, and beverages and can also be used to crush ice. An *immersion blender,* also known as a stick mixer, is a narrow, handheld blender with a rotary blade that is used to purée a product in the container in which it is being prepared. For example, an immersion blender can be inserted into a saucepot to purée a soup.

Blenders and Juicers

Vitamix® Corporation
Blender

Viking Commercial
Immersion Blender

Robot Coupe USA
Juicer

Figure 4-60. *A blender, an immersion blender, and a juicer are each used to blend, purée, or liquefy foods.*

A *juicer* is a device used to extract juice from fruits and vegetables. Common types of juicers include juice extractors and reamers. A *juice extractor* is an electric machine that creates juice by liquefying raw vegetables and fruits and separating the fiber or pulp from the juice. A *reamer* is a manual or electric device used to extract juice from citrus fruits. To use an electric reamer, fruits are cut in half and placed on the juicer screen. The arm is then lowered to squeeze the fruit and extract the juice.

Food Processors. A *food processor* is an appliance with an S-shaped blade and a removable bowl and lid that can be used to quickly chop, purée, blend, or emulsify foods. A food processor also comes with additional blades for shredding, grating, slicing, and julienning foods. *See Figure 4-61.*

Food Choppers. Buffalo choppers and vertical cutter/mixers are used to cut large amounts of foods quickly and efficiently. *See Figure 4-62.* A *buffalo chopper* is an appliance used to process larger amounts of a product into roughly equal-size pieces. Food passes under a hood-like top, which houses a large, S-shaped blade. The coarseness of the cut is dependent on how long the food is left in the machine. The more times the food passes under the blade, the finer the cut. There is a built-in safety switch that prevents the machine from operating if the hood is left open.

Food Choppers

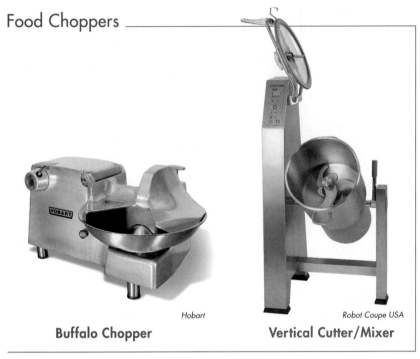

Hobart

Robot Coupe USA

Buffalo Chopper　　　　**Vertical Cutter/Mixer**

Figure 4-62. *Food choppers are used to cut or mix large amounts of foods.*

A *vertical cutter/mixer (VCM)* is an appliance used to cut and mix foods simultaneously using high-speed blades and a mixing baffle, which is used to manually move the product into the blades. For example, a VCM can produce 60 lb of fresh mayonnaise in a matter of minutes. A VCM is usually floor mounted with a 15-80 qt capacity bowl and has a built-in safety switch that prevents the machine from operating if the lid is open.

Cooking Equipment

Cooking equipment is usually placed in a central location in the professional kitchen. When choosing cooking equipment for a professional kitchen, the menu should be considered, as the types of items frequently cooked will help to determine what equipment is needed. For example, if the menu is for a steakhouse, there is a greater need for broilers and grills than for griddles and open-flame ranges. Common cooking equipment includes griddles, ranges, induction cooktops, grills, broilers, steamers, steam-jacketed kettles, tilt skillets, thermal immersion circulators, and fryers.

Griddles

Vulcan-Hart, a division of the ITW Food Equipment Group, LLC

Figure 4-63. *Griddles are solid cooking surfaces made of metal.*

Griddles. A *griddle* is a solid cooking surface made of metal on which foods are cooked. *See Figure 4-63.* Griddles are usually self-standing units commonly used for cooking items such as pancakes, eggs, bacon, and hamburgers. The surface temperature is controlled with a thermostat. Griddles should be seasoned after each cleaning to avoid surface corrosion. To season a griddle, the surface is heated to between 300°F and 350°F and about 1 oz of cooking oil per square foot is spread over the entire surface. Excess oil is wiped off and the procedure is repeated until a slick, mirror-like finish is achieved. A griddle brick is used to clean a griddle. A handle may be attached to the brick for added leverage. *See Figure 4-64.*

Griddle Bricks

Brick   **Brick Handle**

Carlisle FoodService Products

Figure 4-64. *A griddle brick is used to clean a griddle. A handle may be attached to the brick for added leverage.*

Ranges. A variety of open-burner ranges, flat-top ranges, and induction cooktops are used in the professional kitchen. *See Figure 4-65.* A *range* is a large appliance with surface burners. Open ranges have open-flame burners that allow easy regulation of the intensity of heat. Flat-top ranges have a flat top that covers the burners, providing even heat across a larger cooking surface. Although a flat-top range takes longer to initially heat, it allows more cooking vessels to be heated at one time. Some ranges have both open burners and a flat top within the same unit.

Ranges

Vulcan-Hart, a division of the ITW Food Equipment Group LLC

Open-Burner Range

U.S. Range

Flat-Top Range

Edward Don & Company

Induction Cooktops

Figure 4-65. *A variety of open-burner ranges, flat-top ranges, and induction cooktops are used in the professional kitchen.*

Induction Cooktops. An *induction cooktop* is an electromagenetic unit that uses a magnetic coil below a flat surface to heat food rapidly. Induction cooktops interact with cast iron or magnetic cookware, quickly creating heat within the actual cookware. The surface of an induction cooktop never gets hot, so it is always safe to touch. Instead, the cookware itself becomes hot. Professional chefs often use an induction cooktop when giving cooking demonstrations.

Grills. A *grill* is a cooking unit consisting of a large metal grate, also referred to as a grill, placed over a heat source. The heat source may be gas or another burning fuel such as charcoal or wood. Grills are commonly used to cook meats, seafood, poultry, and some fruits and vegetables. *See Figure 4-66.* Food is placed on the preheated metal grate and is turned over halfway through the grilling process to cook the other side.

Broilers. A *broiler* is a large piece of cooking equipment in which the heat source is located above the food instead of below it. The gas, electric, or ceramic heat source cooks the food from above. Broilers can be stand-alone or combined with an oven. The three basic types of broilers include the standard broiler, salamander, and rotisserie. *See Figure 4-67.*

A *salamander* is a small overhead broiler that is usually attached to an open burner range. The heat source is not nearly as intense as a standard broiler. A salamander is primarily used to brown, glaze, melt, or finish cooking foods.

A *rotisserie* is a sideways broiler. Instead of the heat being above the food, it is behind it. Foods are placed on a steel rod or spit that revolves past the heat source to ensure even heating. Rotisseries are most often used to cook poultry and meats.

Grills

Vulcan-Hart, a division of the ITW Food Equipment Group, LLC

Figure 4-66. *Grills are commonly used to cook meats, seafood, poultry, and some fruits and vegetables.*

Broilers

Vulcan-Hart, a division of the ITW Food Equipment Group LLC

Broiler

Vulcan-Hart, a division of the ITW Food Equipment Group LLC

Salamander

Henny Penny Corporation

Rotisserie

Figure 4-67. *Three basic types of broilers include the standard broiler, the salamander, and the rotisserie.*

Steamers. Two common types of steamers are used in the professional kitchen. *See Figure 4-68.* A *convection steamer* is a steamer that generates steam using an internal boiler, which circulates around the food to cook it rapidly. A *pressure steamer* is a steamer that uses water heated within a pressure-controlled, sealed cabinet to cook foods much quicker than a convection steamer.

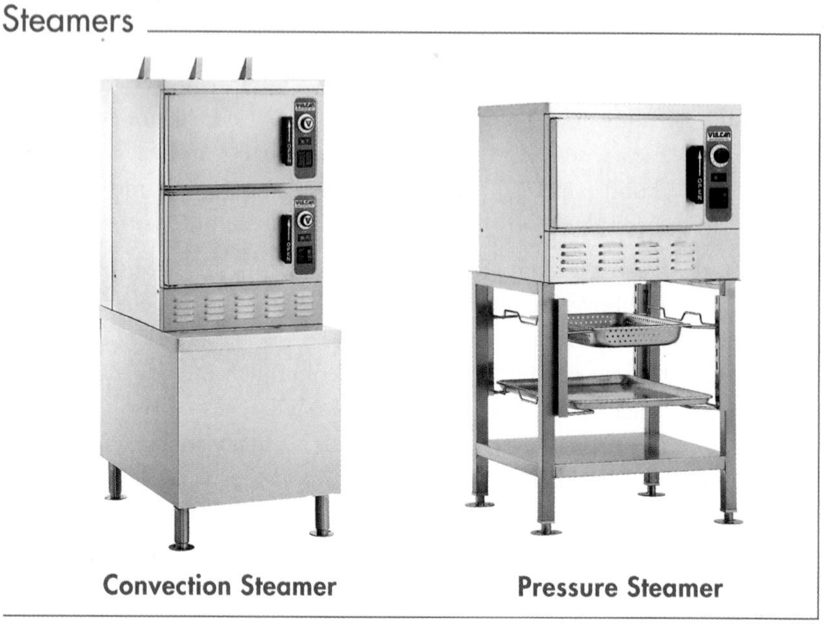

Steamers

Convection Steamer

Pressure Steamer

Vulcan-Hart, a division of the ITW Food Equipment Group LLC

Figure 4-68. *Both convection steamers and pressure steamers are used in the professional kitchen.*

Steam-Jacketed Kettles. A *steam-jacketed kettle,* also known as a steam kettle, is a large cooking kettle that has a hollow lining into which steam is pumped and a spigot at the bottom for easy draining. It cooks foods quickly and evenly while reducing the chance of scorching foods by heating the food without allowing the steam to touch it. *See Figure 4-69.* A *trunnion kettle* is a small steam-jacketed kettle that is tilted by pulling a lever or turning a wheel to empty the kettle. Steam-jacketed kettles can vary greatly in size and style.

Steam-Jacketed Kettles

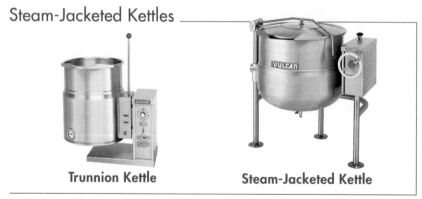

Trunnion Kettle

Steam-Jacketed Kettle

Vulcan-Hart, a division of the ITW Food Equipment Group, LLC

Figure 4-69. *Steam-jacketed kettles and trunnion kettles cook foods quickly and evenly without allowing steam to touch the food. They vary greatly in size.*

Tilt Skillets. A *tilt skillet,* also referred to as an electric braiser, is a versatile piece of cooking equipment with a large-capacity pan, a thermostat, a tilting mechanism, and a cover. The pan of a tilt skillet can hold 30–40 gal. and can be used as an oversize skillet, bain-marie, stockpot, proofing oven, kettle, or evenly heated cooktop. *See Figure 4-70.* Although a tilt skillet can be used to pan-fry or sauté, it should not be used for deep-frying.

Tilt Skillets

Blodgett Oven Company

Figure 4-70. *The pan of a tilt skillet can be used as an oversized skillet, bain-marie, stockpot, proofing oven, kettle, or evenly heated cooktop.*

Sous Vide Equipment. Sous vide, or slow cooking, equipment includes a vacuum sealer, a thermal immersion circulator, and a water bath. *See Figure 4-71.* A *thermal immersion circulator* is a device that is placed in a water bath to keep a uniform temperature for sous vide cooking. Items are vacuum sealed prior to being placed in the water bath.

Sous Vide Equipment

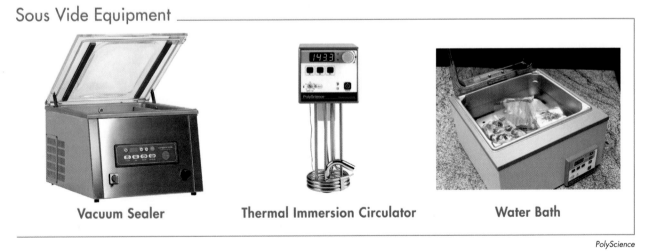

Vacuum Sealer   Thermal Immersion Circulator   Water Bath

PolyScience

Figure 4-71. *Sous vide, or slow cooking, equipment includes a vacuum sealer, a thermal immersion circulator, and a water bath.*

Fryers

Hobart

Figure 4-72. *Fryers are sized by the number of pounds of fat they can hold. For example, a 20 lb fryer holds 20 lb of fat.*

Fryers. A *fryer* is a cooking unit used to cook foods in hot fat. It operates by heating fat to a temperature between 200°F and 400°F. A thermostat regulates the temperature. Fryers are sized by the number of pounds of fat they can hold. For example, a 20 lb fryer holds 20 lb of fat. *See Figure 4-72.*

Ovens

Regardless of the type of food being cooked, using the proper oven is key. An *oven* is an enclosed cabinet where food is baked by being surrounded by hot air. Ovens used in the professional kitchen include convection, combi, deck, infrared, flashbake, wood-fired, smoker, cook-and-hold, revolving tray, impingement conveyor, and microwave ovens. Most professional kitchens have at least one convection oven and combi oven. *See Figure 4-73.*

Convection Ovens and Combi Ovens

Vulcan-Hart, a division of the ITW Food Equipment Group LLC

Convection Oven

Blodgett Oven Company

Combi Oven

Figure 4-73. *Convection ovens and combi ovens are found in most professional kitchens.*

Convection Ovens. A *convection oven* is a gas or electric oven with an interior fan that circulates dry, hot air throughout the cabinet. This fan creates intense and even heating as it keeps the air moving while items are cooking. Cooking temperatures are 50°F lower in a convection oven than in a conventional oven. The airflow leads to more efficient cooking, reduced shrinkage, and cooks more uniformly than a conventional oven.

Combi Ovens. A *combi oven,* also known as a combination oven, is an oven that has both convection and steaming capabilities. Combi ovens can steam and circulate hot air at the same time. They also decrease cooking time, save energy, and enable even roasting and baking.

Deck Ovens. A *deck oven* is a drawer-like oven that is commonly stacked one on top of another, providing multiple-temperature baking shelves. Food items are placed directly on the floor of a deck oven. *See Figure 4-74.* Deck ovens are common in foodservice operations that prepare pizzas.

Infrared and Flashbake Ovens. Two types of ovens involve infrared heat. An *infrared oven* is an oven that uses infrared radiation to evenly and efficiently bake flat foods such as pizza. A *flashbake oven* is an oven that uses both infrared radiation and light waves to cook foods quickly and evenly from above and below. *See Figure 4-75.* Because the heat is so intense, there is no loss of flavor or moisture in the foods cooked in a flashbake oven.

Wood-Fired and Smoker Ovens. A *wood-fired oven* is an oven that is encased with masonry and heated with wood. The extremely hot air inside a wood-fired oven bounces off the inside walls and quickly crisps the outside of the food to seal in moisture. Thermal, convection, radiant, and residual heat all work together to cook foods such as meats, crusty breads, and pizzas.

A *smoker oven* is a gas or electric oven that generates wood smoke and is most often used to smoke or barbeque meats and poultry. Meat can be placed on racks or hung inside a smoker. As the meat cooks, the smoke covers and seasons the outside of the meat. As the temperature inside the meat increases, the seasoning deposited on the outside of the meat is absorbed into the meat, giving it a barbeque flavor.

Hot smoking requires a temperature between 190°F and 225°F, and smokers are insulated to maintain a constant temperature. It is also important for the meat inside a smoker to remain moist. Basting will keep the meat moist, but it also causes the cooking temperature to drop each time the door of the smoker is opened. Often, a pan of water is placed between the heat source and the meat. Smoky flavors are infused in foods that are prepared in wood-fire or smoker ovens. *See Figure 4-76.*

Deck Ovens

Blodgett Oven Company

Figure 4-74. *Deck ovens are often used to bake breads and pizzas.*

Flashbake Ovens

Vulcan-Hart, a division of the ITW Food Equipment Group LLC

Figure 4-75. *Flashbake ovens use infrared radiation and light waves to cook foods.*

Wood-Fired and Smoker Ovens

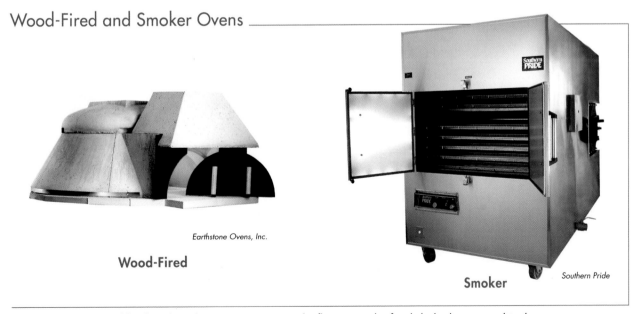

Earthstone Ovens, Inc.

Wood-Fired

Smoker

Southern Pride

Figure 4-76. *Wood-fired and smoker ovens impart smoky flavor into the foods baked or roasted in them.*

Cook-and-Hold Ovens

Figure 4-77. *A cook-and-hold oven can be used to cook, roast, reheat, and hold a variety of foods.*

Cook-and-Hold Ovens. A *cook-and-hold oven,* also known as a retherm oven, is an oven with two separately controlled compartments within one stainless steel cabinet that can be used to cook, roast, reheat, and hold a variety of foods. These ovens use halo heat, a uniform and controlled heat source that gently surrounds foods. Thermostats allow cooking up to 325°F and holding up to 200°F. *See Figure 4-77.*

Rotating Rack Ovens. A *rotating rack oven,* also known as a reel oven, is a large oven that rotates 10–80 pans of food as it cooks. *See Figure 4-78.* Some rotating rack ovens are available with a steam kettle, hood, and trough used for cooking bagels.

Rotating Rack Ovens

Figure 4-78. *A rotating rack oven is a large oven that rotates 10–80 pans of food as it cooks.*

Impinger Conveyor Ovens. An *impinger conveyor oven* is an oven that directs heat from both above and below a food item as it moves along a conveyor belt. Sandwich shops use impinger conveyor ovens to toast bread, melt cheese on bread as it toasts, and to cook pizzas. *See Figure 4-79.*

Impinger Conveyor Ovens

Lincoln FoodService Products, Inc.

Figure 4-79. *Sandwich shops use impinger conveyor ovens to toast bread, melt cheese on bread as it toasts, and to cook pizzas.*

Microwave Ovens

Amana Commercial Products

Figure 4-80. *A microwave oven uses microwaves to heat the water molecules within foods.*

Microwave Ovens. A *microwave oven* is a cooking unit that uses microwaves to heat the water molecules within foods. *See Figure 4-80.* A microwave oven requires no preheating. Metal containers and utensils cannot be put in a microwave because they will damage the oven.

SUMMARY

Foodservice professionals use appropriate tools and equipment for each task in order to obtain optimal efficiency. Professional hand tools include items to cut, shape, measure, strain, sift, lift, mix, blend, turn, and lift food items. Professional cookware includes a variety of skillets, pots, steamware, and ovenware. Equipment used in the professional kitchen must be operated safely to avoid injury.

The professional kitchen includes designated areas that include specific equipment that is used for safety, receiving, storage, sanitation, and preparation tasks. The preparation area of the kitchen is divided into work sections with work stations within each section. It is important to be able to identify and use the variety of tools and equipment used to safely receive, store, sanitize, and prepare foods and beverages.

Refer to DVD for
Quick Quiz® questions

1. Describe six specialized cutting tools that are not knives.
2. Describe three types of shaping tools used in the professional kitchen.
3. Describe seven volume measuring tools used in the professional kitchen.
4. Explain the purpose of a funnel.
5. Describe three types of strainers.
6. Describe four types of sieves.
7. Differentiate between a skimmer and a spider.
8. Describe four mixing and blending tools.
9. Describe 10 types of baking and pastry tools used in the professional kitchen.
10. Describe two types of scraping tools used in the professional kitchen.
11. Describe five turning and lifting tools.
12. Describe six types of pans used in the professional kitchen.
13. Describe four types of pots used in the professional kitchen.
14. Describe three types of warming and steaming cookware used in the professional kitchen.
15. Describe nine types of ovenware used in the professional kitchen.
16. List the specifications that must be met in order for equipment to be NSF certified.
17. List the general safety guidelines for operating and maintaining foodservice equipment.
18. Identify the major areas of the professional kitchen.
19. Describe common safety equipment that is used in the professional kitchen.
20. Differentiate between ventilation systems and fire-suppression systems.
21. Identify three types of fire extinguishers used in the professional kitchen.
22. Describe receiving equipment that is used in the professional kitchen.
23. Describe the types of measurement equipment used in the receiving area.
24. Identify three types of storage areas used in the professional kitchen.
25. Differentiate between the types of shelving units used in dry storage areas.
26. Describe four basic types of cold storage units.
27. Describe the functions of a proofing cabinet.
28. Differentiate between overhead warmers and chafing dishes.
29. Identify common sanitation equipment that is used in the professional kitchen.
30. Differentiate between work sections and work stations.
31. Identify common work sections found in the preparation area.
32. Describe nine different types of preparation equipment used in the professional kitchen.
33. Explain why mixers are one of the most versatile pieces of equipment used in the professional kitchen.
34. Differentiate between blenders and juicers.
35. Differentiate between food processors and food choppers.
36. Describe nine different types of cooking equipment, other than ovens, that are used in the professional kitchen.
37. Explain the purpose of thermal immersion circulators.
38. Describe 11 types of ovens that are used to cook foods in the professional kitchen.

Refer to DVD for
Review Questions

1. Diagram and label each area of the professional kitchen.
2. Diagram the location of each piece of safety equipment used in the professional kitchen.
3. Receive items and document how the receiving area and related equipment are used to perform this task.
4. Store dry, cold, and hot items and document how each storage area and the related equipment are used to perform these tasks.
5. Sanitize a hand tool and document the tools and equipment used.
6. Sanitize a piece of equipment and document the tools and equipment used.
7. Prepare a cold food item and document the tools and equipment used.
8. Prepare a hot food item and document the tools and equipment used.
9. Prepare a cold beverage and document the tools and equipment used.
10. Prepare a hot beverage and document the tools and equipment used.

Refer to DVD for
Application Scoresheets

Chapter 5

Cost Control Fundamentals

Maintaining cost control is essential to a profitable foodservice operation. Earning a profit starts with the planning of the menu and continues with the purchasing, receiving, storing, preparing, and serving of food. Standardized recipes and standard units of measure help ensure that a consistent quality of food is served every day. If customers have a favorite dish at a particular restaurant, the restaurant is likely to see repeat business from those customers. A lack of standardization can lead to a loss of business, which translates into a loss of profit.

Refer to DVD
for **Flash Cards**

Chapter Objectives

1. Identify the common elements of standardized recipes.
2. Differentiate among weight, volume, and count.
3. Explain the difference between ounces and fluid ounces.
4. Convert customary measurements to metric measurements.
5. Convert metric measurements to customary measurements.
6. Scale recipes based on yield, portion size, and product availability.
7. List factors that may have to be adjusted when scaling a recipe.
8. Calculate the as-purchased unit cost of a food item.
9. Calculate the edible-portion cost of a food item.
10. Calculate the yield percentage of a food item.
11. Perform a raw yield test and a cooking-loss yield test.
12. Calculate the as-served cost of a menu item.
13. Calculate three types of food cost percentages.
14. Calculate menu prices using three different methods.
15. Explain the difference between fixed costs and variable costs.
16. Identify the six stages at which costs must be controlled to result in a profit.
17. Explain the difference between gross profit and net profit.
18. Calculate the gross pay and the net pay for a line cook.

Key Terms

- **standardized recipe**
- **yield**
- **portion size**
- **weight**
- **volume**
- **count**
- **measurement equivalent**
- **scaling**
- **as purchased**
- **edible portion**
- **yield percentage**
- **food cost percentage**
- **perceived value pricing**
- **contribution margin**
- **fixed cost**
- **variable cost**
- **purchase specification**
- **par stock**
- **net profit**
- **gross pay**

Viking Commercial

STANDARDIZED RECIPES

One of the most significant challenges for any foodservice operation is to produce food that is consistent. Consistency is important because customers expect food from a particular foodservice operation to look and taste the same every time it is ordered. Since the same cook does not always prepare the same items each day, foodservice operations use standardized recipes. A *standardized recipe* is a list of ingredients, ingredient amounts, and procedural steps for preparing a specific quantity of a food item. Using standardized recipes helps ensure consistent quality, portion size, and cost. A standardized recipe includes the following:

- a list of specific ingredients and quantities listed in the order they are used

- step-by-step instructions for preparation or assembly

- an accurate portion size and total yield

The format of a standardized recipe will vary from one foodservice operation to another. However, most standardized recipes usually contain the following common elements. *See Figure 5-1.*

- **Recipe Name.** The name of a recipe should be descriptive of the dish being prepared and should reflect the name used on the menu. For example, if a restaurant makes two kinds of chili sauce (a spicy version and a mild version) both recipes should not be named "Chili Sauce." Instead, more descriptive names, such as Spicy Red Chili Sauce and Mild Green Chili Sauce, should be used. Differences between the menu name and the standardized recipe name can lead to confusion in the preparation process, which may lead to unnecessary costs.

Standardized Recipe Elements

Meatloaf

Yield: 10 Servings | **Cooking Temperature:** 350°F

Portion Size: 6 oz | **Cooking Time:** 1 hour

Ingredients		Preparation
1 tbsp	vegetable oil	1. Heat oil in small sauté pan and sauté celery and onions until tender. Allow to cool.
3 oz	celery, small dice	
4 oz	onion, small dice	
2 fl oz	milk	2. Combine remaining ingredients and mix well.
2	eggs	
1 c	breadcrumbs	3. Form mixture into a loaf and place in a greased bread pan.
1 tbsp	salt	
1½ tsp	black pepper	4. Bake at 350°F for about 1 hour or until a thermometer inserted in the center of the loaf registers 160°F (to allow for carryover cooking).
1 tsp	thyme	
3 lb	ground beef	
		5. Remove from oven, cover, and let rest for 15 minutes before slicing into 6 oz portions.

Nutrition information (per serving): 465.4 calories; 64% calories from fat; 32.7 g total fat; 149.2 mg cholesterol; 911.5 mg sodium; 503.2 mg potassium; 11.7 g carbohydrates; 0.9 g fiber; 4.0 g sugar; 10.8 g net carbohydrates; 28.6 g protein

Figure 5-1. *Most standardized recipes include the same common elements.*

- **Yield.** A *yield* is the total quantity of a food or beverage item that is made from a standardized recipe. Yield may be given as portions (16 servings, 3 oz each), the size of the item produced (one 8 inch pie), or a measured amount of product (2 gal. of soup).

- **Portion Size.** A *portion size* is the amount of a food or beverage item that is served to an individual person. Portion size is related to yield. For example, a soup recipe that yields 1 gal. can also be said to yield 16 portions of 8 fl oz each because 16×8 fl oz $= 128$ fl oz $= 1$ gal. *See Figure 5-2.*

Yield and Portion Size

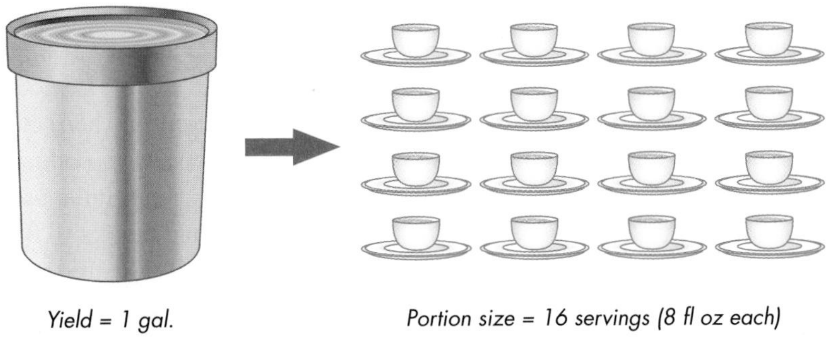

Yield = 1 gal.   Portion size = 16 servings (8 fl oz each)

Figure 5-2. *The yield and the portion size of a recipe are related.*

- **Ingredients.** The amount of each ingredient used in the recipe is listed next to the name of the ingredient. If an ingredient is to be prepared in a certain way prior to being measured, such as minced or chopped, that information is also provided. The ingredients are usually listed in the order that they are incorporated into the recipe to help ensure that no ingredient is left out.

- **Preparation.** Preparation steps are listed in sequential order. Preparation steps direct when and how the ingredients are added to a recipe and are listed in sequential order. Some ingredients are cleaned or cut prior to being measured. For example, "½ cup sliced black olives" requires the olives be sliced before they are measured. In contrast, "½ cup black olives, sliced" requires the olives be measured before they are sliced. Therefore, more olives are required in the first example than in the second.

- **Cooking Temperature.** The cooking temperature may be an exact temperature at which to set an oven, such as 400°F, or a more general indication of temperature such as "low heat" or "high heat." The temperature at which food is cooked greatly affects the outcome of the final product. For example, food that is cooked at too high of a temperature may burn on the outside before it is properly cooked in the center.

- **Cooking Time.** Cooking times provided in standardized recipes are often treated as guidelines. However, professional cooks with experience rely more on the appearance or feel of an item than on a cooking time. They also often rely on exact measurements, such as an internal temperature checked with a thermometer, to determine when a food item is done.

- **Nutrition Information.** While nutrition information is not required to prepare a recipe, it is important information to have for menu planning and for customer inquiries. Due to health or dietary concerns, customers may ask how much fat, carbohydrates, or sodium there is in a menu item.

STANDARD UNITS OF MEASURE

A standardized recipe uses standard units of measure to represent a specific amount of each ingredient. A *unit of measure* is a fixed quantity that is widely accepted as a standard of measurement. Foodservice operations use both customary and metric units of measure to measure food and beverage products. *See Figure 5-3.* Ounces (oz) and pounds (lb) are the customary units used to measure weight. Teaspoons (tsp), tablespoons (tbsp), fluid ounces (fl oz), cups (c), pints (pt), quarts (qt), and gallons (gal.) are the customary units used to measure volume. Inches (in.) and feet (ft) are the customary units used to measure distance, and degrees Fahrenheit (°F) are used to measure temperature.

Grams (g) are the metric unit used to measure weight and liters (L) are used to measure volume. Meters (m) are the metric unit used to measure distance, and degrees Celsius (°C) are used to measure temperature.

Standard Units of Measure

Weight Units		Volume Units		Distance Units		Temperature Units	
Customary System		**Customary System**		**Customary System**		**Customary System**	
Unit	Abbreviation	Unit	Abbreviation	Unit	Abbreviation	Unit	Abbreviation
ounce	oz	teaspoon	tsp	inch	in.	degrees Fahrenheit	°F
pound	lb or #	tablespoon	tbsp	foot	ft		
		fluid ounce	fl oz			**Metric System**	
Metric System		cup	c	**Metric System**		Unit	Abbreviation
Unit	Abbreviation	pint	pt	Unit	Abbreviation	degrees Celsius	°C
milligram	mg	quart	qt	millimeter	mm		
gram	g	gallon	gal.	centimeter	cm		
kilogram	kg			meter	m		
		Metric System					
		Unit	Abbreviation				
		liter	L				
		milliliter	mL				

Figure 5-3. *Professional kitchens use both customary and metric standard units of measure.*

Ingredients are measured by weight, volume, or count. *Weight* is a measurement of the heaviness of a substance. Most dry ingredients are given in weight. *Volume* is a measurement of the physical space a substance occupies. Liquid ingredients are commonly given in volume. *Count* is a measurement of the actual number of items being used. Count is commonly used for whole ingredients, such as 2 whole eggs or 1 medium banana, or for portion sizes, such as 24 servings, 6 oz each of chili. Standardized units of measure are also used to indicate time, temperature, and distance.

In order for a recipe to be successful, the ingredients and portions must be measured accurately. Too much of one ingredient or too little of another can result in a dish that simply does not work. Serving the wrong portion size can result in not having enough servings for a meal service or dissatisfied guests who receive smaller portions than other guests.

Measuring Weight

The most common units for measuring weight in the professional kitchen are the gram (g), ounce (oz), and pound (lb or #). When ingredient weights are listed in customary units, it is common to see pounds and ounces combined. For example, a recipe may call for 1 lb 2 oz of flour. Weight measurements are considered to be more accurate than volume measurements. When measuring by volume, a measurement can be affected by the technique used to fill the measuring tool. However, a scale can only indicate the weight placed on it. *See Figure 5-4.*

Refer to DVD for
Measuring Weight
Culinary Math Tutorial

Weight Measurement Equipment

Edlund Co.

Mechanical Scale

Detecto, A Division of Cardinal
Scale Manufacturing Co.

Digital Scale

Edlund Co.

Balance Scale

Figure 5-4. *Mechanical, digital, and balance scales are used to measure weight.*

To weigh items on a scale, place the empty container that will be used to hold the ingredients on the scale. Next, tare the scale by adjusting the scale to read zero so that the weight of the container is not reflected in the final measurement. Then place the ingredients to be measured in the container until the desired weight is achieved.

Measuring Volume

Volume measures are used to measure liquid ingredients in the professional kitchen. The most common volume measurement units are the milliliter (mL), teaspoon (tsp), tablespoon (tbsp), fluid ounce (fl oz), cup (c), pint (pt), quart (qt), liter (L), and gallon (gal.). The tools used to measure volume are measuring spoons, dry measuring cups, liquid measuring cups, ladles, and portion-controlled scoops. *See Figure 5-5.* Some dry ingredients, such as ground herbs or spices, may also be measured by volume. For example, a recipe may call for 2 tsp of oregano.

Volume Measurement Tools

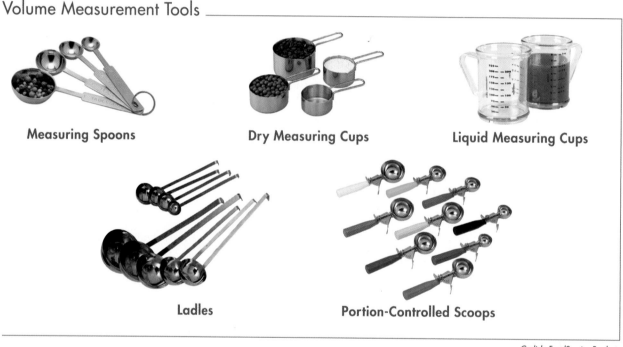

Measuring Spoons

Dry Measuring Cups

Liquid Measuring Cups

Ladles

Portion-Controlled Scoops

Figure 5-5. *A variety of tools for measuring volume are used in the professional kitchen.*

To avoid making costly errors in the professional kitchen, it is important to understand the difference between ounces and fluid ounces. An ounce (oz) is a measurement of weight and a fluid ounce (fl oz) is a measurement of volume. Some ingredients, such as water, weigh the same amount in ounces as they do in fluid ounces. *See Figure 5-6.* However, most ingredients do not weigh the same in ounces and fluid ounces. For example, 8 fl oz of honey actually weighs 12 oz by weight. The reason that 8 fl oz of honey weighs more than 8 fl oz of water is that honey has a higher density than water.

Ounces vs. Fluid Ounces

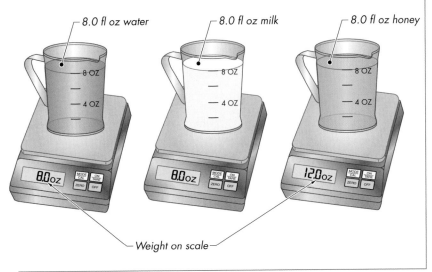

Refer to DVD for
Ounces vs. Fluid Ounces
Culinary Math Tutorial

Figure 5-6. *Depending on the ingredient being measured, a measurement in fluid ounces will not always be the same as a measurement in ounces.*

Density is the measure of how much a given volume of a substance weighs. Only a few ingredients can be measured in either fluid ounces or ounces without affecting a recipe. These ingredients include water and substances with densities very close to water such as alcohol, juices, vinegar, oil, milk, butter, eggs, and granulated sugar.

Most food ingredients have a higher density than water and measure more by weight in ounces than by volume in fluid ounces. For example, 8 fl oz of honey measured by volume will actually weigh about 12 oz. Honey, molasses, and various types of syrup have a different density than water.

Measuring Count

Many food items are measured by count. *See Figure 5-7.* For example, a 90 count package of potatoes contains potatoes of a size that average 90 potatoes per 50 lb case. As the count number gets smaller, the size of the item gets larger. For example, a 60 count package of potatoes contains 60 potatoes per 50 lb case. If a case of 90 count potatoes and a case of 60 count potatoes each weigh 50 lb, there are about 30 fewer potatoes in the 60 count case. Therefore, a 60 count potato must be larger than a 90 count potato.

Measurement Equivalents

A *measurement equivalent* is the amount of one unit of measure that is equal to another unit of measure. For example, 1 gal. is the equivalent of 4 qt. Foodservice employees need to know the basic measurement equivalents

Counts

Cres Cor

Figure 5-7. *Many food items, such as fresh produce, are measured by count.*

for weight and volume and be able to calculate the equivalents between any two volume units of measure. *See Figure 5-8.*

- To convert a customary measurement to a metric measurement, multiply the measurement by the number of metric units that equals one of the customary units. For example, to convert 2 oz into grams, multiply 2 by 28.35 (2 oz × 28.35 = 56.7 g).

- To convert a metric measurement to a customary measurement, multiply the measurement by the number of customary units that equals one of the metric units. For example, to convert 5 L into quarts, multiply 5 by 1.06 (5 L × 1.06 = 5.3 qt).

Customary and Metric Unit Equivalents

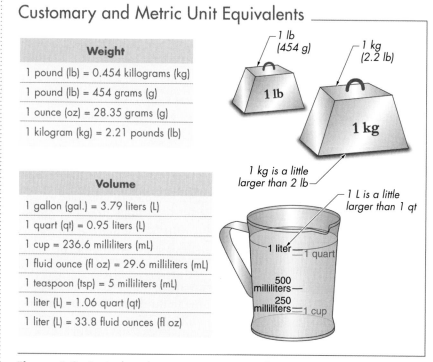

Weight
1 pound (lb) = 0.454 killograms (kg)
1 pound (lb) = 454 grams (g)
1 ounce (oz) = 28.35 grams (g)
1 kilogram (kg) = 2.21 pounds (lb)

Volume
1 gallon (gal.) = 3.79 liters (L)
1 quart (qt) = 0.95 liters (L)
1 cup = 236.6 milliliters (mL)
1 fluid ounce (fl oz) = 29.6 milliliters (mL)
1 teaspoon (tsp) = 5 milliliters (mL)
1 liter (L) = 1.06 quart (qt)
1 liter (L) = 33.8 fluid ounces (fl oz)

1 lb (454 g)

1 kg (2.2 lb)

1 lb

1 kg

1 kg is a little larger than 2 lb

1 L is a little larger than 1 qt

1 liter — 1 quart

500 milliliters —

250 milliliters — 1 cup

Figure 5-8. *Equivalents between customary and metric system units of measure for weight and volume are used to perform calculations in the professional kitchen.*

Weight Equivalents. The only customary weight equivalent used in the professional kitchen is 16 oz = 1 lb. To change a measurement given in pounds to ounces, the number of pounds is multiplied by 16. When an ingredient measurement is provided in pounds and ounces, such as 3 lb 2 oz, the total measurement must be changed to ounces. The first step is to multiply the number of pounds by 16 (3 lb × 16 = 48 oz). Then, that number is added to the amount of ounces in the original measurement (48 oz + 2 oz = 50 oz). In this example, 3 lb 2 oz = 50 oz.

To change an ingredient measurement from ounces only to pounds and ounces, the number of ounces is divided by 16. Then, the quotient is set equal to the number of pounds and any remainder is equal to the number of ounces. For example, 50 oz ÷ 16 = 3 lb with a remainder of 2 oz. This result would be written as 3 lb 2 oz.

Volume Equivalents. It is easy to change a volume measurement with one unit of measure to an equivalent measurement with a different unit of measure by using one of the following rules:

Refer to DVD for
**Volume Measurement
Equivalents**
Culinary Math Tutorial

- To change from a larger to a smaller unit of measure, multiply the number in the original measurement by the number of smaller units that make up the larger unit.

- To change from a smaller to a larger unit, divide the number in the original measurement by the number of smaller units that make up the larger unit.

For example, the measurement equivalent of 4 qt is 1 gal. To change a measurement in gallons to quarts (larger unit to smaller unit), the number of gallons is multiplied by 4 (the number of quarts that make up one gallon). For example, 2 gal. are equivalent to 8 qt (2 gal. × 4 = 8 qt). However, to change a measurement in quarts to gallons (smaller unit to larger unit), the number of quarts is divided by 4. For example, 12 qt are equivalent to 3 gal. (12 qt ÷ 4 = 3 gal.). The same process works using any of the basic volume equivalents. *See Figure 5-9.*

Calculating Volume Equivalents

Volume Equivalents	To Change:	Example Calculations
4 qt = 1 gal.	gallons → quarts (multiply by 4)	2 gal. = 8 qt (2 × 4 = 8)
	quarts → gallons (divide by 4)	12 qt = 3 gal. (12 ÷ 4 = 3)
2 pt = 1 qt	quarts → pints (multiply by 2)	2 qt = 4 pt (2 × 2 = 4)
	pints → quarts (divide by 2)	6 pt = 3 qt (6 ÷ 2 = 3)
2 c = 1 pt	pints → cups (multiply by 2)	8 pt = 16 c (8 × 2 = 16)
	cups → pints (divide by 2)	4 c = 2 pt (4 ÷ 2 = 2)
8 fl oz = 1 c	cups → fluid ounces (multiply by 8)	2 c = 16 fl oz (2 × 8 = 16)
	fluid ounces → cups (divide by 8)	24 fl oz = 3 c (24 ÷ 8 = 3)
2 tbsp = 1 fl oz	fluid ounces → tablespoons (multiply by 2)	5 fl oz = 10 tbsp (5 × 2 = 10)
	tablespoons → fluid ounces (divide by 2)	12 tbsp = 6 fl oz (12 ÷ 2 = 6)
3 tsp = 1 tbsp	tablespoons → teaspoons (multiply by 3)	4 tbsp = 12 tsp (4 × 3 = 12)
	teaspoons → tablespoons (divide by 3)	6 tsp = 2 tbsp (6 ÷ 3 = 2)

Multiply when changing from larger unit to smaller unit

Divide when changing from smaller unit to larger unit

Figure 5-9. *Six volume measurement equivalents can be used to calculate other equivalent volume measurements with different units of measure.*

Other equivalent measurements may need to be calculated in more than one step. For example, if a standardized recipe calls for 8 tbsp of sugar and the chef asks for 4 times the recipe, the cook would need to measure 32 tbsp (8 tbsp × 4 = 32 tbsp). Since it would take too long to measure 32 tablespoons of sugar, the tablespoons are converted to a larger unit of measure. First, since 2 tbsp = 1 fl oz, the measurement in tablespoons is changed to fluid ounces by dividing the number of tablespoons (32) by 2, or 32 tbsp ÷ 2 = 16 fl oz. Then, since 8 fl oz = 1 cup, the measurement in fluid ounces is changed to cups by dividing the number of fluid ounces (16) by 8, or 16 fl oz ÷ 8 = 2 cups.

Converting Measurements

Converting is the process of changing a measurement with one unit of measure to an equivalent measurement with a different unit of measure. Three different types of measurement conversions are used in the profession kitchen.

Converting within the Customary or Metric System. Converting gallons to quarts and grams to kilograms are examples of converting within the customary or metric system. To convert a measurement to a different unit within the same measurement system, the measurement is first written as a fraction. For example, to convert 8 qt to gallons, the first step is to write the original measurement as the numerator in a fraction and 1 as the denominator.

The equivalent is written so that the unit in the denominator is the same as the unit in the original measurement. Then, the original measurement is multiplied by the equivalent. When multiplied, the unit in the numerator of the first part of the calculation cancels the matching unit in the denominator in the next part of the calculation. Cancelling is shown by drawing a line through the matching units. In this example, the quarts cancel each other. *See Figure 5-10.*

When multiplying fractions, the numerators are multiplied by each other and the denominators are multiplied by each other. Then, the numerator is divided by the denominator to obtain the final answer (8 gal. ÷ 4 = 2 gal.).

Converting between Customary and Metric Measurements. Converting quarts to liters and pounds to grams are examples of converting between customary and metric measurements. The equivalents between customary units and metric units are not often whole numbers. For example, 1 L is equivalent to 1.06 qt and 1 kg is equivalent to 2.2 lb. Although the customary to metric equivalents may not be as easy to remember as the equivalents within the same system of measurement, the process for converting these measurements is exactly the same as other conversions. Decimal answers should be rounded based on the degree of accuracy required for that ingredient in the recipe.

Converting between Volume and Weight Measurements. Converting cups to ounces or teaspoons to grams are examples of converting between volume and weight measurements. These conversions are unique because volume-to-weight equivalents are approximations and differ depending on the ingredient being measured. For example, there may be more than one

Cancelling Guide

Convert 4 pints to quarts

Step 1: Write the measurement and the conversion factor as fractions in a calculation using multiplication.

$$\frac{4 \text{ pt}}{1} \times \frac{1 \text{ qt}}{2 \text{ pt}}$$

Step 2: Cancel the matching units in the numerators and denominators and then multiply the resulting fractions.

$$\frac{4 \text{ pt}}{1} \times \frac{1 \text{ qt}}{2 \text{ pt}} = \frac{4}{1} \times \frac{1 \text{ qt}}{2} = \frac{4 \text{ qt}}{2}$$

Step 3: Divide the numerator by the denominator to obtain the final answer.

$$4 \text{ qt} \div 2 = 2 \text{ qt}$$

Figure 5-10. *Cancelling matching units of measure makes converting measurements easier.*

volume-to-weight equivalent listed for grapes depending on whether the grapes are sliced or if the grapes are whole. A cup of sliced grapes will weigh more than a cup of whole grapes because the sliced grapes will fill the cup more efficiently.

Refer to DVD for **Volume-to-Weight Equivalents** Culinary Math Tutorial

SCALING RECIPES

A standardized recipe produces a specific yield. When a larger or smaller yield is needed, the recipe is scaled to produce the desired yield. *Scaling* is the process of calculating new amounts for each ingredient in a recipe when the total amount of food the recipe makes is changed. For example, a recipe that serves 4 people may be scaled for use in a restaurant that plans to make 50 servings of the recipe. Similarly, a recipe used by a banquet hall that normally serves 100 people may need to be scaled to make only 12 servings for a small party.

The scaling process starts by calculating a scaling factor. A *scaling factor* is the number that each ingredient amount in a recipe is multiplied by when the recipe yield is changed. ***See Figure 5-11.*** The formula for calculating a scaling factor is as follows:

SF = DY ÷ OY

where
SF = scaling factor
DY = desired yield
OY = original yield

$$Scaling\ Factor = \frac{Desired\ Yield}{Original\ Yield}$$

- **Scaling Based on Weight.** If a potato salad recipe that makes 4 lb of salad is scaled to make 80 lb of salad, the scaling factor is calculated by dividing the desired yield (80 lb) by the original yield (4 lb).

  SF = DY ÷ OY
  $SF = 80\ lb \div 4\ lb$
  $SF = \textbf{20}$

- **Scaling Based on Volume.** If a soup recipe that makes 8 gal. of soup is scaled to make 3 gal. of soup, the scaling factor is calculated by dividing the desired yield (3 gal.) by the original yield (8 gal.).

  SF = DY ÷ OY
  $SF = 3\ gal. \div 8\ gal.$
  $SF = \textbf{0.375}$

- **Scaling Based on Count.** If a cookie recipe that makes 24 cookies is scaled to make 84 cookies, the scaling factor is calculated by dividing the desired yield (84 cookies) by the original yield (24 cookies).

  SF = DY ÷ OY
  $SF = 84\ cookies \div 24\ cookies$
  $SF = \textbf{3.5}$

- **Scaling Based on Portion Size.** If a fish recipe that makes 12 (8 oz) portions is scaled to make 30 (10 oz) portions, the yields must first be converted to a total number of ounces.

$OY = 12 \times 8 \text{ oz} = 96 \text{ oz}$
$DY = 30 \times 10 \text{ oz} = 300 \text{ oz}$

Then the scaling factor formula can be applied.

SF = DY ÷ OY
$SF = 300 \text{ oz} \div 96 \text{ oz}$
$SF = \textbf{3.125}$

Calculating Scaling Factors

$$\text{Scaling Factor (SF)} = \frac{\text{Desired Yield (DY)}}{\text{Original Yield (OY)}}$$

Refer to DVD for
Calculating Scaling Factors
Culinary Math Tutorial

Type of Yield	Original Yield	Desired Yield	Scaling Factor
Total Weight Yield	4 lb	80 lb	80 ÷ 4 = 20
Total Volume Yield	8 gal.	3 gal.	3 ÷ 8 = 0.375
Count Yield	24 cookies	84 cookies	84 ÷ 24 = 3.5
Portion Yield	12 (8 oz) portions (12 × 8 oz = 96 oz)	30 (10 oz) portions (30 × 10 oz = 300 oz)	300 ÷ 96 = 3.125

Figure 5-11. *When a recipe is scaled based on yield, the scaling factor is calculated by dividing the desired yield by the original yield.*

- **Scaling Based on Product Availability.** In some cases, a recipe is scaled because the amount of one ingredient needs to be changed based on availability. In situations like this, the scaling factor is calculated by dividing the available amount of the ingredient by the original amount of the ingredient. The formula for calculating a scaling factor based on product availability is as follows:

SF = AA ÷ OA
where
SF = scaling factor
AA = available amount
OA = original amount

$$\text{Scaling Factor} = \frac{\text{Available Amount}}{\text{Original Amount}}$$

For example, if a beef stew recipe that calls for 15 lb of beef stew meat needs to be made with only 12 lb of beef stew meat, the scaling factor is calculated by dividing the available ingredient amount by the original ingredient amount.

SF = AA ÷ OA
SF = 12 lb beef stew meat ÷ 15 lb beef stew meat
SF = **0.8**

On the other hand, if there are 18 lb of beef stew meat available and all of the meat is to be used, the scaling factor would be 1.2.

SF = AA ÷ OA
SF = 18 lb beef stew meat ÷ 15 lb beef stew meat
SF = **1.2**

Multiplying Scaling Factors

Once a scaling factor is known, every ingredient amount in the original recipe is multiplied by the scaling factor. *See Figure 5-12.* Using the new ingredient amounts will produce the desired yield of the scaled recipe. For example, a meatloaf recipe that normally yields 10 servings is scaled to make 80 servings. The scaling factor is calculated first.

SF = DY ÷ OY
SF = 80 servings ÷ 10 servings
SF = **8**

Then, the new amount of each ingredient is calculated using the following formula:

NA = OA × SF
where
NA = new amount
OA = original amount
SF = scaling factor

$$\text{New Amount} = \frac{\text{Original}}{\text{Amount}} \times \frac{\text{Scaling}}{\text{Factor}}$$

For example, the original meatloaf recipe requires 3 lb of ground beef. The new amount of ground beef required is calculated as follows:

NA = OA × SF
NA = 3 lb × 8
NA = **24 lb**

The new amounts for the remaining ingredients are then calculated using the same formula. It may be necessary to convert some of the new ingredient amounts to a different unit of measure to make the measuring of the ingredients more efficient.

Chef's Tip

Foodservice staff need to be able to calculate new ingredient amounts when the yield of a recipe is changed. Practicing multiplication, division, addition, and subtraction will help fine tune these needed math skills.

Scaling Recipes

Original yield = 2 gal.
Desired yield = 5 gal.
Scaling factor = 2.5

$$CF = \frac{DY}{OY}$$

$$\frac{5}{2} = \mathbf{2.5}$$

Butternut Squash and Ginger Bisque (Yield: 5 gal.)

Original Quantity	× CF	= Desired Quantity
4 butternut squash, roasted	2.5	10
8 oz onion, diced	2.5	20 oz
4 oz butter	2.5	10 oz
32 fl oz chicken broth	2.5	80 fl oz
1 tsp cinnamon	2.5	2½ tsp
¼ tsp cardamom	2.5	⅝ tsp
¼ tsp ginger	2.5	⅝ tsp
2 tsp cracked black peppercorns	2.5	5 tsp
4 tsp kosher salt	2.5	10 tsp
8 oz brown sugar	2.5	20 oz
16 fl oz heavy cream	2.5	40 fl oz

Figure 5-12. *Once a scaling factor is known, each ingredient amount in the original recipe is multiplied by the scaling factor.*

Scaling Considerations

When recipes are scaled, additional considerations must be taken into account before actually preparing the recipe. Addressing these considerations will help to ensure that the final product is of the same quality as the original recipe.

Measurements. If a new measurement is calculated that is not easily measured, such as 3.7 cups, the measurement will need to be adjusted to make measuring more practical. However, to avoid affecting the final result, the amount should not be adjusted drastically. For example, 3.7 cups should be adjusted to 3.75 cups.

Equipment. Cooking equipment must properly accommodate the adjusted amount of food. It is important that cooking equipment is large enough to handle an increased yield.

Mixing Time. Instructions related to mixing times may need to be adjusted, especially when a recipe yield is significantly increased, to ensure that the appropriate amount of time is given to mix the ingredients adequately.

Cooking Temperature. A professional cook must determine whether the cooking temperature needs to be adjusted. Normally, cooking time is adjusted but in certain circumstances, such as when a recipe yield is significantly decreased, it may be necessary to reduce the cooking temperature to keep food from drying out.

Cooking Time. A professional cook must determine whether the cooking time for a recipe needs to be increased or decreased. If a larger roast is used for a recipe, a longer cooking time will normally be required. However, 10 steaks cooked on a grill will cook in the same amount of time as 2 steaks on a grill.

CALCULATING FOOD COSTS

A foodservice operation needs to continually monitor both the cost of food purchased and the cost of meals served to ensure the operation earns more in food sales than it pays in expenses. Calculating food costs involves much more than simply adding up the cost of ingredients. The terms "as purchased (AP)" and "edible portion (EP)" are commonly used to distinguish between a food item before and after it is trimmed of waste. An *as-purchased (AP) quantity* is the original amount of a food item as it is ordered and received. An *edible-portion (EP) quantity* is the amount of a food item that remains after trimming and is ready to be served or used in a recipe. *See Figure 5-13.*

As Purchased vs. Edible Portion

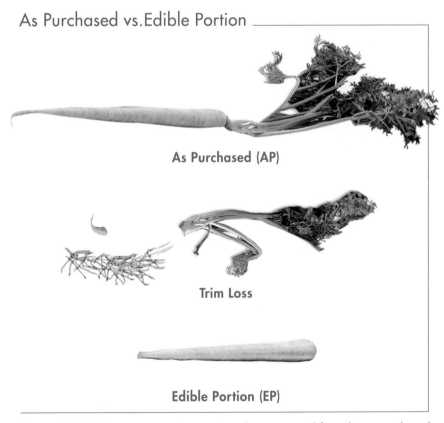

As Purchased (AP)

Trim Loss

Edible Portion (EP)

Figure 5-13. *Carrot greens, tips, and peel are removed from the as-purchased (AP) form prior to use. The edible portion (EP) of the carrot remains.*

Canadian Beef, Beef Information Centre

As-Purchased Costs

An *as-purchased (AP) cost* is the amount paid for a product in the form it was ordered and received. Foodservice operations buy many food products in bulk quantities such as cases of hamburger patties, tubs of ice cream, crates of milk, or pails of pickles. Beverages are also typically purchased in bulk quantities such as 24-bottle cases of iced tea, 10-bottle cases of wine, or half barrels (kegs) of beer. The AP costs of food and beverage products are equal to the prices listed on the invoice. *See Figure 5-14.* An *invoice* is a document provided by a supplier that lists the items delivered to a foodservice operation and the prices of those items.

As-Purchased Costs

DATE	PURCHASE ORDER NO. RA 05353066	PAGE NO. 1 of 1	DELIVERY DATE

SHIP TO: Acme Foods
Silver Spring Rd
Stony Branch, OH 66005

VENDOR Stone Cold Foods

CHARGE & INVOICE TO: Acme Foods
Silver Spring Rd
Stony Branch, OH 66005

QUANTITY	UNIT	ITEM ID—DESCRIPTION	UNIT PRICE	ITEM TOTAL
1	Case (80 ct)	Frozen Hamburger Patties (20 lb)	$30.00	$30.00
1	1 Tub (3 gal.)	Signature Ice Cream	$12.00	$12.00

I certify that sufficient funds are available for this purchase.

Signature

Purchasing Manager

Title

Date

Tax (if applicable)	
Shipping	$6.00
Total	

Type of Payment
☐ Payment Enclosed
☑ COD
☐ Credit

Figure 5-14. *As-purchased (AP) costs are listed on an invoice.*

When a product is purchased in bulk, it is often necessary to calculate a unit cost based on the AP cost. A *unit cost* is the cost of a product per unit of measure. Unit costs can be based on weight (ground turkey at $2.75 per lb), volume (milk at $3.00 per gal.), or count (bread at $2.00 per loaf). The *as-purchased (AP) unit cost* is the unit cost of a food item based on the form in which it is ordered and received. The AP unit cost of an item is calculated by applying the following formula:

$APU = APC \div NU$

where

APU = AP unit cost

APC = AP cost

NU = number of units

$$AP\ Unit\ Cost = \frac{AP\ Cost}{Number\ of\ Units}$$

For example, to calculate the AP unit cost of an egg, divide the AP cost for a case of eggs by the number of eggs in the case. If the price paid for a case of eggs is $14.40 and there are 15 dozen eggs in the case, the AP unit cost would be $0.08 per egg.

APU = APC ÷ NU

APU = $14.40 ÷ 15 dozen

APU = $14.40 ÷ (15 dozen × 12 eggs/dozen)

APU = $14.40 ÷ 180 eggs

APU = **$0.08/egg**

The unit of measure used to calculate a unit cost should be based on how the product is used in recipes. Sometimes it may be helpful to calculate the unit cost of an item in more than one unit of measure. For example, some recipes may call for fluid ounces of heavy cream and other recipes may call for quarts of heavy cream. *See Figure 5-15.*

Calculating As-Purchased Unit Costs

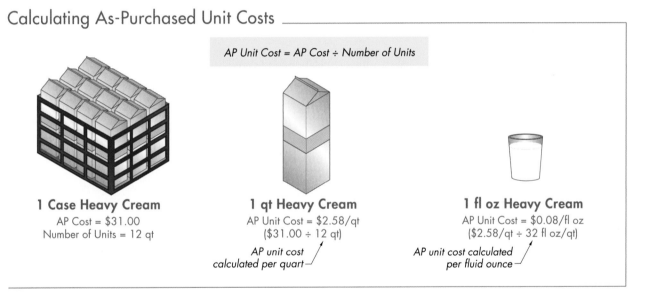

AP Unit Cost = AP Cost ÷ Number of Units

1 Case Heavy Cream
AP Cost = $31.00
Number of Units = 12 qt

1 qt Heavy Cream
AP Unit Cost = $2.58/qt
($31.00 ÷ 12 qt)
AP unit cost calculated per quart

1 fl oz Heavy Cream
AP Unit Cost = $0.08/fl oz
($2.58/qt ÷ 32 fl oz/qt)
AP unit cost calculated per fluid ounce

Figure 5-15. *As-purchased unit costs can be calculated based on more than one unit of measure.*

Edible-Portion Costs

A similar calculation is performed to determine the edible-portion (EP) unit cost based on the AP unit cost. The *edible-portion (EP) unit cost* is the unit cost of a food or beverage item after taking into account the cost of the waste generated by trimming. Unless a food item has a yield percentage of 100%, the EP unit cost will always be higher than the AP unit cost. *See Figure 5-16.* Consider a mashed potato recipe that calls for 10 lb of peeled potatoes. The cost of the potatoes for the recipe will be more than the cost of 10 lb of whole potatoes because more than 10 lb of whole potatoes will need to be peeled in order to yield 10 lb of peeled potatoes.

Edible-Portion Unit Costs

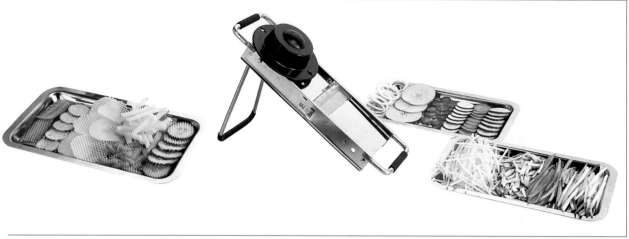

Figure 5-16. *Unless the yield is 100%, the edible-portion (EP) unit cost will always be higher than the as-purchased (AP) unit cost due to the waste generated by trimming the product.*

The EP unit cost of a food item is calculated by applying the following formula:

EPU = APU ÷ YP

$$EP\ Unit\ Cost = \frac{AP\ Unit\ Cost}{Yield\ Percentage}$$

where
EPU = EP unit cost
APU = AP unit cost
YP = yield percentage

For example, if a recipe calls for peeled potatoes and the AP unit cost of whole potatoes is $0.50 per lb and the yield percentage is 80%, what is the EP unit cost of the peeled potatoes?

EPU = APU ÷ YP
EPU = $0.50/lb ÷ 80%
EPU = $0.50/lb ÷ 0.80
EPU = $0.625/lb = **$0.63/lb**

The ingredient amounts provided in a recipe are EP quantities. The total cost of an ingredient in a recipe can be calculated by multiplying the EP quantities provided in the recipe by the EP unit cost for each ingredient. When calculating the EP unit cost of an ingredient, the unit of measure in the EP quantity and the AP unit cost must be the same. For example, if the EP quantity of sugar in a cake recipe is 12 oz and the AP unit cost of sugar is $0.64 per lb, the AP unit cost will need to be converted to a price per ounce.

This conversion is done by multiplying the original cost by the appropriate measurement equivalent. In this case, since 16 oz = 1 lb, $0.64 per lb is equivalent to $0.04 per oz. The total cost of the sugar in the recipe can then be calculated by multiplying the EP amount (12 oz) by the EP unit cost ($0.04 per oz), or 12 oz × $0.04/oz = $0.48.

Yield Percentages

Foodservice operations purchase many products in a form that is different from the way the product will be used. For example, meats may need to be trimmed of excess fat and bone, produce may need to be peeled and seeded, and seafood may need to be scaled, skinned, and have the heads and fins removed. In all of these cases, waste is generated. To account for waste, cost calculations are based on yield percentages.

A *yield percentage* is the edible-portion (EP) quantity of a food item divided by the as-purchased (AP) quantity and is expressed as a percentage. Yield percentages do not apply to food items that are served in the same form as they are purchased, such as premade pastries that are simply plated for service. Three different formulas involving yield percentages are used for the following reasons:

- **Calculating As-Purchased (AP) Quantities.** An AP quantity is calculated when the amount of a food item that is required for a recipe (EP quantity) is known and the amount to be ordered needs to be determined.

- **Calculating Edible-Portion (EP) Quantities.** An EP quantity is calculated when a purchased amount of food (AP quantity) is already on hand and the edible or usable amount of the food needs to be calculated.

- **Calculating Yield Percentages (YP).** A yield percentage is calculated when it cannot be found for a particular food item in any reference material.

The percentage circle can be used to help understand how calculations using yield percentages are performed. The circle is divided into three sections. The top section of the circle represents the EP quantity (EPQ), the part of the food item that is edible. The lower left section of the circle represents the AP quantity (APQ), the whole amount of the item. The lower right section contains the yield percentage (YP). When any two of the three variables are known, the third variable can be calculated. *See Figure 5-17.*

Yield Percentage Circle

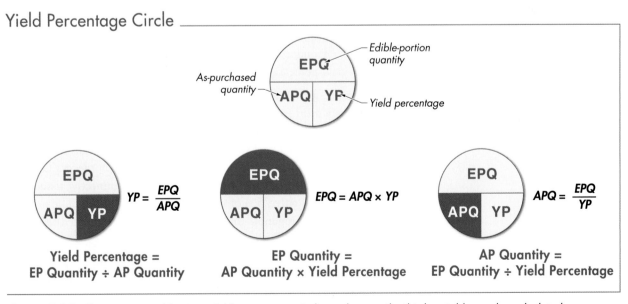

Figure 5-17. *If any two variables in a yield percentage circle are known, the third variable can be calculated.*

Calculating As-Purchased Quantities. Calculating an AP quantity is done to determine the proper amount of an ingredient to order to make a given quantity of food. The AP quantity is calculated by using the following formula:

APQ = EPQ ÷ YP

where

APQ = AP quantity
EPQ = EP quantity
YP = yield percentage

$$\text{AP Quantity} = \frac{\text{EP Quantity}}{\text{Yield Percentage}}$$

Consider a banquet for 100 people that includes grilled salmon on the menu. The recipe for the grilled-salmon dish calls for a ½ lb fillet of salmon per serving. The total EP quantity of salmon can be calculated by multiplying the amount of salmon required per person by the number of people attending the banquet.

If the chef decides to order whole salmon for the banquet, the heads, fins, skin, and bones of the salmon will need to be removed. If the yield percentage for cutting salmon fillets from whole salmon is 50%, the chef can then calculate how much whole salmon to order. The quantity of salmon to be ordered can be calculated by applying the formula for AP quantity as follows:

APQ = EPQ ÷ YP
APQ = 50 lb ÷ 50%
APQ = 50 lb ÷ 0.50
APQ = **100 lb**

Calculating Edible-Portion Quantities. Calculating an EP quantity is done to determine how much food can be prepared using an amount of an ingredient that is already on hand. The EP quantity can be calculated using the following formula:

EPQ = APQ × YP

where

EPQ = EP quantity
APQ = AP quantity
YP = yield percentage

$$\text{EP Quantity} = \text{AP Quantity} \times \text{Yield Percentage}$$

For example, a chef may decide to take advantage of a special price being offered by a meat supplier and purchase 40 lb of whole beef tenderloin. The chef needs to know how many servings of a recipe can be made using the beef tenderloin for a beef stroganoff recipe. The recipe calls for 6 oz per serving of fully trimmed beef tenderloin cut into 1 inch pieces. The chef knows from experience that after trimming the fat and connective tissue from the beef tenderloin, only 75% of the tenderloin will be used. The number of servings of beef stroganoff that can be prepared is calculated using the following formula for EP quantity:

EPQ = APQ × YP
EPQ = 40 lb × 75%
EPQ = 40 lb × 0.75
EPQ = **30 lb**

Calculating Yield Percentages. Common yield percentages are available in various reference tables. However, if a yield percentage for a particular food item is not known and cannot be found in reference material, it can be calculated by performing a raw yield test or a cooking-loss yield test.

- **Raw Yield Tests.** A *raw yield test* is a procedure used to determine the yield percentage of a food item that is trimmed of waste prior to being used in a recipe. Yield percentage is calculated by using the following formula:

YP = EPQ ÷ APQ
where
YP = yield percentage
EPQ = EP quantity
APQ = AP quantity

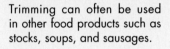

$$\text{Yield Percentage} = \frac{\text{EP Quantity}}{\text{AP Quantity}}$$

To perform the raw yield test, a food item is purchased and weighed to determine the AP quantity. The food item is then trimmed and the edible portion is weighed to determine the EP quantity. For example, to calculate the yield percentage of carrots, the AP quantity of carrots is weighed. Then, the carrots are peeled and the greens and tips are removed as waste (unless the trimmings can be used in another recipe such as a stock). The EP quantity is determined by weighing the cleaned carrots. If 10 lb of carrots weigh 8.5 lb after being trimmed, the yield percentage can be calculated using the following formula:

YP = EPQ ÷ APQ
YP = 8.5 lb ÷ 10 lb
YP = 0.85
YP = **85%**

- **Cooking-Loss Yield Tests.** A *cooking-loss yield test* is a procedure used to determine the yield percentage of a food item that loses weight during the cooking process. For example, meat loses weight as fat is rendered and moisture is lost during cooking. *See Figure 5-18.* The cooking-loss yield test is used when the EP quantity is based on the amount of cooked food to be served as opposed to the amount of raw food to be used in a recipe.

Cooking-Loss Yield Test

Hamburger Before Cooking

Hamburger After Cooking

Figure 5-18. *Meat can lose up to 30% of its weight as moisture is lost and fat is rendered during the cooking process.*

The yield percentage is calculated by dividing the EP quantity by the AP quantity. To perform the test, a food item is weighed to determine the AP quantity. The food item is then weighed again after being trimmed and/ or cooked to determine the EP quantity. For example, a hamburger that weighs 8 oz prior to cooking might weigh 6 oz after cooking. The yield percentage in this case can be calculated by using the following formula:

YP = EPQ ÷ APQ
$YP = 6 ÷ 8$
$YP = 0.75$
$YP =$ **75%**

Calculations involving yield percentages are performed regularly in the professional kitchen. However, variables such as rounding, employee skill level, and product condition affect yield percentages.

- **Rounding.** Special attention should be given to how the results for yield percentage calculations are rounded. *See Figure 5-19.* For example, if it is determined that 30.4 lb of fish should be ordered for a banquet, rounding the amount down to 30 lb using standard math rules would not result in enough fish for the banquet.

Rounding Yield Percentage Calculation Results

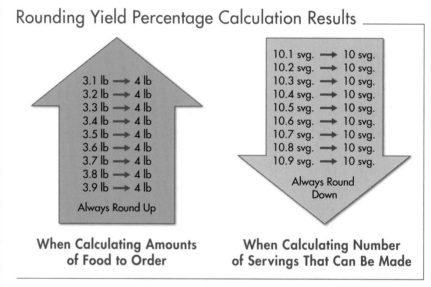

Figure 5-19. *Special rules apply to how the results for yield percentage calculations are rounded depending on the quantity being calculated.*

When ordering food, the yield percentage results should always be rounded up to ensure that enough food is purchased. However, when calculating the number of servings that can be made from a given recipe, such as 14.8, the yield percentage result should be rounded down to 14 because the recipe will not make 15 full servings.

- **Skill Levels.** Employee skill levels can also dramatically affect yield percentages. Butchering meats is a skill that requires practice. Therefore, a butcher will experience higher yield percentages when trimming meats than an entry-level cook assigned to do the same task.

- **Product Condition and Sizes.** Another factor affecting yield percentage is the condition and size of the initial product. For example, fruits that are unblemished will have higher yield percentages than bruised fruits that require the removal of undesirable parts. Likewise, large potatoes have a higher yield percentage than small potatoes. More waste is generated by peeling many small potatoes compared to peeling fewer large potatoes.

As-Served Costs

Once the costs of the individual ingredients in a recipe are calculated, the costs are added to determine the total cost of the recipe. Then, the as-served costs for that recipe can be calculated. An *as-served (AS) cost* is the cost of a menu item as it is served to a customer. It is important to note that an AS cost is the total cost of the ingredients required to prepare one serving of a menu item.

Consider a chicken sandwich that is offered as a stand-alone item on a menu. If the sandwich consists of a grilled chicken breast served on a bun with a slice of tomato, a slice of cheese, lettuce, and mayonnaise, the costs of the ingredients are added to calculate the AS cost of the chicken sandwich. However, if the same chicken sandwich is offered on the menu as a meal that comes with fries and coleslaw, the AS cost of that menu item is equal to the total cost of the sandwich, fries, and coleslaw. On this menu, the chicken sandwich is treated as one ingredient of the chicken sandwich meal. *See Figure 5-20.*

Calculating As-Served Costs

Chicken Sandwich						
Ingredients	EP Quantity	EP Unit of Measure	AP Unit Cost	Yield Percentage	EP Unit Cost	Total Cost
Chicken breast	4	oz	$0.20 per oz	90%	$0.22	$0.89
Bun	1	each	$0.40 per each	100%	$0.40	$0.40
Tomato slice	1	oz	$0.18 per oz	90%	$0.20	$0.20
Cheese slice	1	each	$0.20 per each	100%	$0.20	$0.20
Lettuce, shredded	0.25	oz	$0.10 per oz	80%	$0.13	$0.03
Mayonnaise	0.5	oz	$0.15 per oz	100%	$0.15	$0.08
					Total As-Served Cost:	$1.80

Individual recipe cost becomes part of menu item cost

Chicken Sandwich Meal						
Ingredients	EP Quantity	EP Unit of Measure	AP Unit Cost	Yield Percentage	EP Unit Cost	Total Cost
Chicken sandwich	1	each	$1.80 per each	100%	$1.80	$1.80
French fries	4	oz	$0.20 per oz	100%	$0.20	$0.80
Coleslaw	3	oz	$0.20 per oz	100%	$0.20	$0.60
					Total As-Served Cost:	$3.20

Figure 5-20. *As-served (AS) cost is the total cost of the ingredients and recipes that make up a menu item.*

To calculate the AS cost of a menu item that is prepared in large quantities and then portioned into servings, divide the total cost of the recipe by the number of portions the recipe yields. For example, if the total cost of the ingredients to prepare 2 gal. of tomato soup is $15.00 and the recipe yields 32 (8 fl oz) servings, the AS cost of a serving of tomato soup would be $0.47, or $15.00 ÷ 32 = $0.47/serving.

FOOD COST PERCENTAGES

Another way of examining food costs is through the use of food cost percentages. A *food cost percentage* is a percentage that indicates how the cost of food relates to the menu prices and food sales of a foodservice operation. There are three types of food cost percentages used in the foodservice industry.

- **Menu-Item Food Cost Percentage.** A *menu-item food cost percentage* is the AS cost of a menu item divided by the menu price, written as a percent.

- **Overall Food Cost Percentage.** An *overall food cost percentage* is the total amount of money a foodservice operation spends on food divided by the total food sales over a defined period of time, written as a percent.

- **Target Food Cost Percentage.** A *target food cost percentage* is the percentage of food sales that a foodservice operation plans to spend on purchasing food.

Like a food cost percentage, a *beverage cost percentage* is a percentage that indicates how the cost of beverages relates to menu prices and beverage sales of a foodservice operation. The word beverage can be substituted for the word food in any of these calculations because the math is the same. If a menu item consists of both a food and beverage component that is going to be offered for a single price, the menu price must be based on a combined food and beverage cost percentage.

Carlisle FoodService Products

Menu-Item Food Cost Percentages

A menu-item food cost percentage is equal to the AS cost of a menu item divided by the menu price written as a percent. This percentage shows how the cost of the ingredients relates to the price charged for the menu item. To calculate a menu-item food cost percentage, apply the following formula:

IFC% = ASC ÷ MP

where

IFC% = menu-item food cost percentage

ASC = AS cost

MP = menu price

$$\text{Menu-Item Food Cost Percentage} = \frac{\text{AS Cost}}{\text{Menu Price}}$$

A foodservice operation makes a profit by charging more for the items on a menu than it costs to prepare those items. Therefore, the menu price is always higher than the AS cost of that item. Since a menu-item food cost percentage is equal to the AS cost divided by the menu price, a menu-item food cost percentage will always be less than 100%.

For example, if the menu price for the chicken sandwich is $6.95 and the AS cost is $1.80, the menu-item food cost percentage for the chicken sandwich would be 25.9%.

IFC% = ASC ÷ MP
IFC% = $1.80 ÷ $6.95
IFC% = 0.259
IFC% = **25.9%**

Likewise, if the menu price for the chicken sandwich served with fries and coleslaw is $9.95 and the AS cost is $3.20, the menu-item food cost percentage for the meal would be 32.1%.

IFC% = ASC ÷ MP
IFC% = $3.20 ÷ $9.95
IFC% = 0.321
IFC% = **32.1%**

Carlisle FoodService Products

Overall Food Cost Percentages

In addition to calculating food cost percentages for individual menu items, foodservice operations also calculate overall food cost percentages based on total food sales. An overall food cost percentage is equal to the total amount of money a foodservice operation spends on food divided by its total food sales over a defined period of time. To calculate an overall food cost percentage, use the following formula:

OFC% = FC ÷ FS
where
OFC% = overall food
 cost percentage
FC = total food costs
FS = total food sales

$$\text{Overall Food Cost Percentage} = \frac{\text{Total Food Costs}}{\text{Total Food Sales}}$$

For example, if a restaurant spends $9000 on food in a month when its total food sales were $36,000 for that month, the overall food cost percentage for the restaurant would be 25%.

OFC% = FC ÷ FS
OFC% = $9000 ÷ $36,000
OFC% = 0.25
OFC% = **25%**

The other 75% of the sales would then be available for the foodservice operation to pay for expenses, such as payroll, rent, and utilities, and to contribute to the earnings of the owners of the operation. Lower food cost percentages translate into higher potential earnings for the owners of the operation.

Target Food Cost Percentages

A target food cost percentage is the percentage of total food sales that a foodservice operation plans to spend on purchasing food. Foodservice owners and managers regularly compare overall food cost percentages with target food cost percentages to determine if food costs are higher or lower than planned. *See Figure 5-21.*

When the overall food cost percentage of a foodservice operation is lower than the target food cost percentage, the operation has earned more money selling food than planned. Likewise, if the actual overall food cost percentage is higher than the target food cost percentage, the operation has earned less money selling food than planned. It is the responsibility of management to monitor food costs and make any necessary adjustments.

Food Cost Percentages

Daily Sales Report—Lunch						
Target Food Cost Percentage = 30%						
Menu Item	AS Cost	Menu Price	Menu Item Food Cost % (AS Cost ÷ Menu Price)	Number Sold	Total Food Cost (AS Cost × Number Sold)	Total Food Sales (Menu Price × Number Sold)
Appetizers						
Bruschetta	$1.75	$5.95	29.4%	48	$84.00	$285.60
BBQ shrimp	$4.25	$11.95	35.6%	27	$114.75	$322.65
Entrées						
Fried chicken	$2.65	$12.95	20.5%	43	$113.95	$556.85
Strip steak	$9.00	$25.95	34.7%	32	$288.00	$830.40
Desserts						
Strawberry shortcake	$2.25	$5.95	37.8%	40	$90.00	$238.00
Chocolate cake	$0.50	$3.95	12.7%	30	$15.00	$118.50
Total					**$705.70**	**$2352.00**
Overall Total Food Cost Percentage (Total Food Costs ÷ Total Food Sales)						**30.0%**

Figure 5-21. *The overall food cost percentage is often compared to the target food cost percentage to determine how well a menu item is selling.*

CALCULATING MENU PRICES

There are three primary methods used by foodservice operations to calculate menu prices. These methods include food cost percentage pricing, perceived value pricing, and contribution margin pricing. Regardless of the pricing method used, the AS costs of the menu items must be known first.

Beverage pricing is calculated the same way as food pricing. Simply substitute beverage costs for food costs in the original calculation for menu-item cost percentages.

Food Cost Percentage Pricing

Food cost percentage pricing begins by calculating a target price for each menu item. A *target price* is the price that a foodservice operation needs to charge for a menu item in order to meet its target food cost percentage. Target prices are calculated using the following formula:

$$TP = ASC \div TFC\%$$

where

TP = target price

ASC = AS cost of a menu item

$TFC\%$ = target food cost percentage

$$Target\ Price = \frac{AS\ Cost}{Target\ Food\ Cost\ Percentage}$$

For example, if the AS cost of a sandwich is $1.45 and the target food cost percentage is 30%, the target price would be $4.83.

TP = ASC ÷ TFC%

$TP = \$1.45 \div 30\%$

$TP = \$1.45 \div 0.30$

$TP = \mathbf{\$4.83}$

Some foodservice operations will stop at this point and set the menu price equal to the target price. However, most foodservice operations tend to consider target prices as a starting point and then adjust the price up or down based on how management thinks customers will perceive the price.

Perceived Value Pricing

Perceived value pricing is the process of adjusting a target menu price based on how management thinks a customer will perceive the price. Many foodservice operations use a combination of food cost percentage pricing and perceived value pricing to determine final menu prices. *See Figure 5-22.*

Pricing Menu Items

Food Cost Percentage Pricing

Calculate as-served cost of menu item

Divide as-served menu item cost by target food cost percentage to calculate target menu price

Is target price appropriate for the menu?

Yes → Menu-item food cost percentage is equal to target food cost percentage

No

Is target price too high or too low?

Too Low → Set menu price higher than target price → Menu-item food cost percentage is lower than target food cost percentage

Too High → Set menu price lower than target price → Menu-item food cost percentage is higher than target food cost percentage

Perceived Value Pricing

Figure 5-22. *Many foodservice operations use a combination of food cost percentage pricing and perceived value pricing to calculate menu prices.*

Refer to DVD for
Pricing Menu Items
Culinary Math Tutorial

There are two aspects to perceived value pricing. The first involves adjusting a target price to look like a price customers have become used to seeing on menus. For example, a target price for a menu item is calculated to be $7.03. Most foodservice operations will adjust the price down to $6.95 or $7.00.

Therefore, the actual food cost percentage for the menu item would be slightly higher or lower than the target food cost percentage. For example, if the target price of a bowl of soup is calculated to be $4.67 based on an AS cost of $1.40 and a 30% target food cost, the restaurant might increase the price to $4.95. At $4.95, the actual food cost percentage would be 28.3%, which is slightly lower than the target food cost percentage of 30%.

$$IFC\% = ASC \div MP$$
$$IFC\% = \$1.40 \div \$4.95$$
$$IFC\% = 0.283$$
$$IFC\% = \mathbf{28.3\%}$$

The second aspect of perceived value pricing involves adjusting a target price so that customers will view the price as reasonable relative to other items on the menu and similar items sold elsewhere. For example, a restaurant may offer an appetizer of nachos made with tortillas chips, taco meat, cheese, and tomatoes that has an AS cost of $1.05. If the target price is calculated based on a 28% target food cost percentage, the target price would be $3.75.

$$TP = ASC \div TFC\%$$
$$TP = \$1.05 \div 28\%$$
$$TP = \$1.05 \div 0.28$$
$$TP = \mathbf{\$3.75}$$

If other appetizers on the restaurant menu include Buffalo wings for $8.95 and mozzarella sticks for $5.95, customers may be willing to pay more than $3.75 for the nachos because the perception is that nachos represent a substantial dish and should cost more than the mozzarella sticks but less than the Buffalo wings. In a case like this, the restaurant is likely to increase the price of the nachos to $6.95, which means the menu-item food cost percentage for the nachos would be 15.1%.

$$IFC\% = ASC \div MP$$
$$IFC\% = \$1.05 \div \$6.95$$
$$IFC\% = 0.151$$
$$IFC\% = \mathbf{15.1\%}$$

If the price was left at the target price of $3.75 the restaurant would earn $2.70 ($3.75 −$1.05 = $2.70) on each order of nachos. With the price adjusted to $6.95 the restaurant would earn $5.90 ($6.95 − $1.05 = $5.90) on each order of nachos. A lower menu-item food cost percentage means that the foodservice operation earns more on each order of nachos that are sold. However, just because lower percentages result in higher earnings per order, it does not mean that a foodservice operation can or should base all menu prices on a lower food cost percentage.

To determine the price of other items based on the same menu-item food cost percentage as the nachos, use the following formula:

$$MP = ASC \div IFC\%$$

where
MP = menu price
ASC = AS cost
$IFC\%$ = menu-item food cost percentage

For example, if the same restaurant selling the nachos prices a 24 oz rib-eye steak with an AS cost of $14.00 based on the same 15.1% food cost percentage, the price of the steak would be $92.71.

$$MP = ASC \div IFC\%$$
MP = $14.00 ÷ 15.1%
MP = $14.00 ÷ 0.151
MP = **$92.71**

National Cattlemen's Beef Association

It is unlikely that many customers would pay $92.71 for steak, even at a high-end restaurant. If the target food cost percentage is adjusted to 28%, the price of the steak would still be relatively high at $50.00 ($14.00 ÷ 0.28 = $50.00). If restaurant management thinks most customers will perceive a price of $50.00 as too high, the price may be reduced to be more in line with the perceived value of that restaurant's customers. If the final menu price set for the steak is $36.95, the food cost percentage will increase to 37.9%.

$$IFC\% = ASC \div MP$$
$IFC\%$ = $14.00 ÷ $36.95
$IFC\%$ = 0.379
$IFC\%$ = **37.9%**

It is normal for items on a menu to have different food cost percentages. As long as there is a proper balance between items with lower food cost percentages and items with higher food cost percentages, the foodservice operation can maintain an overall food cost percentage that is close or equal to its target food cost percentage.

Contribution Margin Pricing

Another method used to calculate menu prices is referred to as contribution margin pricing. A *contribution margin* is the amount added to the AS cost of a menu item to determine a menu price. A foodservice operation calculates prices with a contribution margin to ensure that the amount of money made on each serving of food will cover expenses and still result in a profit. This type of pricing is not common in restaurants but is often used when calculating a price charged per person for a special event.

For example, a banquet operation is planning a party for 200 people. The menu includes a salad, an entrée, and a dessert that has a total AS cost of $20.00 per person. Therefore, the total AS cost for all of the food will be $4,000 (200 people × $20/person). The banquet manager calculates that all of the other expenses, such as labor and rental fees, total an additional $5000. It is banquet policy that a party of this size must generate earnings of $2000. Therefore, the contribution margin for this event would be $7000 ($5000 + $2000).

To calculate the contribution margin on a per person basis, the banquet manager simply divides the total contribution margin by the number of people attending the event ($7000 ÷ 200 people = $35/person).

This contribution margin is then added to the AS cost per person to calculate the final menu price for the party, which is $55 per person ($20.00/person + $35.00/person = $55/person).

Pricing Forms

A *pricing form* is a tool often used to help calculate the AS cost of a menu item and establish a menu price. ***See Figure 5-23.*** Pricing forms vary among foodservice operations, but generally contain the following elements:

- **Menu Item Name**—The name on the pricing form should match the name used on the menu.

- **Ingredients**—The list of ingredients required to prepare the menu item.

- **Edible-Portion (EP) Quantity**—The EP quantity is the amount of each ingredient required to prepare the menu item.

- **Number of Portions**—Lists the number of portions that are made with the ingredients listed on the pricing form.

Pricing Forms

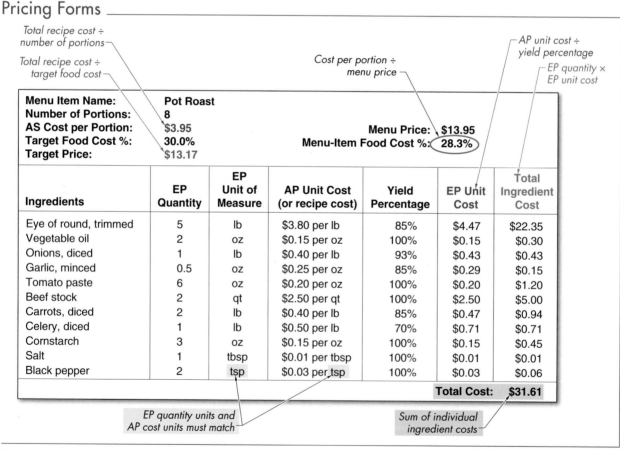

Figure 5-23. *A pricing form is used to help calculate the as-served (AS) cost of a menu item and to establish a menu price.*

- **As-Purchased (AP) Unit Cost**—The unit cost for each ingredient in the same unit of measure as the EP quantity of each ingredient.
- **Yield Percentage**—The yield percentage of each ingredient listed.
- **Edible-Portion (EP) Unit Cost**—Calculated by dividing the AP unit cost of each ingredient by the ingredient's yield percentage.
- **Total Ingredient Cost**—The total cost for each ingredient is calculated by multiplying the EP quantity of each ingredient by the EP unit cost of each ingredient.
- **Total Cost**—The sum of all of the individual ingredient costs.
- **AS Cost per Portion**—Calculated by dividing the total cost by the number of portions.
- **Target Food Cost Percentage**—The target food cost percentage established by the foodservice operation.
- **Target Price**—Calculated by dividing the AS cost per portion by the target food cost percentage.
- **Menu Price**—The price listed on the menu for a single portion of the menu item.
- **Menu-Item Food Cost Percentage**—Calculated by dividing the AS cost per portion by the menu price.

Eloma Combi Ovens

After a pricing form has been completed for all of the items on a menu, the foodservice operation updates the pricing forms when the cost of ingredients change to ensure that menu prices are adjusted as needed. The following steps are typically followed to complete a pricing form.

1. Enter information from the standardized recipe including the menu item name, number of portions, ingredients, and ingredient amounts (EP quantities).
2. Record the AP unit cost of each ingredient. If necessary, convert the unit cost to the same unit of measure as used in the EP quantity.
3. Record the yield percentage for each ingredient.
4. Calculate the EP unit costs by dividing the AP unit cost of each ingredient by the yield percentage of the ingredient.
5. Calculate the total cost of each ingredient by multiplying the EP quantity of each ingredient by the EP unit cost of each ingredient.
6. Calculate the total cost by adding the costs of each ingredient.
7. Calculate the cost per portion by dividing the total recipe cost by the number of portions.
8. Calculate the target price by dividing the total cost by the target food cost percentage.
9. Adjust the target price if desired and record the menu price.
10. Calculate the food cost percentage by dividing the cost per portion by the menu price.

MAKING A PROFIT

There are many costs that must be managed to run a successful foodservice operation. These costs include goods (such as food and beverage products) and services (such as maintenance and repairs). There are two basic types of costs: fixed costs and variable costs. A *fixed cost* is a cost that does not change as sales increase or decrease. Rent, real estate taxes, and insurance are examples of fixed costs.

A *variable cost* is a cost that increases or decreases in proportion to the volume of production. When sales are up variable costs increase. When sales are down variable costs decrease. Food and beverage products and payroll expenses are examples of variable costs because they increase or decrease depending on the sales volume of the foodservice operation.

Determining if a particular foodservice operation is profitable is as simple as comparing revenue to expenses. A *profit* is the amount of money earned by an operation when revenue is greater than expenses. The opposite of a profit is a loss. A *loss* is the amount of money lost by an operation when revenue is less than expenses.

Opportunities to make money or lose money are present in all aspects of a foodservice operation. The focus on earning a profit starts with the planning of the menu and continues with the purchasing, receiving, storing, preparing, and serving of food.

Planning Menus

A menu is a document that markets the foodservice operation to the customer. It lists products that satisfy customers, are profitable, and can compete with the alternatives offered elsewhere. In addition to calculating the appropriate prices to charge for menu items, other considerations must be taken into account when planning a profitable menu.

- **Kitchen Equipment.** The kitchen must be properly equipped to prepare menu items efficiently. When the appropriate equipment is not available, the operation is less efficient. This inefficiency results in higher payroll expenses.

- **Employee Skill Level.** The techniques used to prepare the items on a menu should be in line with the abilities of the kitchen staff. If employees lack the required skills, food products are likely to be wasted due to errors in preparation causing food expenses to increase.

- **Seasonal Ingredients.** Some menu items should only be offered during certain times of the year when key ingredients are readily available at a reasonable price.

- **Customer Demand.** All items on a menu should be popular enough with customers that the ingredients required to prepare them are not wasted.

Purchasing Food

Purchasing involves the selection and ordering of products required to prepare and serve the items on a menu. A tool used to help ensure that the right products are purchased at the best available price is a purchase specification.

Appetizers

baked brie en croute
with sundried tomato pesto and aged balsamic glaze

pan-seared crab cake
with caramelized Vidalia onions, fire-roasted
bell peppers and lemon butter sauce

sautéed black mussels
with garlic, white wine and parsley

Soups

chilled English cucumber soup
with charentais melons, yuzu pearls, salmon ceviche and roe

light lobster bisque
with bomba rice and summer vegetables

creamy yellow spring pea soup
with house-smoked trout

leek and potato potage
with Wellfleet clams and fines herbes

cream of Alsace cabbage soup
with home-smoked sturgeon and caviar

Entrées

molasses & black pepper
basted pork tenderloin
with ginger-scented mashed sweet potatoes, peach-fig chutney

seared duck breast
with cherry reduction sauce, mascarpone polenta cake,
applewood-smoked bacon braised greens

A *purchase specification* is a written form listing the specific characteristics of a product that is to be purchased from a supplier.

Depending on the product, the information on the purchase specification may include the product's quality or grade, variety or place of origin, size, or packaging requirements. It is also common for the purchase specification to include food safety requirements such as the temperature at which products must be transported and received.

A par stock checklist is used to help make sure that the appropriate quantity is ordered. *Par stock* is the maximum amount of a particular product that should be kept in inventory to ensure that an adequate supply is on hand for normal production. **See Figure 5-24.** Par stock values should be set high enough to ensure that the operation does not run out of a product. However, the values should be low enough to avoid disposing of products because they were left in storage too long.

Par Stock				
Item	Par Stock	Stock On Hand	Estimated Use Prior to Delivery	Quantity to Order
Tomatoes	5 cases	2 cases	1 case	4 cases
Lettuce	10 cases	5 cases	5 cases	10 cases
Cucumbers	2 cases	1 case	1 case	2 cases
Onions	1 case	15 cases	5 cases	0 cases

Figure 5-24. *Par stock values affect the amount of product ordered.*

Receiving and Storing Food

All food items need to be checked upon arrival to ensure that the product received actually matches what was on the purchase specification. *See Figure 5-25.* Products should be weighed or counted to make sure that the amount or quantity received is the same as specified on the invoice. Special attention should be paid to perishable foods to make sure they are in good condition and at the proper temperature when received.

Perishable food is food that has a short shelf life and is subject to spoilage and decay. Perishable foods include poultry, fish and shellfish, meats, dairy products, and produce. These items should be fresh, free of damage, and of the appropriate weight or size. Perishable foods should be purchased frequently in the smallest quantities possible.

Nonperishable food items have a much longer shelf life and generally can be kept for six months to a year. These items are normally stored at room temperature in their original packaging. The packaging should be carefully checked upon arrival to avoid accepting dented cans, crushed boxes, and torn or damaged packages. Since these products have a longer shelf life, they can be purchased in bulk.

Receiving Products

Purchase Specification	
Item Name	baking potatoes
Variety	Idaho Russet
Grade	US #1
Count per Case	80
Net Weight per Case	50 lb
Packaging	heavy-duty cardboard box

Figure 5-25. *Products need to be checked against the purchase specification upon arrival.*

Storing foods properly is essential to control costs. Immediately after items have been received and checked-in they should be placed into inventory following the first-in, first-out inventory process. *First-in, first-out (FIFO)* is the process of dating new items as they are placed into inventory and placing them behind or below older items to ensure that older items are used first. ***See Figure 5-26.***

First-In, First-Out (FIFO)

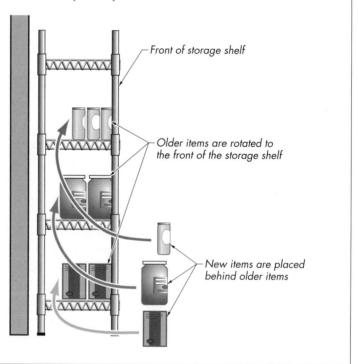

Front of storage shelf

Older items are rotated to the front of the storage shelf

New items are placed behind older items

Refer to DVD for
FIFO Stock Rotation
Media Clip

Figure 5-26. *The first-in, first-out (FIFO) inventory process ensures that older products are used before newer products.*

Many foodservice operations use an issuing procedure, which allows items that are stored to be moved into production only after a requisition is issued. A *requisition* is an internally generated invoice that is used to aid in tracking inventory as it moves from storage to production.

All food and beverage items in dry and cold storage areas must be kept 6 inches off the floor. Thermostats on refrigerators and freezers need to be checked frequently to ensure that the units are operating at proper temperatures to keep food safe. A pest control plan must also be in place to ensure that food and beverage products do not come in contact with pests.

Preparing and Serving Food

Many foods are trimmed or broken down before being served or used in a recipe. Sometimes the parts trimmed from products such as meats and vegetables can be used in other recipes, such as stocks or soups, instead of being disposed of as waste. Unnecessary waste and costs can also be prevented by minimizing overproduction of products.

The FIFO method of stock rotation is the first step in limiting spoilage. Other ways to limit spoilage include ordering the correct amount of an ingredient based on par stock and avoiding obsolescence. *Obsolescence* is the removal of an item from the menu while ingredients for that particular item are still in stock.

One of the most important ways foodservice workers can help minimize expenses is to serve food using the proper portion control techniques. *Portion control* is the process of ensuring that a specific amount of food or beverage is served for a given price. This is accomplished by using portion control equipment such as ladles, portion-controlled scoops, scales, and slicers. *See Figure 5-27.* When the quantity of food served exceeds the amount the menu price was based upon, food expenses increase and profit is reduced.

Portion Control Equipment

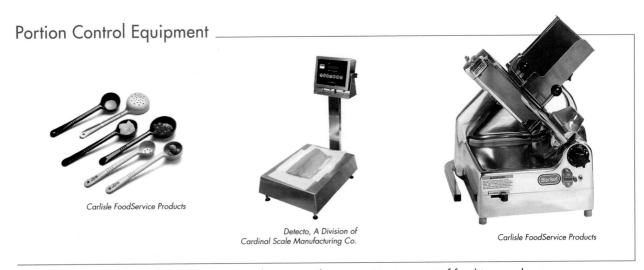

Carlisle FoodService Products

Detecto, A Division of
Cardinal Scale Manufacturing Co.

Carlisle FoodService Products

Figure 5-27. *Portion control equipment is used to ensure that a consistent amount of food is served.*

STANDARD PROFIT AND LOSS

Although every foodservice operation is somewhat unique, the methods for calculating and estimating profits and losses are fairly standard. Foodservice operations have revenues based on the sales of food and beverages. They also have expenses that can be categorized as either fixed or variable based on the amount of those sales. There are two categories of profit associated with any foodservice operation: gross profit and net profit.

Calculating Gross Profit

Gross profit is the calculated difference between total revenue and the cost of goods sold. The formula for calculating gross profit is as follows:

GP = TR − CGS

where

GP = gross profit

TR = total revenue

CGS = cost of goods sold

$$\text{Gross Profit} = \text{Total Revenue} - \text{Cost of Goods Sold}$$

For example, if a restaurant had a total revenue of $31,000 in a month when the cost of goods sold was $10,500, the gross profit would be $20,500.

GP = TR − CGS

$GP = \$31,000 - \$10,500$

$GP = \textbf{\$20,500}$

Calculating Net Profit

Net profit is the calculated difference between the gross profit and operating expenses of a foodservice operation. The formula for calculating net profit is as follows:

NP = GP − OE

where

NP = net profit

GP = gross profit

OE = operating expenses

$$\text{Net Profit} = \text{Gross Profit} - \text{Operating Expenses}$$

If the restaurant with a gross profit of $20,500 had $18,250 in operating expenses, there would be a net profit of $2250.

NP = GP − OE

$NP = \$20,500 - \$18,250$

$NP = \textbf{\$2250}$

However, if the same restaurant had $21,500 in operating expenses, the restaurant would have a net profit of −$1000, which is a loss of $1000.

NP = GP − OE

$NP = \$20,500 - \$21,500$

$NP = \textbf{−\$1000}$ or **($1000)** or **<$1000>**

A loss is not uncommon over certain periods of time. It is especially common in the first several months of a new operation while employees are new and management works to get things operating smoothly.

The relationship between gross profit and net profit is clearly shown on a standard profit and loss statement. A *standard profit and loss statement* is a form that shows the revenue, expenses, and resulting gross and net profit (or loss) over a specific period of time. The purpose of a standard profit and loss statement is to provide management with an indication of the financial status of the foodservice operation. A standard profit and loss statement can be prepared at a detailed level or a summary level depending on the complexity and needs of the foodservice operation. *See Figure 5-28.* The net profit is the last entry on the statement and is referred to as the "bottom line."

National Cancer Institute
Daniel Sone (photographer)

Standard Profit and Loss Statements

Standard Profit and Loss Statement Summary	
Total revenue	$ 34,908.59
Cost of goods sold	$ 10,432.90
Gross profit	$ 24,475.69
Operating expenses	$ 20,768.25
Net profit	$ 3707.44

Standard Profit and Loss Statement Detailed	
Total revenue	$ 34,908.59
Total food revenue	$ 27,801.23
Total beverage revenue	$ 7107.36
Cost of goods sold	$ 10,432.90
Total food costs	$ 7794.38
Total beverage costs	$ 2638.52
Gross profit	$ 24,475.69
Operating expenses	$ 20,768.25
Payroll expenses	$ 13,010.15
Social security taxes	$ 1050.75
Utilities	$ 1124.67
Rent	$ 2800.00
Interest	$ 325.19
Insurance and licenses	$ 356.80
Kitchen supplies	$ 803.21
Sales tax	$ 1047.26
Miscellaneous	$ 250.22
Net profit	$ 3707.44

Figure 5-28. *A standard profit and loss statement can be prepared at a summary level or a detailed level.*

Foodservice Operation

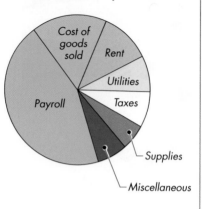

Figure 5-29. *Payroll expenses are frequently the most significant expense of a foodservice operation.*

Payroll Expenses

Payroll expenses are frequently the most significant expense of a foodservice operation. *See Figure 5-29.* A *payroll expense* is any money paid to an employee who performs work for the operation. Employees are paid on either a salaried or hourly basis. A salaried employee is paid based on a fixed amount of money. An hourly employee is paid based on an hourly wage multiplied by the number of hours worked.

Payroll expenses can be fixed or variable depending on how employees are paid. Generally, the payroll expenses of salaried employees are considered fixed expenses and the payroll expenses of hourly employees are considered variable expenses. Hourly employees are required to be paid at least 1.5 times their regular hourly wage for all hours over 40 worked in a week. For example, a cook earning $10 per hour is paid an overtime wage of $15.00 per hour ($10.00 × 1.5 = $15.00). This overtime wage is commonly referred to as "time-and-a-half."

Calculating Gross Pay. Payroll expenses for a foodservice operation are based on the gross pay of its employees. *Gross pay* is the total amount of an employee's pay before any deductions are made. Examples of deductions include federal, state, and local income taxes, Social Security taxes, employee-paid health insurance premiums, and contributions to retirement or savings plans. The gross pay for a salaried employee is the same for every pay period. The gross pay for an hourly employee may vary each pay period depending on the number of hours worked.

To calculate the gross pay of an hourly employee, regular pay is added to any overtime pay. Regular pay is equal to the number of hours worked up to 40 hours each week multiplied by the regular hourly wage. Overtime pay is equal to any hours worked over 40 hours in a week multiplied by the overtime hourly wage. For example, the gross pay for a line cook who earns $12 per hour and works 50 hours in one week would be calculated as follows:

Regular pay = 40 hours × $12.00/hour = $480.00
Overtime pay = 10 hours × ($12.00/hour × 1.5)
= 10 hours × $18.00
= $180.00
Total gross pay = $480.00 + $180.00 = $660.00

To calculate the total payroll expenses for a foodservice operation, the gross pay of each employee is added together. *See Figure 5-30.*

Calculating Net Pay. The net pay of each employee's paycheck will be less than the employee's gross pay. *Net pay* is the actual amount on an employee's paycheck and is equal to the gross pay minus payroll deductions. One of the most common categories of deductions is taxes. All employees are subject to pay Social Security and Medicare taxes. Most employees are required to pay federal income taxes. Depending on the city where the job is located, some employees are also required to pay state and/or local income taxes. Although taxes are not part of the net pay, they are part of the employee's gross pay and are considered foodservice operation expenses. The foodservice operation is required to withhold the amount of taxes owed by the employee and then pay those taxes directly to the government on behalf of the employee. *See Figure 5-31.*

Calculating Payroll Expenses

Payroll Summary Report—Week of Jan 1–7, 20XX

Employee	Regular Hours	Overtime Hours	Pay Rate	Regular Pay	Overtime Pay	Gross Pay
Prep cook	30	0	$9/hr	$270.00	$0.00	$270.00
Salad cook	40	0	$10/hr	$400.00	$0.00	$400.00
Line cook 1	40	10	$12/hr	$480.00	$180.00	$660.00
Line cook 2	40	5	$12/hr	$480.00	$90.00	$570.00
Server 1	40	0	$6/hr	$240.00	$0.00	$240.00
Server 2	40	6	$6/hr	$240.00	$54.00	$294.00
Chef	—	—	$800/wk	$800.00	—	$800.00
General manager	—	—	$1000/wk	$1000.00	—	$1000.00
					Total	**$4234.00**

Figure 5-30. *The total payroll expenses of a foodservice operation are equal to the sum of the gross pay for all employees.*

Net Pay vs. Gross Pay

Company Name
Payroll Account

Check # 12345

Date: 1-14-20XX

Pay to the order of ___ Chris M. Smith ___ $468.23

Four hundred sixty eight and 23/100 dollars

Local Bank

Memo ___ Payroll - 1/14 ___

Signature

Detach before depositing and save for your records.

Employee: Chris M. Smith	**Gross Pay** 660.00
Pay Period: 1/1 – 1/7, 20XX	*Deductions:*

Deductions		
Federal Income Tax	41.15	
State Income Tax	19.80	
Social Security	40.92	
Medicare/Medicaid	9.90	
Insurance	50.00	
Retirement Savings Plan	30.00	
Total Deductions		191.77
Net Pay		468.23

Figure 5-31. *The difference between net pay and gross pay is equal to the sum of the payroll deductions.*

Additional Costs. Foodservice operations are required to pay additional taxes related to payroll that are above and beyond the taxes included in the gross pay. These taxes include the employer's share of Social Security and unemployment taxes. Foodservice operations are also required to carry various types of insurance such as liability insurance and workers compensation insurance.

SUMMARY

Cost control is essential to maintaining a profitable foodservice operation. Standardized recipes and both the customary and metric measurement systems use standardized units to measure weight and volume. When a recipe is scaled, new ingredient amounts must be calculated based on a scaling factor. Adjustments may also need to be made to the recipe due to the change in yield.

As-purchased (AP) costs are broken down into unit costs to calculate the costs of ingredients in a recipe. For food products that are trimmed before being used in a recipe, AP unit costs are converted to edible-portion (EP) unit costs based on the yield percentage of the product. A yield percentage can be found by performing yield tests to ensure that the proper amount of food is ordered.

After the total as-served (AS) cost of a menu item is calculated, menu prices can be established. Most foodservice operations use target food and beverage cost percentages to calculate menu prices. Some operations calculate menu prices by adding contribution margins to the costs of food and beverages. Menu prices are often adjusted based on perceived customer value.

A foodservice operation earns a profit when total revenue is more than total expenses. Expenses must be minimized throughout the processes of planning the menu and purchasing, receiving, storing, preparing, and serving food. Gross profit is the difference between the cost of goods sold and total revenue. Net profit is the difference between the gross profit and operating expenses. The most significant cost of a foodservice operation is payroll expenses. A standard profit and loss statement is used to document financial information.

Refer to DVD for
Quick Quiz® questions

Review

1. Identify the common elements of standardized recipes.
2. Explain the difference between yield and portion size.
3. Identify standard units of measure in the customary system and the metric system.
4. Explain the difference among weight, volume, and count.
5. Name three types of scales used to measure weight.
6. Identify measuring devices used to measure volume.
7. Explain the importance of knowing the difference between ounces and fluid ounces.
8. Explain why some food products are measured by count.
9. Describe the method for converting a customary measurement to a metric measurement.
10. Explain how to determine the scaling factor for a recipe.
11. Identify five factors that may have to be adjusted when scaling a recipe.
12. Explain the difference between as-purchased unit cost and edible-portion cost.

Refer to DVD for
Review Questions

13. Describe the relationship between edible-portion and trim loss.
14. Explain how to calculate yield percentage using the percentage circle.
15. Explain the difference between calculating as-purchased quantities and calculating edible-portion quantities.
16. Explain the difference between a raw yield test and a cooking-loss yield test.
17. Explain how to determine the total as-served cost of a menu item.
18. Identify three types of food cost percentages.
19. Identify three methods used to calculate menu prices.
20. Explain the purpose of a pricing form.
21. Explain the difference between fixed costs and variable costs.
22. Identify four considerations to take into account when planning a profitable menu.
23. Explain the role of purchase specifications and par stock in ordering products.
24. Explain how the process of receiving and storing products is part of cost control.
25. Identify portion-control equipment used to prepare and serve foods and beverages.
26. Explain the difference between gross profit and net profit.
27. Identify the most significant expense of a foodservice operation.
28. Explain the difference between gross pay and net pay.
29. Name the type of statement used to indicate the financial status of a foodservice operation.
30. Identify two additional taxes and types of insurance that foodservice operations are required to pay.

Applications

1. Locate a standardized recipe and identify each of the common elements.
2. Use three types of scales to weigh two different food items.
3. Measure the same volume of two liquids with different densities.
4. Convert six customary measurements to metric measurements.
5. Convert six metric measurements to customary measurements.
6. Scale a recipe based on yield.
7. Scale a recipe based on portion size.
8. Scale a recipe based on product availability.
9. Calculate the as-purchased unit cost of a food item.
10. Calculate the edible-portion cost of a food item.
11. Calculate the yield percentage of a food item.
12. Perform a raw yield test.
13. Perform a cooking-loss yield test.
14. Calculate the as-served price of a menu item.
15. Calculate menu-item food cost percentages, overall food cost percentages, and target food cost percentages for a given menu.
16. Use food cost percentage pricing to set a target price for three menu items.
17. Use perceived value pricing to adjust a target menu price.
18. Use contribution margin pricing to determine a menu price.
19. Complete a pricing form for a given menu item.
20. Calculate the gross pay and the net pay for a line cook.

Refer to DVD for
Application Scoresheets

Menu Planning and Nutrition

Menu planning and nutrition each play an important role in foodservice operations. A menu is the primary way foodservice operations advertise items. Well-planned menus provide choices of enticing dishes that satisfy guests while maximizing profits. Nutrition involves understanding how different foods in the appropriate amounts impact health and well-being. Understanding the roles of menu planning and nutrition is essential for foodservice professionals to be successful.

Refer to DVD for **Flash Cards**

Chapter Objectives

1. Identify the five functions of a menu.
2. Explain the purpose of the truth-in-menu guidelines.
3. Describe the three classifications of menus.
4. Describe six common menu types.
5. Describe the four elements of menu design.
6. Identify three factors that directly impact menu pricing.
7. Describe the dietary considerations that affect menu planning.
8. Describe the function of each of the six nutrients.
9. Contrast complete and incomplete proteins.
10. Explain the role of sugars, starches, and dietary fiber.
11. Explain how different types of fat impact health.
12. Contrast water-soluble and fat-soluble vitamins.
13. Differentiate between macrominerals and microminerals.
14. Describe the role water plays in maintaining health.
15. Summarize the key recommendations in *Dietary Guidelines for Americans*.
16. Explain each of the components on a nutrition facts label.
17. Explain how recipes can be modified to lower fat, sugar, and sodium.
18. Explain the role of portion sizes in meeting nutritional recommendations.

Key Terms

- **fixed menu**
- **market menu**
- **cycle menu**
- **menu mix**
- **food allergy**
- **obesity**
- **nutrient**
- **protein**
- **incomplete protein**
- **carbohydrate**
- **dietary fiber**
- **insoluble fiber**
- **lipid**
- **saturated fat**
- **cholesterol**
- **vitamin**
- **mineral**
- **digestion**
- **calorie**
- **nutrient-dense food**

PLANNING MENUS

A *menu* is a list of items that guests may order from a foodservice operation. Every purchasing and staffing decision relates directly back to the menu. The menu is the primary form of communication that a foodservice operation uses to inform guests of the types of foods and beverages offered and how much each item costs. *See Figure 6-1.* The menu also reflects the target market of the operation. In addition, it dictates the type and amount of ingredients to purchase, the equipment needed, and the level of skill required by the foodservice staff.

Menus

Figure 6-1. *The menu is the primary form of communication that a foodservice operation uses to inform guests of the types of foods and beverages offered and how much each item costs.*

Various individuals may be involved in the planning of menus. For example, a chef and an owner may plan the menu for a family-run restaurant, but a central office might determine the menu for a franchise or chain operation. There are a variety of resources available to assist in the menu planning process, including web sites, design firms, consultants, and dietitians. Regardless of who plans the menu, all menus must follow the truth-in-menu guidelines set by the federal government.

Truth-in-Menu Guidelines

The federal government enacted truth-in-menu guidelines, which require accuracy in statements made on menus. The truth-in-menu guidelines are designed to protect guests from fraudulent food and beverage claims. *See Figure 6-2.* Failure to comply with truth-in-menus guidelines can result in legal claims being made against a foodservice operation for misleading or endangering guests. Given the direct link between food and health, guests rely on menus for accurate information.

Truth-In-Menu Guidelines	
Menu Label	**Example of Misrepresentation**
Portion size of an item	Advertising a 12 oz steak and serving a 10 oz steak
Quality or grade of an item	Listing USDA Prime beef and serving USDA Select beef
Preservation method	Advertising fish as fresh, but serving fish that was previously frozen
Preparation method	Claiming a food is housemade when it is a prepackaged item
Type of product served	Stating the use of extra virgin olive oil when vegetable oil was used
Certified foods	Claiming a food is organic that has not been certified as such
Point of origin	Stating "Florida" oranges when the oranges came from a different location
Nutrition information	Listing a product as low-fat when it does not meet the required criteria
Product brand	Serving a different brand than the one listed

Figure 6-2. *The truth-in-menu guidelines are designed to protect guests from fraudulent food and beverage claims.*

Menu Classifications

Planning menus also involves determining the menu classification. The menu classification is determined by how often menu selections will change. Some menus remain fairly constant, while others change seasonally. Menu classifications include fixed menus, market menus, and cycle menus.

Fixed Menus. A *fixed menu* is a menu that stays the same or rarely changes. Fixed menus are commonly found in chain restaurants where the entire organization has the same menu. They can also be found at entertainment venues, such as ballparks and concert arenas, where the same menu is available throughout the season. Some foodservice operations add to a fixed menu by attaching a list of daily specials.

Market Menus. A *market menu* is a menu that changes frequently to co-incide with changes in the availability of products. Market menus often revolve around seasonal items that are at their peak, which may only be for a short period of time. Market menus allow the chef to be creative and also encourage guests to try new dishes. Farmers' markets are often a local source for seasonal ingredients. Buying locally promotes sustainability and provides guests with foods harvested at the height of freshness.

Daniel NYC

Cycle Menus. A *cycle menu* is a menu written for a specific period and is repeated once that period ends. *See Figure 6-3.* Cycle menus are common in institutional foodservice settings such as hospitals and cafeterias. For example, a cycle menu with predetermined items may repeat each month. Most cycle menus rotate items so that the same meal is not offered on the same day each week. For example, a menu might be written for an 11 day cycle. If lasagna is served on a Monday, the cycle menu may not list lasagna again until 11 days later on a Friday. This rotation helps keep the menu from becoming predictable.

Eleven-Day Cycle Menu				
Monday	**Tuesday**	**Wednesday**	**Thursday**	**Friday**
Cream of Broccoli Soup Broiled Cod Sliced Roasted Pork Glazed Carrots Mashed Potatoes	Vegetable Soup Chicken Tetrazzini Italian Meatloaf Grilled Zucchini Pasta Marinara	Creamy Mushroom Soup Roast Turkey Baked Ham Bread Dressing Broccoli	Chicken Noodle Soup Grilled Chicken Blackened Catfish Rice Pilaf Cauliflower	Cream of Asparagus Soup Fettuccine Alfredo Teriyaki Chicken Fried Rice Carrots and Zucchini
Minestrone Soup Meatballs Baked Scrod Mostaccioli Italian Vegetables	Chicken and Rice Soup Pork Chop Chicken Tacos Spanish Rice Green Beans	Lentil Soup Roast Beef Veggie Lasagna Roasted Potatoes California Blend	Beef Barley Soup Oven-Fried Chicken Spaghetti & Meatballs Sweet Potatoes Peas	Corn Chowder BBQ Pork Stuffed Peppers Boiled Potatoes Crinkle-Cut Carrots
Navy Bean Soup Chicken à la King Pasta Primavera Steamed White Rice Brussels Sprouts	Cream of Broccoli Soup Broiled Cod Sliced Roasted Pork Glazed Carrots Mashed Potatoes	Vegetable Soup Chicken Tetrazzini Italian Meatloaf Grilled Zucchini Pasta Marinara	Creamy Mushroom Soup Roast Turkey Baked Ham Bread Dressing Broccoli	Chicken Noodle Soup Grilled Chicken Blackened Catfish Rice Pilaf Cauliflower
Cream of Asparagus Soup Fettuccine Alfredo Teriyaki Chicken Fried Rice Carrots and Zucchini	Minestrone Soup Meatballs Baked Scrod Mostaccioli Italian Vegetables	Chicken and Rice Soup Pork Chop Chicken Tacos Spanish Rice Green Beans	Lentil Soup Roast Beef Veggie Lasagna Roasted Potatoes California Blend	Beef Barley Soup Oven-Fried Chicken Spaghetti & Meatballs Sweet Potatoes Peas
Corn Chowder BBQ Pork Stuffed Peppers Boiled Potatoes Crinkle-Cut Carrots	Navy Bean Soup Chicken à la King Pasta Primavera Steamed White Rice Brussels Sprouts	Cream of Broccoli Soup Broiled Cod Sliced Roasted Pork Glazed Carrots Mashed Potatoes	Vegetable Soup Chicken Tetrazzini Italian Meatloaf Grilled Zucchini Pasta Marinara	Creamy Mushroom Soup Roast Turkey Baked Ham Bread Dressing Broccoli

Figure 6-3. *A cycle menu is written for a specific period and is repeated once that period ends.*

Menu Types

Planning menus also involves determining the most appropriate menu type for the operation. Menu types are influenced by the menu classification as well as the nature of the foodservice operation. Some restaurants offer multiple courses for a set price, while other restaurants charge separately for each dish. A menu can be displayed in various ways, including on overhead signs, on a single sheet of paper, or in the style of a book. The physical form of the menu often reflects the menu type. Common menu types include à la carte, semi-à la carte, prix fixe, table d'hôte, California, and meal-specific.

À la Carte Menus. An *à la carte menu* is a menu that has all food and beverage items priced separately. À la carte menus can be found in quick service venues where a hamburger is sold separately from a side of fries. À la carte menus are also found in restaurants where a side of grilled zucchini is sold separately from the entrée.

Semi-à la Carte Menus. A *semi-à la carte menu* is a menu that offers entrées along with additional menu items for a set price. The prices on a semi-à la carte menu often include an entrée and a choice of soup, salad, or side. However, appetizers and desserts are commonly sold separately. Some restaurants use semi-à la carte menus to offer value meal combinations that allow a guest to order a main dish, side dish, and beverage for a set price. An à la carte menu can be easily transitioned into a semi-à la carte menu by including side dishes in the price of the entrée instead of pricing each item separately. *See Figure 6-4.*

À la Carte and Semi-à la Carte Menus

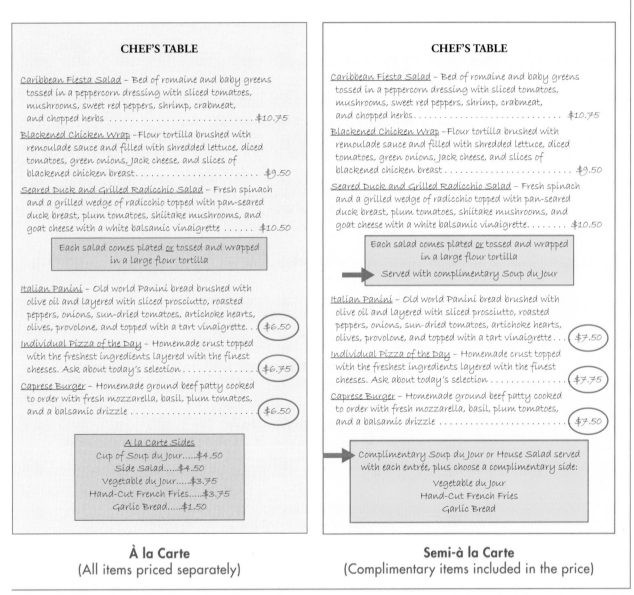

Figure 6-4. *An à la carte menu can be transitioned into a semi-à la carte menu by including side dishes in the price of the entrée instead of pricing each item separately.*

Prix Fixe Menus. A *prix fixe menu* is a menu that offers limited choices within a collection of specific items for a multicourse meal at a set price. There are typically a few options within each course on a prix fix menu. For example, an appetizer, soup, salad, entrée, and dessert may be offered for a set price with three or four choices within each category. The guest is able to choose an item from each category. Prix fixe menus are often used in upscale restaurant settings. *See Figure 6-5.*

Prix Fixe Menus

Prix Fixe Four Course Menu
$75.00

First Course
(Select One)

Lightly smoked duck breast, fresh cheese, and carrot terrine
Shrimp and shaved baby carrot salad with honey coriander vinaigrette

Second Course
(Select One)

Silky cream of peppered butternut squash soup
Roasted baby beet salad
Frisée, green oak, and hazelnut salad

Third Course
(Select One)

Roasted chicken and au jus with ricotta dumplings
Roasted strip loin of beef with red wine sauce
Crispy shrimp, white wine butter sauce, and shellfish foam

Vegetable and Starch Selections
(Select Two)

Slow cooked fennel purée and warm tomato
Bacon, pearl onion, and fresh pea ragout
Leek and potato roulade
Shaved garlic broccolini
Glazed cipollini onions and mushrooms

Fourth Course

A tasting of frozen, warm, and fresh desserts
from our talented pastry department

Figure 6-5. *Prix fixe menus offer limited choices at a set price and are often used in upscale restaurant settings.*

Table d'Hôte Menus. A *table d'hôte menu* is a menu that identifies specific items that will be served for each course at a set price. *See Figure 6-6.* A table d'hôte menu does not offer choices. A plated banquet at which all guests are served the same meal is an example of a table d'hôte menu.

Table d'Hôte Menus

Table d'Hôte Seven Course Menu
$95.00

First Course

Smoked duck breast, fresh cheese, and carrot terrine
Shaved baby carrot salad with honey coriander vinaigrette

Second Course

Crispy shrimp, white wine butter sauce, and shellfish foam
Slow cooked fennel purée and warm tomato

Third Course

Silky cream of peppered butternut squash soup
with vanilla poached pear and orange cream

Fourth Course

Roasted baby beet and goat cheese salad
with frisée, green oak, hazelnuts, cherries, and EVOO

Fifth Course

Roasted chicken and au jus with ricotta dumplings
Bacon, pearl onion, and fresh pea ragout

Sixth Course

Roasted strip loin of beef with red wine sauce
Leek and potato roulade
Shaved garlic broccolini
Glazed cipollini onions and mushrooms

Seventh Course

A tasting of frozen, warm, and fresh desserts from our award-winning pastry chef

Figure 6-6. *A table d'hôte menu identifies specific items that will be served for each course at a set price.*

California Menus. A *California menu* is a menu that offers all food and beverage items for breakfast, lunch, and dinner throughout the entire day. For example, a guest has the option of ordering breakfast items for dinner. California menus are commonly used by restaurants that are open 24 hours a day.

Meal-Specific Menus. A *meal-specific menu* is a menu that only offers a particular meal, such as breakfast, lunch, or dinner. Meal-specific menus feature items that are typical for that meal, such as eggs, pancakes, and waffles for breakfast or sandwiches and salads for lunch. All of the other types of menus, except the California menu, can be meal specific.

Menu Design

Planning menus also involves determining how to promote sales through an effective menu design. A well-designed menu clearly communicates in a manner that promotes sales and increases profitability. To achieve this goal, a menu design should reflect the atmosphere of the foodservice operation and leave a favorable impression on guests. For example, an outdoor café might present a handwritten menu to promote a casual, neighborhood atmosphere. In contrast, a fine dining restaurant may present a menu on linen paper with an elegant script. Even if two restaurants offer the same menu items, the menu design can help create the desired ambiance for each. *See Figure 6-7.*

Menu Design

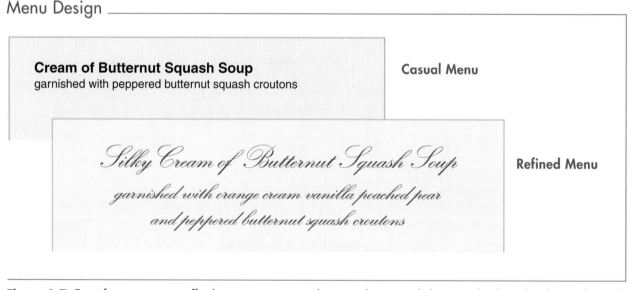

Figure 6-7. *Even if two restaurants offer the same menu items, the menu design can help create the desired ambiance for each.*

A successful menu design includes an effective menu mix. A *menu mix* is the assortment of items that may be ordered from a given menu. For example, a foodservice operation may have a menu mix that includes vegetarian entrées in addition to meat, poultry, and fish entrées. A well-balanced menu mix can promote guest satisfaction, attract more customers, and increase sales.

On average, guests spend approximately two minutes scanning a menu before selecting items. Within this brief time frame, a successful menu design can persuade guests to order items that will generate the greatest profits. For example, an average guest check might be $15.00 for a restaurant selling a large number of sandwiches. However, after redesigning the menu to direct the focus to higher-profit items, the average guest check may rise to $23.00. A good menu design directs guests to the most profitable items by effectively using item descriptions, format, item placement, and graphic elements.

Item Descriptions. Menu item descriptions can entice guests by creating tempting images of menu items in their minds. Some words have more selling power than others and can be used to help direct guests to higher-profit items. For example, "baked in our wood-fire oven" has more appeal than "baked."

In general, menus items that are common or familiar do not require descriptions. For example, New York strip steak, macaroni and cheese, and blueberry pie are menu items that present a clear image without further explanation. However, even if a menu item is familiar to guests, a powerful description can emphasize special components of the dish and elevate it from sounding good to sounding exceptional.

Items that are unclear or unfamiliar to most guests require further description. For example, "stuffed pork tenderloin" does not tell the guest enough about the dish, and some guests may be unfamiliar with an item such as gnocchi. Good menu item descriptions promote clarity, add appeal to the dish, and increase the likelihood that an item will be ordered. *See Figure 6-8.*

Charlie Trotter's

Menu Item Descriptions

Stuffed Pork Tenderloin
Hand-trimmed pork tenderloin filled with a flavorful blend of sweet Italian sausage and Granny Smith apples, pan-seared and glazed with a mandarin orange and honey BBQ sauce

Gnocchi
Delicate potato pasta dumplings simmered in a fresh tomato cream sauce, infused with basil and roasted garlic, and finished with freshly grated pecorino cheese

Figure 6-8. *Good menu item descriptions promote clarity, add appeal to the dish, and increase the likelihood that an item will be ordered.*

American Metalcraft, Inc.

Format. Effective menus require that careful consideration be given to the physical format of the menu and the placement of each menu item within that format. Menus are presented in four basic formats. Menu formats include single-page menus, bifold menus, trifold menus, and multipage menus. *See Figure 6-9.*

- Single-page menus present menu items on one page and may use both the front and back of the sheet.
- Bifold menus open like a book and present items on the left and right sides and possibly the front and back cover.
- Trifold menus have a right and a left panel that fold toward the center and overlap. Trifold menus may list items on the front and back of each panel.
- Multipage menus are similar to bifold menus, but include many more pages.

Menu Formats

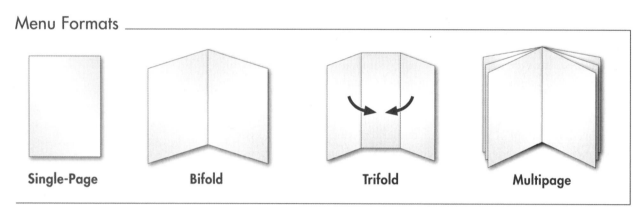

| Single-Page | Bifold | Trifold | Multipage |

Figure 6-9. *Menu formats include single-page, bifold, trifold, and multipage menus.*

Placement of Menu Items

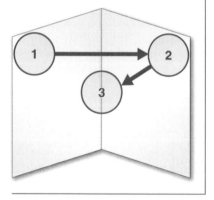

Figure 6-10. *The eye naturally falls on the top left of a bifold menu, then moves to the top right, and finally rests in the center, making these three areas ideal positions for the most profitable items.*

Regardless of the format chosen, menus come in a variety of sizes and shapes. They can be made from materials ranging from laminated card stock to leather-bound booklets. The menu format should reflect the style of the foodservice operation and be influenced by the menu classification and menu type. For example, a fixed, meal-specific menu that offers only breakfast may be formatted as a laminated bifold menu that has been designed for durability. In contrast, a single-page menu on heavy cardstock may be appropriate for a market menu that changes each day.

Item Placement. After the menu format is chosen, the placement of each menu item is the next consideration. Because most guests order food in the order it is eaten, items are typically grouped by type, such as appetizers, soups and salads, entrées, and desserts. The placement of items influences what guests choose because the eye follows a natural path when scanning a menu. For example, the eye naturally falls to the center of the page of a single-page menu. However, the eye naturally falls on the top left of a bifold menu, then moves to the top right, and finally rests in the center, making these three areas ideal positions for the most profitable items. *See Figure 6-10.*

Graphic Elements. Menus should be legible and visually appealing. A legible menu starts with choosing lettering and a background color that make the text easy to read. For example, silver lettering on light-gray paper may strain the eyes and frustrate guests, but simple black letters on off-white paper can be easy to see.

The use of graphic elements, such as borders and images, should create a look that emphasizes high-profit items and reflects the theme of the foodservice operation. *See Figure 6-11.* For example, casual dining operations often use well-positioned photographs to direct guests to featured menu items. In contrast, an upscale restaurant may use an elegant border to highlight signature dishes. Regardless of the graphics, the menu items emphasized should be those that have the greatest revenue potential.

Graphic Elements

Figure 6-11. *The use of graphic elements, such as borders and images, should create a look that emphasizes high-profit items and reflects the theme of the foodservice operation.*

Menu Pricing

Planning menus also involves pricing menu offerings. Menus are used to communicate not only what food and beverage items are available but also the price of those items. Foodservice operations generate profits when they charge more for the items they sell than the amount spent to produce those items. Effective menus can guide guests to specific items and increase the probability that those items will be ordered. Ideally, those items will generate the greatest profits while being perceived as a good value by the guests.

Pricing menu items involves many factors. For example, there are different pricing methods to consider, such as food cost percentage pricing, perceived value pricing, and contribution margin pricing. Foodservice operations might use one pricing method or a combination of pricing methods. Customer demand and market trends also affect menu prices. Finding the right balance of menu prices is a complex process influenced by costs, perceived value, and the placement of prices on the menu.

Refer to DVD for
Pricing Menu Items
Culinary Math Tutorial

Costs. Before a menu price can be established, a foodservice operation must determine the cost of each menu item. *See Figure 6-12.* For example, an operation that serves a quarter-pound cheeseburger needs to know the cost of the bun, the meat, the cheese, and each condiment in order to determine the actual food cost. Knowing how much each menu item costs can help determine how much to charge for that item.

Menu Item Costs

Grilled onions

Bun

Cheese

Beef

Lettuce

Ketchup

Fries

National Cattlemen's Beef Association

Figure 6-12. *Before a menu price can be established, a foodservice operation must determine the cost of each menu item.*

Different menu items have varying food cost percentages in relation to selling price. *See Figure 6-13.* For example, a pasta entrée may have a 25% food cost percentage and a salmon entrée might have a 40% food cost percentage. However, with effective pricing, the salmon can be more profitable even though it has a higher food cost percentage. If the salmon sells for $15.00 with a 40% food cost percentage, the operation makes a $9.00 profit. If the pasta sells for $10.00 with a 25% food cost percentage, the operation makes a $7.50 profit. The salmon generates $1.50 more per order than the pasta despite having a higher food cost percentage.

Food Cost Percentages

Barilla America, Inc.

Lower Food Cost

Tanimura & Antle®

Higher Food Cost

Figure 6-13. *Different menu items have varying food cost percentages in relation to selling price.*

Perceived Value. Perceived value represents the amount that guests are willing to pay for a menu item. Prices should reflect what guests expect. For example, a target price of $4.47 would be rounded up to $4.50. This adjustment reflects a price guests are accustomed to seeing and the three cent difference will not impact the decision to purchase that menu item. Rounding up also adds to profits. However, as a general rule, whole dollars should be rounded down slightly. Most guests perceive an item priced at $9.95 or $9.99 to be a better value than the same item priced at $10.00.

Perceived value is also influenced by the range of prices found on a menu. If a menu prices salads from $6.95–$12.95 with the exception of one priced at $21.95, most guests are not likely to choose the highest-priced salad. Likewise, if most foodservice operations in a specific area sell a shrimp appetizer for $7.95 and one operation charges $14.95, the higher-priced appetizer will not align with guest expectations. *See Figure 6-14.*

Daniel NYC

Perceived Value

Tanimura & Antle®

Shrimp Appetizer

$7.95

Most Area Menus

Shrimp Appetizer

$14.95

One Area Menu

Figure 6-14. *Perceived value is influenced by the range of prices found on a menu.*

Price Placement. The placement of prices on a menu impacts profitability. For example, aligning prices in a column to the right of the item description directs the eye to focus on prices. A different strategy is to position the price after the description using the same font, color, and size as the description. This tactic leads guests to concentrate on the description instead of the price.

Another aspect of placement is the order of prices. If menu prices are listed from least expensive to most expensive or vice versa, guests will tend to focus on the less expensive items. Guests are encouraged to read through all the menu offerings when prices do not reflect a specific order.

Dietary Considerations

Planning menus also involves determining how to accommodate guests who choose meals based on food allergies or food intolerances, weight management issues, or plant-based diets. Dietary considerations often mean that guests have to limit or restrict the consumption of certain foods. Along with well-trained staff, menus that are clear and accurate can help guests make informed decisions.

Nutrition Note

The Food Allergen Labeling and Consumer Protection Act requires food manufacturers to list all major food allergens and any ingredient that contains a protein derived from them.

Food Allergies. Most menus include food allergy warnings. A *food allergy* is a reaction by the immune system to a specific food. Allergic reactions can be minor or extremely severe. *Anaphylaxis* is a severe allergic reaction that causes the airway to narrow and prohibits breathing. An anaphylactic reaction can lead to unconsciousness, coma, or even death. Because of the potential severity involved with food allergies, some operations establish allergy-free work zones in the kitchen that prohibit the use of specific foods known to cause allergic reactions.

There are over 100 foods that can cause allergic reactions. However, eight specific foods are considered major allergens. According to the Food and Drug Administration (FDA), milk, eggs, fish, crustacean shellfish, tree nuts, peanuts, wheat, and soybeans account for 90% of all food allergies. *See Figure 6-15.* In addition to viewing the menu, some guests may request a list of ingredients to avoid having an adverse reaction. This essential information helps individuals with food allergies avoid an allergic reaction by avoiding the food that triggers it.

Sources of Common Food Allergens	
Allergen	**Potential Food Sources**
Milk	• Many nondairy products contain casein (a milk derivative) • Some brands of tuna fish contain casein • Butter melted on top of grilled steaks (often not visible) or foods cooked in butter • Many baked products contain dry milk
Eggs	• Egg whites are often used to create the foam topping on specialty coffee drinks • Egg wash is sometimes used on pretzels before they are dipped in salt • Some brands of egg substitutes contain egg whites • Most processed cooked pastas (including those in prepared soups) contain eggs or are processed on equipment shared with egg-containing pastas
Peanuts	• Chili sauce, hot sauce, mole sauce, marinades, glazes, and salad dressings • Pudding, cookies, crackers, egg rolls, and potato pancakes • Some vegetarian food products advertised as meat substitutes • Foods that contain extruded, cold-pressed, or expelled peanut oil
Tree Nuts	• Salads and salad dressings, barbeque sauce, and honey • Meat-free burgers, fish dishes, pancakes, and pasta • Piecrust and the breading on meat, poultry, and fish
Fish	• Imitation fish or shellfish • Salad dressings, Worcestershire sauce, and sauces made with Worcestershire
Shellfish	• Many Asian dishes include fish sauce • Shellfish protein can become airborne in the steam released during cooking
Soy	• Baked goods, cereals, crackers, infant formula, soups, and sauces • Some brands of tuna and peanut butter • Soybean oil that has not been cold pressed, expeller pressed, or extruded
Wheat	• Some brands of hot dogs, imitation crabmeat, and ice cream contain wheat • Some Asian dishes contain wheat flour shaped to look like beef, pork, or shrimp

Figure 6-15. *Milk, eggs, fish, crustacean shellfish, tree nuts, peanuts, wheat, and soybeans account for 90% of all food allergies.*

Food Intolerances. A *food intolerance* is an adverse reaction to a food that does not involve the immune system. Symptoms of food intolerance usually include bloating, abdominal cramps, nausea, and diarrhea. When the offending food is eliminated by the body or removed from the diet, symptoms usually subside. Some people may be able to tolerate the problem food in small quantities, while other people may need to avoid the food. The most common food intolerances are gluten intolerance and lactose intolerance.

Gluten is a type of protein found in grains such as wheat, rye, barley, and some varieties of oats. Gluten is prevalent in flour and flour-based products such as breads, cereals, pastas, and baked goods. It can also be found in other foods such as bouillon cubes, soy sauce, and deli meats. *See Figure 6-16.* *Celiac disease* is a condition in which gluten damages the small intestine's ability to absorb nutrients. It is essential for a person with celiac disease to abstain from eating gluten. Gluten-free menu offerings can provide guests with safe and healthy alternatives.

Lactose is a sugar found in milk and dairy products. In addition to milk, yogurts, and cheeses, lactose can be found in many processed foods such as salad dressings, soups, and breads. Lactose intolerance is an inability to properly digest lactose. Alternatives such as lactose-free milk, almond milk, or products made from soy can be offered for those who are lactose intolerant.

Weight Management. A sedentary lifestyle and consistent consumption of foods that are high in fat and calories can contribute to weight gain. According to the U.S. Department of Health and Human Services, the number of individuals who are overweight or obese has risen from 13% to 35% over the past several decades. *Obesity* is a medical condition characterized by an excess of body fat. The rise in obesity has paralleled an increase in obesity-related illnesses and diseases, such as high blood pressure, cardiovascular disease, type 2 diabetes, and some cancers. *See Figure 6-17.*

Foods Containing Gluten

- Bouillon cubes
- Brown rice syrup
- Candies
- Cold cuts
- French fries
- Gravies
- Hot dogs
- Matzo
- Potato chips
- Rice mixes
- Salami
- Sauces
- Sausages
- Soups
- Surimi
- Tortilla chips
- Vegetables in sauce

Figure 6-16. *Gluten is found in flour-based products such as breads, cereals, pastas, and baked goods, as well as other types of foods.*

Weight-Related Illnesses and Diseases

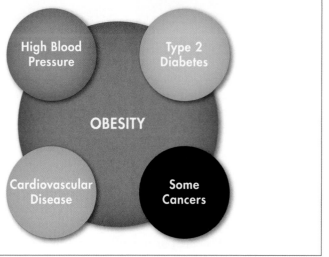

Figure 6-17. *The rise in obesity has paralleled an increase in obesity-related illnesses and diseases, such as high blood pressure, cardiovascular disease, type 2 diabetes, and some cancers.*

To combat obesity, federal initiatives have been developed to promote healthy eating and physical activity. These initiatives encourage the foodservice industry to promote well-balanced and nutritious meals. Menus that include healthy options encourage guests to make more healthy choices. Restaurants that offer appropriate portion sizes can also help individuals manage weight and reduce obesity-related diseases.

Plant-Based Meals. Plant-based meals, also known as vegetarian meals, are being chosen by many people several days per week and every day by others. *See Figure 6-18.* There are many variations of plant-based meals. For example, some individuals exclude all animal-based foods while others may include dairy products or eggs. Menus that include vegetarian dishes, such as broccoli and cashew stir-fry or grilled vegetable risotto, fulfill the needs of this growing segment of the population. Studies have shown that eating plant-based meals can help reduce the risk of cardiovascular disease.

Plant-Based Meals

House Foods

Tofu Salad

Tanimura & Antle®

Broccoli Risotto

Figure 6-18. *Plant-based meals, also known as vegetarian meals, are being chosen by many people several days per week and every day by others.*

NUTRITION FUNDAMENTALS

Nutrition is a key component when planning menus. Nutrition is the study of how food is taken in and utilized by the human body. A nutrient is a substance found in food that is necessary for the body to function properly. Nutrients include proteins, carbohydrates, lipids, vitamins, minerals, and water.

Proteins

A *protein* is a nutrient that consists of one or more chains of amino acids and is essential to living cells. An *amino acid* is the building block of all proteins. The primary function of protein is to build and repair body tissue, but the body can also convert protein to energy when needed. The body requires over 20 different amino acids to function properly. Approximately half of these amino acids are considered essential. An *essential amino acid* is an amino acid that the body cannot manufacture. A *nonessential amino acid* is an amino acid that the body can manufacture. Based upon their amino acid composition, proteins are classified as complete or incomplete.

Complete Proteins. A *complete protein* is a protein that contains all of the essential amino acids. *See Figure 6-19.* An egg is a complete protein. In addition to eggs, complete proteins are primarily found in animal-based foods such as meat, poultry, fish, and dairy foods. Soybeans and quinoa also contain the essential amino acids.

National Cattlemen's Beef Association

Complete Proteins

American Metalcraft, Inc.

Eggs

National Cancer Institute

Meat, Fish, and Poultry

Soybeans

Indian Harvest Specialtifoods, Inc./Rob Yuretich

Quinoa

Figure 6-19. *Complete proteins contain all of the essential amino acids.*

- Hummus and whole wheat pita bread
- Grilled cheddar cheese on whole wheat bread
- Black bean and walnut loaf
- Tofu salad with almond-coconut dressing
- Chili beans with grated cheddar and corn bread
- Stir-fried edamame and brown rice garnished with sesame seeds

Figure 6-20. *Combining incomplete proteins can result in a complete protein.*

Simple Carbohydrates

National Honey Board

Figure 6-21. *Simple carbohydrates are found naturally in honey, fruit, and milk.*

Incomplete Proteins. An *incomplete protein* is a protein that does not contain all of the essential amino acids. Incomplete proteins are found in plant-based foods such as grains, legumes, nuts, and vegetables. Combining grains, nuts, or seeds with lentils, peas, beans, or dairy products results in a complete protein. *See Figure 6-20.* For example, rice is low in the essential amino acid called lysine, and beans are low in the essential amino acid methionine. By eating rice and beans together or in the same day, a complete protein is created. Combining incomplete proteins to form complete proteins is especially important for individuals who only eat plant-based meals.

Carbohydrates

A *carbohydrate* is a nutrient in the form of sugar or starch and is the human body's main source of energy. Carbohydrates are classified as simple or complex. Simple and complex carbohydrates differ in how quickly they are absorbed by the body.

Simple Carbohydrates. A *simple carbohydrate,* also known as a sugar, is a carbohydrate composed of one or two sugar units and is quickly absorbed by the body. Simple carbohydrates are found naturally in honey, fruit, and milk. *See Figure 6-21.* They also exist in refined sugars that are added to foods as sweeteners.

Complex Carbohydrates. A *complex carbohydrate,* also known as a starch, is a carbohydrate composed of three or more sugar units and is slowly absorbed by the body. Because complex carbohydrates are absorbed slowly, the body has a sense of feeling full longer. Complex carbohydrates are found in a variety of grains, legumes, and vegetables. *See Figure 6-22.*

Complex Carbohydrates

Barilla America, Inc.

Figure 6-22. *Complex carbohydrates are found in a variety of grains, legumes, and vegetables.*

Many complex carbohydrates also contain dietary fiber. *Dietary fiber* is the portion of a plant that the body cannot digest. Dietary fiber is classified as soluble or insoluble. *Soluble fiber* is dietary fiber that dissolves in water. Sources of soluble fiber include beans, oats, barley, and many fruits and vegetables. *Insoluble fiber* is dietary fiber that will not dissolve in water. Sources of insoluble fiber include wheat bran, wheat germ, corn, and peas. Insoluble fiber helps remove waste from the digestive tract.

Barilla America, Inc.

Lipids

A *lipid* is a nutrient in the form of fats, oils, and fatty acids. In foodservice operations, lipids are commonly referred to as fats and oils. Lipids help the body absorb nutrients, provide insulation, manufacture hormones, and cushion organs. When the body has used its supply of carbohydrates for energy, lipids are the next energy resource. Unused lipids are stored as body fat. Lipids are found in a variety of foods and are essential in small amounts.

Unsaturated Fats. An *unsaturated fat* is a lipid that is liquid at room temperature. *See Figure 6-23.* Unsaturated fats are derived primarily from plants. They are found in nuts, seeds, olive oil, peanut oil, and canola oil. Fatty fish, such as salmon, also contain unsaturated fats. Unsaturated fats are often referred to as good fats because they have been found to lower the risk of cardiovascular disease and stroke.

Unsaturated Fats

Barilla America, Inc.

Figure 6-23. *An unsaturated fat is a lipid that is liquid at room temperature.*

National Cattlemen's Beef Association

Saturated Fats. A *saturated fat* is a lipid that is solid at room temperature. *See Figure 6-24.* Saturated fats are found in meats, poultry skin, whole milk, butter, and processed foods. Many processed foods contain hydrogenated shortening. *Hydrogenation* is the process that chemically transforms oils into solids to improve shelf life and stabilize flavor. Foods that contain hydrogenated or partially hydrogenated ingredients contain trans fats.

Saturated Fats

National Cancer Institute/Renee Comet

Figure 6-24. *A saturated fat is a lipid that is solid at room temperature.*

Both saturated fats and trans fats contribute to increased levels of cholesterol. *Cholesterol* is a waxy, fat-like substance that is used to form cell membranes, vitamin D, some hormones, and bile acids. The body manufactures all the cholesterol it needs. However, additional cholesterol is supplied through food.

Cholesterol is transported throughout the body by two types of lipoproteins. High-density lipoprotein (HDL) transports cholesterol to the liver where it is used or eliminated. A high level of HDL is considered healthy. Low-density lipoprotein (LDL) transports cholesterol into the blood where it has the potential to build up in the arteries. A high level of LDL has been associated with an increased risk for cardiovascular disease and stroke.

Vitamins

A *vitamin* is a nutrient composed of organic substances and is required in small amounts to help regulate body processes. Vitamins help support cell growth, cell reproduction, and immunity. Vitamins are classified as water-soluble or fat-soluble.

Water-Soluble Vitamins. A *water-soluble vitamin* is a vitamin that dissolves in water. These vitamins need to be replenished daily because they are not stored by the body. Water-soluble vitamins include vitamin C and the vitamin B complex. *See Figure 6-25.* Vitamin C is also known as ascorbic acid. The vitamin B complex functions as a group of vitamins that help activate proteins to build and repair body tissue. Aside from B6 and B12, each B vitamin has a specific name.

Water-Soluble Vitamins

Vitamin	Functions	Food Sources
Vitamin C	Promotes healthy teeth and gums, helps the body absorb iron, and facilitates wound healing	Citrus fruits, broccoli, red bell peppers, tomatoes, and spinach
Thiamin (B1)	Assists nerve processes and aids in metabolism by helping convert carbohydrates, proteins, and lipids into energy	Legumes, pork, pecans, spinach, oranges, cantaloupe, milk, and eggs
Riboflavin (B2)	Aids in the digestion of carbohydrates and proteins, the formation of red blood cells, respiration, antibody production, and growth	Dairy products, dark-green leafy vegetables, liver, yeast, almonds, and soybeans
Niacin (B3)	Assists in the functioning of the digestive system, skin, and nerves	Chicken, fish, milk, eggs, yeast, nuts, fruits, and vegetables
Pantothenic Acid (B5)	Helps convert carbohydrates, proteins, and lipids into energy, defends against infection, and helps stabilize blood sugar	Legumes, eggs, tomatoes, mushrooms, potatoes, and beef
Vitamin B6	Aids protein metabolism, helps form red blood cells, helps maintain the immune system, and stabilizes blood sugar levels	Beans, fish, meat, poultry, and vegetables
Biotin (B7)	Aids the production of cholesterol and amino acids and helps metabolize carbohydrates, proteins, and lipids for energy	Dairy products, soybeans, seafood, chicken, liver, nuts, cauliflower, egg yolks, and yeast
Folate (B9)	Used in cell production and maintenance, particularly during infancy and pregnancy	Dark-green leafy vegetables, legumes, breads, cereals, and orange juice
Vitamin B12	Helps maintain nerve cells and red blood cells	Fish, meat, poultry, eggs, and dairy products

Figure 6-25. *Water-soluble vitamins include vitamin C and the vitamin B complex.*

Fat-Soluble Vitamins. A *fat-soluble vitamin* is a vitamin that dissolves in fat. This means that fat has to be consumed for fat-soluble vitamins to be utilized. Fat-soluble vitamins include vitamins A, D, E, and K. *See Figure 6-26.* Unlike water-soluble vitamins, fat-soluble vitamins are stored by the body in the liver and fatty tissue. Vitamin D is manufactured by the body when the skin is exposed to sunlight.

Fat-Soluble Vitamins

Vitamin	Functions	Food Sources
Vitamin A	Crucial for vision, cell reproduction, and healthy skin	Animal products; red, yellow, and orange fruits and vegetables
Vitamin D	Helps the absorption of calcium and promotes strong teeth and bones	Milk, egg yolks, and enriched cereals and juices
Vitamin E	Aids the immune system, helps build cells, and supports healthy skin	Vegetable oils, nuts, and dark-green leafy vegetables
Vitamin K	Has an essential role in the clotting of blood	Dark-green leafy vegetables, fruits, dairy products, and egg yolks

Figure 6-26. *Fat-soluble vitamins include vitamins A, D, E, and K.*

Minerals

A *mineral* is an inorganic substance that is required in very small amounts to help regulate body processes. Eating a wide variety of foods every day ensures the body receives a sufficient amount of minerals. Minerals are classified as either macrominerals or microminerals.

Macrominerals. A *macromineral* is a mineral that the human body requires at least 100 mg of per day. ***See Figure 6-27.*** Calcium, phosphorus, magnesium, sodium, and potassium are considered macrominerals. A well-balanced diet supplies an adequate amount of macrominerals.

Macrominerals		
Mineral	**Functions**	**Food Sources**
Calcium	Essential for healthy bones and teeth, muscle contraction, and proper nerve function	Dairy products, broccoli, fish with edible bones, and almonds
Phosphorus	Important for healthy bones and teeth and fluid balance	Milk, dairy products, legumes, and nuts
Magnesium	Needed to make proteins and essential for proper muscle and nerve function	Dark-green leafy vegetables, nuts, legumes, and whole grains
Sodium	Helps maintain blood pressure and fluid balance	Cheeses, shellfish, table salt, and processed foods
Potassium	Works with sodium to maintain blood pressure and fluid levels	Fruits, vegetables, dairy products, and seafood

Figure 6-27. *The human body requires at least 100 mg per day of macrominerals.*

Microminerals. A *micromineral* is a mineral that the human body requires less than 100 mg of per day. ***See Figure 6-28.*** Microminerals include iron, iodine, zinc, selenium, copper, chromium, fluoride, manganese, and molybdenum. A well-balanced diet supplies an adequate amount of microminerals.

Microminerals		
Mineral	**Functions**	**Food Sources**
Iron	Helps carry oxygen throughout the body and plays a role in cell growth	Red meat, poultry, legumes, whole grains, and dark-green leafy vegetables
Iodine	Helps make hormones vital to tissue growth and reproduction	Fish, seafood, and iodized salt
Zinc	Needed to make proteins, helps support the immune system, and promotes sense of taste and smell	Meat, shellfish, whole grains, and legumes
Selenium	Required by the heart, immune system, and thyroid gland; also helps make proteins	Fruits, vegetables, some meats, seafood, and nuts
Copper	Helps regulate blood pressure and heart rate; needed to absorb iron and aid in the formation of collagen, which gives strength to bones, teeth, muscles, cartilage, and blood vessels	Seafood, organ meats, legumes, nuts, and seeds
Chromium	Helps produce hormones used to convert carbohydrates, proteins, and lipids into energy; also helps regulate blood sugar	Liver, cheese, yeast, broccoli, and nuts
Fluoride	Necessary for the formation of bones and teeth	Fish with edible bones and most teas
Manganese	Important for proper bone health and nerve function; helps maintain blood sugar levels	Whole grains, spinach, blueberries, pineapples, nuts, and seeds
Molybdenum	Plays a role in energy production and the development of the nervous system; helps the kidneys process waste	Legumes and whole grains

Figure 6-28. *A micromineral is a mineral that the human body requires less than 100 mg of per day.*

Water

Nearly every body function is dependent upon water. Water transports nutrients, carries away waste, provides moisture, and helps normalize body temperature. The human body loses about 2½ qt of water each day through normal processes such as perspiration and waste elimination. *See Figure 6-29.* The best way to replenish the water needed by the body is to drink plain water and eat high-water content foods such as mangoes, watermelon, and spinach each day.

Water Replacement

Figure 6-29. *The human body loses about 2½ qt of water each day through normal processes such as perspiration and waste elimination.*

Carlisle FoodService Products

Sports drinks, energy drinks, sodas, fruit juices, coffee, and tea often quench thirst due to the large amount of water these products contain. However, these products often also contain refined sugars and caffeine. Refined sugars have no nutritional value and have been linked to tooth decay. Caffeine has been found to raise blood pressure and speed the heartbeat.

Digestion

For the body to process nutrients, food must be digested. *Digestion* is the process the human body uses to break down food into a form that can be absorbed and used or excreted. *See Figure 6-30.*

Once food is chewed and swallowed, it travels to the stomach and then the small intestine. Along the way, carbohydrates are broken down into sugars, proteins are broken down into amino acids, and lipids are broken down into fatty acids. Once these nutrients are absorbed by the small intestine, they are carried to the liver along with vitamins and minerals to be processed and sent to cells throughout the body. Inside the cells, nutrients are used for energy or to build new tissue. Any indigestible waste, such as dietary fiber, moves to the large intestine and is excreted.

Digestion

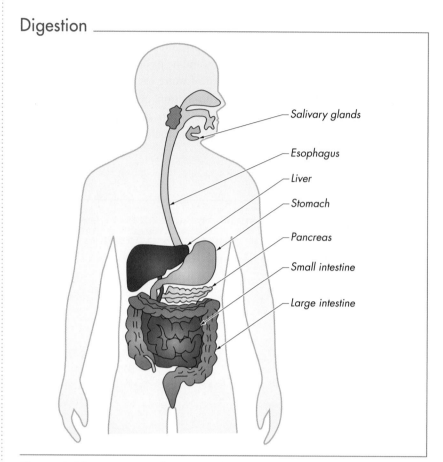

Figure 6-30. *Digestion is the process the human body uses to break down food into a form that can be absorbed and used or excreted.*

DIETARY RECOMMENDATIONS

Every five years, the U.S. Department of Agriculture (USDA) publishes *Dietary Guidelines for Americans* that include key recommendations for healthy eating and physical activity. These recommendations stress the importance of the daily consumption of fruits, vegetables, whole grains, and fat-free or low-fat milk and dairy products. These recommendations also place emphasis on consuming protein from a variety of sources, including lean meats, poultry, fish, eggs, soy, and nuts, and reducing saturated fats, cholesterol, sodium, and sugar.

The 2010 Dietary Guidelines include the following key recommendations:

- Choose foods that provide more potassium, dietary fiber, calcium, and vitamin D. These foods include vegetables, fruits, whole grains, and milk and milk products.

- Increase vegetable and fruit intake.

- Eat a variety of beans, peas, and vegetables, especially dark-green, red, and orange vegetables.

- Consume at least half of all grains as whole grains. Increase whole-grain intake by replacing refined grains with whole grains.

- Choose a variety of protein foods, including seafood, lean meat, poultry, eggs, beans, peas, soy products, and unsalted nuts and seeds.

- Increase the amount and variety of seafood consumed by choosing seafood in place of some meat and poultry.

- Replace protein foods that are high in solid fats with choices that are lower in solid fats, lower in calories, and/or are sources of oils.

- Increase intake of fat-free or low-fat milk and milk products, such as yogurt, cheese, and fortified soy beverages.

- Use oils to replace solid fats where possible.

Photo Courtesy of Lauren Frisch

The Dietary Guidelines are based on a daily 2000 calorie diet. A *calorie* is a unit of measurement that represents the amount of energy in a food. To manage weight and receive an adequate supply of nutrients, most adults require an assortment of nutritious foods each day. The recommended calorie range for an adult male is 2200–3000 per day. For an adult female, the recommended range is 1800–2400 per day. The recommended allowances vary depending on a person's age, height, weight, and level of activity.

The USDA created the ChooseMyPlate web site to help people understand how to meet the Dietary Guidelines and maintain their health and weight. The MyPlate icon is the symbol of the USDA ChooseMyPlate program. *See Figure 6-31.* The ChooseMyPlate program emphasizes the need to eat a variety of vegetables and fruits, whole grains, lean proteins, and low-fat dairy products.

MyPlate Icon

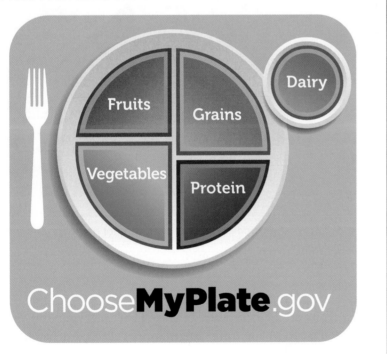

United States Department of Agriculture

Figure 6-31. *The MyPlate icon is the symbol of the USDA ChooseMyPlate program.*

Vegetables and Fruits

Vegetables and fruits, including 100% juice, are an excellent source of vitamins and minerals and a significant source of dietary fiber. The Dietary Guidelines emphasize the importance of eating dark-green, red, and orange vegetables and colorful fruits such as strawberries, blueberries, and pomegranates. Whole vegetables and fruits are considered more nutrient-dense than processed vegetables and fruits. A *nutrient-dense food* is a food that is high in nutrients and low in calories.

Most vegetables are naturally low in fat and calories and do not contain cholesterol. *See Figure 6-32.* The recommended daily allowance (RDA) for vegetables ranges from 1–3 cups depending on a person's age, gender, and level of physical activity. In general, 1 cup of raw vegetables, cooked vegetables, or vegetable juice or 2 cups of raw, leafy greens equals 1 cup of vegetables.

Vegetables

Figure 6-32. *Most vegetables are naturally low in fat and calories and do not contain cholesterol.*

Most fruits are high in water, dietary fiber, and vitamins. *See Figure 6-33.* The RDA for fruits ranges from 1–2 cups depending on a person's age, gender, and level of physical activity. In general, 1 cup of raw fruit, cooked fruit, or 100% fruit juice or ½ cup of dried fruit is equal to 1 cup of fruit.

Fruits

Figure 6-33. *Most fruits are high in water, dietary fiber, and vitamins.*

Vegetables and fruits are important sources of many vitamins, minerals, and dietary fiber. Eating fresh vegetables and fruits can help lower blood pressure and cholesterol levels as well as reduce the risk of cardiovascular disease and some cancers.

Grains

Grains are categorized as whole grains or refined grains. A *whole grain* is a grain that only has the husk removed. The *husk,* also known as the hull, is the inedible, protective outer covering of grain. Beneath the husk, a kernel of grain is composed of a bran, endosperm, and germ. *See Figure 6-34.* The *bran* is the tough outer layer of grain that covers the endosperm. The *endosperm* is the largest component of a grain kernel and consists of carbohydrates and a small amount of protein. The *germ* is the smallest part of a grain kernel and contains a small amount of natural oils as well as vitamins and minerals.

Whole Grain Kernels

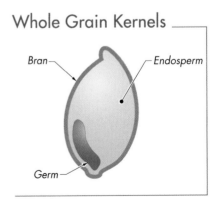

Figure 6-34. *A kernel of grain is composed of a bran, endosperm, and germ.*

Whole grains are an important source of nutrients and dietary fiber. *See Figure 6-35.* Examples of whole grains include brown rice, wild rice, wheat berries, bulgur, oats, barley, rye, quinoa, millet, and spelt. Eating whole grain foods can result in increased energy and a sense of feeling fuller longer. The recommended daily allowance for grains is 3–8 oz, depending on a person's age, gender, and level of physical activity. Of the RDA for grains, 1½–4 oz should be whole grains. A slice of bread, 1 cup of ready-to-eat cereal, or ½ cup of cooked grains or pasta is considered a 1 oz equivalent for grains.

Whole Grains

National Cancer Institute

Figure 6-35. *Whole grains are an important source of nutrients and dietary fiber.*

Grains are important sources of many nutrients, including dietary fiber, several B vitamins, and minerals such as iron, magnesium, and selenium. However, many grains are processed in order to increase shelf life and to offer a softer, moister texture. A *refined grain* is a grain that has been processed to remove the germ, bran, or both. Refined grains lose vitamins, minerals, and dietary fiber during processing. Some refined grains are enriched. An *enriched grain* is a refined grain that has thiamin, riboflavin, niacin, folate, and iron added to it.

Protein Foods

Foods made from meat, poultry, seafood, beans, peas, eggs, processed soy products, nuts, and seeds are considered protein foods. *See Figure 6-36.* Choosing leaner cuts of meat and eating at least 8 oz of cooked seafood per week can help reduce saturated fats and calories. Beans and peas are also considered vegetables but should only be counted in one group or the other. Choosing unsalted nuts and seeds is the best choice.

Protein Foods

Courtesy of The National Pork Board

Meat

Photo Courtesy of Perdue Foodservice, Perdue Farms Incorporated

Poultry

Harbor Seafood., Inc.

Seafood

Courtesy of The National Pork Board

Beans

Eggs

Nuts

Figure 6-36. *Foods made from meat, poultry, seafood, beans, peas, eggs, processed soy products, nuts, and seeds are considered protein foods.*

The RDA for protein foods is 2–6½ oz, depending on a person's age, gender, and level of physical activity. In general, 1 oz of meat, poultry, or fish, ¼ cup cooked beans, 1 egg, 1 tbsp of peanut butter, or ½ oz of nuts or seeds is considered a 1 oz protein equivalent. Protein foods supply many nutrients including protein, several B vitamins, vitamin E, iron, zinc, and magnesium.

Dairy Foods

Dairy foods include all fluid milk products and many foods made from milk that retain their calcium content, such as cheese and yogurt. *See Figure 6-37.* Calcium-fortified soymilk (soy beverage) is also part of this group. Foods made from milk that have little to no calcium, such as butter, cream, and cream cheese, are not considered dairy foods. Low-fat or fat-free dairy products are the best choices for health and wellness.

The RDA for dairy foods is 2–3 cups per day, depending on a person's age, gender, and level of physical activity. In general, 1 cup of dairy is equal to 1½ oz of natural cheese or 1 cup of milk, yogurt, or soymilk. Consuming dairy products provides improved bone health and key nutrients such as calcium, potassium, vitamin D, and protein.

Dairy Foods

Figure 6-37. *Dairy foods include all fluid milk products and many foods made from milk that retain their calcium content, such as cheese and yogurt.*

Physical Activity

In general, it is recommended that individuals participate in 30 minutes of physical activity each day. Walking, hiking, cycling, jogging, dancing, and swimming are all types of physical activity. Regular physical activity can reduce stress, increase bone and muscle strength, lower the risk of certain diseases, and increase the amount of calories used by the body. If the body uses more calories than are consumed, weight loss will occur. However, when more calories are consumed than are used, weight gain will occur. To maintain weight, the amount of calories used should be equal to the amount of calories consumed. *See Figure 6-38.*

Balancing Calories

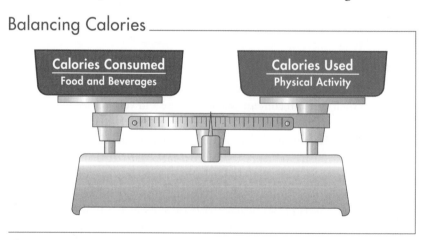

Figure 6-38. *To maintain weight, the amount of calories used should be equal to the amount of calories consumed.*

Nutrition Facts Labels

The Nutrition Labeling and Education Act of 1990 requires food manufacturers to display a nutrition facts label on all packaged food and beverage items. A nutrition facts label lists the nutritional value of the product and provides a standard for comparing foods based on nutrition. The nutrition facts label appears directly above the listing of ingredients and lists the serving size, calories, nutrients, and percent daily values based on a 2000 calorie diet. *See Figure 6-39.*

Nutrition Facts Labels

Figure 6-39. *The nutrition facts label appears directly above the listing of ingredients and lists the (1) serving size, (2) calories, (3) nutrients, and (4) percent daily values based on a 2000 calorie diet.*

Serving Size. The first section of the nutrition facts label includes the serving size and the number of servings per container. Serving sizes are commonly based on one serving and listed in standard measurements such as cups, ounces, or pieces. Calories, nutrient information, and percent daily values are based on the serving size. It is common for a product to have multiple servings. For example, the serving size on a bag of chips may list 15 chips. However, the bag contains 2 servings. If the entire bag is consumed, the calories, nutrients, and percent daily values double.

Calories. The second item on the nutrition facts label identifies the number of calories per serving and the number of those calories that come from fat. This information can be used to compare similar products to determine which is lower in fat and calories.

Nutrients. The next section of the nutrition facts label lists the amount of fat, cholesterol, sodium, carbohydrates, protein, and specific vitamins and minerals. Total fat includes the amount of fat per serving and is listed in grams. Total fat is also broken down into saturated fat and trans fat. Cholesterol and sodium appear next and are listed in milligrams. The USDA recommends limiting cholesterol to 300 mg per day and sodium to less than 2300 mg per day. However, it is best to consume less than 1500 mg of sodium per day.

Total carbohydrates appear next on the label and are listed in grams. Carbohydrates are also broken down into fiber and sugars. Protein is listed next in grams, followed by the percentages of vitamin A, vitamin C, calcium, and iron. The nutrient section of the nutrition facts label can be a valuable tool in helping individuals increase dietary fiber and lower fat, cholesterol, sodium, and sugar.

Percent Daily Values. The percent daily value is established by the recommended daily allowances for each key nutrient. For example, if a food lists calcium as 20%, then one serving provides 20% of the calcium needed per day. The percent daily value can also be used to determine if a food or beverage is high or low in nutrients. A percent daily value of 5% or less is regarded as low. A percent daily value of 20% or more is regarded as high. This information can be used as a guide to choose items high in vitamins, minerals, and dietary fiber.

RECIPE MODIFICATIONS

There are many ways to modify recipes to limit ingredients that are high in fat, sugar, and sodium and offer more nutritious menu items. Recipe management software helps the user easily substitute ingredients and change quantities to create healthier recipes. For example, a fettuccine Alfredo recipe may have 500 calories and 25 g of fat per serving. After modifying the recipe, a nutritional analysis may reveal the new fettuccine Alfredo recipe has 350 calories and 12 g of fat per serving. Using whole grain fettuccine is a simple modification that would increase the amount of dietary fiber, vitamins, and minerals in the dish. Likewise, the portion size of each dish is equally important in maximizing healthy choices.

Lowering Fat

When modifying recipes, a saturated fat, such as butter, can often be replaced with an unsaturated fat, such as olive oil. Fat content may also be lowered by using low-fat proteins such as tofu, poultry without the skin,

Nutrition Note

Instead of sugar, many products are manufactured with sweetening agents such as saccharin, aspartame, acesulfame-K, sucralose, neotame, and rebiana. These sweetening agents can be added to products without increasing the sugar content.

lean cuts of meat, and low-fat or fat-free dairy products. Substituting some or all of the meat in a recipe with vegetables or whole grains can also decrease the amount of fat.

High amounts of fat also can be found in menu items that appear to be healthy. For example, a grilled chicken breast wrap sounds healthy, but the creamy ranch dressing it is often served with is high in fat. Using a low-fat dipping sauce, such as salsa, is an effective way to lower the amount of fat and calories. *See Figure 6-40.*

Low-Fat Dipping Sauces

Figure 6-40. *Using a low-fat dipping sauce, such as salsa, is an effective way to lower the amount of fat and calories.*

The cooking method used also impacts the amount of fat in a recipe. Foods cooked in oil or butter absorb fat, which creates higher-fat menu items. For example, a chicken breast that is broiled, baked, or grilled is lower in fat than chicken fried in oil.

Lowering Sugar

One way to modify recipes by reducing sugar is to avoid processed items and prepare foods in a manner that controls the sugar content. For example, a commercially prepared raspberry vinaigrette may be high in refined sugars. However, when raspberry vinaigrette is made using high-quality ingredients, the raspberries can provide the right balance of sweetness.

Using natural sweeteners, such as honey, molasses, maple syrup, and agave nectar, can help lower the overall sugar content because refined sugars are less sweet than natural sweeteners. *See Figure 6-41.* Additional ways to reduce sugar include using herbs and spices, such as mint or cinnamon, to naturally enhance sweetness, using natural fruit purées as sweeteners in baked goods, and substituting dark chocolate for milk chocolate.

Natural Sweetener Equivalents for ½ Cup of Sugar	
Barley malt	1½ cups
Date sugar	1 cup
Honey	⅓ cup
Maple syrup	½ cup
Molasses	⅓ cup
Rice syrup	1¼ cups
Sorghum syrup	⅓ cup

Figure 6-41. *Using natural sweeteners can help lower the overall sugar content because refined sugars are less sweet than natural sweeteners.*

Barilla America, Inc.

Figure 6-42. *Countless combinations of seasonings can add flavor to foods without using too much salt.*

Lowering Sodium

Research suggests that more than three-fourths of the sodium consumed by the average person comes from processed foods. Using high-quality whole foods can greatly reduce the sodium content of a menu item. When using processed foods, products labeled "low-sodium," "sodium-free," or "salt-free" can help reduce the sodium levels in recipes. Sodium can also be reduced by limiting the amount of salt used to season foods. Instead of salt, flavor can be enhanced with vinegar, peppers, herbs, spices, or an acid, such as lemon juice. Countless combinations of seasonings can add flavor to foods without using too much salt. *See Figure 6-42.*

Portion Sizes

According to the Center for Disease Control (CDC), individuals consume more food when presented with larger portions. This often leads to the consumption of extra fat and calories, resulting in weight gain. Guests have become accustomed to seeing plates overfilled with food and often perceive larger portions as a better value. Presenting food on smaller plates is one way to alter this perception. *See Figure 6-43.* For example, a 12 inch plate can be used instead of a 14 inch plate, and an 8 inch plate can be used in place of a 10 inch plate.

Smaller Plates _____

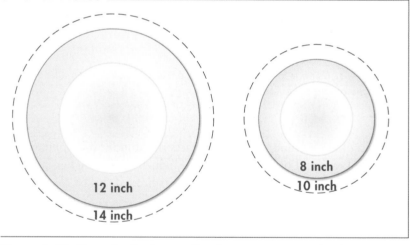

Figure 6-43. *Presenting food on smaller plates is one way to alter the perception that larger portions are a better value.*

In addition to plate size, menus that offer smaller portions, such as two pieces of fried chicken instead of four, can promote smaller portion sizes. For example, instead of offering a 12 oz steak, a 6 oz steak with savory roasted potatoes and a large portion of steamed vegetables can be offered. This type of dish helps guests meet the recommendations of the Dietary Guidelines while still offering a menu item that is full of flavor. Each item on a plate should be examined for appropriate portion size in order to offer guests more healthful options. *See Figure 6-44.*

Smaller Portions

Figure 6-44. *Each item on a plate should be examined for appropriate portion size in order to offer guests more healthful options.*

SUMMARY

Menus are the primary way foodservice operations communicate their products to guests and should follow the truth-in-menu guidelines. Menu planning involves determining which food and beverages a foodservice operation will offer and the prices it will charge for those items. There are several classifications and types of menus from which to choose. A well-designed menu with effective pricing has the potential to positively impact sales, accommodate individuals with dietary considerations, and exceed guest expectations.

The human body needs a variety of nutrients to function properly. Proteins, carbohydrates, lipids, vitamins, minerals, and water are all essential to health. Understanding the function of each nutrient helps foodservice professionals better serve guests. The Dietary Guidelines recommend eating a variety of foods that are low in saturated fat, cholesterol, sugar, and sodium. Changes in weight will occur if the amount of calories consumed is more or less than the amount of calories used during physical activity. The USDA's ChooseMyPlate program and nutrition facts labels found on product packages are tools foodservice operations can use to help prepare menu items that meet nutritional guidelines.

By modifying recipes, foodservice professionals have the opportunity to provide guests with lower fat, sugar, and sodium options that are appealing, nutritious, and flavorful. Foodservice operations that alter the sizes of plates and the amount served for each portion can help guests make better food choices.

Refer to DVD for
Quick Quiz® questions

1. Identify the five functions of a menu.
2. Explain the purpose of the truth-in-menu guidelines.
3. Identify nine types of inaccuracies identified by the truth-in-menu guidelines.
4. Describe the three classifications of menus.
5. Describe six common menu types.
6. Identify the type of menu that cannot be meal specific.
7. Describe the four elements of menu design.
8. Explain three reasons why an effective menu mix is important.
9. Explain why some menu items require a description and other items do not.
10. Describe four common menu formats.
11. Explain how the placement of items on a menu can affect sales.
12. Explain the role that graphics play in menu design.
13. Identify three factors that directly impact menu pricing.
14. Explain how the food cost percentage of an item can affect profits.
15. Explain how perceived value pricing is influenced by the price range on a menu.
16. Describe different strategies for placing menu prices.
17. Describe four types of dietary considerations that affect menu planning.
18. Identify the eight major food allergens.
19. Differentiate between food allergies and food intolerances.
20. Identify illnesses and diseases that have been linked to the rise in obesity.
21. Describe plant-based diets.
22. List the six nutrients.
23. Describe the function of proteins.
24. Contrast complete and incomplete proteins.
25. Describe the function of carbohydrates.
26. Explain the role of sugars, starches, and dietary fiber.
27. Describe the function of lipids.
28. Explain how different types of fat impact health.
29. Explain the link between saturated fats, trans fats, and cholesterol.
30. Explain why a high level of LDL is cause for concern.
31. Describe the function of vitamins.
32. List the water-soluble vitamins.
33. Explain why water-soluble vitamins need to be replenished daily.
34. List the fat-soluble vitamins.
35. Explain why fat-soluble vitamins do not need to be replenished daily.
36. Describe the function of minerals.
37. Differentiate between macrominerals and microminerals.
38. Describe the role water plays in maintaining health.
39. Identify the amount of water lost by the human body each day.
40. Summarize the key recommendations in *Dietary Guidelines for Americans*.
41. Explain the emphasis of the ChooseMyPlate program.

Refer to DVD for
Review Questions

42. Describe a nutrient-dense food.
43. Identify three factors that affect recommended daily allowances (RDA).
44. Identify the RDA range for vegetables.
45. Identify the RDA range for fruits.
46. Differentiate among whole grains, refined grains, and enriched grains.
47. Identify the RDA range for grains.
48. Identify the portion of the RDA for grains that should be whole grains.
49. Explain how to reduce saturated fats and calories when eating protein foods.
50. Identify the RDA range for protein foods.
51. Identify foods made from milk that are not considered dairy foods.
52. Identify the RDA for dairy foods.
53. Name four ways regular physical activity benefits the body.
54. Identify the four main components of a nutrition facts label.
55. Explain how recipes can be modified to lower fat.
56. Explain how recipes can be modified to lower sugar.
57. Explain how recipes can be modified to lower sodium.
58. Explain how recipe management software can help foodservice professionals offer more nutritious menu items.
59. Explain how reducing plate sizes can positively affect guests' health and wellness.
60. Explain how serving smaller portions can positively affect health and wellness.

Applications

1. Develop a fixed menu. Evaluate the menu mix, item descriptions, format, item placement, use of graphics, and price placement.
2. Develop a market menu. Evaluate the menu mix, item descriptions, format, item placement, and use of graphics.
3. Develop a cycle menu. Evaluate the menu mix, item descriptions, format, item placement, and use of graphics.

Refer to DVD for
Application Scoresheets

4. Price the five menu items on the soups menu shown in Figure 6-11. Evaluate the prices based on item costs and perceived value using local information.
5. Evaluate the soups menu for food allergies, food intolerances, weight management, and plant-based meals.
6. Create a dish that combines two incomplete proteins to make a complete protein.
7. Create a dish that offers complex carbohydrates and dietary fiber.
8. Create a dish that is low in saturated fat, does not contain trans fat, and is packed with vitamins and minerals.
9. Create a refreshing beverage that is low in sugar and caffeine.
10. Sketch how food travels through the digestive system.
11. Create a breakfast, lunch, and dinner menu that meets the Dietary Guidelines.
12. Draw or photograph three meals and evaluate them against the MyPlate icon.
13. Record the calories eaten in one day as well as the amount of physical activity. Evaluate the results to see if they balance.
14. Modify a recipe to lower the fat content. Evaluate the nutrition results and portion size.
15. Modify a recipe to lower the sugar content. Evaluate the nutrition results and portion size.
16. Modify a recipe to lower the sodium content. Evaluate the nutrition results and portion size.

Chapter 7

Cooking Techniques

Cooking techniques involve a thorough knowledge of cooking methods and how to develop flavors in foods. The cooking method chosen depends on the type of food being cooked and the flavors the chef is developing in the dish. Flavorings are added to change the natural flavor of a food. Seasonings are added to intensify the flavor of a food. Successfully developing flavors is the key to creating successful dishes that guests will enjoy and want to order again and again.

Refer to DVD
for **Flash Cards**

Chapter Objectives

1. Describe the three methods of heat transfer used to cook food.
2. Identify three types of radiation heat transfer used to cook food.
3. Describe five reactions that change the color or texture of food.
4. Identify the two nutrients most often destroyed by heat.
5. Identify common dry-heat cooking methods.
6. Demonstrate the use of common dry-heat cooking methods.
7. Identify common moist-heat cooking methods.
8. Demonstrate the use of common moist-heat cooking methods.
9. Identify common combination cooking methods.
10. Demonstrate the use of combination cooking methods.
11. Identify four signals the brain receives about food.
12. Explain how flavors are developed in food.
13. Differentiate between flavorings and seasonings.
14. Identify leaf herbs and stem herbs.
15. Identify bark, seed, root, flower, berry, and bean spices.
16. Explain the use of rubs and marinades in flavor development.
17. Identify three common types of condiments.
18. Describe five common types of sauces that may be used as condiments.
19. Describe 12 types of nuts used to flavor food.
20. Identify five categories of seasonings used in the professional kitchen.

Key Terms

- **grilling**
- **barbequing**
- **broiling**
- **roasting**
- **baking**
- **smoking**
- **sautéing**
- **frying**
- **poaching**
- **simmering**
- **blanching**
- **boiling**
- **steaming**
- **braising**
- **stewing**
- **sous vide**
- **flavoring**
- **herb**
- **spice**
- **seasoning**

THE COOKING PROCESS

Cooking is the process of heating foods in order to make them taste better, make them easier to digest, and kill harmful microorganisms that may be present in the food. Heat is energy that is transferred between two objects or substances of different temperatures. Heat typically flows from a warmer material to a cooler material. When heat is transferred, the particles inside an object increase their movement as the temperature increases. The three different methods of heat transfer used with food are conduction, convection, and radiation. Each type of heat transfer cooks food in a different manner and produces different results.

Conduction

Conduction is a type of heat transfer in which heat passes from one object to another through direct contact. *See Figure 7-1.* Heat flows from a warm object to a cool object. When an egg is cracked into an omelet pan and begins to cook, heat transfer occurs as a result of the egg coming in contact with the hot pan, which became hot by direct contact with a heat source. Direct contact enables the heat to be transferred from the heat source to the pan and then to the egg in the pan. Metal is a good conductor of heat, while other materials, such as glass, wood, and plastic, are not.

Convection

Convection is a type of heat transfer that occurs due to the circular movement of a fluid or a gas. *See Figure 7-2.* Heating a pot of water is an example of convection. The pot itself becomes heated by conduction. As the water heats it expands and rises up through the surrounding cooler water, causing the cooler water to sink to the bottom of the pot where it will become heated. This circulating path of the warmer water rising and the cooler water sinking is called a convection current. A convection oven circulates hot air around food items, cooking them more quickly and evenly than a conventional oven.

Convection heat transfer is the reason that fat is a consistent temperature throughout a fryer once it reaches a set temperature. When a fryer is turned on, the fat inside the fryer heats up through direct contact with the heat source in the bottom of the fryer. The cooler fat near the surface of the fryer sinks toward the bottom of the fryer as the hot fat at the bottom moves toward the surface. This constant circulation of warmer and cooler molecules maintains a steady and even temperature.

A convection oven uses a reduced cooking temperature compared to a conventional oven. The difference in cooking temperature is needed because circulating air transfers heat more quickly than still air of the same temperature. An impingement oven also operates on convection heat transfer. Impingement ovens are often used to toast breads, heat sandwiches, and cook pizzas. Impingement ovens circulate hot air above and below food as it moves across a conveyer.

Conduction

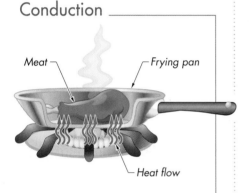

Meat — / — Frying pan

— Heat flow

Figure 7-1. *Conduction is a type of heat transfer in which heat passes from one object to another through direct contact.*

Convection

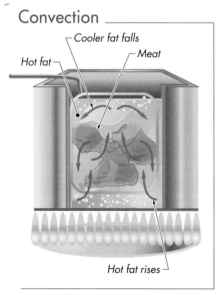

— Cooler fat falls
— Meat
Hot fat —

Hot fat rises —

Figure 7-2. *Convection is a type of heat transfer that occurs due to the circular movement of a fluid or a gas.*

Radiation

Radiation is a type of heat transfer that uses electromagnetic waves to transfer energy. When electromagnetic waves strike an object, the energy carried by the waves is transferred to the object. This is the reason that the sun feels warm. Some materials are better at absorbing radiation than others. Aluminum foil is shiny and therefore reflects rather than absorbs radiation. The three types of radiation heat transfer used to heat foods in the professional kitchen are infrared, microwave, and induction. *See Figure 7-3.*

Radiation

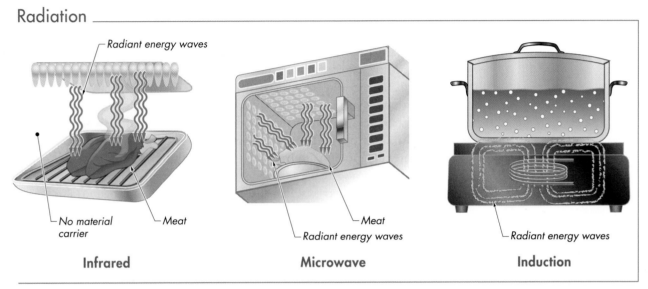

Infrared   Microwave   Induction

Figure 7-3. *The three types of radiation heat transfer used to heat foods in the professional kitchen are infrared, microwave, and induction.*

Infrared. Infrared radiation can be seen in a toaster or a broiler when the heating element glows red. When food is placed in infrared cooking equipment, such as a broiler, the food is cooked as it absorbs the electromagnetic waves. For example, a steak placed in a broiler cooks on top first, even though the bottom of the steak is in direct contact with the broiler rack. The food does not cook from direct contact with the cooking surface but cooks from the heat given off by the heating element.

Microwave. Microwave radiation uses electromagnetic waves to heat the water, fat, and sugar molecules in food. Microwaved food cooks quickly but does not brown. However, it is important to remember that as the food heats up it also conducts heat to the plate through direct contact. Microwaves are generally used to quickly heat items in the professional kitchen.

Induction. Induction radiation uses electromagnetic current to heat magnetic cookware. The cookware heats the food and the cooktop remains cool. Copper coils create an electromagnetic field beneath the smooth ceramic surface of an induction burner. The magnetic pulses cause the molecules in the magnetic cookware to move quickly and generate heat. Induction cooking is rapid, and when the pan is removed from the cooktop, cooking slows quickly.

The cookware used on an induction cooktop must be flat-bottomed and made of a material that will conduct electricity, such as cast iron, stainless steel, copper, or aluminum. Induction radiation gives a chef maximum control over a heated pan.

Changes in Color and Texture

Before learning specific cooking methods, it is important to understand how heat changes the color and texture of food. In general, the longer a food cooks, the more color loss and texture change that occurs. The Maillard reaction, caramelization, reduction, coagulation, and gelatinization are a few ways in which food changes in color and texture.

Maillard Reaction. The *Maillard reaction* is a reaction that occurs when the proteins and sugars in a food are exposed to heat and merge together to form a brown exterior surface. *See Figure 7-4.* For example, the Maillard reaction occurs when meat is seared. *Searing* is the process of using high heat to quickly brown the surface of a food. Searing is done with little or no fat in a very hot sauté pan on a hot grill or griddle. Some foods are seared before cooking, and others may be seared after they are cooked to add color. Searing may also be the entire cooking process. Fish, such as tuna, and some meats are often served seared. Grilling, frying, and toasting foods also causes the Maillard reaction.

Maillard Reaction

Raw Meat   **Seared Meat**

Figure 7-4. *The Maillard reaction is a reaction that occurs when the proteins and sugars in a food are exposed to heat and merge together to form a brown exterior surface.*

Caramelization. Many recipes call for caramelized onions. *Caramelization* is a reaction that occurs when sugars are exposed to high heat and produce browning and a change in flavor. *See Figure 7-5.*

The first step toward caramelization is sweating. *Sweating* is the process of slowly cooking food to soften its texture. Many recipes require the sweating of vegetables such as onions and garlic. In the second step, the sugars begin to darken as the temperature of the food rises, which changes the aroma and color. In the final step, the flavor further develops and completes the caramelization. The top of a crème brûlée and the crust of a freshly baked loaf of bread are examples of caramelization. *See Figure 7-6.*

Caramelization

| Sweating | Browning | Caramelizing |

Figure 7-5. *Caramelization is a reaction that occurs when sugars are exposed to high heat and produce browning and a change in flavor.*

Caramelized Products

Eloma Combi Ovens
Crème Brûlée

Bridgeford Foods Corporation
Freshly Baked Bread

Figure 7-6. *The top of a crème brûlée and the crust of a freshly baked loaf of bread are examples of caramelization.*

Reduction. *Reduction* is the process of gently simmering a liquid until it reduces in volume and results in a thicker liquid with a more concentrated flavor. *See Figure 7-7.* Sauces, stocks, soups and pan drippings are often reduced. Pan drippings left over after cooking meat can be simmered to create a densely flavored sauce. Wine, water, cream, or balsamic vinegar can be added to pan drippings to create a reduction.

Reduction

| Before | After |

Figure 7-7. *Reduction is the process of gently simmering a liquid until it reduces in volume and results in a thicker liquid with a more concentrated flavor.*

Refer to DVD for
Reduction
Media Clip

Coagulation. When proteins are exposed to heat, they begin to firm and coagulate. *Coagulation* is the process of a protein changing from a liquid to a semisolid or a solid state when heat or friction is applied. Coagulation takes foods from a high-moisture state to a low-moisture state, such as when a raw egg begins to coagulate as it is poached. *See Figure 7-8.* Egg whites also coagulate as they are beaten to form a meringue.

Gelatinization. *Gelatinization* is the process of a heated starch absorbing moisture and swelling, which thickens the liquid. As starches (complex carbohydrates) are heated, they begin to absorb liquid and swell. This is why muffin batter is wet in its raw form but expands and dries out as it bakes. The starch molecules in the flour absorb moisture from the wet batter and create steam during the baking process. Starches also absorb moisture when a sauce is made. Adding starch to a stock thickens the stock, and a sauce is formed as the starch absorbs some of the water from the stock. *See Figure 7-9.*

Coagulation

Figure 7-8. *Coagulation takes foods from a high-moisture state to a low-moisture state, such as when a raw egg begins to coagulate as it is poached.*

Gelatinization

Figure 7-9. *Adding starch to a stock thickens the stock, and a sauce forms as the starch absorbs some of the water from the stock.*

Changes in Nutrient Values

Foods are made of various amounts of proteins, carbohydrates, lipids, vitamins, minerals, and water. The nutrient values of food can be altered by the way it is prepared, cooked, and stored. For example, water-soluble vitamins and minerals are easily destroyed by excess heat. However, nutrient-rich blanching water can be used to make soups. The length of cooking time also determines the level of nutrient loss or gain. Cooking bright-green vegetables, such as broccoli, causes them to change color as the chlorophyll is destroyed by heat. *See Figure 7-10.* Likewise, the crisp texture of vegetables decreases as the heat exposure increases.

While heat can destroy vitamins, cooking can also increase the nutrient value of some foods, such as tomatoes. Cooked tomatoes contain more lycopene than uncooked tomatoes. Likewise, heated cinnamon has more antioxidant power than raw cinnamon. It is important to be aware of the impact of heat on nutrient values.

DRY-HEAT COOKING METHODS

The three general categories of cooking methods are dry-heat cooking, moist-heat cooking, and combination cooking. *Dry-heat cooking* is any cooking method that uses hot air, hot metal, a flame, or hot fat to conduct heat and brown food. Grilling, broiling, smoking, barbequing, roasting, baking, sautéing, and frying are dry-heat cooking methods. Tender meats, ground meats, and some fruits and vegetables are often cooked using dry-heat cooking methods.

Grilling

Grilling is a dry-heat cooking method in which food is cooked on open grates above a direct heat source. A grill is scraped clean and then seasoned with fat before use to prevent food from sticking. *See Figure 7-11.*

Changes in Nutrient Values

Steamed

Overcooked

Figure 7-10. *Cooking bright-green vegetables, such as broccoli, causes them to change color as the chlorophyll is destroyed by heat.*

Procedure for Preparing the Grill

1. Preheat the grill. Use a wire brush to scrape clean the grates of the hot grill.

2. Wipe the grates with a towel lightly coated in vegetable oil.

Figure 7-11. *A grill is scraped clean and then seasoned with fat before use to prevent food from sticking.*

Poultry, seafood, meats, and some fruits and vegetables are often grilled. Many foods are cooked directly on the grill, while others are placed on skewers first. Because of the open cooking surface, foods that are grilled have a smoky flavor. Grilled foods have char lines where the food came in contact with the hot grill. Crosshatching the presentation side of grilled food enhances the presentation of the dish. To create a crosshatch marking when grilling, the food is rotated 90° about halfway through the cooking process. *See Figure 7-12.*

Procedure for Grilling

1. Brush a small amount of oil on the items to be grilled and season as desired.

2. Using tongs, place the items on the hot grill with the presentation side facing down.

3. Allow the items to cook until char lines have developed on the presentation side.

4. Use the tongs to rotate the items 90° to create crosshatch markings on the presentation side.

5. Turn the items and cook the other side to the desired degree of doneness. *Note:* Crosshatch markings do not need to be added to this side since it faces the plate.

6. With a gloved finger, gently press the item to determine doneness. *Note:* Fish should flake and a thermometer should be used to check the temperature of thicker meats.

Figure 7-12. *To create a crosshatch marking when grilling, the food is rotated 90° about halfway through the cooking process.*

Grilling and griddling are similar cooking methods. *Griddling* is a dry-heat cooking method in which food is cooked on a solid metal cooking surface called a griddle. The heat comes from below the cooking surface, but items do not come in contact with a flame. Because of the flat surface, a griddle generates less smoke than a grill. A small amount of oil is placed on a hot griddle to prevent foods from sticking. Pancakes, eggs, breakfast meats, and hot sandwich meats are often cooked on a griddle. A *French grill* is a griddle with raised ridges that creates grill marks where the food touches the ridges.

Broiling

Broiling is a dry-heat cooking method in which food is cooked directly under or over a heat source. A broiler is typically scraped clean and then seasoned with fat before use to prevent food from sticking. ***See Figure 7-13.***

Procedure for Preparing a Broiler

1. Preheat the broiler. Use a wire brush to scrape clean the grates of the broiler.

2. Wipe the metal grates with a towel lightly coated in vegetable oil.

Figure 7-13. *A broiler is typically scraped clean and then seasoned with fat before use to prevent food from sticking.*

Poultry, seafood, meats, and some fruits and vegetables are broiled. Other items, such as bruschetta and French onion soup, are often finished in the broiler. Broiled food is exposed to an overhead flame or an overhead electric heating element when it is broiled. ***See Figure 7-14.*** The hot grates on which the food is placed create distinctive markings on broiled items. The grates can be raised or lowered to adjust the intensity of the heat.

Procedure for Broiling

1. Brush a small amount of oil on the items to be broiled and season as desired.

2. Using tongs, place the items in the broiler with the presentation side facing down.

3. Allow the items to cook until char lines have developed on the presentation side.

4. If desired, use the tongs to rotate the items 90° to create crosshatch markings on the presentation side.

5. Use tongs to turn the items and cook the other side to the desired degree of doneness.

6. Fish should flake and a thermometer should be used to check the temperature of thicker meats.

Figure 7-14. *Broiled food is exposed to an overhead flame or an overhead electric heating element when it is broiled.*

Smoking

Smoking is a dry-heat cooking method in which food is flavored, cooked, or preserved by exposing it to the smoke from burning or smoldering plant materials, most often wood. *See Figure 7-15.* Many professional kitchens use gas or electric smokers. Smokers use hardwood shavings or chips to produce the desired smoke flavor. Different types of wood, such as mesquite, hickory, maple, apple, and cherry, are used to impart different smoked flavors. The temperature of a smoker is easy to control, and the chef can easily change the type of wood because it is not the primary heat source.

Refer to DVD for
Smoking
Media Clip

Procedure for Smoking

1. Place soaked wood chips in smoker basket and preheat smoker to desired temperature.

2. Rub seasonings on the item and then place inside the smoker.

3. Smoke item until cooked through or desired level of smokiness has been achieved.

Figure 7-15. *Smoking is a dry-heat cooking method in which food is flavored, cooked, or preserved by exposing it to the smoke from burning or smoldering plant materials, most often wood.*

Smoke roasting is a similar cooking process to smoking, but it is done on a range top. A smoke roaster is a metal or aluminum container lined with wood chips that has a rack and a cover. The wood is heated to a smolder and the food is placed on the rack and then covered. Doneness is determined in the same manner as when roasting food.

Barbequing

Barbequing is a dry-heat cooking method in which food is slowly cooked over hot coals or burning wood. Barbequing is often described as a combination of grilling and roasting, depending on the location of the fire. Direct heat cooks the food quickly, like grilling. If the fire is to the side of the food, the indirect heat has a convection effect. Indirect heat is often used to cook ribs, beef brisket, and delicate fish fillets. It imparts a deeper smoke flavor and a more roasted exterior. The food is often basted with barbeque sauce to keep it moist as it is cooking. *See Figure 7-16.* Some chefs prefer to add the sauce after the food is cooked.

Courtesy of The National Pork Board

Barbequing

Courtesy of The National Pork Board

Figure 7-16. *Barbequed food is often basted with barbeque sauce to keep it moist as it is cooking.*

Roasting

In the past, roasting meant cooking food over an open flame, typically on a spit over a fire. In contemporary cooking, to roast means to cook outside of a liquid. *Roasting* is a dry-heat cooking method in which food that contains fat or that has fat added to it is cooked uncovered at a high temperature in an oven or on a revolving spit over an open flame. Leaving items uncovered while roasting enables caramelization to occur. Roasting produces a well-browned exterior and a moist interior. *See Figure 7-17.* Whole poultry, large cuts of meat, and root vegetables are commonly roasted at lower temperatures to prevent shrinkage and lessen moisture loss.

Cres Cor

1. Trim excess fat from the item and season it. *Note:* Some items, such as turkey, may be brushed with fat before roasting for more even browning.

2. Place the item in a roasting pan on a roasting rack or a bed of mirepoix.

3. Place the uncovered roasting pan in an oven preheated to between 300°F and 350°F, and allow the item to cook.

4. Baste the item as needed.

5. Remove the roasted item from the oven approximately 10 minutes before it reaches the desired degree of doneness. Insert a probe thermometer in the thickest portion to check the internal temperature.

6. Allow the item to rest and carryover cooking to complete the cooking process.

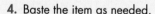

Figure 7-17. *Roasting produces a well-browned exterior and a moist interior.*

Refer to DVD for
Roasting
Media Clip

Allowing roasted items to rest after being removed from the oven permits carryover cooking to occur. *Carryover cooking* is the rise in internal temperature of an item after it is removed from a heat source due to residual heat on the surface of the item. The internal temperature will rise between 5°F and 10°F during this period. Carryover cooking helps a cooked item retain moisture that will escape as steam if the item is cut too soon after cooking. Roasted meat should not be cut immediately after being removed from the oven. If it is, the internal moisture will be released as steam and evaporate into the air. Roasted meat that is allowed to rest before it is cut will retain moisture.

Whole chickens, turkeys, and hogs may be skewered and placed on a rotisserie and roasted. *Spit-roasting* is the process of cooking meat by skewering it and suspending and rotating it above or next to a heat source. Roasting may require periodic basting during the cooking process. *Basting* is the process of using a brush or a ladle to place fat on or pour juices over an item during the cooking process to help retain moisture and enhance flavor.

Baking

Baking is the dry-heat cooking method in which food is cooked uncovered in an oven. Leaving items uncovered while they bake enables caramelization to occur. Like roasting, baking produces a well-browned exterior and a moist interior. Baking time, like roasting time, varies depending upon the type and size of the item, the temperature of the oven, and the ingredients used. Baking is the primary cooking method used to cook yeast breads, quick breads, cookies, cakes, pies, and pastries. *See Figure 7-18.*

Vulcan-Hart, a division of the ITW Food Equipment Group LLC

Procedure for Baking

1. Preheat an oven to the desired temperature.
2. Place items in parchment-lined or oiled pans per recipe directions.
3. Place pans in the oven and bake for the allotted time and until fully cooked.

Figure 7-18. *Baking is the primary cooking method used to cook yeast breads, quick breads, cookies, cakes, pies, and pastries.*

Before placing items in the oven, the oven must be preheated to the correct temperature. When baked items are removed from the oven, the items need to rest before being cut. Carryover cooking helps baked items retain their moisture while the baking process is completed. *See Figure 7-19.* If a loaf of bread is cut as soon as it comes out of the oven, it will immediately begin to dry out and lose much of its freshness.

Carryover Cooking

Paderno World Cuisine

Figure 7-19. *Carryover cooking helps baked items retain their moisture while the baking process is completed.*

Irinox USA

Sautéing

Sautéing is a dry-heat cooking method in which food is cooked quickly in a sauté pan over direct heat using a small amount of fat. Many items are dredged before they are sautéed. *Dredging* is the process of lightly dusting an item in seasoned flour or fine bread crumbs. ***See Figure 7-20.***

Procedure for Dredging

1. Coat items in flour or finely ground crumbs.

2. Shake off excess coating.
3. Immediately cook coated items.

Figure 7-20. *Dredging is the process of lightly dusting an item in seasoned flour or fine bread crumbs for frying.*

Refer to DVD for
Sautéing Large Items
Media Clip

Refer to DVD for
Sautéing Small Items
Media Clip

Sautéing is done over high heat, and caution must be used to not allow the small amount of fat in the pan to burn. Many proteins are sautéed. Some proteins are dredged and others are not. Tongs are typically used to turn large protein items in a sauté pan. ***See Figure 7-21.***

Fresh produce is often sautéed. Items such as vegetables are often flipped using a wrist-flicking motion to ensure even cooking without burning the item. ***See Figure 7-22.***

Stir-frying is the process of quickly cooking items in a heated wok with a very small amount of fat while constantly stirring the items. Foods that are stir-fried are crisp and tender when cooked. The time required to stir-fry an item depends on its thickness. For example, denser vegetables, such as broccoli and carrots, require more cooking time than less dense vegetables, such as bok choy and snow peas. Once the fat is hot, dry herbs and spices are added to the hot wok. Then the proteins are seared before the vegetables and any liquid ingredients are added.

Procedure for Sautéing Large Items

1. Prepare and season items. If desired, dredge items and shake off excess flour.

2. Place sauté pan on burner over high heat and add a small amount of fat to coat the bottom of the pan.

3. When the fat is hot, place the items flat in the pan, using caution to not overfill the pan. *Note:* If the pan is too full, the items will steam or simmer rather than sauté.

4. Once the edges change color and begin to brown, turn the items.

5. Cook the items until golden brown on both sides and cooked through.

Figure 7-21. *Tongs are typically used to turn large items in a sauté pan.*

Procedure for Sautéing Small Items

1. Place sauté pan over high heat and add a small amount of fat to coat the bottom of the pan.

2. When the fat is hot, add the items to the pan, using caution to not overfill the pan. *Note:* If the pan is too full, the items will steam or simmer rather than sauté.

3. Once the items are lightly browned on one side, use a wrist-flicking motion to toss and flip the items. Flip until the items are cooked on all sides. *Note:* Tongs may be used to turn larger items.

Figure 7-22. *Small items such as vegetables are often flipped using a wrist-flicking motion to ensure even cooking without burning the item.*

Frying

Frying is a dry-heat cooking method in which food is cooked in hot fat over moderate to high heat. The fat used for frying can be solid or liquid at room temperature. When fats are exposed to high heat they begin to break down and smoke. A *smoke point* is the temperature at which an oil begins to smoke and give off an odor. Smoke points vary depending on the type, age, and clarity of the oil. The higher the smoke point, the better suited the oil is for frying. The smoke point for butter is around 350°F, whereas vegetable oils begin to smoke around 440°F. The smoke point for olive oil is about 375°F, making it less desirable for frying.

Foods are typically breaded before they are pan-fried or deep-fried. *Breading* is a three-step procedure used to coat and seal an item before it is fried. *See Figure 7-23.* The item is first dredged in flour, then dipped into a mixture of beaten eggs and a liquid such as milk, and finally dipped into a bread crumb mixture. The bread crumbs adhere to the flour and egg mixture, sealing the food item. When breading items, one hand is used exclusively to coat items with dry ingredients and the other hand is used exclusively to coat items in the wet ingredients. By following this guideline, the hands do not become breaded while moving food through the breading procedure.

Procedure for Breading

1. Use the designated dry hand to dredge item in seasoned flour and then shake off the excess.

2. Use the dry hand to gently place the dredged item in an egg wash (a mixture of beaten eggs and milk or water).

3. Use the designated wet hand to coat both sides of the dredged item in the egg wash.

4. Use the wet hand to remove the item from the egg wash and lay it carefully in the bread crumbs.

5. Use the dry hand to coat the surface and edges of the item with bread crumbs.

6. Use the dry hand to gently shake off excess crumbs before setting it aside.

Figure 7-23. *Breading is a three-step procedure used to coat and seal an item before it is fried.*

Pan-Frying. *Pan-frying* is a dry-heat cooking method in which food is cooked in a pan of hot fat. Eggplant parmigiana and country-fried steak are typically pan-fried. Foods that are pan-fried are typically dredged or breaded and then placed in a pan of hot fat. *See Figure 7-24.*

Refer to DVD for
Pan-Frying
Media Clip

Procedure for Pan-Frying

1. Place enough oil in a pan to come halfway up the thickness of the item to be pan-fried. Heat the oil to 350°F.

2. Dredge or bread the item.

3. Gently place the item in the hot oil.

4. Pan-fry the item until cooked halfway up its side.

5. Turn the item over and cook the other side. *Note:* If the item is too thick, it may need to be finished in the oven.

6. Remove the item from the hot oil and drain on a screen.

Figure 7-24. *Foods that are pan-fried are typically dredged or breaded and then placed in a pan of hot fat.*

Chef's Tip

Move fried items away form the fryer before seasoning with salt. Preventing salt from entering the fryer fat prolongs the life of the fat.

Thinner items, such as delicate fish fillets, are often pan-fried. Pan-frying allows items to lie flat in the pan, preventing them from curling. Other foods that are usually pan-fried include boneless chicken breasts and pork chops.

Deep-Frying. *Deep-frying* is a dry-heat cooking method in which food is completely submerged in very hot fat. The fat used for deep-frying must have a high smoke point. Fat heated to the right temperature will produce a crisp exterior and a juicy interior. If the fat is not hot enough, the food will absorb the fat and be greasy. If the fat is too hot, the food will burn. The average temperature for deep-frying is 375°F, but recipes differ depending on the type of food.

Most deep-fried foods are breaded or battered. A coating helps the food brown and become crispy and prevents it from drying out or burning. *Battering* is the process of dipping an item in a wet mixture of flour, liquid, and fat for frying. ***See Figure 7-25.*** Fish and seafood are typically battered before deep-frying. *Tempura* is a very light batter that is used on vegetables, poultry, seafood, and meats served in Asian cuisine.

Procedure for Battering

1. Prepare a batter. *Note:* If the items to be battered are damp, lightly dredge the items in flour before placing them in the batter. If the items are dry, they can be dipped directly in the batter.
2. Dip the items in the batter and slowly add the battered items to the fryer.
3. Fry battered items until crispy and cooked through.
4. Remove items and drain to remove excess fat.

Figure 7-25. *Battering is the process of dipping an item in a wet mixture of flour, liquid, and fat for frying.*

Foods that are deep-fried are added to a preheated fryer by either the swimming method or the basket method. In the swimming method an item is slowly lowered into the hot fat without the use of a fryer basket. The swimming method is used for battered items because battered foods stick together as they fry and can also stick to the basket. By slowly adding each battered item to the fryer, the item first sinks to the bottom of the fryer and then floats back up to the top. As the side that is facing down turns brown, the item is flipped over to brown the other side.

In the basket method, items are added to a fryer basket that is sitting on top of a pan, not over the fryer. Frozen items are often added using the basket method because wet crumbs can fall into the fryer and shorten the usable life of the fat. A fryer basket should never be overfilled. Once the basket is filled, it is submerged in the hot fat. All of the items in the basket should be submerged when the basket is lowered. When the items are fully cooked, they are removed from the fryer and placed in a drain pan to allow excess fat to drain off. ***See Figure 7-26.*** It is important to strain the used fat of all food particles after frying. This process will extend the life of the fat.

Chef's Tip

Because potatoes contain a large amount of water, French fries should be blanched in fat before they are deep-fried. By blanching the French fries in 225°F to 250°F fat for a few minutes, the internal water cooks out. Then the temperature of the fat is raised to 375°F and the French fries are deep-fried until golden brown.

1. Prepare the items to be deep-fried and bread or batter them.

2. Use either the swimming method or the basket method to add the items to the preheated fryer.

3. Fry the items until they are golden brown. *Note:* Some items float on top of the fat when done.

4. Remove the items from the fryer and place them on paper towels or a drain pan to allow excess fat to drain away.

Figure 7-26. *Deep-frying is a dry-heat cooking method in which food is completely submerged in very hot fat.*

MOIST-HEAT COOKING METHODS

Moist-heat cooking is any cooking method that uses liquid or steam as the cooking medium. This involves cooking foods either by submerging them in hot liquid or by exposing them to steam. Because of this, often the natural flavor and aroma of the food is highlighted. It is important to understand the different types of moist-heat cooking to know which method is best for a given preparation. Moist-heat cooking methods include poaching, simmering, blanching, boiling, and steaming. *See Figure 7-27.*

Moist-Heat Cooking Methods			
Method	**Temperature of Cooking Liquid**	**Appearance of Cooking Liquid**	**Common Uses**
Poaching	160°F to 180°F	Very tiny bubbles may have formed on bottom of pot, but they do not break the surface; liquid shows very little movement and emits slight steam	Delicate fish fillets, eggs out of the shell, soft-fleshed fruits, and poultry
Simmering	185°F to 205°F	Small bubbles have formed and come to the surface	Stews, sauces, soups, meats, and poultry
Blanching	212°F (at sea level)	Rapid motion in water as large bubbles break the surface	Vegetables, fruits, and removal of impurities from bones for stocks
Boiling	212°F (at sea level)	Rapid motion in water as large bubbles break the surface	Potatoes, grains, and pasta
Steaming	Greater than 212°F	Moist hot air emitted rapidly from boiling liquid	Vegetables, eggs in the shell, shellfish, some fish, and poultry

Figure 7-27. *Moist-heat cooking methods include poaching, 3, blanching, boiling, and steaming.*

Poaching

Poaching is a moist-heat cooking method in which food is cooked in a liquid that is held between 160°F and 180°F. Poaching imparts some of the flavor of the poaching liquid into the item being cooked. The type and amount of liquid used depends on the food being poached. For example, eggs are poached in lightly salted water and vinegar, and fish are commonly poached in court bouillon. A mirepoix is often added to a poaching liquid. Poaching liquids are often reduced and incorporated into a sauce that is served with the finished dish.

The two methods of poaching are submersion and shallow poaching. Submersion poaching requires the food to be completely covered by the poaching liquid. Eggs are poached using submersion poaching. *See Figure 7-28.*

Refer to DVD for
Poaching Eggs
Media Clip

Procedure for Poaching Eggs

1. Bring the appropriate amount of water, salt and vinegar to between 160°F and 180°F. *Note:* Bubbles should not break the surface of the water.

2. Crack the eggs one at a time into a bowl and then gently lower each one into the poaching water.

3. Cook the eggs to desired doneness.

4. Remove the poached eggs with a perforated spoon and rest them on a clean towel for a few seconds to remove excess water.

5. The finished egg should have a firm white and a soft yolk that is somewhat liquid in the center.

Figure 7-28. *Eggs are poached using submersion poaching.*

Shallow poaching uses only enough poaching liquid to cover the bottom half of the food being cooked. Fish are typically shallow poached. *See Figure 7-29.*

Refer to DVD for **Poaching Fish** Media Clip

Procedure for Poaching Fish

1. Bring the appropriate amount of poaching liquid to 180°F.

2. Gently lower the cleaned and prepared fish into the poaching liquid.

3. Cook the fish until almost cooked through.

4. Remove the pan from the heat and allow the fish to rest in the poaching liquid for a few minutes to finish cooking.

Figure 7-29. *Fish are typically shallow poached.*

Simmering

Simmering is a moist-heat cooking method in which food is gently cooked in a liquid that is between 185°F and 205°F. *See Figure 7-30.* Simmering can be identified by tiny bubbles that reach the surface, but do not break into a full boil. It is important to maintain a constant and even temperature when simmering foods. If a protein-rich food, such as a corned beef brisket, is boiled instead of simmered, the brisket will shrink and become tough. In contrast, simmering a brisket produces a tender product.

Procedure for Simmering

1. Add enough liquid to a pan to completely cover the item.
2. Bring the liquid to between 185°F and 205°F.
3. Add mirepoix, if desired.
4. Carefully place the cleaned and prepared item in the hot liquid and maintain a steady temperature until done. *Note:* Doneness is based on tenderness or cooking time.

Figure 7-30. *Simmering is a moist-heat cooking method in which food is gently cooked in a liquid that is between 185°F and 205°F.*

Blanching

Blanching is a moist-heat cooking method in which food is briefly parcooked and then shocked by placing it in ice-cold water to stop the cooking process. *See Figure 7-31.* Vegetables and fruits are often blanched. For example, asparagus can be blanched to intensify its green color. Blanching may also be used to loosen the skin of a food, such as peaches. Blanching makes items easier to peel, softens harder items, and eliminates bitter or undesirable flavors.

Procedure for Blanching

1. Bring a pot of water to a boil and prepare a separate ice bath. *Note:* Adding salt to the boiling water when blanching green vegetables intensifies their color.

2. Place the cleaned and prepared items in the rapidly boiling water until the desired result is achieved.

3. Remove the items and immediately submerge them in the ice bath to stop the cooking process.

Figure 7-31. *Blanching is a moist-heat cooking method in which food is briefly parcooked and then shocked by placing it in ice-cold water to stop the cooking process.*

Boiling

Boiling is a moist-heat cooking method in which food is cooked by heating a liquid to its boiling point. Different liquids boil at different temperatures. Depending on the liquid, the boiling point may be more or less than 212°F (the boiling point of water at sea level). As the temperature of a liquid rises, it reaches a poaching temperature, then a simmer, and finally a boil. *See Figure 7-32.* When a liquid reaches full boil, large bubbles that have formed at the bottom of the pan rise rapidly and then break on the surface of the liquid. Potatoes, grains, and pasta are often boiled in the professional kitchen.

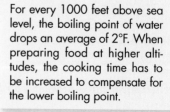

Chef's Tip

For every 1000 feet above sea level, the boiling point of water drops an average of 2°F. When preparing food at higher altitudes, the cooking time has to be increased to compensate for the lower boiling point.

Heating Water

Refer to DVD for
Heating Water
Media Clip

Smooth surface

Water at rest

Normal State 33°F to 155°F

Smooth surface

Small bubbles form on heated surface

Poaching State 160°F to 180°F

Slight surface turbulence

Small bubbles rise to surface at constant speed

Simmering State 185°F to 205°F

Surface turbulence

Rapidly rising large bubbles

Boiling State 212°F (sea level)

Figure 7-32. *As the temperature of a liquid rises, it reaches a poaching temperature, then a simmer, and finally a boil.*

Steaming

Steaming is a moist-heat cooking method in which food is placed in a container that prevents steam from escaping. *See Figure 7-33.* Food is typically placed in perforated pans in a pressure steamer. The movement of the steam cooks the food gently and evenly. Steaming retains the shape, texture, flavor, and many of the vitamins and minerals. Many vegetables, such as broccoli, green beans, carrots, and asparagus, are often steamed. Pot stickers may also be prepared with the steaming method.

Aromatics may be added to flavor the food that is being steamed. An *aromatic* is an ingredient added to a food to enhance its natural flavors and aromas. Aromatics, such as wine, herbs, and spices, release flavors and odors into the steamer that are absorbed by the food as it cooks.

Steaming

Vulcan-Hart, a division of the ITW Food Equipment Group LLC

Figure 7-33. *Steaming is a moist-heat cooking method in which food is placed in a container that prevents steam from escaping.*

Fish are often steamed en papillote. *En papillote* is a technique in which food is steamed in a parchment paper package as it bakes in an oven. ***See Figure 7-34.*** Aromatics and butter are typically added to fish prior to closing the envelope.

Procedure for Steaming en Papillote

1. Fold a 12–16 inch sheet of parchment paper in half and then cut into a half-heart shape.
2. Brush half of the heart with butter and place the fish fillet in the center of the buttered side of the heart.

3. Top fillet with a julienne of fresh vegetables.
4. Place a slice of compound butter on top of vegetables.

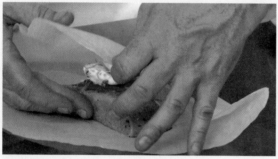

5. Fold the heart in half over the fish.

6. Press the paper down firmly on the fish and align edges of paper to form a half-heart shape.

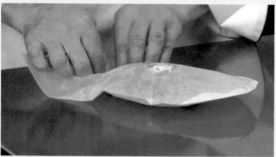

7. Begin folding 1 inch sections of the edge and press tightly to crease and seal.

8. Continue folding and pressing around the entire edge until the package is closed.
9. Place the package on a sheet pan and bake (10–12 minutes) until a thermometer inserted through the side of the package and into the fillet reads 150°F.

Figure 7-34. *En papillote is a technique in which food is steamed in a parchment paper package as it bakes in an oven.*

COMBINATION COOKING METHODS

Combination cooking includes any cooking method that uses both moist and dry heat. Combination cooking methods are often used on tougher cuts of meat to make them more tender and flavorful. The two most common combination cooking methods are braising and stewing. In both methods the protein is seared and then simmered. The main differences between these two methods are the sizes of the proteins cooked and the amount of liquid used to cook them. Poêléing and sous vide are also considered combination cooking methods.

Braising

Braising is a combination cooking method in which food is browned in fat and then cooked, tightly covered, in a small amount of liquid for a long period of time. The slow cooking process tenderizes the food by breaking down fibers. Braising is most often used for larger, roast-sized pieces of meat. *See Figure 7-35.* First, the meat is seared on all sides in a small amount of fat. Aromatic vegetables are often added during the searing process. Then, a flavorful stock or liquid is added about half-way up the side of the seared piece of meat. The seared vegetables intensify the flavor of the braising liquid as it is brought to a simmer. The pot is covered with a lid and left to cook slowly until tender.

Paderno World Cuisine

Stewing

Stewing is a combination cooking method in which bite-sized pieces of food are barely covered with a liquid and simmered for a long period of time in a tightly covered pot. *See Figure 7-36.* Stewing requires more liquid than braising, because the food is completely covered in liquid. The pieces of food need to be very similar in size so they cook evenly.

When stewing, the pot is kept at a simmer on the stove or covered and placed in an oven at a low temperature. Diced vegetables are usually added when the meat is about three-fourths done. Sometimes the vegetables are cooked separately and added at the end. The cooking liquid, usually a stock, develops even more flavor from the caramelized meat and the slow cooking process. The meat and vegetables are removed from the stewing liquid, and a roux is added to thicken the stewing liquid to the desired consistency. The meat and vegetables are then added back to the stewing liquid and the dish is plated for service.

1. Clean, trim, and dry the meat with paper towels.

2. Coat the bottom of a preheated braising pan with a small amount of fat.

3. Add the meat and sear all sides. Remove the meat from the pan. *Note:* If desired, the meat can be dredged in flour prior to searing.

4. Add mirepoix to the pan and sauté until caramelized. Then, add a small amount of tomato product and cook until caramelized.

5. Deglaze the pan with a small amount of stock and add the meat back to the pan.

6. Bring the ingredients to a simmer and cover the pan. Cook until the meat is fork tender.

7. Remove the meat and strain the liquid to remove the mirepoix.

8. Thicken the liquid with brown roux until it reaches a nappe consistency.

9. Slice the meat and serve with the thickened braising liquid.

Figure 7-35. *Braising is most often used for larger roast-sized pieces of meat.*

Refer to DVD for
Braising Media Clip

1. Add a small amount of fat to the bottom of a hot pot. Then, add a single layer of trimmed and cubed meat (1½ – 2 inch cubes).

2. Sear the meat on all sides. *Note:* If desired, the meat can be dredged in flour prior to searing.

3. Add onions or mirepoix and cook until caramelized. Then add a tomato product and stir until it has caramelized.

4. Add stock until it just covers the meat. If desired, add a sachet d'épices.

5. Cover the pot and simmer until the meat is fork tender. Simmer for 30 minutes longer.

6. Add diced potatoes and cover the pot. Simmer for 10 minutes. The potatoes should be fork tender.

7. Remove the meat and vegetables from the liquid.

8. Add a brown roux to thicken the liquid to a nappe consistency.

9. Add the meat and vegetables back to the thickened liquid. Adjust seasonings and serve.

Refer to DVD for **Stewing** Media Clip

Figure 7-36. *Stewing is a combination cooking method in which bite-sized pieces of food are barely covered with a liquid and simmered for a long period of time in a tightly covered pot.*

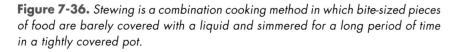

Poêléing

Refer to DVD for
Poêléing
Media Clip

Poêléing is a combination cooking method that is often referred to as butter roasting. Poêléing is similar to roasting and braising but is used for tender meats that do not require a long cooking time. The poultry or meat is placed on top of aromatic vegetables in a pot and brushed with butter. When poêléing an item, the pot is covered and placed in an oven to allow the natural moisture in the food and the butter to steam the food. *See Figure 7-37.* If desired, the poultry or meat can be browned before or after it is poêléed. Doneness is determined in the same manner as roasting. The final dish is served with the pan juices.

Procedure for Poêléing

1. Place mirepoix on the bottom of a roasting pan. Then, place the trussed bird on the mirepoix.

2. Baste the bird with melted butter. Then, cover the pan and place it in the oven.

3. When three-fourths of the cooking time has passed, check to see that the bird has started to brown and that the mirepoix has started to cook down.

4. Remove the lid and return the pan to the oven to finish cooking.

Figure 7-37. *When poêléing an item, the pot is covered and placed in an oven to allow the natural moisture in the food and the butter to steam the food.*

Sous Vide

Sous vide is the process of cooking vacuum-sealed food by maintaining a low temperature and warming food gradually to a set temperature. *See Figure 7-38.* The sous vide method of cooking is based on the chemical reactions that take place in foods at different temperatures. For example, the different proteins in egg whites coagulate at specific temperatures. A few degrees difference in a cooking temperatures will greatly affect how much an egg white coagulates. The texture of an egg yolk also changes as its temperature rises. Temperature affects meat in a similar manner.

Sous Vide Dishes

Seared Tenderloin   **Halibut**   **Salmon**

Figure 7-38. *Sous vide is the process of cooking vacuum-sealed food by maintaining a low temperature and warming food gradually to a set temperature.*

Sous vide is a French term that means "under vacuum." The equipment needed for sous vide cooking includes a vacuum sealer and a thermal circulator or a thermal bath. Sous vide causes very little shrinkage as compared to traditional cooking methods. Foods are allowed to cook slowly and at a low temperature, making them especially tender.

The sous vide process begins by trimming, marinating, or searing the food. If a food is seared it must be thoroughly cooled before being vacuum-sealed to ensure that the food will heat evenly. Flavorings are often added to the pouch before it is vacuum-sealed. The sealed pouch is then placed in the thermal bath and heated to a specific temperature. The thermal bath applies additional heat as needed to maintain the cooking liquid at the desired temperature. Once the food heats to the desired temperature, the food is done.

THE PERCEPTION OF FOOD

Food is presented in many forms, colors, textures, and flavors. *Sensory perception* is the ability of the senses to gather information and evaluate the environment. Signals are sent to the brain regarding the presentation, aroma, taste, and texture of food.

Presentation

A beautifully plated meal is first experienced with the eyes. The plated presentation immediately prompts a judgment on how the meal is likely to taste. *See Figure 7-39.* For example, vegetables that are vibrant shades of green, red, yellow, and orange look more appetizing than drab, overcooked vegetables. If vegetables have lost color, they have also lost flavor and nutrients. Careful thought should be given to making every presentation look appealing to guests. Food should always be properly cooked, neatly plated, and served at the appropriate temperature.

Presentation

MacArthur Place Hotel, Sonoma

Figure 7-39. *The plated presentation immediately prompts a judgment on how the meal is likely to taste.*

Aromas

Garlic

Truffles

Figure 7-40. *Two common aromas that the nose can sense from across a room are garlic and truffles.*

Refer to DVD for
Creaming Garlic
Media Clip

Aromas

The human nose can detect thousands of aromas. The flavor of food is actually sensed by the nose because the nose has more sensory cells than the tongue. The nose can also differentiate between foods that are quite similar in taste. A description such as "a vanilla-flavored coffee" describes the overall aroma of the item rather than its taste. The tongue can taste the sweetness and bitterness of the coffee, but it is the nose that senses the vanilla and the roasted aroma of coffee beans.

Two common aromas the nose can sense from across a room are garlic and truffles. *See Figure 7-40.* Garlic is not an herb, but it is used in the same manner. *Garlic* is a bulb vegetable made up of several small cloves that are enclosed in a thin, husklike skin. White garlic is the most common variety and the most pungent in flavor. Pink garlic, named for its pinkish outer covering, is another strongly flavored variety. Black garlic is made by fermenting bulbs of white garlic at a high temperature to create black cloves. It has a sweet flavor that has hints of vinegar. Elephant garlic contains much larger cloves but is milder in flavor than smaller varieties.

A *truffle* is an edible fungus with a distinct taste. Truffles develop underground, making them hard to locate, and take at least five years to form. Black truffles are bulbous and have dark flesh. They are often used to flavor pâtés and terrines. White truffles are elongated in shape and can weigh up to 1 lb. White truffles have a distinct aroma and are excellent in pastas and sauces. Because their flavor is so strong, a small amount of thinly shaved truffle or truffle oil is quite sufficient in any recipe.

Photo Courtesy of D'Artagnan,
Photography by Doug Adams Studios

Taste and Texture

The taste and texture of food are determining factors in the appeal of a given dish. Even if a food looks and smells appealing, an unappetizing taste or texture can cause a guest to reject the item. The four types of flavor the tongue can distinguish are salty, sweet, bitter, and sour. *See Figure 7-41.* The tongue is covered in papillae. *Papillae* are small bumps found on the tongue that are covered with hundreds of taste buds. A *taste bud* is a cluster of cells that can detect flavor characteristics. As food enters the mouth, flavors are sensed by the taste buds.

Tastes

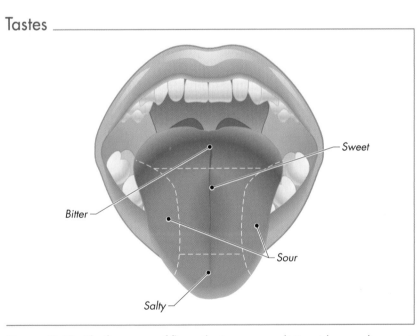

Figure 7-41. *The four types of flavor the tongue can distinguish are salty, sweet, bitter, and sour.*

Great dishes are comprised of foods with contrasting textures and flavors. A soft food should be served with a crisp counterpart, and a chewy item pairs well with a creamy item. Likewise, a sweet flavor paired with a bitter flavor or a salty item served with a sour item excites the taste buds. A variety of flavors and textures make a dish interesting and memorable.

Developing Flavors

Browne-Halco (NJ)

Figure 7-42. *Herbs and spices are used to enhance, not disguise, the natural flavor of food.*

Flavor Development

Chefs are able to blend aromas, tastes, and textures while preserving the individual qualities of the ingredients. Flavor is developed by layering and combining ingredients in different amounts at different times during the preparation process. A critical part of layering flavors is selecting ingredients that are complementary by contrasting one another. These contrasts highlight the flavors present in each item.

Herbs and spices are used to enhance, not disguise, the natural flavor of food. *See Figure 7-42.* Exceptions to this rule include curry and chili dishes, because the defining characteristics of these dishes rely on the use of specific herbs and spices. Moderation is key when using herbs and spices. More can always be added but removal is impossible.

If a dish has a long cooking time, spices should be added early in the process to extract the most flavor. In contrast, herbs should be added near the end of the cooking process. If a dish has a shorter cooking time, herbs are added at the beginning of the process. Herbs or spices should be added several hours before serving uncooked dishes, such as salad dressings, marinades, and fruit beverages, to allow time for flavors to develop. Fresh herbs used in soups, sauces, or gravies should be tied in a sachet d'épices or a bouquet garni for easy removal after use.

Combining flavors offers a better aroma, taste, and texture experience than can be achieved by eating the same foods separately. *See Figure 7-43.* Each ingredient must have a purpose in developing the flavor of the dish or it should be omitted. Flavors are balanced by adding ingredients at various stages of a recipe and taking every opportunity to bring out the unique qualities of each ingredient. Herbs, spices, rubs, marinades, condiments, sauces, and nuts are considered flavorings. A *flavoring* is an item that alters or enhances the natural flavor of food. Salts, peppercorns, citrus zest, vinegars, and oils are considered seasonings because they intensify or improve the flavor of food.

Flavor Combinations			
Food	**Flavoring**	**Food**	**Flavoring**
Beef	Bay leaf, marjoram, nutmeg, sage, thyme	Potatoes	Dill, paprika, parsley, sage
Carrots	Cinnamon, cloves, marjoram, nutmeg, rosemary, sage	Poultry	Ginger, marjoram, oregano, paprika, poultry seasoning, rosemary, sage, tarragon, thyme
Corn	Cayenne, chervil, chives, cumin, curry powder, paprika, parsley	Salads	Basil, celery seed, chervil, chives, cilantro, dill, oregano, rosemary, sage, tarragon, thyme
Eggs	Basil, cilantro, cumin, savory, tarragon, turmeric	Soups	Bay leaves, cayenne, chervil, chili powder, cilantro, cumin, curry, dill, marjoram, nutmeg, oregano, rosemary, sage, savory, thyme
Fish	Curry powder, dill, dry mustard, lemon zest and juice, marjoram, paprika		
Fruits	Cinnamon, ground cloves, ginger, mace, mint	Summer squash	Cloves, curry powder, marjoram, nutmeg, rosemary, sage
Green beans	Dill, curry powder, lemon juice, marjoram, oregano, tarragon, thyme	Tomatoes	Basil, bay leaves, dill, marjoram
Lamb	Curry powder, rosemary, mint	Veal	Bay leaves, curry powder, ginger, marjoram, oregano
Peas	Ginger, marjoram, parsley, sage		
Pork	Sage, oregano	Winter squash	Cinnamon, ginger, nutmeg

Figure 7-43. *Combining flavors offers a better aroma, taste, and texture experience than can be achieved by eating the same foods separately.*

HERBS

An *herb* is a flavoring derived from the leaves or stem of a very aromatic plant. The majority of herbs are grown in temperate climates. Although dried herbs are available throughout the year, the flavors and aromas of herbs change slightly when dried. For this reason, chefs prefer to use fresh herbs. When used and stored effectively, herbs can be incorporated into dishes in ways that gratify the senses with pleasing flavors and aromas.

It is important to add fresh herbs at the appropriate time to ensure maximum flavor and aroma. *See Figure 7-44.* Fresh herbs should be tested for quality when delivered. They should have a bright, rich color and fresh appearance. A small amount of an herb can be rubbed in the palm of the hand and smelled. The scent should be fresh and fairly strong. Freshly cut herbs should be wrapped loosely in damp paper or cloth and placed in a plastic bag to prevent them from drying out and wilting. Properly wrapped fresh herbs should be refrigerated between 35°F and 45°F.

Fresh Herbs

Messermeister

Figure 7-44. *It is important to add fresh herbs at the appropriate time to ensure maximum flavor and aroma.*

When substituting dried herbs for fresh herbs, a smaller amount of dried herbs should be used than the amount of fresh herbs called for in the recipe. Good color, robust flavor, and strong aroma are important points to consider when buying dried herbs. Dried herbs should be stored in airtight containers and out of direct light to help preserve flavors and aromas. Dried herbs tend to lose flavor and aroma the longer they are stored. The two general categories of herbs are leaf herbs and stem herbs.

Chef's Tip

To release the flavor of dried herbs, herbs can be rubbed between the palms of the hands before adding them to a preparation.

Leaf Herbs

A *leaf herb* is a type of herb derived from the leaf portion of a plant. The size, color, and shape of the leaf can vary. Some leaves are delicate and fragile, some have curled or frayed edges, and some are smooth and elongated. In addition to a unique appearance, the leaf of each herb produces an equally unique flavor. There are many leaf herbs. Commonly used leaf herbs include basil, bay leaves, chervil, cilantro, curry leaves, dill, filé powder, lavender, marjoram, mint, oregano, parsley, rosemary, sage, savory, tarragon, tea, and thyme.

Basil. *Basil* is an herb with a pointy green leaf and is a member of the mint family. Sweet basil is the most common type of basil and is traditionally used in tomato sauces and pesto. ***See Figure 7-45.*** It is often used in vinaigrettes, in infused oils, and to add flavor to pasta, vegetable, seafood, chicken, and egg dishes. Opal basil leaves are crinkled and a vibrant plum color. They are slightly firmer yet a little milder in flavor than the leaves of sweet basil. Other varieties of basil have hints of lemon, garlic, cinnamon, clove, and even chocolate flavor.

Basil

Shenandoah Growers

Figure 7-45. *Sweet basil is the most common type of basil and is traditionally used in tomato sauces and pesto.*

Bay Leaves. A *bay leaf* is a thick, aromatic leaf that comes from the evergreen bay laurel tree. Bay leaves are used for flavoring soups, roasts, stews, gravies, and meats and for pickling. ***See Figure 7-46.*** Because bay leaves are tough and can be sharp along the edges, they are always removed and discarded from a dish before serving.

Chervil. *Chervil* is an herb with dark-green, curly leaves with a flavor similar to parsley but more delicate and with a hint of licorice. ***See Figure 7-47.*** Chervil can be used in salad, soup, egg, and cheese dishes. It is also commonly used as a garnish.

Bay Leaves

Shenandoah Growers

Figure 7-46. *Bay leaves are used for flavoring soups, roasts, stews, gravies, and meats and for pickling.*

Chervil

HerbThyme Farms

Figure 7-47. *Chervil has a flavor that is similar to parsley but is more delicate with a hint of licorice.*

Cilantro. *Cilantro,* also known as Chinese parsley, is an herb that comes from the stem and leaves of the coriander plant. Cilantro has a distinct flavor that is slightly lemony. *See Figure 7-48.* It is usually added just before serving because heating cilantro nearly destroys its flavor. It is often used to flavor soups, salads, sandwiches, and salsas.

Curry Leaves. A *curry leaf* is an herb with a small, shiny green leaf from the curry tree. Curry leaves have a pungent aroma and slightly lemony flavor. *See Figure 7-49.* They are often used in Indian and Southeast Asian dishes.

Curry Leaves

The Spice House

Figure 7-49. *Curry leaves have a pungent aroma and slightly lemony flavor.*

Dill. *Dill,* also known as dill weed, is an herb with feathery, blue-green leaves and is a member of the parsley family. *See Figure 7-50.* Dill leaves have a slight licorice flavor. Dill is one of the primary herbs used in pickling foods. It may also be added to salads, vegetables, meats, and sauces.

Dill

Shenandoah Growers

Figure 7-50. *Dill, also known as dill weed, is an herb with feathery, blue-green leaves and is a member of the parsley family.*

Cilantro

Shenandoah Growers

Figure 7-48. *Cilantro has a distinct flavor that is slightly lemony.*

Filé Powder

Figure 7-51. *Filé powder is an herb that is made from the ground leaves of the sassafras tree.*

Filé Powder. *Filé powder* is an herb that is made from the ground leaves of the sassafras tree. *See Figure 7-51.* Its flavor resembles that of root beer and is a common ingredient in Creole cuisine. Filé powder is known for its use as a thickener and flavor enhancer in gumbos. Prolonged cooking can cause filé powder to become stringy and tough so it should be added after a dish is removed from the heat.

Lavender. *Lavender* is an herb with pale-green leaves and purple flowering tops and is a member of the mint family. Both the leaves and flowers of lavender can be used. Lavender is most commonly known for its aromatic qualities and its mild, lemony flavor. *See Figure 7-52.* It is used in classical French cuisine and often added to salads.

Lavender

Fresh Stems **Flower Petals**

Figure 7-52. *Lavender is most commonly known for its aromatic qualities and its mild, lemony flavor.*

Marjoram. *Marjoram* is an herb with short, oval, pale-green leaves and is a member of the mint family. *See Figure 7-53.* Marjoram has a flavor that is similar to a cross between fresh oregano and thyme. Marjoram is used to flavor soups, stews, sausages, cheese dishes, and lamb dishes.

Marjoram

Shenandoah Growers

Fresh **Dried** **Ground**

Figure 7-53. *Marjoram is an herb with short, oval, pale-green leaves.*

Mint. *Mint* is a general term used to describe a family of similar herbs, such as peppermint and spearmint. The bright-green leaves found on the purple-tinted stems of peppermint have a pungent mint flavor that is somewhat peppery. Spearmint has gray-green to pure-green leaves and is mild in aroma and flavor compared to peppermint. *See Figure 7-54.* Mint is widely known for its cool, refreshing flavor and versatility. It can be used in sweet and savory dishes as well as beverages.

Mint

Figure 7-54. *Spearmint has gray-green to pure-green leaves and is mild in aroma and flavor compared to peppermint.*

Oregano. *Oregano,* also known as wild marjoram, is an herb with small, dark-green, slightly curled leaves and is a member of the mint family. Oregano has a pungent, peppery flavor that is slightly stronger than marjoram. *See Figure 7-55.* It is often used in Italian and Greek dishes.

Parsley. *Parsley* is an herb with curly or flat dark-green leaves and is used as both a flavoring and a garnish. *See Figure 7-56.* Italian parsley is called flat-leaf parsley. Parsley has a tangy flavor that is most prominent in the stems. Parsley can be used in a classic bouquet garni. It is also used in soups, salads, stews, sauces, and vegetable dishes. Parsley is often used as a garnish.

Oregano

Shenandoah Growers

Figure 7-55. *Oregano has a pungent, peppery flavor that is slightly stronger than marjoram.*

Parsley

Curly Leaf   **Italian**

Taimura & Antle®

Figure 7-56. *Parsley is an herb with dark-green leaves and is used as both a flavoring and a garnish.*

Rosemary

Shenandoah Growers

Figure 7-57. *Rosemary smells of fresh pine with a hint of mint and is used with grilled or roasted meats.*

Sage

Shenandoah Growers

Figure 7-58. *Sage is often used to flavor stews, sausages, and bean or tomato preparations.*

Rosemary. *Rosemary* is an herb with needlelike leaves and is a member of the evergreen family. Rosemary is one of the two strongest herbs, the other being sage. Fresh rosemary is considered more desirable than dried. The dried form is hard and brittle. Rosemary smells of fresh pine with a hint of mint and is used with grilled or roasted meats. *See Figure 7-57.* Because its needles are secured to a branch, rosemary can easily be removed from a dish when the correct flavor has been achieved.

Sage. *Sage* is an herb with narrow, velvety, green-gray leaves and is a member of the mint family. It can be purchased as whole leaves, crushed, or ground. Sage is often used to flavor stews, sausages, and bean or tomato preparations. *See Figure 7-58.*

Savory. *Savory* is an herb with smooth, slightly narrow leaves and is a member of the mint family. Summer savory is considered to be the best variety because it is harvested when the leaves are at peak quality, producing a delicate, spicy-sweet flavor. Savory is often used to flavor bean, egg, fish, and meat dishes as well as some baked products. *See Figure 7-59.*

Tarragon. *Tarragon* is an herb with smooth, slightly elongated leaves and is best known as the flavoring in béarnaise sauce. Like chervil, tarragon has a flavor that suggests a touch of licorice. Tarragon blends well with seafood, poultry, and tomato dishes and is often added to salads and salad dressings. Tarragon is used in French cuisine more often than any other herb. *See Figure 7-60.*

Savory

Shenandoah Growers

Figure 7-59. *Savory is often used to flavor bean, egg, fish, and meat dishes as well as some baked products.*

Tarragon

Shenandoah Growers

Figure 7-60. *Tarragon is used in French cuisine more often than any other herb.*

Tea. *Tea* is an herb with jagged leathery leaves and comes from an evergreen bush in the magnolia family. The three main types of tea are black tea, green tea, and oolong tea. Black teas are the strongest in aroma and flavor, and green teas are the mildest. Other herbs, spices, and some fruits are often added to tea to create diverse blends. In addition to beverages, tea dust can be used to create sweet and savory dishes. Tea dust can be baked into cookies, cakes, and pies and used to flavor soups, sauces, and meats. *See Figure 7-61.*

Thyme. *Thyme* is an herb with very small gray-green leaves and is a member of the mint family. Thyme is a key ingredient in a classic bouquet garni. *See Figure 7-62.* Some varieties of thyme have a slight taste of nutmeg, lemon, mint, and sage. Thyme is a fragrant herb commonly added to stews, chowders, and poultry and vegetable preparations.

Tea Dust

Figure 7-61. *Tea dust can be baked into cookies, cakes, and pies and used to flavor soups, sauces, and meats.*

Thyme

Shenandoah Growers

Figure 7-62. *Thyme is a key ingredient in a classic bouquet garni.*

Stem Herbs

A *stem herb* is a type of herb that comes from the stem portion of a plant. Stem herbs should be checked for tenderness before use because mature plants tend to have tough stems. Commonly used stem herbs, such as chives, garlic chives, and lemongrass, provide subtle flavors that can enhance a vast array of dishes. *See Figure 7-63.*

Stem Herbs

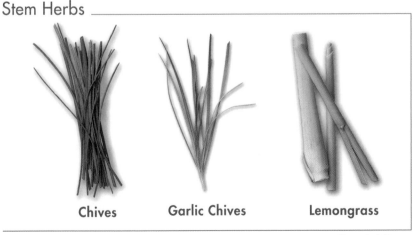

Chives   **Garlic Chives**   **Lemongrass**

Shenandoah Growers

Figure 7-63. *Commonly used stem herbs, such as chives, garlic chives, and lemongrass, provide subtle flavors that can enhance a vast array of dishes.*

Chives. A *chive* is a stem herb with hollow, grass-shaped, green sprouts and a mild onion flavor. Chives add color and flavor to cream cheese, egg dishes, soups, and salads and are a common garnish on top of baked potatoes with sour cream. Although the stem is primarily used, chives have small purple flowers that can be used to flavor salads.

Garlic Chives. A *garlic chive* is a stem herb with a solid, flat, grass-shaped sprout and a mild garlic flavor. Garlic chives come from a different species of plant than regular chives. Their mild flavor and aroma can enrich fresh or cooked dishes, and they are often used in Asian dishes.

Lemongrass. *Lemongrass,* also known as citronella, is a stem herb with long, thin, gray-green leaves, a white scallionlike base, and a lemony flavor. Lemongrass adds fresh lemon flavor without adding the acidity. It is commonly used in Thai dishes and to flavor teas and soups.

Barilla America, Inc.

SPICES

A *spice* is a flavoring derived from the bark, seeds, roots, flowers, berries, or beans of aromatic plants. Unlike herbs, most spices come from tropical climates and are usually available only in dried form. Spices range in flavor from sweet to savory and from mild to hot. Some parts of aromatic plants can be used as herbs while other parts of the same plant can be used as spices. For example, the herb cilantro comes from the leaves of the coriander plant, while its seeds are used to make the spice known as coriander.

Spices can be purchased either whole or ground. Regardless of their form, spices should have powerful, fresh aromas. Ground spices lose flavor and aroma quicker than whole spices. To prevent deterioration, all spices should be stored in tightly closed containers and in a cool, dry location. General categories of spices include bark spices, seed spices, root spices, flower spices, berry spices, and bean spices.

Bark Spices

A *bark spice* is a type of spice derived from the bark portion of a plant. Bark spices are often made from the inner layer of tree bark. Common bark spices used in the professional kitchen include cinnamon and cassia.

Cinnamon. *Cinnamon* is a spice made from the dried, thin, inner bark of a small evergreen tree. After the orange-brown bark has been removed from the tree, it is rolled and cut into 3 inch long quills known as cinnamon sticks. Cinnamon has a distinct flavor and aroma and can be used in both sweet and savory dishes. *See Figure 7-64.* It can also be purchased crushed or ground. Cinnamon is commonly used in baked products, puddings and custards, stewed fruits, and pickled items.

Cinnamon

Sticks　　　　　　　　**Ground**

Figure 7-64. *Cinnamon has a distinct flavor and aroma and can be used in both sweet and savory dishes.*

Cassia. *Cassia,* also known as Chinese cinnamon, is a spice made from the bark of a small evergreen tree. Cassia is thicker and darker in color than cinnamon, and its flavor is stronger and less refined. *See Figure 7-65.* Most spices labeled as cinnamon and sold in the United States are actually cassia, since labeling laws do not require companies to distinguish between the two.

Seed Spices

A *seed spice* is a type of spice derived from the seed portion of a plant. Seed spices are available whole or ground. They are often used for texture as well as flavor. Seed spices include achiote seeds, anise, caraway seeds, celery seeds, coriander, cumin, fennel seeds, fenugreek, mustard seeds, nutmeg, mace, poppy seeds, sesame seeds, and tamarind.

Achiote Seeds. An *achiote seed,* also known as an annatto seed, is a spice made from a red, corn-kernel-shaped seed of the annatto tree. When ground or cooked, achiote seeds give off a yellow-orange color that is used as a natural food coloring for butter, cheese, and smoked fish. *See Figure 7-66.* Achiote seeds are also used in many South American and Mexican dishes.

Anise. *Anise* is a spice made from a comma-shaped seed of the anise plant and is in the parsley family. *See Figure 7-67.* Anise is used in licorice products, sweet rolls, cookies, sweet pickles, and candies.

Star anise, also known as Chinese anise, is not related to the anise spice. It is actually a star-shaped seed fruit from the Chinese magnolia tree. Although similar in flavor to anise, the flavor of star anise is more intense and slightly bitter. It is one of the main components in a Chinese five-spice blend.

Cassia

Figure 7-65. *Cassia is thicker and darker in color than cinnamon, and its flavor is stronger and less refined.*

Achiote Seeds

The Spice House

Figure 7-66. *When ground or cooked, achiote seeds give off a yellow-orange color that is used as a natural food coloring for butter, cheese, and smoked fish.*

Anise

Figure 7-67. *Fresh anise seeds have a green color and a licorice flavor.*

Caraway Seeds

Figure 7-68. *Caraway seeds are commonly used to flavor rye bread.*

Caraway Seeds. A *caraway seed* is a small, crescent-shaped brown seed of the caraway plant and is used as a spice. Caraway seeds are commonly used to flavor rye bread. *See Figure 7-68.* They also can be used to flavor sauerkraut, pork, cabbage, stews, and soups.

Celery Seeds. A *celery seed* is a tiny brown seed of a wild celery plant called lovage and is used as a spice. Celery seeds are commonly used in pickling and to flavor salads, soups, sauces, fish, and some meats. *See Figure 7-69.*

Celery Seeds

Figure 7-69. *Celery seeds are commonly used in pickling and to flavor salads, soups, sauces, fish, and some meats.*

Coriander. *Coriander* is a spice made from a light-brown, ridged seed of the coriander plant. It has a flavor similar to a blend of lemon, sage, and caraway. Although this seed comes from the same plant as the herb cilantro, the flavor of coriander is completely different, and the two should never be substituted. Coriander powder is used in many baked products, hot dogs, sausages, and curry dishes. *See Figure 7-70.* Whole coriander seeds are commonly used in pickling.

Coriander

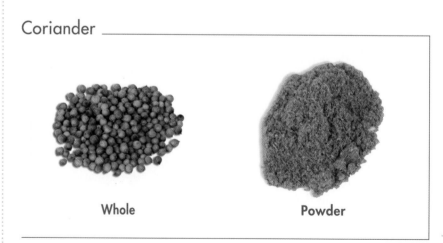

Whole   Powder

Figure 7-70. *Coriander powder is used in many baked products, hot dogs, sausages, and curry dishes.*

Cumin. *Cumin* is a spice made from a crescent-shaped amber, white, or black seed of the cumin plant and is a member of the parsley family. Cumin has a slightly bitter and warm flavor and is an essential ingredient in curry powder and chili powder. *See Figure 7-71.* Cumin is used to flavor chili, soup, tamales, rice and cheese dishes, and many Middle Eastern, Indian, and Mexican dishes.

Cumin

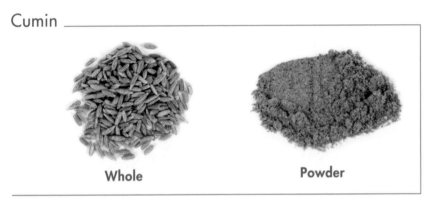

Whole

Powder

Figure 7-71. *Cumin has a slightly bitter and warm flavor and is an essential ingredient in curry powder and chili powder.*

Fennel Seeds. *Fennel seed* is a spice made from an oval, light-brown, and green seed of the fennel plant. Fennel seeds resemble anise in flavor and aroma. *See Figure 7-72.* They are commonly used in Scandinavian dishes. Fennel seeds also enhance the flavor of roast duck, chicken, and pork dishes and are often used in sausage preparation.

Fenugreek. *Fenugreek* is a spice made from the pebble-shaped seed of the fenugreek plant in the pea family. Its flavor is similar to burnt sugar and leaves a bitter aftertaste. The orange color of fenugreek seeps into dishes in which the spice is used. Fenugreek is commonly used in Indian curries, chutneys, and vegetable preparations. *See Figure 7-73.*

Fennel Seeds

Figure 7-72. *Fennel seeds resemble anise in flavor and aroma.*

Fenugreek

The Spice House

Figure 7-73. *Fenugreek is mainly used in Indian curries, chutneys, and vegetable preparations.*

Mustard Seeds. A *mustard seed* is an extremely tiny seed from the mustard plant and is used as a spice. Yellow, brown, and black mustard seeds are available whole and ground. Mustard seeds have virtually no aroma but release a hot, pungent flavor when ground and mixed with hot water. *See Figure 7-74.* Yellow seeds release the mildest flavor, while black seeds are the most intense. Mustard seeds enhance the flavor of pickles, cabbage, beets, sauerkraut, sauces, salad dressings, ham, and cheese.

Mustard Seeds _____

Yellow   Brown

The Spice House

Figure 7-74. *Mustard seeds have virtually no aroma but release a hot, pungent flavor when ground and mixed with hot water.*

Nutmeg and Mace. *Nutmeg* is a spice made from an oval, gray-brown seed found in the yellow, nectarine-shaped fruit of a large tropical evergreen. *See Figure 7-75.* When the nutmeg fruit is ripe, the outer hull splits open, exposing the sister spice mace, which partially covers the nutmeg kernel. The sweet, warm, and spicy flavor of nutmeg can be used to enhance cream soups, custards, puddings, baked products, potato dishes, and sauces.

Nutmeg _____

Whole   Cracked   Ground

Figure 7-75. *Nutmeg is a spice made from an oval, gray-brown seed found in the yellow, nectarine-shaped fruit of a large tropical evergreen.*

Mace is a spice made from the lacy red-orange covering of the nutmeg kernel. Mace turns dark yellow when it is dried. Its flavor resembles that of nutmeg, although it is not as pungent. Mace is often used in pound cake, sweet dough, chocolate dishes, oyster stew, spinach, and pickling.

Poppy Seeds. A *poppy seed* is a spice made from a very small, blue-gray seed of the poppy plant. Poppy seeds have a mild, nutty flavor and are best known for garnishing breads. *See Figure 7-76.* They may also be used in butter sauces for fish, vegetables, and noodles.

Sesame Seeds. A *sesame seed* is a small, flat, white or black seed found inside the pod on a sesame plant and is used as a spice. Often baked on rolls, bread, and buns, sesame seeds supply a nutty flavor. *See Figure 7-77.* The seeds are also toasted and stirred into butter and served over fish, noodles, and vegetables. White sesame seeds are commonly used in Indian and Asian dishes, while black sesame seeds are common in Japanese dishes. Sesame seeds are ground to form tahini, or toasted and pressed to make sesame oil.

Tamarind. *Tamarind* is a spice made from seeds found in the long pods that grow on the tamarind tree. The long pods contain both seeds and pulp that are used as a paste or liquid for flavoring. The flavor is tart and often used in many Middle Eastern and Asian dishes. Tamarind is the main ingredient in Worcestershire sauce. *See Figure 7-78.* It is also commonly used in Indian curries, as well as South American and Asian beverages.

Poppy Seeds

The Spice House

Figure 7-76. *Poppy seeds have a mild, nutty flavor and are best known for garnishing breads.*

Sesame Seeds

Figure 7-77. *Often baked on rolls, bread, and buns, sesame seeds supply a nutty flavor.*

Tamarind

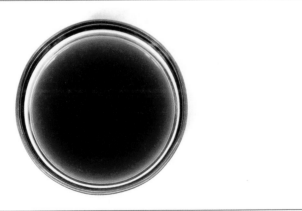

Figure 7-78. *Tamarind is the main ingredient in Worcestershire sauce.*

Root Spices

A *root spice* is a type of spice derived from the root portion of a plant that grows underground. Roots are commonly thick and fibrous. If they are used fresh, roots often have to be peeled and then sliced, grated, or shredded. Roots can also be dried and ground into powders. Root spices include ginger, horseradish, turmeric, and wasabi.

Ginger. *Ginger* is a spice made from the bumpy root of a tropical plant grown in China, India, Jamaica, and the United States. The large root is commonly called a hand, because the thick offshoots resemble fingers. When the plant is about a year old, the roots are dug up and sold fresh, dried, powdered, crystallized, candied, or pickled. Ginger has a warm, pungent, spicy flavor that adds zest to desserts, fruits, and meat preparations. *See Figure 7-79.* Dried ginger is an essential ingredient in gingerbread.

Ginger

| Whole | Powder | Crystallized |

Figure 7-79. *Ginger has a warm, pungent, spicy flavor that adds zest to desserts, fruits, and meat preparations.*

Horseradish. *Horseradish* is a spice made from the large, brown-skinned root of a shrub related to the radish. Horseradish has a pungent, hot, and spicy flavor. *See Figure 7-80.* When horseradish root is cut or grated, enzymes from the damaged root cells break down and produce mustard oil, which irritates the sinuses and eyes. If not used immediately or placed in vinegar, the root darkens and becomes unpleasantly bitter when exposed to air and heat. Grated horseradish is often served with grilled or roasted meats and seafood dishes or incorporated into cream sauces and compound butters.

Horseradish

Figure 7-80. *Horseradish has a pungent, hot, and spicy flavor.*

Turmeric. *Turmeric* is a spice made from the root of a lily-like plant in the ginger family. When the roots are ground, a bright-yellow powder is produced that is sometimes referred to as "Indian saffron." Turmeric has a taste similar to mustard and is an important ingredient in curry powder and some prepared mustards. *See Figure 7-81.*

Turmeric _____

Figure 7-81. *Turmeric has a taste similar to mustard and is an important ingredient in curry powder and some prepared mustards.*

Wasabi. *Wasabi* is a spice made from the light-green root of an Asian plant. Wasabi has a hot and tangy flavor similar to horseradish. It is commonly sold in a dried powder form or as a paste. When wasabi powder is mixed with water, a fiery, green-colored paste that frequently accompanies sushi is produced. *See Figure 7-82.*

Wasabi Powder _____

The Spice House

Figure 7-82. *When wasabi powder is mixed with water, a fiery, green-colored paste that frequently accompanies sushi is produced.*

Capers _____

Figure 7-83. *Capers are commonly used in fish and poultry dishes and are one of the key ingredients in tartar sauce.*

Flower Spices

A *flower spice* is a type of spice derived from the flower of a plant. Capers and cloves are taken from the bud of unopened flowers. Saffron is obtained from the stigma of a flower.

Capers. A *caper* is a spice from the unopened flower bud of a shrub and is only used after being pickled in strongly salted white vinegar. Capers should be stored in their original liquid and will spoil if water or additional vinegar is added. Small, medium, and large caper varieties are available. Capers are commonly used in fish and poultry dishes and are one of the key ingredients in tartar sauce. *See Figure 7-83.*

Cloves. A *clove* is a spice made from the dried, unopened bud of a tropical evergreen tree. Cloves grow in clusters. Cloves are picked when the buds turn red. Dried cloves are dark brown. Cloves possess the most pungent flavor of all the spices and are sometimes referred to as the nail-shaped spice because of their shape. *See Figure 7-84.* They are used in pickling and in the preparation of roast pork, corned beef, baked ham, soups, applesauce, and baked products.

Cloves _____

Figure 7-84. *Cloves are sometimes referred to as the nail-shaped spice because of their shape.*

Saffron. *Saffron* is a spice made from the dried, bright-red stigmas of the purple crocus flower. ***See Figure 7-85.*** It takes over 75,000 flowers to produce 1 lb of saffron. The three stigmas of each flower are removed by hand and dried to produce the delicate, potent, red-orange strands. Saffron is the most expensive spice in the world, but a little goes a long way.

Saffron

Figure 7-85. *Saffron is a spice made from the dried, bright-red stigmas of the purple crocus flower.*

Saffron strands are steeped in warm or hot water before use to release their color, aroma, and flavor. Saffron imparts a perfumelike aroma and deep-yellow color that is desired in certain rice dishes and baked products. The use of saffron is common in Scandinavian and Spanish dishes.

Berry Spices

A *berry spice* is a type of spice made from the berry portion of a plant. Berries used for spices are usually the unripe part of a plant that later develops into a fruit. Berry spices are commonly dried and then crushed or ground. Examples of berry spices include allspice, cardamom, cayenne pepper, crushed red pepper, juniper berries, paprika, and Szechuan pepper.

Allspice. *Allspice,* also known as Jamaican pepper, is a spice made from the dried, unripe fruit of a small pimiento tree. Because its complex flavor suggests a combination of cinnamon, nutmeg, and cloves, this pea-shaped spice is called allspice. ***See Figure 7-86.*** Allspice is used in both whole and ground form in the preparations of pies, cakes, puddings, stews, soups, preserved fruit, curries, relishes, and gravies.

Cardamom. *Cardamom* is a spice made from the dried, immature fruit of a tropical bush in the ginger family. The fruit consists of a yellow pod, about the size of a cranberry, that holds the dark, aromatic cardamom seeds. Cardamom has a lemony flavor and a pleasant aroma. ***See Figure 7-87.*** It is available in whole or ground form. Cardamom is commonly used in Middle Eastern dishes, curries, Danish pastries, and pickling.

Allspice

Whole

Ground

Figure 7-86. *Because its complex flavor suggests a combination of cinnamon, nutmeg, and cloves, this pea-shaped spice is called allspice.*

Cardamom

Whole   **Ground**

Figure 7-87. *Cardamom has a lemony flavor and a pleasant aroma.*

Cayenne Pepper. *Cayenne pepper*, also known as red pepper, is a spice made from dried, ground berries of certain varieties of hot peppers. *See Figure 7-88.* Cayenne pepper is used in soups and meat, fish, and egg dishes. Cayenne pepper should be used in moderation because it is hot and strong.

Crushed Red Pepper. *Crushed red pepper*, also known as crushed chiles or chili flakes, is a spice made from a blend of dried, crushed berries from hot chili peppers. *See Figure 7-89.* Crushed red pepper is often used to flavor soups and pizza.

Juniper Berries. A *juniper berry* is a spice made from a small, purple berry of an evergreen bush. It has an aromatic flavor similar to rosemary but can be slightly tart. Juniper berries are most commonly crushed and used in the preparation of wild game dishes and are also the key ingredient used to make gin alcohol. *See Figure 7-90.*

Cayenne Pepper

Figure 7-88. *Cayenne pepper, also known as red pepper, is a spice made from dried, ground berries of certain varieties of hot peppers.*

Crushed Red Pepper

Figure 7-89. *Crushed red pepper, also known as crushed chiles or chili flakes, is a spice made from a blend of dried, crushed berries from hot chili peppers.*

Juniper Berries

The Spice House

Figure 7-90. *Juniper berries are most commonly crushed and used in the preparation of wild game dishes and are also the key ingredient used to make gin alcohol.*

Paprika. *Paprika* is a spice made from a dried, ground sweet red pepper berry that is often used as a colorful garnish. The two kinds of paprika used in the professional kitchen are Spanish and Hungarian. Spanish paprika is mild in flavor and has a bright-red color. *See Figure 7-91.* Hungarian paprika is darker in color and has a more pungent flavor. Paprika is used to prepare Hungarian goulash, chicken paprika, Newburg sauce, French dressing, and veal paprika.

Paprika

Figure 7-91. *Spanish paprika is mild in flavor and has a bright-red color.*

Szechuan Pepper. *Szechuan pepper* is a spice made from the dried berries of an ash tree. *See Figure 7-92.* Szechuan pepper is a fiery-flavored powder used in many Chinese dishes. It is also a component of Chinese five-spice powder.

Szechuan Pepper

Figure 7-92. *Szechuan pepper is a spice made from the dried berries of an ash tree.*

Bean Spices

A *bean spice* is a type of spice derived from the bean portion of a plant. Bean spices such as vanilla beans, cacao beans, and coffee beans are used to create dishes ranging from appetizers to desserts. *See Figure 7-93.*

Figure 7-93. *Bean spices such as vanilla beans, cacao beans, and coffee beans are used to create dishes ranging from appetizers to desserts.*

Vanilla Beans. A *vanilla bean* is the dark-brown pod of a tropical orchid. Vanilla beans are commonly available fresh, dried and ground, or as a liquid extract. They are known for their warm, sweet aroma and flavor. Vanilla is used to enhance desserts, baked products, and beverages. It may also be used in savory applications, such as salad dressings and sauces, to add complexity and richness to a dish.

Cacao Beans. A *cacao bean* is a bean extracted from the pods of the cacao tree. Once cacao beans are removed from their pod, they are dried, roasted, and commonly used to produce cocoa powder. *Cocoa powder* is a reddish-brown, unsweetened powder extracted from ground cacao beans. Cocoa powder is used to intensify the flavor of chocolate desserts. It is also a primary ingredient in Mexican sauces and can be used to create a savory crust on grilled meats.

Coffee Beans. A *coffee bean* is the unripe bean extracted from the fruit of a coffee tree. Coffee beans are green to yellow in color. The beans are dried and roasted to achieve their well-known dark-brown color. Coffee beans are available whole or ground. They can impart subtle flavors or add a robust and intense flair to a dish. Coffee beans are often used in recipes featuring chocolate because they enhance the flavor of chocolate. They are also used to add flavor and interest to grilled meats, gravies, chili, and some alcoholic beverages.

SPICE AND HERB BLENDS

A variety of spices and herbs can be combined to create complementary blends that add appealing flavors to food. For example, Cajun seasoning is used to impart the bold, spicy flavors associated with Cajun cuisine. Some blends may be used in unintended and creative ways. For example, a blend typically used for pickling can be used to enhance the flavor of a pot roast. Commonly used spice and herb blends include Cajun seasoning, chili powder, Chinese five-spice powder, curry powder, fines herbes, herbes de Provence, jerk seasoning, pickling spice, and poultry seasoning. *See Figure 7-94.*

Spice and Herb Blends		
Name of Blend	**Common Ingredients**	**Common Uses**
Cajun seasoning	Zesty spice blend; ground chiles, fennel seeds, garlic, oregano, paprika, red and black pepper, salt, and thyme	To season blackened poultry, seafood, and meats
Chili powder	Ground chiles, cloves, coriander, cumin, garlic, and oregano	To season chili, tamales, stews, Spanish rice, and gravies
Chinese five-spice powder	Equal proportions of ground cinnamon, cloves, fennel seeds, star anise, and Szechuan pepper	To season Chinese dishes
Curry powder	Mild to spicy blend; can consist of more than 20 spices including cardamom, cinnamon, cloves, coriander, cumin, fenugreek, ginger, mace, nutmeg, red and black pepper, and turmeric	To season Indian and Pakistani dishes featuring poultry, fish, meat, eggs, rice, and soups
Fines herbes	Chopped fresh chervil, chives, parsley, and tarragon; may include marjoram, savory, or thyme	Added to cooked sauces just before service
Herbes de Provence	Dried basil, fennel seed, lavender, marjoram, rosemary, sage, savory, and thyme	To season roasted or grilled poultry and meats and in savory breads
Jerk seasoning	Spicy blend; ground allspice, chiles, cinnamon, cloves, garlic, and ginger	To season poultry, seafood, beef, and pork
Pickling spice	Whole and coarsely ground allspice, bay leaves, cinnamon, cloves, dill, fennel seeds, ginger, mace, mustard, nutmeg, peppercorns, and red pepper	In pickling; a sachet may be added to some stocks and soups
Poultry seasoning	Black pepper, marjoram, nutmeg, rosemary, sage, and thyme; may include celery salt, mustard powder, or oregano	To season poultry, white meat game birds, veal, pork, and lamb

Figure 7-94. *Commonly used spice and herb blends include Cajun seasoning, chili powder, Chinese five-spice powder, curry powder, fines herbes, herbes de Provence, jerk seasoning, pickling spice, and poultry seasoning.*

RUBS AND MARINADES

Rubs and marinades allow food to absorb flavors that can enhance the taste of a finished product. A *rub* is a blend of ingredients that is pressed onto the surface of uncooked foods such as meat, poultry, and fish to impart flavor and sometimes to tenderize. A *marinade* is a flavorful liquid used to soak uncooked foods such as meat, poultry, and fish to impart flavor and sometimes to tenderize.

An endless variety of ingredients can be combined to create rubs and marinades that add flavor without being overbearing. After a rub or marinade has been applied for the appropriate amount of time, the food is often grilled, broiled, baked, or roasted. Rubs are classified as dry rubs or wet rubs. Marinades, including brines, are always liquid based.

Dry Rubs

A *dry rub* is a mixture of finely ground herbs and spices that is rubbed onto the surface of an uncooked food to impart flavor. ***See Figure 7-95.*** A mortar and pestle or an electric spice grinder is often used to grind and combine dry rub ingredients. The rub-coated food is then covered and allowed to rest under refrigeration for a few hours or overnight to enable the complex flavors to be absorbed. Thickly coated rubs can either be wiped off prior to cooking or left on the food to form a crispy crust during cooking.

Dry Rubs

Figure 7-95. *A dry rub is a mixture of finely ground herbs and spices that is rubbed onto the surface of an uncooked food to impart flavor.*

Wet Rubs

A *wet rub,* also known as a paste, is a mixture of wet ingredients and a dry rub and is rubbed onto the surface of uncooked food to impart flavor. Wet ingredients, such as prepared mustards, flavored oils, puréed garlic, honey, lemon juice, ground horseradish, and ground lemongrass, can be added to a dry rub to create a wet rub. The resulting wet rub will have a pastelike consistency. The addition of wet ingredients helps the rub stick to foods and allows flavors to be absorbed more thoroughly than a dry rub. The result is a deeper and more complex finished flavor.

After foods have been coated with a wet rub, they are refrigerated to let the flavors be absorbed. These foods need to be turned occasionally while under refrigeration so the rub does not become more concentrated on one side than on the other side.

Marinades

Figure 7-96. *Unlike rubs, marinades contain a high percentage of liquid ingredients such as vinegars, oils, or fruit juices.*

Marinades

Unlike rubs, marinades contain a high percentage of liquid ingredients such as vinegars, oils, or fruit juices. ***See Figure 7-96.*** Most marinades contain an acidic liquid base, such as lemon juice, wine, or vinegar. Marinades also contain herbs and spices. For example, an Asian marinade may consist of soy sauce, fish sauce, sesame oil, rice wine vinegar, ground lemongrass, ginger, garlic, and scallions.

Marinades are commonly mixed in a deep stainless steel pan or a plastic storage bag. Food items are completely submerged in the marinade to provide a consistent flavor. The intensity of the flavor increases the longer an item sits in a marinade. Once a raw protein, such as poultry, has been removed from a marinade, the marinade should be discarded to prevent cross-contamination.

Brines

Like marinates, brines add flavor and moisture to foods. A *brine* is a salt solution that usually consists of 1 cup of salt per 1 gal. of water. Brines may also contain other flavorful ingredients such as sugar, molasses, fruits, juices, herbs, and spices. Immersing a raw protein, such as poultry or pork, into a brine solution and refrigerating it for several hours or overnight allows the brine to be absorbed. When the brined protein is cooked, it is especially tender because the brine aids in moisture retention. Brines are also used to preserve foods such as herring and pickles.

CONDIMENTS

A *condiment* is a savory, sweet, spicy, or salty accompaniment that is added to or served with a food to impart a particular flavor that will complement the dish. Condiments are sometimes added prior to serving, such as with a sandwich made with Dijon mustard. Some condiments may be used during the cooking process to add flavor. Many condiments are served in a small dish as an accompaniment. Whether it is a creamy, rich mayonnaise or a pungent mustard, condiments add flavor to foods. Common condiments used in the professional kitchen include ketchup, mustards, and mayonnaise. ***See Figure 7-97.*** Many different types of sauces are also served as condiments.

Condiments

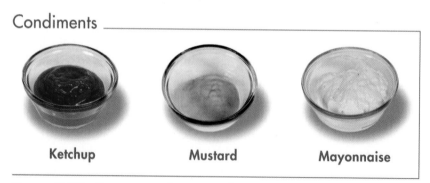

| Ketchup | Mustard | Mayonnaise |

Figure 7-97. *Common condiments used in the professional kitchen include ketchup, mustards, and mayonnaise.*

Ketchup

Ketchup is a thick, tomato-based product that usually includes vinegar, sugar, salt, and spices. Because vinegar and sugar are two of the main ingredients in ketchup, it has a slightly sweet-and-sour acidic flavor. Once opened, ketchup will retain its freshness longer if it is refrigerated, although it is not necessary. Ketchup loses much of its bright-red color as its quality deteriorates with age. It is commonly used as a condiment for sandwiches and French fries.

Carlisle FoodService Products

Mustards

Mustard is a pungent powder or paste made from the seeds of the mustard plant. In addition to mustard seeds, ingredients commonly added to mustards include vinegar or wine, salt, and various spices. Mustard comes in many different varieties. Some mustards are made from ground seeds, resulting in a very smooth product. Other mustards use cracked or whole seeds for added texture and a more rustic appearance. Depending on the type of seed used to prepare mustard, the flavor can range from mild and tangy to hot and spicy. Common mustards used as condiments include prepared yellow mustard, Dijon mustard, and English mustard.

Prepared Yellow Mustard. A *prepared yellow mustard* is a type of bright-yellow mustard that is mild in flavor and made from ground yellow mustard seeds, vinegar, and turmeric. It is the addition of turmeric that gives prepared yellow mustard its color. Because it has a high acid content, prepared yellow mustard does not spoil. However, it can dry out and darken on the surface, and the flavor deteriorates with age. Prepared yellow mustard is often served with sandwiches and is also used in wet rubs.

Dijon Mustard. *Dijon mustard* is a light-tan mustard that has a strong, tangy flavor and is made from brown or black mustard seeds, vinegar, white wine, sugar, and salt. It is named for Dijon, France, where it originated. Because of its ingredient blend, Dijon mustard has a less acidic taste than prepared yellow mustard. In addition to use as a condiment, Dijon mustard is often used in salad dressings, sauces, wet rubs, and marinades.

English Mustard. *English mustard* is a type of yellow mustard that has a hot, spicy flavor and is made from ground yellow and brown or black mustard seeds, wheat flour, and turmeric. Because English mustards are extremely hot and spicy, only a small amount of this condiment is needed.

Mayonnaise

Mayonnaise is a thick, uncooked emulsion formed by combining oil with egg yolks and vinegar or lemon juice. Although often exceptional in quality, house-made mayonnaise only lasts up to four days, while commercially prepared mayonnaise has a shelf life of approximately six months. As a condiment, mayonnaise is often used as a sandwich spread. It is also used to make salad dressings, bound salads, and sauces.

CONDIMENT SAUCES

Some sauces may also be used as condiments. A *sauce* is an accompaniment that is served with a food to complete or enhance the flavor and/or moistness of a dish. Salad dressings, barbeque sauces, pepper sauces, Worcestershire sauce, and Asian sauces are often served as condiment sauces.

Salad Dressings

Salad dressings of the sandwich spread variety are similar in texture and appearance to mayonnaise and may be used in place of mayonnaise in some dishes. A *salad dressing,* also known as a cooked dressing, is a cooked, mayonnaise-like product usually made from distilled vinegar, vegetable oil, water, sugar, mustard, salt, modified corn flour, and emulsifiers. Salad dressings vary in consistency and come in a variety of flavors such as ranch, thousand island, blue cheese, Caesar, French, Italian, and vinaigrette. Salad dressings are commonly served with sandwiches and salads.

National Turkey Federation

Barbeque Sauces

A *barbeque sauce* is a type of sauce used to baste a cooked protein and is often made with tomatoes, onions, mustard, garlic, brown sugar, and vinegar. The flavors of barbeque sauces range from sweet and tangy to smoky and spicy. Some barbeque sauces feature a mayonnaise base instead of a tomato or mustard base, while others may have a vinegar base. Barbeque sauces that contain white or brown sugar, honey, or molasses can burn quickly and should be applied at the very end of the cooking process.

Pepper Sauces

A *pepper sauce* is a type of hot and spicy sauce made from various types of chili peppers and typically includes vinegar. Pepper sauces impart heat or spiciness and flavor. The burn that is felt when pepper sauces are consumed is caused by capsaicin. *Capsaicin* is a potent compound that gives chiles their hot flavor. Capsaicin is found in the seeds and the veins or membranes of a chili pepper. Different chiles contain varying amounts of capsaicin, and Scoville units are used to measure the heat level of hot peppers. *See Figure 7-98.*

Cayenne sauce, also known as Louisiana-style hot sauce, is a type of pepper sauce made from cayenne peppers, vinegar, and salt. This thin, bright-red, pungent sauce adds intense heat to food. Cayenne sauce is commonly used as a condiment but is also used to make marinades, wet rubs, and barbeque sauces.

Frieda's Specialty Produce

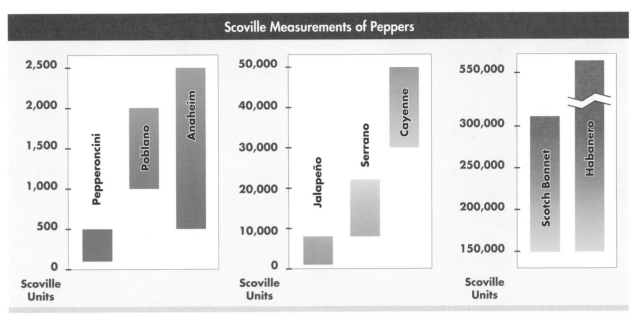

Scoville Measurements of Peppers

Scoville Units (first chart):
- Pepperoncini
- Poblano
- Anaheim
(0 – 2,500)

Scoville Units (second chart):
- Jalapeño
- Serrano
- Cayenne
(0 – 50,000)

Scoville Units (third chart):
- Scotch Bonnet
- Habanero
(150,000 – 550,000)

Figure 7-98. *Different chiles contain varying amounts of capsaicin, and Scoville units are used to measure the heat level of hot peppers.*

Tabasco® sauce is a type of pepper sauce made from Tabasco peppers, vinegar, and salt. Tabasco peppers are named after the Mexican state of Tabasco but are now grown in some areas of Louisiana. Tabasco peppers are used exclusively to make Tabasco® sauce. In addition to being a condiment, Tabasco® sauce is often used to flavor meat sauces, salads, soups, and eggs.

Chipotle sauce is a type of pepper sauce made from dried, smoked jalapeño peppers. The most common sauce featuring chipotle peppers is called adobo sauce. In addition to chipotle peppers, adobo sauce usually includes tomatoes, vinegar, herbs, and spices. This dark-red, spicy sauce is used in marinades, wet rubs, savory stews, soups, and meat dishes.

National Honey Board

Worcestershire Sauce

Worcestershire sauce is a type of sauce traditionally made with anchovies, garlic, onions, lime, molasses, tamarind, and vinegar. It originated in India but takes its name from Worcester, England, where it was first bottled. Worcestershire sauce is a thin, dark-brown sauce commonly served with grilled steak or used as an ingredient in sauces, soups, and stews.

Asian Sauces

Many Asian sauces are commonly used as dipping sauces and as a flavorful ingredient in rice, noodle, and vegetable dishes. Common Asian sauces used in the professional kitchen include soy sauce, fish sauce, oyster sauce, hoisin sauce, Asian chili sauce, and fermented black bean sauce.

Soy Sauce. *Soy sauce* is a type of Asian sauce made from mashed soybeans, wheat, salt, and water. When soy sauce is produced, it is fermented in vats for up to 18 months and is then pressed and strained. Soy sauce comes in several varieties including low-sodium soy sauce and tamari soy sauce, which is prepared with little or no wheat. Tamari is darker, thicker, and milder in taste than regular soy sauce. Soy sauce is often used as a dipping sauce and as an ingredient in savory dishes.

Fish Sauce. *Fish sauce* is a type of Asian sauce made from a liquid drained from fermented, salted fish. Chiles or sugar can be added to the sauce. Fish sauce has a pungent, salty flavor that adds depth of flavor to dishes.

Oyster Sauce. *Oyster sauce* is a type of Asian sauce made from the cooking liquid of boiled oysters, brine, and soy sauce. Oyster sauce is thick and dark brown and has a salty, rich flavor that adds complexity and depth to dishes.

Hoisin Sauce. *Hoisin sauce* is a type of Asian sauce made from fermented soybean paste, garlic, vinegar, chiles, and sugar. Hoisin sauce is a rich amber color and has a sweet and spicy flavor. It is commonly used in meat, poultry, and shellfish dishes or served as a dipping sauce for cooked meats, dumplings, and fried items.

Asian Chili Sauce. *Asian chili sauce* is a red-colored Asian sauce made from puréed red chiles and garlic. Vinegar, sugar, and various spices are sometimes added to Asian chili sauce. Asian chili sauce has a spicy, savory flavor and is marketed as Sriracha, sambal, or sambal oelek depending on where it is produced. It is often used as a condiment as well as an ingredient in a variety of Asian dishes.

Fermented Black Bean Sauce. *Fermented black bean sauce* is a type of Asian sauce made from fermented and salted blackened soybeans mixed with garlic and spices. It is often sold as a smooth puréed paste and is commonly added to Asian dishes near the end of the cooking process.

NUTS

A *nut* is a hard-shelled, dry fruit or seed that contains an inner kernel. Nuts are often added to foods to impart texture or a nutty flavor. For example, a fish can be coated in chopped nuts before being sautéed. The nuts create a flavorful, crispy crust. Likewise, salads are often tossed with toasted nuts for added flavor and texture. Nuts are also used in many baked products and a number of poultry, seafood, and meat dishes.

Nuts commonly used in the professional kitchen include almonds, Brazil nuts, cashews, chestnuts, hazelnuts, macadamia nuts, peanuts, pecans, pine nuts, pistachios, soy nuts, and walnuts. *See Figure 7-99.* Some nuts, such as peanuts and soy nuts, are not nuts. They are actually legumes that are used like nuts.

- An *almond* is a teardrop-shaped fruit that grows on small almond trees. Almonds are marketed whole, chopped, slivered, sliced, ground, or as almond paste.

Nuts _____

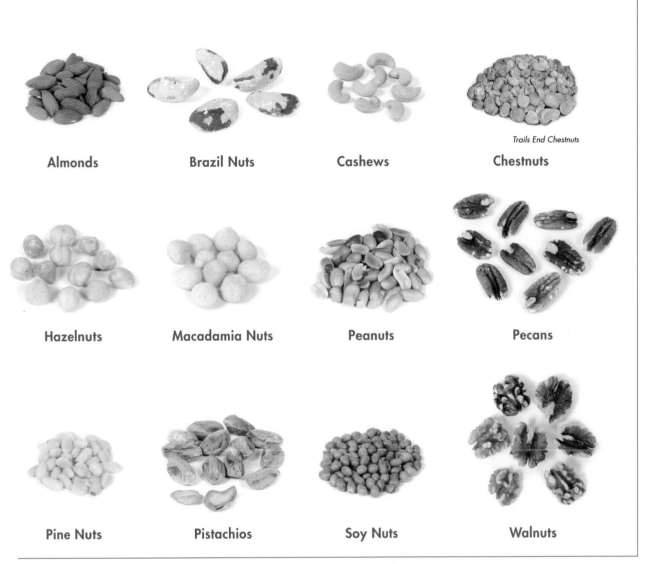

Almonds   Brazil Nuts   Cashews   Chestnuts

Trails End Chestnuts

Hazelnuts   Macadamia Nuts   Peanuts   Pecans

Pine Nuts   Pistachios   Soy Nuts   Walnuts

Figure 7-99. *Nuts commonly used in the professional kitchen include almonds, Brazil nuts, cashews, chestnuts, hazelnuts, macadamia nuts, peanuts, pecans, pine nuts, pistachios, soy nuts, and walnuts.*

- A *Brazil nut* is a 1–2 inch long, white, richly flavored seed of a very large fruit grown on a Brazilian nut tree.

- A *cashew* is a nut from a kidney-shaped kernel from the fruit of the cashew tree and has a buttery flavor. A cashew tree is related to the poison ivy plant. There is a toxin present in the shell, which is why cashews are never sold in shells.

- A *chestnut* is a nut from a rounded-off, triangular-shaped kernel found inside the burrlike fruit of the chestnut tree. Chestnuts are higher in starch and lower in fat than any other nut. When chestnuts are roasted or boiled, they become sweet. Chestnuts are available dried, canned, or as a sweetened paste.

- A *hazelnut,* also known as a filbert, is a grape-sized nut found in the fuzzy outer husks of the hazel tree. Shelled hazelnuts have bitter, brown, papery skins that cover their surface. Skinned hazelnuts have a sweet flavor.

- A *macadamia nut* is a tan, marble-sized nut with a hard shell and is the fruit of the macadamia tree. Macadamia nuts have a very high fat content and a sweet, creamy taste.

- A *peanut* is a legume that is contained in a thin, netted, tan-colored pod that grows underground. A *legume* is the edible seed of a nonwoody plant and grows in multiples within a pod. Unshelled peanuts are often sold roasted or salted. Peanut butter is made from ground peanuts.

- A *pecan* is a nut with a light-brown kernel from the pecan tree. Pecans are 70% fat, making them the fattiest of all nuts and quick to deteriorate. Pecans are often candied or seasoned with spices.

- A *pine nut* is an ivory-colored, torpedo-shaped nut from the pine cone of various types of pine trees.

- A *pistachio* is a pale-green, bean-shaped nut from the pistachio tree.

- A *soy nut* is a legume from the soybean pod that has been soaked in water, drained, and then roasted or baked. Although not technically a nut, soy nuts have a crunchy, nutlike texture.

- A *walnut* is a nut from the fruit of the walnut tree. Two popular varieties of walnuts are English walnuts and black walnuts. The English walnut has a rich sweet flavor. The black walnut has a stronger, somewhat bitter flavor.

SEASONINGS

A *seasoning* is an ingredient used to intensify or improve the natural flavor of foods. When seasonings are used correctly, the flavor of the seasoning should not be noticeable. If a seasoning is used incorrectly or in too great a quantity, it can make an item inedible or unpleasant to eat. Seasoning should be done in moderation. Seasonings can always be added, but they cannot be removed. Common seasonings include salts, peppercorns, citrus zests, vinegars, and oils.

Salts

Salt is the most widely used seasoning in the professional kitchen. *Salt* is a crystalline solid composed mainly of sodium chloride and is used as a seasoning and a preservative. It has a more intense taste flavor on cold foods than it does on hot foods. Salt also has the unique ability to make sweet items taste more full-bodied, sour items taste more pronounced, and bland items taste much more flavorful. Common varieties of salt include kosher salt, sea salt, specialty salt, pickling salt, and curing salt. *See Figure 7-100.*

Salts

Figure 7-100. *Common varieties of salt include kosher salt, sea salt, specialty salt, pickling salt, and curing salt.*

Kosher Salt. *Kosher salt* is a salt used to cure, season, and prepare kosher foods. For example, kosher salt can be rubbed into meat to tenderize it and to draw water out of the meat. Kosher salt contains no iodine or anticaking additives. It is lighter in weight than table salt and sticks to food very well. Kosher salt crystals are larger than other salts and do not contain iodine or chemicals that can give it a metallic taste.

Sea Salts. A *sea salt* is salt produced through the evaporation of seawater. Large, shallow reservoirs (pans) are filled with seawater. As the pans of seawater evaporate, a thick layer of crystallized sea salt is formed. The resulting sea salt is then raked from the salt beds. Sea salt contains calcium, potassium, and magnesium as well as the inherent characteristics of the environment where the body of water is located. These high concentrations of minerals give sea salt a more intense flavor. Sea salt is available in fine and coarse crystals and is used to cook and preserve food. Frequently used sea salts include the following:

- Alaea Hawaiian sea salt is red in color due to the Hawaiian Alaea clay. It is high in minerals and has a very mild ocean flavor.

- Cyprus flake sea salt is harvested from the Mediterranean. Each crystal has a natural pyramid shape, crunchy texture, and mild flavor.

- Flor Blanca sea salt is hand-harvested from salt ponds in Manzanilla, Mexico. It has a white, almost translucent color and a very mild flavor.

- Sel gris is a variety of sea salt that is harvested from the Normandy coast in France. The slightly gray-colored salt is harvested from clay soil and has a moist feel.

- Fleur de sel is a white-colored sea salt also from the Normandy coast. This hand-harvested salt is produced by scraping salt residue from the rims of salt pans as the sea water is evaporating.

- Trapani sea salt is harvested off the coast of Sicily. It is collected from the very first crystals that form on the salt pans. Trapani has a high concentration of minerals and is a prized finishing salt.

Specialty Salts. Specialty salts are mined from oceans and salt water beds that evaporated thousands of years ago. These salts are mined from natural flowing water that runs through natural salt mines. The result is a more intensely flavored salt that carries the characteristics of the soil that is around it as well as the color. Specialty salts include artisan mined salts such as Himalayan pink, Murray River flake, kala namak black, and various smoked salts. *See Figure 7-101.* Smoked salts are produced by placing a specialty salt over a fire to gently impart a subtle smoky flavor of an ingredient such as bamboo, banana leaves, coconut shells, or wood.

Common Specialty Salts		
Name	**Characteristics**	**Attributes**
Himalayan pink	Pink-colored salt mined from deep within the Himalayan Mountains and harvested from ancient sea salt deposits; contains over 84 minerals and trace elements	Mild, pleasant flavor
Murray River flake	Peach-colored or apricot-colored salt harvested from the Murray River in the foothills of the Australian Alps; produced from natural underground brine pools	Crunchy texture; a great finishing salt
Kala namak (also known as Indian black)	Black-colored salt mined from rock salt in an area where the soil is high in minerals such as sulfur	Strong sulfur odor; flavor similar to cooked egg yolks

Figure 7-101. *Specialty salts include artisan mined salts such as Himalayan pink, Murray River flake, kala namak black, and various smoked salts.*

Pickling Salt. *Pickling salt,* also known as canning salt, is a pure form of salt that contains no residual dust, iodine, or other additives. The purity of pickling salt keepings the salt from clouding the brine or settling at the bottom of the jar.

Curing Salts. *Curing salt,* also known TCM, is a pink mixture of table salt and sodium nitrate. Curing salts actually cook proteins through a chemical reaction with the cells. They are died pink to make it easily identifiable.

Peppercorns

Second only to salt, pepper is widely used in the professional kitchen. A *peppercorn* is the dried berry of a climbing vine known as the Piper nigrum and is used whole, ground, or crushed. The Piper nigrum never grows farther than 20° from the equator. Although all peppercorns come from the same plant, their variations in color are a direct result of when they are harvested and how they are processed. *See Figure 7-102.* Green, black, and pink peppercorns are commonly used in the professional kitchen.

The Spice House

Peppercorns

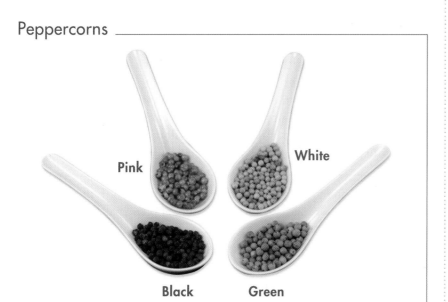

Pepper-Passion Inc.

Figure 7-102. *Although peppercorns are the berries of the same plant, the variations in color are a direct result of when they are harvested and how they are processed.*

Green Peppercorns. Green peppercorns are immature berries that are picked well before they have ripened. They are either pickled in a salted brine or vinegar-based solution, freeze-dried, or air-dried to retain their green color, fresh flavor, and aromatic qualities. Green peppercorns are the mildest of the peppercorn varieties. Pickled green peppercorns are soft and have a taste and texture similar to that of capers. The pickled version is most commonly used in sauces and in condiments. The freeze-dried and air-dried versions are often cracked or ground as a seasoning.

Black Peppercorns. Black peppercorns are the most common variety of peppercorn used in cooking. The berries are harvested when they are just starting to ripen. At this stage they are a yellowish color but still immature. The berries are blanched in boiling water and placed on screens in the sun to ferment and fully dry. The intense heat of the sun causes the berries to shrivel, completely dry out, and turn black.

Black peppercorns have a very strong, spicy-hot, and aromatic quality. Peppercorns grown in different parts of the world will express different characteristics of flavor, aroma, and pungency. Black peppercorns are sold as whole peppercorns, cracked pepper, or ground pepper. The black peppercorn varieties with the best flavor include Tellicherry, Malabar, and Lampong.

- Tellicherry peppercorns are harvested from the Tellicherry Mountain region in India. The berries are left to ripen on the vine a little longer than other varieties, which results in a golden-orange berry that is quite a bit larger than other black pepper varieties. Tellicherry peppercorns are blanched and sun-dried, resulting in a very floral, fruity, and spicy flavor and a color similar to a dark-roasted coffee bean with burnt-red to dark-brown or black undertones.

Cuisine Note

Szechuan peppercorns are actually the product of a small green berry from an ash tree. The flavor is closer to a chile than a peppercorn. The initial floral and pinelike flavor is followed by a slightly lemony and hot finish.

- Malabar peppercorns are named after the coastal area in India where they are harvested. They are very aromatic, have a well-rounded spice flavor, and are burnt green in color.

- Lampong peppercorns are a smaller peppercorn variety harvested in Indonesia on the island of Sumatra. These peppercorns are spicier than other black varieties. Lampong peppercorns have an earthy flavor with a slight hint of smokiness. They are about half the size of Tellicherry peppercorns and are almost black in color.

White pepper is made from black peppercorns. Ripe berries are soaked and fermented in water to soften the outer orange-red husk. The husk is removed when it begins to wrinkle. Removing the husk also removes the spicy characteristic of the black peppercorns. The remaining portion of the berries is white in color. White pepper is sold whole as white peppercorns and ground as white pepper. White pepper is much milder in flavor than black pepper.

Pink Peppercorns. Pink peppercorns are not actual peppercorns. They are the dried berries from the baies rose plant. They are best cracked instead of ground. Their flavor is pungent and slightly sweet, and they are often used with pork and seafood dishes.

Citrus Zests

Zest is the colored, outermost layer of the peel of a citrus fruit and contains a high concentration of oil. When removing the zest from a citrus fruit, it is important to remove only the colored portion of the peel because the white pith beneath the peel is bitter tasting. The zests of lemons and limes are often used in both sweet and savory applications. *See Figure 7-103.*

Citrus Zests

Browne-Halco (NJ)

Figure 7-103. *The zests of lemons and limes are often used in both sweet and savory applications.*

Lemon zest is the most common of the citrus zests. *Lemon zest* is the grated peel of a lemon that is used as a seasoning. Only a small amount is needed in any dish. Lemon zest is added to various sauces, baked products, and sautéed or grilled meats. Lime zest is commonly used for sweet applications such as lime-flavored desserts or savory marinades for grilled meats. Orange zest is commonly used for orange sauces served over roasted or grilled meats and for baked products.

Vinegars

Vinegar is a sour, acidic liquid made from fermented alcohol. Many vinegars are made from wine, but other alcohols can also be used to produce vinegar. High-quality alcohols yield high-quality vinegars. Vinegars can range from strong and pungent to sweet and mild. Vinegar is an essential ingredient in pickles, mustards, and vinaigrettes. Common vinegars used in the professional kitchen include balsamic, flavored, malt, rice, Champagne, sherry, wine, cider, and distilled vinegar. *See Figure 7-104.*

Vinegars

Balsamic   Red Wine   Cider   Distilled

Figure 7-104. *Common vinegars used in the professional kitchen include balsamic, flavored, malt, rice, Champagne, sherry, wine, cider, and distilled vinegars.*

Balsamic Vinegar. *Balsamic vinegar* is a vinegar made by aging red wine vinegar in wooden vats for many years. The vinegar picks up characteristics, such as color and aroma, from the type of wood in which it is stored. After aging, the vinegar darkens in color and becomes sweeter. Although balsamic vinegar has a higher acid content than red wine vinegar, its sweet characteristics mellow its acidic qualities. Balsamic vinegar adds an appealing flavor when drizzled over grilled vegetables or salads.

Flavored Vinegar. A *flavored vinegar* is any vinegar in which other items such as herbs, spices, garlic, fruits, vegetables, or flowers are added. Raspberry, pear, blueberry, and mango are examples of fruit vinegars that add a pleasant sweetness and unique flavor to salads. Tarragon, rosemary, and peppercorns are commonly used to produce savory vinegars.

Malt Vinegar. *Malt vinegar* is vinegar made from malted barley. It has a robust lemony flavor and a lower acidity than other vinegars. Malt vinegar is often served with fish and chips.

Rice Vinegar. *Rice vinegar* is vinegar made from rice wine. Rice vinegar is slightly sweet and mildly acidic. It is most often used in Asian dishes and dipping sauces.

Champagne Vinegar. *Champagne vinegar* is a vinegar made from Champagne grapes. It is a mildly flavored vinegar that pairs well with salads composed of delicate greens.

Sherry Vinegar. *Sherry vinegar* is vinegar made from sherry wine. The strong flavor of sherry comes through in sherry vinegar.

Wine Vinegar. *Wine vinegar* is a vinegar made from red or white wine. Red wine vinegars are aged longer than white wine vinegars and are slightly more pungent.

Cider Vinegar. *Cider vinegar* is honey-colored vinegar made by fermenting unpasteurized apple juice or cider until the sugars are converted into alcohol. It has a mild acidity and a subtle apple taste. Cider vinegar is often used in salad dressings.

Distilled Vinegars. *Distilled vinegar,* also known as white vinegar, is vinegar made by fermenting diluted, distilled grain alcohol. It is most often used in pickling and preserving foods due to its high acidic quality.

Oils

An *oil* is a type of fat that remains in a liquid state at room temperature. Many oils used in the professional kitchen are produced from vegetables, seeds, nuts, or fruit. These oils impart flavors that can be subtle and mild or rich and robust. Because many oils are plant-based products, they do not contain cholesterol and are often high in unsaturated fats. Oils have a limited shelf life and should be stored in a cool, dry place out of direct sunlight in order to prolong freshness.

Different oils are suited for different tasks. For example, some oils have a high smoke point, making them suitable for frying foods. Other oils are more flavorful, yet have low smoke points. For example, olive oil is best used to season foods or for sautéing. Canola, corn, and peanut oil are often used for frying. Oils commonly used in the professional kitchen include olive oil, canola oil, corn oil, peanut oil, walnut oil, hazelnut oil, sesame seed oil, grapeseed oil, and infused oils. *See Figure 7-105.*

Olive Oils. *Olive oil* is a type of oil produced from olives. Olive oil can differ slightly in taste, color, and consistency depending on the region it was produced. The difference in the quality of olive oils is indicated on the label as extra-virgin, virgin, pure, and olive-pomace.

- *Extra-virgin olive oil* is a type of olive oil produced from the first pressing of the olives without the use of heat or chemicals and has an acidic level less than 1%. It is considered the highest quality of olive oil and its rich flavor makes it desirable for salad dressings and garnishing applications.

- *Virgin olive oil* is a type of olive oil produced from the first pressing of the olives without the use of heat or chemicals and has an acid content of as much as 3%.

Oils

Olive Oil   Vegetable Oil

Figure 7-105. *Oils used in the professional kitchen include olive oil, canola oil, corn oil, peanut oil, walnut oil, hazelnut oil, sesame seed oil, grapeseed oil, and infused oils.*

- *Pure olive oil* is a type of olive oil produced using heat, and often chemicals, to extract additional oils from the olive pulp after the first pressing.

- *Olive-pomace oil* is a type of olive oil produced using heat, and often chemicals, to extract additional oils from the olive pulp and olive pits after the first pressing.

Canola Oil. *Canola oil,* also called rapeseed oil, is a type of oil produced from rapeseeds. It is very low in saturated fat making it a healthy choice for dressings, cooking, or baking. Canola oil has a neutral flavor and a high smoke point, making it ideal for frying foods.

Corn Oil. *Corn oil* is a type of oil produced from corn. It has a light golden color, a slight taste of cornmeal, and a high smoke point. Corn oil is often used when a neutral oil is desired.

Peanut Oil. *Peanut oil* is a type of oil produced from peanuts. It has a mild nutty flavor and a high smoke point. It is ideal for frying, sautéing, and stir-frying. Peanut oil has a unique cooking property in that it will not absorb the flavors of other foods.

Walnut and Hazelnut Oils. *Walnut oil* is a type of oil produced from walnuts. *Hazelnut oil* is a type of oil produced from hazelnuts. Due to their distinct flavors and aromas, these oils are used primarily in salad dressings and garnishing applications.

Sesame Oils. *Sesame oil* is a type of oil produced from sesame seeds. Light sesame oil has a mild, nutty flavor. Dark sesame oil is stronger in flavor because the sesame seeds are toasted before the oil is produced. Light sesame oil is often used in salad dressing. Dark sesame oil is often used to accent Asian dishes.

Courtesy of Lauren Frisch

Grapeseed Oil. *Grapeseed oil* is a type of oil produced from grape seeds. It has a delicate flavor with a smoke point suitable for sautéing.

Infused Oils. An *infused oil* is any variety of oil with added herbs, spices, or additional ingredients that increase the flavor. Infused oils are easy to produce in the kitchen. Neutral oils can be heated until warm and then infused with savory ingredients such as garlic, lemongrass, or various spices. Infused oils are used as dipping sauces and in many innovative dishes.

SUMMARY

Cooking refers to the process of transferring heat to food through conduction, convection, or radiation. When heat is introduced to food, the nutrients and properties of the food change in various ways. Proteins become firm and coagulate, starches gelatinize and absorb moisture, simple sugars caramelize, and fats melt.

Food is cooked using a variety of cooking methods. Dry-heat cooking methods include grilling, barbequing, broiling, roasting, baking, smoking, sautéing, and frying. Moist-heat cooking methods include poaching, simmering, blanching, boiling, and steaming. Combination cooking methods include braising, stewing, poêléing, and sous vide. A thorough understanding of how and when to apply various cooking methods helps ensure foods taste good, are easy to digest, and are free of harmful organisms.

Food has a way of awakening the senses through presentation, aromas, tastes, and textures. Flavors are developed in foods by the use of flavorings such as herbs, spices, rubs, marinades, condiments, sauces, and nuts. Flavors are intensified by seasonings such as salts, peppercorns, citrus zest, vinegars, and oils. A chef must have a thorough understanding of flavorings and seasonings in order to fully develop flavors in a variety of foods.

Refer to DVD for
Quick Quiz® questions

1. Describe the three methods of heat transfer used to cook food.
2. Identify three types of radiation heat transfer used to cook food.
3. Describe five reactions that change the color or texture of food.
4. Identify the two nutrients most often destroyed by heat.
5. Identify the eight dry-heat cooking methods.
6. Describe how to grill food.
7. Describe how to broil food.
8. Contrast smoking and barbequing.
9. Contrast roasting and baking.
10. Describe how to sauté food.
11. Define frying.
12. Explain why it is important to know the smoke point of oils used for frying.
13. Describe the standard breading procedure.
14. Contrast pan-frying and deep-frying.
15. Identify five moist-heat cooking methods.
16. Contrast submersion poaching and shallow poaching.
17. Describe how to simmer food.
18. Describe how to blanch food.
19. Describe how to know when a liquid has reached a full boil.
20. Describe how to steam food.
21. Describe how to steam food en papillote.
22. Identify four combination cooking methods.
23. Contrast braising and stewing.
24. Describe how to poêlé food.
25. Define sous vide.
26. Identify four signals the brain receives about food.
27. Explain how flavors are developed in food.
28. Differentiate between flavorings and seasonings.
29. Identify 18 common leaf herbs.
30. Identify three common stem herbs.
31. Identify two common bark spices.
32. Identify 14 common seed spices.
33. Identify four common root spices.
34. Identify three common flower spices.
35. Identify seven common berry spices.
36. Identify three common bean spices.
37. Identify nine common spice and herb blends.
38. Explain the use of rubs and marinades in flavor development.
39. Contrast dry rubs and wet rubs.

Refer to DVD for
Review Questions

40. Differentiate between marinades and brines.
41. Identify three common condiments.
42. Identify five common types of condiment sauces.
43. Identify 12 common nuts used to flavor food.
44. Identify five categories of seasonings used in the professional kitchen.
45. List two reasons salt is used in the professional kitchen.
46. Explain why peppercorns vary in color.
47. Define zest.
48. Identify nine commonly used vinegars.
49. Explain how to select the appropriate oil to use for a given preparation.
50. Identify nine commonly used oils.

Applications

1. Grill a food and evaluate the final product for preparation, color, texture, and flavor.
2. Broil a food and evaluate the final product for preparation, color, texture, and flavor.
3. Smoke a food and evaluate the final product for preparation, color, texture, and flavor.
4. Barbeque a food and evaluate the final product for preparation, color, texture, and flavor.
5. Roast a food and evaluate the final product for preparation, color, texture, and flavor.
6. Bake a food and evaluate the final product for preparation, color, texture, and flavor.
7. Sauté a protein and evaluate the final product for preparation, color, texture, and flavor.
8. Sauté a vegetable and evaluate the final product for preparation, color, texture, and flavor.
9. Stir-fry a dish and evaluate the final product for preparation, color, texture, and flavor.
10. Bread and pan-fry a protein and evaluate the final product for preparation, color, texture, and flavor.
11. Batter and deep-fry a potato and evaluate the final product for preparation, color, texture, and flavor.
12. Poach an egg and evaluate the final product for preparation, color, texture, and flavor.
13. Poach a piece of fish and evaluate the final product for preparation, color, texture, and flavor.
14. Simmer a grain and evaluate the final product for preparation, color, texture, and flavor.
15. Blanch a vegetable and evaluate the final product for preparation, color, texture, and flavor.
16. Boil a potato and evaluate the final product for preparation, color, texture, and flavor.
17. Steam a vegetable and evaluate the final product for preparation, color, texture, and flavor.
18. Steam fish en papillote and evaluate the final product for preparation, color, texture, and flavor.
19. Braise a meat and evaluate the final product for preparation, color, texture, and flavor.
20. Stew a meat and evaluate the final product for preparation, color, texture, and flavor.
21. Poêlé poultry and evaluate the final product for preparation, color, texture, and flavor.
22. Prepare a food using the sous vide method and evaluate the final product for preparation, color, texture, and flavor.
23. Use herbs to flavor a food. Evaluate the food for contrast, balance, and flavor.

Refer to DVD for
Application Scoresheets

24. Use spices to flavor a food. Evaluate the food for contrast, balance, and flavor.
25. Use a dry rub or a wet rub to flavor a food. Evaluate the food for contrast, balance, and flavor.
26. Use a marinade to flavor a food. Evaluate the food for contrast, balance, and flavor.
27. Use a brine to flavor a food. Evaluate the food for contrast, balance, and flavor.
28. Use a condiment to add flavor to a dish. Evaluate the dish for contrast, balance, and flavor.
29. Use nuts to add flavor to a dish. Evaluate the dish for contrast, balance, and flavor.
30. Use salts, peppercorns, citrus zest, vinegar, and oil to flavor a salad. Evaluate the salad for contrast, balance, and flavor.

Chapter 8

Stocks and Sauces

Stocks and sauces are fundamentally important items produced in the professional kitchen. The French word for stock is "fond," which means "foundation." Stocks serve as the foundation for countless sauces and soups. A wide variety of sauces complement other foods by adding flavor and visual appeal. A skilled chef is a master at preparing quality stocks and sauces. This key skill demonstrates knowledge of the ingredients, procedures, and storage methods used to prepare quality stocks and sauces.

Refer to DVD for **Flash Cards**

Chapter Objectives

1. Describe the basic composition of stocks.
2. Describe the general guidelines for preparing stocks.
3. Contrast two common methods for cooling stocks.
4. Prepare a brown stock, a white stock, a fish stock, and a vegetable stock.
5. Contrast an essence and a fumet.
6. Contrast a glace, a remouillage, and a bouillon.
7. Explain the process of reduction.
8. Describe common thickening agents used to prepare sauces.
9. Prepare a roux and a beurre manié.
10. Demonstrate how to add a liaison to a liquid.
11. Describe the five classical mother sauces.
12. Prepare a hollandaise sauce.
13. Describe three types of butter sauces.
14. Prepare a beurre blanc sauce.
15. Contrast common contemporary sauces.
16. Prepare flavored oils.

Key Terms

- **stock**
- **fumet**
- **glace**
- **remouillage**
- **sauce**
- **nappe**
- **reduction**
- **thickening agent**
- **gelatinization**
- **roux**
- **beurre manié**
- **slurry**
- **liaison**
- **coagulation**
- **mother sauce**
- **emulsification**
- **beurre blanc**
- **coulis**
- **nage**

Dashi is a stock used as the base for many Japanese soups and sauces. There are different kinds of dashi. For example, dashi can be made from dried kelp, dried shiitake mushrooms, dried bonito, or dried sardines. Dashi powder can also be used to make a stock, but the seasonings have to be adjusted because dashi powder contains salt.

STOCKS

A *stock* is an unthickened liquid that is flavored by simmering seasonings with vegetables, and often, the bones of meat, poultry, or fish. The ingredients in a fresh stock are simmered to extract flavors, nutrients, and color. However, many foodservice operations use commercially prepared bases to save time and money. Bases are available in highly reduced pastes or dehydrated forms. Quality bases are made from the meat, bones, and juices of the product they represent. Other bases rely on high-sodium ingredients for flavor. Bases may be used to add flavor to a fresh stock. Likewise, fresh vegetables and bones may be added to a stock made from a base to make the stock taste fresher.

Stock Composition

Stocks are composed of water, a flavoring component, mirepoix, and aromatics. The composition of a stock can vary, but is generally 10 parts water, 5 parts flavoring component, and 1 part mirepoix, plus aromatics. *See Figure 8-1.* A stock is often reduced to further concentrate its flavor, especially when used to make sauces. Salt is never added to stock in the early stages because the resulting flavor is too intense. If desired, salt is added at the final stage of preparation.

Stock Composition

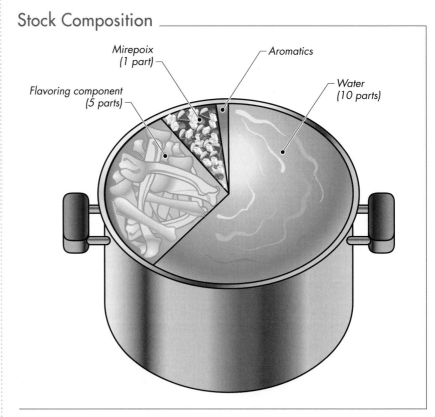

Figure 8-1. *The initial composition of a stock is approximately 10 parts water, five parts flavoring component, and 1 part mirepoix, plus aromatics.*

Water. Cold water is used to start a stock. As the water heats, it slowly draws out impurities and blood from the bones which then coagulate and float to the surface where they can be removed by skimming. If hot water is used, the impurities will coagulate too quickly and stick to the bones, causing the stock to become cloudy as it simmers.

Flavoring Ingredients. Each flavoring ingredient adds to the flavor and nutritional makeup of a stock. Bones or trimmings are the primary flavoring ingredients used to make meat, poultry, or fish stocks. Animal bones and trimmings contain collagen, a fibrous protein that is converted to gelatin when heated. In order to more effectively expose the collagen, bones are cut into smaller pieces or split in half lengthwise. *See Figure 8-2.* The resulting gelatin that forms when the collagen cooks out of the bones and trimmings provides the stock with nutrients and body. A stock that is rich in gelatin will thicken naturally as it is reduced.

Stock Bones

Roasted Beef Bones

Fish Bones

Figure 8-2. *Animal bones are cut into smaller pieces or split in half lengthwise in order to more effectively expose the collagen, which helps flavor and color the stock.*

Bones can also be roasted or blanched before being used in a stock. The blanching process removes any loose blood proteins and impurities from the bones that could cloud the stock. *See Figure 8-3.* After the bones have been blanched, the blanching water is discarded and the bones are covered with fresh cold water to start a stock.

Procedure for Blanching Bones

1. Place the bones in a pot of cold water.
2. Bring water to a boil over high heat.
3. Reduce heat to a simmer and cook for a few minutes.
4. Remove the bones and plunge them into cold water.
5. Discard the blanching water.

Figure 8-3. *Animal bones are often blanched to remove blood proteins and impurities.*

Vegetables must be thoroughly washed, cleaned, and trimmed prior to being used in a stock. Most chefs remove the peelings from carrots because the peels will break down and discolor the stock. Celery leaves are not used in stock because the leaves are often bitter. Onion peels are sometimes used in brown stocks, but never in white stocks because the peels may darken the stock. Stock should always be relatively clear, not cloudy. Care should be taken to remove any impurities that could cloud the stock, such as overcooked vegetables.

Mirepoix. *Mirepoix* is a mixture of 50% onions, 25% celery, and 25% carrots, roughly cut into the appropriate size for the stock being produced. *See Figure 8-4.* Because the mirepoix is strained from the finished stock, a perfect dice is not required. A white mirepoix is used to make a white stock. A *white mirepoix* is a mirepoix made of onions, celery, and leeks or parsnips instead of carrots.

Mirepoix

25% celery ⌐   25% carrots ⌐   50% onions ⌐

Figure 8-4. *Mirepoix is a mixture of 50% onions, 25% celery, and 25% carrots cut into the appropriate size for the stock being produced.*

A stock made from large bones, such as beef stock, requires large cuts of mirepoix, usually 1–2 inch square pieces. Fish and chicken stocks that cook for a shorter time and contain smaller, thinner bones require smaller cuts of mirepoix, typically ½ inch square pieces. The smaller the cut, the quicker the mirepoix cooks. If a small-cut mirepoix is added to a beef stock, the vegetables will overcook before the stock is done. Overcooked mirepoix can make the stock cloudy.

Sometimes a matignon is used in place of mirepoix. A *matignon* is a uniformly cut mixture of onions or leeks, carrots, and celery and may also contain smoked bacon or ham. A white matignon includes parsnips instead of carrots. Matignon is often referred to as "edible mirepoix" because it is cut uniformly and left in the completed dish.

Aromatics. An *aromatic* is an ingredient such as an herb, spice, or vegetable added to a food to enhance its natural flavors and aromas. There are two standard methods for adding aromatics to a stock. ***See Figure 8-5.*** A *bouquet garni* is a bundle of herbs and vegetables tied together with twine that is used to flavor stocks and soups. It usually consists of a sprig of thyme, several parsley stems, a dried bay leaf, and leek leaves or a celery stalk cut in half. The bundle is tied to the handle of the stockpot so that it can be easily retrieved. A *sachet d'épices*, or sachet, is a mixture of spices and herbs placed in a piece of cheesecloth and tied with butcher's twine. A sachet often contains several parsley stems, a dried bay leaf, a sprig of thyme, a teaspoon of cracked peppercorns, and a clove of garlic. Sometimes spices such as cardamom, ginger, or cinnamon are added. The sachet is tied to the handle of the stockpot for easy retrieval. Aromatics infuse their flavor and aroma, adding depth to the flavor of the stock.

Basic Guidelines for Preparing Stocks

Although there are many different types of stocks used in the professional kitchen, there are basic guidelines that can be applied to any stock. For example, stock is always prepared in a pot that is taller than it is wide to allow for slower evaporation. Following these general guidelines yields consistent stocks that are easy to produce.

1. **Begin with cold water.** Cold water added to bones at the beginning stages of stock preparation helps remove any impurities and blood proteins that could later cloud a stock. As the stock is heated, these impurities coagulate or thicken and float to the surface. The coagulated impurities are commonly referred to as "scum," which can be skimmed from the surface and discarded.

2. **Cover bones completely.** Bones should be completely submerged in water during preparation. A bone that is above the surface of the water does not add any flavor or nutrients to the stock. Stocks made with larger and thicker bones, such as a beef stock, require more cooking time than stocks made with small, thin bones. A longer cooking time is needed to extract the nutrients of dense bones than is needed for small bones. Beef stocks usually simmer for about 8 hours. A chicken stock made with many small, less dense bones usually cooks for about 4 hours. A fish stock made from thin, fragile fish bones usually requires a cooking time of only 30–45 minutes. *See Figure 8-6.*

3. **Simmer stock.** Simmering stock at a temperature between 180°F and 185°F helps draw flavor and nutrients from the bones. Boiling a stock (even if only for a short time) breaks up coagulated impurities as they are released from bones and causes the impurities to break up and cloud the stock.

4. **Do not stir while simmering.** The solid components of a stock should not be disturbed. Stirring the stock during cooking causes impurities in and around the solid ingredients to be broken up throughout the stock and results in cloudiness.

Aromatics _____

Bouquet Garni

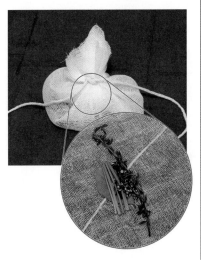

Sachet d'Épices

Figure 8-5. *A bouquet garni and a sachet d'épices are made with aromatics that enhance the flavor and aroma of a stock.*

Cooking Times for Stocks	
Type	**Cooking Time***
Beef	8 hours
Chicken	4 hours
Fish	45 minutes
Vegetable	30 minutes

* approximate

Figure 8-6. *Required cooking times for stocks vary depending upon the size of the bones used.*

Cooling Paddles

San Jamar

Figure 8-7. *A cooling paddle may be used to cool stocks quickly for storage.*

Roasting Bones

Figure 8-8. *The process of roasting bones creates a rich, brown color, flavor, and high-gelatin content that are characteristic of a quality brown stock.*

5. **Remove impurities from the top.** The solid material that rises to the surface of a stock should be skimmed off continuously during the cooking process. Care should be taken not to disturb any material remaining on the surface while draining the stock. Many stockpots have a spigot drain that makes it easy to drain all liquid out through the bottom of the stockpot without needing to tip the pot. This gentle draining of unstrained stock helps ensure that impurities remain in the pot and are not disturbed, which would cause them to break up and cloud the stock. If a stockpot with a drain spigot is not available, a large ladle is used to gently remove the stock.

6. **Strain stock thoroughly.** A chinois or a cheesecloth-lined china cap is used to strain stock. The cheesecloth helps remove small particles that can cloud the stock.

7. **Cool stock rapidly.** Stock is first cooled to 70°F and then to below 41°F before being stored. The most common methods for cooling stocks are in an ice bath and with a cooling paddle. When using an ice bath, the hot stockpot is placed on top of an inverted perforated hotel pan inside a clean sink filled with ice cubes and cold water. Then, the stock is stirred to speed the cooling process.

   Another method for cooling a stock is to use a cooling paddle. A *cooling paddle* is a heavy-gauge, hollow plastic paddle with a screw-on cap at the top of the handle that is filled with water and frozen prior to use. *See Figure 8-7.* The cooling paddle is simply inserted into the hot stock and used to stir the stock until it is cooled.

8. **Follow proper storage procedures.** When a stock has cooled, any remaining fat will have risen to the surface as a solid and can be easily removed. Many chefs use this solidified fat in place of butter or oil when sautéing meats to add additional flavor to various dishes. Once the fat is removed, the stock is covered, labeled, dated, and stored in a refrigeration or freezer unit.

The way a stock is prepared determines the outcome of the stock. Some stocks are made by simply simmering uncooked ingredients together. Other stocks include ingredients that are roasted prior to use, which produces a rich, brown color and roasted flavor. The taste of a stock should be identifiable, but should not overpower the taste of the dish that it is used in, such as a sauce or a soup.

Brown Stocks

A *brown stock* is a stock produced by simmering roasted meat, poultry, or game bones with mirepoix and an optional tomato product. When roasting the bones, they must be placed in a single layer in an oiled roasting pan. The bones should be roasted in a 400°F oven for 45–60 minutes and stirred occasionally to produce even browning. *See Figure 8-8.* The roasted bones are then transferred to a stockpot, covered with cold water, and brought to a simmer. The mirepoix is caramelized in the same roasting pan and then added to the pot. The roasted bones and vegetables add the rich, brown color, flavor, and high-gelatin content that are characteristic of a well-made brown stock. *See Figure 8-9.*

1. Roast bones in a roasting pan until evenly brown. Transfer the roasted bones to a stockpot and cover with cold water. Reserve the rendered fat in the roasting pan.

2. Begin heating the contents of the stockpot. Then, sauté the mirepoix in the reserved rendered fat until it is well caramelized. Stir the mirepoix continuously to avoid burning.

3. Pour off excess fat and reserve for later use.

4. If desired, add a small amount of tomato sauce or paste to the mirepoix and cook until the tomato product caramelizes.

5. Deglaze the roasting pan.

6. Once the water in the stockpot has reached a simmer, skim the impurities from the surface and then add the contents of the roasting pan to the stockpot.

7. Return the contents of the stockpot to a simmer and continue cooking and skimming impurities from the surface until done. Do not let the stock boil.

8. Strain the stock with a chinois or cheesecloth-lined china cap.

9. Quickly cool the strained stock in an ice bath or with a cooling paddle and refrigerate or freeze.

10. Label and date the stock and refrigerate or freeze until needed.

Refer to DVD for
Preparing Brown Stock
Media Clip

Figure 8-9. *A brown stock is prepared by simmering roasted meat, poultry, or game bones with mirepoix and an optional tomato product.*

White Stocks

A *white stock* is a light-colored stock produced by gently simmering poultry, veal, or fish bones in water with vegetables and herbs. A white mirepoix is used to avoid a possible color change in the stock. Chefs often disagree about whether or not bones should be blanched prior to being used in a white stock. Some chefs think that blanching removes many impurities from the bones. *See Figure 8-10.* Other chefs think that blanching bones wastes flavor and nutrients. It is common practice to simply rinse the bones in clean, cold water to remove any loose impurities that are present on the surface of the bones before using them to prepare a white stock.

1. Rinse the bones in cold water and place in a stockpot. Completely cover the bones with cold water.

2. Bring water to a simmer. Continually skim off impurities as they rise to the surface.

3. Add a white mirepoix and a sachet to the pot. Maintain a simmer and continually skim off impurities. *Note:* A stock with smaller bones such as poultry will only need to simmer 4 hours, while large veal or beef bones will require 8 hours of simmering.

4. Carefully drain the stock to minimize disturbing the cooked ingredients.

5. Strain the stock with a chinois or cheesecloth-lined china cap.

6. Quickly cool the strained stock in an ice bath or with a cooling wand.

7. Label the stock with the correct product name and date before storing in the refrigerator or freezer.

Figure 8-10. *White stocks can be prepared using poultry, veal, or fish bones and a white mirepoix.*

Fish Stocks

A *fish stock* is a basic stock prepared by adding fish bones or shellfish shells, vegetables, a sachet, and cold water to a stockpot and bringing it to a simmer over medium heat. The resulting stock is clear and mild in flavor. The cooking time for fish stocks is much shorter than white or brown stocks because fish bones and shellfish shells are thinner and more delicate than the bones used for other stocks. Bones used for a fish stock should be those from a lean fish, such as sole or snapper, as opposed to an oily fish, such as salmon. *See Figure 8-11.* Fish heads with the eyes and gills removed may also be used.

Fish Bones

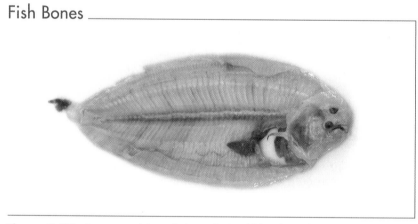

Figure 8-11. *Bones used for a fish stock should be those from a lean fish, such as sole.*

Fish bones and shellfish shells must be completely rinsed prior to use in a fish stock in order to make a clear stock. Vegetables used in a mirepoix for a fish stock are cut very small due to the quick cooking time, usually no more than 45 minutes. Often mushroom stems or trimmings are added to fish stocks to increase the flavor and aromatic characteristics. *See Figure 8-12.*

1. Rinse fish bones or shellfish shells in cold water to remove any surface debris.
2. Place the bones or shells, a small diced mirepoix (including mushroom pieces if desired), a sachet, and cold water in a stockpot over medium heat and bring to a simmer.
3. Simmer for 30–45 minutes, using caution to never let the stock reach a boil. Continually skim the surface to remove impurities.
4. Strain the stock carefully with a chinois or cheesecloth-lined china cap.
5. Quickly cool the strained stock in an ice bath or with a cooling wand.
6. Label the stock with the correct product name and date before storing in the refrigerator or freezer.

Figure 8-12. *Fish stocks require less cooking time due to the delicate fish bones and shellfish shells used as flavoring ingredients.*

Fumets

A *fumet* is a concentrated stock made from fish bones or shellfish shells and vegetables. A fumet differs from a fish stock in that the mirepoix, bones, and shells are sweated in a stockpot over low-to-medium heat prior to use. Care is taken to not discolor the vegetables, bones, or shells in the sweating process in order to produce a flavorful yet relatively colorless stock. Fish stock or water and a sachet are added to the sweated ingredients, and wine and/or lemon juice is added for flavor. ***See Figure 8-13.***

Refer to DVD for
Preparing Fumets
Media Clip

Procedure for Preparing Fumets

1. Rinse fish bones and shellfish shells in cold water to remove any surface debris.
2. Melt a small amount of butter in a stockpot over low-to-medium heat. Add onions, parsley stems, thyme, mushroom pieces, garlic, trimmings, bones, and shells and slowly sweat them, being careful not to allow them to gain color.

3. When the onions are translucent, deglaze the pot with a small amount of lemon juice and white wine.

4. Add fish stock or water and a few lemon slices and bring to a simmer.
5. Simmer for about 30 minutes. Continually skim the surface to remove impurities.
6. Quickly cool the stock in an ice bath or with a cooling wand.
7. Label and date the stock.
8. Refrigerate or freeze the stock until needed.

Figure 8-13. *When preparing a fumet, the small-diced vegetables, fish bones, and shellfish shells are sweated over low-to-medium heat prior to adding the fish stock.*

Essences

An *essence* is a concentrated fish stock that includes a large amount of aromatic ingredients such as celery, morels, a bouquet garni, a sachet, and fennel root. The flavor of an essence is more concentrated than a fumet and extremely aromatic.

Vegetable Stocks

A *vegetable stock* is a clear, light-colored stock produced by gently simmering vegetables with a white mirepoix and a sachet. A vegetable stock can be used in place of a meat-based stock in many recipes. This substitution option is helpful when a chef chooses to offer a more nutritious sauce or soup. Almost any combination of vegetables may be used, but strongly flavored vegetables such as asparagus and spinach should be avoided when making an all-purpose vegetable stock. Starchy vegetables such as potatoes should also be avoided unless a cloudy stock is not a concern.

Glaces

A *glace* is a highly reduced stock that results in an intense flavor. For example, one gallon of a stock can be slowly reduced to one-eighth or one-tenth of its original volume to yield just one or two cups of glace. *See Figure 8-14.*

Procedure for Preparing Glaces

1. Simmer the stock slowly to allow a slow reduction. Continually check the bottom of the saucepan to ensure that the reduction does not begin to scorch or burn.
2. Continually skim any surface impurities.
3. As the stock reduces, strain and transfer it into progressively smaller pans. Straining the stock each time it is transferred helps ensure the final glace is as free from impurities as possible.
4. When the reduction is finished, strain the glace carefully with a chinois or cheesecloth-lined china cap.
5. Quickly cool the strained glace in an ice bath.
6. Label and date the glace before storing in the refrigerator or freezer.

Figure 8-14. *Preparing a glace requires the reduction of a stock to one-eighth or one-tenth of its original volume.*

Once completely cooled, glace has a rubbery texture due to the concentrated amount of gelatin that results from the reduction process. The glace can then be cut into cubes, which are easy to store. *See Figure 8-15.* It is important to avoid handling glace with bare hands as this will cause the glace to spoil quickly. Improper storage and handling can greatly shorten the amount of time that glace can be stored. When properly prepared, stored, and handled, a glace will last for several months. Classic glaces are added to soups or sauces to intensify their flavor. *See Figure 8-16.*

Glaces

Figure 8-15. *Cold glace has a rubbery texture that allows it to be easily cut into cubes and stored for future use.*

Glaces	
Glace	**Description**
Glace de poisson (fish glaze)	Made from a reduced fish stock
Glace de veau (veal glaze)	Made from a veal stock reduction
Glace de viande (beef glaze)	Made from caramelized beef bones and mirepoix
Glace de volaille (chicken glaze)	Made from a concentrated reduction of chicken stock

Figure 8-16. *Glaces can be easily flavored to suit different dishes.*

Remouillages

Remouillage means "rewetting" in French. A *remouillage* is a stock made from using bones that have already been used once to make a stock. Used bones are placed in water with vegetables and aromatics and simmered to extract any additional flavor they may contain. The resulting stock is less flavorful than the original stock made from the same bones. A remouillage is often used in place of water when starting a new stock as it has more flavor than plain water.

Bouillon

Bouillon means "broth" in French. *Bouillon* is the liquid that is strained off after cooking vegetables, poultry, meat, or seafood in water. It is not a stock, but bouillon can form the base for sauces and soups.

Likewise, court bouillon is also not a stock. *Court bouillon* is a highly flavored, aromatic vegetable broth made from simmering vegetables with herbs and a small amount of an acidic liquid (usually vinegar or wine). Court bouillon is used to poach fish, seafood, and vegetables.

Sauces

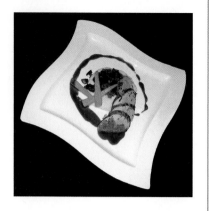

Figure 8-17. *A sauce completes or enhances the flavor, moistness, or texture of a dish.*

SAUCE BASICS

Many chefs consider a sauce to be the unifying component of a dish. A *sauce* is an accompaniment that is served with a food to complete or enhance the flavor and/or moistness of a dish. Sauces also add texture to a dish and can be the ingredient that pulls a dish together. *See Figure 8-17.* Sauces are meant to complement, not overpower. A sauce is similar to the icing on a cake. The cake may taste good by itself, but when icing is added, the cake becomes delectable. Likewise, adding an excellent sauce to a good dish creates something extraordinary.

Proper sauce preparation yields a smooth sauce with a nappe consistency. *Nappe* is the consistency of a liquid that thinly coats the back of a spoon and ensures that a sauce will cling lightly to another food. To achieve a nappe, sauces are either thickened by reduction or the addition of a thickening agent. *Reduction* is the process of gently simmering a liquid until it lessens in volume and results in a thicker liquid with a more concentrated flavor. Simmering can reduce a sauce by up to three-fourths of its original volume. This reduction lessens or eliminates the need for a thickening agent.

Thickening Agents

A *thickening agent* is a substance that adds body to a hot liquid. For example, a thickening agent is added to chicken stock to create chicken gravy. Common thickening agents include flour, cornstarch, arrowroot, roux, beurre manié, and a liaison. The first three of these thickeners are starches, which undergo gelatinization when added to a liquid. *Gelatinization* is the process of a heated starch absorbing moisture and swelling, which thickens the liquid. *See Figure 8-18.* When using a thickening agent, the final sauce should be strained prior to serving. It is always possible that some of the thickening agent did not dissolve completely. Straining removes any unappetizing lumps from the finished sauce.

Gelatinization

Thickening agent

Thickened liquid

Figure 8-18. *Starches undergo gelatinization when added to a liquid. The starch granules absorb water and swell, thickening the liquid.*

Flour. Flour is a primary thickening agent and is often combined with fat before being added to a liquid. Pastry and cake flour are optimal choices when making products that are used to thicken sauces and soups, such as roux and beurre manié.

Cornstarch. *Cornstarch* is the white, powdery, pure starch derived from corn. Unlike roux, which is only used to thicken hot foods, cornstarch can be used to thicken hot or cold foods. Cornstarch has double the thickening power of flour, so less is needed to thicken a liquid. However, cornstarch-thickened sauces have a tendency to break down if held over heat for long periods of time. It is not recommended to reheat a sauce thickened with cornstarch.

Cornstarch must be made into a slurry before it is used to thicken a liquid. A *slurry* is a mixture of equal parts of cool liquid and a starch that is used to thicken other liquids. *See Figure 8-19.* Making a slurry allows the starch granules to absorb liquid most efficiently. If cornstarch alone is sprinkled into a hot liquid, the starch granules will clump together and form lumps. These lumps cannot be whisked or completely dissolved and have to be removed by straining.

The slurry is then added to either a hot or cold liquid with a whisk and heated thoroughly. Once the mixture reaches a boil, the slurry thickens to its full potential. After reaching a boil, it takes about 5 minutes of simmering to completely cook out the taste of the cornstarch. A cornstarch slurry is often used to thicken sauces and gravies because a slurry does not require fat and is therefore less expensive.

Arrowroot. *Arrowroot* is a thickening agent that is the edible starch from the rootstock of the arrowroot plant. Like cornstarch, it is mixed with a cool liquid before being heated or added to hot liquids. However, a sauce thickened with arrowroot is clearer than a sauce made with cornstarch. Arrowroot is flavorless, but more expensive than cornstarch.

Roux. A *roux* is a thickening agent made by cooking a mixture of equal amounts, by weight, of flour and fat and is used to thicken sauces and soups. Pastry flour and cake flour are the most effective types of flour to use in a roux as they both contain more starch than bread flour. All-purpose flour, which is a combination of pastry flour and bread flour, may also be used.

Cooking flour in fat coats the starch granules in fat. This coating of fat helps prevent the starch from lumping when added to a hot liquid. If flour is added by itself to a liquid such as a hot stock, clumps of flour will stick together as the starch goes though the gelatinization process. Once flour begins to clump, these improperly gelatinized flour pieces will never be broken down, no matter how much the mixture is whisked. When a roux is made properly, by cooking the flour and fat together first, the fat-coated starch granules dissolve smoothly in the hot stock. *See Figure 8-20.*

Figure 8-19. *A slurry is a mixture of equal parts of cool water and a starch such as cornstarch or arrowroot.*

Procedure for Preparing Roux

1. Using a heavy-bottomed sauce-pan to prevent scorching, heat fat until hot.

2. Add an equal amount, by weight, of sifted pastry or cake flour to the hot fat.

3. Stir to form a pasty consistency while cooking over medium heat.

4. Continue stirring until desired color of roux is achieved.

Figure 8-20. A roux is made by cooking equal parts of flour and fat. Rouxs are used to thicken sauces and soups.

Refer to DVD for **Preparing Roux** Media Clip

Chef's Tip

As a roux is cooked, and turns from white to brown, it will lose some gelatinizing strength. Therefore, a larger quantity of brown roux is required than the quantity of blonde or white roux needed to thicken a sauce or a soup.

There are three types of roux made in the professional kitchen: white, blonde, and brown. *See Figure 8-21.*

- White roux is made by cooking equal amounts, by weight, of fat and flour until a slightly foamy appearance is achieved. It is then immediately removed from the heat and allowed to rest for 5–10 minutes before it is added to a liquid. A white roux has a faint smell of freshly baked bread. White roux is used in white sauces when no color is desired.

- Blonde roux is made by cooking the same mixture of fat and flour until the roux exhibits a slightly golden or tan color. A blonde roux develops an aroma similar to freshly popped popcorn. Blonde roux is used in sauces that have a slight golden or tan color such as those made with chicken or light veal stock. Blonde roux not only adds some of its color to a sauce but also adds a slightly toasted flavor from starches in the flour beginning to caramelize.

- Brown roux is made by cooking the same mixture of fat and flour for an even longer period of time until the roux changes to a deep brown color and develops a nutty aroma. Due to its intense flavor and color, a brown roux is used with sauces that are made with a rich veal or beef stock. It is important to use caution when making a brown roux to avoid overcooking it. A roux that is too dark has an unpleasant, bitter taste and will leave black flecks in a prepared dish.

Roux

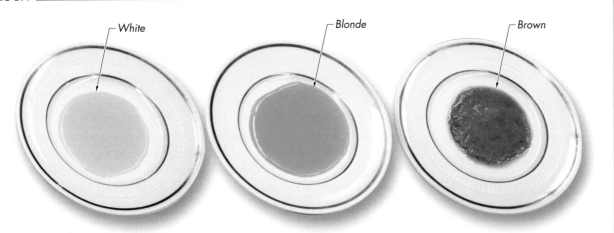

Type	Cooking Time	Characteristics	Uses
White roux	2 minutes	White, bubbly, slightly foamy, fresh bread smell	White sauces
Blonde roux	4 – 7 minutes	Golden tan color, slightly nutty popcorn smell	Light-colored sauces, such as chicken-based or veal-based sauces
Brown roux	10 –12 minutes	Deep brown color, very nutty smell	Dark sauces, such as beef-based sauces

Figure 8-21. *White, blonde, and brown roux are used in the professional kitchen, each with specific characteristics and uses.*

Adding a roux to a liquid can be a challenging task. Adding the roux correctly results in a smooth and silky sauce that has a nappe consistency. Using the proper proportions of flour and fat is the first step. *See Figure 8-22.* When adding roux to a liquid, the roux and liquid must be at different temperatures. If they are both either extremely hot or extremely cold, the roux may develop lumps and may have a floury taste.

Proportions of Roux Ingredients to Liquid					
Sauce Consistency	= (Flour*	+ Fat	= Roux)	+ Liquid	
Light	6 oz	6 oz	12 oz	1 gal.	
Medium	8 oz	8 oz	16 oz	1 gal.	
Heavy	10 oz	10 oz	20 oz	1 gal.	

Light refers to the consistency of most fine and delicate sauces.

Medium refers to the consistency of a cream soup or standard smooth gravy.

Heavy refers to the consistency of a binding sauce, such as the sauce used in a turkey pot pie.

* The above proportions ratios are for use with pastry flour or cake flour. If using all-purpose or bread flour, more flour may be needed to achieve the same results.

Figure 8-22. *A sauce can be made light, medium, or heavy depending on the amount of roux added in proportion to the amount of liquid.*

Refer to DVD for
Preparing a Buerre Manié
Media Clip

There are two accepted methods for adding roux, and the appropriate method depends upon the temperature of the roux. With a hot roux, use a cold or room-temperature stock. The roux should be in a pot large enough to incorporate the cold stock. The cold stock is carefully added to the hot roux while whisking vigorously. If using a roux that is at room temperature or colder, a hot stock should be used. Carefully add room-temperature roux to the hot stock while whisking vigorously. Many restaurants make large batches of roux and store it for future use, rather than making individual batches frequently.

Beurre Manié. Beurre is the French word for butter. A *beurre manié* is a thickening agent made by kneading equal amounts, by weight, of pastry or cake flour and softened butter and can be whisked into a sauce just before service. Unlike roux, a beurre manié is not cooked. Instead, the flour and butter are kneaded together until a smooth texture is achieved. The mixture is then separated into small balls and whisked into a sauce to finish it. The butter component adds additional sheen and richness to the sauce. The process of kneading butter and flour together coats each starch granule with fat, allowing the starch to better absorb moisture without lumping. *See Figure 8-23.* A beurre manié is often used to thicken a small amount of sauce, as when making a pan sauce, while a roux is more often used to thicken larger quantities of sauce.

Procedure for Preparing a Beurre Manié

1. Add equal parts softened butter to pastry or cake flour.

2. Mix ingredients thoroughly.

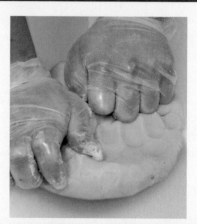

3. Knead mixture until all of the flour is incorporated.

4. Separate the mixture into small balls and store for later use. *Note:* One or more balls can be whisked into a sauce to finish it.

Figure 8-23. *A beurre manié is prepared by kneading together equal amounts of pastry or cake flour and softened butter.*

Liaison. A *liaison* is a thickening agent that is a mixture of egg yolks and heavy cream used to thicken sauces. Because a liaison does not contain a starch ingredient, it does not use the process of gelatinization. A liaison is typically used for cream- or milk-based dishes such as custards, puddings, or rich cream sauces.

A liaison is made using a 1:3 ratio, composed of 1 part egg yolk and 3 parts heavy cream. The yolk and cream are whisked together after the liquid to be thickened has been heated to a simmer and removed from the heat. A small amount of the heated liquid is slowly added to the yolk mixture, causing coagulation. *Coagulation* is the process of a protein changing from a liquid to a semisolid or a solid state when heat or friction is applied. Egg yolks begin to coagulate at temperatures above 145°F, but the addition of cream raises the coagulation point of the liaison to approximately 180°F. The difference in temperature between the liaison and a hot liquid can cause the yolks to coagulate. Therefore, only a small amount of the hot liquid to be thickened is whisked into the yolk mixture. *See Figure 8-24.*

Refer to DVD for
Adding a Liaison
Media Clip

Procedure for Adding a Liaison

1. Heat liquid to be thickened to a simmer. Do not overheat the liquid.

2. Remove the liquid from the heat.

3. Whisk together 1 part egg yolk and 3 parts heavy cream.

4. Add a small amount of the heated liquid to the liaison while whisking rapidly. Continue adding the hot liquid while whisking until the liaison is fairly warm.

5. Add the warmed liaison into the remainder of the heated liquid while whisking vigorously.

6. Carefully heat the thickened sauce to a temperature between 140°F and 185°F.

Note: A thickened sauce should have a nappe consistency.

Figure 8-24. *A liaison is a thickening agent made from a mixture of egg yolks and heavy cream that is used to thicken sauces.*

After the liaison has been whisked well, it is added to the hot liquid while whisking vigorously. If the liaison-thickened sauce needs to be heated, the temperature should be kept between 140°F and 185°F. The sauce must be heated to a minimum of 140°F to cook the egg yolks to a safe temperature for consumption. The liaison should not be heated above 185°F as the yolks will coagulate to the point of appearing curdled. If the yolks coagulate, the liaison must not be used or the sauce will also appear curdled.

CLASSICAL SAUCES

Most classical sauces are composed of a flavorful liquid, a thickening agent, and additional flavorings and seasonings. Classical sauces are often referred to as mother sauces and small sauces. A *mother sauce,* also known as a leading or grand sauce, is one of five sauces from which the small classical sauces described by Escoffier are produced. The five mother sauces are béchamel, velouté, espagnole, tomato, and hollandaise. With the exception of hollandaise, these sauces are rarely used by themselves.

Béchamel

Béchamel, also known as a cream sauce or a white sauce, is a mother sauce that is made by thickening milk with a white roux and seasonings. Traditionally, a béchamel was made by adding hot, heavy cream to a thick, reduced white veal stock. Today, the simpler version of béchamel prepared with scalded milk and a white roux is used in many vegetable and gratin dishes.

A well-made béchamel should be silky smooth and have a nappe consistency. It has a very mild flavor with a subtle taste of onion. The onion taste comes from the onion piquet that is added to the milk and then removed after the milk is scalded. An *onion piquet* is half of an onion studded with cloves and a bay leaf. *See Figure 8-25.* Then, a white roux is whisked in. The sauce is finished with a little nutmeg and seasoned to taste. A béchamel can be made lighter or heavier by adding slightly less or slightly more roux. Béchamel is strained prior to use in a dish.

Onion Piquet

Figure 8-25. *An onion piquet is half of an onion studded with cloves and a bay leaf.*

With the addition of different seasonings, ingredients, and garnishes, a simple but well-made béchamel can be transformed into countless small sauces. For example, a Mornay sauce is made by adding grated cheese to a béchamel, while a soubise sauce is made by adding puréed cooked onions to a béchamel.

Velouté

Velouté is a mother sauce made from a flavorful white stock (either veal, chicken, or fish stock) and a blonde roux. Traditionally, only veal stock was used. Today, chicken stock is typically used instead of veal stock because of its versatility and low cost. Fish stock is not often used because the flavor may overpower a dish. No matter what type of stock is used, a velouté should have a very smooth and silky texture similar to a béchamel. A velouté is typically made into a middle, or intermediate, sauce before being used for a small sauce. There are two middle sauces, suprême and allemande.

- A *suprême sauce* is a sauce made by adding cream to a chicken velouté. A suprême sauce can be wonderful by itself. It also is used as a binding sauce in many casseroles.

- An *allemande sauce* is a sauce made by adding fresh lemon juice and a yolk-and-cream liaison to a velouté. The liaison makes allemande a rich, wonderful sauce.

Daniel NYC

Suprême and allemande sauces can be used to make a large number of small sauces. Because both middle sauces are derived from velouté, almost any small sauce made from one type of middle sauce can also be made with the other type. For example, an aurora sauce is made from a velouté with a tomato purée, a bercy sauce is made from a fish velouté, and a poulette sauce is made from an allemande sauce flavored with mushroom essence.

Espagnole

Espagnole is a mother sauce made from a full-bodied brown stock, brown roux, tomato purée, and a hearty caramelized mirepoix. An espagnole is typically made into a middle sauce such as a demi-glace or jus lié.

- A *demi-glace* is a sauce made by adding equal parts of espagnole and brown stock together and reducing the mixture by half. A demi-glace is the most widely used middle sauce. This richly flavored reduction is often finished with a small amount of Madeira or sherry to further boost its flavor. A properly made demi-glace should be a nappe consistency when hot and somewhat rubbery when cold. Espagnole is converted into a demi-glace before being made into one of many small sauces. ***See Figure 8-26.***

- A *jus lié,* also known as a fond lié, is a sauce that is made by thickening a brown stock either by adding a cornstarch or arrowroot slurry or simply by a slow reduction.

Espagnole Sauces	
Sauce	**Description**
Bordelaise	Demi-glace flavored with red wine, herbs, and shallots
Châteaubriand	Demi-glace flavored with white wine, shallots, tarragon, cayenne, and lemon juice
Madeira	Demi-glace flavored with shallots, Madeira wine, and butter
Périgueux	Demi-glace flavored with Madeira wine and truffles
Robert	Demi-glace flavored with onions, white wine, vinegar, and Dijon mustard

Figure 8-26. *An espagnole is typically made into a middle sauce called demi-glace which can then be used to make many small sauces.*

Some chefs prefer a jus lié to a demi-glace because it does not have the taste of roux and is easier to make. A disadvantage of a jus lié is that it will only be as rich and flavorful as the brown stock from which it was produced because it is thickened with only a cornstarch or arrowroot slurry. In comparison, an espagnole has a richer taste because of the butter in the roux. When completed, a jus lié is not quite as thick as a demi-glace.

Tomato Sauce

A *tomato sauce* is a mother sauce made by sautéing mirepoix and tomatoes, adding white stock, and thickening with a roux. A well-made tomato sauce is thick, flavorful, and tastes like ripe tomatoes. It is coarser than the other classical sauces, but should have similarly rich characteristics that allow it to cling to the foods it is served with. Typically, plum (Roma) tomatoes are used to make tomato sauce, as they have less water, fewer seeds, and more meat than other varieties. Tomatoes are most often peeled and seeded before use in the sauce.

Some chefs do not use a roux to thicken a tomato sauce. Instead, tomato sauces are either thickened with a slurry, or are puréed or ground in a food mill, resulting in a hearty sauce. A frequent problem with modern sauces, however, is that water can separate from the puréed pulp and form an unattractive pool beneath pasta on a prepared plate. This can be avoided by using a thickening agent such as a roux or a slurry, which helps by causing the water that separates from the tomato pulp to bind with the starch. A properly thickened sauce makes an attractive presentation on a pasta dish, such as spaghetti. *See Figure 8-27.*

Many chefs rely heavily on the flavor of seasonal, ripe tomatoes cooked for a relatively short period of time to produce a rich and fresh tomato sauce. Others use canned tomato products sautéed with mirepoix, roasted pork or veal bones, and rendered bacon and cook the sauce for a long time over low heat. Some tomato sauces are chunky, while others are puréed smooth.

Tomato Sauce

Barilla America Inc.

Figure 8-27. *A well-made tomato sauce is thick, flavorful, and tastes like ripe tomatoes.*

Many chefs also add a small amount of gastrique to a tomato sauce to mask the acidic qualities of the sauce. *Gastrique* is a sugar syrup made by caramelizing a small amount of granulated sugar in a saucepan and deglazing the pan with a small amount of vinegar. The gastrique does not reduce the acid in a tomato sauce, but merely softens the harshness of the acidic qualities. To reduce acid in a tomato sauce, baking soda can be added while the sauce simmers. Tomato sauce immediately foams when baking soda is added, because of chemical reactions taking place as the soda neutralizes some of the acid. When the foam disappears, the acid in the sauce has been reduced.

Tomato sauce can be made into many small sauces, the majority of which are still called tomato sauce. Tomato sauce variations have an ethnic description attached to identify the sauce, such as Creole, Milanaise, Portuguese, or Spanish.

Hollandaise

Hollandaise is a mother sauce made by thickening melted butter with egg yolks. A classical hollandaise is started with a flavorful reduction of dry white wine, white wine vinegar, minced shallots, and freshly cracked peppercorns, reduced over medium heat. Hollandaise relies on an emulsification rather than coagulation and is therefore different from other sauces thickened with yolks. *Emulsification* is the process of temporarily binding two liquids that do not combine easily, such as oil and vinegar.

For example, emulsification occurs when warm, clarified butter is rapidly whisked into heat-tempered egg yolks. Egg yolks contain a natural emulsifier called lecithin, which coats each fat droplet, suspending the fat in the mixture. When hollandaise is made properly, each yolk should completely emulsify 4 oz of clarified butter. The resulting sauce should be light and silky, a very pale lemon color, and should have a well-rounded flavor that is more complex than that of just butter. *See Figure 8-28.*

When making a hollandaise for the first time, the sauce may break and appear thin, curdled, or separated. This most often happens either because the yolks are heated at too high of a temperature, resulting in overcooked yolks, or because butter is added too quickly and does not emulsify properly. Sometimes if a hollandaise is kept too cold it becomes thick but does not break until it is served over a warm menu item, such as grilled asparagus.

Hollandaise should be held for no more than 2 hours at 145°F. If the sauce is held above 150°F, the emulsification breaks down because egg yolks curdle and solidify at higher temperatures. If the sauce temperature falls below 100°F, the butter will solidify and the sauce will break.

To repair a broken hollandaise, the first step is to determine whether it is broken because of overcooked yolks or from holding a completed sauce at too low a temperature. If the hollandaise is broken because of overcooked yolks, add 1 yolk to a clean mixing bowl and whisk with 1 tsp of warm water until thick. Slowly add the broken hollandaise to the yolk while whisking rapidly to incorporate. Continue adding broken sauce to the mixture until all of the broken hollandaise has been incorporated.

National Turkey Federation

1. Add dry white wine and white wine vinegar, minced shallots, and freshly cracked peppercorns to a saucepan. Reduce over medium heat until almost all of the liquid has evaporated.

2. Strain and slightly cool the reduction.

3. Transfer the reduction to a stainless steel bowl. Add egg yolks and whisk vigorously until light and foamy.

4. Add a small amount of warm water (1 oz for every 6 yolks) and whisk to incorporate.

5. Place the mixture over a gently simmering water bath and whisk rapidly until the mixture reaches approximately 145°F and has at least doubled in volume. *Note:* The bowl may be momentarily removed from the water bath to allow a few seconds of thorough whisking without overcooking the yolks.

6. Remove the bowl from the water bath and place on a moistened towel. Begin gradually whisking in warm clarified butter very slowly in a fine, steady stream. As the clarified butter is incorporated, the mixture will begin to thicken. If the sauce begins to get too thick, add a small amount (1–2 tsp) of warm water to thin the sauce slightly. Continue whisking in the clarified butter until about 4 oz of clarified butter per yolk has been added.

7. Add a small amount of lemon juice, salt, white pepper, cayenne pepper, and Worcestershire to finish the sauce.

8. Serve immediately.

Figure 8-28. *Hollandaise sauce is thickened by an emulsification that occurs when warm, clarified butter is rapidly whisked into heat-tempered egg yolks.*

If the hollandaise breaks from being held at too low of a temperature, reheat the sauce over a double boiler while whisking continuously until the sauce is soft and warm. In a separate stainless steel bowl, add 2 tsp of warm water and a few tablespoons of the broken sauce. Whisk rapidly to emulsify and then slowly whisk in the remainder of the broken hollandaise sauce until all of the sauce has been added.

A béarnaise sauce is very similar to a hollandaise, except for the tarragon flavor. Béarnaise sauce can also be used to make small sauces such as a choron sauce, which is made by blending a béarnaise with a tomato purée.

Hollandaise sauce can also be used to make small sauces, such as a mousseline sauce, by blending in unsweetened heavy cream. A Maltaise sauce can be made by blending in Maltese orange juice and grated orange rind.

BUTTER SAUCES

Not all sauces are derived from the classical mother sauces. There are also many sauces made from butter. Butter sauces may be as simple as browned butter or as complex as an emulsion. Three common types of butter sauces are compound butters, beurre blanc, and broken butters.

Compound Butters

A *compound butter* is a flavorful butter sauce made by mixing cold, softened butter with flavoring ingredients such as fresh herbs, garlic, vegetable purées, dried fruits, preserves, or wine reductions. Typically, compound butters are prepared, rolled into a cylinder shape on a sheet of plastic wrap or parchment paper, and refrigerated until needed. When needed, a ¼–½ inch thick disk is cut and then placed on top of a freshly cooked item. The dish is served immediately so that the butter finishes melting in front of the customer. *See Figure 8-29.*

Compound butters are often prepared in advance and held in the refrigerator for a few days or in the freezer for longer periods. *See Figure 8-30.* Compound butters may be served alone or over grilled meats, fish, or vegetable dishes. They can also be added to other sauces for additional flavor. Although the variations of compound butters are virtually endless, there are several common preparations. *See Figure 8-31.*

Compound Butters

Figure 8-29. *A compound butter is chilled and served immediately after being placed on top of a freshly cooked item so that the butter finishes melting in front of the customer.*

Procedure for Preparing Compound Butters

1. In a food processor, combine and process the flavoring ingredients.
2. Add whole butter, salt, and pepper, and purée until all ingredients are well incorporated.
3. Remove butter and place on a sheet of parchment or plastic wrap. Roll the mixture into a cylinder, approximately 1–1½ inch thick.
4. Place in refrigerator until needed, or freeze for later use.

Figure 8-30. *Compound butters are often prepared in advance and refrigerated or frozen until needed.*

Compound Butters	
Butter	**Preparation**
Almond	Finely grind 4 oz of almonds until very smooth; add 8 oz of whole butter and mix well
Anchovy	Soak 1 oz of anchovies in cold water for 15 minutes to remove excess oil and salt; pat anchovies with paper towels to dry; place anchovies in food processor with 8 oz of whole butter; purée until smooth
Bercy	Reduce 1 cup of white wine and 2 minced shallots until only about 2 tbsp of liquid remain; cool mixture; add 8 oz of whole butter and 2 tbsp of diced parsley; purée until smooth
Colbert	Place 1 tbsp of chopped fresh tarragon, 3 tbsp of glace de viande, 3 tbsp of chopped fresh parsley, and 1 tsp of lemon juice in a food processor and purée until smooth
Herb	Add 2–4 tbsp of desired chopped fresh herb to 8 oz of whole butter and purée until smooth
Lobster, shrimp, or crayfish	In a meat grinder, grind 6 oz of lobster, shrimp, or crayfish shells with 8 oz of whole butter; place in small saucepan over low heat until butter melts; keep on very low heat for 30 minutes and strain; chill the melted butter until solid; season to taste and purée until smooth
Maître d'hôtel	Place 3 tbsp of freshly chopped parsley, 1 tsp of fresh lemon juice, and 8 oz of whole butter in a food processor and purée until smooth (*Note:* this is the most commonly used compound butter.)
Marchand de Vin	Combine 1 cup of dry red wine, 1 tbsp of glace de viande, 1 minced shallot, 1 clove garlic, and 6 cracked black peppercorns; reduce until about 2 tbsp of liquid remain and chill; blend in a food processor with 8 oz of whole butter

Figure 8-31. *Compound butters can be made with flavoring ingredients such as fresh herbs, garlic, vegetable purées, dried fruits, preserves, or wine reductions.*

Beurre Blanc

A *beurre blanc,* also known as white butter sauce, is a butter-based emulsified sauce made by whisking cold, softened butter into a wine, white-wine vinegar, shallot, and peppercorn reduction. The butter becomes the sauce, while the shallots and wine are the primary flavoring ingredients. Beurre blanc sauce is lighter and thinner than hollandaise but slightly thicker than heavy cream. It is often served with fish or seafood.

In the traditional preparation, after a fish is poached, a small amount of the poaching liquid is reduced with shallots, white wine, and a small amount of white-wine vinegar. The reduction is removed from the stove and small pieces of whole butter are whisked into the reduction. Trace amounts of lecithin and other emulsifiers naturally found in butter help to create an emulsification. The key to a successful beurre blanc sauce is constantly whisking while adding the butter to the reduction. *See Figure 8-32.*

The simplicity of a beurre blanc makes it a popular sauce for many different dishes. A beurre blanc sauce can be easily transformed into similar sauces by substituting or adding different flavoring ingredients. *See Figure 8-33.*

Broken Butters

A *broken butter* is a butter sauce made by heating butter until the fat, milk solids, and water separate or "break." Broken butters are finished with a small amount of an acid, such as lemon juice or wine vinegar. Broken butter sauces are easy to prepare and are intended to be made à la minute, reusing the pan that an item was cooked in. By reusing the pan, flavors from the dish are incorporated into the sauce. A classic dish served with a broken butter is Dover sole à la Munière.

1. In a small saucepan, combine dry white wine, white wine vinegar, shallots, and a few whole peppercorns.

2. Over medium heat, reduce the mixture until only a few tablespoons of liquid remain. If too much of the reduction remains, the resulting sauce will be too thin.

3. Add tablespoon-size pieces of cold butter over a very low heat while whisking constantly. *Note:* Cold butter is whisked into the mixture to keep the sauce between 100°F and 120°F. At temperatures above 120°F, the emulsification can break down. At temperatures below 100°F, the sauce becomes too thick.

4. Strain the sauce with a chinois or cheesecloth-lined china cap.

5. If desired, finish with a small amount of warm heavy cream.

Figure 8-32. *The three main components of a beurre blanc sauce are butter, shallots, and wine.*

Refer to DVD for
Preparing a Beurre Blanc
Media Clip

Beurre Blanc Sauces	
Sauce	**Description**
Beurre citron	The juice of 1 lemon per pound of butter is substituted for the wine and vinegar; a small amount of lemon zest is added to finish the sauce
Beurre nantais	Muscadet wine is substituted for the dry white wine
Beurre rouge	Dry red wine and red wine vinegar are substituted for the white wine and white wine vinegar
Dill beurre blanc	Fresh dill is added to the beurre blanc just prior to service; this sauce is often finished with a touch of lemon juice or zest
Peppercorn beurre blanc	The strained beurre blanc is finished with cracked pink, green, or black peppercorns

Figure 8-33. *A beurre blanc sauce can be easily transformed by substituting or adding different flavoring ingredients.*

Almost any flavoring ingredient can be added to a broken butter sauce to change the flavor profile. For example, chopped herbs, julienned leeks, crushed garlic, or sliced truffles can each be added to give a broken butter sauce an exciting flavor. Two common broken butters are beurre noisette and beurre noir.

- Beurre noisette is commonly referred to as "nut-brown butter" because of its color and nutty aroma. To prepare a beurre noisette, whole butter is cooked over medium heat until the white milk solids begin to turn brown. As the solids change color, they add a toasted nut flavor to the butter. The butter is then allowed to cool slightly and fresh lemon juice is added. The juice from one lemon is sufficient for one cup of butter.

- Beurre noir is commonly referred to as "black butter." In a beurre noir, the lemon juice is replaced with wine vinegar and the butter is cooked a bit longer than a beurre noisette, allowing the milk solids to turn a rich, dark brown. Different kinds of vinegar may be used, however, sherry vinegar, champagne vinegar, and balsamic vinegar provide the best overall flavors.

CONTEMPORARY SAUCES

Items traditionally called condiments are now viewed as sauces. Although classical sauces and butter sauces remain popular, many of these contemporary sauces offer a more healthful alternative and have gained acceptance on menus around the globe. Common contemporary sauces include salsas, relishes, pestos, chutneys, coulis, nages, flavored oils, and foams.

Salsas, Relishes, and Pestos

Some sauces are more often thought of as condiments. Salsas and relishes are simply made by mixing diced vegetables or fruits, herbs, and spices together. Salsas offer low fat content and high nutrient values. *See Figure 8-34.* Both salsas and relishes have textures that range from coarse to puréed. Similarly, a pesto is made from fresh ingredients that have been either crushed with a mortar and pestle or finely chopped in a food processor before being mixed with olive oil. Cilantro, basil, mint, pine nuts, and parmesan cheese are common pesto ingredients.

Chutneys and Coulis

The ingredients in chutneys are cooked with sugar and spices to yield a sweet-and-sour flavor. Like jams and jellies, chutneys can be smooth or a chunky. Popular chutney ingredients include apples, cilantro, cinnamon, cloves, coconuts, garlic, jalapenos, lemons, limes, mangoes, peaches, plums, tamarind, and tomatoes.

A *coulis* is a sauce typically made from either raw or cooked puréed fruits or vegetables. A coulis can be served either warm or cold over grilled or sautéed items, as well as desserts. *See Figure 8-35.* The texture of a coulis can range from fairly smooth to slightly coarse. A fruit coulis is usually made from frozen fruit that is puréed with simple syrup and a hint of lemon or vanilla.

Salsas

Figure 8-34. *Salsas offer nutritious options that are low in fat.*

Coulis

Coulis

Figure 8-35. *A coulis can be served either warm or cold over grilled or sautéed items, as well as desserts.*

Nages

A *nage* is an aromatic court bouillon that is often used as a finishing sauce. A nage is prepared by reserving and straining the liquid in which the main item was cooked. Vegetables and aromatics are added to the broth, and the broth is reduced by up to half of the original volume. The resulting liquid is an aromatic broth that is often served to accompany fish and shellfish. Some chefs choose to clarify the broth, while others may whisk in whole butter or a bit of heavy cream to add richness. Only a small amount of a nage is needed to add moisture to a dish. *See Figure 8-36.*

Nages

Nage

Figure 8-36. *As a finishing sauce, a nage adds an aromatic quality to a dish.*

Flavored Oils

Flavored oils can be used to add flavor, moisture, color, and aroma to a dish. Typically, a neutral-flavored oil such as canola or grape seed oil is used, although olive oil can be used if the flavor pairs well with the flavoring ingredients. Blanching the flavoring ingredients in salt water gives them a more vibrant color, resulting in a more colorful oil. *See Figure 8-37.* Since flavored oils are generally used in small amounts, only a small amount of fat and calories are added to the finished dishes.

Procedure for Preparing Flavored Oils

1. In salted water, quickly blanch any herbs or vegetables to be used to retain their color. *Note:* Spices are not blanched.
2. Remove flavoring ingredients from the hot water and immediately shock in ice water to stop the cooking process.
3. After the ingredients are cool, remove them from the ice water and squeeze out any excess water.
4. Place all herbs, spices, and vegetables in a blender with the desired oil and purée until no portions of the flavoring ingredients remain as visible solids.
5. Strain the oil through a chinois and reserve until needed.

Figure 8-37. *Flavored oils are often made from blanched flavoring ingredients in order to enhance the color of the resulting oil.*

Foams

Foams are produced by making a reduction of a flavoring ingredient, shallots, garlic, and wine. A small amount of lecithin powder is added as a stabilizer and then the mixture is puréed. This reduced sauce has a very light texture. A touch of cream can be added if desired. The sauce is then whipped with a stick mixer until foamy. The resulting foam is spooned over an item or drizzled on the plate. *See Figure 8-38.*

Foams

Foam

Daniel NYC

Figure 8-38. *Foams are spooned over an item or drizzled on the plate.*

SUMMARY

Flavorful and intense stocks are produced from simple flavoring ingredients and water, yet they form the foundation of classical cuisine. Stocks are either reduced or thickened to make sauces or soups. Whether it is a classical sauce such as hollandaise or a contemporary salsa, the primary purpose of a sauce is to complement and enhance the flavor of a dish. With practice and quality ingredients a chef can produce extraordinary stocks and sauces to complement a vast array of foods.

Refer to DVD for
Quick Quiz® questions

Review

Refer to DVD for
Review Questions

1. Identify the components of a stock and their respective proportions.
2. List the eight general guidelines for preparing stocks.
3. Describe the procedure for preparing brown stock.
4. List the main ingredients of a white stock.
5. Describe the procedure for preparing fish stock.
6. Explain how an essence differs from a fumet.
7. Describe the procedure for preparing vegetable stock.
8. Explain how to store a glace.
9. Explain the primary use of a remouillage.
10. Describe a bouillon and a court bouillon.
11. Explain why it is important that sauces have a nappe consistency.
12. Explain the process of reduction.
13. Define thickening agent.
14. Define gelatinization.
15. Contrast six thickening agents used to prepare sauces.
16. Explain the purpose of a slurry.
17. Describe the procedure for preparing a roux.
18. Contrast the different uses for white, blonde, and brown roux.
19. Contrast a beurre manié and a roux.
20. Describe the procedure for adding a liaison to a sauce.
21. Describe the five classical mother (leading or grand) sauces.
22. Identify the main ingredient in a béchamel sauce.
23. Identify two middle sauces made from a velouté.
24. Name the middle sauce most often made from an espagnole.
25. Contrast a jus lié and a demi-glace.
26. Explain why plum (Roma) tomatoes are often used to make tomato sauce.
27. Explain how a gastrique is made.
28. Explain the role of emulsification in making a hollandaise sauce.
29. Describe the procedure for preparing a hollandaise sauce.
30. Explain how to repair a broken hollandaise.
31. Identify the ingredient in a béarnaise sauce that is not in a hollandaise sauce.
32. Identify the three common types of butter sauces.

33. Describe the procedure for preparing a compound butter.
34. Explain how a beurre blanc is made.
35. Explain what is meant by "broken" butter.
36. Identify eight contemporary sauces.
37. Identify the types of ingredients used to make a coulis.
38. Describe how a nage is prepared.
39. Explain the advantages of using flavored oils.
40. Describe how foams are produced.

Applications

1. Prepare a brown stock and a white stock. Compare the appearance and flavor of each stock.
2. Prepare a fish stock and a vegetable stock. Compare the appearance and flavor of each stock.

Refer to DVD for
Application Scoresheets

3. Prepare an essence and a fumet using the same main ingredient. Compare the appearance, flavor, and texture of the two stocks.
4. Prepare a glace and a remouillage. Compare the appearance and flavor of the two stocks.
5. Contrast the appearance and flavor of a bouillon and a court bouillon.
6. Reduce a stock. Evaluate the appearance and flavor of the reduction.
7. Prepare a white, blonde, and brown roux. Evaluate each roux for appearance, flavor, and texture.
8. Prepare a slurry using cornstarch and another slurry using arrowroot. Compare the slurries.
9. Demonstrate how to add a liaison to a liquid.
10. Prepare a hollandaise sauce. Evaluate the sauce for appearance, flavor, and texture of the sauce.
11. Prepare a compound butter, a beurre blanc, and a broken butter. Compare the appearance, flavor, and texture of each butter sauce.
12. Prepare a fruit or vegetable salsa, relish, and chutney. Contrast the appearance and flavor of these contemporary sauces.
13. Prepare a nage. Evaluate the appearance, flavor, and texture of the sauce.
14. Prepare two flavored oils. Compare the appearance and flavor of the oils.
15. Prepare two foams. Compare the appearance and flavor of the foams.

Clear Vegetable Stock

Yield: *32 servings, 1 fl oz each*

Ingredients

vegetable oil	2 tsp
mirepoix	
onions, medium dice	4 oz
carrots, medium dice	2 oz
celery, medium dice	2 oz
leek	2 oz
turnip, cut small dice	2 oz
fennel, cut small dice	2 oz
parsnips, cut small dice	2 oz
mushroom stems, sliced thin	1 oz
garlic cloves, peeled and	
rough chopped	1 tsp
tomato, rough chopped	1 oz
water	36 fl oz
sachet d'épices	
bay leaf	1 ea
parsley stems	3 ea
dried thyme	⅛ tsp
peppercorns, cracked	2 ea

Preparation

1. Place oil in saucepot over medium heat.
2. Add mirepoix, leek, turnip, fennel, parsnips, mushrooms, and garlic and sweat until onions are translucent.
3. Add tomatoes and cook for 1 minute more.
4. Add water and sachet d'épices and simmer 30–45 minutes until vegetables are tender.
5. Strain through a cheesecloth-lined chinois.
6. Chill completely and refrigerate until needed.

> **NUTRITION FACTS**
> **Per serving:** 9 calories, 3 calories from fat, <1 g total fat, 0 mg cholesterol, 6.5 mg sodium, 41.6 mg potassium, 1.5 g carbohydrates, <1 g fiber, <1 g sugar, <1 g protein

Brown Vegetable Stock

Yield: *32 servings, 1 fl oz each*

Ingredients

vegetable oil	2 tsp
mirepoix	
onions, medium dice	4 oz
carrots, medium dice	2 oz
celery, medium dice	2 oz
leek, sliced	2 oz
turnip, cut small dice	2 oz
fennel, cut small dice	2 oz
mushroom stems, sliced thin	1 oz
garlic cloves, peeled and rough	
chopped	1 tsp
tomato, rough chopped	2 oz
tomato paste	1 oz
onion brûlé	1 ea
water	36 fl oz
sachet d'épices	
bay leaf	1 ea
parsley stems	3 ea
dried thyme	⅛ tsp
peppercorns, cracked	2 ea

Preparation

1. Place oil in saucepot over medium heat.
2. Add mirepoix, leeks, turnips, fennel, mushrooms, and garlic and sweat until onions are translucent.
3. Continue to cook over medium-high heat until vegetables are caramelized.
4. Add tomato and tomato paste and cook until slightly caramelized.
5. Add onion brûlé, water, and sachet d'épices and simmer 45 minutes until vegetables are tender.
6. Strain through a cheesecloth-lined chinois.
7. Chill completely and refrigerate until needed.

> **NUTRITION FACTS**
> **Per serving:** 20 calories, 3 calories from fat, <1 g total fat, 0 mg cholesterol, 29.6 mg sodium, 137.9 mg potassium, 3.9 g carbohydrates, 1.1 g fiber, 1.3 g sugar, <1 g protein

Basic Dashi Stock

Yield: *5 servings, 6 fl oz each*

Ingredients

water	1 qt
kombu	6-inch piece
katsuobushi	1 oz
soy sauce	½ fl oz
mirin	1 tsp
rice vinegar	1 tsp

Preparation

1. Soak the kombu overnight in cold water. Drain.
2. Add the kombu to clean water and simmer over medium heat for 15 minutes.
3. Strain the liquid into a bowl. Discard the kombu.
4. Immediately add the katsuobushi to the liquid and stir until the flakes settle at the bottom (approximately 1–2 minutes).
5. Strain the liquid into another bowl through a chinois lined with a coffee filter.
6. Immediately stir in soy sauce, mirin, and rice vinegar.
7. Cover and refrigerate until needed. Note: Stock should be used within 24 hours.

Variation: To make vegan dashi stock, omit the katsuobushi and use a 9-inch piece of kombu.

NUTRITION FACTS
Per serving: 27 calories, 1 calories from fat, <1 g total fat, 8.6 mg cholesterol, 529 mg sodium, 101.4 mg potassium, 1.2 g carbohydrates, <1 g fiber, <1 g sugar, 3.8 g protein

Court Bouillon

Yield: *5 servings, 6 fl oz each*

Ingredients

butter	1 tbsp
mirepoix	
onions, sliced thin	4 oz
celery, sliced thin	2 oz
carrots, sliced thin	2 oz
white leeks, sliced thin	1 oz
fennel, sliced thin	1 oz
garlic clove, peeled	1 ea
kosher salt	½ oz
peppercorns, crushed	2 ea
bay leaf	1 ea
fresh thyme	1 sprig
parsley stems	4 ea
water, cold	1 qt
dry white wine	1 fl oz
white wine vinegar	1 fl oz
lemon juice	1 fl oz

Preparation

1. Add butter to saucepot over low heat and sweat mirepoix, leeks, fennel, and garlic for about 10 minutes (without browning).
2. Add salt, pepper, bay leaf, thyme, parsley stems, and cold water and simmer for 10 minutes.
3. Add the wine, vinegar, and lemon juice and simmer for 20 additional minutes.
4. Remove from heat and let cool slightly.
5. Strain court bouillon through a chinois lined with a coffee filter.
6. Refrigerate bouillon until needed.

NUTRITION FACTS
Per serving: 90 calories, 22 calories from fat, 2.6 g total fat, 6.1 mg cholesterol, 1117.3 mg sodium, 355.7 mg potassium, 15.5 g carbohydrates, 3.2 g fiber, 6.4 g sugar, 1.8 g protein

Béchamel Sauce

Yield: *16 servings, 2 fl oz each*

Ingredients

milk	1 qt
onion piquet	1 ea
clarified butter	2 fl oz
pastry or cake flour	2 oz
salt and white pepper	TT

Preparation

1. Add milk and onion piquet to a heavy-bottomed stainless steel saucepan. Scald milk.
2. In a separate pan, heat clarified butter and flour together to make a white roux. Allow the roux to cool slightly.
3. Remove onion piquet from the scalded milk. Add white roux to the milk and whisk.
4. Bring sauce to a boil. Lower heat and simmer the sauce for at least 30 minutes, stirring occasionally to prevent scorching.
5. Strain sauce through a chinois or cheesecloth-lined china cap.
6. Season to taste.

NUTRITION FACTS
Per serving: 80 calories, 46 calories from fat, 5.2 g total fat, 14.3 mg cholesterol, 44.7 mg sodium, 90.1 mg potassium, 6.1 g carbohydrates, <1 g fiber, 3.3 g sugar, 2.3 g protein

Cream Sauce

Yield: *24 servings, 2 fl oz each*

Ingredients

shallots, minced	2 oz
clarified butter	2 tbsp
white wine	4 fl oz
heavy cream	8 fl oz
béchamel sauce	1 qt
salt and white pepper	TT

Preparation

1. Sauté shallots in clarified butter in a heavy-bottomed stainless steel saucepan until translucent.
2. Deglaze with white wine and reduce by half.
3. Add heavy cream and bring to a simmer.
4. Add prepared béchamel sauce and return to a simmer.
5. Season to taste.

NUTRITION FACTS
Per serving: 103 calories, 72 calories from fat, 8.2 g total fat, 25.8 mg cholesterol, 46.2 mg sodium, 79 mg potassium, 4.9 g carbohydrates, <1 g fiber, 2.2 g sugar, 1.8 g protein

Caper Sauce

Yield: *20 servings, 2 fl oz each*

Ingredients

shallots, minced	2 oz
capers	2 tbsp
clarified butter	2 tbsp
white wine	3 fl oz
heavy cream	4 fl oz
béchamel sauce	1 qt
salt and white pepper	TT

Preparation

1. Sauté shallots and capers in clarified butter in a heavy-bottomed stainless steel saucepan until shallots are translucent.
2. Deglaze with white wine and reduce by half.
3. Add heavy cream and prepared béchamel sauce and bring to a simmer.
4. Season to taste.

NUTRITION FACTS
Per serving: 102 calories, 68 calories from fat, 7.7 g total fat, 22.9 mg cholesterol, 78.6 mg sodium, 89.6 mg potassium, 5.7 g carbohydrates, <1 g fiber, 2.7 g sugar, 2 g protein

Cheddar Cheese Sauce

Yield: *20 servings, 2 fl oz each*

Ingredients

béchamel sauce	1 qt
cheddar cheese, grated	8 oz
Worcestershire sauce	1 tsp
dry mustard	1 tsp
processed cheese food (optional)	2 oz
salt and white pepper	TT

Preparation

1. Bring béchamel sauce to a simmer in a heavy-bottomed stainless steel saucepan.
2. Remove from heat and add cheddar cheese, Worcestershire sauce, and dry mustard. Whisk until smooth.
3. If desired, add processed cheese food for extra smoothness.
4. Season to taste.

NUTRITION FACTS
Per serving: 114 calories, 71 calories from fat, 8 g total fat, 23.5 mg cholesterol, 158.3 mg sodium, 95.4 mg potassium, 5.4 g carbohydrates, <1 g fiber, 2.9 g sugar, 5.3 g protein

Mornay Sauce

Yield: *9 servings, 4 fl oz each*

Ingredients

béchamel sauce	1 qt
Gruyère cheese, grated	4 oz
Parmesan cheese, grated	2 oz
heavy cream	2 tbsp
salt and white pepper	TT

Preparation

1. Bring béchamel sauce to a simmer in a heavy-bottomed stainless steel saucepan.
2. Remove from heat and whisk in Gruyère and Parmesan until smooth. *Note:* If necessary, thin the sauce by adding a little scalded heavy cream.
3. Season to taste.

NUTRITION FACTS
Per serving: 233 calories, 145 calories from fat, 16.4 g total fat, 49.4 mg cholesterol, 251.7 mg sodium, 180.8 mg potassium, 11.3 g carbohydrates, <1 g fiber, 5.9 g sugar, 10.3 g protein

Creamy Dijon Sauce

Yield: *18 servings, 2 fl oz each*

Ingredients

shallots, minced	2 oz
clarified butter	2 tbsp
white wine	3 fl oz
béchamel sauce	1 qt
Dijon mustard	2 oz
salt and white pepper	TT

Preparation

1. Sauté shallots in clarified butter in a heavy-bottomed stainless steel saucepan until shallots are translucent.
2. Deglaze with white wine and reduce by half.
3. Add béchamel sauce and Dijon mustard and bring to a simmer.
4. Season to taste.

> **NUTRITION FACTS**
> **Per serving:** 92 calories, 55 calories from fat, 6.2 g total fat, 16.3 mg cholesterol, 96 mg sodium, 98.3 mg potassium, 6.3 g carbohydrates, <1 g fiber, 2.9 g sugar, 2.3 g protein

Soubise Sauce

Yield: *16 servings, 2 fl oz each*

Ingredients

onions, finely diced	1 lb
clarified butter	2 tbsp
béchamel sauce	1 qt
salt and white pepper	TT

Preparation

1. Sweat onions in clarified butter in a heavy-bottomed stainless steel saucepan until translucent.
2. Add prepared béchamel sauce and simmer the mixture until the onions are tender.
3. Strain sauce through a chinois or cheesecloth-lined china cap.
4. Season to taste.

> **NUTRITION FACTS**
> **Per serving:** 105 calories, 60 calories from fat, 6.8 g total fat, 18.4 mg cholesterol, 64 mg sodium, 131.6 mg potassium, 8.8 g carbohydrates, <1 g fiber, 4.5 g sugar, 2.6 g protein

Tomato Soubise Sauce

Yield: *18 servings, 2 fl oz each*

Ingredients

onions, finely diced	1 lb
clarified butter	2 tbsp
tomato paste	4 oz
béchamel sauce	1 qt
salt and white pepper	TT

Preparation

1. Sweat onions in clarified butter in a heavy-bottomed stainless steel saucepan until translucent.
2. Add tomato paste to sweated onions and cook for 3 minutes.
3. Add prepared béchamel sauce and simmer the mixture until the onions are tender.
4. Strain sauce through a chinois or cheesecloth-lined china cap.
5. Season to taste.

> **NUTRITION FACTS**
> **Per serving:** 99 calories, 54 calories from fat, 6.1 g total fat, 16.3 mg cholesterol, 63.1 mg sodium, 180.8 mg potassium, 9 g carbohydrates, <1 g fiber, 4.7 g sugar, 2.6 g protein

Velouté Sauce

Yield: *16 servings, 2 fl oz each*

Ingredients

clarified butter	2 fl oz
pastry or cake flour	2 oz
light stock	1 qt
salt and white pepper	TT

Preparation

1. Add clarified butter and flour to a heavy-bottomed stainless steel saucepan. Cook to make a blonde roux. Remove from heat and allow to rest for 10 minutes.
2. Return roux to medium heat and slowly add light stock to roux while whisking.
3. Bring sauce to a boil. Lower heat and simmer the sauce for at least 30 minutes, stirring occasionally to prevent scorching.
4. Strain sauce through a chinois or cheesecloth-lined china cap.
5. Season to taste.

> **NUTRITION FACTS**
> **Per serving:** 63 calories, 35 calories from fat, 3.9 g total fat, 10 mg cholesterol, 104.1 mg sodium, 66.9 mg potassium, 4.9 g carbohydrates, <1 g fiber, <1 g sugar, 1.8 g protein

Suprême Sauce

Yield: *16 servings, 2 fl oz each*

Ingredients

velouté sauce	1 qt
heavy cream	1 c
salt and white pepper	TT

Preparation

1. Add velouté to a heavy-bottomed stainless steel saucepan and simmer until slightly reduced.
2. Slowly whisk in heavy cream until well incorporated.
3. Strain sauce through a chinois or cheesecloth-lined china cap.
4. Season to taste.

> **NUTRITION FACTS**
> **Per serving:** 114 calories, 83 calories from fat, 9.4 g total fat, 30.4 mg cholesterol, 127.9 mg sodium, 78.1 mg potassium, 5.4 g carbohydrates, <1 g fiber, <1 g sugar, 2.1 g protein

Allemande Sauce

Yield: *18 servings, 2 fl oz each*

Ingredients

veal or chicken velouté	1 qt
egg yolks	2 ea
heavy cream	6 fl oz
lemon juice	1 ½ tsp
salt and white pepper	TT

Preparation

1. Add veal or chicken velouté to a heavy-bottomed saucepan and simmer.
2. In a stainless steel mixing bowl, add yolks and heavy cream. Whisk to blend thoroughly.
3. Temper the yolks by adding a small amount of hot velouté to the liaison mixture while whisking vigorously to prevent yolks from coagulating.
4. Pour the tempered liaison into the hot velouté slowly while whisking continuously to prevent coagulation.
5. Bring the sauce to a simmer and remove from heat. Do not heat over 180°F.
6. Strain the sauce through a chinois or cheesecloth-lined china cap.
7. Add lemon juice and season to taste.

> **NUTRITION FACTS**
> **Per serving:** 96 calories, 68 calories from fat, 7.7 g total fat, 42.5 mg cholesterol, 113.3 mg sodium, 69.4 mg potassium, 4.8 g carbohydrates, <1 g fiber, <1 g sugar, 2.1 g protein

Aurore Sauce

Yield: *18 servings, 2 fl oz each*

Ingredients

tomato paste	6 oz
clarified butter	2 tbsp
allemande or suprême sauce	1 qt
salt and white pepper	TT

Preparation

1. Sweat tomato paste in clarified butter.
2. Add allemande or suprême sauce and bring to a simmer.
3. Season to taste.

> **NUTRITION FACTS**
> **Per serving:** 106 calories, 73 calories from fat, 8.3 g total fat, 41.4 mg cholesterol, 126.2 mg sodium, 157.6 mg potassium, 6 g carbohydrates, <1 g fiber, 1.9 g sugar, 2.3 g protein

Bercy Sauce

Yield: *18 servings, 2 fl oz each*

Ingredients

shallots, minced	2 oz
clarified butter	2 tbsp
white wine	½ c
allemande or fish velouté sauce	1 qt
butter	3 tbsp
fresh parsley, chopped	1 tbsp
lemon juice	1 tbsp
salt and white pepper	TT

Preparation

1. Sweat shallots in clarified butter.
2. Deglaze with white wine and reduce by half.
3. Add allemande or fish velouté sauce and bring to a simmer.
4. Whisk in the butter, parsley, and lemon juice.
5. Season to taste.

> **NUTRITION FACTS**
> **Per serving:** 123 calories, 90 calories from fat, 10.1 g total fat, 46.5 mg cholesterol, 118 mg sodium, 79.6 mg potassium, 5 g carbohydrates, <1 g fiber, <1 g sugar, 2 g protein

Poulette Sauce

Yield: *20 servings, 2 fl oz each*

Ingredients

mushrooms, diced	1 c
water	1 c
allemande sauce	1 qt
lemon juice	1 tbsp
fresh parsley, chopped	2 tbsp
whole butter	2 tbsp
salt and white pepper	TT

Preparation

1. Simmer mushrooms in water until mixture is reduced by one-fourth.
2. Add allemande sauce and bring to a simmer.
3. Whisk in the lemon juice, parsley, and whole butter.
4. Season to taste.

> **NUTRITION FACTS**
> **Per serving:** 88 calories, 64 calories from fat, 7.3 g total fat, 37 mg cholesterol, 106.1 mg sodium, 70 mg potassium, 4 g carbohydrates, <1 g fiber, <1 g sugar, 1.8 g protein

Dugléré Sauce

Yield: *20 servings, 2 fl oz each*

Ingredients

mushrooms, small dice	4 oz
tomatoes, diced	2 oz
clarified butter	2 tbsp
white wine	3 fl oz
suprême sauce	1 qt
fresh parsley, chopped	1 tbsp
salt and white pepper	TT

Preparation

1. Sauté wild mushrooms and tomatoes in clarified butter.
2. Deglaze with white wine and reduce by half.
3. Add suprême sauce and bring to a simmer.
4. Stir in fresh parsley.
5. Season to taste.

> **NUTRITION FACTS**
> **Per serving:** 97 calories, 75 calories from fat, 9.3 g total fat, 35.3 mg cholesterol, 63.3 mg sodium, 29 mg potassium, 3.6 g carbohydrates, <1 g fiber, <1 g sugar, 1.8 g protein

Curry Sauce

Yield: *18 servings, 2 fl oz each*

Ingredients

onions, brunoise	2 oz
garlic, minced	1 tbsp
clarified butter	2 tbsp
curry powder	2 tbsp
fresh thyme, minced	½ tsp
white wine	3 fl oz
allemande or suprême sauce	1 qt
fresh parsley, chopped	1 tbsp
salt and white pepper	TT

Preparation

1. Sauté onions and garlic in clarified butter until onions are translucent.
2. Add curry powder and fresh thyme. Sauté for 1 minute.
3. Deglaze with white wine and reduce by half.
4. Add allemande or suprême sauce and bring to a simmer.
5. Stir in fresh parsley.
6. Season to taste.

NUTRITION FACTS
Per serving: 106 calories, 74 calories from fat, 8.3 g total fat, 41.4 mg cholesterol, 117.9 mg sodium, 84.3 mg potassium, 5.3 g carbohydrates, <1 g fiber, 1 g sugar, 2.1 g protein

Espagnole (Brown) Sauce

Yield: *16 servings, 2 fl oz each*

Ingredients

mirepoix	
onions, small dice	4 oz
celery, small dice	2 oz
carrots, small dice	2 oz
clarified butter	2 oz
flour	2 oz
brown stock	1½ qt
tomato purée	2 oz
sachet d'épices	
bay leaf	½ leaf
thyme	⅛ tsp
parsley	2 sprigs
peppercorns, cracked	2 ea

Preparation

1. Sauté mirepoix in clarified butter until well browned.
2. Add flour and heat to make a brown roux.
3. Carefully add hot brown stock and the tomato purée. Whisk to incorporate and dissolve the roux.
4. Add the sachet d'épices and bring to a boil. Lower heat to a simmer.
5. Move saucepot over to one side of the burner, forcing the sauce to simmer only on one side of the pot. *Note:* This causes any impurities floating to the surface to be forced to one side, making them much easier to skim.
6. Simmer for 1½ hours until reduced to approximately 1 qt.
7. Strain through a chinois or cheesecloth-lined china cap.
8. Season to taste.

NUTRITION FACTS
Per serving: 59 calories, 29 calories from fat, 3.3 g total fat, 8.2 mg cholesterol, 185 mg sodium, 218.4 mg potassium, 5.3 g carbohydrates, <1 g fiber, 1.2 g sugar, 2.4 g protein

Jus Lié

Yield: *16 servings, 2 fl oz each*

Ingredients

brown veal stock	1 qt
cornstarch	1 oz
water	1 fl oz

Preparation

1. Bring brown veal stock to a simmer.
2. Mix cornstarch with water to form a slurry.
3. Whisk cornstarch slurry into stock and return stock to a simmer.
4. Simmer for 15 minutes, until sauce is completely thickened and starch flavor has disappeared.
5. Strain with a chinois or cheesecloth-lined china cap.

NUTRITION FACTS
Per serving: 15 calories, <1 calorie from fat, <1 g total fat, 0 mg cholesterol, 119 mg sodium, 111.1 mg potassium, 2.3 g carbohydrates, <1 g fiber, <1 g sugar, 1.2 g protein

Demi-Glace

Yield: *16 servings, 2 fl oz each*

Ingredients

brown stock	1 qt
espagnole sauce	1 qt

Preparation

1. Combine brown stock and espagnole sauce in saucepot over medium heat. *Note:* Offset the pot to allow simmering on only one side of the pot.
2. Continue to simmer and skim until sauce is reduced by half.
3. Strain with a chinois or cheesecloth-lined china cap.

NUTRITION FACTS
Per serving: 67 calories, 30 calories from fat, 3.4 g total fat, 8.2 mg cholesterol, 303.8 mg sodium, 329.4 mg potassium, 6 g carbohydrates, <1 g fiber, 1.5 g sugar, 3.5 g protein

Bordelaise Sauce

Yield: *18 servings, 2 fl oz each*

Ingredients

dry red wine	12 fl oz
shallots, minced	2 oz
fresh thyme	1 sprig
bay leaf	1 ea
peppercorns, cracked	½ tsp
demi-glace	1 qt
whole butter	2 oz
salt and white pepper	TT

Preparation

1. Combine dry red wine, shallots, fresh thyme, bay leaf, and peppercorns in a medium saucepan over medium heat.
2. Simmer the mixture until reduced by three-fourths.
3. Add demi-glace and simmer for 15–20 minutes.
4. Strain through a chinois or cheesecloth-lined china cap.
5. Whisk in whole butter.
6. Season to taste. *Note:* A classical bordelaise is served with a slice of poached beef marrow.

NUTRITION FACTS
Per serving: 143 calories, 57 calories from fat, 6.5 g total fat, 14.1 mg cholesterol, 208.4 mg sodium, 549.8 mg potassium, 16 g carbohydrates, 3.4 g fiber, 4.6 g sugar, 3.2 g protein

Madeira Sauce

Yield: *16 servings, 2 fl oz each*

Ingredients

Madeira wine	6 fl oz
shallots, minced	2 oz
fresh thyme	1 sprig
bay leaf	1 ea
peppercorns, cracked	½ tsp
demi-glace	1 qt
whole butter	2 oz
salt and white pepper	TT

Preparation

1. Combine the Madeira wine, shallots, fresh thyme, bay leaf, and peppercorns in a medium saucepan over medium heat.
2. Simmer the mixture until reduced by three-fourths.
3. Add demi-glace and simmer for 15–20 minutes.
4. Strain through a fine chinois.
5. Whisk in whole butter.
6. Season to taste.

NUTRITION FACTS
Per serving: 168 calories, 64 calories from fat, 7.3 g total fat, 15.8 mg cholesterol, 233.6 mg sodium, 590.7 mg potassium, 17.4 g carbohydrates, 3.8 g fiber, 5 g sugar, 3.6 g protein

Périgueux Sauce

Yield: *16 servings, 2 fl oz each*

Ingredients

Madeira wine	6 fl oz
shallots, minced	2 oz
fresh thyme	1 sprig
bay leaf	1 ea
peppercorns, cracked	½ tsp
demi-glace	1 qt
whole butter	2 oz
truffles, finely diced	½ oz
salt and white pepper	TT

Preparation

1. Combine the Madeira wine, shallots, fresh thyme, bay leaf, and peppercorns in a medium saucepan over medium heat.
2. Simmer the mixture until reduced by three-fourths.
3. Add demi-glace and simmer for 15–20 minutes.
4. Strain through a fine chinois.
5. Whisk in whole butter.
6. Stir in truffles.
7. Season to taste.

NUTRITION FACTS
Per serving: 182 calories, 87 calories from fat, 9.9 g total fat, 23 mg cholesterol, 627.9 mg sodium, 681 mg potassium, 14.2 g carbohydrates, <1 g fiber, 3.9 g sugar, 7.2 g protein

Chasseur (Hunter's) Sauce

Yield: *24 servings, 2 fl oz each*

Ingredients

mushrooms, sliced	6 oz
shallots, finely diced	2 oz
whole butter	2 tbsp
white wine	1 c
demi-glace	1 qt
tomatoes, diced	6 oz
fresh parsley, chopped	1 tbsp
salt and white pepper	TT

Preparation

1. Sauté mushrooms and shallots in whole butter.
2. Add white wine and reduce by three-fourths.
3. Add demi-glace and diced tomatoes and simmer for 15 minutes.
4. Stir in fresh parsley. Season to taste.

NUTRITION FACTS
Per serving: 107 calories, 45 calories from fat, 5.1 g total fat, 12.8 mg cholesterol, 419.1 mg sodium, 493.3 mg potassium, 9.2 g carbohydrates, <1 g fiber, 2.5 g sugar, 5.1 g protein

Champignon (Mushroom) Sauce

Yield: *24 servings, 2 fl oz each*

Ingredients

mushrooms, sliced	8 oz
clarified butter	2 tbsp
shallots, finely diced	2 oz
white wine	6 fl oz
demi-glace	1 qt
lemon juice	½ tsp
sherry	1 tbsp
whole butter	2 oz
salt and white pepper	TT

Preparation

1. Sauté sliced mushrooms in clarified butter until browned.
2. Add the shallots and cook for 1 minute.
3. Deglaze with white wine and reduce by three-fourths.
4. Add the demi-glace and simmer for 15 minutes.
5. Whisk in lemon juice, sherry, and whole butter.
6. Season to taste.

NUTRITION FACTS
Per serving: 123 calories, 63 calories from fat, 7.1 g total fat, 18.1 mg cholesterol, 418.8 mg sodium, 482.4 mg potassium, 9 g carbohydrates, <1 g fiber, 2.3 g sugar, 5.1 g protein

Châteaubriand Sauce

Yield: *18 servings, 2 fl oz each*

Ingredients

dry white wine	12 fl oz
shallots, minced	2 oz
demi-glace	1 qt
lemon juice	2 tsp
cayenne pepper	dash
fresh tarragon, chopped	1 tbsp
whole butter	2 oz
salt and white pepper	TT

Preparation

1. Combine dry white wine and shallots in a medium saucepan over medium heat.
2. Simmer the mixture until reduced by three-fourths.
3. Add demi-glace and simmer for 30 minutes to reduce slightly.
4. Add lemon juice, cayenne pepper, and fresh tarragon.
5. Whisk in whole butter.
6. Season to taste.

NUTRITION FACTS
Per serving: 143 calories, 57 calories from fat, 6.5 g total fat, 14.1 mg cholesterol, 208.7 mg sodium, 546.6 mg potassium, 16.1 g carbohydrates, 3.4 g fiber, 4.7 g sugar, 3.3 g protein

Marchand de Vin (Wine Merchant) Sauce

Yield: *16 servings, 2 fl oz each*

Ingredients

whole butter	2 oz
dry red wine	8 fl oz
shallots, minced	2 oz
demi-glace	1 qt
salt and white pepper	TT

Preparation

1. Combine whole butter, dry red wine, and shallots in a medium saucepan over medium heat.
2. Simmer the mixture until reduced by two-thirds.
3. Add demi-glace and bring to a simmer.
4. Strain through a fine chinois or a cheesecloth-lined china cap.
5. Season to taste.

NUTRITION FACTS
Per serving: 105 calories, 52 calories from fat, 6 g total fat, 15.3 mg cholesterol, 323.5 mg sodium, 359.9 mg potassium, 7 g carbohydrates, <1 g fiber, 1.6 g sugar, 3.7 g protein

Piquant Sauce

Yield: *18 servings, 2 fl oz each*

Ingredients

shallots, diced	2 oz
white wine	4 fl oz
white wine vinegar	4 fl oz
demi-glace	1 qt
cornichons, diced	2 oz
fresh parsley, chopped	1 tbsp
fresh tarragon, chopped	2 tsp
salt and white pepper	TT

Preparation

1. Combine shallots, white wine, and vinegar in a medium saucepan over medium heat.
2. Simmer the mixture until reduced by two-thirds.
3. Add demi-glace and simmer for 15 minutes.
4. Stir in cornichons, fresh parsley, and fresh tarragon.
5. Season to taste.

NUTRITION FACTS
Per serving: 114 calories, 35 calories from fat, 4 g total fat, 7.3 mg cholesterol, 237.4 mg sodium, 547.2 mg potassium, 17.1 g carbohydrates, 3.4 g fiber, 4.5 g sugar, 3.2 g protein

Robert Sauce

Yield: *16 servings, 2 fl oz each*

Ingredients

onions, diced	2 oz
whole butter	2 tbsp
dry white wine	1 c
demi-glace	1 qt
dry mustard	2 tsp
sugar	½ tbsp
lemon juice	1 tsp
salt and white pepper	TT

Preparation

1. Sauté onions in whole butter. Do not brown the onions.
2. Deglaze with dry white wine and reduce by two-thirds.
3. Add demi-glace and simmer for 15 minutes.
4. Strain the sauce with a chinois or cheesecloth-lined china cap.
5. Stir in dry mustard and sugar dissolved in lemon juice.
6. Season to taste.

NUTRITION FACTS
Per serving: 157 calories, 67 calories from fat, 7.6 g total fat, 19.2 mg cholesterol, 627.2 mg sodium, 675.1 mg potassium, 13.2 g carbohydrates, <1 g fiber, 3.7 g sugar, 7.2 g protein

Roquefort Sauce

Yield: *18 servings, 2 fl oz each*

Ingredients

shallots, minced	2 oz
whole butter	2 tbsp
garlic, minced	1 tsp
dry white wine	1 c
Roquefort cheese, crumbled	3 oz
demi-glace	1 qt
salt and white pepper	TT

Preparation

1. Sauté shallots in whole butter until slightly golden brown.
2. Add garlic and cook for 1 minute.
3. Deglaze with dry white wine and reduce by three-fourths.
4. Add Roquefort cheese.
5. Whisk in demi-glace.
6. Season to taste.

NUTRITION FACTS
Per serving: 143 calories, 58 calories from fat, 6.7 g total fat, 14.9 mg cholesterol, 293.6 mg sodium, 537.6 mg potassium, 15.9 g carbohydrates, 3.3 g fiber, 4.6 g sugar, 4.2 g protein

Stroganoff Sauce

Yield: *16 servings, 3 fl oz each*

Ingredients

shallots, minced	3 oz
mushrooms, sliced	6 oz
whole butter	2 tbsp
dry white wine	1 c
demi-glace	1 qt
sour cream	4 oz
fresh parsley, chopped	1 tbsp
dry mustard	½ tsp
salt and white pepper	TT

Preparation

1. Sauté shallots and mushrooms in whole butter until slightly golden brown.
2. Deglaze with dry white wine and reduce by three-fourths.
3. Add demi-glace and bring to a simmer.
4. Whisk in sour cream, fresh parsley, and dry mustard.
5. Season to taste.

NUTRITION FACTS
Per serving: 159 calories, 64 calories from fat, 7.3 g total fat, 15.7 mg cholesterol, 240.6 mg sodium, 650.8 mg potassium, 18.6 g carbohydrates, 3.9 g fiber, 5.6 g sugar, 4.1 g protein

Tomato Sauce

Yield: *8 servings, 5 fl oz each*

Ingredients

salt pork or bacon, diced	1 oz
mirepoix	
onions, small dice	4 oz
celery, small dice	2 oz
carrots, small dice	2 oz
garlic, crushed	1 clove
canned or fresh tomatoes	1 qt
canned tomato purée	1 c
granulated sugar	1 tsp
light stock	1 c
sachet d'épices	
bay leaf	1 leaf
thyme, dried	¼ tsp
parsley	3 sprigs
peppercorns, cracked	¼ tsp
baking soda	1 tsp
cornstarch	1 oz
water	1 fl oz
salt and white pepper	TT

Preparation

1. Render salt pork or bacon in heavy-bottomed saucepot until it is slightly crisp, but not browned.
2. Add mirepoix and sauté until tender but not browned.
3. Add garlic and cook for 1–2 minutes without browning.
4. Add tomatoes, tomato purée, sugar, light stock, and sachet d'épices.
5. Bring sauce to a boil and immediately lower heat. Simmer sauce for 1 hour.
6. Add baking soda and stir to incorporate. Simmer until no longer foaming.
7. Mix cornstarch with water to form a slurry.
8. Whisk in cornstarch slurry and simmer for 10 minutes.
9. Remove sachet and either pass the sauce through a food mill or purée to reach desired consistency.
10. Season to taste.

NUTRITION FACTS
Per serving: 89 calories, 31 calories from fat, 3.5 g total fat, 4 mg cholesterol, 311.4 mg sodium, 414.6 mg potassium, 13 g carbohydrates, 2.2 g fiber, 5.5 g sugar, 2.4 g protein

---

Vegetarian Tomato Sauce

Yield: *8 servings, 5 fl oz each*

Ingredients

mirepoix	
onions, small dice	4 oz
celery, small dice	2 oz
carrots, small dice	2 oz
olive oil	1 tbsp
garlic, crushed	1 clove
canned or fresh tomatoes	1 qt
canned tomato purée	1 c
granulated sugar	1 tsp
vegetable stock	1 c
sachet d'épices	
bay leaf	1 leaf
thyme, dried	¼ tsp
parsley	3 sprigs
peppercorns, cracked	¼ tsp
baking soda	1 tsp
cornstarch	1 oz
water	1 fl oz
salt and white pepper	TT

Preparation

1. Sauté mirepoix in olive oil until tender but not browned
2. Add garlic and cook for 1–2 minutes without browning.
3. Add tomatoes, tomato purée, sugar, vegetable stock, and sachet d'épices.
4. Bring sauce to a boil and immediately lower heat. Simmer sauce for 1 hour.
5. Add baking soda and stir to incorporate. Simmer until no longer foaming.
6. Mix cornstarch with water to form a slurry.
7. Whisk in cornstarch slurry and simmer for 10 minutes.
8. Remove sachet and either pass sauce through a food mill or purée to reach desired consistency.
9. Season to taste.

NUTRITION FACTS
Per serving: 87 calories, 22 calories from fat, 2.4 g total fat, <1 mg cholesterol, 421.3 mg sodium, 429.1 mg potassium, 15.3 g carbohydrates, 2.6 g fiber, 5.1 g sugar, 2.3 g protein

Creole Sauce

Yield: *10 servings, 4 fl oz each*

Ingredients

onion, small dice	4 oz
bell pepper, small dice	4 oz
celery, thinly sliced	2 oz
olive oil	1 fl oz
garlic, crushed	1 tsp
chili powder	½ tsp
cayenne pepper	¼ tsp
bay leaf	1 ea
thyme	½ tsp
hot pepper sauce	dash
prepared tomato sauce	1 qt
salt and white pepper	TT

Preparation

1. Sauté onions, peppers, and celery in olive oil until tender.
2. Add garlic and cook for 1–2 minutes without browning.
3. Add chili powder, cayenne pepper, bay leaf, thyme, and hot pepper sauce.
4. Add prepared tomato sauce and simmer for 15 minutes.
5. Season to taste.

NUTRITION FACTS
Per serving: 70 calories, 38 calories from fat, 4.3 g total fat, 1.6 mg cholesterol, 161.5 mg sodium, 222.4 mg potassium, 7.3 g carbohydrates, 1.5 g fiber, 3.1 g sugar, 1.3 g protein

Marinara Sauce

Yield: *10 servings, 5 fl oz each*

Ingredients

onion, small dice	4 oz
whole butter	3 oz
garlic, crushed	3 cloves
fresh parsley, chopped	½ c
tomatoes, diced	1 c
vegetarian tomato sauce	1 qt
salt and white pepper	TT

Preparation

1. Sauté onions in whole butter until softened.
2. Add garlic and cook for 1–2 minutes without browning.
3. Add fresh parsley and diced tomatoes and simmer for 1 minute.
4. Add vegetarian tomato sauce and simmer for 15 minutes. Season to taste.

NUTRITION FACTS
Per serving: 211 calories, 96 calories from fat, 10.9 g total fat, 18.8 mg cholesterol, 707.3 mg sodium, 768.1 mg potassium, 26.7 g carbohydrates, 4.6 g fiber, 9.1 g sugar, 4.1 g protein

Mexican Tomato Sauce

Yield: *8 servings, 5 fl oz each*

Ingredients

onion, small dice	4 oz
bell peppers, small dice	4 oz
jalapeño pepper, small dice	1 ea
vegetable oil	1 fl oz
garlic, crushed	1 tsp
chili powder	½ tsp
cumin	½ tsp
cayenne pepper	¼ tsp
coriander	¼ tsp
bay leaf	1 ea
prepared tomato sauce	1 qt
salt and white pepper	TT

Preparation

1. Sauté onions, bell peppers, and jalapeño pepper in vegetable oil until tender.
2. Add garlic and cook for 1–2 minutes without browning.
3. Add chili powder, cumin, cayenne pepper, coriander, and bay leaf.
4. Add prepared tomato sauce and simmer for 15 minutes. Season to taste.

NUTRITION FACTS
Per serving: 72 calories, 34 calories from fat, 3.8 g total fat, 0 mg cholesterol, 682.4 mg sodium, 464.4 mg potassium, 9.1 g carbohydrates, 2.5 g fiber, 6.2 g sugar, 2 g protein

Milanaise Sauce

Yield: *8 servings, 5 fl oz each*

Ingredients

mushrooms, sliced	1 c
mushrooms, diced	3 oz
garlic, crushed	2 cloves
smoked ham, cooked	5 oz
whole butter	2 tbsp
prepared tomato sauce	1 qt
fresh parsley, chopped	2 tbsp
salt and white pepper	TT

Preparation

1. Sauté both sliced and diced mushrooms, garlic, and smoked ham in whole butter until mushrooms are tender.
2. Add prepared tomato sauce and simmer for 15 minutes.
3. Stir in parsley.
4. Season to taste.

NUTRITION FACTS
Per serving: 164 calories, 87 calories from fat, 9.7 g total fat, 21.5 mg cholesterol, 577.3 mg sodium, 540.4 mg potassium, 14 g carbohydrates, 2.4 g fiber, 5.9 g sugar, 6.4 g protein

Portuguese Tomato Sauce

Yield: *8 servings, 5 fl oz each*

Ingredients

onion, small dice	4 oz
olive oil	1 tbsp
garlic, crushed	2 cloves
tomatoes, crushed or concassé	8 oz
prepared tomato sauce	1 qt
fresh parsley, chopped	2 tbsp
salt and white pepper	TT

Preparation

1. Sauté onions in olive oil until tender.
2. Add garlic and tomatoes. Simmer for 20–30 minutes.
3. Add prepared tomato sauce and simmer for 15 minutes.
4. Stir in parsley.
5. Season to taste.

NUTRITION FACTS
Per serving: 117 calories, 47 calories from fat, 5.3 g total fat, 4 mg cholesterol, 350.4 mg sodium, 510.9 mg potassium, 15.8 g carbohydrates, 2.8 g fiber, 6.9 g sugar, 2.9 g protein

Vera Cruz Tomato Sauce

Yield: *8 servings, 4 fl oz each*

Ingredients

onion, small dice	4 oz
bell pepper, small dice	4 oz
celery, thinly sliced	2 oz
mushrooms, sliced	4 oz
garlic, crushed	1 tsp
olive oil	2 fl oz
thyme	½ tsp
cayenne pepper	¼ tsp
bay leaf	1 ea
hot pepper sauce	dash
prepared tomato sauce	1 qt
green olives, sliced	3 oz
salt and white pepper	TT

Preparation

1. Sauté onions, peppers, celery, mushrooms, and garlic in olive oil until tender.
2. Add thyme, cayenne pepper, bay leaf, and hot pepper sauce.
3. Add prepared tomato sauce and simmer for 15 minutes.
4. Add green olives.
5. Season to taste.

NUTRITION FACTS
Per serving: 118 calories, 74 calories from fat, 8.5 g total fat, 0 mg cholesterol, 779.4 mg sodium, 518.8 mg potassium, 10.2 g carbohydrates, 3 g fiber, 6.6 g sugar, 2.5 g protein

Meat Sauce

Yield: *18 servings, 6 fl oz each*

Ingredients

salt pork or bacon, small dice	14 oz
ground pork or beef (or a mixture)	2 lbs
white mirepoix	
onions, small dice	8 oz
celery, small dice	4 oz
leek stems, small dice	4 oz
garlic cloves, crushed	4 ea
red wine	8 oz
canned tomatoes	2 qt
tomato puree	2 c
granulated sugar	1 tsp
white stock	2 c
roasted veal bones	10 oz
sachet d'épices	
bay leaf	2 ea
thyme, dried	½ tsp
dried oregano	3 tbsp
fresh parsley, minced	¼ c
peppercorns, cracked	½ tsp
baking soda	1 tsp
cornstarch	1 oz
water	1 fl oz
salt and white pepper	TT

Preparation

1. Render salt pork or bacon in heavy bottomed saucepot until slightly crisp. Drain excess fat.
2. Add ground meat and cook through. Drain excess fat.
3. Add mirepoix and sweat until cooked but not browned.
4. Add garlic and cook 1–2 minutes without browning.
5. Add wine and reduce by half.
6. Add tomatoes, tomato puree, sugar, stock, bones, and sachet d'épices.
7. Bring to a boil and immediately lower heat.
8. Simmer for 1 hour.
9. Add baking soda and stir to incorporate. Cook until all foam has subsided.
10. Mix cornstarch with water to form a slurry.
11. Before foam has completely disappeared, whisk in cornstarch slurry.
12. Simmer for 10 minutes.
13. Remove sachet and bones.
14. Season to taste.

NUTRITION FACTS
Per serving: 283 calories, 144 calories from fat, 16 g total fat, 56.3 mg cholesterol, 829.6 mg sodium, 686.2 mg potassium, 13.2 g carbohydrates, 2.4 g fiber, 5.6 g sugar, 19.6 g protein

Hollandaise Sauce

Yield: *16 servings, 2 fl oz each*

Ingredients

peppercorns, cracked	½ tsp
white wine vinegar	2 fl oz
water	2½ fl oz
shallot, minced	1 ea
egg yolks	4 ea
lemon juice	2 tsp
clarified butter	1 qt
Tabasco® sauce	2 shakes
Worcestershire sauce	½ tsp
salt and white pepper	TT

Preparation

1. Combine cracked peppercorns, vinegar, four-fifths of the water, and minced shallot in small saucepan and reduce by half.
2. Place yolks in a stainless steel mixing bowl and whisk until creamy in color and the volume has almost doubled.
3. Strain the vinegar reduction into the yolks and whisk rapidly to incorporate.
4. Place the mixture in the mixing bowl over a double boiler to begin heating the sauce while whisking rapidly. As the mixture cooks, it will begin to thicken. When the mixture has thickened so that the whisk leaves a thick trail behind it, the yolks have cooked enough. *Note:* Do not allow the yolks to cook too rapidly and curdle. It is acceptable to remove the bowl from the double boiler to allow thorough mixing and to prevent overcooking.
5. Remove the mixing bowl from the double boiler and place on top of a saucepot lined with a kitchen towel to stabilize the bowl.
6. Quickly whisk half of the lemon juice and the remaining lukewarm water into the cooked yolk mixture.
7. While whisking rapidly, slowly begin to drizzle the warm clarified butter in a narrow, steady stream until all of the butter has been added. Do not stop whisking until all the butter has been incorporated, or the hollandaise could separate and break. *Note:* If the mixture gets too thick at any point when adding the butter, add a small amount of lukewarm water to thin the mixture slightly.
8. Whisk in the remaining lemon juice and all of the Tabasco® and Worcestershire.
9. Season to taste.
10. Hold for service over a warm bain marie.

> **NUTRITION FACTS**
> Per serving: 481 calories, 459 calories from fat, 52.1 g total fat, 176.2 mg cholesterol, 26.3 mg sodium, 96.4 mg potassium, 4.7 g carbohydrates, <1 g fiber, <1 g sugar, 1.4 g protein

Béarnaise Sauce

Yield: *16 servings, 2 fl oz each*

Ingredients

shallots, minced	2 oz
fresh tarragon, chopped	5 tbsp
peppercorns, cracked	1 tsp
white wine vinegar	8 fl oz
hollandaise sauce	1 qt
salt and white pepper	TT

Preparation

1. Combine shallots, peppercorns, and white wine vinegar in a small saucepan and reduce by three-fourths.
2. Strain the reduction and then add the tarragon.
3. Add the vinegar reduction to hollandaise sauce, whisking rapidly to incorporate.
4. Season to taste.
5. Hold for service over a warm bain marie.

> **NUTRITION FACTS**
> **Per serving:** 169 calories, 155 calories from fat, 17.6 g total fat, 130.7 mg cholesterol, 25.3 mg sodium, 53.3 mg potassium, 2.4 g carbohydrates, <1 g fiber, <1 g sugar, 1.6 g protein

Grimrod Sauce

Yield: *16 servings, 2 fl oz each*

Ingredients

saffron	¼ tsp
hollandaise sauce	1 qt
salt and white pepper	TT

Preparation

1. Whisk saffron into warm hollandaise sauce.
2. Season to taste.
3. Hold for service over a warm bain marie.

> **NUTRITION FACTS**
> **Per serving:** 481 calories, 459 calories from fat, 52.1 g total fat, 176.2 mg cholesterol, 44.4 mg sodium, 96.7 mg potassium, 4.7 g carbohydrates, <1 g fiber, <1 g sugar, 1.4 g protein

Choron Sauce

Yield: *16 servings, 2 fl oz each*

Ingredients

tomato puree, stewed	2 oz
béarnaise sauce	1 qt
salt and white pepper	TT

Preparation

1. Stir stewed tomato puree into warm béarnaise sauce and thoroughly incorporate.
2. Season to taste.
3. Hold for service over a warm bain marie.

> **NUTRITION FACTS**
> **Per serving:** 271 calories, 232 calories from fat, 26.3 g total fat, 196.1 mg cholesterol, 327.7 mg sodium, 63.1 mg potassium, 2 g carbohydrates, <1 g fiber, <1 g sugar, 2.4 g protein

Maltaise Sauce

Yield: *16 servings, 2 fl oz each*

Ingredients

orange juice, Maltese	2 fl oz
orange zest	2 tsp
hollandaise sauce	1 qt
salt and white pepper	TT

Preparation

1. Whisk concentrated orange juice and orange zest into warm hollandaise sauce.
2. Season to taste.
3. Hold for service over a warm bain marie.

> **NUTRITION FACTS**
> **Per serving:** 489 calories, 459 calories from fat, 52.1 g total fat, 176.2 mg cholesterol, 44.6 mg sodium, 126.9 mg potassium, 6.5 g carbohydrates, <1 g fiber, 1.7 g sugar, 1.5 g protein

Mousseline Sauce

Yield: *20 servings, 2 fl oz each*

Ingredients

heavy cream	8 fl oz
hollandaise sauce	1 qt

Preparation

1. Whip the heavy cream to soft peaks.
2. Fold the whipped cream into warm hollandaise sauce.
   Note: Items dressed with mousseline sauce may be placed under the broiler, resulting in a very light golden glaze known as a glaçage.

NUTRITION FACTS
Per serving: 426 calories, 406 calories from fat, 46.1 g total fat, 157.3 mg cholesterol, 25.5 mg sodium, 86.1 mg potassium, 4.1 g carbohydrates, <1 g fiber, <1 g sugar, 1.4 g protein

Noissette Sauce

Yield: *16 servings, 2 fl oz each*

Ingredients

hazelnut butter	2 oz
hollandaise sauce	1 qt
salt and white pepper	TT

Preparation

1. Whisk hazelnut butter into warm hollandaise sauce.
2. Season to taste.
3. Hold for service over a warm bain marie.

NUTRITION FACTS
Per serving: 26 calories, 25 calories from fat, 2.9 g total fat, 7.6 mg cholesterol, 18.6 mg sodium, <1 mg potassium, <1 g carbohydrates, <1 g fiber, <1 g sugar, <1 g protein

Valois Sauce

Yield: *20 servings, 2 fl oz each*

Ingredients

glace de viande	8 fl oz
béarnaise sauce	1 qt
salt and white pepper	TT

Preparation

1. Stir glace de viande into warm béarnaise sauce.
2. Season to taste.
3. Hold for service over a warm bain marie.

NUTRITION FACTS
Per serving: 229 calories, 191 calories from fat, 21.7 g total fat, 158.4 mg cholesterol, 316.7 mg sodium, 110.3 mg potassium, 2.6 g carbohydrates, <1 g fiber, <1 g sugar, 2.6 g protein

Beurre Blanc

Yield: *8 servings, 2 fl oz each*

Ingredients

dry white wine	8 fl oz
white wine vinegar	1 fl oz
shallots, minced	1 oz
peppercorns, whole	3 ea
cold butter, 1 inch chunks	1 lb
heavy cream	2 tbsp
salt and white pepper	TT

Preparation

1. Combine dry white wine, vinegar, shallots, and peppercorns over medium heat. Reduce to a few tablespoons.
2. Whisk in chunks of cold butter. Do not let the sauce boil.
3. Strain sauce with a chinois or cheesecloth-lined china cap.
4. Whisk in a warm heavy cream.
5. Season to taste.

NUTRITION FACTS
Per serving: 447 calories, 417 calories from fat, 47.4 g total fat, 127 mg cholesterol, 46 mg sodium, 54.4 mg potassium, 1.8 g carbohydrates, <1 g fiber, <1 g sugar, <1 g protein

Beurre Rouge

Yield: *8 servings, 2 fl oz each*

Ingredients

dry red wine	8 fl oz
white wine vinegar	1 fl oz
shallots, minced	1 oz
peppercorns, whole	3 ea
cold butter, 1 inch chunks	1 lb
heavy cream	2 tbsp
salt and white pepper	TT

Preparation

1. Combine dry red wine, vinegar, shallots, and peppercorns over medium heat. Reduce to a few tablespoons.
2. Whisk in chunks of cold butter. Do not let the sauce boil.
3. Strain sauce with a chinois or cheesecloth-lined china cap.
4. Whisk in warm heavy cream.
5. Season to taste.

NUTRITION FACTS
Per serving: 448 calories, 417 calories from fat, 47.4 g total fat, 127 mg cholesterol, 45.7 mg sodium, 70.9 mg potassium, 1.8 g carbohydrates, <1 g fiber, <1 g sugar, <1 g protein

---

Mango and Roasted Red Pepper Salsa

Yield: *8 servings, 2 fl oz each*

Ingredients

red bell pepper	1 ea
jalapeno, seeded	½ ea
mango, small dice	2 ea
shallot, minced	1 ea
cilantro, chopped fine	1 tbsp
sherry vinegar	1 tbsp
sugar	½ tsp
salt and pepper	TT

Preparation

1. Flame roast the red bell pepper and jalapeno until charred on all sides.
2. Remove peppers from flame and wrap in plastic wrap. Allow peppers to steam for 10 minutes.
3. Dice the mango. Mince the shallot. Chop the cilantro.
4. Place mango, shallot, cilantro, vinegar, sugar, salt, and pepper in mixing bowl and mix well.
5. After 10 minutes, gently rinse peppers under cold water to remove burnt peel and seeds.
6. Small dice the red pepper and add to mango mixture.
7. Mince jalapeno and add to mixture as well.
8. Toss all ingredients to incorporate.
9. Let rest in refrigerator for at least 30 minutes prior to serving to develop flavors.

NUTRITION FACTS
Per serving: 74 calories, 3 calories from fat, <1 g total fat, 0 mg cholesterol, 43.9 mg sodium, 299 mg potassium, 17.7 g carbohydrates, 1.3 g fiber, 8.1 g sugar, 1.9 g protein

---

Pesto

Yield: *8 servings, 1 fl oz each*

Ingredients

fresh basil leaves, stems removed	½ lb
garlic cloves, fresh peeled	3 ea
pine nuts, toasted	2 tbsp
parmesan, fresh grated	4 oz
kosher salt	½ tbsp
olive oil	4 fl oz

Preparation

1. Place all ingredients except the oil in a food processor and pulse until finely chopped.
2. Slowly add olive oil in a fine stream while pulsing to incorporate until pesto is fairly smooth.

NUTRITION FACTS
Per serving: 208 calories, 173 calories from fat, 19.7 g total fat, 12.5 mg cholesterol, 570.9 mg sodium, 118.8 mg potassium, 2 g carbohydrates, <1 g fiber, <1 g sugar, 6.7 g protein

Cooked Asparagus and Rosemary Coulis

Yield: *16 servings, 2 fl oz each*

Ingredients

garlic, crushed	1 clove
shallot, minced	1 ea
vegetable oil	2 tbsp
asparagus	2 lb
white wine	¼ c
fresh rosemary	2 sprigs
white stock	2 fl oz
salt and white pepper	TT

Preparation

1. In a saucepan, sweat crushed garlic and minced shallot in vegetable oil.
2. Add tender portion of asparagus and sweat until bright green in color.
3. Deglaze pan with white wine.
4. Place ingredients in a blender with freshly chopped rosemary and purée.
5. Strain sauce with a chinois or cheesecloth-lined china cap. *Note:* If coulis is too thick, it can be thinned with white stock.
6. Season to taste.

NUTRITION FACTS
Per serving: 50 calories, 17 calories from fat, 1.9 g total fat, <1 mg cholesterol, 27.9 mg sodium, 205.1 mg potassium, 6.7 g carbohydrates, 1.2 g fiber, 1.2 g sugar, 2 g protein

Red Bell Pepper Coulis

Yield: *16 servings, 1 fl oz each*

Ingredients

olive oil	1 tbsp
shallots, minced	1 ½ oz
garlic, minced	1 tsp
red bell peppers, small dice	2 lbs
white wine	4 fl oz
chicken or vegetable stock	1 c
salt and pepper	TT

Preparation

1. Add oil to large sauté pan over low heat and sweat shallots and garlic until translucent.
2. Add diced red peppers and sweat until tender.
3. Deglaze sauté pan with white wine and reduce by half.
4. Add stock and simmer uncovered for 20 minutes.
5. Add mixture to food processor and season with salt and pepper.
6. Puree until smooth.
7. Strain puree through a china cap and adjust seasonings if necessary. Serve warm.

Variation: A vegetable coulis can be served cold as a sandwich spread or to accompany a charcuterie dish. For this application, finish the cold red bell pepper coulis with 1 tsp of sherry vinegar and whisk to incorporate.

NUTRITION FACTS
Per serving: 39 calories, 11 calories from fat, 1.2 g total fat, <1 mg cholesterol, 42.6 mg sodium, 150.2 mg potassium, 4.6 g carbohydrates, 1.2 g fiber, 2.7 g sugar, 1 g protein

Nage

Yield: *6 servings, 6 fl oz each*

Ingredients

butter	1 tbsp
mirepoix	
onion, sliced thin	1 oz
celery, sliced thin	½ oz
carrots, sliced thin	½ oz
white leeks, sliced thin	½ oz
fennel branches, sliced thin	½ oz
court bouillon	1 qt
softened butter	3 oz

Preparation

1. Add butter to saucepot over low heat and sweat mirepoix, leeks, and fennel for about 10 minutes (without browning).
2. Add the court bouillon and reduce slightly.
3. Remove from heat and strain using a china cap lined with a coffee filter.
4. Whisk in softened butter until fully incorporated. Serve warm.

NUTRITION FACTS
Per serving: 163 calories, 90 calories from fat, 10.3 g total fat, 26.4 mg cholesterol, 1123.7 mg sodium, 392.6 mg potassium, 16.8 g carbohydrates, 3.5 g fiber, 6.9 g sugar, 2 g protein

Szechuan Chile Garlic Oil

Yield: *8 servings, 1 fl oz each*

Ingredients

Szechuan dried chiles, chopped fine	10 ea
garlic cloves, peeled	4 ea
ginger root, chopped	1 tbsp
peanut oil	7 fl oz
sesame oil	1 fl oz

Preparation

1. Place all ingredients except the sesame oil in small saucepan over medium heat. Cook until garlic and ginger begin to sizzle.
2. Turn burner to low and let the mixture simmer gently for about 7–8 minutes, being careful not to burn it.
3. Remove pan from burner and add the sesame oil. Allow to cool until just warm.
4. Strain the mixture through a chinois and refrigerate until needed. *Note:* Oil can be used to give sautéed or stir-fry foods a smoky, spicy flavor.

NUTRITION FACTS
Per serving: 87 calories, 83 calories from fat, 9.4 g total fat, 0 mg cholesterol, <1 mg sodium, 20.8 mg potassium, 1.1 g carbohydrates, <1 g fiber, <1 g sugar, <1 g protein

Eggplant Foam

Yield: *10 servings, ½ fl oz each*

Ingredients

shallot, minced	1 ea
fresh garlic, crushed	½ tsp
whole butter	2 tbsp
eggplant, small diced	2 oz
white wine	1 tbsp
white stock	2 fl oz
heavy cream	1 tbsp
lecithin powder	⅛ tsp

Preparation

1. Sweat shallot and garlic in butter.
2. When shallots are tender, add eggplant and sweat until tender.
3. Deglaze the pan with white wine and white stock. Reduce liquid by half.
4. Add heavy cream and cook gently for 2 minutes.
5. Add lecithin powder and whisk to incorporate.
6. Using a stick mixer, purée all ingredients until very smooth and foamy.
7. Serve immediately. *Note:* Foam can be remixed if more volume is needed.

NUTRITION FACTS
Per serving: 59 calories, 27 calories from fat, 3 g total fat, 8.3 mg cholesterol, 14.4 mg sodium, 155.6 mg potassium, 7.3 g carbohydrates, <1 g fiber, <1 g sugar, 1.3 g protein

Culinary Arts
PRINCIPLES AND APPLICATIONS

United States Potato Board

Soups

*S*oups are versatile foods that can range from flavorful broths to hearty chowders. A variety of soups can be found in every culture. Soup can be a nutritious meal in itself, served as an appetizer prior to the main course, or as an accompaniment to a salad or sandwich. A wonderfully prepared soup can set the stage for an unforgettable meal, while a poorly made soup may suggest that the courses to follow will be mediocre at best. Every pot of soup can be customized to personal taste by adding or substituting one or more ingredients.

Refer to DVD
for **Flash Cards**

Chapter Objectives

1. Identify soup varieties from around the world.
2. Describe clear soups.
3. Prepare broths.
4. Clarify a consommé.
5. Describe thick soups.
6. Prepare cream soups.
7. Prepare purée soups.
8. Contrast the three varieties of specialty soups.
9. Prepare bisques.
10. Prepare chowders.

Key Terms

- **clear soup**
- **broth**
- **consommé**
- **clarify**
- **clearmeat**
- **oignon brûlé**
- **raft**
- **thick soup**
- **bisque**
- **chowder**

SOUP VARIETIES

Soup is one of the most versatile categories of food and allows chefs to be creative while demonstrating classical preparation techniques. Many soups are produced by using a high-quality stock as a base and adding other ingredients to make the dish unique. Soups can be made using fresh, seasonal ingredients or high-quality leftovers such as the meat removed from baked chicken. Countless ingredient combinations are used to create soups in every culture around the world. *See Figure 9-1.*

Soup Varieties Around the World		
Soup	**Origin**	**Key Ingredients**
Asopao de Gandules	Puerto Rico	Pigeon peas and rice
Avgolemono	Greece	Chicken, egg, and lemon
Borscht	Ukraine	Beets, onions, beef stock or water, red-wine vinegar, dill, sugar, sour cream, and optional vegetables such as cabbage, beans, and celery root
Bouillabaisse	France	Fish stock, cooked fish and shellfish, garlic, orange peel, basil, bay leaf, fennel, and saffron
Bredie	South Africa	Mutton, tomatoes, potatoes, peppercorns, and onions
Caldo Verde	Portugal	Mashed potatoes, minced collards, savoy cabbage, kale, onions, and chorizo
Callaloo	Trinidad and Tobago	Callaloo leaves or spinach, okra, crab meat, chicken stock, onions, thyme, chile pepper, and salt beef
Erwtensoep	Netherlands	Peas and sliced sausage
Fanesca	Ecuador	Figleaf gourd, pumpkin, twelve grains, and salt cod
Fruktsuppe	Norway	Grape juice, pitted prunes, raisins, lemons, sugar, water, and sabo
Fufu and Egusi	Ghana	Vegetables, meat, fish, and balls of wheat gluten
Gazpacho	Spain	Tomatoes, peppers, cucumbers, garlic, oil, and vinegar
Ginataan	Philippines	Coconut milk, rice flour, jackfruit, yams, taro root, saba, sugar, and water
Goulash	Hungary	Beef, onions, red peppers, and paprika
Harira	North Africa	Chunks of lamb, tomatoes, chickpeas, spices, and herbs
Menudo	Mexico	Beef tripe, white hominy, and onions
Minestrone	Italy	Beans, onions, celery, carrots, stock, tomatoes, and optional pasta
Miso	Japan	Fish broth, fermented soy, and dashi
Phở	Vietnam	Beef or chicken, scallions, Welsh onions, charred ginger, wild coriander, basil, cinnamon, star anise, cloves, and black cardamom
Pozole	Columbia	Hominy, pork, chilies, cabbage, oregano, cilantro, avocado, radish, and lime juice
Sambar	India	Yellow lentils, tamarind-flavored water, a variety of vegetables, sautéed curry leaves, and sambar powder
Shchi	Russia	Cabbage, beef brisket, and a variety of root vegetables
Shark Fin	China	Shark fins
Tarator	Bulgaria	Yogurt, cucumbers, garlic, nuts, dill, oil, and water
Tom Yam	Thailand	Stock, lemon grass, kaffir lime leaves, galangal, shallots, lime juice, fish sauce, tamarind, and crushed chilies
Trahana	Turkey	Cracked wheat, yogurt, and dried fermented vegetables
Vatapá	Brazil	Seafood and coconut milk
Vichyssoise	France	Potatoes, chicken stock, and heavy cream
Zurek	Poland	Soured rye flour and meat

Figure 9-1. *A wide variety of ingredient combinations are used to create soups in every culture around the world.*

Different varieties of soups are characterized by the overall finished appearance and consistency of the soup and the preparation and cooking methods used to make the soup. Most soups can be classified as clear, thick, or specialty. Specialty soups include bisques, chowders, and cold soups, as well as unique regional varieties. Although many specialty soups are thick in consistency, they are produced using different methods than those used to make traditional thick soups.

CLEAR SOUPS

A *clear soup* is a stock-based soup with a thin, watery consistency. Clear soups include broths and consommés. Broths, which are produced from well-made stocks, can be made from meat, poultry, seafood, or vegetables. When meat is used, a broth will have a much more concentrated flavor and color than a stock made from bones. Consommés are made from high-quality broths that have been further clarified to remove all impurities and surface fat. *See Figure 9-2.*

Components of Clear Soups		
Soup	**Liquid**	**Ingredients**
Broth	Prepared stock	Main flavoring ingredients, mirepoix, and sachet d'épices or bouquet garni
Consommé	High-quality prepared broth	Clearmeat, mirepoix, sachet d'épices or bouquet garni, and oignon brûlé

Figure 9-2. *Broths are prepared from well-made stocks, and consommés are made from high-quality broths.*

Broths

A *broth* is a flavorful liquid made by simmering stock along with meat, poultry, seafood, or vegetables and seasonings. A stock is used as the liquid ingredient in a broth, as opposed to water. Meat or vegetables may be added to the stock and simmered to create a more intense flavor, color, and overall body. A broth can be served as a finished product or made into a broth-based soup or consommé.

Tougher cuts of meat or poultry are typically used to produce a full-flavored, rich broth. In addition, not allowing the broth to boil vigorously, continually skimming impurities from the surface, and carefully straining the cooked broth produces a broth of excellent quality. *See Figure 9-3.*

National Chicken Council

1. Cut the main flavoring ingredient to desired size or truss if needed.

2. If a dark color and roasted flavor are desired, carefully brown the meat and the mirepoix. If no color is desired, simply sweat the mirepoix.

3. Completely cover the main ingredient and mirepoix with cold stock. Add a sachet d'épices.

4. Bring to a simmer and continually skim off impurities as they rise to the surface.

5. Reduce heat to a low simmer and cook until the main ingredient is tender and the desired flavor has developed. Occasionally taste the broth to monitor progress.

6. Carefully drain the broth to minimize disturbing the cooked ingredients.

7. Strain the broth with a chinois or cheesecloth-lined china cap.

8. Quickly cool the strained broth using an ice bath or a cooling wand.

9. When the broth is cold, remove excess fat and grease.

Figure 9-3. A broth is prepared by simmering a flavoring ingredient, such as meat, in a well-prepared stock and then removing excess fat and grease from the surface.

The mirepoix used to make a broth may be cut larger or smaller depending on the time required to cook the main flavoring ingredient. For example, if the broth contains a cut of meat that requires a lengthy cooking time, a large-cut mirepoix can be added halfway through the cooking process. Beef broth, for example, typically simmers for 2 or more hours. However, if the main flavoring ingredient cooks more quickly, a smaller mirepoix can be added at the beginning of the cooking process. Chicken broth, for example, is typically done within 30–40 minutes. The size of the mirepoix should be such that it is cooked thoroughly in the allotted time, but not overcooked. *See Figure 9-4.*

Mirepoix

Figure 9-4. *The mirepoix used to make broths may be cut larger or smaller depending on the time required to cook the main flavoring ingredient.*

If a broth is needed for service immediately, the completed broth is brought to a boil. Seasonings are adjusted to taste. Fat droplets on the surface of the broth can be removed with a clean paper towel. *See Figure 9-5.* The broth is then garnished and served.

Consommés

A *consommé* is a very rich and flavorful broth that has been further clarified to remove any impurities or particles that could cloud the finished product. The flavor of the main ingredient should be pronounced, and the color of the consommé should be very rich. Consommés made from poultry should have a deep golden or amber color, while consommés made from red meat should have a dark, roasted meat color. It is very important that a consommé be made with the highest-quality ingredients and be properly prepared in a nonreactive pot, such as stainless steel, that is taller than it is wide. A finished consommé is only as good as the broth and ingredients used.

Consommés were once used as a way to display a chef's skill to people of nobility or wealth. A perfectly clear consommé showed that the chef had refined soup-making skills. Classical consommés can be made with any meat, poultry, or seafood and are often named after their main flavoring ingredient. *See Figure 9-6.* There are countless appropriate garnishes for a consommé. It is important the garnish be precisely prepared, as it can be seen through a clear consommé.

Removing Fat Droplets

Figure 9-5. *A clean paper towel can be used to absorb fat droplets from the surface of a broth.*

Classical Consommé Presentations	
Consommé	**Description**
Bouquetière	Garnished with a petite bouquet of assorted fresh vegetables tied with a string of leek
Brunoise	Garnished with a brunoise of carrot, celery, leek, and turnip and finished with minced chervil
Célestine	Slightly thickened with tapioca and garnished with a julienne of crêpe, chopped truffles, and savory herbs
Chasseur	Garnished game consommé with quenelles (poached puréed meat dumplings) of game meat or petite profiteroles (cream puffs) studded with game forcemeat
Diplomate	Chicken consommé slightly thickened with tapioca and garnished with julienne of truffles and quenelles of chicken and crawfish butter
Grimaldi	Clarified consommé with a greater quantity of tomato purée; garnished with cooked custard (royale) cut into various small shapes and a julienne of blanched celery
Julienne	Garnished with a blanched julienne of leeks, carrots, turnips, celery, and cabbage and finished with a chiffonade of chervil or sorrel
Madrilène	Served as a tomato-based consommé in cold jelly (aspic)
Mikado	Garnished chicken consommé that has a slight tomato flavor, with diced, peeled, and seeded tomato and cooked chicken meat
Printanièr	Garnished with petite blanched balls of turnip, carrots, peas, and chiffonade of chervil
Royale	Garnished with large cubes or batonnet of cooked custard (royale)

Figure 9-6. *Classical consommé presentations are often named for their main flavoring ingredient.*

Oignon Brûlé

Figure 9-7. *A burnt onion, or oignon brûlé, gives intense flavor and color to a consommé.*

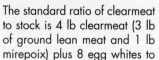

Chef's Tip

The standard ratio of clearmeat to stock is 4 lb clearmeat (3 lb of ground lean meat and 1 lb mirepoix) plus 8 egg whites to every gallon of cold stock.

Clarifying Consommés. Although the clarification process adds a small bit of flavor, the bulk of a consommé's flavor is derived from the quality of the broth. To *clarify* is to remove impurities, sediment, cloudiness, and particles to leave a clear, pure liquid. For this to occur, very cold stock or broth is combined with a mixture known as a clearmeat. A *clearmeat* is a cold, lean ground meat, fish, or poultry that is combined with an acid (such as wine, lemon juice, or a tomato product), ground mirepoix, egg whites, and an oignon brûlé. An *oignon brûlé* is half an onion that is charred on the cut side. The term oignon brûlé is French for "burnt onion." *See Figure 9-7.* An oignon brûlé is used to give an intense roasted flavor and deeper color to a darker-colored consommé.

To clarify a consommé, the clearmeat is mixed well and then whisked into the cold broth and brought to a low simmer. As the temperature increases, the mixture is occasionally stirred until the clearmeat begins to coagulate. Once the clearmeat begins to form a solid mass, the mixture cannot be disturbed or stirred as the clearmeat could break apart and cloud the consommé. As the clearmeat gently simmers, it rises to the surface and, in doing so, strains the impurities from the consommé. When the clearmeat rises, it is known as a raft. A *raft* is the clearmeat that has risen to the surface. In addition to clarifying the consommé, the raft enhances the flavor, color, and richness of the final consommé. *See Figure 9-8.*

1. Prepare the clearmeat by combining freshly ground lean meat, mirepoix, and egg whites with an acid (such as a tomato product or lemon juice) and an oignon brûlé.

2. Blend the cold clearmeat well with a small amount of cold broth to loosen the mixture. Both the clearmeat and broth should be between 32°F and 35°F. A small amount of crushed ice can be added to bring down the temperature, if needed.

3. In a heavy-bottomed stockpot (preferably with a drain spigot), combine the clarification mixture and cold broth. Stir well.

4. While occasionally stirring, bring the mixture to a simmer over medium heat. The clearmeat will begin to form a solid mass, or raft. Do not allow to boil.

5. Once a solid mass forms, stop stirring and gently poke a hole in the top of the newly formed raft. This hole allows the progress of the consommé to be monitored as it cooks.

6. Allow the consommé to gently simmer for 1–2 hours until a desired flavor and color are achieved. Occasionally baste the raft with consommé to prevent the raft from drying or breaking apart.

7. Carefully strain the consommé through a chinois or china cap lined with multiple layers of moistened cheesecloth. Degrease completely.

Refer to DVD for
Clarifying Consommés
Media Clip

Figure 9-8. *Consommé is prepared by using clearmeat to clarify a high-quality broth.*

Once the raft forms, a hole is gently poked into the top of it. This hole allows steam to escape without disrupting the raft. It also allows the progress of the consommé to be monitored as it cooks. After gently simmering for 1–2 hours, the consommé is removed through the hole in the raft using a ladle or from the bottom of a stockpot through the drain spigot. The consommé is then strained through multiple layers of dampened cheesecloth to remove any small impurities that may not have been trapped by the raft. The consommé should be cooled immediately and stored properly. Once completely cooled, any remaining fat will solidify on the surface and can be easily removed. The resulting liquid should be a clear and intensely flavored consommé. *See Figure 9-9.*

Garnished Consommé

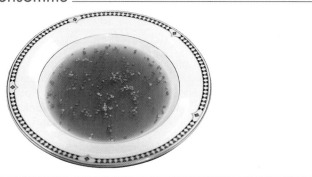

Figure 9-9. *A well-prepared consommé is clear and intensely flavored.*

Rescuing Cloudy Consommés. On rare occasions, factors such as accidentally boiling the consommé instead of simmering, stirring the consommé after the raft has formed, and starting the consommé with clearmeat and broth that are not cold enough can cause the final consommé to be cloudy. When this happens, it is necessary to conduct a secondary clarification.

To perform a secondary clarification on a gallon of consommé, a mixture consisting of 4 lightly beaten egg whites, ¼ cup of finely ground mirepoix, and either 1 tbsp tomato purée or 2 tsp lemon juice is added to the consommé. The mixture is added to the consommé and the same clarification procedure is repeated. The mixture is brought to a simmer and forms a raft. After the consommé is removed through the hole in the raft or the drain spigot, it is strained to remove any remaining impurities.

THICK SOUPS

A *thick soup* is a soup having a thick texture and consistency. The two basic types of thick soups are cream soups and purée soups. The main difference between a cream soup and a purée soup is that most cream soups are thickened by an added starch, such as the flour in a roux. Purée soups, on the other hand, are thickened by puréeing all of the ingredients. Additionally, purée soups are generally thicker and coarser than cream soups and are seldom strained. Due to the slightly lower starch content of vegetable purée soups, rice or potatoes are sometimes added to increase the texture and add body to the soup.

Cream Soups

Cream soups are traditionally made using one of two preparation methods. The first method uses velouté, and the second uses a roux. These two preparation methods work in much the same way, and the choice of which to use is a matter of preference. Regardless of the method chosen, the main ingredients need to be simmered until tender in order to be puréed smooth.

Velouté Preparation Method. The velouté preparation method consists of sweating a white matignon and the flavoring ingredient, such as broccoli for a cream of broccoli soup, in fat. Next, a velouté (a white stock thickened with a roux) is added and the mixture is simmered until the flavoring ingredient is tender. Then the mixture is puréed. Finally, the soup is seasoned to taste and finished with hot cream. *See Figure 9-10.*

Chef's Tip

When making a cream soup or a purée soup, always use a heavy-bottomed saucepot to avoid burning or scorching.

Refer to DVD for
Preparing Cream Soups (Velouté Method)
Media Clip

Procedure for Preparing Cream Soups (Velouté Method)

1. In a heavy-bottomed saucepot, sweat a white matignon and the flavoring ingredient in fat until slightly tender.

2. Add prepared velouté sauce and simmer until the flavoring ingredient is completely tender, stirring often.

3. Carefully purée the soup until smooth.

4. Strain through a chinois or cheesecloth-lined china cap.

5. Return strained soup to a simmer and thin with stock if needed.

6. Season to taste and finish with hot cream.

7. Serve immediately.

Figure 9-10. *To prepare cream soups using the velouté method, the main ingredient is simmered in a velouté and then puréed.*

Refer to DVD for
**Preparing Cream Soups
(Roux Method)**
Media Clip

Roux Preparation Method. The roux preparation method consists of sweating a white matignon and the flavoring ingredient in fat. Next, flour is added to the mixture to make a roux. Then hot stock is added to the roux, and the mixture is allowed to simmer and thicken. Finally, the soup is puréed until smooth, seasoned to taste, and finished with hot cream. *See Figure 9-11.*

Procedure for Preparing Cream Soups (Roux Method)

1. In a heavy-bottomed saucepot, sweat a white matignon and the main flavoring ingredient in enough fat to also make the roux.

2. When the aromatics and main ingredient are slightly tender, add enough flour to make a roux.

3. Cook the roux just long enough to achieve a pale golden color and then add warm white stock while whisking vigorously. Bring to a slight simmer and cook until the taste of flour is gone and the flavoring ingredient is completely tender.

4. Carefully purée the soup until smooth.

5. Strain through a chinois or cheesecloth-lined china cap.

6. Return strained soup to a simmer and thin with stock if needed.

7. Finish with hot cream and season to taste.

8. Serve immediately.

Figure 9-11. *To prepare cream soups using the roux method, a roux is made of the main ingredient, matignon, and flour, then stock is added and the soup is puréed.*

Purée Soups

Purée soups are made by cooking vegetables such as potatoes, squash, turnips, or carrots or dried legumes such as lentils, beans, or split peas in a broth until tender. Preparing vegetables in a relatively consistent size ensures uniform cooking. The cooked ingredients are puréed. Often a portion of the main ingredient is reserved and added into the purée to add texture to the soup. Purée soups can have a completely smooth texture for a sophisticated look or a coarse texture for a more rustic look. Purée soups can contain many different ingredients, from roasted vegetables and chilies to bacon, salt pork, and smoked ham. *See Figure 9-12.*

Refer to DVD for
Preparing Purée Soups
Media Clip

Procedure for Preparing Purée Soups

1. Per the recipe, either sweat or lightly brown the mirepoix in fat. Add the cooking liquid and bring to a simmer.

2. Add the main ingredients and sachet d'épices. Simmer until the main ingredients are tender enough to purée. Do not allow the soup to scorch or burn on the bottom.

3. Remove the sachet d'épices and reserve about 20% of the liquid.

4. Purée the remainder of the ingredients until the desired texture is achieved. Use the reserved liquid to thin the soup as needed.

5. Return the soup to a simmer and season to taste.

6. Garnish and serve immediately.

Figure 9-12. *Purée soups are thicker and coarser than cream soups.*

Purée soups tend to further thicken if they are made in advance and stored. When it is time to reheat, a small amount of hot water, stock, broth, milk, or cream may be added to thin the soup to the desired consistency. If a purée soup is too thin, a small amount of roux, beurre manié, or cornstarch slurry may be whisked in and the soup brought to a simmer.

SPECIALTY SOUPS

Many soups do not fall into the traditional categories of clear or thick soups. These other soups are considered specialty soups and fall into one of three categories: bisques, chowders, or cold soups.

Bisques

A *bisque* is a form of cream soup that is typically made from shellfish. The name bisque traditionally applies to soup made from shellfish and thickened with cooked, puréed rice. Today bisques are commonly thickened with a roux instead of rice because roux achieves a smoother and richer consistency. In modern cuisine, many puréed soups are referred to as bisques. Bisques are typically prepared using either a modern preparation method or a classical preparation method.

Modern Preparation Method. The modern preparation method consists of caramelizing an aromatic mirepoix in a small amount of fat. Then flavoring ingredients are added and cooked until they caramelize. Next, tomato paste is added to the pan and lightly sautéed. The pan is deglazed with white wine, brandy, or cognac. A sachet d'épices and a white stock are added before simmering until a strong flavor and color are achieved. Roux is then whisked in to thicken the bisque. Finally, the bisque simmers for 15–20 minutes or until the taste of flour is gone. *See Figure 9-13.*

Classical Preparation Method. The classical preparation method involves sweating a mirepoix in a small amount of fat. Then flavoring ingredients are added and cooked until they caramelize. Next, tomato paste is added to the pan and lightly sautéed. The pan is deglazed with white wine, brandy, or cognac. A sachet d'épices and small amount of white stock are added before allowing the bisque to simmer until reduced by half. Heavy cream is added before simmering until the soup is reduced to a nappe consistency. The sachet d'épices is removed and the soup is puréed slightly. The puréed soup is strained and then brought to a simmer. *See Figure 9-14.*

1. In a heavy-bottomed saucepot, caramelize the mirepoix in a small amount of fat.

2. Add flavoring ingredients and cook until caramelized.

3. Add a small amount of tomato paste to the pan and sauté lightly.

4. Deglaze the pan with a small amount of white wine, brandy, or cognac.

5. Add sachet d'épices and a white stock. Simmer for 20–25 minutes or until a strong flavor and color are achieved.

6. Add roux to thicken while whisking vigorously.

7. Simmer gently for 15–20 minutes or until the taste of flour is gone.

8. Remove the sachet and pureé the bisque.

9. Strain the bisque through a chinois or cheesecloth-lined china cap.

10. Return strained soup to saucepot and bring to a simmer. If needed, thin the bisque with hot stock and bring to a simmer again.

11. Finish with hot cream and season to taste.

12. Garnish and serve immediately.

Figure 9-13. *Modern bisques are prepared using roux and should be simmered until the flour taste is gone.*

Refer to DVD for
Preparing Bisques (Modern Method)
Media Clip

1. In a heavy-bottomed saucepot, carmelize the mirepoix in a small amount of fat.

2. Add flavoring ingredients and cook until caramelized.

3. Add a small amount of tomato paste to the pan and sauté lightly.

4. Deglaze the pan with a small amount of white wine, brandy, or cognac.

5. Add sachet d'épices and a small amount of white stock. Simmer 15–20 minutes or until reduced by half.

6. Add heavy cream. Simmer gently until reduced to a nappe consistency.

7. Remove the sachet and purée the bisque.

8. Strain the bisque through a chinois or cheesecloth-lined china cap.

9. Return strained soup to saucepot and bring to a simmer. If needed, thin the bisque with hot stock and bring to a simmer again. Season to taste.

10. Garnish and serve immediately.

Figure 9-14. *Bisques prepared using the heavy cream reduction method are simmered until the liquid is reduced by half.*

Refer to DVD for
Preparing Bisques (Classical Method)
Media Clip

Chowders

A *chowder* is a hearty soup that contains visibly large chunks of the main ingredients. Although most cream-based chowders are thick, a few chowders are broth-based and therefore thinner. The primary difference between cream soups and chowders is that chowders are not puréed. The majority of cream-based chowders are thickened with a roux. *See Figure 9-15.*

Florida Department of Agriculture and Consumer Services, Bureau of Seafood and Aquaculture Marketing

Refer to DVD for
Preparing Chowders
Media Clip

Procedure for Preparing Chowders

1. Render diced bacon or salt pork until it begins to caramelize.

2. Add mirepoix and sweat until onions are translucent.

3. Add flour to make a roux.

4. Add stock or flavoring liquid.
5. Add the main flavoring ingredients and a sachet d'épices.

6. Simmer until main ingredients are fully cooked.
7. Remove the sachet.

8. Finish with hot cream.
9. Serve immediately.

Figure 9-15. *Chowders are hearty soups that are not puréed.*

Cold Soups

There are two basic types of cold soups, those that require cooking the main ingredients and those that use raw ingredients that have been puréed. Cold soups include traditional cold cream soups such as vichyssoise, a cold potato and leek soup. Cold soups can also be made from a purée of fruit or vegetables and yogurt. For example, a refreshing cantaloupe soup can be made by puréeing fresh cantaloupe with mint and yogurt. Many cold soups are made using fruit or vegetable juice. A popular cold soup made using juice is gazpacho. To prepare gazpacho, fresh raw vegetables are puréed with tomato juice and spices and served cold. *See Figure 9-16.*

Gazpacho

Figure 9-16. *Gazpacho is a cold puréed soup made from raw vegetables and tomato juice.*

SUMMARY

Every culture around the world eats a variety of soups. Soups can set the stage for a wonderful meal and can also be a good way to use high-quality leftovers such as chicken. However, a soup is only as good as the ingredients used to make it, so only quality ingredients should be used. Soups are generally classified as clear, thick, or specialty soups. Clear soups include broths and consommés. Cream soups and purée soups are known as thick soups. Specialty soups include bisques, chowders, and cold soups. No matter what type of soup is being served, it is important to serve hot soup very hot and cold soup very cold.

Refer to DVD for
Quick Quiz® questions

1. Identify soup varieties from around the world.
2. Describe two types of clear soups.
3. Describe the procedure for preparing broths.
4. Explain what determines the size of the mirepoix used to make a broth.
5. Define consommé.
6. Describe the function of clearmeat.
7. Describe the function of an oignon brûlé.
8. Describe the procedure for clarifying consommés.
9. Identify factors that contribute to an unsuccessfully clarified consommé.
10. Identify two basic types of thick soups.
11. Differentiate between the velouté method and the roux method of preparing cream soups.
12. Explain how to prepare purée soups.
13. Identify the three categories of specialty soups.
14. Differentiate between the modern method and the classical method of preparing bisques.
15. Describe the procedure for preparing chowders.
16. Identify common types of cold soups.

Refer to DVD for
Review Questions

1. Find recipes for at least four of the soups listed in Figure 9-1, Soup Varieties Around the World.
2. Prepare a broth. Evaluate the appearance and flavor of the broth.
3. Clarify a consommé. Evaluate the clarity and flavor of the consommé.
4. Find a recipe for one of the consommés listed in Figure 9-6, Classical Consommés.
5. Prepare two cream soups using the same main ingredient. Use the velouté method to prepare one cream soup and the roux method to prepare the other cream soup. Evaluate and compare the appearance, flavor, and texture of the two cream soups.
6. Prepare a purée soup. Evaluate the appearance, flavor, and texture of the puréed soup.
7. Prepare two bisques using the same main ingredient. Use the roux method to prepare one bisque and the heavy cream reduction method to prepare the other bisque. Evaluate and compare the appearance, flavor, and texture of the two bisques.
8. Prepare a chowder. Evaluate the appearance, flavor, and texture of the chowder.
9. Prepare a cold soup. Evaluate the appearance, flavor, and texture of the cold soup.

Refer to DVD for
Application Scoresheets

Chicken Broth

Yield: *16 servings, 8 fl oz each*

Ingredients

mirepoix	
onions, medium dice	12 oz
carrots, medium dice	6 oz
celery, medium dice	6 oz
vegetable oil	1 fl oz
stewing hen	3 lb
chicken stock or water	1 gal.
sachet d'épices	
bay leaves	2 ea
thyme	1 sprig
parsley	3 stems
peppercorns, cracked	¼ tsp
garlic, crushed	1 small clove

Preparation

1. Sweat mirepoix in vegetable oil over medium heat without browning.
2. Place chicken in stockpot with mirepoix.
3. Cover chicken and mirepoix with cold chicken stock or water.
4. Add sachet d'épices and bring to a full simmer.
5. Reduce to a low simmer and skim occasionally.
6. Continue cooking until the desired flavor and color have developed and the chicken is completely cooked (approximately 30–45 minutes).
7. When fully cooked, remove chicken and sachet. Cool and store chicken for later use.
8. Carefully strain the chicken broth using a chinois or cheesecloth-lined china cap.
9. Garnish for service or cool and store for later use.

NUTRITION FACTS
Per serving: 248 calories, 136 calories from fat, 15 g total fat, 43.3 mg cholesterol, 396.1 mg sodium, 451 mg potassium, 11.9 g carbohydrates, <1 g fiber, 5.4 g sugar, 15.4 g protein

---

Beef Broth

Yield: *16 servings, 8 fl oz each*

Ingredients

beef shank or oxtail	6 lb
vegetable oil	1½ fl oz + 1 fl oz
beef stock or water	5 qt
mirepoix	
onions, medium dice	12 oz
carrots, medium dice	6 oz
celery, medium dice	6 oz
leeks, medium dice	4 oz
tomatoes, seeded and diced	4 oz
sachet d'épices	
bay leaf	1 ea
thyme, diced	⅛ tsp
black peppercorns, cracked	¼ tsp
garlic, crushed	1 clove
salt	TT

Preparation

1. Cut meat into 1–2 inch pieces. Brown in medium stockpot in 1½ fl oz of oil.
2. Completely cover meat with cold beef stock or water. Bring to a simmer while continually skimming the surface.
3. Simmer for 2 hours.
4. In another pan, brown the mirepoix in 1 fl oz of oil over medium-high heat. Then add leeks and tomatoes and cook a minute longer.
5. Deglaze pan with a small amount of broth. Add mirepoix and deglazing liquor to stockpot along with sachet d'épices.
6. Simmer for an additional hour while skimming the surface as needed.
7. Strain the broth using a chinois or cheesecloth-lined china cap. Discard sachet and vegetables. Reserve all usable pieces of meat.
8. Season to taste. Garnish and serve or cool and store for later use.

NUTRITION FACTS
Per serving: 399 calories, 198 calories from fat, 21.6 g total fat, 74.8 mg cholesterol, 732.7 mg sodium, 1287.5 mg potassium, 8.3 g carbohydrates, 1.1 g fiber, 3.7 g sugar, 41.4 g protein

---

Vegetable Broth

Yield: *4 servings, 8 fl oz each*

Ingredients

vegetable oil	1 tbsp
mirepoix	
onions, medium dice	8 oz
carrots, medium dice	4 oz
celery, medium dice	4 oz
leeks, medium dice	4 oz
turnips, cut small dice	4 oz
fennel, cut small dice	4 oz
parsnips, cut small dice	4 oz
mushroom stems, sliced thin	4 oz
garlic cloves, peeled and	
rough chopped	2 tsp
tomato, rough chopped	2 oz
water	36 fl oz
sachet d'épices	
bay leaf	1 ea
parsley stems	3 ea
dried thyme	⅛ tsp
peppercorns, cracked	2 ea

Preparation

1. Place oil in saucepot over medium heat.
2. Add mirepoix, turnips, fennel, parsnips, mushrooms, and garlic and sweat until onions are translucent.
3. Add tomatoes and cook for 1 minute more.
4. Add water and sachet d'épices and simmer 45–60 minutes until vegetables are tender.
5. Strain through a cheesecloth-lined chinois.
6. Chill completely and refrigerate until needed.

NUTRITION FACTS
Per serving: 115 calories, 15 calories from fat, 1.7 g total fat, 0 mg cholesterol, 97.1 mg sodium, 708.1 mg potassium, 23.8 g carbohydrates, 6 g fiber, 8.8 g sugar, 3.5 g protein

Miso Soup

Yield: *8 servings, 4 fl oz each*

Ingredients

dashi	1 qt
aka miso	2 oz
tofu, rinsed, patted dry, and cut	
medium dice	4 oz
scallions, sliced very thin	2 ea
mushrooms, sliced very thin	2 oz

Preparation

1. Heat dashi in medium saucepot until hot.
2. Place miso in small bowl with one ladle of hot dashi. Mix until miso is completely dissolved.
3. Add the dashi and miso mixture back to the saucepot.
4. To serve, add tofu and a small amount of sliced scallions and mushrooms to each bowl and top with hot broth.

NUTRITION FACTS
Per serving: 51 calories, 12 calories from fat, 1.5 g total fat, 5.9 mg cholesterol, 686.1 mg sodium, 139.8 mg potassium, 3.9 g carbohydrates, <1 g fiber, <1 g sugar, 5.3 g protein

Beef Consommé

Yield: *16 servings, 8 fl oz each*

Ingredients

egg whites	8 ea	beef stock or broth	5 qt
ground beef, lean		oignon brûlé	½ onion
(shank, shoulder, or neck)	3 lb	sachet d'épices	
mirepoix, ground fine		bay leaves	3 ea
onions, small dice	8 oz	thyme	2 sprigs
carrots, small dice	4 oz	parsley	4 stems
celery, small dice	4 oz	peppercorns, cracked	½ tsp
tomatoes, seeded and diced	2 ea	salt	TT

Preparation

1. Slightly whip egg whites until frothy.
2. Prepare clearmeat by combining ground beef, mirepoix, tomato, and whipped egg whites in mixing bowl.
3. Mix in a small amount of cold beef stock or broth to loosen the clearmeat.
4. Place clearmeat in stockpot along with remaining stock or broth. Mix well.
5. Add oignon brûlé and sachet d'épices.
6. Over medium heat, bring mixture to a gentle simmer while stirring occasionally.
7. Stop stirring when mixture reaches a full simmer. The clearmeat will rise to the surface and form a raft.
8. Carefully poke a 2 inch hole in the top of the raft.
9. Simmer gently for 1–2 hours until a full-bodied and rich-colored consommé develops.
10. Carefully strain the consommé through multiple layers of moistened cheesecloth.
11. Season to taste.
12. Garnish and serve or completely chill and remove any remaining surface fat.

NUTRITION FACTS
Per serving: 179 calories, 42 calories from fat, 4.6 g total fat, 52.7 mg cholesterol, 708.3 mg sodium, 986.4 mg potassium, 7 g carbohydrates, <1 g fiber, 3.4 g sugar, 26.4 g protein

---

Variations:

Consommé Bouquetière: Garnish with a petite bouquet of assorted fresh vegetables tied with a string of leek.

Consommé Brunoise: Garnish with a brunoise cut of carrot, celery, leek, and turnip and finish with minced chervil.

Consommé Célestine: Thicken slightly with tapioca and garnish with a julienne of crêpe, chopped truffles, and savory herbs.

Consommé Chasseur: Garnish game consommé with quenelles (poached puréed meat dumplings) of game meat or petite profiteroles (cream puffs) studded with game forcemeat.

Consommé Diplomate: Thicken chicken consommé slightly with tapioca and garnish with a julienne of truffles and quenelles (poached puréed meat dumplings) of chicken and crawfish butter.

Consommé Grimaldi: Clarify consommé with a greater quantity of tomato purée. Garnish with cooked custard (royale) cut into various small shapes and a julienne of blanched celery.

Consommé Julienne: Garnish with a blanched julienne of leeks, carrots, turnips, celery, and cabbage and finish with a chiffonade of chervil or sorrel.

Consommé Mikado: Garnish chicken consommé that has a slight tomato flavor with peeled, seeded, diced tomato and cooked chicken meat.

Consommé Madrilène: Serve a tomato-based consommé in cold jelly (aspic) form.

Consommé Printanier (Parisienne): Garnish with petite blanched balls of turnip, carrots, peas, and a chiffonade of chervil.

Consommé Royale: Garnish with large cubes or batonnet of cooked custard (royale).

Cream of Broccoli Soup (Velouté Method)

Yield: *24 servings, 8 fl oz each*

Ingredients

broccoli, diced	2¼ lb
white matignon	
onions, small dice	8 oz
celery, small dice	4 oz
leek stems, small dice	4 oz
whole butter	3 oz
chicken velouté	1 gal.
heavy cream	8 fl oz
salt and white pepper	TT

Preparation

1. Sweat the white matignon and broccoli in butter until slightly tender.
2. Add hot velouté and whisk.
3. Simmer for approximately 30 minutes or until vegetables are tender, stirring often.
4. Purée until smooth.
5. Place over medium heat and bring to a simmer. Add heated cream. If soup is too thick, add a small amount of stock.
6. Season to taste.

NUTRITION FACTS
Per serving: 238 calories, 144 calories from fat, 17.4 g total fat, 47.9 mg cholesterol, 85.5 mg sodium, 172.7 mg potassium, 20.1 g carbohydrates, <1 g fiber, <1 g sugar, 7 g protein

Cream of Broccoli Soup (Roux Method)

Yield: *16 servings, 8 fl oz each*

Ingredients

broccoli, diced	2¼ lb
white matignon	
onions, small dice	8 oz
celery, small dice	4 oz
leek stems, small dice	4 oz
butter, unsalted	3 oz
flour	8 oz
chicken stock	1 gal.
heavy cream	8 oz
salt and white pepper	TT

Preparation

1. Sweat the white matignon and asparagus in butter until slightly tender.
2. Add flour and cook until golden brown.
3. Remove from heat and add warm chicken stock, whisking continuously.
4. Purée until smooth.
5. Place over medium heat and bring to a simmer. Add heated cream. If soup is too thick, add a small amount of stock.
6. Season to taste.

NUTRITION FACTS
Per serving: 213 calories, 120 calories from fat, 13.5 g total fat, 38.7 mg cholesterol, 261.8 mg sodium, 352.2 mg potassium, 17 g carbohydrates, <1 g fiber, 3.2 g sugar, 6.7 g protein

Purée of Navy Bean Soup

Yield: *22 servings, 8 fl oz each*

Ingredients

mirepoix	
onions, medium dice	8 oz
carrots, medium dice	4 oz
celery, medium dice	4 oz
leeks, diced	2 oz
garlic, chopped	1 tbsp
vegetable stock	1 gal.
navy beans, soaked overnight	1 lb
sachet d'épices	
bay leaves	2 ea
thyme	2 sprigs or ½ tsp
peppercorns, cracked	1 tsp
tomatoes, canned (drained), diced	8 oz
salt and pepper	TT

Preparation

1. In a heavy-bottomed saucepot, cook mirepoix, leeks, and garlic over low heat until slightly tender.
2. Add the stock and bring to a simmer.
3. Add drained beans and sachet d'épices.
4. Cover the pot and simmer until the beans are tender (approximately 1½–2 hours).
5. Add tomatoes and cook an additional 15 minutes.
6. Remove sachet and reserve 20% of the soup.
7. Purée remaining soup until desired texture is achieved. Add reserved soup as needed to reach desired consistency.
8. Bring to a simmer over medium heat.
9. Season to taste.

NUTRITION FACTS
Per serving: 289 calories, 31 calories from fat, 3.6 g total fat, 1.8 mg cholesterol, 1221.6 mg sodium, 913.9 mg potassium, 50.5 g carbohydrates, 14.5 g fiber, 3 g sugar, 15.3 g protein

---

Minestrone Soup

Yield: *11 servings, 6 fl oz each*

Ingredients

dried white beans, soaked in cold water overnight	2 oz
olive oil	2 tsp
onions, medium dice	1 oz
celery, medium dice	2 oz
carrot, medium dice	2 oz
garlic, minced	2 tsp
zucchini, medium dice	3 oz
green beans, cut into 1 inch sections	2 oz
cabbage, medium diced	2 oz
kale, stems removed, chopped	1 oz
vegetable or chicken stock	1 qt
canned tomatoes, diced	8 fl oz
tomato paste	3 tbsp
fresh basil, chopped	2 tsp
fresh parsley, chopped	2 tsp
fresh oregano, chopped	1 tsp
elbow macaroni, cooked	4 oz
salt and pepper	TT
parmesan, grated	3 tbsp

Preparation

1. Drain beans and place in saucepan covered with water at least 2 inches above the beans.
2. Simmer approximately 45 minutes to 1 hour or until tender. Drain beans and set aside.
3. In saucepot, add olive oil over medium-high heat and sweat onions until translucent.
4. Add celery and carrots and cook for 4 minutes.
5. Add garlic and cook for 1 minute more.
6. Add zucchini, green beans, cabbage, and kale and cook for 2 minutes.
7. Add stock, tomatoes, and tomato paste. Cover soup and simmer for 1 hour.
8. Add fresh chopped herbs and stir well.
9. Add white beans and macaroni and stir well.
10. Bring soup back to a simmer and cook for 5 minutes.
11. Season to taste.
12. Serve soup garnished with parmesan cheese.

NUTRITION FACTS
Per serving: 78 calories, 22 calories from fat, 2.5 g total fat, 3.8 mg cholesterol, 254 mg sodium, 291 mg potassium, 10 g carbohydrates, 1.5 g fiber, 3.5 g sugar, 4.3 g protein

---

Shrimp Bisque

Yield: *16 servings, 8 fl oz each*

Ingredients

mirepoix	
onions, small dice	8 oz
carrots, small dice	4 oz
celery, small dice	4 oz
clarified butter	3 oz
shrimp shells	2 lb
garlic, minced	3 cloves
tomato paste or purée	3 oz
flour	½ c
brandy	2 fl oz
dry white wine	1 c
seafood velouté	1 gal.
sachet d'épices	
bay leaves	2 ea
thyme	4 sprigs
peppercorns, cracked	4 ea
heavy cream	1 pt
shrimp, diced and sautéed	1 lb
chives, finely diced	2 oz
salt and pepper	TT

Preparation

1. In a heavy-bottomed saucepot over medium heat, caramelize mirepoix in clarified butter.
2. Add shrimp shells and cook until pink and caramelized.
3. Add garlic and cook for 1 minute.
4. Add tomato paste or purée. Cook until slightly caramelized, stirring frequently.
5. Deglaze the pan with white wine and brandy. Reduce liquid by half.
6. Add velouté and sachet d'épices. Gently simmer for 30–45 minutes.
7. Remove sachet. Carefully strain stock, reserving the solids and liquids separately.
8. Purée solids with a small amount of liquid.
9. Place all ingredients back in pot and gently simmer for 15 minutes.
10. Whisk in heated cream.
11. Strain soup through a chinois or cheesecloth-lined china cap. Season to taste.
12. Place 2 oz of shrimp and 1½ tsp of chives in each bowl. Fill bowls with bisque.
13. Season to taste.

NUTRITION FACTS
Per serving: 457 calories, 228 calories from fat, 27.5 g total fat, 200.2 mg cholesterol, 693.8 mg sodium, 267 mg potassium, 32.4 g carbohydrates, 1 g fiber, 2 g sugar, 23.2 g protein

New England-Style Clam Chowder

Yield: *20 servings, 8 fl oz each*

Ingredients

salt pork or bacon, finely minced	6 oz
white matignon	
onions, small dice	8 oz
celery, small dice	4 oz
leek stems, small dice	4 oz
flour	4 oz
canned clams (with juice)	2 qt
clam stock or fish stock	1 qt
potatoes, small dice	1 lb
sachet d'épices	
bay leaves	2 ea
thyme	2 sprigs
parsley	4 stems
peppercorns, crushed	4 ea
heavy cream	1 qt
Tabasco® sauce	½ tsp
Worcestershire sauce	½ tsp
salt and white pepper	TT

Preparation

1. In a heavy-bottomed saucepot, render fat until pork begins to caramelize.
2. Add white matignon and sweat until onions are translucent.
3. Add flour and stir well while cooking slowly to make a roux. Cook roux for 5–7 minutes.
4. Strain clams from juice and reserve clams until needed.
5. Add clam juice and clam or fish stock while whisking continuously. Bring to a simmer.
6. Add potatoes and sachet d'épices. Simmer until potatoes are slightly tender.
7. Remove sachet. Add reserved clams, heated cream, Tabasco®, and Worcestershire. Season to taste.

NUTRITION FACTS
Per serving: 204 calories, 144 calories from fat, 16.2 g total fat, 43.4 mg cholesterol, 433.5 mg sodium, 381 mg potassium, 11.2 g carbohydrates, 1.1 g fiber, 1 g sugar, 3.7 g protein

Manhattan-Style Clam Chowder

Yield: *20 servings, 8 fl oz each*

Ingredients

salt pork or bacon, small dice	4 oz
celery, medium dice	8 oz
carrots, medium dice	4 oz
leeks, medium dice	4 oz
onions, medium dice	8 oz
green bell peppers, medium dice	4 oz
garlic, finely minced	3 cloves
medium-dice canned	
tomatoes (with juice),	1 qt
canned clams (with juice)	2 qt
clam stock or fish stock	1 qt
potatoes, small dice	1 lb
sachet d'épices	
bay leaves	2 ea
thyme	2 sprigs
parsley	4 stems
peppercorns, cracked	4 ea
Tabasco® sauce	½ tsp
Old Bay® seasoning	½ tsp
Worcestershire sauce	½ tsp
salt and white pepper	TT

Preparation

1. In a heavy-bottomed saucepot, render fat until pork begins to caramelize.
2. Add celery, carrots, leeks, and onions. Sweat until onions are translucent.
3. Add bell peppers and garlic. Sauté for 1–2 minutes.
4. Add tomatoes with juice and deglaze pan.
5. Strain clams from juice and reserve clams until needed.
6. Add clam juice and clam or fish stock to pot.
7. Add potatoes and sachet d'épices. Simmer until potatoes are slightly tender.
8. Remove sachet. Add clams, Tabasco®, Old Bay®, and Worcestershire. Season to taste.

NUTRITION FACTS
Per serving: 98 calories, 46 calories from fat, 5.2 g total fat, 8.2 mg cholesterol, 507.6 mg sodium, 530.8 mg potassium, 10.7 g carbohydrates, 2 g fiber, 3.5 g sugar, 3 g protein

Pho Bo Soup

Yield: *12 servings, 6 fl oz each*

Ingredients

oxtail, cut into 2 inch thick pieces	1½ lb
beef stock	1½ qt
onions, thinly sliced	6 oz
vegetable oil	1 tsp
fresh ginger root, peeled, and sliced in half lengthwise	2 inch long piece (2 oz)
garlic cloves, peeled and sliced in half lengthwise	2 ea
daikon radish, cut medium dice	4 oz
carrot, cut medium dice	2 oz
green onion, minced	2 ea
cinnamon stick	1 ea
star anise	1 ea
black pepper, cracked	½ tsp
fish sauce	2 fl oz

Garnishes:

rice stick noodles (bahn pho)	6 oz
bean sprouts	1 c
green onions, sliced thin on a bias	4 ea
green chiles, jalapeno chiles, or Thai chiles, sliced very thin	2 ea
Thai basil leaves, cut chiffonade	½ c
cilantro leaves, cut chiffonade	½ c
mint leaves, cut chiffonade	¼ c
lime wedges	12 ea
Thai garlic chili paste	1 tbsp
sirloin, partially frozen and then sliced thin	9 oz

Preparation

1. Rinse the oxtails under cold water and then place in saucepot and cover with cold beef stock.
2. Slowly bring stock to simmer, skimming often for the first 30–40 minutes to remove any impurities that rise to the surface.
3. In a saucepan, sauté sliced onions in vegetable oil over medium-high heat until onions are caramelized.
4. Add ginger root and garlic and cook for a few minutes to caramelize slightly. Remove from saucepan to drain excess oil.
5. When oxtails no longer need frequent skimming, add onions, ginger, garlic, daikon, carrot, green onion, cinnamon stick, star anise, black pepper, and fish sauce.
6. Allow stock to simmer until oxtails are tender, approximately 2–3 hours. Skim as necessary when foam or impurities are present.
7. Soak noodles in a bowl of cool water for 20–30 minutes.
8. Remove oxtails from the pot and set aside to cool slightly.
9. Shred the oxtail meat and set aside.
10. Drain the stock through a cheesecloth-lined chinois and adjust seasoning as needed.
11. Bring a medium saucepot of water to a boil and blanch bean sprouts for 3–4 seconds. Remove bean sprouts from the hot water with a strainer (do not discard water).
12. Rinse the sprouts under cold water and set aside until garnishes are needed.
13. Bring the water back to a boil. Drain noodles, add to the boiling water, and cook for 1–2 minutes until tender.
14. Remove, drain, and set aside the noodles if needed immediately or rinse and store in cold water if not needed until later.
15. To serve, place approximately 1 oz of noodles in each bowl and top with approximately ½ oz of the cooked oxtail meat and 3 small slices (¾ oz total) of raw beef. *Note:* If the noodles were held in cold water simply dip them in boiling water for a few seconds to reheat.
16. Garnish with remaining ingredients as desired and pour hot broth over the raw sirloin to cook the meat.

Variation: Use 1 lb beef chuck in place of the oxtail.

NUTRITION FACTS
Per serving: 284 calories, 192 calories from fat, 21.3 g total fat, 43.1 mg cholesterol, 674 mg sodium, 518.3 mg potassium, 11.8 g carbohydrates, 2.4 g fiber, 2.5 g sugar, 12 g protein

Vichyssoise

Yield: *16 servings, 8 fl oz each*

Ingredients

whole butter	6 oz
white matignon	
onions, small dice	8 oz
celery, small dice	4 oz
leek stems, small dice	4 oz
chicken stock	3 qt
potatoes, medium dice	2 lb
sachet d'épices	
bay leaves	2 ea
thyme	2 sprigs
parsley	4 stems
peppercorns, cracked	4 ea
heavy cream	1 qt
salt and white pepper	TT
sour cream	4 oz
chives, chopped	2 tbsp

Preparation

1. Melt butter in a heavy-bottomed saucepot over medium heat.
2. Add white matignon. Sweat until onions are translucent and leeks have wilted.
3. Add stock, potatoes, and sachet d'épices. Simmer 30–40 minutes or until potatoes are almost tender.
4. Remove sachet. Purée soup until smooth.
5. Strain mixture through a chinois or cheesecloth-lined china cap.
6. Chill soup in an ice bath and reserve until needed.
7. When ready for service, stir in cold heavy cream. Season to taste.
8. Garnish with sour cream and chives.

NUTRITION FACTS
Per serving: 327 calories, 206 calories from fat, 23.3 g total fat, 72.9 mg cholesterol, 305.3 mg sodium, 672.7 mg potassium, 23.6 g carbohydrates, 1.5 g fiber, 4.1 g sugar, 7.1 g protein

Cold Curried Carrot and Coconut Milk Soup

Yield: *16 servings, 8 fl oz each*

Ingredients

whole butter	4 oz
onions, small dice	8 oz
carrots, small dice	3 lb
celery, small dice	6 oz
ginger, small dice	3 tbsp
green onions, small dice	8 oz
curry powder	3 tbsp
chicken stock	2 qt
coconut milk, unsweetened	3 c
lime juice, fresh	1½ fl oz

Preparation

1. In a heavy-bottomed saucepot over medium heat, sweat onions, carrots, celery, and ginger in butter until onions are translucent.
2. Add green onions and curry powder. Sauté 1 minute.
3. Add stock. Cover and simmer 20–30 minutes or until carrots are soft.
4. Add coconut milk and bring to a boil.
5. Purée in small batches until very smooth. Transfer puréed soup to a clean bowl.
6. Stir in lime juice. Chill.

NUTRITION FACTS
Per serving: 230 calories, 143 calories from fat, 16.7 g total fat, 18.9 mg cholesterol, 248.7 mg sodium, 606.8 mg potassium, 17.4 g carbohydrates, 3.6 g fiber, 7.2 g sugar, 5.4 g protein

Gazpacho

Yield: *16 servings, 8 fl oz each*

Ingredients

tomatoes concassé	2 lb
onions, medium dice	4 oz
green onions, medium dice	3 oz
jalapenos, finely minced	1 tbsp
green bell pepper, medium dice	1 ea
red bell pepper, medium dice	1 ea
cucumbers, peeled and seeded, medium dice	1 lb
celery, medium dice	½ lb
garlic, minced	1 tbsp
basil, chiffonade	2 tbsp
tarragon, chopped	2 tsp
balsamic vinegar	3 fl oz
olive oil	3 fl oz
lemon juice	2 fl oz
Worcestershire sauce	1 tbsp
Tabasco® sauce	2 tsp
tomato juice	2 qt
vegetable stock	as needed
cayenne pepper	TT
salt and pepper	TT
cilantro	1 tbsp

Preparation

1. Purée all ingredients except tomato juice and vegetable stock, leaving some texture to ingredients.
2. Add tomato juice and pulse to incorporate.
3. Thin with vegetable stock as needed. Season to taste. Chill.
4. Garnish with cilantro.

NUTRITION FACTS
Per serving: 121 calories, 53 calories from fat, 6.1 g total fat, <1 mg cholesterol, 266.1 mg sodium, 623.4 mg potassium, 15.5 g carbohydrates, 2.7 g fiber, 8.6 g sugar, 2.9 g protein

Sandwiches

*P*reparing sandwiches is a basic skill in the professional kitchen and is often the first kitchen position a person may be hired for when entering the foodservice industry. An operation that prepares great tasting and creatively composed sandwiches quickly will keep customers coming back for more. Sandwiches can be very simple and casual or they can be quite elaborate and require a number of gourmet ingredients. The combinations of sandwich components are only limited by the imagination of the sandwich maker. It is important to master the art, speed, and skill of sandwich making to be a successful food-service employee.

Refer to DVD
for **Flash Cards**

Chapter Objectives

1. Explain why there is a large variety of sandwiches.
2. Explain how to lower the fat and calorie contents of sandwiches.
3. Identify the four main sandwich components.
4. Identify common types of sandwich bases.
5. Prepare sandwich bases for use at a sandwich station.
6. Identify five common types of sandwich spreads.
7. Prepare a variety of sandwich spreads for use at a sandwich station.
8. Identify seven common types of sandwich fillings.
9. Prepare a variety of sandwich fillings for use at a sandwich station.
10. Identify common sandwich garnishes.
11. Prepare a variety of sandwich garnishes for use at a sandwich station.
12. Prepare common types of hot sandwiches.
13. Prepare common types of cold sandwiches.
14. Explain the importance of range of motion at a sandwich station.
15. Prepare large quantities of sandwiches.
16. Identify common side dishes served with plated sandwiches.

Key Terms

- **sandwich base**
- **sandwich spread**
- **sandwich filling**
- **bound salad**
- **sandwich garnish**
- **hot open-faced sandwich**
- **hot closed sandwich**
- **hot wrap sandwich**
- **grilled sandwich**
- **fried sandwich**
- **cold open-faced sandwich**
- **cold closed sandwich**
- **multidecker sandwich**
- **cold wrap sandwich**
- **tea sandwich**

SANDWICH VARIETIES AND NUTRITION

Sandwiches have become the symbol of a quick and tasty meal on the go. They can be made from many combinations of ingredients and can be served hot or cold. Sandwiches can be simple or elaborate. Sandwiches are one of the most versatile categories of food and allow chefs to use their imagination. Special ingredient combinations are used to create signature sandwiches around the world. *See Figure 10-1.*

Sandwiches can be nutritious meals when the ingredients and preparation method used are carefully selected. Changing the chosen cooking method or the ingredient(s) can positively impact the nutritional composition. For example, grilling a chicken breast instead of frying it would provide a healthier chicken sandwich that is lower in fat and calories. Other simple changes such as making a tuna salad sandwich with light or fat-free mayonnaise can also decrease the fat and calorie content. The use of flatbreads or thinly sliced whole-grain breads, fat-free or low-fat spreads such as mustard, lower fat meats such as turkey, and high-fiber fruits and vegetables adds nutrients and flavor without adding a lot of fat and calories.

Sandwich Varieties Around the World		
Sandwich	**Origin**	**Key Ingredients**
Arepa	Venezuela	A corn cake filled with chicken, avocado, cheese, shredded beef, eggs, or other ingredients; an arepa is often toasted on a griddle
Cha Gio	Vietnam	A crispy rice paper wrap filled with julienned fresh vegetables and shrimp; served with a dipping sauce
Cubano	Cuba	Cuban bread filled with ham and roast pork marinated in mojo sauce (garlic, salt, pepper, oregano, and citrus juice), dill pickles, and swiss cheese
Falafel	Middle East	A pita filled with a fried patty or ball of spiced chickpeas with lettuce, tomatoes, cucumbers, and onions, along with a tahini or yogurt-based sauce
Gyro	Greece	A pita filled with roasted lamb or beef cut from a vertical spit; garnishes include lettuce, tomatoes, onions, cucumbers, and a yogurt-based tzatziki sauce
Muffuletta	New Orleans	A Sicilian round bread filled with a marinated olive salad, mortadella or smoked ham, salami, and provolone cheese; the olive salad contains olives, assorted pickled vegetables, celery, pimentos, garlic, and cocktail onions
Pan Bagnat	France	A French roll filled with cooked tuna, red onions, hard cooked eggs, olives, capers, tomatoes, anchovies, and red wine vinegar
Po' Boy	Louisiana	A baguette typically filled with fried shrimp, oysters, or catfish; ham, turkey, or roast beef are considered alternate fillings; may also include lettuce, tomatoes, and remoulade or tartar sauce
Roti	India	An Indian flatbread filled with curried meats, potatoes, and vegetables
Shawarma	Middle East	A pita filled with lamb or chicken flavored with vinegar and spices such as cardamom, cinnamon, and nutmeg; garnishes include onions, tomatoes, lettuce, pickled turnips, and cucumbers; a shawarma is often topped with hummus, tahini, or hot sauce
Torta	Mexico	A crusty or a soft roll filled with marinated steak, fried pork, or marinated pork; garnishes include avocados, lettuce, jalapeños, tomatoes, and cheese; a torta is often grilled in a sandwich press or on a griddle

Figure 10-1. *Signature sandwiches are created by combining ingredients that reflect flavor combinations from particular regions or cultures.*

SANDWICH COMPONENTS

Most sandwiches consist of four main components: a base, a spread, one or more fillings, and one or more garnishes. *See Figure 10-2.* Although some sandwiches may have all four of these components, it is not mandatory for a sandwich to include every one of them. Understanding each of the main sandwich components helps a chef to choose interesting combinations that complement each other and deliver satisfying flavors and an enticing presentation. Due to the vast variety of bases, spreads, fillings, and garnishes available, the potential number of sandwich combinations is endless.

Sandwich Components

National Turkey Federation

Figure 10-2. *Most sandwiches have four main components: a base, a spread, one or more fillings, and one or more garnishes.*

Sandwich Bases

A *sandwich base* is the edible packaging that holds the contents of a sandwich. Bread is the most common and easy-to-handle sandwich base. Other bases include rice paper wrappers, nori, and various types of leaves such as lettuce, grape leaves, and banana leaves. The base determines the shape of the sandwich and adds flavor, color, and nutritional value. *See Figure 10-3.*

Breads commonly used for sandwiches include various rolls such as kaiser, French, and Italian, as well as tortillas, pitas, croissants, bagels, baguettes, sourdough bread, rye bread, foccacia bread, ciabatta bread, whole-grain breads, dried cherry or walnut breads, fruit breads, and savory breads such as onion rye or rosemary garlic. *See Figure 10-4.* Whole-grain and multi-grain breads provide more nutrients than refined breads. Bread also quickly satisfies the appetite.

Sandwich Bases

Baguettes

Pitas

Tortillas

Lettuce Leaves

Courtesy of The National Pork Board

Figure 10-3. *Sandwich bases come in many forms, such as baguettes, pitas, tortillas, and lettuce leaves.*

Sandwich Breads

Figure 10-4. *Breads are available in a wide range of textures, flavors, and colors.*

Pullman Loaves

Alpha Baking Co., Inc.

Figure 10-5. *A Pullman loaf is a long, rectangular loaf of bread with four square sides and a consistently fine, dry texture.*

Some bread varieties have a very soft texture while others, such as sourdough, are more firm and chewy. Baguettes and other artisan breads such as ciabatta and pane campagnolo have a hard crusty surface and a chewy interior. When choosing a sandwich bread, consider the following points:

- The flavor of a sandwich bread should complement the other sandwich components, not overpower them.
- The thickness of the bread should not make the fillings appear skimpy.
- The bread should not tear when a spread is applied nor be too tough to bite or chew.

Years ago the Pullman loaf was the most popular bread used to make sandwiches in professional kitchens due to its consistent shape and durability. A *Pullman loaf* is a long, rectangular loaf of sandwich bread with four square sides and a fine, dry texture. *See Figure 10-5.* Its dry texture keeps the bread from becoming soggy after a spread is applied. In years past, white or wheat Pullman loaves were used as the base for many sandwiches. However, use of Pullman loaves has diminished in favor of a wide variety of whole-grain, artisan, and flavored breads.

Breads should only be ordered as needed to offer the freshest bread possible. Other than crusty varieties, breads should be wrapped airtight to prevent them from becoming stale. Crusty varieties are not wrapped airtight as this causes the crust to become soft and lose its crispness. It is important to remember that crusty breads stale quickly and should be used within a day of purchase.

Sandwich Spreads

A *sandwich spread* is a slightly moist, flavorful substance that seals the pores of the bread and creates a thin moisture barrier. *See Figure 10-6.* However, because spreads are slightly moist, they also can make breads soggy if added too far in advance of service. Common types of sandwich spreads include mayonnaise, butters, salad dressings, purées, and variety spreads. A spread should complement the flavor of the main sandwich ingredient.

Spreads as Moisture Barriers

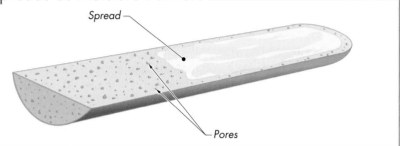

Spread

Pores

Figure 10-6. *A sandwich spread seals the pores of the bread, creating a thin moisture barrier.*

Mayonnaise. Mayonnaise is the most commonly used spread. Mayonnaise is an uncooked emulsion made by combining oil, egg yolks, and lemon juice or vinegar. An *emulsion* is a combination of two unlike liquids that have been forced to bond with each other. Rapid mixing separates tiny droplets of each ingredient, allowing them to combine. As an oil-based spread, mayonnaise acts as a moisture blocker, helping to prevent bread from becoming soggy. However, butter is a better moisture blocker than mayonnaise.

Mayonnaise is often used as a binding agent to hold sandwich fillings together, such as with egg salad, chicken salad, tuna salad, and potato salad. *See Figure 10-7.* Mayonnaise also can be flavored in the same manner as a compound butter by adding garlic, mustard, hot chili paste, curry, lemon, dill, capers, wasabi, or horseradish. Because mayonnaise is much less dense than butter, it does not have to be whipped to incorporate flavoring ingredients.

Butters. Butter is one of the most common sandwich spreads. It adds richness and flavor to a sandwich and also acts as a barrier to prevent moisture from saturating the bread. Butters used on sandwiches should be somewhat soft or whipped before use to make the butter spread more evenly without tearing the bread. Chefs will often whip a small amount of water into butter to lighten its texture and density before using it as a sandwich spread. This technique also increases the volume of the butter as air is incorporated during the whipping process.

Salad Dressings. Salad dressing refers to sandwich spreads similar in flavor, texture, and appearance to mayonnaise. The primary difference between salad dressing and mayonnaise is that salad dressing does not contain egg yolks and it is typically sweeter. Like mayonnaise, salad dressing can also be flavored with other ingredients.

Spreads as Binding Agents

Alpha Baking Co., Inc.

Figure 10-7. *Mayonnaise is used as a binding agent for sandwich fillings, such as with egg salad.*

Chef's Tip

Compound butters are used as sandwich spreads and are simply made by adding flavoring ingredients such as peppers, herbs, or avocado. For example, a spicy avocado butter complements a southwest-style grilled chicken breast sandwich.

Purées. Purées of fruits, vegetables, and legumes can also be used as flavorful sandwich spreads. Purées offer lower fat, cholesterol, and calorie alternatives to other types of spreads. *See Figure 10-8.* For example, a cilantro hummus purée made with garbanzo beans, cilantro, lemon, tahini paste, and olive oil enhances the flavor of a grilled southwest chicken sandwich. A sundried tomato, roasted garlic, and pine nut purée complements a grilled portobello mushroom sandwich. Likewise, a spicy jalapeño and basil pesto purée spread pairs well with a grilled cumin-rubbed steak sandwich. A roasted pineapple honey mustard purée goes nicely with a grilled mahi mahi sandwich. Other types of purées include guacamole, refried beans, baba ghanoush (eggplant purée), and spreadable cooked pâtés.

Nutritional Comparison of Spreads (1 oz each)					
Nutrition Facts	**Butter**	**Mayonnaise**	**Cream Cheese**	**Fresh Avocado**	**Hummus**
Calories	200	110	100	50	50
Total Fat	25 g	10 g	10 g	5 g	5 g
Saturated Fat	14 g	1 g	6 g	1 g	0 g
Cholesterol	60 mg	10 mg	30 mg	0 mg	0 mg

Figure 10-8. *Purées of fruits, such as avocado, and legumes, such as chickpeas (hummus), offer lower fat, cholesterol, and calorie spreads compared to butter, mayonnaise, or cream cheese.*

Variety Spreads. A *variety spread* is any other food mixture of a spreadable consistency that can be added to a sandwich to complement or increase flavor and moisture. Spreadable cheeses, an olive tapenade, a spicy green olive jardinière, and jams are examples of variety spreads that enhance sandwiches.

Sandwich Fillings

A *sandwich filling* is the main ingredient in a sandwich and is stacked, layered, or folded on top of the base to form a sandwich. The filling is often the reason people order a particular sandwich. Many sandwiches are named for their filling, such as hot dogs, hamburgers, and Italian sausage and peppers. *See Figure 10-9.* Types of sandwich fillings include poultry, seafood, meats, cheeses, bound salads, vegetables, or a combination of these items.

Some sandwiches may be composed of several fillings. For example, the Reuben contains corned beef, sauerkraut, and swiss cheese. When multiple fillings are used, it is important to select flavors that complement one another.

Since the filling is the main attraction of a sandwich, it essential that the filling be prepared and served properly. The filling served should never be at room temperature. The temperature of the filling should be appropriately hot or cold, depending on the type of sandwich. For example, when serving a hot fish sandwich, the fish should be hot, moist, flaky, and not overcooked or undercooked. It is also important to place the driest fillings next to the bread to increase the time it takes for the bread to become soggy from the moisture-rich fillings.

Filling-Named Sandwiches

Alpha Baking Co., Inc.

Mini Hot Dogs

*Photo Courtesy of Perdue Foodservice,
Perdue Farms Incorporated*

Thai Turkey Wrap

Courtesy of The National Pork Board

Italian Sausage and Peppers

Figure 10-9. *Many sandwiches are named for their filling, such as mini hot dogs, Thai turkey wraps, and Italian sausage and peppers.*

Cheeses. Cheese is a wonderful sandwich filling that adds color, flavor, and nutrients. Cheese complements almost any other type of filling. Almost any type of cheese can be added to hot or cold sandwiches. *See Figure 10-10.* On a hot sandwich, cheese helps hold the sandwich together as it melts between the bread and the fillings. Sliced cheese dries out easily so it is important to keep it covered in an airtight container or in plastic wrap. Gloves should always be worn when handling cheese to prevent mold growth.

Cream cheese can be used as a filling or a spread. The neutral flavor of cream cheese lends itself to the addition of a variety of flavoring ingredients. Common cream cheese variations include veggie cream cheese (includes puréed or grated vegetables such as carrots, zucchini, and onions), fruit cream cheese (includes puréed fruit such as strawberries or pineapple), and caper cream cheese, which often accompanies lox and bagels.

Cheese Varieties

Wisconsin Milk Marketing Board, Inc.

Figure 10-10. *Almost any type of cheese can be added to hot or cold sandwiches.*

Poultry. Poultry used on sandwiches can be served either precooked and chilled or cooked to order. Precooked poultry includes sliced or diced breast meat, grilled strips, smoked breast meat, and deli-style meat. A chicken breast is grilled, broiled, or fried before being added to a sandwich. *See Figure 10-11.* Chicken breasts also may be marinated, seasoned, or barbequed to add flavor and moisture during the cooking process.

Turkey-based meat substitution products that are lower in fat and in cholesterol than other meats are also popular sandwich fillings. Examples of turkey-based meat substitution products include turkey sausage, turkey bacon, turkey bologna, turkey pastrami, turkey ham, ground turkey burgers, and turkey hot dogs.

Poultry Fillings

— Portabello mushrooms

— Cheese

— Grilled chicken breast

Photo courtesy of Perdue Foodservice, Perdue Farms Incorporated

Figure 10-11. *A chicken breast can be grilled, broiled, or fried before being used as a sandwich filling.*

Fish and Shellfish. Fish and shellfish are also common protein fillings that are used to make sandwiches. Flake tuna and salmon, crab meat, surimi (imitation crab meat), and cooked shrimp are commonly used to make bound salad fillings. *See Figure 10-12.* These items are either fresh or canned and have been fully cooked before being added to a bound salad.

Fish fillets are also common sandwich fillings. The fillets can be grilled, broiled, sautéed, or fried. Other seafood products may be cured or smoked, such as smoked salmon or lox, and simply sliced and served.

Meats. Many meat varieties used in sandwiches are precooked and chilled prior to slicing and serving. This is because attempting to cook them to order in a quick amount of time would not be possible as they come from larger cuts that take hours to prepare properly. Some of them are referred to as deli meats or processed meats. Common examples of fully cooked and cold sandwich style meats include ham, bacon, roast beef, pastrami, salami, and corned beef. Some precooked meats can also be reheated and served hot such as roast beef, hot dogs, pulled or barbecued pork, sausages, corned beef, and bologna.

Surimi filling

Alpha Baking Co., Inc.

Figure 10-12. *Seafood fillings include crab meat and surimi.*

A common mistake is slicing sandwich meats thick instead of thin. A few thick slices of meat stacked on a sandwich will have far less volume and therefore look like less than an equal amount of thinly sliced meat. *See Figure 10-13.* Also, thicker cut meats tend to be chewier, especially when cold. Thinly sliced sandwich meats make the sandwich portion appear larger, are more tender, and are generally easier to eat.

Some sandwich meats are cooked to order. These meat fillings are individually portioned and have a short cooking time. Examples of à la carte sandwich meats include hamburgers, sausages, steaks, and fish fillets.

Meat Fillings

Photo Courtesy of Perdue Foodservice, Perdue Farms Incorporated

Figure 10-13. *Thinly sliced meat gives a sandwich more volume than thick slices of meat.*

Meatless Fillings

California Strawberry Commission

Figure 10-14. *Grilled fruit or vegetable sandwiches, such as a strawberry panini, are a meatless option full of flavor, color, and nutrients.*

Vegetables and Fruits. Vegetables and fruits can also be used as the main filling for a sandwich. Grilled vegetable or fruit sandwiches provide a meatless option and are full of flavor, color, and nutrients. *See Figure 10-14.* Vegetables and fruits also add texture, flavor, nutrients, and moisture to meat-filled sandwiches.

Vegetables and fruits can also serve as a garnish for a sandwich. Sliced tomatoes, lettuce, and pickles are commonly served on the side of a hot sandwich so they do not get limp from the heat. These same vegetables add color and contrast to cold sandwich fillings. Diced vegetables and fruits such as celery and apples are often folded into bound salads to add crunch and flavor. Fresh vegetables and fruits add a fresh, crisp taste to any sandwich.

Eggs. Eggs are a versatile filling for many types of sandwiches and can be served at any mealtime. *See Figure 10-15.* Hard-cooked eggs are cooled, peeled, chopped, and then tossed with mayonnaise, seasonings, and diced vegetables to make a bound egg salad that is perfect on toasted or fresh sliced bread. Hard-cooked eggs can also be sliced and used as a sandwich garnish.

Egg Fillings

Photo Courtesy of Perdue Foodservice, Perdue Farms Incorporated

Figure 10-15. *Eggs are a versatile filling for many types of sandwiches, such as a pastrami egg muffin.*

Bound Salads. A *bound salad* is a salad made by combining a main ingredient, often a protein, with a binding agent such as mayonnaise or yogurt and other flavoring ingredients. Popular bound salads include egg, chicken, ham, and tuna salad. *See Figure 10-16.* A vegetable bound salad can be used to create a meatless sandwich.

Bound Salad Fillings

Figure 10-16. *Popular bound salads include egg, chicken, ham, and tuna salad.*

Sandwich Garnishes

Garnishes add color, texture, and nutrition to sandwiches and should be carefully chosen. A *sandwich garnish* is a complementary food item that is served on or with a sandwich. *See Figure 10-17.* Examples of garnishes include lettuces, tomato and onion slices, pickle slices or spears, olives, raw vegetables (crudités), grilled peppers, crumbled or shredded cheese, sliced fresh fruit, pasta salad, and coleslaw. The number of garnishes that can be used alone or in various combinations to complement sandwiches is endless. It is important to choose garnishes that do not overpower or overshadow the main filling.

Sandwich Garnishes

Hot peppers — Carrots — Coleslaw — Lettuce — Tomatoes

Florida Department of Agriculture and Consumer Services

Figure 10-17. *Garnishes complement sandwiches by adding color, texture, and nutrition.*

HOT SANDWICHES

Hot sandwiches feature cooked fillings and are served hot. Most hot sandwiches are served on warm toasted or grilled bread. Care must be taken when preparing hot sandwiches with uncooked garnishes, such as lettuce, tomatoes, or pickle slices, to avoid wilting the garnish. Fresh garnishes are placed on the sandwich after the hot filling and bread are plated. Hot sandwiches can be grouped into five basic types: hot open-faced sandwiches, hot closed sandwiches, hot wrap sandwiches, grilled sandwiches, and fried sandwiches.

Hot Open-Faced Sandwiches

A *hot open-faced sandwich* is a sandwich consisting of one or two slices of fresh, toasted, or grilled bread, topped with one or more hot fillings, and covered with a sauce, gravy, or a melted cheese topping. *See Figure 10-18.* Often, the hot open-faced sandwich is browned under a broiler just prior to serving. Due to the fact that this variety of sandwich is often covered with a sauce, it is usually eaten with a knife and fork. Popular hot open-faced sandwiches include hot turkey, hot roast beef, and various "melts" such as the patty melt, tuna melt, crab melt, and turkey melt.

Hot Open-Faced Sandwiches

Figure 10-18. *A hot open-faced sandwich is often covered with a sauce and is usually eaten with a knife and fork.*

The most popular hot open-faced sandwich is one that is usually not thought of as a sandwich: the pizza. A pizza consists of bread dough topped with a spread (tomato sauce), a filling (sausage, vegetables, etc.), and a garnish (cheese or spices) all baked until golden brown. In the case of a plain cheese pizza, cheese is the filling and there is typically no garnish. Although once considered exclusively a casual food, gourmet pizzas can be found in restaurants around the world. *See Figure 10-19.* Many chefs use flatbreads and organic flours and toppings to prepare a wide variety of pizzas.

Pizzas

Courtesy of The National Pork Board

Figure 10-19. *Pizza is the most popular type of hot open-faced sandwich.*

Hot Closed Sandwiches

A *hot closed sandwich* is a sandwich made by placing one or more hot fillings between two pieces of bread or a split roll or bun. Hot closed sandwiches are often served with cold garnishes, such as lettuce and slices of tomato and onion. The most common hot closed sandwich is the hamburger. Other hot closed sandwiches include the fish fillet sandwich, grilled chicken sandwich, pastrami sandwich, chicken parmesan sandwich, hot roast beef sandwich, meatball marinara sandwich, steak sandwich, quesadillas, and the veggie burger. *See Figure 10-20.*

Hot Closed Sandwiches

National Turkey Federation

Turkey Pastrami Reuben

Courtesy of The National Pork Board

Pork and Mushroom Quesadilla

Figure 10-20. *A turkey pastrami Reuben and a pork and mushroom quesadilla are both examples of hot closed sandwiches.*

Hot Wrap Sandwiches

A *hot wrap sandwich* is a sandwich made by adding a spread and precooked fillings to a flatbread and then cooking it. Hot wraps can be made using one of two methods. In the first method, flatbread is covered with a spread, topped with a cold precooked filling, and rolled up. The resulting wrap sandwich is then baked, fried, or grilled. An alternate method of making a hot wrap is to place flatbread on a heated griddle, cover it with a spread, top it with a hot precooked filling, and then roll it tightly and serve. Common examples of hot wrap sandwiches are burritos, tacos, fajitas, and enchiladas. *See Figure 10-21.*

Hot Wrap Sandwiches

USA Rice Federation

Burrito

The National Pork Board

Hard Tacos

The National Pork Board

Soft Tacos

Figure 10-21. *Burritos and tacos are popular hot wrap sandwiches.*

Grilled Sandwiches

A *grilled sandwich* is a hot sandwich made by adding a precooked filling or cheese to bread that has been buttered on the exterior and then heated on a griddle, in a sauté pan, or on a panini grill after assembly. A *panini grill* is an Italian clamshell-style grill made specifically to cook grilled sandwiches. *See Figure 10-22.* To grill a sandwich on a panini grill, the bread is first buttered on the exterior sides. The filled sandwich is then placed in the grill and the hinged lid is closed over the sandwich, holding it in place while grilling both the top and the bottom of the bread.

Panini Sandwiches

Panini Sandwiches

Panini Grill

Edward Don & Company

Figure 10-22. *A panini sandwich is prepared on an Italian clamshell-style panini grill.*

Grilled sandwiches are sometimes referred to as toasted sandwiches because the bread turns a toasty golden-brown color during cooking. Grilled sandwiches typically include cheese as a filling, as it melts during cooking and holds the bread to the filling. All grilled sandwich fillings, such as bacon, chicken, or beef, must be thoroughly cooked prior to assembly. Two common grilled sandwiches are the grilled cheese and the Reuben.

Roasted root vegetables can be combined with various cheeses and artisan breads to create tasty grilled sandwiches. Likewise, strawberries and bananas paired with cheese and a hearty bread make great grilled sandwiches. Some people even enjoy grilled peanut butter and jelly sandwiches.

Fried Sandwiches

A *fried sandwich* is a hot sandwich that consists of precooked fillings placed within a closed or wrapped sandwich and then fried. *See Figure 10-23.* Fried sandwiches are prepared in a few different ways. Some fried sandwiches are prepared by dipping a hot closed sandwich into beaten eggs and sometimes bread crumbs before gently placing it in a deep fryer and cooking it until a safe internal temperature is reached to properly cook the raw egg. Fried sandwiches can also be cooked on a griddle or on a sheet pan in the oven.

The Monte Cristo is a popular fried sandwich that consists of two pieces of bread filled with cooked ham, turkey, and swiss cheese that is then soaked in beaten eggs before it is pan-fried. It is often served with fruit or preserves on the side and may be dusted with powdered sugar.

Another popular fried sandwich is the chimichanga. A *chimichanga* is a variety of hot sandwich that consists of a tortilla wrap filled with precooked meat and beans that is then fried. The tortilla is secured with toothpicks before being lowered into the fryer. The tortilla becomes very crisp and the fillings heat to a safe internal temperature. The toothpicks are removed before the sandwich is served.

Fried Sandwiches

Figure 10-23. *A fried sandwich is made by placing precooked fillings within a closed or wrapped sandwich and then frying it.*

COLD SANDWICHES

Cold sandwiches are so called because their fillings are served below room temperature. Often the fillings are cooked ahead of time and then refrigerated, such as with roast beef, turkey, or ham luncheon meats. Cold sandwiches can be grouped into four distinct types: cold open-faced sandwiches, cold closed sandwiches, cold wrap sandwiches, and tea sandwiches. Although each type has some or all of the four basic sandwich components, they are quite different in presentation.

Cold Open-Faced Sandwiches

A *cold open-faced sandwich* is a variety of cold sandwich that consists of a single slice of bread that is often toasted or grilled and then coated with a spread and topped with thin slices of poultry, seafood, meat, partially cooked or raw vegetables, or a thin layer of a bound salad and a garnish. Cold open-faced sandwiches have a more upscale appearance than closed sandwiches. A lox and cream cheese sandwich is a popular cold open-faced sandwich. ***See Figure 10-24.***

Cold Closed Sandwiches

Cold closed sandwiches are the quickest to prepare and the most commonly served. A *cold closed sandwich* is a variety of cold sandwich that consists of two pieces of bread, or the top and bottom of a bun or roll, coated with a spread and one or more fillings and garnishes. Examples of cold closed sandwiches include hero, submarine, and liverwurst sandwiches. ***See Figure 10-25.***

Cold Open-Faced Sandwiches

Alpha Baking Co., Inc.

Figure 10-24. *A lox and cream cheese sandwich is a popular cold open-faced sandwich.*

Cold Closed Sandwiches

Photo Courtesy of Perdue Foodservice, Perdue Farms Incorporated

Figure 10-25. *A cold closed sandwich, such as an Italian submarine, consists of two pieces of bread, or the top and bottom of a bun or roll, coated with a spread and loaded with one or more fillings and garnishes.*

Multidecker sandwiches, such as the turkey club, are a special type of cold closed sandwich. A *multidecker sandwich* consists of three pieces of bread, a spread, and at least two layers of garnishes and fillings. *See Figure 10-26.*

Refer to DVD for
Preparing Multidecker Sandwiches
Media Clip

Procedure for Preparing Multidecker Sandwiches

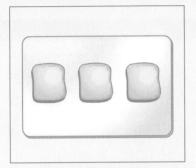

1. Coat one side of three toasted slices of bread with a spread.

2. Layer the garnishes (e.g., lettuce and tomato slices) on two of the slices of bread.

3. Place the fillings (e.g., cheese, turkey, and bacon slices) on top of the garnishes.

4. Pick up one stack of bread, garnishes, and fillings and place it on top of the other stack with the garnishes facing up.

5. Place the third slice of bread, with the spread side down, on top of the double stacked sandwich.

6. Secure the center of each of the four sides of the sandwich with frill picks.

7. Cut the sandwich from corner to corner to produce four triangular sections.

8. Arrange the four frill picked triangles on a plate in an appealing format.

Figure 10-26. *Multidecker sandwiches consist of three pieces of bread, a spread, and at least two layers of garnishes and fillings.*

Cold Wrap Sandwiches

A *cold wrap sandwich* is a variety of cold sandwich in which a flat bread or tortilla is coated with a spread, topped with one or more fillings and garnishes, and rolled tightly. It is typically wrapped in parchment or waxed paper and cut in half on the bias to reveal the filling. *See Figure 10-27.* A cold wrap sandwich is one of the most convenient sandwiches to eat, because the person eating it can hold on to the paper rather than the sandwich.

Procedure for Preparing Cold Wraps

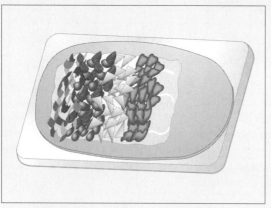

1. Lightly coat ⅔ of a piece of flatbread (such as a tortilla) with a spread, staying ¾ inch from the outside edge.

2. Arrange desired fillings and garnishes on top of the spread.

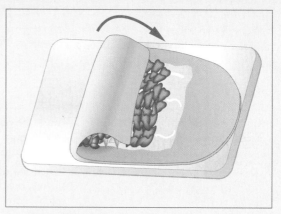

3. Roll the flatbread around the fillings about ⅓ of the way up the flatbread.

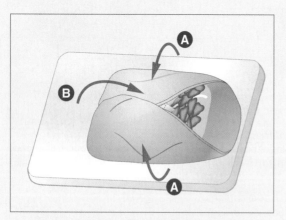

4. Fold in both sides of the flatbread around the rolled portion to seal the ends of the wrap.

5. Continue to roll the flatbread tightly around the fillings.

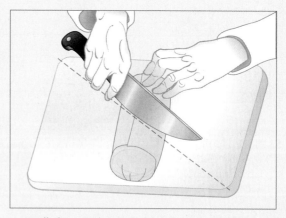

6. Roll the sandwich in deli wrap or parchment paper in the same manner that the flatbread was rolled.

7. Cut the wrap in half on the bias and serve.

Figure 10-27. *A cold wrap sandwich is typically cut in half on the bias to reveal the filling.*

An aram sandwich, also known as a levant, is another type of a cold wrap. An aram sandwich is created by spreading cream cheese on softened cracker bread called lahvosh and then layering thin slices of meats, cheeses, lettuce, and pickles before rolling it up like a jelly roll, wrapping it tightly in plastic wrap, and refrigerating it for several hours. This cold wrap sandwich is cut into 1 inch slices before being served.

Tea Sandwiches

A *tea sandwich* is a petite and delicate cold sandwich with a trimmed crust and a soft filling. The tea sandwich originated in England as a snack served with afternoon tea. The breads used for tea sandwiches are light and soft in texture. The crusts are trimmed off to make them easy to eat without creating a lot of crumbs. The spreads used in tea sandwiches are usually flavorful, and the fillings are usually delicate in flavor, such as with cucumber or watercress fillings. Tea sandwiches are often cut into rectangles, squares, triangles, or circles or can be rolled and sliced into pinwheels. *See Figure 10-28.*

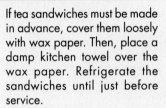
Tea Sandwich Cutting Patterns

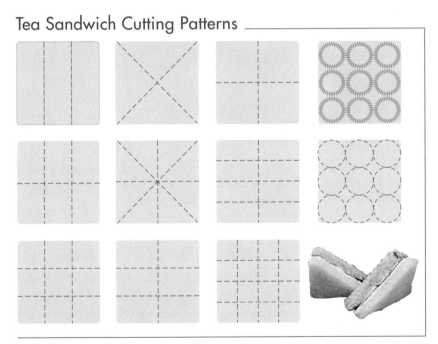

Figure 10-28. *Tea sandwiches are often cut into rectangles, squares, triangles, or circles or can be rolled and sliced into pinwheels.*

SANDWICH STATION SANITATION

Practicing proper sanitation is extremely important at the sandwich station because many sandwiches are served cold and undergo quite a bit of handling by the sandwich maker. All ingredients must be kept out of the temperature danger zone at all times. Likewise, sandwich makers must wear hair restraints. Most foodservice operations and local health departments also require foodservice workers to wear gloves when preparing sandwiches. *See Figure 10-29.* When gloves are used, it is imperative that the gloves be changed frequently to avoid cross-contamination. It is important to wash hands both before and after wearing gloves. Some foodservice operations do not require gloves, but do require frequent handwashing between sandwiches to avoid cross-contamination.

Sandwich Station Sanitation

Edward Don & Company

Figure 10-29. *Many foodservice operations and local health departments require foodservice workers to wear gloves when preparing sandwiches.*

SANDWICH STATION MISE EN PLACE

Preparing sandwiches is a skill that relies on organization, consistency, and speed. Whether preparing sandwiches in large quantities, such as for a banquet, or individually to order, the sandwich station must be well organized to reduce the amount of movement needed and maximize efficiency. If needed, each ingredient in the sandwich station must be washed, dried, sliced, cut, mixed, and portioned in advance.

The equipment needs of a sandwich station are determined by the menu. For example, if deli meats and cheeses are bought sliced, there is no need for an electric slicer at the sandwich station. If there are hot items, such as a hot roast beef sandwich with au jus, the station needs to have a steam table to keep the items hot. Equipment found at a sandwich station includes storage equipment, hand tools, portioning equipment, and cooking equipment. *See Figure 10-30.*

- Storage equipment at a sandwich station typically includes refrigerators, steam tables, and holding cabinets.

- Hand tools such as knives, cutting boards, spatulas, and spreaders are basic necessities for a sandwich station.
- Portioning equipment such as digital or mechanical scales and portion control scoops are essential for making consistent sandwiches.
- Cooking equipment at a sandwich station may include a griddle, grill, broiler, oven, toaster, and a fryer.

Sandwich Station Equipment

Barker Company

Figure 10-30. *A sandwich station includes storage equipment, hand tools, portioning equipment, and cooking equipment.*

Component Preparation

The primary responsibility of a sandwich cook, also known as a pantry cook, is to assemble sandwiches quickly, neatly, consistently, and efficiently. A typical sandwich station might be very busy for 3–4 hours at a time with no opportunity to do additional prep work or refill the sandwich station. All of the sandwich spreads must already be prepared, the meats and cheeses must be sliced and portioned, and the garnishes must be washed and ready to use. *See Figure 10-31.*

Every component needs to be ready for each sandwich combination prior to the first order coming into the kitchen so that there is nothing left to prepare except the sandwiches. Flavored spreads need to be mixed, seasoned, and tasted for flavor in advance. The fillings and garnishes for each sandwich also must be prepared in advance. All produce needs to be washed and dried before starting sandwich preparation. Whole produce such as tomatoes and onions need to be sliced to the proper thickness. Side dishes such as coleslaw, potato salad, fruit, and crudités must also be ready to serve.

Component Preparation

Paderno World Cuisine

Figure 10-31. *All sandwich components must be prepared, portioned, and ready to use prior to filling a sandwich order.*

Range of Motion

Because much of sandwich making is assembly work, it is important that the individual making the sandwiches use both hands for efficient production. A sandwich maker can determine a comfortable range of motion by extending both arms directly out in front and then sweeping each arm in an arc to each side. *See Figure 10-32.* All of the ingredients in a sandwich station should be within this range of motion to maximize efficiency and speed. This also allows one hand to place the filling on the bread while the other hand adds the garnishes.

Range of Motion

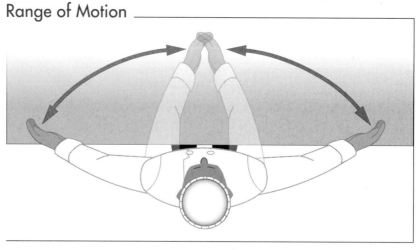

Figure 10-32. *A comfortable range of motion can be determined by extending both arms directly out in front and then sweeping each arm in an arc to each side.*

Right-handed sandwich makers should place the bread supply to their left, and left-handed sandwich makers should place the bread supply to their right. Spreads and filling ingredients should be placed directly in front of the sandwich maker. An organized workspace enables efficient production and reduces the time that ingredients are exposed to the temperature danger zone.

Although a foodservice operation may have different procedures for making each sandwich on the menu, quantities of sandwiches can be made using an assembly procedure. *See Figure 10-33.* This procedure includes grilling or toasting the bread (if desired) and coating one side of each slice with a spread, placing garnishes and hot or cold fillings on top of the spread, adding the top piece of bread, and then securing the sandwiches before cutting them for service.

Portion Control

All items for sandwich making should be portioned prior to service. All meats should be sliced and portioned by weight. Individual portions can either be placed in small plastic bags or in small piles separated by individual squares of waxed paper and then stacked or layered in a storage container. Items that are served by the slice, such as cheese, should be sliced to an exact thickness to ensure a consistent portion.

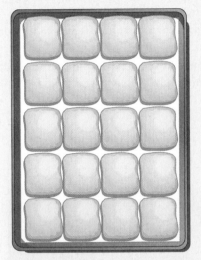

1. Arrange bread or toast slices in rows on a sheet pan lined with parchment paper.

2. Use a spatula to evenly coat each slice of bread or toast with a spread.

3. Arrange garnishes on top of the bread or toast in the two center rows.

4. Place portioned fillings on top of the garnishes, making sure the bread is well covered.

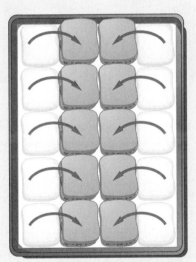

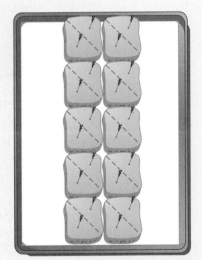

5. Place the remaining slices of bread on the top of slices holding the fillings and garnishes.

6. Secure each sandwich with frill picks and then cut each sandwich in half. *Note:* Sandwiches may be stacked so that several may be cut at the same time.

Refer to DVD for
Preparing Sandwiches in Quantity
Media Clip

Figure 10-33. *Large quantities of sandwiches can be made by using a basic assembly procedure.*

Practicing proper portion control helps maintain accurate food costs, as the menu price of an item is based on that item's portion size. If portions are not consistent from one sandwich to the next, the actual cost of making each sandwich will also not be consistent. *See Figure 10-34.*

Proper portion control also ensures that customers receive the same value for their money. For example, if two customers at the same table order corned beef sandwiches and one customer receives more meat than the other customer, the customer who receives less meat will not be happy with the sandwich.

Portion Control

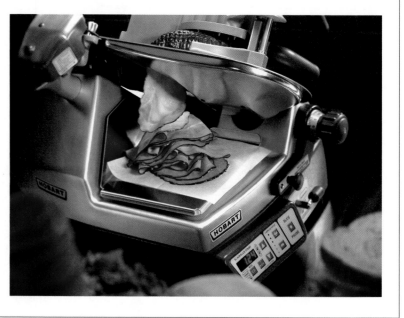

Hobart

Figure 10-34. *If portions are not consistent from one sandwich to the next, the actual cost of making each sandwich will also not be consistent.*

Sandwich Stabilizers

An assembled sandwich should have a neat appearance and should be easy to eat. While different types of bread, buns, and flatbreads make handling sandwiches easier, a sandwich often needs to be secured to prevent the fillings from falling out while the sandwich is being held. Cheeses, wrappers, and frill picks serve as stabilizers that secure sandwich components together. *See Figure 10-35.*

Cheeses. In addition to being a sandwich filling, cheese also serves as a stabilizer that makes a sandwich easier to hold and eat. The cheeses used on many hot closed sandwiches bind the bread to the fillings due to the heat of the main filling.

Wrappers. One of the easiest ways to secure a sandwich is with a wrapper. Sandwiches on a bun can be served in a paper sleeve to make them easier and less messy to handle. Likewise, sandwich wraps can be rolled in a piece of waxed paper or parchment paper. Having a paper wrapper around a sandwich not only keeps the hands cleaner but is a fast, safe, and convenient way to hold the entire sandwich together.

Frill Picks. Cold closed sandwiches are commonly held together with frill picks, which are slightly longer toothpicks that have a frilly plastic decoration at one end and a sharp point on the other end. The pointed end is inserted into the top of a sandwich to secure the top and bottom breads to the contents inside. The frill picks are quite visible to prevent a choking accident. A frill pick may also be used to secure the breads to the fillings of some hot closed sandwiches.

PLATING SANDWICHES

Just as the presentation is important for every other dish that comes out of the kitchen, consistent sandwich presentation is equally as important. Hot sandwiches such as a cheeseburger, grilled chicken breast sandwich, or pulled pork sandwich are often served open-faced. *See Figure 10-36.* The top bun and garnish are not placed on top of the hot meat as it would cause the cold garnishes to wilt. Cold sandwiches are typically cut into halves or quarters to expose the interior of the sandwich and to make it easier to eat. Cutting sandwiches in half or into quarters reveals the layers of color and texture.

Sandwich Stabilizers

Alpha Baking Co., Inc.

Cheeses

Photo Courtesy of Perdue Foodservice, Perdue Farms Incorporated

Wrappers

Photo Courtesy of Perdue Foodservice, Perdue Farms Incorporated

Frill Picks

Figure 10-35. *Cheeses, wrappers, and frill picks serve as stabilizers that secure sandwich components together.*

Plating Hot Sandwiches

Courtesy of The National Pork Board

Figure 10-36. *Hot sandwiches such as pulled pork sandwiches are often served open-faced.*

Sandwiches are usually served with a side dish that complements that particular sandwich. Common side dishes such as coleslaw, pasta salad, potato salad, french fries, sweet potato fries, potato or vegetable chips, crudités, cut fruit, or a side salad add visual appeal, flavor, and nutrition to plated sandwiches. *See Figure 10-37.*

Sandwich Sides

Photo Courtesy of McCain Foods USA

Figure 10-37. *Common side dishes such as sweet potato fries add visual appeal and flavor to a plated sandwich.*

SUMMARY

The sandwich is one of the most basic and easy-to-prepare food items. However, success at making sandwiches requires organizational skills, speed, consistency, and coordination. A sandwich is typically composed of a base, a spread, and one or more fillings and garnishes. The possible combinations of sandwich ingredients are limited only by a sandwich maker's imagination.

Some sandwiches are named for their fillings such as a hot roast beef sandwich, fillet of fish sandwich, or cheeseburger. Sandwiches are divided into hot and cold categories, depending on the serving temperature of the fillings. They are served open-faced, closed, or as wraps. Sandwiches also may be grilled, fried, or cut into various shapes before service.

All sandwich components must be prepared and ready for assembly prior to filling the first sandwich order. Sanitation is also critical at a sandwich station. Components must be kept out of the temperature danger zone at all times. Placing items within the sandwich maker's range of motion will increase efficiency and speed. The appropriate stabilizer is added to each sandwich before it is plated. Hot sandwiches must be plated and served immediately. If cold sandwiches are not served promptly, they must be tightly sealed in plastic wrap and refrigerated.

Refer to DVD for
Quick Quiz® questions

1. Explain why there is a large variety of sandwiches.
2. Explain how to lower the fat and calorie contents of sandwiches.
3. Identify the four main sandwich components.
4. Identify common types of sandwich bases.
5. Describe the primary purpose that a sandwich base serves.
6. Identify five common types of sandwich spreads.
7. Identify seven common types of sandwich fillings.
8. List three sandwiches named for their filling.
9. Identify common sandwich garnishes.
10. Identify five types of hot sandwiches.
11. Identify four types of cold sandwiches.
12. Describe how to prepare multidecker sandwiches.
13. Describe how to prepare cold wraps.
14. Identify the items that should be worn at a sandwich station for sanitation purposes.
15. Explain the importance of range of motion at a sandwich station.
16. Describe how to prepare large quantities of sandwiches.
17. Describe common ways to portion meats and cheeses at a sandwich station.
18. Identify common side dishes served with plated sandwiches.

Refer to DVD for
Review Questions

Applications

1. Research one sandwich from the "Sandwiches Around the World" chart and prepare it to order.
2. Prepare a variety of sandwich bases for use at a sandwich station.
3. Prepare a flavored butter. Evaluate the texture and flavor of the spread.
4. Prepare a flavored mayonnaise or salad dressing. Evaluate the texture and flavor of the spread.
5. Prepare a purée sandwich spread. Evaluate the texture and flavor of the spread.
6. Prepare a variety spread. Evaluate the texture and flavor of the spread.
7. Prepare a variety of sandwich fillings for use at a sandwich station.
8. Prepare a bound salad for use as a sandwich filling. Evaluate the texture and flavor of the filling.
9. Prepare a variety of sandwich garnishes for use at a sandwich station.
10. Prepare and plate a hot open-faced sandwich and a hot closed sandwich using the same filling(s). Evaluate the appearance and flavor of the two sandwiches.
11. Prepare and plate a hot wrap. Evaluate the appearance and flavor of the sandwich.
12. Prepare and plate a grilled sandwich. Evaluate the appearance and flavor of the sandwich.
13. Prepare and plate a fried sandwich. Evaluate the appearance and flavor of the sandwich.
14. Prepare and plate a cold open-faced sandwich and a cold closed sandwich. Evaluate the appearance and flavor of the two sandwiches.
15. Prepare and plate a multidecker sandwich. Evaluate the appearance and flavor of the sandwich.
16. Prepare and plate a cold wrap. Evaluate the appearance and flavor of the sandwich.
17. Prepare and plate tea sandwiches using two different cutting patterns. Evaluate the appearance of the two cutting patterns.
18. Prepare and plate a quantity of sandwiches using a bound salad filling. Evaluate the appearance and flavor of the sandwiches.

Refer to DVD for
Application Scoresheets

Spicy Avocado Butter

Yield: *8 oz*

Ingredients

unsalted butter, softened	8 oz
avocado, sliced	½ ea
jalapeño, minced	1 ea
cilantro, minced	1 tbsp
lemon juice	2 tsp
salt and white pepper	TT

Preparation

1. Place all ingredients in a food processor and purée until almost smooth.
2. Lay out a 6 × 12 inch piece of parchment paper. Scoop the butter mixture into a log shape on the parchment paper.
3. Roll the butter into a 10–12 inch log, approximately 1½ inches in diameter.
4. Refrigerate the log until needed.
5. When needed, slice a ½ inch thick slice of spicy avocado butter and remove the wrapping.
6. Place the butter slice on freshly grilled or cooked food just before service.

> **NUTRITION FACTS**
> **Per serving:** 74 calories, 72 calories from fat, 8.2 g total fat, 20.3 mg cholesterol, 13.6 mg sodium, 23.4 mg potassium, <1 g carbohydrates, <1 g fiber, <1 g sugar, <1 g protein

Herbed Chicken Salad Sandwiches

Yield: *6 servings, 1 sandwich each*

Ingredients

boneless, skinless chicken breasts	24 oz
mayonnaise	¾ c
celery, small dice	2 tbsp
onion, minced	1 tbsp
salt and pepper	TT
fresh tarragon, minced	1 tsp
fresh chives, minced	1 tsp
sandwich bread	12 slices

Preparation

1. Steam or lightly sauté chicken breasts until cooked through. Let cool.
2. When cool, dice chicken into medium dice and place in mixing bowl.
3. Add mayonnaise, celery, onions, salt, pepper, and herbs. Mix the ingredients well.
4. Refrigerate chicken salad for several hours or overnight.
5. Place 5 ounces of chicken salad on a slice of bread. Top with a second slice of bread.
6. Serve sandwich immediately or wrap tightly in plastic wrap and refrigerate.

> **NUTRITION FACTS**
> **Per serving:** 378 calories, 128 calories from fat, 14.4 g total fat, 80.2 mg cholesterol, 646.4 mg sodium, 482.4 mg potassium, 32.6 g carbohydrates, 1.3 g fiber, 4.2 g sugar, 28.2 g protein

Roasted Tomato, Prosciutto, and Garlic Pizza

Yield: *4 servings, 2 slices each*

Ingredients

pizza dough	8 oz
corn meal	2 tbsp
tomato sauce	1 c
fresh basil, chopped	1 tbsp
olive oil	1 tbsp
mozzarella cheese, shredded	1 c
medium tomato, sliced	1 ea
roasted garlic, chopped	5 cloves
prosciutto, thinly sliced and cut into small pieces	6 oz

Preparation

1. On a lightly floured surface, roll out pizza dough into a 12 inch circle.
2. Spread corn meal on the pizza pan and then place the dough on top.
3. Spread tomato sauce evenly over the dough, leaving a ⅛ inch border around the edge.
4. Top tomato sauce with fresh basil.
5. Brush the outer edge with olive oil.
6. Layer top of pizza with shredded mozzarella, sliced tomatoes, roasted garlic cloves, and prosciutto.
7. Bake in a 475°F oven for 10 minutes, or until cheese melts and crust is golden brown.

NUTRITION FACTS
Per serving: 361 calories, 125 calories from fat, 13.9 g total fat, 50.9 mg cholesterol, 1677 mg sodium, 607 mg potassium, 33.8 g carbohydrates, 2 g fiber, 3.9 g sugar, 25 g protein

Philly Beef Sandwiches

Yield: *4 servings, 1 sandwich each*

Ingredients

medium onions, sliced	2 ea
sweet peppers (red or green), julienned	2 ea
garlic, minced	1 clove
olive oil	1 tbsp
pepper	TT
beef tenderloin, sliced very thin	1 lb
Italian rolls	4 ea
mozzarella cheese	4 slices

Preparation

1. Sauté onions, peppers, and garlic in olive oil until tender. Season with pepper and set aside.
2. Pan-fry beef until brown, but not crispy.
3. Slice each roll in half lengthwise, leaving the two halves attached.
4. Layer beef, onions, and peppers inside each roll. Top with a slice of mozzarella cheese.
5. Wrap sandwich in foil to hold in the heat and allow the cheese to melt. Serve immediately.

NUTRITION FACTS
Per serving: 529 calories, 196 calories from fat, 22 g total fat, 112.7 mg cholesterol, 554.2 mg sodium, 639.7 mg potassium, 37.3 g carbohydrates, 3.1 g fiber, 4.3 g sugar, 43.7 g protein

Bistro Burgers

Yield: *4 servings, 1 hamburger each*

Ingredients

ground beef	1 lb
garlic, minced	1 clove
fresh parsley, chopped	2 tbsp
pepper	TT
sesame rolls	4 ea
Boston lettuce	4 leaves
onion, sliced thin	4 oz
tomato, sliced	4 oz
dill pickle, sliced	2 oz

Preparation

1. Combine ground beef, garlic, parsley, and pepper. Blend mixture well.
2. Divide into four equal balls. Flatten each ball into a patty with a 6 inch diameter.
3. Grill patties to desired doneness.
4. Slice sesame rolls in half.
5. Place a cooked hamburger on the bottom half of each roll.
6. Layer the top half of each roll with lettuce, onion, tomato, and then pickles.
7. Immediately plate the open-faced hamburgers.

NUTRITION FACTS
Per serving: 381 calories, 130 calories from fat, 14.5 g total fat, 69.3 mg cholesterol, 493.7 mg sodium, 462.1 mg potassium, 34.8 g carbohydrates, 2.5 g fiber, 3.2 g sugar, 26.5 g protein

Reuben Sandwiches

Yield: *4 servings, 1 sandwich each*

Ingredients

butter	2 tbsp
dark pumpernickel rye bread	8 slices
corned beef, shredded	1 lb
sauerkraut	8 oz
swiss cheese	4 slices

Preparation

1. Spread butter on one side of each slice of bread.
2. Place the bread butter-side down on a cutting board.
3. Layer corned beef, sauerkraut, and then swiss cheese on four of the slices.
4. Top each sandwich with remaining slices of bread, butter side up.
5. Grill sandwiches on both sides until browned and cheese has melted. Serve immediately.

NUTRITION FACTS
Per serving: 700 calories, 408 calories from fat, 45.6 g total fat, 197.8 mg cholesterol, 2592.6 mg sodium, 459.8 mg potassium, 29.4 g carbohydrates, 5 g fiber, 1.7 g sugar, 41.7 g protein

Monte Cristo Sandwiches

Yield: *2 servings, 1 sandwich each*

Ingredients

white sandwich bread	4 slices
Dijon mustard	3 tsp
gruyère cheese	4 slices
ham, sliced thin	4 slices
roasted turkey, sliced thin	4 slices
eggs	2 ea
water	2 tbsp
butter	4 tbsp

Preparation

1. Arrange bread on cutting board. Spread mustard on each slice.
2. Place two slices of cheese, two slices of ham, and two slices of turkey on two of the slices of bread. Top each sandwich with the remaining slices of bread.
3. Whisk eggs and water together in a shallow bowl until well blended.
4. Melt butter in a skillet over medium-low heat.
5. Dip both sides of each sandwich in the egg batter.
6. Cook the egg-battered sandwiches in the hot butter until the bread is golden (approximately 4 minutes per side), flipping as needed.
7. Cut fried sandwiches in half diagonally and serve immediately.

> **NUTRITION FACTS**
> **Per serving:** 763 calories, 443 calories from fat, 50.3 g total fat, 357.9 mg cholesterol, 1774.1 mg sodium, 713.6 mg potassium, 29.1 g carbohydrates, 1.7 g fiber, 4.5 g sugar, 47.5 g protein

Classic Italian Submarine Sandwiches

Yield: *4 servings, 1 sandwich each*

Ingredients

Italian rolls	4 ea
extra virgin olive oil	2 tbsp
red wine vinegar	1 tbsp
garlic, minced	1 clove
fresh oregano, finely chopped	1 tsp
pepper	TT
capicola, thinly sliced	8 oz
Genoa salami, thinly sliced	8 oz
provolone cheese, sliced	4 oz
mozzarella cheese, sliced	4 oz
romaine lettuce, shredded	2 c
medium tomato, sliced	1 ea
medium red onion, thinly sliced	1 ea
red pepper, roasted and julienned	1 ea
black olives, sliced	4 tbsp

Preparation

1. Partially slice each roll in half lengthwise, leaving the two halves attached.
2. Combine olive oil, vinegar, garlic, oregano, and pepper in a small bowl. Mix well to create vinaigrette.
3. Brush a small amount of the vinaigrette on the cut side of each roll.
4. Layer the meats, cheeses, and vegetables, placing the meats on the bottom and vegetables on top.
5. Drizzle the remaining vinaigrette on the vegetables.
6. Serve sandwich immediately or wrap tightly in plastic wrap and refrigerate.

> **NUTRITION FACTS**
> **Per serving:** 731 calories, 388 calories from fat, 43.7 g total fat, 124.3 mg cholesterol, 2366.8 mg sodium, 650 mg potassium, 41.4 g carbohydrates, 4.3 g fiber, 3.2 g sugar, 42.1 g protein

Triple-Decker Turkey Club Sandwiches

Yield: *2 servings, 1 sandwich each*

Ingredients

sandwich bread	6 slices
mayonnaise	3 tbsp
romaine lettuce	6 leaves
medium tomato, sliced into	
4 or 8 slices	1 ea
turkey breast, cooked	4 oz
bacon, cooked and cut into	
4 inch lengths	4 slices

Preparation

1. Lightly toast bread.
2. Lightly coat one side of 6 slices of bread with mayonnaise. Set aside 2 coated slices.
3. Top 4 of the coated slices with lettuce, tomato, turkey, and then bacon.
4. Doublestack the filled slices on top of each other to form two sandwiches.
5. Place the two reserved slices of bread, spread side down, on top of the doublestacks to form multidecker sandwiches.
6. Insert frill picks into the top center of each side of the two sandwiches.
7. Cut each sandwich into four triangular portions before plating.

NUTRITION FACTS
Per serving: 446 calories, 157 calories from fat, 17.6 g total fat, 47.7 mg cholesterol, 1489.9 mg sodium, 543.2 mg potassium, 49.1 g carbohydrates, 3.3 g fiber, 8.6 g sugar, 22.3 g protein

Chipotle Turkey Wraps

Yield: *4 servings, 1 wrap each*

Ingredients

mayonnaise	½ c
fresh cilantro, chopped	3 tbsp
red onion, minced	2 tbsp
chipotle hot sauce	2 tsp
lime	½ ea
salt	TT
flour tortillas, 8–10 inches	
in diameter	4 ea
smoked turkey breast, thinly sliced	1 lb
romaine lettuce, shredded	2 leaves
tomato, diced	4 oz
cheddar cheese, shredded	4 oz

Preparation

1. In a bowl, combine mayonnaise, cilantro, onion, hot sauce, and juice from half a lime to create chipotle mayonnaise. Season with salt to taste.
2. Spread chipotle mayonnaise on each tortilla.
3. Layer turkey, lettuce, tomato, and then cheese on each tortilla.
4. Roll up tortilla tightly.
5. Cut each wrap in half and serve immediately.

NUTRITION FACTS
Per serving: 576 calories, 235 calories from fat, 26.6 g total fat, 86.2 mg cholesterol, 2119.8 mg sodium, 586.3 mg potassium, 50.8 g carbohydrates, 3.6 g fiber, 8.3 g sugar, 33 g protein

Smoked Salmon Wasabi Tea Sandwiches

Yield: *4 servings, 3 tea sandwiches each*

Ingredients

sandwich bread	6 slices
wasabi paste	1 tbsp
cream cheese	4 oz
smoked salmon, thinly sliced	4 oz

Preparation

1. Place bread slices on a cutting board. If slices are more than ¼ inch thick, use a rolling pin to gently flatten bread.
2. Whisk wasabi paste and cream cheese together until well blended.
3. Spread wasabi cream cheese on each slice of bread.
4. Evenly divide salmon and place on half of the slices of bread.
5. Top salmon with remaining slices of bread, spread-side down.
6. Cut off the crusts and cut each sandwich into quarters.
7. Serve immediately or wrap individually and refrigerate.

NUTRITION FACTS
Per serving: 232 calories, 107 calories from fat, 12.2 g total fat, 37.7 mg cholesterol, 516.7 mg sodium, 135.5 mg potassium, 20.6 g carbohydrates, 1 g fiber, 2.8 g sugar, 9.8 g protein

National Turkey Federation

the fragile yolk and white, it is porous and fragile. This is why eggs should be stored away from strong smelling foods such as garlic and onions. The porous shell also allows the moisture inside the egg to evaporate over time.

In addition to the shell, the fragile yolk and white are also protected by a shell membrane. The *shell membrane* is a thin, skinlike material located directly under the eggshell. The *yolk* is the yellow portion of the egg. Although the yolk is only about one-third the weight of the egg, it contains more than three-fourths of the calories and all of the cholesterol found in the egg.

The *albumen,* also known as the white, is the clear portion of the raw egg, which makes up two-thirds of the egg and consists mostly of ovalbumin protein. It is the primary food source for the developing embryo and provides a cushion from the hard shell covering. The albumen contains the majority of protein, no cholesterol, a trace amount of fat, and only about 17 calories per large egg. When cooked, the albumen turns white, which is why it is commonly referred to as the egg "white."

Another component of the egg found in the albumin is the chalazae. The *chalazae* are twisted cordlike strands that anchor the yolk to the center of the albumen. The chalazae maintain the internal structure of the egg.

Egg Nutrition

Eggs are a nutrient-dense food, which means they have a high proportion of nutrients to calories. *See Figure 11-3.* A large egg provides varying amounts of 13 essential nutrients. Eggs are an affordable source of high-quality protein that includes all nine essential amino acids and two antioxidants, lutein and zeaxanthin, which help prevent common causes of age-related blindness.

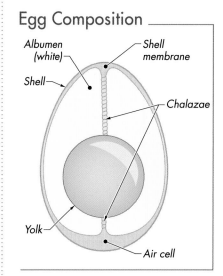

Egg Composition

Figure 11-2. *An egg is composed of four main parts: the shell, shell membrane, yolk, and albumen (white).*

Nutritional Data for a Large (Raw) Egg						
Total Calories	72	**Vitamins**	**Amount**	**Minerals**	**Amount**	
From Fat (63%)	46.03	Vitamin A	270 IU	Calcium	28 mg	
From Protein (35%)	25.89	Vitamin D	41 IU	Copper	0.04 mg	
From Carbohydrates (2%)	1.58	Vitamin E	0.53 mg	Fluoride	0.6 mcg	
Total Fat	4.75 g	Vitamin K	0.10 mcg	Iodine	23.70 mcg	
Saturated Fat	1.50 g	Vitamin C	0 IU	Iron	0.88 mg	
Monounsaturated Fat	1.91 g	Thiamin (Vitamin B1)	0.02 mg	Manganese	0.01 mg	
Polyunsaturated Fat	0.68 g	Riboflavin (Vitamin B2)	0.23 mg	Magnesium	6 mg	
Cholesterol	211.50 mg	Niacin (Vitamin B3)	0.04 mg	Phosphorus	99 mg	
Sodium	71 mg	Pantothenic Acid (Vitamin B5)	0.77 mg	Potassium	69 mg	
Potassium	67 mg	Vitamin B6	0.08 mg	Selenium	15.30 mcg	
Total Carbohydrates	0.36 g	Choline	125.50 mg	Zinc	0.65 mg	
Fiber	0 g	Folate (Vitamin B9)	24 mcg			
Sugar	0.18 g	Vitamin B12	0.45 mcg			
Protein	6.28 g					

Source: USDA National Nutrient Database, SR 23

Figure 11-3. *Eggs are a nutrient-dense food, which means they have a high proportion of nutrients to calories.*

Egg yolks contain a lot of cholesterol, but they are also packed with vitamins, minerals, and essential fatty acids. Although the yolk is only one third of the egg, it is the nutrition powerhouse. In addition to cholesterol and fat, the yolk contains the majority of the thiamin, pantothenic acid, folate, vitamin B6 and B12, calcium, iron, phosphorous, and zinc. The egg white contains most of the sodium, magnesium, and niacin.

Egg Substitutes

Eggs (more specifically egg yolks) are naturally high in fat and cholesterol, which causes concern for people with high cholesterol, heart disease, and related health concerns. Egg whites, egg substitutes, and eggless egg substitutes each offer alternatives to whole eggs and egg yolks. For example, egg whites do not contain any cholesterol but can be used to make a fluffy omelet.

An *egg substitute* is a liquid product that is typically made from a blend of egg whites, vegetable oil, food starch, powdered milk, artificial colorings, and additives. One-fourth cup of liquid egg substitute contains only 1 mg of cholesterol as compared to 211.5 mg in one large egg yolk. Egg substitutes are used to prepare omelets, scrambled eggs, quiches, and custards. Eggs and egg substitutes are sold in cartons and often sold frozen for a longer shelf life. *See Figure 11-4.*

Eggs and Egg Substitutes

Liquid egg whites

Pasteurized whole eggs

Liquid egg yolks

American Egg Board

Figure 11-4. *Eggs and egg substitutes are also sold frozen for a longer shelf life.*

An *eggless egg substitute* is a yellow-colored liquid composed of soy, vegetable gums, and starches derived from corn or flour. Eggless egg substitutes do not contain any egg product. They have the appearance of beaten eggs but not the flavor. For this reason, eggless egg substitutes are not used to prepare egg dishes such as omelets, scrambled eggs, or custards. Instead, they are almost exclusively used to produce cholesterol free or lower cholesterol baked goods.

Egg Allergies

Some customers have an allergic reaction when they eat eggs or egg-derived ingredients. As one of the eight most common food allergens, current FDA labeling regulations require the presence of eggs be marked on food labels in bold print. *See Figure 11-5.* It is important to note that eggs and egg-derived ingredients may be listed as the following:

- eggs, powdered eggs, egg whites, or egg yolks
- any ingredient with the prefix ovo- or ova- (meaning "egg" in Latin)
- albumin, globulin, livetin, or lysozyme (which are all egg proteins)
- Simplesse® (a fat substitute made from egg protein)
- mayonnaise

Eggs or egg-derived ingredients may also be present in many menu items served in restaurants, such as the following:

- aioli, hollandaise, béarnaise, and rémoulade sauce
- noodles, pasta, and breads with shiny crusts (brushed with egg wash)
- cakes, cookies, doughnuts, pancakes, and waffles
- meringues, puddings, divinity, jelly beans, and filled chocolates
- ice cream, gelato, frozen custards, and sherbets
- pretzels, zwieback, and crackers
- breaded vegetables, poultry, seafood, and meats
- soups and casseroles
- meatloaf, meatballs, croquettes, and sausages

Egg Sizes and Grades

Eggs are classified according to size based on the minimum weight per dozen eggs. Each named size weighs about 3 oz per dozen more than the previous size. *See Figure 11-6.* The size does not refer to the dimensions of an egg or how big it appears. For example, some eggs may be slightly smaller or larger than others in the same carton. Size refers to the total weight of the dozen, which places the eggs in one of the following categories:

- pee wee—15 oz to 18 oz
- small—18 oz to 21 oz
- medium—21 oz to 24 oz
- large—24 oz to 27 oz
- extra large—27 oz to 30 oz
- jumbo—30 oz and above

Food Labels

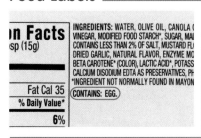

Figure 11-5. *U.S. Food and Drug Administration (FDA) labeling regulations require the presence of eggs be marked on food labels in bold print.*

Chef's Tip

Large eggs are the most common size of egg used in foodservice operations, and most recipes use large eggs as the standard.

Egg Sizes

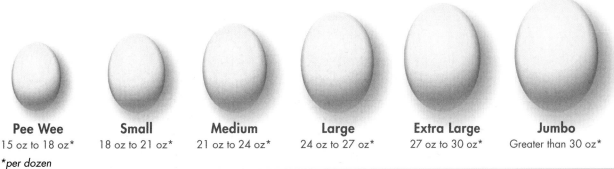

| Pee Wee | Small | Medium | Large | Extra Large | Jumbo |
| 15 oz to 18 oz* | 18 oz to 21 oz* | 21 oz to 24 oz* | 24 oz to 27 oz* | 27 oz to 30 oz* | Greater than 30 oz* |

*per dozen

Figure 11-6. *Eggs are classified according to size based on the minimum weight per dozen eggs.*

The United States Department of Agriculture (USDA) oversees the inspection of eggs for wholesomeness. Inspection is mandatory, but grading for quality is voluntary. The USDA Agricultural Marketing Service administers the voluntary egg-quality grading program for shell eggs, which is paid for by the egg processing plants. Egg cartons bearing the USDA shield indicate they have been graded for quality by a trained USDA grader. Egg cartons marked "Grade A" that do not bear the USDA shield have not been graded. State agencies monitor egg packers that do not use the grading service.

The size of an egg, the color of its shell, and its grade have no bearing on its nutritional value. Eggs are graded as AA, A, or B. The highest quality egg, Grade AA, has a firm yolk and white that both stand tall when the egg is broken onto a flat surface. As an egg ages or if it is of a lower grade, the yolk and the white will lie flatter and spread out farther on a flat surface. *See Figure 11-7.* The white will also be looser and more watery, and the air pocket inside the egg will be larger.

Egg Grades

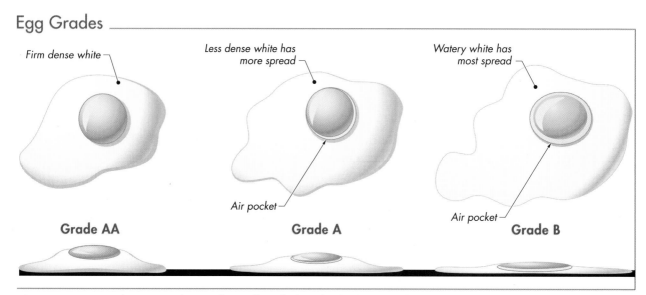

Figure 11-7. *Grade AA eggs have a firm yolk and white that both stand tall, whereas the yolk and white of lower grade eggs lie flatter and spread out farther.*

Egg Safety and Sanitation

Fresh eggs are used more than any other form of egg in the professional kitchen. Like all protein-rich foods, eggs are a potentially hazardous when not handled and stored properly. Protein-rich foods are excellent breeding grounds for harmful bacteria. Salmonella enteritidis is the bacteria found inside eggs that have become contaminated from infected laying hens. *See Figure 11-8.* The potential presence of Salmonella enteritidis is why it is best to thoroughly cook eggs prior to eating them.

Salmonella Enteritidis

Figure 11-8. *Eggs are tested for Salmonella enteritidis, the bacteria found inside eggs that are contaminated from infected laying hens.*

To avoid the contamination issue, some foodservice operations use pasteurized eggs and egg products. A *pasteurized egg* is an egg that has been heated to a specific temperature for a specific period of time to kill bacteria that can cause foodborne illnesses. The pasteurization process takes place before the egg is packaged, either in the shell or as a liquid egg product. The use of pasteurized eggs and egg products reduces the potential of a foodborne illness being contracted from eating foods that contain uncooked or partially cooked eggs. Pasteurized eggs are a good option for preparing egg dishes that are not fully cooked, such as sunny-side up eggs and over-easy eggs.

Eggs may also be purchased in liquid, frozen, and dried forms. These alternative forms of eggs are convenient in volume food preparation settings, especially for preparing baked products. Dried eggs are not often used in foodservice operations. However, dried egg whites are occasionally used in bakeshops to make meringues.

Cooked eggs perish rapidly so it is best to cook eggs in small quantities and as close to serving time as possible. Holding cooked eggs for even 10 minutes will show signs of quality loss.

Egg Storage

Shell eggs should always be refrigerated at 40°F or below and may be kept refrigerated for about a month past the Julian date. The Julian date stamped on the egg carton indicates the day the eggs were packed. Julian dates begin with January 1 as day 001 and number successively through December 31 as day 365 to represent the days of the calendar year.

Leftover egg whites and egg yolks should be tightly covered with plastic wrap before being refrigerated and should be used within 2–4 days. *See Figure 11-9.* Yolks that are left uncovered will form a surface crust and dry out.

Opened liquid-egg substitutes can be stored for three days. Frozen egg products should always be thawed in the refrigerator and then stirred thoroughly before use. Dried eggs should always be stored in a cool, dry place and then refrigerated once they have been reconstituted.

Leftover Eggs

Figure 11-9. *Leftover eggs should be tightly covered with plastic wrap before being refrigerated and should be used within two to four days.*

EGG PREPARATION

The most common mistake made when cooking eggs is overcooking or under-cooking them. Using very high heat or too long a cooking time will cause eggs to become rubbery. It can also change their color and flavor. For example, the yolk of an egg cooked in the shell may form a green outer ring when overcooked due to the sulfur in the egg white reacting with the iron in the yolk. Likewise, overcooked scrambled eggs that sit in a steam table too long often develop a green cast.

Eggs can be prepared many different ways using dry-heat and moist-heat cooking methods. Eggs can be fried, scrambled, made into omelets and frittatas, baked, poached, and simmered in the shell.

Fried Eggs

Sunny-side up, basted, over-easy, over-medium, and over-hard eggs are cooked to order. Although they are all referred to as fried eggs, these egg dishes are technically sautéed and pan-fried. *See Figure 11-10.* The main difference among each dish is the variation in cooking time and technique.

Fried Eggs

Figure 11-10. *Fried eggs are technically sautéed and pan-fried.*

Sunny-Side Up Eggs. A *sunny-side up egg* is a lightly fried egg with an unbroken yolk that is not flipped over to cook the other side. The yolk of a sunny-side up egg is still liquid inside and bright yellow on top.

Basted Eggs. A *basted egg* is a fried egg with an unbroken yolk that is cooked like a sunny-side up egg, yet the tops are slightly cooked by tilting the pan and basting the top of the egg with hot butter. The yolk is still liquid inside, but is covered with a thin white film.

Over-Easy Eggs. An *over-easy egg* is a fried egg with an unbroken yolk that is cooked until the egg white has gone from translucent to white on the bottom and then flipped over and cooked on the other side just until the white is no longer translucent. The yolk of an over-easy egg is still liquid inside and covered with a white film.

Over-Medium Eggs. An *over-medium egg* is a fried egg with a completely cooked white and a yolk that is cooked nearly all the way through. Over-medium eggs are simply cooked a little longer than over-easy eggs. The yolk of an over-medium egg is fairly set around the edges, has an almost fully cooked appearance, and yet is soft in the very center.

Over-Hard Eggs. An *over-hard egg* is a fried egg with a bright, firm white that is not rubbery and a pale yellow yolk that looks almost fluffy. Over-hard eggs are cooked a little longer than over-medium eggs.

Fried eggs should always be cooked to order using fresh Grade AA eggs. The fresher the egg, the more the yolk stands up and the less runny the white will be. If fried eggs are cooked in butter, only clarified butter should be used due to it having a higher smoke point than whole butter, which burns easily. Many foodservice operations use nonstick cooking spray instead of using butter.

Selecting the correct size sauté pan always yields good results. For a single egg, the sauté pan should be 4 inches in diameter at the bottom. For two eggs, the pan should be 6 inches in diameter. A nonstick sauté pan is often used to help prevent fried eggs from sticking and potentially breaking the yolks. *See Figure 11-11.*

Nonstick Sauté Pans

The Vollrath Company, LLC

Figure 11-11. *A nonstick sauté pan is often used to help prevent fried eggs from sticking and potentially breaking the yolks.*

Prior to cooking, eggs should be cracked into a small dish so that the raw eggs can be inspected for quality, as well as to help ensure that the yolks do not break when added to the pan. Then, the sauté pan is heated over moderate heat and approximately 2 tsp of fat or a light coating of nonstick cooking spray is added to the bottom of the pan. Next, the eggs are gently poured into the pan. The hot oil immediately solidifies the eggs so the whites will not spread. The eggs are cooked to the desired doneness. *See Figure 11-12.*

Procedure for Frying Eggs

1. Crack eggs into a small dish or bowl.
2. Place a small amount of fat into a hot sauté pan or on a hot griddle.
3. Gently slide the eggs into the pan or onto the griddle.
4. When whites are firm, gently flip the egg over to cook the other side. The egg yolk should always be on the far side of the pan when flipping an egg. *Note:* Remember that sunny-side up eggs and basted eggs are not turned over.
5. Cook eggs to desired degree of doneness (i.e., over-easy, over-medium, or over-hard).
6. When eggs are done, gently flip them back over, slide them onto a plate, and serve immediately.

Figure 11-12. *Sunny-side up, basted, over-easy, over-medium, and over-hard eggs are cooked to order.*

- Use a conditioned pan or griddle. A poorly conditioned pan or griddle can cause eggs to break and burn.
- Use the correct amount of fat. Too much fat results in greasy eggs that may also burn. Too little fat results in eggs that stick to the pan and break when trying to remove them from the pan.
- Use the correct temperature. Too high a temperature can cause eggs to burn and overcooking may result in a tough exterior and a runny interior. Too low a temperature can cause egg whites to spread too rapidly.

National Turkey Federation

Scrambled Eggs

Scrambled eggs are a popular egg dish that is often served on buffets. Scrambling is the easiest method to use when preparing eggs in quantity. First, the eggs are whisked using a tilted wheel-like motion to combine the yolks and whites. *See Figure 11-13.* Then the eggs are sautéed in a steam-jacketed kettle, in tilt skillet, on a griddle, in a double boiler, or in a sauté pan to produce fluffy curds. When scrambling small amounts of eggs, it is best to use a sauté pan.

To prepare scrambled eggs, the eggs are broken into a nonreactive bowl and seasoned with salt and pepper. An aluminum bowl should never be used because it will discolor the eggs as they are whisked. The eggs are lightly beaten with a whisk or kitchen fork. If desired, about 1 tsp of milk, cream, or water per egg may be added. Too much added liquid will cause the eggs to weep (give off water) after they are cooked.

The beaten eggs are poured into a heated, oiled, or buttered sauté pan, where they start to coagulate immediately. After the eggs are added to the pan, the heat is reduced and the egg mixture is carefully stirred from the bottom with a high-heat spatula. Scrambled eggs should always be slightly undercooked, because they become firm as they are held for service. If scrambled eggs are overcooked, they will become dry, hard, and unpalatable. Scrambled eggs are properly cooked when they are fluffy, moist, and not runny. *See Figure 11-14.*

Chef's Tip

When scrambled eggs are to be held longer than 5 minutes, a small amount of béchamel sauce or heavy whipping cream can be added to extend the holding time and prevent the eggs from drying out and discoloring.

Whisking Technique

Figure 11-13. *Scrambled eggs are whisked using a tilted wheel-like motion to combine the yolks and whites.*

Procedure for Preparing Scrambled Eggs

1. Break the eggs into a stainless steel mixing bowl, season with salt and pepper, and, if desired, add a small amount of milk, cream, or water.
2. Beat the eggs with a wire whisk until well-mixed and fluffy.
3. Heat a sauté pan or griddle with a small amount of fat.
4. Pour the eggs into the pan or onto the griddle and stir slowly with a high-heat spatula until the eggs are set but still moist.
5. Serve immediately.

Figure 11-14. *Scrambled eggs are properly cooked when they are fluffy, moist, and not runny.*

Tips for Scrambling Eggs

- Use the correct amount of fat. Too much fat yields greasy eggs. Too little fat can cause eggs to stick, become tough and dry, or burn.
- Cook eggs at the correct temperature. Too high a temperature yields overcooked, dry, brown eggs that may also burn.
- Do not stir eggs excessively. Excessive stirring breaks the egg into very fine particles, resulting in a poor appearance once they are cooked.
- Do not hold scrambled eggs too long. Holding eggs too long results in an unappetizing flavor and a green cast.

Omelets

An *omelet* is an egg dish made with beaten eggs and cooked into a solid form. Omelets are one of the most commonly ordered egg dishes in restaurants. Omelets may be filled with a variety of cheeses, vegetables, and meats. ***See Figure 11-15.*** A filled omelet is often named for the filling ingredient, such as a bacon omelet or a mushroom omelet.

Omelets are prepared to order and usually consist of two or three eggs per order. If omelets are held for even a short period of time, they lose their fluffiness and become rubbery. There are four common varieties of omelets: rolled omelets, folded omelets, soufflé omelets, and frittatas.

Omelet Fillings

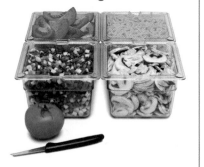

Carlisle FoodService Products

Figure 11-15. *Omelets may be filled with a variety of cheeses, vegetables, and meats.*

Rolled Omelets. A *rolled omelet,* also known as a French omelet, is an omelet that is cooked and then rolled onto a plate and cooked filling ingredients are added through a slit cut into the top. To prepare a rolled omelet, cook an omelet until it is slightly set but still very moist. Then, tilt the pan about 60° and use a high-heat spatula to roll the omelet. The eggs should be pale yellow and browned as little as possible or not at all. Once the egg portion of the omelet is cooked, roll the omelet tightly on the plate and add cooked ingredients and cheeses through a slit cut into the top of the omelet. ***See Figure 11-16.***

1. Add fat to a heated omelet pan over medium heat.

2. Whisk eggs in a nonreactive bowl. If desired, season with salt and pepper.

3. Pour the egg mixture in the pan and gently shake the pan while stirring with a heat-resistant spatula.

4. When the mixture is almost set, gently lift one side of the omelet to allow any liquid to run underneath and cook. Then, sweep the sides of the pan with the spatula to round out the omelet.

5. Fold one side of the omelet a third of the way toward the center.

6. Place the omelet pan over the serving plate with the unfolded end of the omelet nearest the plate and gently roll the omelet onto the plate, seam side down. Serve immediately.

7. If desired, cooked filling ingredients can be added through a slit cut in the top of the omelet.

Figure 11-16. *A rolled (French) omelet is cooked and then rolled onto a plate. Cooked filling ingredients can be added through a slit cut in the top, if desired.*

Tips for Preparing Omelets

- Use a conditioned sauté pan. A poorly conditioned pan can cause omelets to stick as well as break when rolled or folded.
- Use the correct amount of fat. Too much fat yields greasy eggs. Too little fat can cause eggs to stick, become tough and dry, or burn.
- Cook eggs at the correct temperature. Too high a temperature yields brown omelets that may also crack when rolled or folded.
- Do not stir the eggs once the omelet begins to set or the end product will resemble scrambled eggs.
- Do not hold omelets. Held omelets are tough and rubbery.

Refer to DVD for
Preparing Rolled (French) Omelets
Media Clip

Folded Omelets. A *folded omelet* is an omelet that is cooked until nearly done and then folded before serving. Any vegetables or meats need to be precooked, or at least heated, prior to adding them to the beaten eggs. The exception to the rule is cheese, which is added just before the omelet is folded near the end of the cooking process.

After the egg has been cooked almost all of the way through, the omelet is typically flipped over and allowed to cook for approximately 15 seconds longer. During this time, the cheese is added to the center of the omelet. The omelet is then folded in half and served. ***See Figure 11-17.***

1. Add fat to a heated omelet pan over medium heat.

2. Add fillings (except cheese) to the omelet pan and cook until done. *Note:* If preferred, precooked fillings may be added just prior to folding the omelet.

3. Whisk eggs in a nonreactive bowl. If desired, season with salt and pepper.

4. Pour beaten eggs over the cooked fillings, and stir with a heat-resistant spatula until the eggs are somewhat set.

5. When the mixture is almost set, gently lift one side of the omelet to allow any liquid to run underneath and cook. Then, sweep the sides of the pan with the spatula to round out the omelet.

6. Let the omelet sit undisturbed for about 15 seconds.

7. If desired, add cheese to the center of the omelet.

8. Fold the omelet in half. *Note:* If a tri-fold is desired, fold a third toward the center and then fold the other third toward the center. Turn the omelet over onto the plate, seam side down.

9. Slide the unfolded end of the omelet gently onto a plate. Serve immediately.

Figure 11-17. *A folded omelet is cooked, then folded in half, and served.*

Refer to DVD for
Preparing Folded Omelets
Media Clip

Soufflé Omelets. A *soufflé omelet,* also known as a fluffy omelet, is a lighter variation of a folded omelet. To make a soufflé omelet, the eggs are beaten to incorporate air, which increases the volume. Air can be incorporated into a soufflé omelet two different ways. One method is to separate the yolks from the whites and then whisk each into a soft foam. Next, the whisked whites and the whisked yolks are folded together and then poured into a hot, oiled pan and cooked in the same manner as a folded omelet.

An immersion blender can also be used to whip air into the egg mixture. When this technique is used, the egg mixture will cook up even lighter than an egg mixture that was whisked by hand. The omelet is then placed under a broiler for 10–15 seconds to crisp. A soufflé omelet will puff slightly due to the air incorporated from using the blender and the addition of whipped egg whites.

Frittatas. A *frittata* is basically a traditional folded omelet served open-faced after being browned in a broiler or hot oven. The ingredients are all precooked and mixed with the eggs, but instead of folding the omelet, it is left open and cheese (if desired) is added on top. The omelet is then placed under a broiler or in a very hot oven to melt and brown the cheese. *See Figure 11-18.*

Refer to DVD for
Preparing Frittatas
Media Clip

Procedure for Preparing Frittatas

1. Add fat to a heated omelet pan over medium heat.
2. Add fillings to the omelet pan and cook until done.
3. Whisk eggs in a nonreactive bowl. If desired, season with salt and pepper.
4. Pour beaten eggs over the cooked fillings and stir with a heat-resistant spatula until the eggs are somewhat set.
5. When the mixture is almost set, gently lift one side of the frittata to allow any liquid to run underneath and cook.
6. Top the frittata with cheese, if desired.
7. Place the frittata under a broiler or in a hot oven to melt and brown the cheese. Serve immediately.

Figure 11-18. *A frittata is basically a traditional folded omelet served open-faced after being browned in a broiler or hot oven.*

Egg Sandwiches

A fried egg, scrambled eggs, or a mini omelet can be added to a hot biscuit, bagel, English muffin, tortilla, or croissant to create a delicious handheld breakfast. *See Figure 11-19.* Cheese, vegetables, or meats may also be added to enhance the flavor and the size of the sandwich.

Breakfast Sandwiches

American Egg Board

Courtesy of the National Pork Board

Figure 11-19. *A handheld breakfast can be created by adding a fried egg, scrambled eggs, or a mini omelet to a hot biscuit, bagel, English muffin, tortilla, or croissant.*

Shirred Eggs

A *shirred egg,* also known as a baked egg, is an egg that is baked on top of other ingredients in a shallow dish in the oven. After the dish has been coated with butter, any type of ingredients may be placed into the dish. Then, raw eggs are removed from the shell and gently poured on top of the ingredients. The dish is then placed in an oven and baked at 350°F until the eggs set and cook until the desired degree of doneness. *See Figure 11-20.* Shirred eggs often top ingredients such as ham, sausage, Canadian bacon, cheese, asparagus, mushrooms, or artichoke bottoms.

Procedure for Preparing Shirred Eggs

1. Butter a ramekin or other small casserole dish with butter.
2. Place a small amount of cooked garnish ingredients in the buttered dish.
3. Remove two eggs from their shells and pour over ingredients.
4. Season the eggs and bake in a 350°F oven until the whites are completely set and yolk is cooked to the desired degree of doneness.

Figure 11-20. *Shirred eggs are baked at 350°F until the eggs set and the desired degree of doneness is achieved.*

It is important to not overcook shirred eggs. When cooked at too high a temperature, shirred eggs become tough and usually burn. When cooked at too low a temperature, the whites spread and the yolk has a tendency to break.

Quiches

A *quiche* is a baked egg dish composed of a savory custard baked in a piecrust. *See Figure 11-21.* A savory custard consists of beaten eggs mixed with cream, milk, and seasonings. The cream and milk added to the beaten eggs cause the egg protein in the custard to be more delicate and tender when cooked. To prepare a quiche, a piecrust can be filled with a variety of cooked vegetables, meat, and cheeses. Then, the savory custard mixture is poured on top of the chosen ingredients and the quiche is baked until set. *See Figure 11-22.* It is important to note that the custard should not be cooked to an internal temperature above 185°F or the egg proteins will begin to curdle and separate from the water molecules in the egg.

Quiches

Courtesy of The National Pork Board

Figure 11-21. A quiche is a baked egg dish composed of a savory custard baked in a piecrust.

Procedure for Preparing Quiche

1. Roll 8 oz of pie dough into a 10 inch circle.
2. Place the dough into an 8 inch pie shell and press gently to set. Place in refrigerator until well chilled.
3. Once the crust has chilled completely, place a second pie tin inside the first pie tin and bake upside down in a 450°F preheated oven for 10 minutes to prebake the crust. *Note:* Prebaking will ensure the crust will be cooked by the time the quiche is finished baking and that the crust will not get soggy when the custard is added.
4. While the crust is prebaking, prepare the filling ingredients (e.g., diced bacon or shredded cheese).
5. Remove the piecrust from the oven and add the prepared filling ingredients.
6. Whisk milk, eggs, cream, salt, and pepper together to create the custard.
7. Pour the custard on top of the prepared filling ingredients.
8. Place the quiche on a sheet pan to catch any spillover during baking.
9. Bake quiche in a 375°F oven for about 35–45 minutes or until the custard is firm and set. *Note:* When a toothpick inserted into the center of the quiche comes out clean, the quiche is done.

Figure 11-22. To prepare quiche, piecrust is filled with cooked vegetables, cooked meats, cheeses, and a savory custard mixture.

Poached Eggs

Poached eggs are very popular on breakfast menus for dishes such as eggs à la Florentine and eggs Benedict. Poached eggs are removed from the shell prior to cooking. Vinegar is used to set the white firmly around the yolk when the egg is placed in the water. The acid from the vinegar actually toughens the protein in the egg white and does not affect the flavor of the egg.

Although poached eggs are very easy to prepare, the challenge is in determining doneness as well as transporting eggs from the poaching water to the plate without breaking the yolks. Eggs cooked at too high a temperature are usually overcooked and tough. Eggs cooked at too low a temperature are too tender and very difficult to handle when they are served.

About 8 eggs can be poached at a time in a gallon of water. The water may be used for three different batches before being discarded. Eggs that are poached in quantity should be slightly undercooked and then immediately placed in cold water to stop further cooking and held until service. To serve, the precooked eggs are reheated in hot salted water. This quantity method is often used for banquet or buffet preparations. For individual breakfast preparations, eggs should be poached to order. *See Figure 11-23.*

Refer to DVD for
Poaching Eggs
Media Clip

Procedure for Poaching Eggs

1. Fill a pan with enough water to cover the eggs.

2. Add 1 tsp of salt and 1 tbsp of distilled white vinegar per quart of water.

3. Bring the water to a boil and then lower the heat to approximately 180°F.

4. Break each egg into a nonreactive bowl and then gently slide the eggs into the simmering liquid so the yolk remains in the center of the white and does not break.

5. Cook to desired doneness (3–5 minutes).

6. Remove poached eggs with a skimmer or slotted spoon and drain well. Serve immediately.

Figure 11-23. *Eggs can be poached to order or partially poached in quantity and held for later service.*

Traditional eggs Benedict, known as the most lavish of egg dishes, consists of two toasted English muffin halves, each topped with a slice of Canadian bacon and a poached egg. The poached eggs are then topped with a coating of fresh hollandaise sauce. *See Figure 11-24.*

1. Prepare hollandaise sauce and set aside until needed.
2. Prepare poaching liquid for eggs and hold between 160°F and 180°F.
3. Heat two slices of Canadian bacon on a griddle or in a sauté pan until warmed through. Set aside until needed.
4. Crack two eggs into a small bowl and pour them gently into the poaching water.
5. Split an English muffin and toast in toaster until done. Set on serving plate until eggs are ready.
6. Top each muffin half with one slice of heated Canadian bacon.
7. Using a slotted or perforated spoon, remove eggs from poaching water at desired degree of doneness (soft, medium, or hard cooked). Place the spoon with eggs on a folded paper towel to draw any excess water from the eggs.
8. Place well drained egg on top of each piece of Canadian bacon and muffin half.
9. Ladle an ample portion of hollandaise sauce over each poached egg and serve immediately.

Figure 11-24. *Eggs Benedict is a lavish egg dish in which hollandaise sauce is poured over poached eggs atop Canadian bacon and an English muffin.*

The richness of the poached eggs and hollandaise sauce complement the saltiness of the Canadian bacon and the crunch of a toasted English muffin. Variations of traditional eggs Benedict are offered by many foodservice operations. For example, avocado and turkey can be placed under the egg and hollandaise sauce and atop an English muffin to offer a unique flavor combination. *See Figure 11-25.*

National Turkey Federation

Variations of Eggs Benedict	
Variation	**Description**
Country Benedict	Replace English muffin and Canadian bacon with a buttermilk biscuit and a sausage patty; then top eggs with country-style white gravy
Eggs Blackstone	Replace Canadian bacon with a slice of regular smoked ham topped with a tomato slice
Eggs Bombay	Serve poached eggs on a bed of rice pilaf and top with a creamy curry sauce
Eggs Florentine	Replace Canadian bacon with sautéed spinach and top eggs with hollandaise or Mornay sauce
Eggs Sardou	Replace English muffins with cooked artichoke bottoms and the Canadian bacon with two crossed anchovies; then top eggs with a slice of truffle
Irish Benedict	Replace Canadian bacon with corned beef hash
Lobster Benedict	Replace Canadian bacon with slices of sautéed lobster tail or claw meat

Figure 11-25. *Popular variations of eggs Benedict are offered by many foodservice operations.*

Eggs in the Shell

Although the common term for cooking eggs in the shell is "boiled eggs," the fact is that eggs in the shell should be simmered, never boiled. Boiling tends to toughen the texture of the egg and can create a green sulfur ring around the outside of the yolk. Simmering at a temperature of 195°F is recommended, as higher temperatures yield tough, rubbery eggs. Temperatures below 195°F may yield undercooked eggs. *See Figure 11-26.*

Procedure for Preparing Eggs in the Shell

1. Place the room temperature eggs in a pot and cover them with cold water.
2. Bring the water to a gentle boil over medium heat.
3. Reduce the heat to a simmer and begin timing. Simmer eggs to desired doneness (i.e., soft, medium, or hard).

Figure 11-26. *Eggs in the shell are simmered to the desired degree of doneness.*

Hard-Cooked Eggs

Egg white is bright white and has solid texture

Yolk is pale yellow with a crumbly texture

American Metalcraft, Inc.

Figure 11-27. *Hard-cooked eggs in the shell have a bright, solid white and a pale, crumbly yolk.*

Eggs in the shell that are cooked for 3 minutes are considered soft-cooked. Medium-cooked eggs are cooked for 4–6 minutes. Hard-cooked eggs are cooked for 9–10 minutes and have a bright, solid white and a pale, crumbly yolk. *See Figure 11-27.*

Hard-cooked eggs should be plunged into cold water to stop the cooking process. If the eggs are to be held for use in a cold preparation, such as egg salad, they should be cooled in ice-cold water for about 5 minutes immediately after cooking, peeled and placed in a bain-marie, covered with cold water, and refrigerated for up to a day. This process will prevent the eggs from drying out. Cooked eggs in the shell can be stored in the refrigerator for up to a week.

Always begin peeling eggs in the shell at the large end of the egg and move toward the narrow end. Keeping the egg submerged in cold water while peeling will help loosen the shell.

PANCAKES AND WAFFLES

Pancakes and waffles are both popular breakfast items that can be served with a variety of different toppings. Pancakes and waffles are always served piping hot. Berries, apples, bananas, or pecans are often added to pancake batter or served on top of waffles. Both may also be served with breakfast meats such as sausage or bacon.

Pancakes are also known as hotcakes, griddle cakes, or flapjacks. The basic recipe for pancakes consists of flour, sugar, salt, baking powder, baking soda, milk or buttermilk, eggs, and butter or oil. The dry ingredients are sifted together, and the milk, eggs, and butter are whisked together in a separate

Cuisine Note

Potato pancakes are commonly served with sour cream or applesauce. Shredded potatoes are mixed with flour, eggs, onions, and baking powder and then cooked on a griddle.

bowl. Then the liquid mixture is slowly combined with the dry mixture and the batter is ladled on a hot oiled griddle or into a hot sauté pan. The pancake is ready to be flipped over when the surface is nearly covered in bubbles. When the second side turns golden brown, the pancakes are ready to be plated and served. *See Figure 11-28.*

Pancakes

Pancake Preparation

Plated Pancakes

Figure 11-28. *When the surface of a pancake is nearly covered in bubbles, it is ready to be flipped over. When the second side turns golden brown, the pancakes are ready to be plated and served.*

The procedure for making waffles is very similar to pancakes except that a waffle iron is used in place of a griddle or a pan. Waffle batter is poured onto a well-oiled or nonstick waffle iron and the lid is closed. The batter is simultaneously cooked for a few minutes on both sides. *See Figure 11-29.* Waffles are most often served with a fruit topping and whipped cream.

Pancakes and waffles may also be prepared in advance, although the quality is not nearly as good as when they are freshly made. Both are completely cooked and then shingle-layered on sheet pans. The pancakes and waffles are then warmed in the oven before plating.

FRENCH TOAST

French toast is a popular breakfast item prepared by dipping slices of bread into a batter made from eggs, milk, sugar, and vanilla. The batter-dipped bread is then placed on a lightly oiled griddle or pan and cooked on each side until golden brown. *See Figure 11-30.*

Crunchy French toast is prepared the same way as traditional French toast except that the batter dipped bread is also dipped in a crunchy coating such as crumbled Corn Flakes™ or sliced nuts. The toast is then cooked on a griddle or in a pan until it is golden on both sides. It is then placed on a sheet pan and placed in a 350°F oven for 10 minutes until crispy and cooked through. Both traditional and crunchy French toast can be served with a fruit topping, syrup, or confectioner's sugar.

Waffles

Chef's Choice® by EdgeCraft Corporation

Figure 11-29. *A waffle iron simultaneously cooks both sides of a waffle.*

French Toast

National Honey Board

Figure 11-30. *French toast is batter-dipped bread that is cooked until each side is golden brown.*

CRÊPES AND BLINTZES

A crêpe is a French pancake that is light and very thin. Crêpe batter does not contain leavening agents and therefore remains very thin when cooked. Crêpes are cooked either in a very hot crêpe pan, in a small sauté pan, or on a crêpe maker. The surface of a crêpe is cooked to a light golden-brown color. Typically, crêpes are filled with fruit or cheese, rolled up, and topped with confectioners' sugar. *See Figure 11-31.* Crêpes may be served for breakfast, as a dessert, or filled with savory items for lunch or dinner. Although crêpes are delicate, they can be frozen.

Refer to DVD for
Preparing Crêpes
Media Clip

Procedure for Preparing Crêpes

1. Add oil, clarified butter, or non-stick cooking spray to a small omelet pan over medium heat.

2. Add just enough batter to lightly coat the pan.

3. Spread the batter across the diameter of the pan and cook over medium-low heat.

4. Before the crêpe begins to turn golden, flip with a rubber spatula.

5. When the batter is set, add filling if desired.

6. Fold an edge of the crêpe toward the center and then position the pan over the plate.

7. Roll the crêpe onto the plate and serve immediately.

Figure 11-31. *Crêpes are often filled with fruit or cheese, rolled up, and topped with confectioners' sugar.*

A *blintz* is a crêpe that is only cooked on one side and not flipped over to cook the other side. Like crêpes, blintzes can be filled with sweet or savory fillings. Sweet blintzes are often filled with a sweetened ricotta or a farmer's cheese mixture and then rolled or folded into small pillow-shaped parcels. *See Figure 11-32.* Then, the blintz is put back into a hot sauté pan with clarified butter and browned on the top and bottom. Blintzes are garnished with any variety of fruit compote, warmed fruit preserves, or sour cream. A savory blintz can be filled with any variety of sautéed vegetables, but the most common filling is potato and onion garnished with sour cream.

BREAKFAST MEATS

Breakfast meats commonly include a variety of sausages, bacon, ham and Canadian bacon, steaks and pork chops, herring and salmon, and hash. Most breakfast meats may be precooked and then reheated for service when the breakfast demand is large. When the demand is smaller, breakfast meats are usually cooked to order.

Sausages

Breakfast pork sausages often have a strong flavor of sage and crushed red pepper flakes. Turkey sausages offer a lower fat and calorie alternative to pork sausage. *See Figure 11-33.* For example, a fresh cooked turkey sausage patty contains 55 calories, 3 g of fat, 26 mg of cholesterol, and 7 g of protein. In contrast, a fresh cooked pork sausage patty contains 95 calories, 8 g of fat, 24 mg of cholesterol, and 5 g of protein. However, turkey sausage does contain more cholesterol than pork sausage.

Sausages come in many flavors, which often connect to their cuisine of origin. For example, chorizo con huevos mixes chopped and fried chorizo sausage with scrambled eggs. *Chorizo* is a Mexican pork sausage flavored with paprika that can be sweet or spicy. Likewise, *kielbasa* is a Polish pork or beef sausage flavored with garlic, pimento, and cloves. *Boudin* is a highly seasoned Creole link sausage made of pork, pork liver, and rice. In the midwestern United States, bratwurst is a common accompaniment for eggs. *Bratwurst* is a German sausage composed of veal, pork, or beef.

Breakfast sausages are served in the form of patties or links. Sausage patties may be purchased in bulk or prepared in the professional kitchen. Patties are usually portioned into 1 oz or 2 oz servings. Sausage patties can be precooked on sheet pans in a 350°F oven, under a broiler, on a griddle, or in a skillet.

Sausage links typically range in size from about 1 oz each (16 per lb) to 2 oz each (8 per lb). The portion served for breakfast is generally three or four links if they are smaller and one or two links if they are larger. Sausage links are cooked by separating the links, placing them on sheet pans, and baking at 350°F until done. *See Figure 11-34.* After cooking, sausage links are placed in a hotel pan and held for service.

Blintzes _____

Figure 11-32. *Blintzes are often filled with a sweetened cheese mixture.*

Turkey Sausages _____

Photo Courtesy of Perdue Foodservice, Perdue Farms Incorporated

Figure 11-33. *Turkey sausages offer a lower fat alternative to pork sausages.*

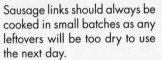

Chef's Tip

Sausage links should always be cooked in small batches as any leftovers will be too dry to use the next day.

Sausage Links

Figure 11-34. *Sausage links are cooked by separating the links, placing them on a sheet pan, and baking at 350°F until done.*

Bacon

Different types of bacon are served at breakfast. Pork bacon comes from the belly of a pig and is the most common breakfast bacon. Hickory smoked, applewood smoked, peppered, and turkey bacon frequently appear on breakfast menus. Italian breakfasts often feature strips of pancetta. *Pancetta* is a dry, salt-cured pork belly that has been spiced and then dried for about three months. Unlike bacon, pancetta is traditionally not smoked.

Sliced bacon is packaged in a few different forms. Slab-packed bacon is presliced but the user has to pull the slices apart. Shingled bacon consists of slightly overlapped sliced strips of bacon. Layer-packed bacon is packaged with each slice separated and laid out on sheets of parchment paper. *See Figure 11-35.* Layer-packed bacon is the type commonly used in foodservice operations, because it does not require someone to spend time taking the slices apart. Many restaurateurs are willing to pay for the convenience.

The most common thickness of sliced bacon is an 18–22 count, meaning there will be between 18 and 22 slices per pound. Bacon slices are typically placed side by side, but not touching, on a sheet pan and baked at 350°F until they are three quarters done. The bacon is then removed from the oven and placed in a bread-lined hotel pan to absorb excess grease. The bacon is finished just before service. This method of preparation also reduces shrinkage and curling. Bacon slices may also be cooked in a broiler or on a griddle.

Layer-Packed Bacon

Figure 11-35. *Layer-packed bacon is packaged with each slice separated and laid out on sheets of parchment paper.*

Ham and Canadian Bacon

Ham and Canadian bacon are popular breakfast meats. ***See Figure 11-36.*** Both meats are precooked and only need to be heated on a griddle, under a broiler, or in a hot pan before service. Breakfast ham is typically purchased in a boneless or boned and rolled form, providing 3 oz or 4 oz portions when sliced. *Canadian bacon* is a hamlike breakfast meat made from boneless, smoked, pressed pork loin. The perfectly round shape and size of Canadian bacon serves as a great base for a breakfast sandwich or eggs Benedict.

Ham and Canadian Bacon

Breakfast Ham Focaccia

Canadian Bacon Omelet

Figure 11-36. *Ham and Canadian bacon are popular breakfast meats.*

Courtesy of The National Pork Board

Steaks and Pork Chops

Steak and eggs or pork chops and eggs are often featured on breakfast menus. These dishes typically include eggs cooked to order and a side of breakfast potatoes. *See Figure 11-37.* Steaks served for breakfast include the T-bone, New York strip, and butt steak. Breakfast steaks and pork chops are usually cut thin for a shorter cooking time on a griddle or under a broiler.

Chicken-fried steak, also known as country-fried steak, is another popular breakfast steak. Chicken-fried steak is prepared by breading a thin, tenderized cut of cubed or round beef steak and pan-frying it in oil. The name of this dish reflects its similarity to the preparation of fried chicken.

Steak and Eggs

Figure 11-37. *A steak and eggs or chops and eggs dish typically includes eggs cooked to order and a side of breakfast potatoes.*

Herring and Salmon

Herring and smoked salmon are often served at breakfast in the form of kippers and lox. A *kipper* is a whole herring that has been split from tail to head, gutted, salted or pickled, and cold smoked. *Lox* is salmon that has been brine cured and then cold smoked. Kippers and lox may be served with scrambled eggs, on a bagel or rye bread with cream cheese and slices of onion, or with dry toast, sliced tomatoes, and ripe avocado.

Hash

Hash dishes are a creative way to use up the trimmings left behind when cutting corned beef or roast beef on a slicer. *Hash* is shredded and chopped meat that has been mixed and cooked with diced potatoes, onions, and seasonings. *See Figure 11-38.* Hash is cooked on a griddle or in a sauté pan until heated through and commonly topped with two eggs cooked to order.

Hash

Figure 11-38. *Hash is composed of shredded and chopped meat and cooked with diced potatoes, onions, and seasonings.*

BREAKFAST SIDES

Eggs and omelets are rarely served alone. They are most commonly served with some sort of accompaniment or side dish. Common side dishes include breakfast potatoes, breads, pastries, fritters, fruit, yogurts, and cheeses.

Breakfast Potatoes

Breakfast potatoes are usually served fried or griddled. Potatoes may be served à la carte or included in featured breakfast combinations, such as two eggs with ham and hash brown potatoes. Two common breakfast potato preparations are hash browns and home fries. *See Figure 11-39.*

Breakfast Potatoes

Basic American Foods

Hash Browns

Idaho Potato Commison

Home Fries

Figure 11-39. *Two popular breakfast potato preparations are hash browns and home fries.*

Hash Browns. Hash browns are a common breakfast side dish. Although the name "hash" brown refers to a chopped, recooked potato, most people refer to the shredded potato dish as hash browns. Most foodservice operations purchase already prepared hash browns rather than producing them from scratch to save time and labor costs. To prepare hash browns from scratch, potatoes are peeled and then boiled until tender, yet slightly al dente. The potatoes are cooled and then shredded. Vegetable oil or clarified butter is added to a very hot sauté pan or griddle and the potatoes are cooked until golden brown on both sides. Hash browns need to be served immediately.

Home Fries. Home fries are another simple breakfast potato dish that may or may not be made from scratch, depending on the foodservice operation. Home fries are made by peeling and boiling red mealy potatoes until tender, yet al dente. The potatoes are then cooled and diced or sliced and browned on all sides in a hot sauté pan with butter. They are seasoned with salt and pepper and served immediately.

Breakfast Breads

Common breakfast breads include toast, biscuits, bagels, English muffins, and sweet muffins. *See Figure 11-40.* Butter, honey, jellies, jams, and preserves are common accompaniments for breakfast breads.

Breakfast Breads

Biscuits

Bagels

Photo Courtesy of Perdue Foodservice, Perdue Farms Incorporated

Figure 11-40. *Biscuits and bagels are two common breakfast breads.*

Toast. The majority of restaurants serve a variety of loaf breads as toast. They include, but are not limited to, wheat, white, rye, multigrain, and Texas toast. These presliced breads are simply toasted immediately before being served.

Biscuits. A *biscuit* is a quick bread made by mixing solid fat, baking powder or baking soda, salt, and milk with flour. At breakfast, biscuits are often topped with sausage gravy or served as a side for egg dishes and with grits.

Bagels. Bagels are usually sliced in half horizontally and then toasted. Typical condiments include butter, plain or flavored cream cheese, jams, preserves, and peanut butter. Bagels also are used as a base for breakfast sandwiches. The flavor of a bagel is usually reflected in the toppings it wears since most bagels are produced from standard bagel dough. Cinnamon raisin and pumpernickel bagels are exceptions.

Bagels begin as dough that is shaped into a small cigar shaped log. The log is then looped over and pinched to create a circle or bracelet shape. The raw bagels are then boiled in a malt solution before being baked. Boiling the dough in a malt solution prior to baking is what gives bagels their tough, chewy exterior. Warm bagels are dipped in a variety of toppings such as poppy seeds, sesame seeds, garlic and herbs, or coarse salt.

English Muffins. English muffins are often served toasted and also may be used as the base for breakfast sandwiches. They are usually heated and buttered. English muffins are made from a sticky dough with the addition of sugar and nonfat milk solids. After the dough is kneaded, it is divided into 1½ oz portions that are formed into circles. The circles of dough are precooked on a griddle until done on both sides. Once cooked and cooled slightly, they are split horizontally to produce a top and bottom half.

Sweet Muffins. Sweet muffins are cup-shaped quick breads that may be served with any meal. A variety of sweet muffins including, but not limited to, bran, blueberry, lemon poppy seed, and pumpkin are often served as breakfast breads. Sweet muffins complement a steamy cup of coffee or hot tea and commonly appear on room service menus as a breakfast item.

Breakfast Pastries

Danishes, sweet rolls, coffee cakes, doughnuts, and fritters are common breakfast pastries. *See Figure 11-41.* Many foodservice operations purchase these items from distributors or commercial bakeries to reduce labor costs, storage space, and spoilage. However, some foodservice operations have their own bakeshops and take great pride in producing their own pastries.

A *fritter,* also known as a beignet, is a fried donutlike item that may or may not be filled with fruit. Fritters are made from a flavored donutlike batter that is dropped into a fryer in dumpling-shaped portions. Fritters can be sweet or savory. For example, a sweet caramel apple fritter can be made by cooking a frozen apple wedge in cinnamon and sugar and then dipping it in a batter and frying it. Likewise, a savory fritter can be made by mixing roasted jalapeños and kernels of sweet corn into a sweet cornbread batter and frying it.

Fruits

Any type of fruit may be served for breakfast. Fruit that is in season is most often served as an accompaniment. *See Figure 11-42.* A combination of melons, such as honeydew and cantaloupe, are often sliced or served in chunks. Fresh strawberries are washed and served whole or sliced and fanned as a garnish on breakfast plates. Oranges are often cut into slices or wedges to add color and freshness to a hot breakfast plate. Frozen or canned fruit is typically used to fill crêpes and blintzes and top pancakes, waffles, or French toast.

Fresh Fruits

Wisconsin Milk Marketing Board

Figure 11-42. *Fruit that is in season is most often served as an accompaniment.*

Danishes

Figure 11-41. *Danishes are a common breakfast pastry.*

Yogurts and Granolas

Yogurt is a lower fat, protein rich food that is not as filling as an omelet or pancakes. A *yogurt* is a tangy, custardlike cultured dairy product produced by adding a safe bacteria and an acid to whole, low-fat, or fat-free milk. It is important to note that yogurt has the same fat content as the type of milk from which it is produced. Many yogurts often have added sugar, flavorings, or bits of fruit added. Greek-style yogurt is creamier and thicker than traditional yogurt. It is made by straining most of the waterlike whey from the yogurt and adding milk solids during production. Yogurts are typically served with fresh fruit, granolas, or both.

Granola is a baked mixture of rolled oats, nuts, dried fruit, and honey. Granolas may also contain other grains, cinnamon, coconut, chocolate bits, and other tasty items. Granolas and muesli are often served with yogurts, milks, fresh fruit, or porridge. *See Figure 11-43. Muesli* is an unbaked mixture of rolled oats, wheat flakes, oat bran, raisins, dates, sunflower seeds, hazelnuts, and wheat germ. Fresh cut fruit, such as bananas, may be added to granola or muesli along with milk. Some prepackaged breakfast cereals and cereal bars are made from granola and muesli. Granolas typically contain more calories than muesli.

Granolas

Alpha Baking Co., Inc.

Figure 11-43. *Granolas are often served with yogurts, milks, fresh fruit, or porridge.*

BREAKFAST CEREALS

Foodservice operations commonly serve hot cereals and ready-to-eat cereals for breakfast. Hot cereals such as oatmeal, porridge, grits, and farina are made of whole, cracked, ground, or flaked grains. Hot cereals need to be cooked on the stovetop or in a steam-jacketed kettle and may be served with milk, cream, nuts, fruit, cinnamon, brown sugar, butter, or cheese.

Hot Cereals

Hot cereals are often served with butter or cream and brown sugar. Porridge, oatmeal, farina, and grits are popular hot breakfast cereals made from ground grains. *See Figure 11-44. Porridge* is a hot breakfast dish made by heating a cereal grain in milk, water, or both. Oatmeal can be made from crushed, rolled, steel cut, or coarsely ground oats that are slowly simmered. Farina is made from a ground soft wheat called semolina and commonly sold under the brand name of Cream of Wheat®. Farina is boiled until the semolina becomes tender and creamy. Farina is often served alone or with a side of toast or fruit.

Grits are made from ground corn called hominy. Grits are simmered in water or milk until the grain becomes tender and creamy. Both white and yellow grits have an earthy flavor and are often served with eggs and toast or biscuits. These hearty cereals are often served only during the morning hours due to how quickly they coagulate when held too long.

Hot Cereals

Oatmeal   **Grits**

Figure 11-44. *Oatmeal and grits are popular hot breakfast cereals.*

Cold Cereals

Prepackaged breakfast cereals such as Corn Flakes® simply require the addition of milk. These cereals come in a variety of flavors and textures from very sweet to whole grain and high fiber. Cold cereals are typically served with a side of whole, low-fat, or fat-free milk, a side of fruit or toast, and juice. Foodservice operations typically purchase prepackaged cereals in bulk or in single-serving sized boxes.

BREAKFAST BEVERAGES

Almost as important as the breakfast menu items, are the breakfast beverages that accompany the meal. A wide variety of coffees, teas, juices, milks, and smoothies may be served.

Coffees

Coffee is prepared from the fruit of a coffee tree, which is found in tropical and subtropical climates around the world. The small red fruit, known as a cherry, contains two small green seeds called coffee beans. Ripened coffee cherries are harvested, fermented, and hulled to remove the two green coffee beans inside each cherry. During processing, the pulp, parchment, and silverskin are removed. *See Figure 11-45.* The green coffee beans are then dried and roasted to produce a rich, intensely flavored, and aromatic beverage called coffee.

Coffee Beans

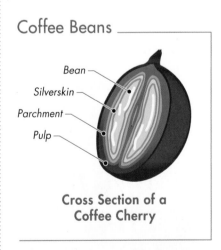

Cross Section of a Coffee Cherry

Figure 11-45. *A coffee cherry contains two small green seeds called coffee beans. The pulp, parchment, and silverskin are removed during processing. The beans are then roasted.*

Coffee beans are roasted to varying levels of darkness to produce different levels of flavor and aromas. The lighter the beans and the shorter the time period that they are roasted, the lighter the flavor and less intense the aroma. The darker the beans and the longer time period that they are roasted, the darker the flavor and the more intense the aroma. Most coffees are roasted to at least a medium roast because most coffee drinkers prefer a robust flavor. *See Figure 11-46.* Once the coffee beans have been roasted, coffee artisans blend different beans to produce a wide variety of desired coffee flavor profiles.

Types of Coffee Roasts		
Roast	**Common Name**	**Description**
Medium	City roast	Coffee beans are roasted to a medium-brown color, yielding a medium body and medium-flavored coffee
Medium dark	Viennese roast	Coffee beans are roasted to a rich, dark-brown color, yielding a rich and intensely flavored coffee
Dark	French roast	Coffee beans are roasted to a very dark-brown color, yielding a very intensely flavored coffee with a strong aroma and a slightly bitter taste
Espresso	Italian roast	Coffee beans are roasted until they almost reach a burnt state, yielding a bitter, full-flavored, and intensely strong aromatic coffee

Figure 11-46. *Coffee beans are roasted to varying levels of darkness to produce different levels of flavor and aromas.*

Although there are numerous species of coffee beans grown around the world, arabica beans and canephora beans are used in approximately 90% of coffee production. Arabica beans are typically used to produce higher-quality and more expensive coffees and are the best beans for brewing. Canephora beans, commonly known as robusta beans, produce a less flavorful coffee but are often blended with arabica beans to produce a less bitter and smoother tasting coffee.

Espresso. *Espresso* is an intensely flavored coffee made from beans that have been roasted to the very dark or espresso-roasted stage. The almost burnt beans are then ground to a very fine powder, allowing a high extraction of flavor from the dark beans. Espresso is used as the base for many specialty coffee drinks such as cappuccinos, lattes, and mochas. *See Figure 11-47.*

Common Coffee Drinks	
Coffee Drink	**Description**
Americano	Equal ratio of espresso to very hot water
Cappuccino	Equal ratio of espresso to frothy steamed milk
Frappuccino™	Iced coffee blended to make a coffee milkshake
Iced coffee	Espresso with milk and sugar served over ice
Latte	Espresso with double the ratio of steamed milk to espresso
Macchiato	Espresso served with a small amount of frothy steamed milk
Mocha	Espresso mixed with cocoa or hot chocolate and served with a dollop of whipped cream

Figure 11-47. *Espresso is used as the base for many specialty coffee drinks such as cappuccinos, lattes, and mochas.*

Decaffeinated Coffees. Decaffeinated coffee undergoes a process that removes most of the caffeine from the coffee beans. Removing the caffeine allows a coffee drinker to consume coffee without the effects of the stimulant.

Instant Coffees. Instant coffee is made by freeze-drying coffee to remove the water. The resulting powdered coffee can then be rehydrated quickly by adding hot water. However, instant coffee lacks the quality of a naturally brewed pot of coffee.

Teas

Tea is the primary breakfast drink for a majority of the world's population. In some locations this beverage is also enjoyed as an afternoon snack or after an evening meal. *See Figure 11-48.* Tea can be served hot or cold and is generally less expensive to serve than coffee. Tea comes in many varieties and flavors, depending on the growing region and length of processing. Common teas include green, black, and oolong. Tea-like beverages such as chamomile are often listed on menus as teas.

Teas

Rishi Tea

Figure 11-48. *Tea is the primary breakfast drink for a majority of the world's population.*

Green Teas. Green teas are dried without fermentation. It is a slightly bitter flavored tea and is most commonly served hot, without the addition of milk, lemon, or any sweetener. One cup of green tea contains approximately 20–22 mg of caffeine. Sencha is the most common variety of green tea. Green tea loses its flavor more quickly than black tea and should not be stored for extended periods of time.

Rishi Tea

Black Teas. Black tea is a strongly flavored tea resulting from the fermentation of tea leaves. Black tea varieties vary from bold and smoky to light and fruity. To produce black tea, tea leaves are left to dry and then cracked to begin the oxidation process. When oxidation is complete, the once green leaves are left either deep brown or black. Unlike green tea, black tea does not lose its flavor for several years. The various types of black tea derive their names from the places where they are produced. China, India, and Sri Lanka are the primary producers of black tea, but it is also produced in Kenya, Vietnam, Nepal, Turkey, Thailand, Azerbaijan, Georgia, Russia, Indonesia, Sumatra, and Malaysia.

Black tea is graded according to the shape and size of the tea leaves. The smaller the tea leaf, the shorter the brewing time needed to release the maximum amount of flavor. Likewise, tea leaves rolled one way adopt a particular flavor that differs from the same tea leaves rolled a different way. For example, orange pekoe is a tea industry term that refers to leaf size, not a type of tea. *See Figure 11-49.*

	Common Tea Grades	
Tea Grade	**Grade Translation**	**Description**
S.	Souchong	Whole twisting leaf that is often light in color; China is the main producer
F.O.P.	Flowery orange pekoe	Long whole leaf with a crushed flower appearance; India is the main producer
O.P.	Orange pekoe	Thin, wiry leaf with a tighter roll than F.O.P.
T. and G.	Tippy and golden	Modifiers used to describe both whole leaf and broken grades; indicates colorful tips on the leaf
P.	Pekoe	Curly, large broken leaf without a visible tip; Sri Lanka is the main producer
B.O.P.	Broken orange pekoe	Small, squarish broken leaf with good body and strength; India is the main producer
F.	Fannings	Smaller than B.O.P. with less keeping quality; used for commercial tea bags
D.	Dust	Smallest grade produced; quick liquoring

Figure 11-49. *Black tea is graded according to the shape and size of the tea leaves.*

It is important to note that quality is determined by the taste of the tea, not its grade. The purpose of grading tea leaves is to create teas that have a uniform leaf density and appearance and that will extract at the same rate when brewed. The more broken the leaf, the more full-bodied and pungent the tea and the shorter the brewing time. Black teas are often served hot and cold with milk, lemon, and sweetener. A cup of black tea contains approximately 40–42 mg of caffeine.

Oolong Teas. Oolong tea is partially fermented, meaning that it goes through a short period of fermentation that turns the tea leaves a red-brown color. Oolong tea is pale yellow and has a fruity, floral quality with a hint of smoke. Considered the champagne of teas, oolong is commonly served hot and without milk, lemon, or sweeteners. Formosa is the most common variety of oolong tea.

Herbal Beverages. Tea is only made from the leaves of an evergreen plant named Camellia sinensis or one of its subspecies. Tea-like beverages are called tisanes or herbal infusions. A *tisane* is an herbal beverage created by steeping herbs, spices, flowers, dried fruits, or roots in boiling water. The act of steeping creates an infusion. Common tisanes or herbal infusions include chamomile, rose hip, peppermint, and lemon verbena.

Juices

Both fruit and vegetable juices are common breakfast menu items. Juices may be purchased fresh, frozen, or canned. Breakfast juices commonly include orange, grapefruit, cranberry, pineapple, tomato, prune, and mixed or blended juices such as cranapple. *See Figure 11-50.* A standard serving of juice is a 4 fl oz glass.

Milks

Milk is a nutritional beverage from the mammary glands of cows, goats, sheep, or water buffalo. Milk is available in whole, low-fat, and skim and can be flavored, such as with chocolate and strawberry milk. Milk is served as a stand-alone beverage as well as an accompaniment to coffee and hot tea. Milk is extremely perishable, so proper storage and handling procedures must always be followed.

Milks are classified according to the percentage of milk fat that remains after processing. Each milk product contains a different percentage of milk fat depending on the amount of milk fat removed. *See Figure 11-51.* Milk classifications are determined based on the following characteristics:

- Whole milk must contain at least 3.5% milk fat in order to be labeled as whole milk.
- Low-fat milk is commonly labeled as 1% or 2% because the majority of the milk fat has been removed. The percentage of milk fat in low-fat milk can vary between 0.5% and 2%.
- Skim milk must contain less than 0.5% milk fat in order to be labeled as skim milk. Nonfat milk is another name for skim milk.
- Flavored milks contain flavoring ingredients such as chocolate or strawberry syrup plus added sugars or sweeteners. Flavored milk is typically higher in calories than plain milk.

Fruit Juices

Florida Department of Citrus

Figure 11-50. *Grapefruit juice is just one of the many fruit juices commonly offered on breakfast menus.*

Milks

CROPP Cooperative

Figure 11-51. *Milks are classified according to the percentage of milk fat that remains after processing.*

Smoothies

Frieda's Specialty Produce

Figure 11-52. *Smoothies are blended drinks made of fruit, yogurt, and/or milklike beverages.*

Simple Garnishes

Idaho Potato Commission

Figure 11-53. *Common breakfast garnishes include strawberries and orange slices.*

Skillet Dishes

Figure 11-54. *Hot skillet breakfasts often incorporate eggs, potatoes, and cheeses.*

Milklike beverages made from plants such as soy, rice, almond, hemp, and oat milk are also available. However, these products are not typically served in commercial foodservice operations.

Smoothies

A *smoothie* is a blended drink made of fruit, such as berries or bananas, yogurt, and/or milklike beverages. *See Figure 11-52.* All of the ingredients are blended together to the desired thickness and served in a glass with a straw. Using frozen fruit will make the consistency of the smoothie similar to that of a milkshake. The use of frozen fruit also means the smoothie will have less ice crystals than a smoothie made with ice cubes.

PLATING BREAKFAST

Common plating of breakfast includes the use of round or oval plates for the entrée and placement of the main protein as the focal point of the dish. Accompaniments such as toast are commonly served on side plates that are about half the size of the entrée plate. However, many restaurants and hotels serve pancakes on larger side plates. Simple garnishes add color and freshness to breakfast plates. Common breakfast garnishes include sliced or fanned strawberries and orange slices. *See Figure 11-53.*

Beverage cups and glasses are chosen based on the guest-to-waitstaff ratio. If a restaurant has a large number of guests per waitstaff, the staff may not be able to keep up with demand if smaller cups and glasses are used. Slightly larger cups and glasses mean fewer trips to the table for refills. However, it is important to not serve hot coffee or hot tea in too large of a cup as these beverages may become cold prior to the guest consuming them.

When planning table layout for breakfast service, it is important to consider table size in proportion to the number of breakfast plates used in an average order. Guests perceive more value if their meal is provided on multiple plates, but the table must be able to accommodate all of the plates without overcrowding the guests' eating space.

Skillet Dishes

Many restaurants utilize hot skillets to express their creativity with the breakfast menu. Hot skillet breakfasts often incorporate multiple items such as eggs, breakfast meats, potatoes, cheeses, and vegetables. *See Figure 11-54.* The eggs are generally placed on top of the other ingredients and covered with cheese, hollandaise sauce, or some other sort of finished sauce. Skillet ingredients are typically cooked prior to being placed in the skillet. After the ingredients are placed in the skillet, it is finished in an oven or under a broiler.

Continental Breakfasts

A *continental breakfast* is a light breakfast consisting of fruit or juice, toast or pastries, and coffee or tea. Continental breakfasts are common in hotels offering a complementary light breakfast and at conference centers prior to business meetings or conferences.

Breakfast Buffets

A successful breakfast buffet offers a variety of foods that are well prepared and arranged in a manner that stimulates the appetite. *See Figure 11-55.* Many foodservice operations display their awareness of health and wellness by placing tags near their high protein, low-fat, low-carb, or gluten-free breakfast items, while still offering traditional favorites.

Breakfast Buffets

MacArthur Place Hotel, Sonoma

Figure 11-55. *A successful breakfast buffet offers a variety of foods that are well prepared and arranged to showcase a wide variety of colors.*

Breakfast buffets allow guests to eat at their own pace. Offering a breakfast buffet also takes pressure off the kitchen to quickly turn orders. Breakfast buffets also allow a foodservice operation to charge a higher price for breakfast. Most people perceive buffets as a better value because there are more choices and no limit to how much a person can consume. However, it is important to note that most people do not eat that much more at a buffet than they would have eaten if they had ordered an á la carte breakfast.

SUMMARY

Although many items can be served for breakfast, eggs have long been a popular choice due to their versatility, availability, and reasonable cost. Knowledge of egg composition, various uses of eggs, and proper cooking methods is essential to creating egg-based preparations in the professional kitchen. Eggs can be fried, scrambled, made into omelets, baked, poached, and cooked in the shell. Eggs are also used to create many other food preparations due to their properties as an adhesive agent, an emulsifier, a clarifying agent, a thickening agent, and a leavening agent.

Breakfast dishes also include pancakes, waffles, French toast, crêpes and blintzes, assorted meats, various potato dishes, a variety of breads and pastries, fruits, yogurts, granolas, and hot and cold cereals. All of these breakfast dishes can be complemented with a variety of beverages including coffees, teas, juices, milks, and smoothies. Plating breakfast requires a fair number of plates, bowls, cups, and glasses of various sizes. Some breakfast items are plated in skillets, served on buffets, or make up a continental breakfast. Although most breakfast items are easy to prepare, the position of breakfast cook is a great place to develop speed, organization, and timing.

Refer to DVD for
Quick Quiz® questions

Review

1. Identify the six main uses of eggs in food preparation.
2. Describe the four main parts of the egg.
3. Identify the part of the egg that contains most of the calories and all of the cholesterol.
4. Describe the purpose of egg substitutes.
5. Describe the six classifications of eggs according to size.
6. Explain how eggs are graded.
7. Explain the advantage of using pasteurized eggs.
8. Explain why eggs should be cooked in small quantities and very close to serving time.
9. Describe the storage requirements for eggs.
10. Identify five types of fried eggs.
11. Identify the type of egg preparation most often used when preparing eggs in quantity.
12. Contrast the preparation of a rolled omelet, a folded omelet, a soufflé omelet, and a frittata.
13. Identify five bread products often used to create egg sandwiches.
14. Explain how to prepare shirred eggs.
15. Explain how to prepare a quiche.
16. Explain how to poach eggs.
17. Explain how to prepare traditional eggs Benedict.
18. Identify the cooking times for preparing soft-cooked, medium-cooked, and hard-cooked eggs in the shell.
19. Contrast the preparation of pancakes, waffles, and French toast.
20. Contrast the preparation of crêpes and blintzes.
21. Identify nine types of breakfast meats.

Refer to DVD for
Review Questions

22. Identify the two forms of sausages.

23. Identify the form of bacon commonly used in food service.

24. Identify the two breakfast meats that are purchased precooked.

25. Explain how steaks, pork chops, herring, and salmon are served at breakfast.

26. Describe how hash is prepared.

27. Differentiate between hash browns and home fries.

28. Identify five types of common breakfast breads.

29. Identify five types of common breakfast pastries.

30. Explain how fruit is served at breakfast.

31. Explain why it matters which type of milk is used to make yogurts.

32. Contrast granolas and muesli.

33. Identify three common hot cereals served at breakfast.

34. Describe various types of breakfast beverages.

35. Explain how to plate breakfast items.

36. Contrast skillet dishes, a continental breakfast, and a breakfast buffet.

Applications

1. Prepare and plate five types of fried eggs. Compare the appearance, flavor, and texture of the five dishes.

2. Prepare and plate scrambled eggs. Evaluate the appearance, flavor, and texture of the eggs.

3. Prepare and plate a rolled omelet, a folded omelet, a soufflé omelet, and a frittata. Compare the appearance, flavor, and texture of the four dishes.

Refer to DVD for
Application Scoresheets

4. Prepare and plate two different types of egg sandwiches. Compare the appearance and flavor of the two sandwiches.

5. Prepare and plate shirred eggs. Evaluate the appearance, flavor, and texture of the eggs.

6. Prepare a quiche. Evaluate the appearance, flavor, and texture of the quiche.

7. Prepare and plate eggs Benedict. Evaluate the appearance, flavor, and texture of the eggs.

8. Prepare and plate eggs in the shell. Evaluate the appearance, flavor, and texture of the eggs.

9. Prepare and plate pancakes. Evaluate the appearance, flavor, and texture of the pancakes.

10. Prepare and plate waffles. Evaluate the appearance, flavor, and texture of the waffles.

11. Prepare and plate French toast. Evaluate the appearance, flavor, and texture of the French toast.

12. Prepare and plate crêpes and blintzes. Compare the appearance, flavor, and texture of the two dishes.

13. Prepare and plate three breakfast meats. Compare the appearance, flavor, and texture of the three dishes.

14. Prepare and plate hash browns and home fries. Compare the appearance and flavor of the two dishes.

15. Plate breakfast breads and pastries. Evaluate the appearance of the dishes.

16. Plate breakfast fruits, yogurts, and granolas. Evaluate the appearance of the dishes.

17. Prepare and plate a hot breakfast cereal. Evaluate the appearance, flavor, and texture of the cereal.

18. Prepare and present a hot and a cold breakfast beverage. Compare the appearance, flavor, and texture of the beverages.

19. Prepare and plate a breakfast skillet dish. Evaluate the appearance, flavor, and texture of the dish.

20. Plan a breakfast buffet. Evaluate the planned appearance, composition, and pricing of the buffet.

Scrambled Eggs

Yield: *12 servings, 2 eggs each*

Ingredients

eggs, beaten	24 ea
whole milk	4 fl oz
salt and pepper	TT
clarified butter	2 fl oz

Preparation

1. Whisk eggs, milk, salt, and pepper together in a stainless steel bowl. *Note:* If eggs are to be served in a chafing dish over high heat, substitute heavy cream or béchamel for the milk to prevent the scrambled yolks from turning green due to overcooking.
2. Add butter to a large, preheated sauté pan over medium heat.
3. Pour egg mixture into the sauté pan and stir until eggs are fully cooked, but still creamy.

NUTRITION FACTS
Per serving: 191 calories, 130 calories from fat, 14.5 g total fat, 385.1 mg cholesterol, 170.7 mg sodium, 151.6 mg potassium, 1.2 g carbohydrates, 0 g fiber, <1 g sugar, 12.9 g protein

Spanish Omelet

Yield: *1 serving, 1 omelet each*

Ingredients

vegetable oil	1 tsp
onion, diced	1 tsp
red pepper, diced	1 tsp
green pepper, diced	1 tsp
eggs, beaten	2 ea
cheddar cheese (or other desired cheese)	1 tbsp
salsa	1 tbsp
sour cream	1 tsp

Preparation

1. Add vegetable oil to a nonstick omelet pan over medium heat.
2. When the oil is hot, add the vegetables and sauté for 1 minute.
3. Add the beaten eggs and stir until almost set.
4. Gently flip the omelet over to cook the other side.
5. Place cheese in the center of omelet and fold one side over to form a half circle.
6. Cook until cheese is slightly melted.
7. Slide omelet onto a warm plate.
8. Garnish omelet with salsa and sour cream.

NUTRITION FACTS
Per serving: 667 calories, 548 calories from fat, 62.2 g total fat, 500.3 mg cholesterol, 473.6 mg sodium, 534.7 mg potassium, 9.1 g carbohydrates, <1 g fiber, 9.3 g sugar, 19.7 g protein

Spinach and Feta Frittata

Yield: *2 servings, half a frittata each*

Ingredients

clarified butter	1 tsp
fresh spinach, stems removed	¼ c
eggs	3 ea
salt and pepper	TT
feta cheese, crumbled	2 oz

Preparation

1. Add butter to a 6 inch well-seasoned or nonstick omelet pan over medium heat.
2. Add spinach and sauté for 30 seconds.
3. Whisk eggs in a stainless steel bowl and then pour into the pan.
4. Add salt and pepper and stir until the eggs begin to set.
5. Top with crumbled feta and place the frittata under a broiler or salamander for about 5 minutes to brown the top and finish cooking.

NUTRITION FACTS
Per serving: 202 calories, 136 calories from fat, 15.3 g total fat, 309.7 mg cholesterol, 571.2 mg sodium, 142.1 mg potassium, 1.8 g carbohydrates, <1 g fiber, 1.5 g sugar, 13.6 g protein

Shirred Eggs with Canadian Bacon

Yield: *1 serving*

Ingredients

unsalted butter, melted	1 tsp
Canadian bacon, sliced thin	2 slices
eggs	2 ea
salt and pepper	TT

Preparation

1. Brush the inside of a casserole dish with melted butter.
2. Lay 2 slices of Canadian bacon in the bottom of the casserole.
3. Break 2 eggs into a small bowl and then pour them over the bacon.
4. Add salt and pepper to taste.
5. Bake eggs in a 350°F oven for 15–20 minutes, until the egg whites have set. Serve immediately.

NUTRITION FACTS
Per serving: 356 calories, 191 calories from fat, 21.3 g total fat, 439.2 mg cholesterol, 2039.5 mg sodium, 531.4 mg potassium, 2.6 g carbohydrates, 0 g fiber, <1 g sugar, 36.1 g protein

Quiche Lorraine

Yield: *6 servings (one 8-inch quiche)*

Ingredients

mealy pie dough	8 oz
bacon, medium dice	4 oz
onions, small dice	4 oz
Gruyère cheese, grated	4 oz
eggs, beaten	5 ea
heavy cream	4 fl oz
milk	4 fl oz
salt	½ tsp
white pepper	pinch
nutmeg, ground	pinch

Preparation

1. Roll mealy pie dough into a 10 inch circle.
2. Place dough in an 8 inch pie shell and press gently to set. Chill in the refrigerator.
3. Once the crust has chilled completely, place another pie tin inside the first pie tin and bake upside down in a 450°F preheated oven for 10 minutes to prebake the crust. *Note:* Prebaking will ensure that the crust will be cooked by the time the quiche is finished baking and that the crust will not get soggy when the custard is added.
4. While crust is prebaking, sauté diced bacon until crisp.
5. Add diced onions and cook until translucent. Remove bacon mixture from the heat.
6. Add grated cheese to the mixture and stir well.
7. Fill prebaked crust with mixture.
8. In a stainless steel bowl, whisk eggs, cream, milk, salt, pepper, and nutmeg to make a custard.
9. Pour custard on top of the bacon mixture. Place quiche on a sheet pan to catch any spillover during baking.
10. Bake quiche at 375°F for approximately 40 minutes or until the custard is firm and set. *Note:* Doneness is determined by inserting a toothpick in the center of the quiche and checking to see if it comes out clean.

> **NUTRITION FACTS**
> **Per serving:** 483 calories, 324 calories from fat, 36.3 g total fat, 217.8 mg cholesterol, 647 mg sodium, 209.6 mg potassium, 23.5 g carbohydrates, 1.1 g fiber, 2.1 g sugar, 15.5 g protein

Eggs Benedict

Yield: *1 serving*

Ingredients

white vinegar	1 tsp
salt	pinch
eggs	2 ea
English muffin, separated into 2 halves	1 ea
Canadian bacon, ¼ inch thick	2 slices
prepared hollandaise sauce	2 fl oz

Preparation

1. Place vinegar and salt in small saucepan of water and heat between 160°F and 170°F.
2. Crack eggs into a small bowl. Then carefully slide the eggs into the hot water.
3. While the eggs are poaching, toast the muffins.
4. Heat Canadian bacon in sauté pan or on a griddle until warmed through.
5. Place a piece of Canadian bacon on top of each muffin half.
6. When eggs are done, remove them from the water with a slotted spoon and drain well.
7. Place a poached egg on top of each piece of Canadian bacon.
8. Top the eggs with hollandaise sauce and serve immediately.

> **NUTRITION FACTS**
> **Per serving:** 848 calories, 589 calories from fat, 66.6 g total fat, 576.7 mg cholesterol, 1526.6 mg sodium, 510.2 mg potassium, 32.9 g carbohydrates, 1.6 g fiber, <1 g sugar, 30.1 g protein

Variations:

Eggs Benedict Florentine: Add 1 tbsp sautéed spinach between the Canadian bacon and the poached eggs.

Smoked Salmon Benedict: Use 1 oz thinly sliced smoked salmon instead of Canadian bacon and garnish with ¼ tsp capers.

Buttermilk Pancakes

Yield: *8 servings, 3 pancakes each*

Ingredients

all-purpose or pastry flour	1 lb
granulated sugar	3 tbsp
salt	1 tsp
baking powder	1 tbsp
baking soda	1 tsp
buttermilk	1 qt
whole eggs, beaten	4 ea
unsalted butter, melted	3 fl oz
vegetable oil	2 tbsp

Preparation

1. Sift all dry ingredients together in a large bowl.
2. In a separate bowl, whisk together buttermilk, eggs, and butter.
3. Add wet ingredients to dry ingredients and mix slightly. *Note:* If the batter is too thick, add a small amount of water.
4. Add oil to a preheated griddle.
5. Pour batter onto the griddle in equal amounts using a ladle.
6. Cook until small bubbles appear on the tops of the pancakes and the bottoms are golden brown.
7. When the bubbles begin to pop, flip the pancakes and cook until golden brown.
8. Plate the pancakes and serve immediately.

NUTRITION FACTS
Per serving: 418 calories, 142 calories from fat, 16.1 g total fat, 120.8 mg cholesterol, 797.3 mg sodium, 283.2 mg potassium, 54.5 g carbohydrates, 1.5 g fiber, 10.8 g sugar, 13.1 g protein

Waffles

Yield: *12 servings, 2 waffles each*

Ingredients

all-purpose or pastry flour	22 oz
salt	1 tsp
baking powder	2 tbsp
granulated sugar	2½ oz
eggs, beaten	5 ea
milk	24 fl oz
butter, melted	6 fl oz
vanilla	2 tsp
vegetable oil	1 tsp

Preparation

1. Sift dry ingredients together in large bowl.
2. In a separate bowl, mix the beaten eggs, milk, butter, and vanilla together.
3. Add the wet ingredients to the dry ingredients and mix well. *Note:* The batter can be made and stored a day ahead of service.
3. Brush preheated waffle iron with vegetable oil or spray with nonstick cooking spray.
4. Pour enough batter onto the waffle grid to barely cover the surface.
5. Cook according to manufacturer's directions.
6. Plate and serve waffles immediately.

NUTRITION FACTS
Per serving: 387 calories, 144 calories from fat, 16.3 g total fat, 114.1 mg cholesterol, 496 mg sodium, 169.7 mg potassium, 49.4 g carbohydrates, 1.4 g fiber, 9.3 g sugar, 10 g protein

Whole-Grain French Toast

Yield: *4 servings, 2 pieces each*

Ingredients

eggs	3 ea
cinnamon	⅛ tsp
sugar	1 tbsp
milk	⅔ c
vanilla	½ tsp
whole-grain bread	8 slices
confectioners' sugar	½ oz

Preparation

1. Break the eggs into a stainless steel bowl and beat with a wire whisk.
2. Sift together cinnamon and sugar.
3. Add milk, vanilla, and eggs to sugar mixture and blend well.
4. Dip each slice into the batter to coat both sides of the bread.
5. Remove the bread slices from the batter and let drain slightly.
6. Brown the bread on both sides in a skillet or on a griddle.
7. Place two pieces of French toast on each plate and sprinkle with confectioners' sugar.

NUTRITION FACTS
Per serving: 244 calories, 64 calories from fat, 7.1 g total fat, 143.6 mg cholesterol, 289.3 mg sodium, 226.3 mg potassium, 31.6 g carbohydrates, 3.9 g fiber, 12.2 g sugar, 12.9 g protein

Crunchy Cinnamon-Raisin French Toast

Yield: *6 servings, 2 slices each*

Ingredients

eggs, beaten	8 ea
half and half	6 fl oz
milk	2 fl oz
vanilla extract	1 tsp
cinnamon	1 tsp
nutmeg	¼ tsp
salt	⅛ tsp
Corn Flakes®, slightly crushed	3 c
almond slivers, slightly crushed	½ c
raisin bread slices	12 ea
vegetable oil	1 tbsp
confectioners' sugar	½ oz

Preparation

1. Whisk eggs, half and half, milk, vanilla extract, cinnamon, nutmeg, and salt in a stainless steel bowl to make a batter.
2. Mix the crushed Corn Flakes® with the almond slivers in a separate bowl.
3. Dip each slice of bread into the batter and coat both sides of the bread.
4. Remove the bread from the batter and let drain slightly.
5. Dip each battered slice into the Corn Flake® mixture to coat well.
6. Add vegetable oil to a hot skillet or griddle.
7. Place each slice of bread in a hot skillet or on a griddle and cook until golden brown on each side.
8. Place cooked slices on a sheet pan in a 350°F oven for 5 minutes to finish cooking.
9. Dust with confectioners' sugar and serve immediately.

NUTRITION FACTS
Per serving: 443 calories, 173 calories from fat, 19.8 g total fat, 260.1 mg cholesterol, 493.2 mg sodium, 372.7 mg potassium, 50.5 g carbohydrates, 4.7 g fiber, 6 g sugar, 17.5 g protein

Crêpes

Yield: *8 servings, 3 crêpes each*

Ingredients

eggs	6 ea
milk	1 qt
flour	10 oz
sugar	3 oz
salt	½ tsp
butter, melted	3 fl oz
vegetable oil	1 tbsp
fruit preserves	1½ c
confectioners' sugar	4 oz

Preparation

1. Whisk eggs in a mixing bowl.
2. Add milk and mix until blended.
3. Combine the flour, sugar, and salt in a separate bowl.
4. Gradually add the dry mixture to the wet mixture and mix until all ingredients are well blended.
5. Add the melted butter and mix until well blended.
6. Pour the batter into a bain-marie and refrigerate until ready for use.
7. Coat a crêpe or omelet pan with vegetable oil and heat slightly.
8. Use a 2 oz ladle to thinly coat the bottom of the skillet with batter while rotating the skillet clockwise.
9. Place the skillet over medium-low heat until the batter sets.
10. Flip or turn the crêpe by hand and cook on the other side.
11. Remove the crêpe from the skillet and place on a sheet pan covered with parchment paper. *Note:* If crêpes are being stacked, place a sheet of wax paper between each one.
12. Spread 1 tbsp of fruit preserves on each crêpe. *Note:* Jelly, jam, marmalade, or applesauce may be used in place of preserves.
13. Roll up each crêpe and place three on each plate, seam side down.
14. Dust crêpes with confectioners' sugar and serve immediately.

NUTRITION FACTS
Per serving: 590 calories, 163 calories from fat, 18.4 g total fat, 174.6 mg cholesterol, 277.3 mg sodium, 300 mg potassium, 96.6 g carbohydrates, 1.1 g fiber, 56.9 g sugar, 12.7 g protein

Sweet Cheese Filled Blintzes

Yield: *8 servings, 2 blintzes each*

Ingredients

prepared crêpe batter recipe	1 ea
ricotta cheese	16 oz
egg yolk	1 ea
vanilla extract	1½ tsp
lemon juice	1 tsp
salt	⅛ tsp
butter	3 tbsp

Preparation

1. Mix one recipe of crêpe batter and cook crêpes on one side only in crêpe pan. Set aside until needed. Repeat until all batter has been used.
2. Place the ricotta in a cheesecloth-lined china cap and drain well.
3. In a stainless steel mixing bowl, whisk the egg yolk. Then, whisk in the ricotta, vanilla, lemon juice, and salt to create the sweet cheese filling.
4. Place crêpes on a clean surface, cooked side up.
5. Spoon 2 tbsp of the filling into the center of each crêpe.
6. Fold one end of each crêpe toward the center to cover the filling.
7. Fold the two sides in and then roll each crêpe to form a pillow-shaped package.
8. Melt butter in a sauté pan over medium heat and sauté blintzes until warmed through. Serve immediately.

NUTRITION FACTS
Per serving: 258 calories, 122 calories from fat, 13.8 g total fat, 113.1 mg cholesterol, 181 mg sodium, 164.9 mg potassium, 23.5 g carbohydrates, <1 g fiber, 8.7 g sugar, 9.6 g protein

Savory Potato Filled Blintzes

Yield: *16 blintzes, 2 per serving*

Ingredients

red potatoes, large	1 lb
onions, small dice	½ lb
oil or clarified butter	1 tbsp
salt	TT
white pepper	TT
prepared crêpe batter recipe	1 ea
vegetable oil	1 tbsp
butter	3 tbsp

Preparation

1. Peel and boil potatoes until tender. Drain and set aside.
2. Sauté onions in oil or butter over medium heat until slightly golden brown.
3. Run cooked potatoes through a ricer or mash by hand.
4. Add cooked onions and seasonings to the potatoes and mix well.
5. Mix one recipe of crêpe batter, and cook crêpes on one side only in crêpe pan. Set aside until needed. Repeat until all batter has been used.
6. Place crêpes on a clean surface, cooked side up.
7. Spoon 2 tbsp of the filling into the center of each crêpe.
8. Fold one end of each crêpe toward the center to cover the filling.
9. Fold the two sides in and then roll each crêpe to form a pillow-shaped package.
10. Melt butter in a sauté pan over medium heat and sauté blintzes until warmed through. Serve immediately.

NUTRITION FACTS
Per serving: 257 calories, 115 calories from fat, 13 g total fat, 95.1 mg cholesterol, 147.3 mg sodium, 302.1 mg potassium, 28.4 g carbohydrates, 1.2 g fiber, 9.1 g sugar, 6.9 g protein

Breakfast Sausage Patties

Yield: *20 servings, two 2-oz patties each*

Ingredients

salt	¾ tsp
sage	½ oz
white pepper	¼ tsp
red pepper flakes, crushed	⅛ oz
pork butt, ground (70% lean; 30% fat)	5 lb
ice water	1 c

Preparation

1. Combine all seasonings in a small bowl and mix well.
2. In an appropriately sized bowl, add pork and toss well in seasonings.
3. Grind pork and seasonings together through the medium plate of a meat grinder and into a large bowl placed over ice.
4. Place the ground meat and seasoning mixture in a stand mixer.
5. Mix on low and slowly add ice water until the mixture becomes sticky to the touch.
6. Form the sausage into patties and pan-fry or broil.

NUTRITION FACTS
Per serving: 301 calories, 218 calories from fat, 24.2 g total fat, 81.6 mg cholesterol, 151.2 mg sodium, 336.8 mg potassium, <1 g carbohydrates, <1 g fiber, <1 g sugar, 19.2 g protein

Corned Beef Hash

Yield: *12 servings, two 3-oz patties each*

Ingredients

potatoes, waxy peeled and quartered	1¾ lb
salt	1 tsp
onions	½ lb
oil	2 tsp
chicken stock	4 fl oz
eggs, beaten	2 ea
parsley, minced	1 tbsp
whole-grain mustard	2 tsp
black pepper, ground	1 tsp
thyme, dried	¼ tsp
nutmeg, ground	pinch
corned beef, shredded or medium dice	2 lb
butter	2 tbsp

Preparation

1. Boil the potatoes in salted water until tender. Strain and allow to cool slightly.
2. Grate onions and drain in a strainer.
3. Sauté onions in oil until slightly caramelized and set aside.
4. Coarsely mash half of the potatoes and mix with the remainder of the potatoes and all the remaining ingredients except the butter.
5. Refrigerate the mixture for at least 4 hours to cool completely.
6. Remove hash mixture from refrigerator and shape into 3 oz patties about ¾ inch thick.
7. Heat butter in a heavy bottomed sauté pan and add the hash patties.
8. Cook until golden brown on both sides and heated through.

NUTRITION FACTS
Per serving: 383 calories, 215 calories from fat, 23.9 g total fat, 140.9 mg cholesterol, 1446.5 mg sodium, 667.5 mg potassium, 18.6 g carbohydrates, 1.5 g fiber, 1 g sugar, 22.6 g protein

Savory Breakfast Potatoes

Yield: *16 servings, 4 oz each*

Ingredients

red potatoes	4 lb
clarified butter	2 fl oz
paprika	½ tsp
garlic powder	⅛ tsp
salt and pepper	TT
scallions, minced	4 ea

Preparation

1. In a large pot, completely cover potatoes with water. Simmer until slightly tender but still firm.
2. Remove potatoes and cool completely.
3. Cut potatoes into 1 inch chunks, leaving the skin intact.
4. Add butter to a large sauté pan over medium-high heat and cook potatoes until golden brown on all sides.
5. Season potatoes and add scallions. Serve immediately.

NUTRITION FACTS
Per serving: 119 calories, 29 calories from fat, 3.3 g total fat, 8.2 mg cholesterol, 25.7 mg sodium, 628.6 mg potassium, 20.7 g carbohydrates, 1.9 g fiber, <1 g sugar, 2.4 g protein

Culinary Arts
PRINCIPLES AND APPLICATIONS

Frieda's Specialty Produce

Fruits

*F*ruits are a satisfying way to get vitamins, minerals, and fiber without consuming a lot of calories. Fruits are most flavorful and of the highest quality at the peak of their growing seasons. A chef must possess a solid understanding of available fruit varieties, how to judge freshness and quality, and when different fruits are in season. Having this important information helps a chef choose the best fruits to offer on menus at different times of the year.

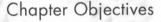

Refer to DVD for **Flash Cards**

Chapter Objectives

1. Identify the nutritional benefits of eating fruit.
2. Describe berries that are used in the professional kitchen.
3. Describe grapes that are used in the professional kitchen.
4. Describe pomes that are used in the professional kitchen.
5. Demonstrate how to core apples.
6. Describe drupes that are used in the professional kitchen.
7. Demonstrate how to prepare avocados.
8. Demonstrate how to seed melons.
9. Describe melons that are used in the professional kitchen.
10. Demonstrate how to cut citrus supremes.
11. Describe citrus fruits that are used in the professional kitchen.
12. Describe tropical fruits that are used in the professional kitchen.
13. Demonstrate how to prepare mangoes.
14. Demonstrate how to core pineapples.
15. Describe exotic fruits that are used in the professional kitchen.
16. Describe ways to accelerate and delay the ripening of fruits.
17. Explain how the pectin level of a fruit affects the cooking process.
18. Cook fruits and evaluate the quality of the prepared dishes.

Key Terms

- fruit
- variety
- hybrid
- berry
- aggregate fruit
- grape
- pome
- drupe
- melon
- citrus
- peel
- pith
- zest
- ethylene gas

Strawberries

California Strawberry Commission

Figure 12-1. *Strawberries are covered with tiny seeds.*

Blueberries

Figure 12-2. *Blueberries are small, dark-blue berries.*

Cranberries

Melissa's Produce

Figure 12-3. *Cranberries have a tart flavor.*

FRUIT CLASSIFICATIONS

Fruit is classified into several major categories, including berries, grapes, pomes, drupes, melons, citrus fruits, tropical fruits, exotic fruits, and fruit-vegetables. A *fruit* is the edible, ripened ovary of a flowering plant that usually contains one or more seeds. Fruits are nutritious because they are high in water, dietary fiber, vitamins, fructose, and antioxidants. Fruit adds color, texture, and flavor to a meal. The vast assortment of available fruits is primarily due to the large number of fruit varieties and hybrid fruits created by fruit growers over the last 2000 years. A *variety* is a fruit that is the result of breeding two or more fruits of the same species that have different characteristics. For example, a Jonagold apple is a variety of apple created by breeding a Jonathan apple with a Golden Delicious apple. A *hybrid* is a fruit that is the result of crossbreeding two or more fruits of different species to obtain a completely new fruit. For example, the loganberry was created by crossbreeding a raspberry and a blackberry.

Berries

A *berry* is a type of fruit that is small and has many tiny, edible seeds. Quality berries are sweet and evenly colored. Berries are harvested ripe because they do not continue to ripen after harvest. Frequently used berries in the professional kitchen include strawberries, blueberries, cranberries, gooseberries, currants, blackberries, raspberries, loganberries, and boysenberries.

Strawberries. A *strawberry* is a bright-red, heart-shaped berry covered with tiny black seeds. *See Figure 12-1.* A ripe strawberry should be evenly red and free of brown or soft spots. Strawberries should not be stored with any berries that are molding. A moldy berry can quickly spread mold and spoil other berries. Strawberries are widely used in desserts, sauces, and salads.

Blueberries. A *blueberry* is a small, dark-blue berry that grows on a shrub. *See Figure 12-2.* Blueberries are in peak season from mid-June to August. Blueberries are often used in jams, jellies, assorted desserts, salads, and quick breads such as muffins.

Cranberries. A *cranberry* is a small, red, round berry that has a tart flavor. The skin of a cranberry is white when young and turns a deep red when ripe. Because of their strong, sour flavor, cranberries are most often used in desserts, breads, sauces, and jellies. Fresh or frozen cranberries are always cooked with sugar or simple syrup before being added to a dish in order to soften their tart flavor. *See Figure 12-3.*

Gooseberries. A *gooseberry* is a smooth-skinned berry that can be green, golden, red, purple, or white and has many tiny seeds on the inside. *See Figure 12-4.* Gooseberries are used to make a classic English dessert called a fool, which has puréed gooseberries mixed with sweetened whipped cream. They are also used to make jams, jellies, and chutneys.

Currants. A *currant* is a small red, black, or golden-white berry that grows in grape-like clusters. **See Figure 12-5.** Currants are used in jams, jellies, sauces, and in some savory dishes. Dried currants used in baking are most often Zante currants. The Zante currant is not a currant at all, but is actually a small, sweet variety of seedless grape.

Aggregate Fruits

Several types of berries are actually aggregate fruits. An *aggregate fruit* is a cluster of very tiny fruits. Blackberries, raspberries, loganberries, and boysenberries are aggregate fruits. **See Figure 12-6.**

Blackberries. A *blackberry* is a sweet, dark-purple to black, aggregate fruit that grows on a bramble bush. Blackberries have a shiny outer layer and are in peak season during the summer months. Blackberries are used in many desserts and sauces.

Raspberries. A *raspberry* is a slightly tart, red fruit that grows in clusters. Raspberries are an aggregate fruit. The velvety soft texture of raspberries makes them one of the most fragile fruits. Raspberries are used in a variety of desserts and to make bright red sauces to complement desserts. The sweet yet tart flavor of raspberries also makes them ideal for use in glazes for savory dishes such as roasted meats or poultry as well as for raspberry vinaigrette salad dressing.

Loganberries. A *loganberry* is a red-purple hybrid berry made by crossbreeding a raspberry and a blackberry. Loganberries are an aggregate fruit similar in size and shape to blackberries but are redder in color. Loganberries are eaten raw and can be made into jams, preserves, and syrup. Loganberry juice is also used to make beverages.

Boysenberries. A *boysenberry* is a deep-maroon, hybrid berry made by crossbreeding a raspberry, a blackberry, and a loganberry. Boysenberries are similar in size and shape to blackberries. This aggregate fruit can be eaten raw and made into jams, preserves, syrup, and wine.

Gooseberries

Frieda's Specialty Produce

Figure 12-4. *Gooseberries have many tiny seeds inside.*

Currants

Frieda's Specialty Produce

Figure 12-5. *Currants grow in grape-like clusters.*

Aggregate Fruits

Oregon Raspberry & Blackberry Commission

Oregon Raspberry & Blackberry Commission

Blackberries   **Raspberries**   **Loganberries**   **Boysenberries**

Figure 12-6. *Blackberries, raspberries, loganberries, and boysenberries are all aggregate fruits.*

Grapes

A *grape* is an oval fruit that has a smooth skin and grows on woody vines in large clusters. Grapes are the most widely grown fruit because of their use in winemaking. Table grapes are the varieties suitable for eating, as opposed to varieties grown specifically for winemaking. The two main classifications of table grapes are white grapes, which are usually green in color, and black grapes, which are usually red to dark blue in color. Most of the flavor of a grape comes from its skin. It is important to choose grapes that are firm with no discoloration. Grapes frequently used in the professional kitchen include Thompson, red flame, and Concord grapes. Dried grapes are called raisins. *See Figure 12-7.*

Grapes

Figure 12-7. *Common varieties of table grapes include Concord, red flame, and Thompson grapes.*

Thompson Grapes. A *Thompson grape* is a seedless grape that is pale to light green in color. Thompson grapes are available year-round, but are in peak season from June to November. Thompson grapes may be used in salads or served with a cheese platter. Dried seedless grapes are called sultanas.

Red Flame Grapes. A *red flame grape* is a seedless grape that ranges from a light purple-red color to a dark-purple color. Red flame grapes are typically sweeter and crisper in texture than Thompson grapes. Red flame grapes are often used as an accompaniment to salads, cheese trays, or platters.

Concord Grapes. A *Concord grape* is a seeded grape with a deep black color. It is sometimes called a slipskin grape, as its skin is very easily separated from the fruit. Concord grapes are available in the market as a table grape but, because they have seeds, they are used less often than the Thompson or red flame varieties. Concord grapes are commonly used in jams, jellies, and grape juice due to their high sugar content and distinguishable flavor.

Pomes

A *pome* is a fleshy fruit that contains a core of seeds and has an edible skin. Pomes have thin skin, grow on trees or bushes, and are an excellent source of antioxidants and dietary fiber. Quality pomes are free of blemishes and bruises and have no soft spots. Pomes often used in the professional kitchen include apples, pears, and quinces. *See Figure 12-8.*

Pomes

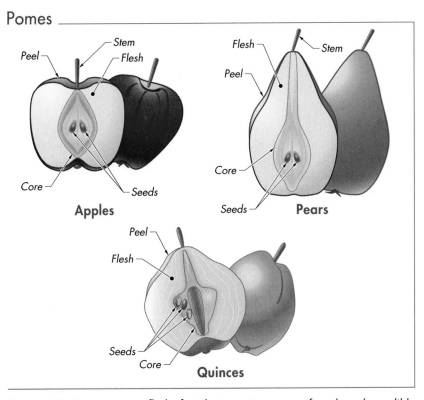

Figure 12-8. *A pome is a fleshy fruit that contains a core of seeds and an edible peel. Apples, pears, and quinces are pomes.*

Apples. An *apple* is a hard, round pome that can range in flavor from sweet to tart and in color from pale yellow to dark red. There are more than 7500 cultivated varieties of apples. Some varieties are not suitable for cooking purposes, and others become bitter when baked. In general, tart apples, or apples that have a high acid content, are desirable for cooking because of their intense flavor. Apples are available year-round, but are in peak season between late August and early November. *See Figure 12-9.*

When cooking apples for use in sauces or pies, the yield percentage is 50%. For example, 1 lb of raw apples yields ½ lb of cooked apples.

Common Apple Varieties		
Name	**Description**	**Availability**
 Cortland	Sweet, hint of tartness, juicy tender, white flesh	Sept. to Apr.
 Crispin	Sweet, juicy, crisp	Oct. to Sept.
 Empire	Sweet, tart, juicy, creamy-white flesh	Sept. to July
 Fuji	Spicy, sweet, juicy, firm cream-colored flesh, tender skin	Oct. to June
 Gala	Yellow to red, sweet, juicy, crisp yellow flesh	Sept. to June
 Golden Delicious	Sweet, crisp, light yellow flesh	Sept. to June
 Granny Smith	Tart, crisp, juicy	Sept. to June
 Idared	Sweet, tart, juicy, firm, pale yellow-green flesh, sometimes rosy pink flesh	Oct. to Aug.
 Jonagold	Tangy, sweet	Oct. to May
 McIntosh	Sweet, tangy, juicy, tender, white flesh	Sept. to June
 Rome	Mildly tart, firm, greenish-white flesh	Oct. to Sept.

U.S. Apple Association

Figure 12-9. *Common varieties of apples are used to prepare a vast array of dishes, including salads, sauces, purées, desserts, and savory dishes.*

When preparing an apple for use in a dish, the inner core must be removed. The core contains inedible seeds and tough, inedible fibers. A fruit corer or a chef's knife can be used to remove the core. A fruit corer is inserted into the stem end of the apple to cut out the cylindrical portion of the apple that contains the core. The cylinder is removed and the remaining edible fruit is prepared as desired. A chef's knife can also be used to remove the core. First, the apple is cut in half and then into quarters. Then, each quarter is cut at an angle to remove the core. *See Figure 12-10.*

Refer to DVD for
Coring Apples
Media Clip

Procedure for Coring Apples

Using a Fruit Corer

1. After removing the stem, grip the apple firmly and place the fruit corer at the top center of the apple.

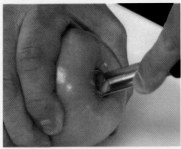

2. Push the corer straight through to the other end of the apple, twisting as necessary to make sure that the entire core is removed.

3. Pull the corer out of the apple and discard the core.

Using a Chef's Knife

1. After removing the stem, hold the top of the apple with tucked fingers while cutting down the middle to split the apple in half.

2. Lay each apple half on its cut side and cut it in half again.

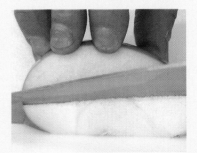

3. Cut each apple quarter at an angle to remove the core.

Figure 12-10. *The inedible core must be removed when preparing an apple.*

Chef's Tip

Adding pear slices to grilled chicken sandwiches or diced pears to chicken or tuna salad balances the flavor profile. Likewise, thin pear wedges go nicely with cheese or smoked turkey.

Pears. A *pear* is a bell-shaped pome with a thin peel and sweet flesh. Pears should be harvested before they are ripe, but if picked too early they will not develop their full flavor. Conversely, pears that are picked too late turn brown and become watery inside. If left to ripen on the tree, pears develop concentrations of cellulose, resulting in a grainy texture. There are thousands of different kinds of pears. Commonly available varieties of pears include the Anjou, Asian, Bartlett, Bosc, Comice, and Seckel pears. ***See Figure 12-11.***

Common Pear Varieties

Name	Description	Availability
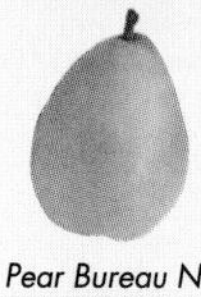 **Anjou** *Pear Bureau Northwest*	Sweet, juicy	Sept. to July
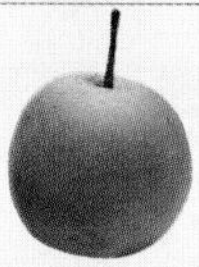 **Asian** *Frieda's Specialty Produce*	Texture of apple, taste of pear	Aug. to Sept.
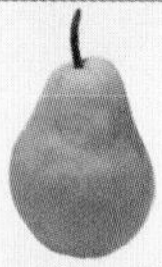 **Bartlett** *Pear Bureau Northwest*	Aromatic, sweet, juicy	Aug. to Jan.
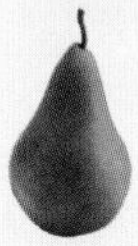 **Bosc** *Pear Bureau Northwest*	Dense flesh, highly aromatic, spicy sweet	Sept. to Apr.
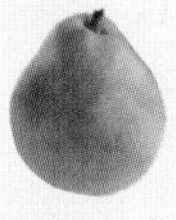 **Comice** *Pear Bureau Northwest*	Very sweet, juicy	Sept. to May
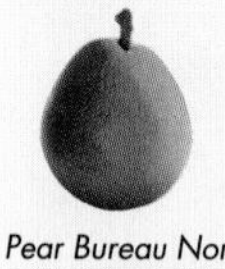 **Seckel** *Pear Bureau Northwest*	Extremely sweet	Sept. to Feb.

Figure 12-11. *Common varieties of pears can be used to prepare salads, sauces, purées, desserts, and savory dishes.*

- An *Anjou pear* is a plump, lopsided pear that is green in color. The Anjou pear is a standard pear in the professional kitchen because of its almost year-round availability and low cost. Anjou pears have a juicy, firm, and smooth flesh that is not as sweet as some other varieties. Ripe Anjou pears are best used in salads. Anjou pears that are not fully ripe can be baked, poached, or roasted.

- An *Asian pear,* also known as an apple-pear, is a round pear with the texture of an apple and yellow-colored skin. Asian pears are ready to eat when harvested instead of being allowed to ripen and soften off the tree like other pears. The peak season for Asian pears is late summer. In addition to being eaten raw, Asian pears are used in salads and slaws.

- A *Bartlett pear* is a large, golden pear with a bell shape. Bartlett pears have a creamy, juicy flesh and a musky flavor and aroma. Bartlett pears are delicious raw, can be used in sauces and sorbets, and can be baked in pies and tarts.

- A *Bosc pear* is a pear with a gourd-like shape and a brown-colored to bronze-colored peel. The peel of a quality Bosc pear has a yellow rather than a green undertone. The flesh of a Bosc pear is rich, syrupy, and sweet. Because of their firm texture, Bosc pears hold their shape when cooked. Bosc pears are winter pears and need cold storage to ripen fully.

- A *Comice pear* is a fairly large pear with a rotund body and a very short, well-defined neck. They are most often green-yellow in color with a red blush covering part of the surface, although some varieties are almost entirely red in color. Comice pears are very sweet and juicy, making them a favorite complement for soft-ripening cheeses like Brie and Camembert. They are most often eaten raw, as their juiciness makes them a poor choice for cooking. Comice pears are usually available from September through May.

- A *Seckel pear* is a small pear that is sometimes called a honey pear or sugar pear because of its syrupy, fine-grained flesh and complex sweetness. Seckel pears mature in late summer and early fall and are superb for use in desserts.

Quinces. A *quince* is a hard yellow pome that grows in warm climates. *See Figure 12-12.* Quinces are not eaten raw because they have a bitter taste. This characteristic disappears during cooking. Quinces are typically cooked in sugar syrup, which turns the fruit slightly darker with a pinkish tint. Quinces are often used in jams and jellies.

Quinces

Frieda's Specialty Produce

Figure 12-12. *Quinces are hard yellow pomes that grow in warm climates.*

Drupes

A *drupe,* also known as a stone fruit, is a type of fruit that contains one hard seed or pit. Drupes grow on shrubs and trees and are usually harvested before they are ripe. High-quality drupes are free of blemishes or bruises. Peaches, nectarines, apricots, avocados, dates, plums, cherries, and olives are all drupes that are used in the professional kitchen. Peaches, nectarines, and apricots are similar in color, but differ in taste. *See Figure 12-13.*

Peaches. A *peach* is a sweet, orange to yellow fruit with downy skin. The flesh of a peach is juicy, yet firm enough to hold its shape. The skin is edible, but the large oval pit inside the peach is not. Peaches should be ripened at room temperature and then refrigerated to prevent them from becoming mealy in texture. Peaches are excellent raw, are used in salsas or chutneys, and are baked into pies and other pastries.

Nectarines. A *nectarine* is a sweet, slightly tart, orange to yellow drupe with a firm, yellow flesh and a large oval pit. Nectarines share many characteristics with peaches. In fact, the juicy nectarine is a result of a peach mutation. Like peaches, nectarines need to ripen at room temperature to prevent them from becoming mealy. Virtually any recipe that calls for peaches can also be made with nectarines.

Apricots. An *apricot* is a drupe that has pale orange-yellow skin with a fine, downy texture and a sweet and aromatic flesh. Apricots are quite delicate and may be harvested before they are ripe to avoid damage in shipping. Apricots can be enjoyed raw or cooked. They are a popular choice for fruit tarts and are commonly puréed to make dessert sauces, jams, and pastry fillings. Apricots can also be used to make savory sauces for meat or poultry.

Peaches, Nectarines, and Apricots

| Peaches | Nectarines | Apricots |

Figure 12-13. *Peaches, nectarines, and apricots are similar in color, but differ in taste.*

Avocados. An *avocado*, also known as an alligator pear, is a pear-shaped fruit with a rough green skin and a large pit surrounded by yellow-green flesh. *See Figure 12-14.* The inedible skin and pit must be removed before the avocado can be eaten. The color of the skin varies depending on the variety. For example, the skin of a Hass avocado, a common variety, turns black when ripe. Other varieties remain green when ripe. Avocados are harvested from tall trees that grow in groves like citrus. They are native to Central America but are also grown in other warm climates such as California. Avocados have a very short period of peak ripeness.

Avocados are full of vitamins and minerals and contain both omega 3 and 6 fatty acids. The majority of the fat in avocados is monounsaturated, with less than 20% saturated. They do not contain cholesterol. Avocados are also a good source of protein and are often used as a meat substitute in vegetarian diets.

Avocados are often eaten raw or made into guacamole. *See Figure 12-15.* They are also used in salads and as an accompaniment to savory dishes and as a sandwich garnish. The rich, smooth, buttery flesh also makes avocados a good sandwich spread that can be used in place of mayonnaise, butter, or salad dressing.

Avocados _____

Figure 12-14. *Avocados are often called alligator pears due to their rough outer skin and their pear-like shape.*

Refer to DVD for
Preparing Avocados
Media Clip

Procedure for Preparing Avocados

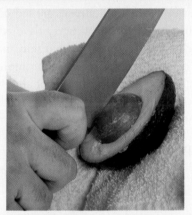

1. Holding the avocado with the guiding hand, use a chef's knife to cut the avocado in half lengthwise, cutting all the way around the pit.

2. Use a gentle twisting motion to separate the two halves.

3. Hold the half containing the pit with the guiding hand and gently hit the pit with the blade of a chef's knife.

4. Gently twist the knife to lift and discard the pit.

5. Use a spoon to gently scoop out the flesh to be mashed for guacamole or sandwich spread. *Note:* Use a dinner knife to cut the flesh into slices for a more elegant presentation.

Figure 12-15. *To prepare avocados for making guacamole or salads, the pit must be removed.*

Dates

Frieda's Specialty Produce

Figure 12-16. *Dates grow on a date palm tree.*

Plums

Figure 12-17. *A plum is an oval drupe that comes in a variety of colors.*

Cherries

National Cherry Growers and Industries Foundation

Figure 12-18. *Cherries have a long, thin stem that holds them on the tree.*

Dates. A *date* is a plump, juicy, and meaty drupe that grows on a date palm tree. *See Figure 12-16.* Dates are high in dietary fiber, carbohydrates, and contain more potassium than bananas. They are also low in fat, cholesterol, and sodium. In addition to being eaten fresh, dates may be dried or stuffed with savory or sweet fillings after the pit has been removed.

Plums. A *plum* is an oval-shaped drupe that grows on trees in warm climates and comes in a variety of colors such as blue-purple, red, yellow, or green. *See Figure 12-17.* There are more than 2000 varieties of plums. Some types of plums are sweet, juicy, and fragrant, while others can be sour, crisp, or mealy. The flesh can be red-orange, yellow, or green-yellow. A Damson plum is a blue-skinned plum with an acidic, tart flavor used to make jams, jellies, and preserves. A Japanese plum is a slightly heart-shaped plum with a crimson to green-yellow or purple to green-yellow skin. The sweet, juicy flesh is pale green or golden yellow in color. Plums can be eaten raw or baked in cobblers and tarts. Dried plums are known as prunes.

Cherries. A *cherry* is a small, smooth-skinned drupe that grows in a cluster on a cherry tree. Cherries have a long, thin stem that holds them on the tree. *See Figure 12-18.* The skin of cherries typically ranges in color from a bright red to a deep red that is nearly black. There are also golden-skinned varieties. The flesh of a cherry is pulpy and juicy and ranges in color from a dark yellow-orange to a deep reddish black. The small pit of a cherry is easily removed with a cherry pitter. *See Figure 12-19.* Cherries are used in a wide variety of recipes in both raw and cooked form and are classified as sweet or sour.

Sweet cherries include Bing cherries, Gean cherries, and Rainier cherries. Bing cherries have a thin skin with a dark red hue and are known for their sweetness. Gean cherries are red or black, soft-fleshed cherries used in the making of Kirsch, a cherry-flavored brandy. Rainier cherries are a premium cherry with yellow and red skin and a sweet yellow flesh.

Sour cherries include Montmorency cherries and Morello cherries. Sour cherries are dark red in color and, because of their sourness, are more often cooked to make jams, jellies, pie fillings, and liqueurs.

Pitting Cherries

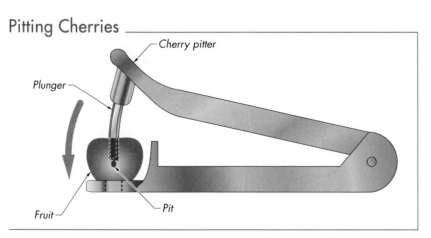

Figure 12-19. *A cherry pitter is used to remove the pit from cherries.*

Olives. An *olive* is a small, green or black drupe that is grown for both the fruit and its oil. *See Figure 12-20.* Unripe olives are green. Olives that are tree-ripened naturally turn dark brown to black. Fresh olives are bitter and their flavor reflects how ripe they were at harvest and how they were processed. A majority of olive varieties are brined or salt cured and then packed in olive oil or vinegar. Black olives, also known as Mission olives, get their color and flavor from lye curing and oxygenation. Greek Kalamata olives and French Niçoise olives are also popular varieties. Pitted and unpitted olives are used in both raw and cooked dishes.

Tree-ripened olives are pressed to extract olive oil. Olive oils are classified by their level of acidity. Extra virgin olive oil is oil from the first cold pressing of tree-ripened olives and is only 1% acid, whereas virgin olive oil is 3% acid.

Melons

A *melon* is a type of fruit that has a hard outer rind (skin) and a soft inner flesh that contains many seeds. The hard outer rind can be netted, ribbed, or smooth in texture. Most melons are picked just before they are ripe. Characteristics of a good melon include firmness and a good aroma. Melons are a good source of vitamin C and potassium and are relatively low in calories. Cantaloupes, muskmelons, honeydew melons, canary melons, watermelons, casaba melons, and Crenshaw melons are all used in the professional kitchen. Cantaloupes and muskmelons are often mistaken for one another in the marketplace. *See Figure 12-21.*

Cantaloupes. A *cantaloupe* is an orange-fleshed melon with a rough, deeply grooved rind. The grooves run lengthwise from the stem. The most widely cultivated variety of the cantaloupe is the Charentais cantaloupe, which has pale-green, lightly ridged rind and is native to Europe. American cantaloupe is actually a type of muskmelon.

Olives

Green Olives

Black Olives

Olive Oil

Figure 12-20. *Olives are grown for both their fruit and their oil.*

Cantaloupes and Muskmelons

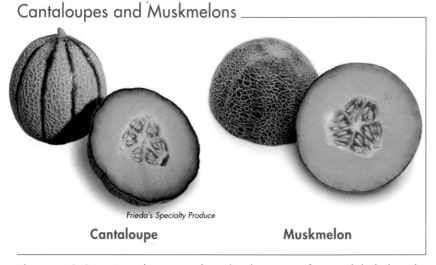

Frieda's Specialty Produce

Cantaloupe   **Muskmelon**

Figure 12-21. *Cantaloupes and muskmelons are often mislabeled in the marketplace.*

Muskmelons. A *muskmelon* is a round, orange-fleshed melon with a beige or brown, netted rind. The inside flesh can range in color from salmon to an orange-yellow color. There are many hybrids derived from muskmelon, such as the American cantaloupe. A *Santa Claus melon* is a large, mottled yellow and green variety of muskmelon that has a slightly waxy skin and soft stem end when ripe. Santa Claus melons are typically used in fruit salads and platters. A *Persian melon,* also known as a patelquat, is a green muskmelon with finely textured net on the rind. The green coloring lightens and the netting turns brown as the melon ripens. Persian melons are often sliced and served with breakfast.

Honeydew Melons. A *honeydew melon* is a melon with a smooth outer rind that changes from a pale-green color to a creamy-yellow color as it ripens. Honeydews have a mild, sweet flavor. They are available almost year-round, but at peak of season from June through October. Honeydew melons are used in the same manner as cantaloupes. Honeydew melons, canary melons, and watermelons have a smooth outer rind. *See Figure 12-22.*

Canary Melons. A *canary melon* is a fairly large, canary yellow melon with a smooth rind that is slightly waxy when ripe. Canary melons are available from June through September and are served raw in fruit salads and platters.

Watermelons. A *watermelon* is a sweet, extremely juicy melon that is round or oblong in shape, with pink, red, or golden flesh and green skin. The watermelon is named for its high water content—watermelons are over 90% water. The weight of a watermelon varies by variety, but can weigh up to 30 pounds. The thick rind ranges from light green to dark green in color and is often striped or solid. Some watermelons are seedless while others contain a lot of seeds. Watermelon seeds thicken and turn black when mature. Immature watermelon seeds are thin and white. Watermelon is most often enjoyed fresh or puréed into sauces and cold fruit soups.

Cuisine Note

In some cultures, watermelon rind is eaten as a vegetable. In China the rind is stir-fried, and in Russia the rind is pickled.

Smooth Melons

Melissa's Produce

National Watermelon Promotion Board

| Honeydew Melon | Canary Melon | Watermelon |

Figure 12-22. *Honeydew melons, canary melons, and watermelons have a smooth outer rind.*

Casaba Melons. A *casaba melon* is a teardrop-shaped melon with a thick, bright-yellow, ridged rind and white flesh. Casaba melons are unique in that they do not have an aroma so the quality must be judged by looking for a deep yellow color and an absence of blemishes. Casaba melons are at peak of season from late summer to early fall. Virtually any recipe that calls for cantaloupe can also be made with casaba melon. Casaba melons and Crenshaw melons have a rough outer rind. *See Figure 12-23.*

Crenshaw Melons. A *Crenshaw melon* is a large, pear-shaped melon with a yellow-green, slightly ribbed rind and an orange or salmon-colored flesh. The peak season for Crenshaw melons is late summer. They are used much like cantaloupes but have a sweet and spicy flavor. Some Crenshaw melons weigh as much as 10 pounds.

With the exception of watermelon, the seeds are typically removed from fresh melons before service. *See Figure 12-24.* The rind may or may not be removed depending on the desired presentation. Melons can be used as garnishes and to make cold soups, sorbets, ice creams, and parfaits.

Casabas and Crenshaws

Casaba Melon

Crenshaw Melon

Melissa's Produce

Figure 12-23. *Casaba and Crenshaw melons have a rough outer rind.*

Procedure for Seeding Melons

1. Cut off the rind from the top and bottom of the melon.

2. Place one cut end of the melon on the cutting board. Cut off even strips of the rind, starting at the top edge and moving down while carefully following the contour of the melon. Do not cut deeply into the fruit.

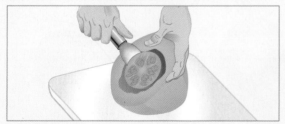

3. Cut the melon in half. If the melon contains seeds, remove the seeds by scraping them out with a spoon.

4. Place a melon half on the cutting board with the flat side facing down. Slice the melon for presentation on a fruit platter, or dice the melon for use in a fruit salad or other recipe.

Figure 12-24. *The rind and seeds must be removed prior to slicing or dicing melons.*

Oranges

Figure 12-25. *Oranges are orange-colored citrus fruits that grow on trees.*

Blood Oranges

Frieda's Specialty Produce

Figure 12-26. *Blood oranges are often used to add color to salads and sauces.*

Citrus Fruits

A *citrus fruit* is a type of fruit with a brightly colored, thick rind and pulpy, segmented flesh that grows on trees in warm climates. The peel and the pith are both removed during preparation because of their bitter taste. The *peel* is the thick outer rind of a citrus fruit. The *pith* is the white layer just beneath the peel of a citrus fruit. Citrus fruits are harvested fully ripe, as they do not continue to ripen after they are picked. Fresh citrus should be kept refrigerated to extend storage life. Citrus fruits are an excellent source of vitamin C and are quite acidic. Citrus fruits used in the professional kitchen include oranges, mandarins, tangerines, lemons, limes, grapefruits, and ugli fruit.

Oranges. An *orange* is a round, orange-colored citrus fruit. *See Figure 12-25.* Valencia and navel oranges are two popular varieties of sweet oranges and are available year-round, but are at peak of season from December to April. Blood oranges are another variety of sweet orange. Blood oranges are smaller than the Valencia and navel varieties and have a deep-red flesh. They are often used to make beautifully colored sauces and to add color to salads. *See Figure 12-26.* Some recipes call for bitter oranges, also known as Seville oranges, in which both the flesh and the peel are sour tasting. Oranges are used in sauces, juices, jams, jellies, and desserts.

Mandarins. A *mandarin* is a small, intensely sweet citrus fruit that is closely related to the orange, but is more fragrant. The peel of a mandarin is thinner than that of an orange and is more easily separated from the flesh. Mandarins are usually eaten raw because of their intense sweetness. A *Satsuma* is a small, seedless variety of mandarin. Mandarins and tangerines are close relatives of oranges. *See Figure 12-27.* The peels of mandarins and tangerines are thinner than that of an orange and are more easily separated from the flesh.

Tangerines. A *tangerine* is a small citrus fruit with a slightly red-orange peel that can be easily removed without a knife. Tangerines are a hybrid of a mandarin and a bitter variety of orange. A *Clementine* is similar to a tangerine, but has a rougher skin. A *tangor* is hybrid of a tangerine and a sweet orange. A *tangelo* is a hybrid of a tangerine and a grapefruit. Tangerines are typically eaten raw.

Mandarins and Tangerines

Frieda's Specialty Produce
Mandarins

Florida Department of Citrus
Tangerines

Figure 12-27. *The peels of mandarins and tangerines are thinner than that of an orange and are more easily separated from the flesh.*

Lemons. A *lemon* is a tart yellow citrus fruit with high acidity levels. The juice and rind of lemons are used in many dishes. Lemon juice is used in desserts, as well as many types of sauces that flavor poultry, fish, and shellfish. Lemon juice can also be used as a salad dressing or as a flavoring in marinades and beverages.

A *Meyer lemon* is cross between a lemon and an orange. Meyer lemons are round and have a smooth, dark-yellow peel. The flesh has a yellow-orange color and the juice is sweeter and less acidic than a regular lemon. The zest and juice of a Meyer lemon can be used just like a regular lemon but will contribute a sweeter and more floral flavor. Lemons and limes are small, tart citrus fruits that are often used to flavor many dishes. *See Figure 12-28.*

Limes. A *lime* is a small citrus fruit that can range in color from dark green to yellow-green. A *key lime* is a variety of lime that is smaller, more acidic, and more strongly flavored than other limes. Key limes have a juicy green-yellow flesh. Limes are available year-round but are at peak of season in the summer. Lime juice adds a fresh, crisp flavor to many marinades, sauces, ethnic dishes, desserts, and beverages.

Carlisle FoodService Products

Lemons and Limes

Figure 12-28. *Lemons and limes are tart citrus fruits with high acidity levels.*

Many recipes call for the zest of a citrus fruit, such as a lemon or a lime. *Zest* is the colored, outermost layer of the peel of a citrus fruit that contains a high concentration of oil. Zest is used to add flavor to a dish. Fine shreds of zest are made by gently rubbing the fruit against a zester or a rasp grater. *See Figure 12-29.* Thin strips of zest can be obtained by running a five-hole zester along the surface of the rind. Large strips of zest can be obtained by using a vegetable peeler to peel off the colored surface of the rind and then be julienned. Zest may also be candied for use as a confection or a decoration.

Grapefruits. A *grapefruit* is a round citrus fruit with a thick, yellow outer rind and tart flesh. *See Figure 12-30.* Common varieties of grapefruit include white grapefruit, pink grapefruit, and ruby red grapefruit. White grapefruit has pale-yellow flesh and pink grapefruit has pink-colored flesh. Ruby red grapefruit has red-colored flesh. Fresh grapefruit and grapefruit juice are commonly found on breakfast menus.

Zesting

Browne-Halco (NJ)

Figure 12-29. *Zesting is the process of removing the thin, colored peel of a citrus fruit.*

Grapefruits

Flesh
Pith
Peel

Figure 12-30. *A grapefruit is a round citrus fruit with a thick, yellow outer rind and tart flesh.*

Ugli Fruit. An *ugli fruit*, also known as a uniq fruit, is a large, teardrop-shaped, seedless citrus fruit. Ugli fruit is similar in size to a large grapefruit but is sweeter. *See Figure 12-31.* The peel is lumpy with large pores and ranges from green to orange when fully ripe. The juicy flesh is yellow-orange and tastes like grapefruit with a hint of sweet orange. Ugli fruit is a hybrid made by crossbreeding a grapefruit and a tangerine and is available from December through April. Ugli fruit is eaten fresh and the juice can be used in salad dressings and marinades.

Ugli Fruit

Flesh
Pith
Peel

Frieda's Specialty Produce

Figure 12-31. *Ugli fruit is a large, teardrop-shaped, seedless citrus fruit similar in size to grapefruit.*

The flesh of a citrus fruit is naturally divided into wedge-shaped segments that are separated by a thin membrane. These segments are often separated or cut to form "supremes" when preparing citrus fruits for use in a dish or as a garnish. A *supreme* is the flesh from a segment of a citrus fruit that has been cut away from the membrane. *See Figure 12-32.*

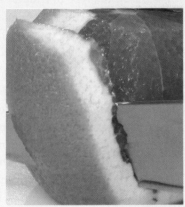

1. Using a chef's knife, carefully slice off the top and bottom of the fruit. Place the fruit on the cutting board so the bottom is lying flat. *Note:* The top of the fruit can be identified by a small circle where the stem has been removed.

2. Place the tip of the knife at the top edge of the fruit and cut from top to bottom using a slight sawing motion to remove a strip of the peel and the pith. *Note:* The pith can be used as a guide to gage the depth of the peel and pith.

3. Following the contour of the fruit, use a slight sawing motion to cut away a small slice of peel and pith. Rotate the fruit clockwise and removed another small slice. Repeat until the entire peel and pith have been removed. *Note:* Try not to remove too much of the flesh.

4. When the peel has been removed, check to make sure that none of the pith remains. Carefully cut off any remaining pith from the fruit.

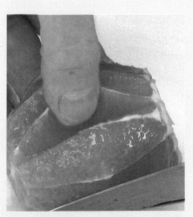

5. Hold the fruit against the cutting board and cut along the membrane in a "V" to remove a supreme. Hold back the membrane as each cut is made.

6. Continue to cut each segment until all the supremes have been separated from the fruit. *Note:* The remainder of the fruit can be juiced to flavor a dressing or a sauce.

The supremes are ready for use.

Figure 12-32. *Cutting citrus supremes involves separating each segment of flesh from the membrane within the fruit.*

Refer to DVD for
Cutting Citrus Supremes
Media Clip

Bananas and Plantains

Bananas

Plantains

Figure 12-33. *Bananas and plantains are close relatives that both grow in tiered bunches called hands.*

Tropical Fruits

A *tropical fruit* is a type of fruit that comes from a hot, humid location but is readily available. Tropical fruits can range in flavor from sweet to tangy and in texture from soft to crisp. Bananas, plantains, coconuts, kiwifruit, pineapples, papayas, figs, pomegranates, prickly pears, guavas, persimmons, and mangoes are considered tropical fruits and are used frequently in the professional kitchen. Bananas and plantains are relatives. *See Figure 12-33.*

Bananas. A *banana* is a yellow, elongated tropical fruit that grows in hanging bunches on a banana plant. Banana bunches grow in tiers called hands, with each hand having up to 20 bananas. There are typically 5–20 hands of bananas in a bunch. The most common banana is the Cavendish. Bananas are picked and shipped while still green. At their shipping destination, bananas are often placed in airtight rooms filled with ethylene gas to accelerate ripening. *Ethylene gas* is an odorless gas that a fruit emits as it ripens. If bananas are purchased green and allowed to ripen naturally, their flavor is notably richer. Bananas at peak ripeness have a bright yellow skin that is speckled with brown spots. Bananas can be eaten raw, grilled, fried, or baked.

Plantains. A *plantain* is a tropical fruit that is a close relative of the banana, but is larger and has a dark brown skin when ripening. When extremely ripe, the skin of a plantain turns black and the flesh is a soft, deep yellow. Unripe plantains are firm and starchy, similar to potatoes. Plantains are usually fried and are sometimes served with a sweet sauce to reduce their starchy flavor. Plantains can also be dried and ground into banana meal.

Coconuts. A *coconut* is a drupe with a white meat that is housed within a hard, fibrous brown husk. Coconut meat contains less sugar and more protein than apples, bananas, and oranges. *See Figure 12-34.* It is also a good source of iron, phosphorus, and zinc. However, 90% of the fat in a coconut is saturated. Young coconuts are used for coconut water, which is sweet and aerated, whereas the meat of a young coconut is so gelatinous that it is called coconut jelly. Coconuts are used in desserts, salads, soups, and sauces.

Coconuts

Frieda's Specialty Produce

Figure 12-34. *Coconut meat contains less sugar and more protein than apples, bananas, or oranges.*

Kiwifruit. A *kiwifruit* is a small, barrel-shaped tropical fruit, approximately 3 inches long and weighing between 2–4 ounces. Kiwifruit has a very thin, brown, fuzzy skin. Its bright green flesh has a white core that is surrounded by hundreds of tiny black seeds. ***See Figure 12-35.*** Kiwifruit is native to China and is also known as the Chinese gooseberry. It is harvested when ripe but still firm. Kiwifruit contains twice as much vitamin C as an orange. It is most often eaten raw. Kiwifruit also makes a colorful garnish.

Kiwifruit

Figure 12-35. *Kiwifruit has a bright green flesh with a white core surrounded by hundreds of tiny black seeds.*

Pineapples. A *pineapple* is a sweet, acidic tropical fruit with a prickly, pinecone-like exterior and juicy, yellow flesh. ***See Figure 12-36.*** Pineapples grow on a shrub-like perennial plant that grows to be about 3 feet tall. The pineapple plant bears hundreds of small purple flowers that grow in a spiral pattern around a central axis. These flowers join to form a single fruit, know as the pineapple. The leaves at the top of the fruit are called the crown. Pineapples take 18–20 months to fully mature, and must be fully ripe when harvested because they will not ripen after harvesting. Pineapples are eaten raw, in salads, baked with ham or in desserts, grilled as an accompaniment, or made into a juice. Before pineapple can be used in a dish or served fresh, the spiny outer covering and tough inner core must be removed. ***See Figure 12-37.***

Pineapples

Figure 12-36. *Pineapples have a prickly, pinecone-like exterior and a juicy, yellow flesh.*

1. Use a chef's knife to cut off the top and bottom of the pineapple. Place the pineapple on the cutting board so the bottom is lying flat.

2. Place the tip of the knife at the top edge of the pineapple and cut from top to bottom using a slight sawing motion to remove a strip of the peel.

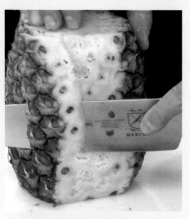

3. Rotate the pineapple clockwise and continue to remove the peel in slices, cutting from top to bottom while following the contour of the fruit and using the next row of eyes to guide each cut. *Note:* Try not to remove too much of the flesh, but cut deep enough to remove the eyes.

4. Once the peel is removed, cut the pineapple in half lengthwise.

5. Place each half so the cut side is facing down on the cutting board and cut each half lengthwise again to quarter the pineapple.

6. Make a final lengthwise cut along the corner of each wedge to remove the core. *Note:* The peeled and cored pineapple can be further cut for a fruit tray or use in a salsa, salad, or other recipe.

Figure 12-37. *Coring pineapple involves the removal of the spiny outer covering and the tough inner core.*

Refer to DVD for
Coring Pineapples
Media Clip

Papayas. A *papaya* is a pear- or cylinder-shaped tropical fruit weighing 1–2 pounds with flesh that ranges in color from orange to red-yellow. The center cavity of a papaya is filled with numerous edible seeds that resemble peppercorns and have a sharp taste similar to pepper. *See Figure 12-38.* Peak papaya season is from April through June. Papayas are ready to be harvested as soon as they are streaked with yellow. Green papayas will not ripen but can be pickled or cooked for use in savory applications. Papayas contain the enzyme papain, which is used as a meat tenderizer.

Papayas

Figure 12-38. *The center cavity of a papaya is filled with numerous edible seeds that resemble peppercorns and have a sharp taste similar to pepper.*

Figs. A *fig* is the small pear-shaped fruit of the fig tree. There are more than 150 varieties, including black figs, green figs, white figs, and purple figs. Figs have a sweet, rich flavor and are highly perishable. *See Figure 12-39.* The texture is somewhat tough and gritty due to the massive amount of tiny seeds inside. Fresh figs are an excellent accompaniment to a cheese platter. Figs can be stuffed and are also used in compotes and as pastry fillings.

Pomegranates. The *pomegranate* is a round, bright-red tropical fruit with a hard, thick outer skin. Pomegranates measure about 3 inches in diameter and contain a thick white membrane that encloses hundreds of red seeds that are the edible portion of the pomegranate. The sweet and juicy seeds are used in salads, soups, sauces, and beverages. Pomegranates and prickly pears both contain edible seeds. *See Figure 12-40.*

Prickly Pears. The *prickly pear* is a pear-shaped tropical fruit with protruding prickly fibers that is a member of the cactus family. The prickly pear is 2–4 inches long and has a very thick, coarse outer skin that can be green, yellow, orange, red, or deep purple, depending on the variety. The flesh is sweet and juicy and contains sweet, crisp, edible seeds. Peeled and sliced prickly pears are typically served cold.

Figs

Figure 12-39. *Figs are small pear-shaped fruits that are highly perishable.*

Edible Seeds

Frieda's Specialty Produce

Pomegranate   **Prickly Pear**

Figure 12-40. *Pomegranates and prickly pears both contain edible seeds.*

Guavas. A *guava* is a small oval-shaped fruit, usually 2–3 inches in diameter, with thin edible skin that can be yellow, red, or green. The flesh can be white, yellow, or pink in color. Guava juice is often blended into tropical fruit drinks. Guavas are used in jams, jellies, and chutneys. ***See Figure 12-41.***

Guavas

Florida Department of Agriculture and Consumer Services

Figure 12-41. *Guavas are tropical fruits that make great jams, jellies, and chutneys.*

Persimmons

Frieda's Specialty Produce

Figure 12-42. *A persimmon is similar in shape to a tomato.*

Persimmons. A *persimmon* is a bright-orange tropical fruit that grows on trees and is similar in shape to a tomato. ***See Figure 12-42.*** Ripe persimmons are very sweet. Unripe persimmons are tannic in flavor and have a chalky aftertaste. Persimmons grow on trees that are native to China. They are a winter fruit that remains on the tree even after the leaves have fallen. Persimmons are at peak of season from October through January. Persimmons are eaten raw or are baked into breads, cakes, and desserts.

Mangoes. A *mango* is an oval or kidney-shaped drupe with orange to orange-yellow flesh. The thin outer skin of the mango can include shades of light yellow, red, green, and pink. There are over 1000 varieties of mangoes grown around the world. Peak mango season is from May through August. The flesh of a mango clings to a large flat stone in the center of the fruit. The peel and stone are removed when preparing mangoes for use in a dish. ***See Figure 12-43.*** Mangoes can be used in a variety of ways, such as in sauces, desserts, drinks, salads, and salsas.

1. Using a chef's knife, carefully slice off the top and bottom of the mango. Place the mango on the cutting board so the bottom is lying flat. *Note:* This should slightly expose the stone inside the mango.

2. Using the tip of the chef's knife, start at the top edge of the mango and cut down in a slight sawing motion to remove a strip of the peel. *Note:* Try not to remove too much of the flesh.

3. Following the contour of the mango, use a slight sawing motion to cut away another small slice of peel. Rotate the mango clockwise and repeat the cutting motion until the entire peel has been removed.

4. Once the peel been removed, position the knife about ¼ inch on one side of the stone. *Note:* The stone is a flat oval about ½ inch thick and about 1–1½ inches wide.

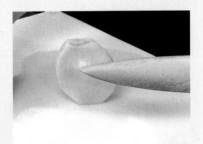

5. Slice down from the top to the bottom of the mango to remove the flesh on one side of the stone. Repeat this same cut along the other side of the stone.

6. Reserve the mango halves and place the piece containing the stone on the cutting board.

7. Cut the flesh away from the stone by slicing at angles. Discard the stone.

8. Slice or dice the flesh of the mango.

Use the diced mango in a recipe of choice.

Figure 12-43. *Dicing mangoes involves the removal of the peel and stone prior to use in a dish.*

Refer to DVD for
Preparing Mangoes
Media Clip

Star Fruit

Frieda's Specialty Produce

Figure 12-44. *Star fruit is an exotic fruit that is shaped like a star when cut perpendicular to the stem.*

Exotic Fruits

An *exotic fruit* is type of fruit that comes from a hot, humid location but is not as readily available as a tropical fruit. Many exotic fruits are used in the professional kitchen. Star fruit, dragon fruit, mangosteens, lychees, rambutans, durians, passion fruit, breadfruit, kiwanos, kumquats, and jackfruit are considered exotic fruits due to their limited availability.

Star Fruit. A *star fruit,* also known as a carambola, is an exotic fruit that is shaped like a star when cut perpendicular to the stem. *See Figure 12-44.* It has a thin, edible skin enclosing translucent flesh that is crisp and juicy. Star fruit is native to Asia and ranges from 2–5 inches in length, 1–1½ inches in diameter, and 3–4 ounces in weight. Its flavor ranges from very tart and acidic to somewhat sweet. Star fruit is ripe when the tips of the "star" begin to turn brown. Star fruit may be added to salads or used as a garnish.

Dragon Fruit. A *dragon fruit* is an exotic fruit of a cactus with an inedible pink skin, green scales, and white or red flesh speckled with small crunchy black seeds. The flesh is mildly sweet and similar in texture to kiwifruit. *See Figure 12-45.* The flesh makes an interesting addition to a tropical fruit salad and can be juiced or blended into beverages.

Dragon Fruit

Frieda's Specialty Produce

Figure 12-45. *Dragon fruit flesh is mildly sweet and similar in texture to kiwifruit.*

Mangosteens

Frieda's Specialty Produce

Figure 12-46. *Mangosteens have a juicy, cream-colored flesh that is segmented, similar to that of a tangerine.*

Mangosteens. A *mangosteen* is a round, sweet and juicy exotic fruit about the size of an orange with a hard, thick, dark-purple rind that is inedible. Underneath the hard rind is an inedible red, membrane that encloses the juicy, cream-colored flesh. *See Figure 12-46.* The sweet-tart flesh is segmented similarly to a tangerine. Mangosteens must be ripe before being harvested as they spoil quickly. This prized fruit from Asia is eaten raw.

Lychees. A *lychee* is an exotic drupe covered with a thin, red, inedible shell and has a light-pink to white flesh that is refreshing, juicy, and sweet. The lychee also contains an inedible seed. Lychees are native to China and are grown on evergreen trees that can reach heights of 50–60 feet and bear an average of 300 pounds of fruit annually. Lychees must be harvested ripe, as they do not continue to ripen after being harvested. With the shell and seed removed, lychees can be eaten raw or in salads and desserts. Lychees and rambutans are exotic fruits with similar flesh surrounding a single, inedible seed. *See Figure 12-47.*

Rambutans. A *rambutan* is fragrant and sweet exotic fruit covered on the outside with soft, hair-like spikes. The rambutan is native to Malaysia and grows in clusters on an evergreen tree. Rambutans are easily opened revealing flesh, similar to that of a lychee, that surrounds a single, inedible seed.

Durians. A *durian* is an exotic fruit that contains several pods of sweet, yellow flesh and has a custard-like texture. *See Figure 12-48.* Durians are about the size of a large cantaloupe, have a thick thorny skin that is golden to green in color, and can weigh up to 10 pounds. This Malaysian fruit is known for having a strong, unpleasant odor but a very appealing taste.

Refreshing Fruits

United States Department of Agriculture

Lychees

Frieda's Specialty Produce

Rambutans

Figure 12-47. *Lychees and rambutans both have a refreshing, sweet flesh.*

Durians

Frieda's Specialty Produce

Figure 12-48. *Durians contain several pods of sweet yellow flesh that have a custard-like texture.*

Nutrition Note

Passion fruit is low in calories and a good source of potassium, iron, and vitamins A and C. When eaten with the seeds, passion fruit is also a good source of fiber.

Passion Fruit. A *passion fruit* is a small, oval-shaped exotic fruit that typically weighs 2–3 ounces and has firm, inedible skin that can be either yellow or purple. As the fruit ripens, the skin becomes thinner and wrinkles. The orange-yellow or pink-green pulp of the passion fruit is sweet and juicy, with a gelatin-like texture and a citrus flavor. Its small black seeds are edible. ***See Figure 12-49.*** Passion fruit may be used in ice jams, jellies, ice creams, and a variety of desserts.

Passion Fruit

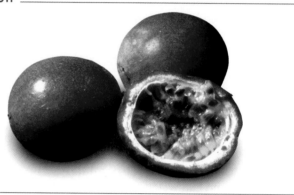

Florida Department of Agriculture and Consumer Services

Figure 12-49. *Passion fruit has a sweet and juicy orange-yellow or pink-green pulp with a gelatin-like texture, a citrus flavor, and small, edible black seeds.*

Cuisine Note

In Polynesia breadfruit is baked, boiled, or fried as a potato-like vegetable and also used to make breads, pies, and puddings.

Breadfruit. A *breadfruit* is an exotic fruit that is native to Polynesia and has bumpy green skin and a white starchy flesh. Breadfruit is about the size of a small cantaloupe. When unripe, the flesh is cooked and used much like a potato and has a flavor similar to fresh bread. ***See Figure 12-50.*** As the fruit ripens, the flesh becomes softer and slightly sweeter and can be filled with butter and brown sugar and roasted.

Breadfruit

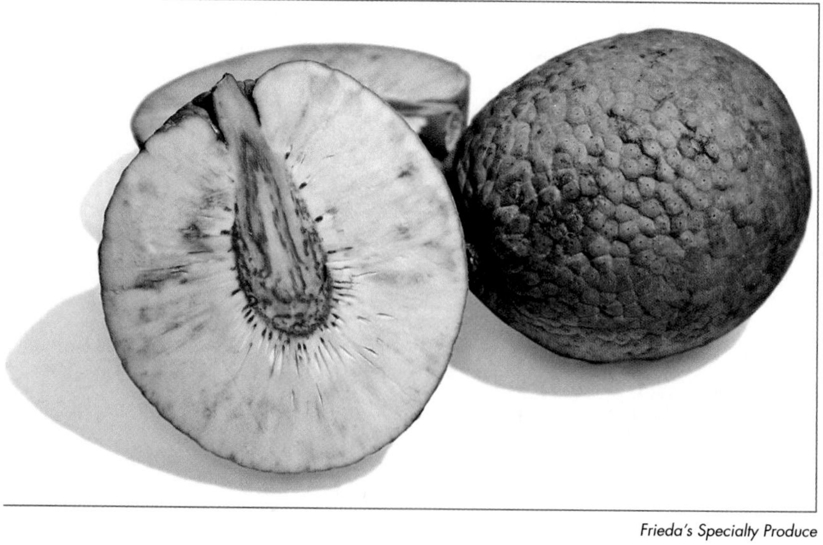

Frieda's Specialty Produce

Figure 12-50. *The flesh of breadfruit is cooked and used much like a potato and has a flavor similar to fresh bread.*

Kiwanos. A *kiwano,* also known as horned melon, is an exotic fruit with jagged peaks rising from an orange and red-ringed rind that is native to Africa. Its flesh is a brilliant lime-green color with cucumber-like seeds and a tart, refreshing kiwi-cucumber flavor. *See Figure 12-51.* Kiwano melons can be cut in half lengthwise and eaten right out of the rind or sliced and added to fruit plates. The flesh can be strained to make a tangy-sweet sauce, salad dressing, or a refreshing beverage.

Kiwanos

Figure 12-51. *Kiwanos, also known as horned melons, have a brilliant lime-green flesh with cucumber-like seeds and a tart, refreshing kiwi-cucumber flavor.*

Kumquats. A *kumquat* is a small, golden, oval-shaped fruit with a thin, sweet peel and tart center. A kumquat is similar in appearance to a small citrus fruit but is eaten peel and all. *See Figure 12-52.* Kumquats are available from November to April. Kumquats are used to make jellies, marmalades, and chutneys. They are also an attractive garnish, especially when displayed with their leaves.

Jackfruit. A *jackfruit* is an enormous, spiny, oval exotic fruit with yellow flesh that tastes like a banana and has seeds that can be boiled or roasted and then eaten. Jackfruit is the largest tree-borne fruit in the world. *See Figure 12-53.* A jackfruit can weigh up to 100 pounds and be up to 36 inches long and approximately 20 inches in diameter. There are 100–500 edible seeds in a single jackfruit. Immature jackfruit is similar in texture to chicken, making it a vegetarian substitute known as "vegetable meat." The flesh can be eaten raw or preserved in syrup. Like pineapple, jackfruit is a multiple fruit derived from the convergence of many individual flowers and a fleshy stem axis.

Fruit-Vegetables

Fruit-vegetables are botanical fruits that are usually sold, prepared, and served like vegetables. Fruit-vegetables are typically more tart than sweet. Tomatoes, cucumbers, eggplants, okra, edible bean and pea pods, sweet and hot peppers, summer and winter squashes, and pumpkins are fruit-vegetables commonly used in the professional kitchen. Conversely, rhubarb is a vegetable that is often prepared and served like a fruit. See Chapter 13—Vegetables for more information.

Kumquats

Figure 12-52. *Kumquats are small, golden, oval-shaped fruits that are eaten peel and all.*

Jackfruit

Florida Department of Citrus   Melissa's Produce

Figure 12-53. *Jackfruit is an enormous, exotic fruit that tastes like banana and has seeds that can be boiled or roasted and then eaten. Jackfruit is the largest tree-borne fruit in the world.*

PURCHASING FRESH FRUIT

Fresh fruit is packed in cartons, lugs, flats, crates, or bushels and is sold by weight or count. *See Figure 12-54.* Lugs can hold between 25–50 pounds of fruit, whereas flats are usually used to ship pint-size or quart-size containers of fruits such as blueberries or strawberries. Individual states have diverse regulations on the weight the containers in a flat can hold. Fruit size may also need to be specified when ordering from a vendor. For example, a 25 pound case of apples could contain 90, 110, or 125 apples, depending on the size of the fruit.

Some fruits can be purchased in a prepared form—cleaned, peeled, and cut. These convenience products save time but cost more than purchasing the whole fruit. Prepared fruit also may not taste as fresh as fruit bought whole due to processing.

Fresh Fruit Grades

The USDA has a voluntary grading program for fresh fruit. Grades are based on a variety of characteristics including uniformity of shape, size, color, texture, and the absence of defects. The USDA grades for fresh fruit include U.S. Fancy, U.S. No. 1, U.S. No. 2, and U.S. No. 3. Some fruit varieties have additional grades, such as U.S. Extra Fancy, U.S. Utility, or U.S. Commercial, that are specific to that particular variety. Most fruit used in restaurants is either U.S. Fancy or U.S. No. 1.

Fresh Fruit

Figure 12-54. *Fresh fruit is packed in cartons, lugs, flats, crates, or bushels and is sold by weight or count.*

Ripening Fresh Fruit

It is important to know when fresh fruit should be purchased in order to purchase fruit with the best flavor and quality. As fruit begins to ripen, it changes in color and size and its flesh becomes soft and succulent. Left on the plant, fruit does not stop ripening when it reaches full maturity. Rather, it continues to ripen, breaking down in texture and flavor and eventually spoiling. Some fruits continue to ripen and mature after harvest, while other fruits do not. Fruits that continue to ripen after being harvested, such as bananas and pears, are often purchased before they are fully ripe. Other fruits, such as pineapples, have to be harvested fully ripe because they do not continue to mature after they are picked.

The ripening process can also be accelerated by storing fruit at room temperature or with other fruits that emit a large amount of ethylene gas. Apples, melons, and bananas give off ethylene gas and should be stored away from delicate fruits that could quickly ripen and spoil. Ripening can be delayed by chilling.

Irradiated Fruit

Some fruits are irradiated prior to being sold. *Irradiation* is the process of exposing food to low doses of gamma rays in order to destroy deadly organisms such as E. coli O157:H7, campylobacter, and salmonella. Irradiation also reduces spoilage bacteria, reduces insects and parasites such as fruit flies and the mango seed weevil, and in certain fruits and vegetables it inhibits sprouting and delays ripening. Alternatives to irradiation include cold treatment and fumigation. A statement indicating that the fruit was "treated with irradiation" and a radura, the international symbol of irradiation, appear on the label of irradiated foods. *See Figure 12-55.*

Canned Fruit

Almost any type of fruit can be canned. Peaches, pears, and pineapples are commonly canned fruits. Fruit is canned at its peak. Depending on how long after harvest fresh fruit is eaten, canned fruit can have similar nutritional value. While the heating process does destroy some vitamins, canned fruits are adequate substitutes. Canned fruits that are packed in their own juice and those that have low amounts of added sugar and salt are the best choices.

Fruits are canned in water, light syrup, medium syrup, or heavy syrup. The packing method should be considered when using canned fruit in recipes, as the canning liquid can affect the flavor, texture, and nutritional value of the fruit. After opening canned fruit, any unused portions should be transferred to an airtight storage container, dated and labeled, and then refrigerated.

During the process of canning, raw fruit is subjected to very high temperatures to ensure that all of the bacteria that could cause spoilage or illness are destroyed. Canned fruit can be stored for indefinite periods as long as it is kept in a cool, dry place. *See Figure 12-56.* Cans that are dented or bulging should be disposed of as they may contain harmful bacteria.

Radura Symbol

Figure 12-55. *The radura symbol is required to be on the label of any food that has been irradiated.*

Canned Fruit

InterMetro Industries Corporation

Figure 12-56. *Canned fruits can be stored for indefinite periods when kept in a cool, dry place.*

Frozen Fruit

Freezing fruit inhibits the growth of microorganisms without affecting the nutritional value of the fruit. Many fruits are individually quick-frozen. Liquid nitrogen is commonly used to produce a quick chill, which speeds the freezing process. Fruits, such as berries, are frozen whole, while other fruits, such as peaches, are cleaned, sliced, and frozen. *See Figure 12-57.* Individually quick-frozen fruits are convenient because the entire content need not be thawed to use a small portion. Some frozen fruits packaged in heavy syrup are also sold as purées that are convenient for making ice creams, sorbets, and some sauces. The USDA grades for frozen fruit include U.S. Grade A or U.S. Fancy, U.S. Grade B or U.S. Choice, and U.S. Grade C or U.S. Standard.

Frozen Fruit

Figure 12-57. *Frozen fruits, such as berries, are frozen whole and are convenient because the entire package need not be thawed in order to use a small portion.*

Dried Fruit

Dried fruit is fruit that has had most of the moisture removed either naturally or through use of a machine, such as a food dehydrator. Raisins, prunes, and dates are popular dried fruits. Dried apples, apricots, cherries, cranberries, figs, kiwifruit, mangoes, peaches, pears, persimmons, and strawberries are also available. Dried fruit has a much sweeter taste than fresh fruit due to the sugars being more concentrated. *See Figure 12-58.* Dried fruit is widely used in chutneys and compotes. It can also be added to breads, desserts, or eaten as a snack. For maximum shelf life, dried fruit is stored in an airtight container in a cool, dry place.

Dried Fruit

Frieda's Specialty Produce

Figure 12-58. *Dried fruits, such as cranberries, have a much sweeter taste than fresh fruit due to the sugars being more concentrated.*

COOKING FRUIT

Although many dishes involve raw fruit, fruit can also be prepared using various cooking methods such as simmering and poaching, grilling and broiling, baking and roasting, sautéing, and frying. Regardless of the cooking method used, it is important to remember that fruit is delicate and can become soft or mushy very quickly. Adding sugar to fruit can help prevent it from becoming mushy in the cooking process. The sugar is absorbed by the cells of the fruit, helping the fruit to plump up and stay firm. *See Figure 12-59.* Adding lemon juice or another acid has a similar effect. However, any sort of an alkali, such as baking soda, quickly breaks down cells in the fruit, turning the fruit to mush.

Cooking Fruit

Figure 12-59. *Adding sugar to fruit helps the fruit stay plump.*

Another characteristic of fruit to understand is the varying levels of pectin that different fruits contain. *Pectin* is a chemical present in all fruits that acts as a thickening agent when it is cooked in the presence of sugar and an acid. For example, cranberry sauce can be made by simmering cranberries (which contain acid naturally) with sugar. When the cranberries break down during the cooking process pectin is released. As the mixture is allowed to cool it thickens to a jelly-like consistency due to the pectin. Fruits that are high in pectin include apples, blackberries, cranberries, quinces, gooseberries, grapes, and plums. The peels of citrus fruits also contain a lot of pectin. Some fruits that have a smaller amount of pectin include strawberries, raspberries, peaches, apricots, and pears. When making jams, jellies, or marmalades using fruits that are naturally low in pectin, such as strawberries and apricots, packaged pectin is typically added to the recipe to achieve the desired consistency.

National Honey Board

Simmering and Poaching

The simmering method is often used to make fruit compotes and stewed fruit. Fresh, frozen, canned, or dried fruit can be simmered. Simmering tenderizes and sweetens fruit. Simmered fruit can be served hot or cold and can accompany a dessert or entrée.

Fruit is poached in various liquids, such as water, liquor, wine, or syrup. Apples, pears, peaches, and plums are often poached. Poaching is done at 185°F. This low temperature ensures that the fruit retains its shape while cooking.

Grilling and Broiling

When grilling or broiling fruit, the sugars must be allowed time to caramelize. This happens quickly, as broiling and grilling occur at very high temperatures. Fruits that are good to broil or grill include pineapples, peaches, grapefruits, bananas, and apples. These fruits can be cut into slices or chunks and soaked in liquor or coated with sugar, honey, or liqueur for extra flavor before cooking.

When broiling fruit, the fruit should be placed on a sheet pan lined with parchment paper. Any grilling of fruit should be done on a clean grill without any residue from previously grilled foods. Fruit can be placed directly on the grill or cooked on skewers to make fruit kabobs. *See Figure 12-60.* Grilled or broiled fruit can be eaten alone or added as an accompaniment.

Baking and Roasting

Most berries, pomes, and drupes are well suited for baking. The inner cavity of an apple, fig, or pear can be stuffed with a flavorful filling. Pies, tarts, cobblers, strudels, and turnovers are typically filled with fruits such as apples, blueberries, cherries, or peaches. *See Figure 12-61.*

Fruit can also be added to meats that are being roasted. For example, ham is often covered with pineapple rings while roasting to add extra sweetness to the meat. Placing peach halves atop chicken pieces during the final stages of roasting adds flavor and beauty to the dish.

Grilling Fruit

National Chicken Council

Figure 12-60. *Fruit can be added to skewers and grilled as kabobs.*

Baked Fruit Pies and Tarts

National Cherry Growers and Industries Foundation

Fruit Pie   **Fruit Tart**

Figure 12-61. *Baked fruit pies and tarts are typically filled with fruits such as cherries, apples, blueberries, or peaches.*

Sautéing and Frying

Fruit is often sautéed in butter, sugar, spices, or liquor. The fruit develops a sweet, rich flavor and a syrupy, caramelized glaze. Sautéed fruit can be used in dessert dishes such as in crêpes or as toppings for ice cream. It can also be incorporated into savory mixtures that include garlic, onions, or shallots. Savory fruit mixtures pair well with entrées such as pork and poultry. *See Figure 12-62.*

Apples, bananas, pears, and peaches are suitable fruits for frying because they do not break down when exposed to very high temperatures. Before frying, the fruit is sliced into uniformly sized pieces so that it cooks evenly. The fruit is patted dry with a paper towel to help the batter adhere to the fruit. The fruit is dipped in the batter and then fried in fat. When the batter turns golden brown, the fruit is removed from the hot fat. Cooling fruit on a rack allows the excess fat to drain. Fried fruits may be garnished with powdered sugar or melted chocolate.

Sautéed Fruit

Courtesy of The National Pork Board

Figure 12-62. *Sautéed fruits pair well with entrées such as pork.*

SUMMARY

Fruits are packed with vitamins, minerals, and fiber and provide a sense of fullness without fat or excess calories. There are many varieties of fruit, including berries, grapes, pomes, drupes, melons, citrus fruits, tropical fruits, exotic fruits, and fruit-vegetables. Fruit is always most flavorful at its peak season. Fresh fruit should be purchased when in season and used upon ripening. Fruit can be simmered, poached, grilled, broiled, baked, roasted, sautéed, and fried.

Refer to DVD for
Quick Quiz® questions

Review

1. List the nutritional benefits of eating fruit.
2. Differentiate between variety fruits and hybrid fruits.
3. Identify types of berries used in the professional kitchen.
4. Identify four types of berries that are also aggregate fruits.
5. Identify three type of grapes used in the professional kitchen.
6. Explain why Concord grapes are most often used to make jams, jellies, and grape juice.
7. Identify three types of pomes used in the professional kitchen.
8. Explain two methods that can be used to core and peel apples.
9. Identify six types of pears used in the professional kitchen.
10. Describe how quinces are typically cooked.
11. Explain why drupes are often referred to as stone fruits.
12. Identify seven types of drupes used in the professional kitchen.
13. Explain how to prepare avocados.
14. Identify the drupe that is harvested both for its fruit and its oil.
15. Identify the three types of rind that can be found on melons.
16. Identify seven types of melons used in the professional kitchen.

Refer to DVD for
Review Questions

17. Explain how to seed and cut a melon.
18. Explain the difference between the peel and pith of citrus fruits.
19. Identify seven types of citrus fruits used in the professional kitchen.
20. Explain the purpose of zesting a citrus fruit.
21. Describe how to cut citrus fruit into supremes.
22. Identify 12 tropical fruits used in the professional kitchen.
23. Differentiate between bananas and plantains.
24. Describe how to core a pineapple.
25. Describe how to prepare a mango.
26. Identify 11 exotic fruits used in the professional kitchen.
27. Describe ways to accelerate and delay the ripening of fruit.
28. Explain how the pectin level in a fruit affects the cooking process.
29. List common methods of cooking fruit.
30. Explain why fruit is poached at 185°F.

Applications

1. Core and peel an apple using a fruit corer and a vegetable peeler. Core and peel an apple using a chef's knife and a paring knife. Compare the two methods.
2. Demonstrate how to prepare avocados.
3. Demonstrate how to seed and slice a melon.
4. Cut an orange or a grapefruit into supremes.
5. Zest a lemon or a lime.
6. Peel, core, and slice a pineapple using a chef's knife.
7. Peel and dice a mango using a chef's knife.
8. Simmer or poach a fruit and evaluate the quality of the prepared fruit.
9. Grill or broil a fruit and evaluate the quality of the prepared fruit.
10. Bake or roast a fruit and evaluate the quality of the prepared fruit.
11. Sauté or fry a fruit and evaluate the quality of the prepared fruit.

Refer to DVD for
Application Scoresheets

Grilled Fruit Kebabs

Yield: *4 servings, 1 kebab each*

Ingredients

wooden skewers	8 ea
pineapple, 1 inch cubes	4 oz
honeydew, 1 inch cubes	4 oz
cantaloupe, 1 inch cubes	4 oz
strawberries	4 ea
powdered sugar	2 tsp

Preparation

1. Soak wooden skewers in water for 10 minutes. Cut fruit into cubes.
2. Run 2 skewers through each piece of fruit about ¼ inch apart. This will prevent the fruit from spinning around on a single skewer.
3. Dust the fruit skewers lightly with powdered sugar and grill on each side for 2–3 minutes or until marked.

NUTRITION FACTS
Per serving: 45 calories, 2 calories from fat, <1 g total fat, 0 mg cholesterol, 10.1 mg sodium, 198.8 mg potassium, 11.2 g carbohydrates, 1.2 g fiber, 9.4 g sugar, <1 g protein

Poached Pears

Yield: *4 servings, 1 pear each*

Ingredients

ripe Anjou pears	4 ea
juice oranges	3 ea
lemon	1 ea
red wine	½ c
pear nectar	½ c
cinnamon	⅛ tsp
vanilla extract	4 drops
cornstarch	½ tsp

Preparation

1. Peel pears and remove the core from each pear, leaving the stem intact and the fruit otherwise whole.
2. Juice the oranges and lemon.
3. Combine juices, red wine, pear nectar, cinnamon, and vanilla in a saucepot that is just large enough to hold the 4 pears standing upright.
4. Add pears to the poaching liquid. The liquid should cover the pears halfway. Bring the liquid to a simmer and cook over low heat until tender, turning pears halfway through the cooking time.
5. When done, remove pears from poaching liquid and cool completely.
6. Use a few tablespoons of the poaching liquid to make a slurry with the cornstarch.
7. Slowly add the cornstarch mixture to the poaching liquid to create a sauce. Bring sauce to a simmer and stir until thickened.
8. Serve sauce with pears.

NUTRITION FACTS
Per serving: 235 calories, 4 calories from fat, <1 g total fat, 0 mg cholesterol, 4.7 mg sodium, 556 mg potassium, 55.6 g carbohydrates, 10.1 g fiber, 34 g sugar, 2.2 g protein

Baked Apples

Yield: *4 servings, 1 apple each*

Ingredients

apples, peeled and cored	4 ea
butter	4 tbsp
brown sugar	½ c
cinnamon and nutmeg	TT

Preparation

1. Place peeled and cored apples upright in a roasting pan so that the hole from each core can be filled.
2. Place 1 tbsp of butter in the center hole of each apple.
3. Add 2 tbsp of brown sugar in the center hole of each apple.
4. Lightly dust apples with cinnamon and nutmeg to taste.
5. Bake in a 400°F oven, basting often with melted butter, until the apples are soft (approximately 35–40 minutes).

NUTRITION FACTS
Per serving: 269 calories, 103 calories from fat, 11.7 g total fat, 30.5 mg cholesterol, 9.3 mg sodium, 156.8 mg potassium, 43.6 g carbohydrates, 1.9 g fiber, 39.6 g sugar, <1 g protein

Mango-Peach Cobbler

Yield: *6 servings, 3 oz each*

Ingredients

Filling

mangoes, diced	2 ea
peaches, sliced thin	4 ea
granulated sugar	¼ c
lemon juice	1 tbsp
cornstarch	1 tsp
nonstick cooking spray	1 tsp

Biscuit Topping

all-purpose flour	1 c
granulated sugar	½ c
baking powder	1 tsp
nutmeg	½ tsp
salt	½ tsp
unsalted butter	6 tbsp
water	¼ c

Preparation

1. Dice the mangos and slice the peaches.
2. Mix mangoes, peaches, ¼ c of sugar, lemon juice, and cornstarch in a mixing bowl until well combined.
3. Spray bottom and sides of a 2-qt baking dish with nonstick cooking spray.
4. Place fruit mixture in an even layer in baking dish.
5. Heat fruit mixture in a 400°F oven for 8 minutes.
6. Mix flour, ½ c of sugar, baking powder, nutmeg, and salt in a large bowl.
7. Using a pastry blender, blend cold butter into the dry mixture until butter is pea sized.
8. Boil water and add to dry mixture. Stir until just combined.
9. Spoon the biscuit topping over the heated fruit.
10. Bake in 400°F oven until topping is golden.

Variation: To make Peach Cobbler, omit the mangoes and use 9 peaches, sliced thin.

Variation: To make Apricot Cobbler, omit the mangoes and the peaches. Use 12 apricots, sliced thin, and ½ c granulated sugar in the filling.

NUTRITION FACTS
Per serving: 385 calories, 132 calories from fat, 14.8 g total fat, 30.5 mg cholesterol, 278.4 mg sodium, 344.6 mg potassium, 62.1 g carbohydrates, 3.3 g fiber, 43.3 g sugar, 3.8 g protein

Blueberry Cobbler

Yield: *12 servings, 3 oz each*

Ingredients

Filling

blueberries	6 c
granulated sugar	½ c
lemon juice	1½ tbsp
cornstarch	2 tbsp
nonstick cooking spray	1 tsp

Biscuit Topping

all-purpose flour	1 c
granulated sugar	½ c
baking powder	1 tsp
nutmeg	½ tsp
salt	½ tsp
unsalted butter	6 tbsp
water	¼ c

Preparation

1. Mix blueberries, ½ c of sugar, lemon juice, and cornstarch in bowl until well combined.
2. Spray bottom and sides of a 2-qt baking dish with nonstick cooking spray.
3. Place fruit mixture in an even layer in baking dish.
4. Heat fruit mixture in a 400°F oven for 8 minutes.
5. Mix flour, ½ c of sugar, baking powder, nutmeg, and salt in a large bowl.
6. Using a pastry blender, blend butter into the dry mixture until butter is pea sized.
7. Boil water and add to dry mixture. Stir until just combined.
8. Spoon the biscuit topping over the heated fruit.
9. Bake in 400°F oven until topping is golden.

NUTRITION FACTS
Per serving: 214 calories, 66 calories from fat, 7.4 g total fat, 15.3 mg cholesterol, 139.7 mg sodium, 72.6 mg potassium, 36.8 g carbohydrates, 2.1 g fiber, 24.1 g sugar, 1.7 g protein

Apple Fritters

Yield: *16 servings, 1 fritter each*

Ingredients

all-purpose flour, sifted	1 c
granulated sugar	1 tbsp
baking powder	1½ tsp
salt	¼ tsp
milk	½ c
egg, beaten	1 ea
butter, melted	1 tbsp
apples, peeled and sliced	2 ea
vegetable oil	as needed
confectioners' sugar	½ c

Preparation

1. Mix flour, sugar, baking powder, and salt together in a large bowl.
2. Mix milk, egg, and melted butter together in a small bowl while whisking rapidly. *Note:* Butter that is too hot can cause the egg to curdle.
3. Add the wet mixture to the dry mixture and slowly mix until incorporated.
4. Dip the peeled apple slices in the batter to coat.
5. Fry the battered apples in vegetable oil at 350°F until golden brown.
6. Remove from fat and allow the cooked fritters to drain.
7. Dust the hot fritters with confectioners' sugar just prior to serving.

NUTRITION FACTS
Per serving: 70 calories, 15 calories from fat, 1.6 g total fat, 14.3 mg cholesterol, 90.1 mg sodium, 37.5 mg potassium, 12.4 g carbohydrates, <1 g fiber, 5.9 g sugar, 1.5 g protein

Apple and Walnut Stuffing

Yield: *12 servings, 3 oz each*

Ingredients

whole butter	2 oz
onions, small dice	4 oz
celery stalks, small dice	4 ea
apple, cored and seeded, medium dice	2 ea
walnuts, chopped	4 oz
sage, chiffonade	2 tbsp
wheat or multigrain toast, 1 inch cubes	10 slices
chicken stock	¼ c
whole egg, beaten	1 ea
salt and pepper	TT

Preparation

1. In medium sauté pan over medium heat, add butter and cook until melted taking precaution to not burn.
2. Add diced onions and celery to sauté pan and sweat until onions are almost translucent.
3. Add diced apples and chopped walnuts. Sauté until apples begin to turn golden brown.
4. Place apple mixture in mixing bowl along with remaining ingredients and stir well to moisten.
5. Stuff apple walnut mixture into meat or poultry and cook until it reaches an internal temperature of 165°F.

NUTRITION FACTS
Per serving: 178 calories, 97 calories from fat, 11.3 g total fat, 25.7 mg cholesterol, 165.8 mg sodium, 176.1 mg potassium, 15.5 g carbohydrates, 3.1 g fiber, 4.6 g sugar, 5.4 g protein

Raspberry and Apple Compote

Yield: *12 servings, 3 oz each*

Ingredients

tart apples, medium dice	2 large
brandy	1 c
sugar	¼ c
red or black raspberries	4 c
unsalted butter	1 tbsp
spearmint, chiffonade	1 tsp

Preparation

1. Add diced apples, brandy, and sugar to a saucepan and bring to simmer. Reduce liquid to ¼ c.
2. Stir in berries.
3. Gently stir in unsalted butter and spearmint until the butter is melted and thoroughly incorporated.
4. Serve compote warm over an entrée or side dish.

NUTRITION FACTS
Per serving: 111 calories, 11 calories from fat, 1.3 g total fat, 2.5 mg cholesterol, 1.2 mg sodium, 104.5 mg potassium, 14.4 g carbohydrates, 3.6 g fiber, 10 g sugar, <1 g protein

Greek Yogurt with Minted Pineapple, Candied Ginger, and Sweet Granola

Yield: *8 servings, 6 oz each*

Ingredients

pineapple, medium dice	2¼ c
clover honey	¼ c
spearmint, chiffonade	1 tbsp
candied or crystallized ginger root, minced	2 tbsp
Greek-style plain yogurt	2½ c
granola	1 c

Preparation

1. Dice the pineapples. Chiffonade the spearmint. Mince the ginger root.
2. Mix pineapple, honey, spearmint, and ginger in bowl.
3. Divide yogurt between the serving dishes.
4. Top yogurt with fruit mixture.
5. Top fruit mixture with an equal portion of granola and serve.

NUTRITION FACTS
Per serving: 186 calories, 44 calories from fat, 4.9 g total fat, 4.6 mg cholesterol, 58.5 mg sodium, 317.8 mg potassium, 30.3 g carbohydrates, 2.1 g fiber, 21.4 g sugar, 6.6 g protein

Watermelon and Arugula Salad with Crumbled Feta

Yield: *4 servings, 6 oz each*

Ingredients

Dressing

olive oil	2 fl oz
balsamic vinegar	1 fl oz
granulated sugar	¼ tsp
salt and freshly ground black pepper	TT

Salad

seedless watermelon, large dice	3 c
feta cheese, crumbled	2 oz
baby arugula, washed and spun dry	4 oz

Preparation

1. Mix dressing ingredients in large bowl and season to taste.
2. Add diced watermelon, feta, and arugula.
3. Toss well to incorporate.

NUTRITION FACTS
Per serving: 205 calories, 152 calories from fat, 17.2 g total fat, 12.6 mg cholesterol, 234.9 mg sodium, 157.8 mg potassium, 11.1 g carbohydrates, <1 g fiber, 9.2 g sugar, 2.8 g protein

Raspberry Coulis

Yield: *8 oz*

Ingredients

fresh raspberries	2 c
simple syrup	1 tbsp

Preparation

1. Place raspberries and simple syrup in food processor and purée until smooth.
2. Strain through a chinois to remove all of the seeds.

NUTRITION FACTS
Per serving: 152 calories, 13 calories from fat, 1.6 g total fat, 0 mg cholesterol, 2.5 mg sodium, 371.6 mg potassium, 35.7 g carbohydrates, 16 g fiber, 17.2 g sugar, 3 g protein

Raspberry Vinaigrette

Yield: *6 servings, 1½ fl oz each*

Ingredients

vegetable oil	6 fl oz
raspberry vinegar	2 fl oz
fresh raspberries	½ c
sugar	½ tsp
lemon juice, fresh	⅛ tsp
salt and pepper	TT

Preparation

1. Place all ingredients in blender.
2. Purée until smooth.
3. Strain out seeds and discard.

> **NUTRITION FACTS**
> **Per serving:** 256 calories, 248 calories from fat, 28.1 g total fat,
> 0 mg cholesterol, 48.7 mg sodium, 27.4 mg potassium,
> 2.3 g carbohydrates, <1 g fiber, <1 g sugar, <1 g protein

Citrus, Radicchio, and Fresh Fennel Salad with Tarragon Sherry Vinaigrette

Yield: *8 servings, 4 oz each*

Ingredients

lime juice	½ tbsp
sherry vinegar	1 fl oz
olive oil	3 fl oz
sugar	½ tsp
salt and pepper	TT
ruby red grapefruit, supremed	2 ea
juice oranges, supremed	2 ea
fennel bulb, sliced very thin	1 ea
radicchio, torn into bite-sized pieces	1 small head
tarragon leaves, shredded fine	1 tsp

Preparation

1. Supreme grapefruit and oranges.
2. Prepare fennel, radicchio, and tarragon.
3. Place lime juice, vinegar, oil, and sugar in bowl and mix well. Season to taste.
4. Add remaining ingredients and toss well to coat. Serve immediately.

> **NUTRITION FACTS**
> **Per serving:** 146 calories, 95 calories from fat, 10.8 g total fat,
> 0 mg cholesterol, 52.9 mg sodium, 280.2 mg potassium,
> 13.1 g carbohydrates, 2.7 g fiber, 4.5 g sugar, 1.2 g protein

Cold Strawberry Soup with Prosecco and Fresh Basil

Yield: *6 servings, 5 oz each*

Ingredients

strawberries, leaves and hull removed	1 pt
confectioners' sugar	½ tbsp
basil, stems removed, chiffonade	1 tbsp
black pepper, freshly ground	pinch
salt	pinch
Prosecco wine	1 c
plain yogurt	3 oz
clover honey	2 tsp

Preparation

1. Clean the strawberries and chiffonade the basil.
2. Place all ingredients in blender except the Prosecco, yogurt, and honey. Purée until smooth.
3. Add Prosecco and pulse to incorporate.
4. In a separate bowl, mix yogurt and honey and reserve.
5. Plate the soup in chilled bowl and garnish with a dollop of sweetened yogurt.

NUTRITION FACTS
Per serving: 122 calories, 6 calories from fat, <1 g total fat, <1 mg cholesterol, 64.2 mg sodium, 284.3 mg potassium, 18.6 g carbohydrates, 2.8 g fiber, 10.6 g sugar, 1.8 g protein

Pineapple, Banana, and Fresh Spinach Smoothie

Yield: *5 servings, 8 fl oz each*

Ingredients

pineapple chunks	1 lb
apple, peeled, cored, and seeds removed	1 ea
banana, peeled	1 ea
spinach leaves, stems removed	4 oz
apple juice	5 fl oz
ice cubes	1 c

Preparation

1. Prepare fruit and spinach.
2. Place all ingredients in blender.
3. Purée until very smooth and serve immediately.

NUTRITION FACTS
Per serving: 98 calories, 3 calories from fat, <1 g total fat, 0 mg cholesterol, 21.7 mg sodium, 364.8 mg potassium, 24.9 g carbohydrates, 2.8 g fiber, 17.5 g sugar, 1.5 g protein

Culinary Arts
PRINCIPLES AND APPLICATIONS

Barilla America, Inc.

Vegetables

Vegetables come in many colors, shapes, sizes, and flavors and are an essential part of a healthy diet. It is important to be able to identify the array of available vegetables and to select quality vegetables at their peak. Vegetables can be prepared and served many ways. Adding vegetables to a plate adds color, texture, flavor, and nutrition. Knowing the proper selection and preparation of a wide variety of vegetables enables culinary professionals to offer delicious, healthy dishes that are pleasing to the eye and the palate.

Chapter Objectives

1. Describe types of edible roots used in the professional kitchen.
2. Describe types of edible bulbs used in the professional kitchen.
3. Demonstrate how to clean leeks.
4. Describe types of edible tubers used in the professional kitchen.
5. Describe types of edible stems used in the professional kitchen.
6. Describe types of edible leaves used in the professional kitchen.
7. Describe types of edible flowers used in the professional kitchen.
8. Demonstrate how to prepare artichokes.
9. Describe types of edible seeds used in the professional kitchen.
10. Demonstrate how to rehydrate pulses.
11. Describe types of fruit-vegetables used in the professional kitchen.
12. Demonstrate how to prepare tomato concassé.
13. Demonstrate how to core bell peppers.
14. Describe types of sea vegetables used in the professional kitchen.
15. Describe types of edible mushrooms used in the professional kitchen.
16. Identify factors to consider when purchasing vegetables.
17. Explain how acidic and alkaline ingredients affect cooked vegetables.
18. Cook a variety of vegetables and evaluate the quality of the prepared dishes.
19. Demonstrate how to fire-roast peppers.

Refer to DVD for **Flash Cards**

Key Terms

- **vegetable**
- **edible root**
- **edible bulb**
- **edible tuber**
- **edible stem**
- **edible leaf**
- **edible flower**
- **edible seed**
- **legume**
- **pulse**
- **lentil**
- **fruit-vegetable**
- **sea vegetable**
- **edible mushroom**
- **chlorophyll**
- **carotenoid**
- **flavonoid**

VEGETABLE CLASSIFICATIONS

A *vegetable* is an edible root, bulb, tuber, stem, leaf, flower, or seed of a non-woody plant. Fruit-vegetables, sea vegetables, and mushrooms are typically prepared and served like vegetables. Each vegetable offers culinary professionals unique opportunities to enhance menus with healthy, earthy flavors. Vegetables are an excellent source of vitamins, minerals, and fiber. They are also low in fat and calories. Vegetables have the best flavor and color when purchased in season.

Edible Roots

An *edible root vegetable* is an earthy-flavored vegetable that grows underground and has leaves that extend above ground. Edible roots include carrots, parsnips, salsify, radishes, turnips, rutabagas, beets, celeriac, jicamas, lotus roots, and bamboo shoots. There are also edible roots that are not classified as vegetables, such as ginger and horseradish.

Carrots. A *carrot* is an elongated root vegetable that is rich in vitamin A. Carrots come in many colors and are sold with and without their green tops. *See Figure 13-1.* Carrots are available year-round. When purchasing carrots, those that are firm and bright in color should be chosen. Carrots can be eaten raw, sautéed, broiled, blanched, or steamed.

Carrots

Orange

Maroon

Melissa's Produce

Figure 13-1. *Carrots are elongated root vegetables that are rich in vitamin A and come in many colors.*

Parsnips. A *parsnip* is an off-white root vegetable, similar in shape to a carrot, that ranges from 5–10 inches in length. *See Figure 13-2.* Parsnips are available from late fall through winter. Parsnips that are harvested later in the season are sweeter, as the cold converts some of the starch into sugar. Parsnips can be eaten raw, blanched, steamed, broiled, or roasted.

Parsnips

Frieda's Specialty Produce

Figure 13-2. *Parsnips are off-white root vegetables, similar in shape to carrots, that range from 5–10 inches in length.*

Salsify. *Salsify* is a white or black root vegetable, similar in shape to a carrot, and can grow up to 12 inches in length. When cooked, white salsify tastes like an artichoke heart. The more preferred black salsify, also known as scorzonera or black oyster plant, has a savory fish flavor. ***See Figure 13-3.*** The thin inedible skin is usually removed after the root is boiled. Salsify is often added to other vegetables such as peas and carrots and served with a béchamel or mustard sauce. Salsify can also be battered and fried.

Chef's Tip

Peeled salsify should be immediately immersed in water mixed with vinegar or lemon juice to prevent discoloration.

Salsify

Melissa's Produce

Figure 13-3. *Salsify is a white or black root vegetable that resembles the shape of a carrot and can grow up to 12 inches in length.*

Radishes. A *radish* is a root vegetable that is small in diameter with a white flesh and a peppery taste that comes in many colors and shapes. Radishes can be eaten raw, added to salads, used as garnishes, and added to stir-fries and soups. Radishes are a good source of vitamin C, potassium, and folic acid. Common varieties include the red radish, black radish, and daikon.

- A red radish is about 1 inch in diameter. The exterior is red or pinkish red. The white flesh is crisp and juicy with a sharp, peppery flavor.
- A black radish can be round or nearly pear-shaped and about 4 inches in diameter. The exterior is rough and black. The flesh is white with a hot, peppery flavor.
- A daikon radish is usually 8–12 inches in length with an elongated shape, similar to that of a carrot. *See Figure 13-4.* Daikons are milder in flavor than red radishes. Most of the pungent flavor of radishes is found in the skin, which is commonly removed prior to cooking.

Nutrition Note

Radishes are high in fiber and water content and low in calories. Radish greens contain more vitamin C, calcium, and iron than the radish root.

Radishes

Red   Daikon

Melissa's Produce

Figure 13-4. *Common varieties of radishes include the red and daikon radishes.*

Turnips. A *turnip* is a round, fleshy root vegetable that is purple and white in color. Turnips have a peppery flavor, similar to that of a radish, and are a good source of vitamin C and potassium. They need to be washed and peeled prior to being cooked. Turnips can be simmered and then puréed, mashed like potatoes, or diced and then sautéed or blanched. When roasted, they develop a buttery taste.

Rutabagas. A *rutabaga* is a round root vegetable derived from a cross between a Savoy cabbage and a turnip. Rutabagas are often confused with turnips, but rutabagas are longer and rounder by comparison. ***See Figure 13-5.*** The flesh has a yellow tint and a more distinct flavor than that of a turnip. Rutabagas are high in potassium and vitamin C. They are often added to soups or puréed like potatoes. Rutabagas can also be prepared in the same ways as turnips.

Turnips and Rutabagas

Turnips   Rutabagas

Melissa's Produce

Figure 13-5. *Turnips and rutabagas are two root vegetables that are often confused for one another, but rutabagas are longer and rounder than turnips.*

Nutrition Note

Betacyanin is the phytochemical that gives a beet its rich color and significantly reduces the homocysteine levels in the blood.

Beets. A *beet* is a round root vegetable with a deep reddish purple or gold color. ***See Figure 13-6.*** Beets are rich in nutrients, including vitamins A and C and potassium. Quality beets are firm with a smooth skin and no spots or bruising. Beets can be eaten raw, cooked, or pickled. Borscht is a type of soup that gets its deep red color from beets.

Beets

Red   Gold

Melissa's Produce

Figure 13-6. *Beets are round root vegetables that are a deep reddish purple color.*

Celeriac. *Celeriac,* also known as celery root, is a knobby, brown root vegetable cultivated from a type of celery grown for its root rather than its stalk. *See Figure 13-7.* Celery root measures about 4–5 inches in diameter and can weigh 2–4 lb. Celeriac tastes like a cross between strong celery and parsley. Dried celeriac is ground to make celery salt. Celeriac should be washed thoroughly and peeled before use. Since it oxidizes quickly, lemon juice or vinegar should be added to the water prior to cooking to avoid discoloration. Raw celeriac can be diced, grated, or shredded for use in salads. It also can be boiled, braised, sautéed, baked, or puréed.

Celeriac

Frieda's Specialty Produce

Figure 13-7. *Celeriac, also known as celery root, is a knobby, brown root vegetable cultivated from a type of celery grown for its root rather than its stalk.*

Jicamas. A *jicama* is a large, brown root vegetable that ranges in size from 4 oz to 6 lb and is a good source of vitamin C and potassium. Jicama, also referred to as the Mexican potato, is native to Central America. The thin brown skin must be removed before use. The crisp, white flesh has a delicate, sweet flavor. Small jicamas are more flavorful than larger ones. *See Figure 13-8.* Jicama can be eaten raw in slaws and salads. It also can be steamed, boiled, puréed, baked, or fried like potatoes. When cooked briefly, jicama retains a crisp, water chestnut texture.

Jicamas

Melissa's Produce

Figure 13-8. *A jicama is a large, brown root vegetable that ranges in size from 4 oz to 6 lb and is referred to as the Mexican potato.*

Lotus Roots. A *lotus root* is the underwater root vegetable of an Asian water lily that looks like a solid-link chain about 3 inches in diameter and up to 4 feet in length. *See Figure 13-9.* Removal of the reddish-brown skin reveals a creamy white flesh that tastes like fresh coconut and has the texture of a raw potato. Lotus root can be purchased fresh, canned, dried, or candied. It can be served as a vegetable and as a dessert. Water chestnuts and sunchokes may be substituted in recipes calling for lotus root.

Bamboo Shoots. A *bamboo shoot* is a root vegetable that is the immature shoot of the bamboo plant. Bamboo shoots are harvested when they reach approximately 6 inches in length. Bamboo shoots cannot be eaten raw because they contain a toxic substance. The cooking process removes the toxin, making them safe to eat. Bamboo shoots are primarily available canned or dried. *See Figure 13-10.* They are an excellent addition to a stir-fry or salad.

Edible Bulbs

An *edible bulb vegetable* is a strongly flavored vegetable that grows underground and consists of a short stem base with one or more buds that are enclosed in overlapping membranes or leaves. Examples of edible bulbs include garlic, shallots, onions, scallions, and leeks. Bulbs are very fragrant and are used for their aromatic qualities as well as their flavor.

Garlic. *Garlic* is a bulb vegetable made up of several small cloves that are enclosed in a thin, husklike skin. White garlic is the most common

Lotus Roots

Melissa's Produce

Figure 13-9. *A lotus root is the underwater root vegetable of an Asian water lily that looks like a solid-link chain.*

Bamboo Shoots

Figure 13-10. *A bamboo shoot is a root vegetable that is the immature shoot of the bamboo plant.*

variety and the most pungent in flavor. Pink garlic, named for its pinkish outer covering, is another strongly flavored variety. Black garlic is made by fermenting bulbs of white garlic at a high temperature to create black cloves. It has a sweet flavor that has hints of vinegar. Elephant garlic contains much larger cloves but is milder in flavor than smaller varieties. *See Figure 13-11.*

Garlic

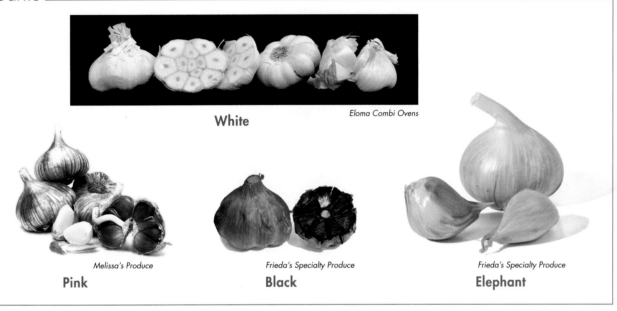

White

Eloma Combi Ovens

Melissa's Produce
Pink

Frieda's Specialty Produce
Black

Frieda's Specialty Produce
Elephant

Figure 13-11. *Garlic is a bulb vegetable made up of several small cloves that are enclosed in a thin, husklike skin that comes in many colors.*

The flavor of garlic is released when a clove is cut, crushed, or minced and increases the more finely the clove is cut. Crushing garlic with the side of a chef's knife is an easy way to remove the peel. Fresh garlic should be stored in a cool, dry place. Cut garlic must be refrigerated.

Shallots. A *shallot* is a very small bulb vegetable that is similar in shape to garlic and has two or three cloves inside. The outer covering can be bronze-colored, rose-colored, or pale gray. *See Figure 13-12.* Shallots have a pink-tinged ivory flesh and a more subtle flavor than onions. When purchasing shallots, it is important to choose those that are firm and dry-skinned and to avoid any that are sprouting.

Onions. An *onion* is a bulb vegetable made up of many concentric layers of fleshy leaves. Onion varieties include white, yellow, red, and pearl onions. *See Figure 13-13.* The variety of the onion and the climate where it was grown determine how strong its flavor is. White onions have a slightly sweet flavor. Yellow onions, also known as Spanish onions, are very mild in flavor. Red onions are the sweetest variety and are commonly added to salads and sandwiches for color. Onions are a key flavoring ingredient in many dishes. They can be sautéed, grilled, roasted, stir-fried, deep-fried, or eaten raw.

Shallots

Melissa's Produce
Bronze

Frieda's Specialty Produce
Rose-Colored

Figure 13-12. *Shallots are small bulb vegetables that can be bronze, rose, or pale gray in color.*

Cuisine Note

Cipollinis, also known as Italian pearl onions, have a flat top and are squatty in shape. They are mild in flavor and easy to roast whole or grill on skewers.

Onions

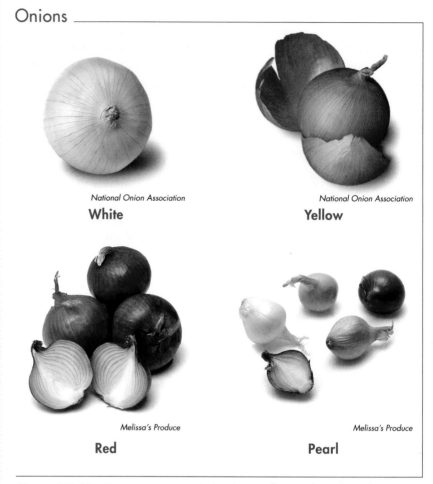

National Onion Association
White

National Onion Association
Yellow

Melissa's Produce
Red

Melissa's Produce
Pearl

Figure 13-13. *Onion varieties include white, yellow, red, and pearl onions.*

Scallions. A *scallion*, also known as a green onion, is a small bulb vegetable with a slightly swollen base and long, slender, green leaves that are hollow. Scallions are mildly flavored compared to onions. The best scallions have a pleasant aroma and brightly colored leaves. Scallions are often added to salads or used as a garnish.

Leeks. A *leek* is a long, white bulb vegetable, with long, wide, flat leaves. Leeks are similar in appearance to scallions, but leeks are much larger. Leeks are milder and sweeter than onions. The white portion of the leek is used most often in a variety of recipes. The green leaves are most often used to flavor soups and stocks. A *ramp* is a wild leek with a flavor similar to scallions, yet with more zing. *See Figure 13-14.* Ramps are often diced for use in salads or on sandwiches. They can also be sautéed for use in egg or potato dishes.

When purchasing leeks or ramps, it is important to choose those with firm bulbs and bright green leaves. It is also important to clean leeks very well because soil and grit often become trapped between the layers of the bulb. *See Figure 13-15.*

Scallions, Leeks, and Ramps

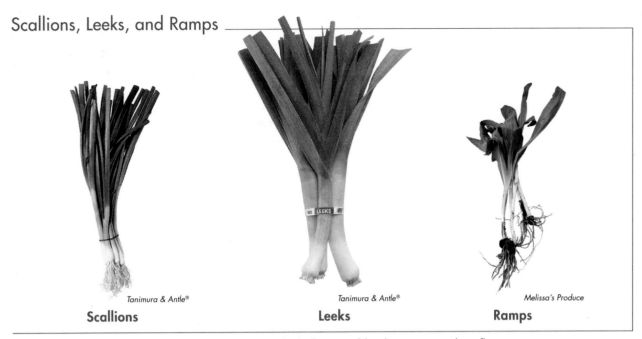

Scallions — *Tanimura & Antle®*

Leeks — *Tanimura & Antle®*

Ramps — *Melissa's Produce*

Figure 13-14. *Scallions, leeks, and ramps are similar bulb vegetables that vary greatly in flavor.*

Procedure for Cleaning Leeks

1. Cut off the root end of the leek just above the root.

2. Split the leek lengthwise down the center from top to bottom.

3. Cut off the top portion of the dark-green end and remove any white portions that look old.

4. Rinse the leek thoroughly to remove any soil or grit that may have settled between the layers.

Figure 13-15. *It is important to clean leeks very well because soil and grit often become trapped between the layers of the bulb.*

Edible Tubers

An *edible tuber* is a short, fleshy vegetable that grows underground and bears buds capable of producing new plants. Examples of edible tubers include potatoes, sweet potatoes, yams, ocas, sunchokes, and water chestnuts.

Potatoes. A *potato* is a round, oval, or elongated tuber that is the only edible part of the potato plant. The color of potato skin differs among varieties and can be brown, red, yellow, white, orange, blue, or purple. *See Figure 13-16.* Potato flesh can be creamy white to yellow-gold or purple in color. When purchasing potatoes, firm, undamaged potatoes with no signs of sprouting should be chosen. Potatoes must be stored in a dry, cool, dark place that allows them to breathe. If potatoes do not have adequate ventilation they quickly rot.

Potatoes

White

Purple

Red

Gold

Frieda's Specialty Produce

Figure 13-16. *The color of potato skin differs among varieties and can be brown, red, yellow-gold, white, orange, blue, or purple.*

Potatoes begin to turn brown when they are peeled or cut. Placing peeled or cut potatoes in cold water prevents discoloration. Potatoes are often added to soups and stews. They also can be baked, sautéed, broiled, grilled, or fried.

Sweet Potatoes. A *sweet potato* is a tuber that grows on a vine and has a paper-thin skin and flesh that ranges in color from ivory to dark orange. Sweet potatoes are an excellent source of vitamin A and potassium. The skin is edible, although it is often removed before cooking. Peeled or cut sweet potatoes oxidize, so it is important to place them in cold water until they are used. Sweet potatoes can be prepared in the same manner as potatoes. They also can be incorporated into breads, cookies, pies, and cakes. Sweet potatoes are often puréed with cinnamon, butter, nutmeg, or brown sugar to enhance their sweetness.

Yams. A *yam* is a large tuber that has thick, barklike skin and a flesh that varies in color from ivory to purple. The skin is inedible. Yams are commonly confused with sweet potatoes because they are often labeled as sweet potatoes in the United States, but they are different vegetables. *See Figure 13-17.* Common varieties of yams include the tropical yam, garnet yam, and jewel yam. Yams can be used in soups, pies, breads, and casseroles. They are low in fat and a good source of carbohydrates, protein, and vitamins A and C.

Sweet Potatoes and Yams

Thin skin

Sweet Potatoes

Barklike skin

Yams

Melissa's Produce

Figure 13-17. *Sweet potatoes have a thin skin and yams have a bark-like skin.*

Ocas. An *oca,* also known as a New Zealand yam, is a small, knobby tuber that has a potato-like flesh and ranges in flavor from very sweet to slightly acidic. Ocas are native to South America and are white, pink, or red in color. *See Figure 13-18.* They are a good source of carbohydrates, calcium, dietary fiber, and iron. Ocas must be kept in a cool, dark place at room temperature or refrigerated in a crisper. Ocas can be used raw in salads or cooked like potatoes. They also may be pickled.

Sunchokes. A *sunchoke,* also known as a Jerusalem artichoke, is a tuber with thin, brown, knobby-looking skin. *See Figure 13-19.* The skin is edible but is often removed before cooking. The white flesh is crisp and sweet and can be used in salads. Sunchokes are related to sunflowers. They can be blanched, steamed, puréed, or used to flavor soups.

Ocas

Melissa's Produce

Figure 13-18. *Ocas are small, knobby tubers that have a potato-like flesh.*

Sunchokes

Frieda's Specialty Produce

Figure 13-19. *Sunchokes are tubers with thin, brown, knobby-looking skin.*

Water Chestnuts. A *water chestnut,* also known as a water caltrop, is a small tuber with brownish-black skin and white flesh. A water chestnut is crunchy and juicy and resembles a chestnut in exterior color and shape. ***See Figure 13-20.*** Water chestnuts have a mild, sweet flavor. Native to Southeast Asia, water chestnuts must be peeled before use in both raw and cooked dishes. Water chestnuts may be refrigerated for up to one week if they are tightly wrapped. Canned water chestnuts are available but inferior in quality.

Water Chestnuts

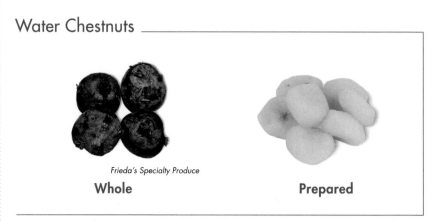

Frieda's Specialty Produce

Whole   **Prepared**

Figure 13-20. *Water chestnuts are small tubers with brownish-black skin and white, crunchy flesh.*

Edible Stems

An *edible stem vegetable* is the main trunk of a plant that develops buds and shoots instead of roots. Stems contain a lot of cellulose and become tougher as they continue to develop. Therefore, stems are usually harvested while tender. Examples of edible stems include asparagus, celery, fennel, rhubarb, kohlrabi, and hearts of palm.

Asparagus. *Asparagus* is a green, white, or purple edible stem that is referred to as a spear. Asparagus is harvested in the spring while it is young. The longer asparagus grows, the woodier it becomes, making it less palatable. Raw asparagus is excellent in salads and is a popular ingredient in omelets, quiches, and pasta dishes. Asparagus can be broiled, grilled, steamed, or puréed.

Green asparagus is the most common variety of asparagus. White asparagus is grown covered in soil to prevent photosynthesis from taking place and harvested as soon as the spears begin to emerge. White asparagus is more tender than green asparagus but is less flavorful. Purple asparagus is sweeter than green asparagus because it has a higher sugar content. A cancer-fighting phytochemical called anthocyanin gives it a purple hue. ***See Figure 13-21.***

Celery. *Celery* is a green stem vegetable that has multiple stalks measuring 12–20 inches in length. The inner stalks are sweeter and more tender than the outer stalks. Celery should be purchased when it is shiny, firm, and crisp. Stalks that have brown or yellow leaves should be avoided. Celery is often eaten raw. It can be sautéed, stir-fried, roasted, or used in stocks and soups.

Tanimura & Antle®

Asparagus

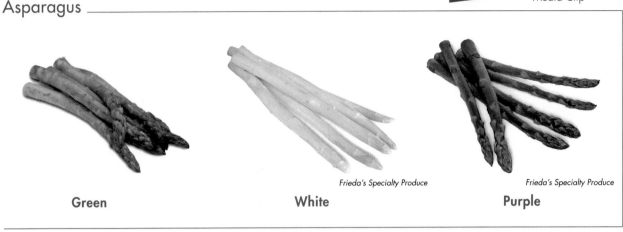

Green

White

Frieda's Specialty Produce

Purple

Frieda's Specialty Produce

Figure 13-21. *Asparagus is a green, white, or purple edible stem that can be broiled, grilled, steamed, or puréed.*

Fennel. *Fennel* is a celery-like stem vegetable with overlapping leaves that grow out of a large bulb at its base. *See Figure 13-22.* Fennel has a mild, sweet flavor that is often associated with licorice or anise. When purchasing fennel, it is important to choose stalks that are firm and unblemished with healthy-looking, bright green leaves. Fennel can be eaten raw but is usually cooked. It can be diced or sliced and sautéed, broiled, blanched, or steamed. It can also be puréed into a soup or side dish or made into an au gratin similar to potatoes.

Celery and Fennel

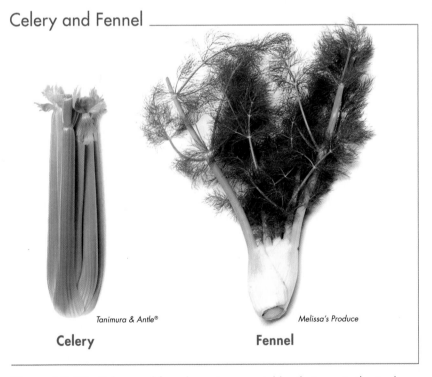

Tanimura & Antle®

Celery

Melissa's Produce

Fennel

Figure 13-22. *Celery and fennel are stem vegetables that are similar in shape with very different flavors and uses.*

Rhubarb

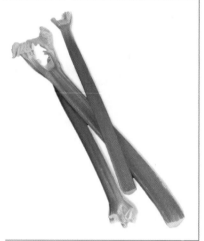

Frieda's Specialty Produce

Figure 13-23. *Rhubarb is a tart stem vegetable that is most often prepared like a fruit.*

Kohlrabi

Frieda's Specialty Produce

Figure 13-24. *Kohlrabi is a sweet, crisp, stem vegetable that has a pale-green or purple, bulbous stem.*

Hearts of Palm

Melissa's Produce

Figure 13-25. *A heart of palm is a slender, white, stem vegetable.*

Rhubarb. *Rhubarb* is a tart stem vegetable that ranges in color from pink to red and is most often prepared like a fruit. *See Figure 13-23.* It may be peeled or left with the skin intact, depending on the use. Rhubarb is best purchased uncut to prevent drying out. Never use rhubarb leaves because they contain a poisonous toxin called oxalate. Rhubarb can be sweetened and stewed to make sauces for meats or poultry, but it is most commonly used to make pies, tarts, and other desserts.

Kohlrabi. *Kohlrabi* is a sweet, crisp, stem vegetable that has a pale-green or purple, bulbous stem and dark-green leaves. Kohlrabi is created by crossbreeding a cabbage and a turnip. Although the entire kohlrabi is edible, the bulbous stem is the portion primarily used in professional cooking. *See Figure 13-24.* The inner part of the stem base may be removed to produce a cavity that can be stuffed. Kohlrabi can be eaten raw, blanched, sautéed, or stir-fried. It is available year-round, but it is at its peak from June through September.

Hearts of Palm. A *heart of palm* is a slender, white, stem vegetable that is surrounded by a tough husk. Hearts of palm are about 4 inches long and can be up to 1½ inches thick, although most are very thin. *See Figure 13-25.* They are good sources of fiber and do not contain any cholesterol. Once the husks have been removed, hearts of palm can be served raw, steamed, or fried. They are a good side dish and a nice complement to salads and pastas.

Edible Leaves

Edible leaves, also known as greens, are plant leaves that are often accompanied by edible leafstalks and shoots. Although edible leaves or greens can be eaten raw, they are often cooked to decrease their bitterness and increase their palatability. Examples of edible leaves or greens include cabbages, bok choy, Brussels sprouts, lettuces, chicory, watercress, spinach, sorrel, chard, kale, collards, mustard greens, turnip greens, beet greens, dandelion greens, nopales, tatsoi, and fiddlehead ferns.

Cabbages. Varieties of cabbages used in the professional kitchen include head cabbage, Napa cabbage, and Savoy cabbage. *See Figure 13-26.* *Head cabbage* is a tightly packed, round head of overlapping edible leaves that can be green, purple, red, or white in color. The inner leaves are usually lighter in color than the outer leaves because they have been exposed to less sunlight. The base of the head where the leaves attach to the stalk is known as the heart. The inedible heart is removed during preparation. Head cabbage usually ranges from 2–8 pounds and from 4–10 inches in diameter. The best heads are heavy and compact, with shiny, unblemished leaves. Head cabbage can be eaten raw, steamed, braised, roasted, or stir-fried.

Napa cabbage, also known as celery cabbage, is an elongated head of crinkly and overlapping edible leaves that are a pale yellow-green color with a white vein. In many parts of the world, Napa cabbage is referred to as Chinese cabbage. Its leaves are more tender than those of head cabbage and it has a very delicate flavor due to its high water content. Napa cabbage is most often used raw in salads or stir-fried.

Cabbages

Green Head

Purple Head

Frieda's Specialty Produce

Napa

Melissa's Produce

Savoy

Figure 13-26. *Varieties of cabbages used in the professional kitchen include head cabbage, Napa cabbage, and Savoy cabbage.*

Savoy cabbage is a conical-shaped head of tender, crinkly, edible leaves that are blue-green on the exterior and pale green on the interior. The leaves are very pliable and have a distinct sweet flavor. Savoy cabbages also lack the sulfur-like odor often associated with cooking other cabbage varieties. They are available year-round, but reach their peak in the winter months. Heavier cabbages should be selected and should not be refrigerated for more than one week. Savoy cabbage can be stir-fried, stuffed, or used raw in salads.

Bok Choy. *Bok choy* is an edible leaf that has tender white ribs, bright-green leaves, and a more subtle flavor than head cabbage. There are many varieties of bok choy, some having short ribs and others having long ribs. *See Figure 13-27.* Bok choy is readily available year-round. Bok choy can be eaten raw, but it is often sautéed, stir-fried, or added to soups.

> **Nutrition Note**
>
> Cabbage is an excellent source of fiber, vitamins, and minerals and has been proven to have cancer-fighting properties.

Cuisine Note

There are over 20 varieties of bok choy. In China, the smallest varieties are valued for their tenderness and often cooked whole. Choy sum is considered the highest quality. It has light green leaves and tiny yellow flowers. Choy sum is also known as Chinese flowering cabbage.

Bok Choy

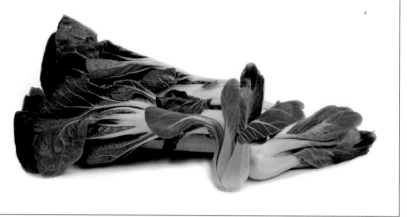

Frieda's Specialty Produce

Figure 13-27. *Bok choy is an edible leaf that has tender white ribs and bright-green leaves.*

Brussels Sprouts. A *Brussels sprout* is a very small round head of tightly packed leaves that looks like a tiny cabbage. ***See Figure 13-28.*** Brussels sprouts grow along an upright stalk and are ready to be harvested when they reach a diameter of about 1 inch. The best sprouts are bright green and have no yellowing leaves. Their peak season is from September through February. Brussels sprouts can be steamed, broiled, grilled, or sautéed.

Brussels Sprouts

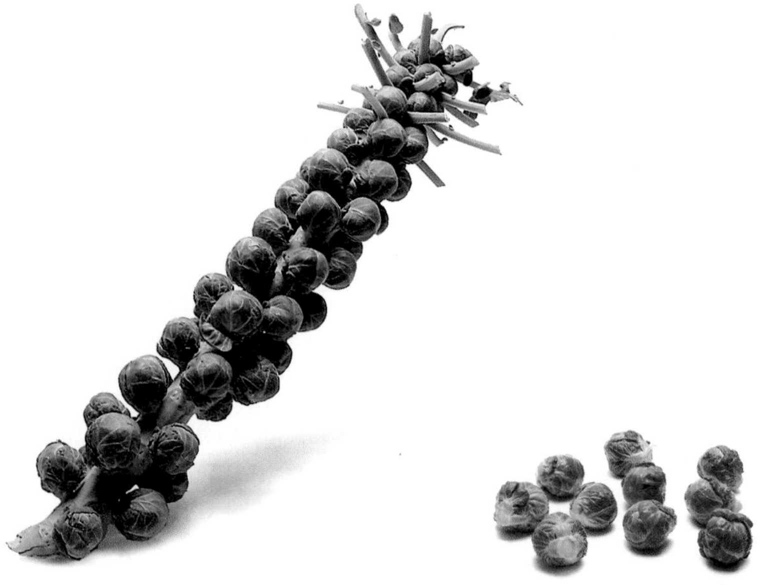

Stalk of Brussels Sprouts **Brussels Sprouts**

Melissa's Produce

Figure 13-28. *Brussels sprouts are very small round heads of tightly packed leaves that look like tiny cabbages.*

Lettuces. *Lettuce* is an edible leaf that is almost exclusively used in salads or as a garnish. *See Figure 13-29.* The four main types of lettuce used in the professional kitchen are looseleaf, romaine, butterhead, and crisphead. These lettuce varieties are discussed in Chapter 15–Garde Manger.

Lettuces _____

Tanimura & Antle®

Figure 13-29. *Lettuce is an edible leaf that is almost exclusively used in salads or as a garnish.*

Chicory. *Chicory,* also known as escarole, is a curly, edible leaf with a slightly bitter-tasting flavor. Chicory should be crisp and brightly colored when purchased. One variety of chicory is frisee. *See Figure 13-30.* Endive is another variety of chicory. Chicory can be used in a salad or served as cooked greens. Chicory roots can be roasted and ground for use in or as a substitute for coffee.

Chicory _____

Melissa's Produce

Figure 13-30. *Frisee is a type of chicory, which has a slightly bitter flavor and an interesting texture.*

Watercress

Tanimura & Antle®

Figure 13-31. *Watercress is a small, crisp, dark-green edible leaf that is typically sold in bouquets.*

Sorrel

Melissa's Produce

Figure 13-33. *Sorrel is a large, green edible leaf that ranges in color from pale green to dark green.*

Watercress. *Watercress* is a small, crisp, dark-green, edible leaf that is a member of the mustard family. Watercress has a pungent, yet slightly peppery flavor. It is typically sold in bouquets and can be refrigerated for up to five days if the stems are in water. It should be crisp and brightly colored when purchased. *See Figure 13-31.* Watercress is a popular garnish that is used on sandwiches and in salads, soups, and stir-fries.

Spinach. *Spinach* is a dark-green, edible leaf with a slightly bitter flavor that may have flat or curly leaves, depending on the variety. *See Figure 13-32.* Fresh spinach is available year-round and is rich in vitamins A and C, folate, potassium, iron, and magnesium. Fresh spinach may be refrigerated in plastic for up to three days. Spinach is usually very gritty and must be thoroughly rinsed. The stems are usually removed before it is cooked. Spinach can be sautéed or added to soups or creamed dishes. It is also served raw in salads or on sandwiches.

Spinach

Tanimura & Antle®

Figure 13-32. *Spinach is a dark-green, edible leaf with a slightly bitter flavor that may have flat or curly leaves, depending on the variety.*

Sorrel. *Sorrel* is a large, green, edible leaf that ranges in color from pale green to dark green and from 2–12 inches in length. Sorrel is quite acidic in flavor. The most strongly flavored sorrel is called sour dock or sour grass. *See Figure 13-33.* The acidic flavor comes from the presence of oxalic acid. Like spinach, sorrel can be eaten raw, used in salads, or cooked. It also is used to flavor cream soups and sauces. Sorrel should not be refrigerated more than three days. Sorrel is a good source of vitamin A.

Chard. *Chard* is a large, dark-green, edible leaf with white or reddish stalks. Swiss chard is grown for its silvery stalks and crinkly leaves. It is sometimes called rhubarb chard and has a strong flavor. Ruby chard has deep-red leaves tinged with green and bright-red stalks. The flavor of ruby chard is milder than Swiss chard. *See Figure 13-34.* Chard is a good source of vitamins A and C and iron. It is available year-round but is in peak season in the summer. The tender greens and crisp stalks can be wrapped in plastic and refrigerated for up to three days. Chard can be prepared in the same manner as spinach.

Chard

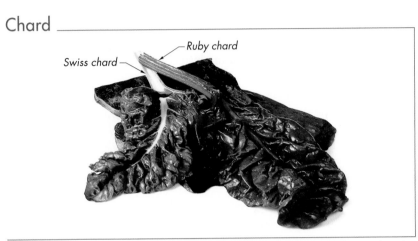

Ruby chard

Swiss chard

Frieda's Specialty Produce

Figure 13-34. *Swiss chard is grown for its silvery stalks and crinkly leaves. Ruby chard has deep-red leaves tinged with green and bright-red stalks.*

Kale. *Kale* is a large, frilly, edible leaf that varies in color from green and white to shades of purple. Although all varieties of kale are edible, the green varieties are better for cooking and the other varieties are used as garnishes. *See Figure 13-35.*

Kale

Melissa's Produce

Figure 13-35. *Kale is a large, frilly, edible leaf that varies in color from green and white to shades of purple.*

Collards

Figure 13-36. *Collards, also known as collard greens, are large, dark-green, edible leaves with a thick, white vein.*

Melissa's Produce

Kale is available year-round and can be refrigerated for two to three days. Longer storage times yield limp leaves and a stronger flavor. Because of its bitterness, kale is rarely eaten raw. The center stalk is often removed before kale is cooked. It may be prepared in the same manner as spinach and is often added to soups or sautéed in flavorful oil and served as a side dish. Kale is an ample source of vitamins A and C, folic acid, calcium, and iron.

Collards. A *collard,* also known as a collard green, is a large, dark-green, edible leaf with a thick, white vein that resembles kale. The flavor of collards is a cross between kale and cabbage. Collards are a variety of cabbage that does not form a head, but grows in clusters at the top of a tall stem. *See Figure 13-36.*

Collards are an excellent source of vitamins A and C, calcium, and iron. They are available year-round, but reach their peak from January through April. Collards are often gritty and must be thoroughly washed before being prepared. They are prepared in the same manner as cabbage or spinach and can be served as a side dish or added to soups. Collards should not be refrigerated for more than five days.

Mustard Greens. A *mustard green* is a large, dark-green, edible leaf from the mustard plant that has a strong peppery flavor. *See Figure 13-37.* Mustard greens are an excellent source of vitamins A and C, thiamin, and riboflavin. Mustard greens must be thoroughly washed before cooking and may be refrigerated up to one week. They can be found year-round, but are at their peak from December through early March. They may be steamed, braised, sautéed, or stir-fried.

Mustard Greens

Figure 13-37. *Mustard greens are large, dark-green, edible leaves from the mustard plant.*

Turnip Greens. A *turnip green* is a dark-green, edible leaf that grows out of the top of the turnip root vegetable. *See Figure 13-38.* Young turnip greens have a sweet flavor. As the plant ages, the leaves become bitter. Turnip greens are an excellent source of vitamins A and C and a good source of riboflavin, calcium, and iron. They can be found year-round, but are at their peak from October through February. Fresh turnip greens should be crisp and have even coloring without any wilted or off-colored leaves. Turnip greens must be thoroughly washed before being cooked. They may be refrigerated for up to three days. The removal of the veins (ribs) from the leaves will yield a more tender batch of greens. Turnip greens may be steamed, braised, sautéed, or stir-fried.

Turnip Greens

Figure 13-38. *Turnip greens are dark-green, edible leaves from the turnip plant.*

Beet Greens. A *beet green* is the green, edible leaf that grows out of the top of the beet root vegetable. *See Figure 13-39.* Beet greens are full of vitamins and minerals and contain more iron than spinach. They have a bitter taste similar to chard. Beet greens are often braised or sautéed. They should be cooked within seven days of purchase.

Beet Greens

Tanimura & Antle®

Figure 13-39. *Beet greens are the green, edible leaves that grow out of the top of the beet root vegetable.*

Dandelion Greens. A *dandelion green* is the dark-green, edible leaf of the dandelion plant. *See Figure 13-40.* Dandelion greens are quite bitter. Often considered a salad green, young dandelion greens are added to salads with a vinaigrette dressing that cuts the harsh bitterness of the green. Dandelion greens are also sautéed and served as a side dish.

Dandelion Greens

Frieda's Specialty Produce

Figure 13-40. *Dandelion greens are the dark-green, edible leaves of the dandelion plant.*

Nopales

Figure 13-41. *Nopales are the green, edible leaves of the prickly pear cactus.*

Tatsoi

Figure 13-42. *Tatsoi is a spoon-shaped, emerald-colored leaf vegetable native to Japan.*

Nopales. A *nopal* is the green, edible leaf of the prickly pear cactus. Nopales measure about 5 inches in length and 3–4 inches in width. *See Figure 13-41.* They are crunchy and slippery, and they have a slight tartness. Nopales are a great source of calcium and vitamins A and C. When preparing nopales the eyes, prickles, and all fibrous or dry areas are removed. Nopales may be steamed, sautéed, or eaten raw in salads.

Tatsoi. *Tatsoi* is a spoon-shaped, emerald-colored leaf vegetable native to Japan. *See Figure 13-42.* Tatsoi has a mild flavor and is a good source of vitamins, minerals, and antioxidants. It may be served raw in salads, steamed, sautéed, or boiled. It is often served in soups, served as a side dish, or used to create a pesto-like sauce.

Fiddlehead Ferns. A *fiddlehead fern* is the curled tip of an ostrich fern frond with a nutty and slightly bitter flavor similar to asparagus and artichokes. Fiddlehead ferns are only available for a few weeks in spring. *See Figure 13-43.* They are a great source of vitamin A, vitamin C, and antioxidants. Before use, fiddlehead ferns should be rubbed between the hands in order to remove the brown scales and silk. They can be boiled, roasted, grilled, sautéed, or used as a garnish.

Fiddlehead Ferns

Fern frond

Fiddlehead ferns

Figure 13-43. *A fiddlehead fern is the curled tip of an ostrich fern frond.*

Edible Flowers

Edible flowers are the flowers of nonwoody plants that are prepared as vegetables. Edible flowers can be eaten raw or cooked. Examples of edible flowers include squash blossoms, broccoli, cauliflower, and artichokes.

Squash Blossoms. A *squash blossom* is the edible flower of a summer or a winter squash. They come in varying shades of yellow and orange and often taste a bit like the parent squash. ***See Figure 13-44.*** Ideal squash blossoms are closed buds. They are extremely perishable and cannot be refrigerated more than a day. Squash blossoms are used as a garnish and to flavor salads and soups. They also can be lightly battered and sautéed, stuffed with soft cheese and baked, or batter dipped and fried.

Broccoli. *Broccoli* is an edible flower that is a member of the cabbage family and has tight clusters of dark-green florets on top of a pale-green stalk with dark-green leaves. When buying broccoli, it is important the broccoli is firm and evenly colored. Broccoli can be eaten raw or can be steamed, blanched, broiled, sautéed, or stir-fried. Broccoli is available year-round.

Cauliflower. *Cauliflower* is an edible flower that is a member of the cabbage family and has tightly packed white florets on a short, white-green stalk with large, pale-green leaves. Some varieties of cauliflower have a purple or greenish tinge to the florets. Cauliflower grows covered with numerous layers of leaves attached to the stalk and surrounding the head. These leaves protect the head from sunlight and preserve its white color. When purchasing cauliflower, it is important to choose heads that are firm and compact. Cauliflower can be eaten raw or can be steamed, blanched, sautéed, stir-fried, or broiled.

A hybrid created from broccoli and cauliflower is called broccoflower®. It has a mild and sweet, nutty flavor and a tender, yet firm texture that is less crumbly than cauliflower. ***See Figure 13-45.*** It can be cooked in the same manner as cauliflower and is high in vitamin C, folic acid, and copper. Like cauliflower, the head should be firm with compact florets and must be washed before use. Orange and purple cauliflower do not differ in taste or use from white cauliflower.

Squash Blossoms

Figure 13-44. *A squash blossom is the edible flower of a summer or a winter squash.*

Broccoli and Cauliflower

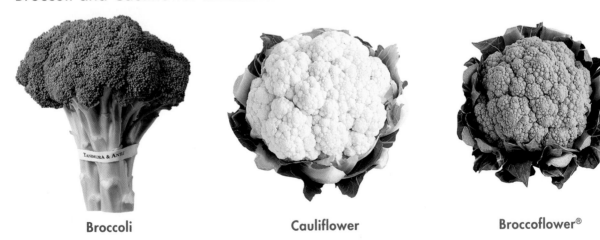

Broccoli   Cauliflower   Broccoflower®

Tanimura & Antle®

Figure 13-45. *Broccoli has tight clusters of dark-green florets on top of a pale-green stalk. Cauliflower has tightly packed white florets on a short, white-green stalk. Broccoflower® is a hybrid created from broccoli and cauliflower.*

Artichokes. An *artichoke* is the edible flower bud of a large, thistle-family plant that comes in many varieties. *See Figure 13-46.* Globe artichokes are available year-round, but reach their peak from March through May. Artichokes that have a tight leaf formation, are deep-green in color, and are heavy for their size should be purchased. In general, the smaller the artichoke, the more tender it will be, and the rounder it is, the larger the heart. Artichokes must be washed before cooking. They also should be prepared in nonreactive bowls and pans to prevent off-flavors.

Artichokes _____

Globe

Purple

Melissa's Produce

Figure 13-46. *An artichoke is the flower bud of a large, thistle-family plant that comes in many varieties.*

The petals of an artichoke, called leaves, are the edible part of the bud. The tips of the larger leaves can be very sharp with a spiny thorn that must be removed during preparation. The dense area where the leaves attach to the stem is known as the heart. As the flower matures, a fuzzy, sometimes thorny, center called the choke develops just above the center of the heart. The choke must be removed before the artichoke is served. *See Figure 13-47.*

1. With a French knife, remove the entire top half of the artichoke, exposing the lighter-colored choke in the center. Discard the top half.

2. With kitchen shears, cut off the top third (the thorn) of each of the larger exposed leaves. Discard the removed tips.

3. Grasp the bottom of the artichoke, and use both thumbs to spread open the center and expose the choke.

4. With a spoon, thoroughly scrape the choke from the solid heart. Discard the choke.

5. Cut off any extra stem.

6. A small amount of lemon juice can be squeezed onto the exposed heart to deter browning.

Figure 13-47. *The leaves of an artichoke must be trimmed and the choke must be removed before the artichoke is served.*

Once the choke has been removed, an artichoke can be simmered, steamed, or baked until the heart becomes tender. Artichokes are commonly filled with a mixture of lemon, garlic, and bread crumbs and baked until the filling is golden. Cooked artichokes may be refrigerated for up to three days.

Refer to DVD for **Preparing Artichokes** Media Clip

Edible Seeds

An *edible seed* is the seed of a nonwoody plant. Edible seeds are prepared as vegetables. Edible seeds include some of the oldest recorded forms of food. Many edible seeds can be eaten raw, and all of them can be cooked. Examples of edible seeds include all varieties of legumes and sprouts. A *legume* is the edible seed of a nonwoody plant and grows in multiples within a pod. There are thousands of varieties of legumes, but the most popular varieties include beans, peas, pulses, and lentils. In some cases, the pods are eaten along with the seeds. Legumes are rich in fiber and protein and contain little or no fat. They are often used as the protein component of a dish.

Beans and Peas. Beans and peas are usually kidney-shaped or round. They can be purchased fresh, canned, frozen, or dried. Popular varieties of beans include limas, cannellinis, anasazis, peruanos, calypsos, flageolets, pintos, kidney beans, great northern beans, and black beans. *See Figure 13-48.*

Beans

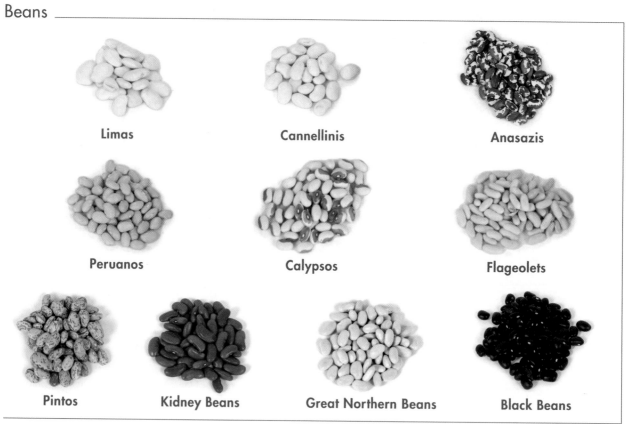

Figure 13-48. *Popular varieties of beans include limas, cannellinis, anasazis, peruanos, calypsos, flageolets, pintos, kidney beans, great northern beans, and black beans.*

Beans and peas can be eaten hot or cold and are used to make hearty soups. Some beans, such as pintos, can be puréed to make refried beans. When purchasing fresh beans and peas, smooth, shiny seeds or pods should be chosen. Some peas, called split peas, are harvested fully mature, left to dry, and then split. Split peas can be puréed into tasty soups.

Some fresh bean and pea varieties are called edible pods, meaning that both the exterior skin and the interior seeds are edible. For example, fresh green beans and fresh wax beans are actually immature beans with under-developed pods that are therefore edible. Likewise, snow peas have a flat pod that is entirely edible. Sugar snap peas have a more rounded pod, but are still tender enough to eat. These types of pod vegetables can be eaten raw, steamed, sautéed, or fried. *See Figure 13-49.*

Other types of fresh beans, such as edamame, do not have edible pods. *Edamame* are green soybeans housed within a fibrous, inedible pod. Edamame can be purchased fresh or frozen. Most fresh beans can be eaten raw, steamed, sautéed, grilled, or fried.

Edible Pods

Green Beans   **Wax Beans**

Snow Peas   **Sugar Snap Peas**

Melissa's Produce

Figure 13-49. *Green beans, wax beans, snow peas, and sugar snap peas are beans and peas with edible pods.*

Pulses. A *pulse* is a dried seed of a legume. Dried beans and peas, such as cannellini beans and black-eyed peas, are shelled and then left to dry until they become rock hard. Pulses must be rehydrated by soaking them overnight or by using a quick-soaking process. Rehydration decrease the total cooking time and result in an even texture throughout. *See Figure 13-50.* Changing the water once or twice while soaking pulses will also help remove impurities that can cause gas during digestion.

Indian Harvest Specialtifoods, Inc./Rob Yuretich

Soaking Overnight

1. Remove any cracked pulses and any debris.
2. Rinse the pulses several times in cold water.
3. Transfer the pulses to a bowl large enough to hold 3 parts water and 1 part pulses.
4. Let the pulses soak overnight in the refrigerator. *Note:* The soaking liquid contains nutrients and can be used to make stock or soup.

Quick-Soaking

1. Remove any cracked pulses and any debris.
2. Rinse the pulses several times in cold water.
3. In a large stockpot, add 4 parts water to 1 part pulses.
4. Slowly bring the pulses to a boil. Reduce the heat and simmer for 2 minutes.
5. Remove the pot from the heat, cover it and let it stand for 1–2 hours or until the pulses swell.
6. Strain the pulses and proceed with the recipe. *Note:* The soaking liquid contains nutrients and can be used to make stock or soup.

Figure 13-50. *Pulses such as cannellini beans and black-eyed peas must be rehydrated by soaking them overnight or by using a quick-soaking process.*

Lentils. A *lentil* is a very small, dried pulse that has been split in half. There are many varieties of lentils, with colors ranging from white to green. ***See Figure 13-51.*** Unlike dried beans and peas, lentils do not have to be soaked because they are smaller and already split in half. However, lentils must be thoroughly washed before cooking because they often contain small stones. Lentils are used to make soups, added to salads, combined and served with other vegetables, and served as sides. Lentils turn mushy when overcooked. They can be stored in airtight containers and held at room temperature for up to one year.

Nutrition Note

Lentils contain calcium and vitamins A and B. They are also a good source of iron and phosphorus.

Lentils

Split White

Petite Crimson

Black Beluga

French Green

Figure 13-51. *Varieties of lentils range in color from white to green.*

Sprouts. A *sprout* is an edible strand with an attached bud that comes from a germinated bean or seed. *See Figure 13-52.* Varieties of sprouts include mung bean, soybean, alfalfa, and radish sprouts. Depending on the plant of origin, sprouts range in taste from mild to spicy. When purchasing sprouts, crisp sprouts that are not wilted and that have attached buds should be chosen. They should be used within a day or two of purchase as they expire quickly. Sprouts are most often used in salads and on sandwiches. They also may be sautéed or stir-fried for less than 30 seconds.

Sprouts _____

Bean **Alfalfa**

Figure 13-52. *A sprout is an edible strand with an attached bud that comes from a germinated bean or seed.*

Fruit-Vegetables

A *fruit-vegetable* is a botanical fruit that is sold, prepared, and served as a vegetable. Fruit-vegetables are typically more tart than sweet. Fruit-vegetables used in the professional kitchen include tomatoes, cucumbers, eggplants, sweet peppers, hot peppers, okra, sweet corn, summer squashes, winter squashes, and pumpkins.

Tomatoes. A *tomato* is a juicy fruit-vegetable that contains edible seeds. Beefsteak, cherry, yellow, pear, and plum tomatoes are just a few of the thousand plus varieties. Tomatoes are available in many different colors, sizes, and shapes. Most tomatoes are red, but some tomatoes are yellow, orange, pink, purple, green, black, white, multicolored, or striped. *See Figure 13-53.*

A *tomatillo,* also known as a Mexican husk tomato, is a small tomato with a thin, papery husk covering a pale-green skin that encases a pale-green flesh. The husk is always removed before cooking. Green tomatillos are tart in flavor, but become sweeter as they ripen and turn yellow. Tomatillos are primarily used to make salsa verde.

Tomatoes are highly perishable and should be used within a few days of purchase. The number of ways tomatoes can be incorporated into a recipe is virtually endless. Tomatoes can be eaten raw, added to salads or sandwiches, and made into soups, sauces, and juice. Tomatoes are also used in many prepared dishes. *Concassé* is a preparation method where a tomato is peeled, seeded, and then chopped or diced. *See Figure 13-54.*

Tomatoes _____

Florida Tomato Committee

Figure 13-53. *Tomatoes are juicy fruit-vegetables that contain edible seeds and come in a thousand plus varieties.*

Chef's Tip

Plum tomatoes are elongated tomatoes about 3–4 inches long. They have a much lower water content than other varieties, making them a perfect choice for sauces.

1. With a paring knife, make a ½ inch wide "X" in the bottom of a tomato, just slightly deeper than the surface of the skin.

2. Place the tomato in a pot of boiling water and blanch for 20–30 seconds or until the skin near the "X" begins to wrinkle or come free from the tomato.

3. Remove the tomato with a strainer or slotted spoon and shock in ice water until cold.

4. Remove the tomato from the ice water. With a paring knife, make a circular cut around the core, and remove the core.

5. Using the tip of a paring knife grab the loose tomato skin and peel it away.

6. Cut the tomato in half horizontally and gently squeeze each half to remove the seeds and juice.

7. Slice, chop, or dice the peeled and seeded tomato.

Figure 13-54. *Preparing tomato concassé requires blanching and shocking the fruit-vegetable prior to removing the skin and seeds and slicing or dicing the flesh.*

Cucumbers. A *cucumber* is a green, cylindrical fruit-vegetable that has an edible skin, edible seeds, and a moist flesh. Cucumbers are often eaten raw or pickled. *See Figure 13-55.* The cucumber is a widely cultivated member of the gourd family. The most common varieties of cucumbers are English (burpless), Japanese, Mediterranean, and dosakai, which are yellow, round cucumbers. Pickling cucumbers have a bumpy, light-green skin and are smaller and thicker than cucumber varieties that are eaten fresh. Cucumbers are used in salads and soups.

Cucumbers

Melissa's Produce

Seeded

Frieda's Specialty Produce

Seedless

Figure 13-55. *Cucumbers are fruit-vegetables that are often eaten raw or pickled.*

Eggplants. An *eggplant* is a deep-purple, white, or variegated fruit-vegetable with edible skin and a yellow to white, spongy flesh that contains small, brown, edible seeds. Although eggplants are available year-round, their peak of season is from August to September. There are many varieties of eggplant. *See Figure 13-56.* The black beauty variety is a large, dark, glossy eggplant commonly found in the United States. Japanese eggplants are long, slender, and a lighter purple than the black beauty variety.

Eggplants

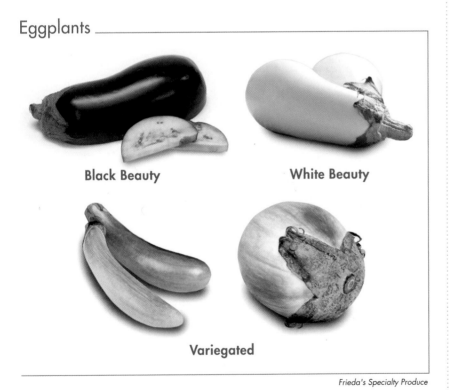

Black Beauty

White Beauty

Variegated

Frieda's Specialty Produce

Figure 13-56. *Varieties of eggplant include black beauty, white beauty, and variegated eggplants that can be oblong or round.*

Barilla America, Inc.

An eggplant begins to discolor as soon as it is cut, so it is important either to cook the eggplant immediately after it is sliced or to sprinkle it with lemon juice. Sliced, raw eggplant can be lightly salted and left on paper towels to drain some of the moisture from the eggplant before cooking. Eggplant can be roasted, grilled, fried, puréed, and baked. Eggplant lasagna is the vegetarian alternative to meat lasagna. Moussaka is a famous Mediterranean dish made with eggplant.

Sweet Peppers. Sweet peppers are called bell peppers. A *bell pepper* is a fruit-vegetable with three or more lobes of crisp flesh that surround hundreds of seeds in an inner cavity. Bell peppers turn yellow and ultimately red if left to ripen on the vine. *See Figure 13-57.* The longer the pepper stays on the vine to ripen, the sweeter it becomes. Red peppers are the sweetest peppers because they are the ripest. The core and seeds can be easily removed from bell peppers using the rolling method. Once the seeds have been removed, bell peppers are typically julienned or diced before use. *See Figure 13-58.*

Sweet Peppers

Barilla America, Inc.

Figure 13-57. *Sweet peppers, also known as bell peppers, turn from green to yellow and then red if left to ripen on the vine.*

Refer to DVD for
Coring Peppers
Media Clip

Procedure for Coring Peppers

1. Use a chef's knife to cut off the top of the pepper just below the stem. Remove the stem and reserve the top piece.

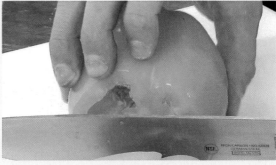

2. Cut off the bottom of the pepper, so that the pepper sits flat on the cutting board. Reserve the bottom piece.

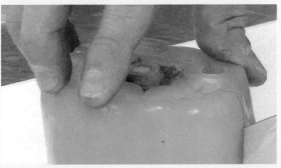

3. Cut a vertical slice to create an opening in the exterior of the pepper.

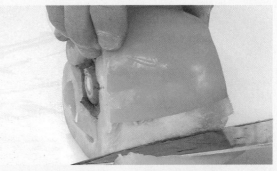

4. Turn the pepper on its side and insert the top of the knife blade between the outer skin and the internal ribs.

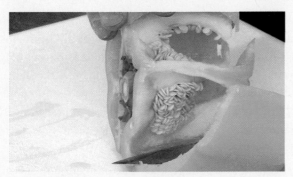

5. Slowly move the knife along the inside of the skin while carefully rolling the pepper away from the knife blade in one continuous motion.

6. Discard the center portion of the pepper. Julienne or dice the cored pepper.

Figure 13-58. *The core and seeds can be easily removed from bell peppers using the rolling method. Once the seeds have been removed, bell peppers are typically julienned or diced before use.*

Hot Peppers. Hot peppers are called chiles. A *chile* is a brightly colored fruit-vegetable pod with distinct mild to hot flavors. Chiles come in many colors, shapes, and sizes. There are more than 200 varieties, including the jalapeño, habanero, poblano, and serrano. *See Figure 13-59.*

Hot Peppers

Jalapeño   Habanero   Poblano   Serrano

Figure 13-59. *Hot peppers, also known as chiles, come in many colors, shapes, and sizes. There are more than 200 varieties, including the jalapeño, habanero, poblano, and serrano.*

Okra

Figure 13-60. *Okra is a green fruit-vegetable pod containing round, white seeds and a gelatinous, slimy liquid.*

Sweet Corn

Figure 13-61. *Sweet corn has kernels that grow in rows on a cob.*

Chiles range in color from yellow to green and fire red to black. They also range from ¼–4 inches in size. The seeds and veins, or membranes, located inside the chile pod contain capsaicin. *Capsaicin* is a potent compound that gives chiles their hot flavor. Larger chiles, such as the poblano, are milder than smaller chiles, such as the serrano. Smaller chiles typically contain more seeds and membranes and are therefore hotter in flavor. Chiles are native to Mexican cuisine and Central American cuisine and are used to flavor sauces and soups. Some chiles, like the poblano, can be stuffed and made into chile relleno.

Okra. *Okra* is a green fruit-vegetable pod that contains small, round, white seeds and a gelatinous liquid. *See Figure 13-60.* Originally from Africa, okra is available year-round in the southern United States and from May to October in other areas of the country. When preparing okra, the stem end is trimmed, and the pod is thoroughly rinsed in cold water before cooking. Okra should be cooked only in stainless steel cookware to prevent it from turning dark. Okra may be steamed, boiled, fried, or pickled. It also can be blanched whole and frozen for up to one year.

Sweet Corn. *Sweet corn* is a fruit-vegetable that has edible seeds called kernels that grow in rows on a spongy cob encased by thin leaves (husks), forming what is referred to as an ear of corn. Beneath the husks, the kernels are covered with fine, hairlike material called silk. An ear of corn can contain 200–400 kernels and be 4–10 inches in length. *See Figure 13-61.*

Sweet corn reaches its peak during July and August. Sweet corn can be grilled, roasted, steamed, or boiled as corn on the cob or as kernels that have been cut off the cob. When grilling or roasting sweet corn, the outer leaves of the husk are removed, while the inner leaves are left on the cob to retain moisture and prevent the kernels from drying out. The thin strands of corn silk are removed prior to eating.

Summer Squashes. A *summer squash* is a fruit-vegetable that grows on a vine and has edible skin, flesh, and seeds. Summer squashes are harvested as immature vegetables two to eight days after flowering. They come in many shapes, sizes, and colors and can be eaten raw or cooked. The tender flesh of a summer squash has high water content and a mild flavor.

Summer squashes are highly perishable, so they should be used soon after they are purchased. They can be sautéed, grilled, stir-fried, baked, or deep-fried. Summer squash varieties include zucchini, straightneck, crookneck, and pattypan squash. *See Figure 13-62.*

- *Zucchini* is an elongated squash that resembles a cucumber and is available in green or yellow varieties.

- *Straightneck squash* is a yellow squash that resembles a bowling pin.

- *Crookneck squash* is a yellow squash that resembles a bowling pin with a bent neck.

- *Pattypan* is a round, shallow squash with scalloped edges and is best harvested when it is no larger than 2–3 inches in diameter.

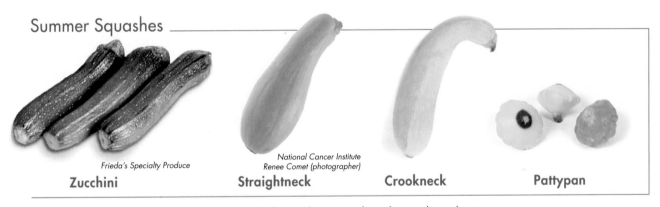

Summer Squashes

Zucchini — *Frieda's Specialty Produce*

Straightneck — *National Cancer Institute Renee Comet (photographer)*

Crookneck

Pattypan

Figure 13-62. *Summer squash varieties include zucchini, straightneck squash, and pattypan.*

Winter Squashes. A *winter squash* is a fruit-vegetable that grows on a vine and has a thick, hard, inedible skin and firm flesh surrounding a cavity filled with seeds. The firm flesh of a winter squash is deep yellow to orange in color, varies in color and taste by variety, and requires longer cooking times than summer squash.

Most winter squashes are roasted and made into soups or puréed and served as a side dish. Winter squashes are harvested fully ripe and come in a variety of shapes and sizes. Acorn, butternut, spaghetti, turban, Hubbard, and kabocha squash are all varieties of winter squash. *See Figure 13-63.*

- *Acorn squash* is a winter squash that looks like a large, dark-green acorn and can be baked, steamed, or puréed.

- *Butternut squash* is a large, bottom-heavy, tan-colored winter squash.

- *Spaghetti squash* is a dark-yellow winter squash with pale-yellow flesh that can be separated into spaghetti-like strands after it is cooked. Spaghetti squash can be used as a substitute for regular spaghetti noodles.

- *Turban squash* is a turban-shaped winter squash. Turban squash has a striped rind and a thick, dry flesh that is orange or golden yellow. The flavor is mild and sweet with nutty overtones. Turban squash measures 6–8 inches in diameter and weighs about 3 lb. Turban squash adds nutritional value to muffins, puddings, and cookies.

- *Hubbard squash* is a large, oval winter squash with a bumpy, pale-green, blue-gray, or orange skin and a sweet orange flesh. It is a relatively dry squash, which means it does not lose much volume when cooked. Hubbard squash weighs around 20 lb, so the flesh is cubed before it is roasted or baked.

- *Kabocha squash* is a winter squash with a jade-green and celadon-streaked rind and a pale-orange flesh that is sweet. A kabocha can range from 2–8 lb. Before cooking, kabochas must be halved and seeded. They can be cooked in the same way as acorn squash.

Winter Squashes

Melissa's Produce
Acorn

Melissa's Produce
Butternut

Melissa's Produce
Spaghetti

Frieda's Specialty Produce
Turban

Melissa's Produce
Hubbard

Melissa's Produce
Kabocha

Figure 13-63. *Winter squash varieties include acorn, butternut, spaghetti, turban, Hubbard, and kabocha squash.*

Pumpkins. A *pumpkin* is a round fruit-vegetable with a hard orange skin and a firm flesh that surrounds a cavity filled with seeds. Pumpkins vary in size and weight and peak in the fall and winter. The flesh has a mild, sweet flavor and can be prepared in the same manner as winter squash. Pumpkin seeds are often roasted for use as a garnish or to add flavor to salads. *See Figure 13-64.*

Pumpkins

Figure 13-64. *Pumpkins have a mild, sweet flesh that can be prepared in the same manner as winter squash. Roasted pumpkin seeds are often used as a garnish.*

Sea Vegetables

Sea vegetables are edible saltwater plants that contain high amounts of dietary fiber, vitamins, and minerals. Sea vegetables lend a salty flavor to food because of the minerals they absorb from the ocean. When adding sea vegetables to a dish, no salt should be added. Many sea vegetables also contain alginic acid, which is used as a stabilizer and a thickener when making processed foods such as ice creams, puddings, and pie fillings. Sea vegetables can be roasted along with other vegetables or crumbled and added to soups, sauces, salads, pastas, and rice dishes. Nori, kombu, arame, wakame, and dulse are types of sea vegetables commonly used in the professional kitchen. *See Figure 13-65.*

Cuisine Note

Sea vegetables are consumed by many cultures around the world. People living in coastal Asian countries, New Zealand, the Pacific Islands, coastal South American countries, Ireland, Scotland, Norway, and Iceland have been harvesting and eating sea vegetables for thousands of years.

Sea Vegetables

Nori

Kombu

Wakame

Figure 13-65. *Sea vegetables such as nori, kombu, and wakame lend a salty flavor to food because of the minerals they absorb from the ocean.*

Nori. *Nori* is a thin, purple-black sea vegetable that turns green when it is toasted. Nori is often used as a wrapper for sushi. It contains iodine and vitamin C. Nori is often toasted and crushed before being used as a condiment for grain dishes, soups, and salads.

Kombu. *Kombu* is a long, dark-brown to purple sea vegetable that is used to flavor dashi stock. It is sold in dried strips or sheets. Quality kombu is almost black with a white residue on its surface that is quite flavorful. It contains iodine, calcium, magnesium, and iron. Kombu may be substituted for monosodium glutamate (MSG). It is often added to the cooking liquid for rice, soups, and beans. When kombu is added to these dishes, extra liquid is required because it soaks up liquid and doubles in volume.

Arame. *Arame* is a thin and wiry, shredded black sea vegetable. It contains calcium, iodine, potassium, vitamin A, and dietary fiber. Arame must be soaked in warm water 10–15 minutes before it is cooked. It is often added to quiches, omelets, stir-fries, and cold pasta salads.

Wakame. *Wakame* is a long, tender, grayish-green sea vegetable that expands to seven times its size when soaked in water. Wakame is high in dietary fiber and potassium. It can be eaten raw, added to soups and stir-fries, or roasted and sprinkled on salads.

Dulse. *Dulse* is a stringy, reddish-brown sea vegetable with a fishy odor that is rich in iron, iodine, potassium, and vitamin A. It has a mild flavor and is sold in whole, crushed, flaked, and powdered form. Dulse powder is often used in soups and stews as a salt substitute. Whole dulse can be pan-fried until it is crispy to add crunch to sandwiches and salads. It has a chewy texture and a savory flavor that is similar to bacon.

Edible Mushrooms

Mushrooms are not vegetables, but they are prepared and used in the same manner. An *edible mushroom* is the fleshy, spore-bearing body of an edible fungus that grows above the ground. Many mushrooms are commercially raised, but others are harvested from the wild. Mushrooms can be purchased fresh, dried, or canned.

Fresh mushrooms should be firm and not spotted or slimy. Fresh mushrooms should be cleaned with a damp towel and lightly rinsed, if necessary, before being used in a recipe. Fresh mushrooms must be stored in a cool, dry place. They also require air circulation to stay fresh, so storing them in paper bags works best. Dried mushrooms must be rehydrated before they are used in a recipe. Canned mushrooms do not have the same quality as fresh or dried mushrooms.

Mushrooms are used to add flavor to many dishes and can be eaten raw, sautéed, stir-fried, or deep-fried. There are many varieties of edible mushrooms. Common varieties of mushrooms used in the professional kitchen include button, portobello, enokitake, wood ear, shiitake, oyster, chanterelles, morel, and porcini.

Irinox USA

Button Mushrooms. A *button mushroom,* also known as a white mushroom, is a cultivated mushroom with a very smooth, rounded cap and completely closed gills atop a short stem. *See Figure 13-66.* Button mushrooms are one of the most widely consumed mushrooms. Button mushrooms have a strong flavor that complements most foods. They can be eaten raw, sautéed, fried, or added to omelets, soups, meat dishes, poultry dishes, seafood dishes, and pasta dishes.

Portobello Mushrooms. A *portobello mushroom* is a very large and mature, brown cremini mushroom that has a flat cap measuring up to 6 inches in diameter. *See Figure 13-67.* The gills of the portobello mushroom are fully exposed, leaving the mushroom without much moisture and creating a dense, meaty texture. Their woody stems are removed and used to flavor stocks and soups. The caps are typically used whole, but can be diced for use in a wide variety of dishes. Portobello mushrooms are popular grilled, on sandwiches, or cut into thick slices and added to salads. They have a meaty, savory flavor.

Button Mushrooms

Mushroom Council

Figure 13-66. *A button mushroom has a very smooth, rounded cap and completely closed gills atop a short stem.*

Portobello Mushrooms

Mushroom Council

Figure 13-67. *A portobello mushroom is a very large and mature, brown cremini mushroom that has a flat cap measuring up to 6 inches in diameter.*

Enokitake Mushrooms. An *enokitake mushroom,* also known as an enoki or a snow puff mushroom, is a crisp, delicate mushroom that has spaghetti-like stems topped with white caps. *See Figure 13-68.* Enokitake mushrooms have a crunchy texture and an almost fruity flavor. Enokitake mushrooms are available almost year-round and can also be purchased canned. Fresh enokitake mushrooms should be wrapped in a paper towel and then a plastic bag before being refrigerated. They should be used within five days.

Enokitake Mushrooms

Mushroom Council

Figure 13-68. *An enokitake mushroom has spaghetti-like stems topped with white caps.*

Wood Ear Mushrooms. A *wood ear mushroom,* also known as a cloud ear or a tree ear mushroom, is a brownish-black, ear-shaped mushroom that has a slightly crunchy texture. Wood ear mushrooms have a delicate flavor and often absorb the taste of other ingredients in a dish. *See Figure 13-69.* When dried wood ear mushrooms are reconstituted they increase 5–6 times in size and are a popular addition to stir-fries and soups. Wood ear mushrooms are often combined with tiger lily buds. The albino variety of wood ear mushrooms is white in color.

Wood Ear Mushrooms

Figure 13-69. *A wood ear mushroom is a brownish-black, ear-shaped mushroom that has a slightly crunchy texture.*

Shitake Mushrooms

Figure 13-70. *A shiitake mushroom has an umbrella shape and curled edges.*

Nutrition Note

Oyster mushrooms contain small amounts of arabitol, a sugar alcohol, which may cause gastrointestinal upset.

Shiitake Mushrooms. A *shiitake mushroom,* also known as a forest mushroom, is an amber, tan, brown, or dark-brown mushroom with an umbrella shape and curled edges. Shiitake mushroom caps range in size from 3–10 inches in diameter. Cooked shiitakes release a pinelike aroma and have a rich, earthy, savory flavor. *See Figure 13-70.* The tough stems are usually removed and used to flavor stocks and soups. Spring and fall are their peak seasons, but fresh and dried shiitakes are typically available year-round. Shiitakes can be sautéed, broiled, or baked.

Oyster Mushrooms. An *oyster mushroom* is a broad, fanlike or oyster-shaped mushroom that varies in color from white to gray or tan to dark brown. The white flesh is firm and varies in thickness. Oyster mushrooms span 2–10 inches. The gills of the mushroom are white to cream and descend toward the stalk, which is often not present at the point of sale. *See Figure 13-71.*

Oyster mushrooms are often made into a soup. They also may be stuffed or stir-fried. They are sometimes made into an oyster mushroom sauce used in Asian cooking. Their flavor is mild and they have a slight odor that resembles anise. Oyster mushrooms are often picked when they are young. As the mushroom ages, the flesh toughens and the flavor becomes unpleasant.

Oyster Mushrooms

Figure 13-71. *An oyster mushroom is a broad, fanlike or oyster-shaped mushroom that varies in color from white to gray or tan to dark brown.*

Chanterelle Mushrooms. A *chanterelle mushroom* is a trumpet-shaped mushroom that ranges in color from bright yellow to orange and has a nutty flavor and a chewy texture. *See Figure 13-72.* Chanterelle mushrooms are not widely cultivated, but are found in some markets during the summer and winter months. When purchasing, mushrooms with plump and spongy caps should be chosen. Chanterelles tend to toughen when overcooked, so it is best to add them to a dish near the end of the cooking time. They are also available dried and canned.

Morel Mushrooms. A *morel mushroom* is an uncultivated mushroom with a cone-shaped cap that ranges in height from 2–4 inches and in color from tan to very dark brown. *See Figure 13-73.* Morels belong to the same fungus species as the truffle and are favored for their earthy, nutty flavor. Darker-colored morels tend to have a stronger flavor. Fresh morels can be found in specialty produce markets from April through June. Dried and canned morels are available year-round. Dried morels have a more intense and smoky flavor. Morels are typically sautéed in butter.

Chanterelle Mushrooms

Mushroom Council

Figure 13-72. *A chanterelle mushroom has a nutty flavor and a chewy texture.*

Morel Mushrooms

Melissa's Produce

Figure 13-73. *A morel mushroom is an uncultivated mushroom with a cone-shaped cap.*

Porcini Mushrooms. A *porcini mushroom,* also known as a cèpe, is an uncultivated, pale-brown mushroom with a smooth, meaty texture and a pungent flavor. Porcini mushrooms can weigh anywhere from 1 oz to 1 lb and have a cap that ranges from 1–10 inches in diameter. *See Figure 13-74.* Porcini mushrooms can be found in specialty produce markets in late spring and fall. When purchasing porcini mushrooms, firm large caps with pale undersides should be chosen. Dried porcini mushrooms must be softened in hot water for about 20 minutes before use. They can be substituted for cultivated mushrooms in most recipes.

Porcini Mushrooms

Figure 13-74. *A porcini mushroom is an uncultivated, pale-brown mushroom with a smooth, meaty texture and a pungent flavor.*

PURCHASING VEGETABLES

Vegetables are available fresh, canned, frozen, and dried. Understanding how to purchase and store vegetables is very important in maintaining the highest-quality produce. Because most fresh vegetables have a short shelf life, it is important to know how to maximize their use. Fresh vegetables are also less expensive during their peak season.

Fresh Vegetables

Fresh vegetables are packed in cartons, lugs, flats, crates, or bushels and sold by weight or count. The weight or count of a packed container depends on the size and type of vegetable. *See Figure 13-75.* The USDA voluntary grading system for fresh vegetables includes U.S. Extra Fancy, U.S. Fancy, U.S. Extra No. 1, U.S. No. 1, U.S. No. 2, and U.S. No. 3. The grade of vegetable to purchase depends on how the vegetable will be used. Recipes using fresh or slightly cooked vegetables require premium ingredients, while lesser grades are acceptable in soup recipes.

Fresh Vegetables

Barilla America, Inc.

Figure 13-75. *Fresh vegetables are packed in cartons, lugs, flats, crates, or bushels and sold by weight or count.*

Most vegetables should be stored in a produce cooler at a temperature of 41°F or below. Vegetables should always be stored away from fruits that emit ethylene gas, such as apples and bananas, as the gas can cause the vegetables to overripen and spoil. It is also important to store vegetables away from poultry, meat, seafood, and dairy products.

Potatoes, onions, garlic, and squash should be stored in a cool, dry location that is between 60°F and 70°F. Storing these vegetables in a refrigerator causes their starches to convert to sugars, which negatively alters their texture and flavor.

Canned Vegetables

Canned vegetables are a staple in the professional kitchen. They are sold in a variety of commercial sizes and packed by weight. *See Figure 13-76.* Canned vegetables are USDA graded as U.S. Grade A or U.S. Fancy, U.S. Grade B, and U.S. Grade C or U.S. Standard. Canned vegetables have already been cleaned, cut, peeled, cooked, and treated with heat to kill any harmful microorganisms. However, the canning process often softens vegetables and can sometimes cause nutrient loss.

Canned vegetables can be stored for long periods if they are kept in a cool, dry place. Dented or bulging cans should be discarded as they may contain harmful bacteria. After opening canned vegetables, the unused portion should be placed in an airtight storage container, dated, labeled, and refrigerated.

Cuisine Note

Many fresh vegetables have been treated with irradiation to destroy deadly bacteria such as E. coli O157:H7, campyloacter, and salmonella. A radura symbol appears on the label of irradiated vegetables.

Sullivan University

Vegetable Can Sizes		
Can Size	**Weight**	**Cans Per Case**
No. 2	20 oz	24
No. 2½	28 oz	24
No. 5	46–51 oz	12
No. 10	6 lb 10 oz	6
No. 300	14–15 oz	36
No. 303	16–17 oz	36

Figure 13-76. *Canned vegetables are a staple in the professional kitchen and come in standard sizes.*

Frozen Vegetables

Frozen vegetables offer the same convenience as canned vegetables, with an additional advantage. Frozen vegetables often retain their color and nutrients better than canned vegetables. *See Figure 13-77.* Like fruits, some vegetables are individually quick-frozen to preserve their texture and appearance. Some vegetables are blanched before being frozen, which reduces overall cooking time. Other frozen vegetables are already fully cooked and need only to be heated for service. The USDA grading system used for canned vegetables also applies to frozen vegetables. Frozen vegetables are usually packed in 1 lb or 2 lb bags.

Frozen Vegetables

Figure 13-77. *Frozen vegetables often retain their color and nutrients better than canned vegetables.*

Dried Vegetables

Dried vegetables have had most of their moisture removed by a food dehydrator or the freeze drying process. *See Figure 13-78. Freeze drying* is the process of removing the water content from a food and replacing it with a gas. Freeze dried vegetables retain more color, texture, and shape than dehydrated vegetables.

Dried onions and legumes are commonly used for convenience. For example, 1 lb of dried onions can be used instead of 8 lb of fresh onions. Dried legumes need to be soaked before cooking. For maximum shelf life, dried vegetables should be stored in an airtight container in a cool, dry place.

Dried Vegetables

Frieda's Specialty Produce

Figure 13-78. *Dried vegetables such as sun-dried tomatoes, have had most of their moisture removed by the dehydration process.*

COOKING VEGETABLES

Vegetables should be cooked until they are just tender enough to be easily digested. At this stage of cooking, most vegetables retain the majority of their nutritional value, flavor, and color. Overcooked vegetables often lose their bright colors, may become mushy in texture, and lose nutrients, as vitamins and minerals are destroyed by excess heat. Common methods for cooking vegetables include steaming, blanching, grilling, broiling, baking, roasting, sautéing, and frying.

Regardless of the cooking method used, it is important to understand that the addition of acidic or alkaline ingredients when cooking vegetables causes chemical reactions that affect the color and texture of the vegetables. An acidic ingredient, such as lemon juice or vinegar, is often added to a cooking liquid to contribute flavor to the dish. An alkali, such as baking soda, is often added to tough vegetables, such as dried beans, to speed up the softening process. Adding acidic or alkaline ingredients also alters the natural pigments present in vegetables. The types of natural pigments in vegetables include chlorophyll, carotenoids, and flavonoids. *See Figure 13-79.*

- *Chlorophyll* is an organic pigment found in green vegetables. When an acidic ingredient is added to the cooking liquid, green vegetables turn a drab olive color, but retain their naturally firm texture. If an alkali is added to the cooking liquid, green vegetables become brighter in color but mushy in texture.

- A *carotenoid* is an organic pigment found in orange or yellow vegetables. Acidic ingredients have little to no affect on carotenoids. Alkaline ingredients do not affect the color of carotenoids, but do cause orange and yellow vegetables to become mushy.

Acid and Alkali Reactions				
Pigment	**Cooked Vegetables***		**Acid Added**	**Alkali Added**
Chlorophyll	Broccoli		Color loss	 Mushy texture 
Carotenoids	Carrots		Little or no effect	 Mushy texture 
Flavonoids	Beets		Brighter red	 Turns blue; mushy texture 

* No acidic or alkaline ingredients used

Figure 13-79. *The addition of acidic or alkaline ingredients when cooking vegetables causes chemical reactions that affect the color and texture of the vegetables.*

- A *flavonoid* is an organic pigment found in purple, dark-red, and white vegetables. Acidic ingredients cause purple and dark-red vegetables to turn bright red. In contrast, alkaline ingredients cause purple and dark-red vegetables to turn blue and white vegetables to turn yellow. Alkaline ingredients also cause purple, dark-red, and white vegetables to have a mushy texture.

Steaming and Blanching

Steamers that can accommodate hotel pans allow large quantities of fresh or frozen vegetables to be steamed in a short amount of time. Steamed vegetables can be finished by sautéing to add flavor.

Blanching is accomplished by placing fresh vegetables in boiling water and then quickly removing and placing them in an ice bath to stop the cooking process. *See Figure 13-80.* Some vegetables, such as asparagus, are often blanched and then finished in a broiler or on a grill. Other vegetables, such as tomatoes, may be blanched to make it easier to remove their skin.

Grilling and Broiling

Grilling and broiling are both fast and easy ways to prepare vegetables. Fresh vegetables are seasoned as desired, drizzled with a little oil, and then placed under the broiler or directly on the grill. The size of the vegetables determines how long they need to cook. Grilling and broiling caramelize the sugars in vegetables, giving them a sweeter flavor. *See Figure 13-81.*

Steamed Vegetables

Carlisle FoodService Products

Figure 13-80. *Steamed vegetables can be finished by sautéing to add flavor.*

Grilled Vegetables

Figure 13-81. *Grilling vegetables caramelizes their sugars giving them a sweeter flavor.*

Baking and Roasting

Baking and roasting are excellent ways to prepare vegetables. Dishes such as a broccoli and cheese casserole are typically baked. Many root vegetables such as onions, carrots, turnips, parsnips, and various types of potatoes are often roasted together or placed alongside a large cut of meat. *See Figure 13-82.* When baking and roasting vegetables, it is important to cut vegetables into uniform sizes to ensure doneness.

Roasted Vegetables

Figure 13-82. *Many root vegetables, such as onions, are often roasted.*

Fire-roasting vegetables is another way to roast vegetables. Fire-roasting a vegetable over an open flame allows the item to be cooked whole. For example, peppers are often fire roasted. *See Figure 13-83.*

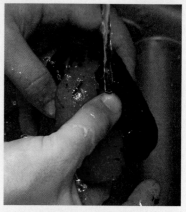

1. Place the washed pepper over an open flame. An alternate method is to place the pepper in a broiler, directly under the flame.

2. Roast the pepper until charred on all sides, turning continuously. *Note:* The pepper should not directly touch the flame.

3. When the pepper is completely charred on all sides, remove it and place it in a paper bag or wrap it in plastic and allow to rest for 10 minutes. *Note:* This allows the pepper to steam itself and loosens its skin. If a paper bag is used, check the pepper for doneness after 5 minutes.

4. Remove the pepper and peel away the loosened skin.

5. Rinse the pepper under cold water to remove any remaining pieces of charred skin.

Figure 13-83. *Fire-roasting a vegetable over an open flame allows the vegetable to be cooked whole.*

Sautéing and Frying

Vegetables prepared for sautéing or stir-frying are usually diced small or thinly sliced. They are sautéed and stir-fried very quickly in a hot pan with a small amount of oil. The finished vegetables are firm.

Deep-fried vegetables such as onions, mushrooms, cauliflower, zucchini, and eggplant are popular appetizers. *See Figure 13-84.* These vegetables are batter-coated and fried until crisp. French fries are also a popular fried vegetable.

Deep-Fried Vegetables

Sullivan University

Figure 13-84. *Vegetables may be dipped into a batter and deep-fried, such as this asparagus tempura.*

SUMMARY

Vegetables are rich in vitamins, minerals, and fiber and low in fat and calories. Vegetables include edible roots, bulbs, tubers, stems, leaves, flowers, and seeds. In addition, fruit-vegetables, sea vegetables, and mushrooms are prepared and served as vegetables. Each vegetable and vegetable-like food has the best color and flavor when purchased in season.

Vegetables offer culinary professionals ways to enhance menus with nutritious, earthy flavors. Common methods for preparing vegetables include steaming, blanching, grilling, broiling, baking, roasting, sautéing, and frying. They can be used to create appetizers, soups, salads, sandwiches, pasta dishes, sides, entrées, and desserts. Vegetables only need to be cooked until they become tender. Cutting vegetables into uniform sizes prevents uneven cooking.

Refer to DVD for
Quick Quiz® questions

Review

Refer to DVD for
Review Questions

1. Describe 11 edible roots that are used in the professional kitchen.
2. Describe five types of edible bulbs that are used in the professional kitchen.
3. Explain how to clean leeks.
4. Describe six types of edible tubers that are used in the professional kitchen.
5. Describe six types of edible stems that are used in the professional kitchen.
6. Describe 18 types of edible leaves that are used in the professional kitchen.
7. Describe four types of edible flowers that are used in the professional kitchen.
8. Explain how to prepare artichokes.
9. Describe five types of edible seeds that are used in the professional kitchen.
10. Explain how to rehydrate pulses.
11. Describe 10 types of fruit-vegetables that are used in the professional kitchen.
12. Explain how to prepare tomato concassé.
13. Explain how to core bell peppers.
14. Describe five types of sea vegetables used in the professional kitchen.
15. Describe nine types of edible mushrooms used in the professional kitchen.
16. Describe factors to consider when purchasing fresh vegetables.
17. Explain the role of canned and frozen vegetables in the professional kitchen.
18. Describe the quality of dried vegetables and their use.
19. Explain how acidic and alkaline ingredients affect cooked vegetables.
20. Identify eight methods that can be used to cook vegetables.
21. Explain how to fire-roast peppers.

1. Prepare an edible root and evaluate the quality of the prepared dish.
2. Clean and prepare leeks. Evaluate the quality of the prepared dish.
3. Prepare an edible tuber and evaluate the quality of the prepared dish.
4. Prepare an edible stem and evaluate the quality of the prepared dish.
5. Prepare an edible leaf and evaluate the quality of the prepared dish.
6. Prepare and cook artichokes. Evaluate the quality of the prepared dish.
7. Prepare two legume dishes and compare the quality of the two prepared dishes.
8. Soak a pulse overnight. Then quick-soak the same type of pulse. Prepare both pulses, and compare the quality of the two prepared dishes.
9. Prepare lentils and evaluate the quality of the prepared dish.
10. Prepare two fruit-vegetable dishes. Compare the quality of the two prepared dishes.
11. Prepare a tomato concassé.
12. Core a bell pepper.
13. Prepare a sea vegetable dish and evaluate the quality of the prepared dish.
14. Prepare a mushroom dish and evaluate the quality of the prepared dish.
15. Steam or blanch a vegetable and evaluate the quality of the prepared dish.
16. Grill or broil a vegetable and evaluate the quality of the prepared dish.
17. Bake or roast a vegetable and evaluate the quality of the prepared dish.
18. Fire-roast a vegetable and evaluate the quality of the prepared dish.
19. Sauté a vegetable and evaluate the quality of the prepared dish.
20. Stir-fry a vegetable and evaluate the quality of the prepared dish.
21. Deep-fry a vegetable and evaluate the quality of the prepared dish.

Refer to DVD for
Application Scoresheets

Steamed Vegetable Salad with Asian Dressing

Yield: *6 servings, 4 oz each*

Ingredients _____

romaine lettuce, cut into	
1 inch pieces	3 oz
broccoli florets	4 oz
cauliflower florets	4 oz
carrots, cut into thick diagonal slices	4 oz
yellow squash, cut in ¼ inch rounds	4 oz
asparagus	4 oz
scallions, cut into thirds	1 oz

Asian Dressing

yellow miso	1½ tbsp
water	1 tbsp
rice vinegar	½ tbsp
soy sauce	½ tsp
fresh ginger, grated	½ tbsp
red chile paste	½ tsp
peanut oil	2 tbsp
sesame oil	½ tsp

Preparation _____

1. Place romaine lettuce in a large mixing bowl and set aside.
2. Pour about 1 inch of water in a wok or sauté pan and bring to a boil over medium-high heat.
3. Add the vegetables and steam for 4–6 minutes or until the vegetables are al dente.
4. While the vegetables are cooking, combine the miso, water, vinegar, soy sauce, ginger, and chile paste into a small mixing bowl, and whisk until well blended.
5. Gradually whisk in the peanut oil. Start with a few drops and then add the remainder in a steady stream to make a smooth, slightly thick dressing.
6. Gradually whisk in the sesame oil. Start with a few drops and then add the remainder in a steady stream.
7. Add the steamed vegetables to the romaine lettuce in the large mixing bowl.
8. Pour the dressing over the salad and toss. Serve immediately.

NUTRITION FACTS
Per serving: 79 calories, 48 calories from fat, 5.4 g total fat, 0 mg cholesterol, 212.8 mg sodium, 305.9 mg potassium, 7.2 g carbohydrates, 2 g fiber, 2.5 g sugar, 2.4 g protein

Roasted Beet, Mandarin Orange, and Feta Salad

Yield: *4 servings, 4 oz each*

Ingredients _____

fresh beet, peeled and cut	
medium dice	1 ea
vegetable oil	1 tbsp
salt and pepper	TT
romaine lettuce, cut into	
1 inch pieces	6 oz
Mandarin orange segments, drained	2 oz
feta cheese, crumbled	2 oz
fresh mint, minced	1 tsp
red wine vinaigrette	4 fl oz

Preparation _____

1. Place cubed beets in a roasting pan, toss with vegetable oil, and season with salt and pepper.
2. Roast beets in a 350°F oven until they are slightly tender, but not overcooked.
3. Remove the beets from the oven and cool in the refrigerator until needed. *Note:* Beets can be roasted a day ahead.
4. Place romaine lettuce in a large mixing bowl and set aside.
5. Add the roasted beets, Mandarin oranges, cheese, and mint to the romaine lettuce.
6. Pour the vinaigrette over the ingredients and toss gently to coat thoroughly.

NUTRITION FACTS
Per serving: 96 calories, 59 calories from fat, 6.7 g total fat, 12.6 mg cholesterol, 254.3 mg sodium, 224.5 mg potassium, 5.4 g carbohydrates, 1.8 g fiber, 3.4 g sugar, 3.1 g protein

Stir-Fried Vegetables

Yield: *6 servings, 5 oz each*

Ingredients

peanut oil or vegetable oil	3 tbsp
fresh ginger, minced	1 tbsp
fresh garlic, minced	1 tbsp
red onion, quartered with layers separated	½ ea
bok choy, cut into 1 inch pieces	1 head
Chinese broccoli, cut into 1 inch pieces	¼ lb
Chinese long beans, cut into 1 inch pieces	½ lb
red bell pepper, cut into 1 inch pieces	1 ea
scallions, cut diagonally into 1 inch pieces	5 ea
Napa cabbage, cut crosswise into 1 inch strips	¼ head
chicken stock or vegetable broth, heated	⅔ c
cornstarch	1 tbsp
soy sauce	1 tbsp
sesame oil	1 tbsp
scallions, thinly sliced	1 ea
sesame seeds, toasted	2 tsp

Preparation

1. Place a large wok or sauté pan over high heat and add two-thirds of the oil.
2. When the oil is hot, add the garlic and stir-fry for about 30 seconds. Scoop out the garlic and discard.
3. Add the remaining peanut or vegetable oil. When the oil is hot, add the ginger and sauté for about 30 seconds.
4. Add the red onion and stir-fry until the onion turns glossy and bright (1–2 minutes).
5. Add the bok choy and Chinese broccoli. Stir-fry for 1–2 minutes.
6. Add the beans, bell peppers, and the scallions. Continue stir-frying the vegetables until they are glossy (1–2 minutes).
7. Add the Napa cabbage and one half of the hot stock or broth.
8. Continue stir-frying the vegetables until all are tender and crisp (about 2 minutes).
9. Dissolve the cornstarch in the soy sauce to create a slurry.
10. Add the slurry and the remaining stock or broth to the pan. Stir-fry the vegetables for 1 minute or until they are lightly glazed with sauce.
11. Add sesame oil and stir to coat.
12. Transfer the vegetables to a heated serving dish and garnish them with scallions and sesame seeds. Serve immediately.

NUTRITION FACTS
Per serving: 252 calories, 159 calories from fat, 18.2 g total fat, 1.6 mg cholesterol, 355.4 mg sodium, 746.5 mg potassium, 18.3 g carbohydrates, 6.4 g fiber, 6.5 g sugar, 7.9 g protein

Roasted Turnips and Fingerling Potatoes

Yield: *6 servings, 4 oz each*

Ingredient

extra virgin olive oil	2 tbsp
fresh rosemary	2 sprigs
fresh tarragon, minced	1 sprig
fresh parsley, minced	1 tbsp
salt	½ tsp
black pepper	⅛ tsp
fingerling potatoes	1 lb
medium turnips, peeled and cut into ½ inch wedges	¾ lb

Preparation

1. Whisk together the olive oil, rosemary, tarragon, parsley, salt, and pepper.
2. Add the potatoes and turnips to the mixture and toss until they are well coated.
3. Pour the vegetables into a roasting pan and roast at 350°F for 45–60 minutes or until the vegetables are tender.

NUTRITION FACTS
Per serving: 115 calories, 41 calories from fat, 4.7 g total fat, 0 mg cholesterol, 237 mg sodium, 438.3 mg potassium, 17.1 g carbohydrates, 2.8 g fiber, 2.7 g sugar, 2.1 g protein

Grilled Asparagus with Garlic, Rosemary, and Lemon

Yield: *6 servings, 5 oz each*

Ingredients

asparagus	2 lb
olive oil	1½ tbsp
garlic, finely minced	2 cloves
fresh rosemary	2 tsp
lemon zest	1 tsp
salt and black pepper	TT

Preparation

1. Trim asparagus and discard the woody ends.
2. Combine oil, garlic, rosemary, and lemon zest in a small bowl. Mix well.
3. Add asparagus and toss to coat.
4. Place asparagus spears in a grill pan on a heated grill.
5. Grill asparagus to desired tenderness.
6. Season to taste and serve immediately.

NUTRITION FACTS
Per serving: 62 calories, 32 calories from fat, 3.6 g total fat, 0 mg cholesterol, 51.8 mg sodium, 311.6 mg potassium, 6.3 g carbohydrates, 3.3 g fiber, 2.9 g sugar, 3.4 g protein

Cabbage and White Bean Soup

Yield: *11 servings, 6 fl oz each*

Ingredients

dried white beans, rinsed	4 oz
onion, diced	2 oz
unsalted butter	1 oz
garlic, finely chopped	1 clove
ham hocks, smoked	1¼ lb
white stock	1½ qt
bay leaf	1 ea
fresh thyme	¾ tsp
Yukon Gold potatoes	8 oz
cabbage, cut into 1/2 inch pieces	8 oz
fresh parsley, chopped	1 tbsp

Preparation

1. Place beans in container. Cover the beans with cold water at least 2 inches over the beans and let soak overnight. Drain in a colander the next day. *Note:* The beans can be quick-soaked by placing them in a pot with cold water, bringing the water to a boil, and simmering for 2 minutes. After 2 minutes, the beans are removed from the heat and allowed to rest, uncovered, for 1 hour.

2. In medium stockpot, sweat the onion in butter over medium heat. Add garlic and sweat an additional 1 minute.

3. Add the ham hocks, stock, bay leaf, and thyme. Bring to a simmer, cover, and cook for 1 hour. Skim off impurities as they rise to the surface.

4. Add the beans and simmer uncovered, stirring occasionally for 40–50 minutes or until the beans are almost tender.

5. Peel and dice the potatoes into 1 inch cubes and cover with water until needed.

6. Add the potatoes and the cabbage. Return to a simmer and cook uncovered for 20–25 minutes or until the potatoes are tender.

7. Remove and discard the bay leaf.

8. Remove the ham hocks. When they are cool enough to handle, discard the skin and bones and then cut the ham into bite-size pieces and add it back to the soup.

9. Garnish with fresh parsley and serve immediately.

NUTRITION FACTS
Per serving: 154 calories, 57 calories from fat, 6.4 g total fat, 26.7 mg cholesterol, 203.9 mg sodium, 419.6 mg potassium, 14.8 g carbohydrates, 2.2 g fiber, 2.9 g sugar, 9.6 g protein

Caramelized Brussels Sprouts, Roasted New Potatoes, and Smoked Bacon

Yield: *6 servings, 4 oz each*

Ingredients

smoked bacon, diced medium	4 slices
Brussels sprouts, trimmed and quartered	8 oz
garlic cloves	3 ea
new potatoes, washed and cut in half	8 oz
olive oil	1 tbsp
salt and pepper	TT

Preparation

1. Cook the bacon in a roasting pan until slightly golden in color but not cooked through.

2. Remove the bacon from pan and set aside. Pour off most of the bacon fat from the pan.

3. Add the bacon back to the roasting pan along with the Brussels sprouts, garlic cloves, and new potatoes.

4. Add the olive oil, salt, and pepper and toss well.

5. Place the roasting pan in a 400°F oven and roast until the potatoes are tender, stirring occasionally.

6. Serve immediately.

NUTRITION FACTS
Per serving: 236 calories, 42 calories from fat, 4.8 g total fat, 5.9 mg cholesterol, 191.9 mg sodium, 1089 mg potassium, 42.4 g carbohydrates, 6 g fiber, 2.3 g sugar, 7.5 g protein

Sautéed Brussels Sprouts and Pistachios

Yield: *4 servings, 3 oz each*

Ingredients

Brussels sprouts	1½ lb
salt	1 tsp
shallot, minced	1 ea
olive oil	1 tbsp
unsalted pistachios	2 oz
fresh lemon juice	1½ tsp
salt and pepper	TT

Preparation

1. Remove the stem end from each Brussels sprout and cut ¾ of the sprouts into quarters.
2. Remove the leaves from the remaining sprouts and set aside.
3. In lightly salted water, blanch the quartered sprouts until they turn bright green. Immediately remove the sprouts and shock in ice water to stop the cooking process. Drain the sprouts and reserve until needed.
4. In a sauté pan, cook shallots in olive oil until slightly brown.
5. Add the blanched sprouts and the pistachios to the sauté pan. Cook until the sprouts are heated through.
6. Add the reserved sprout leaves, the lemon juice, and seasonings to taste.
7. Toss to warm the leaves, but not wilt them completely. Serve immediately.

NUTRITION FACTS
Per serving: 260 calories, 89 calories from fat, 10.4 g total fat, 0 mg cholesterol, 709.6 mg sodium, 1158.1 mg potassium, 37.6 g carbohydrates, 7.9 g fiber, 5.4 g sugar, 11.3 g protein

Stuffed Artichokes

Yield: *6 servings, 1 artichoke each*

Ingredients

lemon juice	3 tbsp
large artichokes, with chokes removed	6 ea
onion, minced	2 oz
garlic, minced	2 cloves
olive oil	3 fl oz
white wine	2 fl oz
parsley, chopped	1 tbsp
bread crumbs	4 oz
scallion greens, minced	2 ea
Parmesan cheese, grated	2 oz
salt and pepper	TT

Preparation

1. Pour 1–2 inches of water and one-third of lemon juice in a sauté pan. Place prepared artichokes in pan, top side up.
2. Pour one-third of the lemon juice in the center of all of the artichoke to coat the heart.
3. Bring the artichokes to a boil over medium-high heat.
4. Cover the pan and lower the heat to a simmer to steam the artichokes for about 8 minutes or until the stem ends are tender when pierced with a paring knife.
5. Remove the covered pan from the heat for 5 minutes. Meanwhile, sauté the onion and garlic in one-third of olive oil until they are soft.
6. Deglaze the sauté pan with white wine and reduce by half.
7. Remove the onion mixture from the heat and add the parsley, bread crumbs, scallion greens, Parmesan cheese, remaining lemon juice, salt, and pepper. Mix well.
8. Move the artichokes into a small baking dish just large enough to hold them.
9. Add a small amount of water to cover the bottom of the pan.
10. Fill the center of each artichoke with the bread crumb mixture.
11. Drizzle the remaining olive oil over the stuffed artichokes.
12. Bake at 375°F for 12 minutes or until golden brown.

NUTRITION FACTS
Per serving: 316 calories, 158 calories from fat, 17.9 g total fat, 8.3 mg cholesterol, 454.2 mg sodium, 573.1 mg potassium, 29.9 g carbohydrates, 8.1 g fiber, 3.4 g sugar, 10.7 g protein

Tempura-Battered Zucchini

Yield: *4 servings, 5 oz each*

Ingredients

club soda, cold	8 fl oz
all-purpose flour	5 oz
cornstarch	1 oz
salt	1 tsp
medium zucchini, cut on the bias into ⅓ inch slices	2 ea
vegetable oil	1 c

Preparation

1. Pour club soda into a medium-sized mixing bowl. Sift in four-fifths of the flour, the cornstarch, and the salt. Whisk until evenly blended.
2. Cover and refrigerate for 45–60 minutes.
3. Dust the zucchini slices with the remaining flour. Shake off any excess flour.
4. Dip the floured slices into the batter and coat completely.
5. Using the swimming method, fry the zucchini in batches in 375°F oil until they are golden.
6. Use a slotted spoon to transfer the cooked zucchini onto paper towels to drain.
7. Serve immediately.

NUTRITION FACTS
Per serving: 655 calories, 488 calories from fat, 55.2 g total fat, 0 mg cholesterol, 603.4 mg sodium, 308.7 mg potassium, 36.7 g carbohydrates, 2.1 g fiber, 2.7 g sugar, 4.9 g protein

Beer-Battered Zucchini

Yield: *4 servings, 5 oz each*

Ingredients

all-purpose flour	5 oz
baking powder	½ tsp
salt	1 tsp
beer	8 fl oz
medium zucchini, cut on the bias into ½ inch slices	2 ea
vegetable oil	1 c

Preparation

1. Mix four-fifths of flour, the baking powder, and salt in a large mixing bowl.
2. Make a well in the center of the flour. Pour the beer into the well and whisk until the mixture is combined. Cover and let it rest for 1 hour.
3. Dust the zucchini slices with the remaining flour. Shake off any excess flour.
4. Dip the floured slices into the batter and coat completely.
5. Using the swimming method, fry the zucchini in batches in 375°F oil until they are golden.
6. Use a slotted spoon to transfer the cooked zucchini onto paper towels to drain.
7. Serve immediately.

NUTRITION FACTS
Per serving: 654 calories, 487 calories from fat, 55.2 g total fat, 0 mg cholesterol, 653.7 mg sodium, 323.5 mg potassium, 32.5 g carbohydrates, 2 g fiber, 2.7 g sugar, 5.2 g protein

Curried Pumpkin Soup

Yield: *16 servings, 8 oz each*

Ingredients

onions, diced	6 oz
butter, unsalted	2 oz
garlic, minced	2 cloves
ginger, fresh, minced	1½ tbsp
cumin, ground	3 tsp
coriander, ground	2 tsp
cardamom	⅛ tsp
pumpkin or butternut squash, roasted	4 lb
vegetable or chicken stock	2 qt
coconut milk	14 fl oz
red curry paste	1 tbsp
salt and pepper	TT

Procedures

1. Sauté onions in butter over moderate heat, stirring occasionally until softened and slightly caramelized.
2. Add garlic and ginger. Cook for 1 minute while stirring.
3. Add cumin, coriander, and cardamom. Cook for 1 additional minute.
4. Add pumpkin, stock, coconut milk, and curry paste.
5. Simmer uncovered for 20 minutes.
6. Purée until smooth and light. Season to taste.
7. Garnish with crème fraiche and pumpkin seed oil.

NUTRITION FACTS
Per serving: 139 calories, 79 calories from fat, 9.2 g total fat, 10.8 mg cholesterol, 170.8 mg sodium, 504 mg potassium, 11.8 g carbohydrates, <1 g fiber, 3.3 g sugar, 4.3 g protein

Sesame Kale with Sautéed Shiitake Mushrooms

Yield: *4 servings, 4 oz each*

Ingredients

sesame oil	½ tsp
olive oil	1 tsp
shiitake mushrooms, sliced thin	4 oz
garlic clove, crushed	1 ea
fresh kale, cut chiffonade	12 oz
soy sauce	½ tbsp
salt and pepper	TT

Preparation

1. Add sesame oil and olive oil to a hot sauté pan.
2. Add the shiitake mushrooms and cook until they are a rich golden brown on one side. Toss to cook on the opposite side.
3. Add the garlic and sauté quickly until the aroma is dominant, but garlic has not browned.
4. Add the kale and sauté for about 30 seconds.
5. Add 1 tbsp water and soy sauce and stir until the kale is well coated and the moisture has evaporated.
6. Salt and pepper to taste. Serve immediately.

NUTRITION FACTS
Per serving: 66 calories, 21 calories from fat, 2.4 g total fat, 0 mg cholesterol, 236.8 mg sodium, 445.7 mg potassium, 10.2 g carbohydrates, 2.2 g fiber, <1 g sugar, 3.5 g protein

Barilla America, Inc.

Potatoes, Grains, and Pastas

*P*otatoes, grains, and pastas are often referred to as starches. They are relatively easy to prepare and complement foods they are paired with. Potatoes used in the professional kitchen include mealy, waxy, new, and sweet potatoes. A wide variety of grains from rice to quinoa offer additional flavor and nutrition to many dishes. Whole grains and processed grains are both used extensively. Pastas can be made by hand or purchased in dried or frozen form. Pastas and Asian noodles are served as appetizers, entrées, salads, sides, and desserts.

Refer to DVD
for **Flash Cards**

Chapter Objectives

1. Describe the four major classifications of potatoes.
2. Identify five market forms of potatoes.
3. Describe the guidelines for receiving and storing potatoes.
4. Explain how to determine the doneness of potatoes.
5. Prepare potatoes using six different cooking methods.
6. Identify the four parts of a whole grain.
7. Differentiate between whole grains and refined grains.
8. Differentiate among milled, pearled, and flaked grains.
9. Describe the three major classifications of rice.
10. Identify forms of corn, wheat, and oats used in the professional kitchen.
11. Describe barley, quinoa, rye, buckwheat, farro, millet, and spelt.
12. Explain the importance of storing grains in an airtight container and in a cool, dry place.
13. Prepare grains using the risotto method and the pilaf method
14. Explain how pasta dough is formed.
15. Identify three forms of pasta used in the professional kitchen.
16. Describe tube, ribbon, shaped, and formed pastas.
17. Prepare pasta dough, ravioli, and tortellini.
18. Explain how to determine if pasta is cooked al dente.
19. Describe 10 types of Asian noodles and how they are prepared.

Key Terms

- **potato**
- **mealy potato**
- **waxy potato**
- **fingerling potato**
- **new potato**
- **solanine**
- **gratinée**
- **grain**
- **husk**
- **bran**
- **endosperm**
- **germ**
- **whole grain**
- **refined grain**
- **pasta**
- **gluten**
- **tube pasta**
- **ribbon pasta**
- **shaped pasta**
- **formed pasta**

POTATOES

A *potato* is a round, oval, or elongated tuber that is the only edible part of the potato plant. The color of potato skin differs among varieties and can be brown, white, purple, red, or yellow. Depending on the variety, potato flesh can be creamy white, yellow-gold, or purple in color. A potato eaten with the skin on provides dietary fiber, vitamin C, vitamin B6, and potassium. Mealy, waxy, and new potatoes, as well as sweet potatoes and yams, are used in the professional kitchen.

Mealy Potatoes

A *mealy potato* is a type of potato that is higher in starch and lower in moisture than other types of potatoes. After cooking, these potatoes become light and fluffy inside. They are the preferred type of potato for baking, frying, mashing, puréeing, and casseroles. Mealy potatoes include russets, white, and purple potatoes. ***See Figure 14-1.***

Mealy Potatoes

Russet

White

Purple

United States Potato Board

Figure 14-1. *Mealy potatoes include russets, white, and purple potatoes.*

Russet Potatoes. A *russet potato* is a mealy potato with thin brown skin, an elongated shape, and shallow eyes. Russets have white flesh that is high in starch. They are commonly baked, mashed, or fried. Russet potatoes are sometimes referred to as bakers or baking potatoes. Common varieties of russets include Idaho and Burbank potatoes.

White Potatoes. A *white potato* is an oblong mealy potato with a thin, white or light-brown skin and tender, white flesh. White potatoes contain less starch than russet potatoes and can be prepared using almost any cooking method. They are often used to make scalloped and au gratin potatoes.

Purple Potatoes. A *purple potato,* also known as a blue potato, is a mealy potato with smooth, thin, bluish-purple skin and purple flesh. Purple potatoes have a fine texture and a mild, nutty flavor. They are typically roasted, baked, mashed, fried, or steamed. Purple potatoes can be used to make mashed potatoes and potato salad.

United States Potato Board

Waxy Potatoes

A *waxy potato* is a type of potato with a thin skin and slightly waxy flesh that is lower in starch and higher in moisture than mealy potatoes. Waxy potatoes include red potatoes, yellow potatoes, and fingerling potatoes. *See Figure 14-2.* Compared to mealy potatoes, waxy potatoes stay much firmer in the center when fully cooked and also retain their shape better. Waxy potatoes can be roasted, sautéed, steamed, or simmered. Waxy potatoes are the common choice for cooking except when baking or deep-frying.

Waxy Potatoes

Idaho Potato Commission
Red

Idaho Potato Commission
Yellow

United States Potato Board
Fingerling

Figure 14-2. *Waxy potatoes include red potatoes, yellow potatoes, and fingerling potatoes.*

Red Potatoes. A *red potato* is a round, waxy, red-skinned potato with white flesh. Red potatoes are often grilled, roasted, or simmered. Popular varieties of red potatoes include red bliss, chieftain, and Norland red potatoes.

Yellow Potatoes. A *yellow potato* is an oval, waxy potato with thin, yellowish skin and flesh and pink eyes. The flesh has a buttery, nutty flavor and remains a yellowish color after cooking. Popular varieties include Yukon gold and yellow finn potatoes. Yellow potatoes are typically roasted, mashed, or simmered. Their sweet flavor pairs well with citrus, parsnips, and cauliflower.

Fingerling Potatoes. A *fingerling potato* is a small, tapered, waxy potato with butter-colored flesh and tan, red, or purple skin. Popular varieties of fingerlings include Russian banana fingerlings and French fingerlings. Russian banana fingerlings have tan or yellow skin. French fingerlings have light-purple skin. Fingerlings can be roasted, baked, or steamed. They are often cooked whole or halved lengthwise and used to add visual interest to salads and entrées.

New Potatoes

United States Potato Board

Figure 14-3. *Because they are harvested so early, new potatoes are small and relatively uniform in size.*

Potatoes and Yams

Sweet Potato

Yam

Melissa's Produce

Figure 14-4. *Sweet potatoes and yams have a similar outer appearance but differ from each other.*

New Potatoes

A *new potato,* also known as an early crop potato, refers to any variety of potato that is harvested before the sugar is converted to starch. Because they are harvested so early, new potatoes are small and relatively uniform in size. *See Figure 14-3.* They have a thin, delicate skin and tender flesh. New potatoes hold their shape after cooking. They are often roasted, steamed, or simmered. Due to their high moisture content, new potatoes spoil more quickly than other types of potatoes.

Sweet Potatoes and Yams

Sweet potatoes and yams have a similar outer appearance but differ from each other. *See Figure 14-4.* A *sweet potato* is a tuber that grows on a vine and has a paper-thin skin and flesh that ranges in color from ivory to dark orange. Sweet potatoes are members of the morning glory family. Different varieties can be yellow, red, or brown in skin color with yellow to orange-red flesh.

Firm varieties of sweet potatoes remain firm when cooked. The edible skin is often removed before cooking. Sweet potatoes can be roasted, baked, sautéed, fried, or simmered. They complement pork and poultry dishes and are often incorporated into breads and desserts. Roasting or baking sweet potatoes caramelizes the sugars and releases a sweet flavor. Sweet potatoes are often served topped with butter and brown sugar or cinnamon. Sweet potatoes are an excellent source of vitamin A and potassium. They are often puréed.

A *yam* is a large tuber that has thick, barklike skin and flesh that varies in color from ivory to purple. Like sweet potatoes, yams can be round or oblong in shape and vary in color. The flavor of a yam is somewhat dry and starchier than a sweet potato. Common varieties include tropical yams, garnet yams, and jewel yams. They are low in fat and a good source of carbohydrates, protein, vitamin A, and vitamin C. The USDA requires labels using the term "yam" to be accompanied by the term "sweet potato" since yams are not readily available in the United States.

MARKET FORMS OF POTATOES

Potatoes can be purchased fresh, frozen, canned, dehydrated, or as processed items such as potato chips. *See Figure 14-5.* Various market forms of potatoes can be used to reduce time and labor costs. For example, frozen precooked French fries, canned cooked whole potatoes, and dehydrated mashed potatoes are often used in the professional kitchen.

Market Forms of Potatoes

United States Department of Agriculture

Figure 14-5. *Market forms of potatoes include processed, frozen, dehydrated, and fresh potatoes. Potatoes are also available canned.*

RECEIVING AND STORING FRESH POTATOES

Potatoes should be firm, undamaged, and not show any signs of sprouting. *See Figure 14-6.* Quality potatoes are firm when pressed. They should be clean and have shallow eyes. Potato eyes are the undeveloped buds on a potato. Eyes, sprouts, and any green-colored surfaces should be removed from potatoes prior to cooking. Potatoes that are sprouting or that have a green-colored flesh or skin should not be eaten because they contain a toxin called solanine. Solanine tastes bitter and can be harmful if eaten in large quantities.

USDA grading of potatoes is voluntary. Most graded potatoes are U.S. No. 1. If the external appearance is not important, U.S. Commercial or U.S. No. 2 potatoes are often used.

Quality Potatoes

Firm when pressed

Clean

No sprouts

Shallow eyes

Undamaged

Idaho Potato Commission

Figure 14-6. *Potatoes should be firm, undamaged, and not show any signs of sprouting.*

Potato Packaging

Idaho Potato Commission

Figure 14-7. *Potatoes are best stored in cardboard boxes or mesh bags to allow adequate ventilation.*

Potatoes are generally sold in 50 lb cases and are packed according to the size of the potato. They may be packaged as 80 count potatoes, 90 count potatoes, or 100 count potatoes. In a 50 lb box of 90 count potatoes, there are approximately 90 potatoes weighing around ½ lb each. Potatoes must be inspected upon receipt to ensure they are firm and undamaged and show no signs of sprouting or green patches. Potatoes that are purchased cleaned have shorter storage life because cleaning removes a protective outer coating that deters bacteria.

Potatoes are best stored in cardboard boxes or mesh bags to allow adequate ventilation. *See Figure 14-7.* Potatoes must be kept in a dry, cool, dark place that allows them to breathe. If potatoes do not have adequate ventilation, they quickly rot. The conditions under which potatoes are stored determine the length of time the potatoes will keep. The temperature of the storage area should be between 45°F and 55°F. Higher temperatures will shorten the lifespan of the potatoes because potatoes begin to sprout and dehydrate under higher temperatures.

Refrigeration is not recommended, as colder temperatures increase the conversion of starch to sugar. Potatoes that are exposed to too much light will turn green and begin to sprout. Sprouts and any green areas should be cut off potatoes before they are cooked.

COOKING POTATOES

Potatoes can be prepared in a variety of ways for breakfast, lunch, and dinner. They can be grilled, roasted, baked, sautéed, fried, or simmered. Potatoes are often added to soups, braised dishes, and stewed dishes for additional texture, flavor, and nutrients. Potatoes also help thicken sauces by releasing starch during the cooking process. Some potato preparations require more than one cooking method.

It is important to choose the appropriate potato size for the cooking time. For example, a shorter cooking time requires smaller pieces of potatoes than a dish that cooks for a longer period. If a dish with a longer cooking time calls for small-dice potatoes, the potatoes should be added late in the cooking process to avoid overcooking.

Uncooked potatoes begin to oxidize, causing them to discolor immediately after they are cut or peeled. A freshly peeled potato quickly turns brown or gray and eventually black. *See Figure 14-8.* To prevent discoloration, potatoes should be placed in cold water as soon as they are cut or peeled and left there until they are ready to be cooked.

Determining Doneness

Fully cooked potatoes are fluffy and have little resistance when pierced by a fork. If a potato is hard in the center, it needs to be cooked longer. Roasted and fried potatoes should be crunchy on the outside and soft on the inside. Cutting open these types of cooked potatoes is the only way to determine if they are done. Lightly squeezing a baked potato will indicate whether it is done. The skin of a fully roasted potato will look slightly shriveled, and the skin of a simmered potato will begin to separate from the flesh when the potato is done.

Courtesy of The National Pork Board

Oxidation of Potatoes

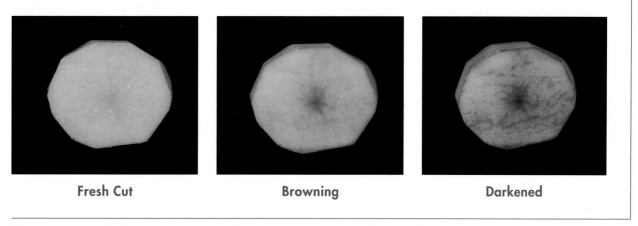

| Fresh Cut | Browning | Darkened |

Figure 14-8. *A freshly peeled potato quickly turns brown or gray and eventually black.*

Grilling Potatoes

Grilled potatoes have a golden color and smoky flavor. ***See Figure 14-9.*** When grilling potatoes, they are sliced into ¼ inch thick slices or ½ inch wedges, coated with oil and seasonings, and then placed on a preheated grill. Grilled potatoes should be cooked al dente. Overcooking potatoes makes them soft and difficult to remove from the grill without breaking.

Grilled Potatoes

United States Potato Board

Figure 14-9. *Grilled potatoes have a golden color and smoky flavor.*

Roasted Potatoes

United States Potato Board

Figure 14-10. *Waxy potatoes can be tossed lightly in oil, seasoned, and then roasted until golden brown.*

Roasting Potatoes

Potatoes can be roasted or baked. Roasted potatoes are usually cut up, tossed with oil and other seasonings, and then cooked in an oven. Waxy potatoes, such as red, yellow, and fingerling potatoes, can be tossed lightly in oil, seasoned, and then roasted until golden brown on the outside and soft and creamy on the inside. *See Figure 14-10.* Mealy potatoes can also be roasted in this way, but will have a slightly tough exterior. Roasted potatoes are done when a fork can be inserted into the potato with little resistance. Roasted potatoes are typically cooked between 350°F and 400°F for 20–45 minutes.

Baking Potatoes

Baked potatoes are cooked whole. Russet potatoes make the best baked potatoes. When baked, these low-moisture potatoes produce fluffy and light flesh that readily absorbs butter, sour cream, and a wide variety of other toppings. *See Figure 14-11.* Baked potatoes are typically cooked between 350°F and 400°F until fork tender.

Baked Potatoes

United States Potato Board

Plain Baked Potato

Idaho Potato Commission

Toppings

Figure 14-11. *When baked, low-moisture potatoes produce fluffy and light flesh that readily absorbs butter, sour cream, and a wide variety of other toppings.*

Chef's Tip

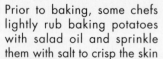

Prior to baking, some chefs lightly rub baking potatoes with salad oil and sprinkle them with salt to crisp the skin and add flavor.

Baked potatoes should be served as quickly as possible because their quality deteriorates if they are held too long. They can be baked directly on the oven rack or can be placed on sheet pans. The skin of a potato is washed thoroughly and pricked with a fork prior to baking. The tiny holes left by the fork allow moisture to escape during the baking process. *See Figure 14-12.*

1. Wash potatoes thoroughly and pierce several times with a fork.

2. Place the potatoes on a sheet pan and bake for approximately 45–60 minutes in a 400°F oven until the potatoes become slightly soft when gently squeezed.

Figure 14-12. *The skin of a potato is washed thoroughly and pricked with a fork prior to baking.*

Potato casseroles are also baked. A *casserole* is a baked dish containing a starch (such as potatoes, grains, or pasta), other ingredients (such as meat or vegetables), and a sauce. Mealy potatoes are the best choice for potato casseroles because they are low in moisture and can absorb the most flavor from a sauce. A potato casserole, such as scalloped potatoes or potatoes au gratin, can be baked with or without toppings, such as bread crumbs or cheese. *See Figure 14-13*. *Gratinée* is the process of topping a dish with a thick sauce, cheese, or bread crumbs and then browning it in a broiler or high-temperature oven. *Gratin* is any dish prepared using the gratinée method.

Potato Casseroles

Scalloped

Au Gratin

Basic American Foods

Figure 14-13. *A potato casserole, such as scalloped potatoes or potatoes au gratin, can be baked with or without toppings, such as bread crumbs or cheese.*

Potatoes used in casseroles may be sliced raw or partially cooked prior to slicing. *See Figure 14-14.* The sauce used for a casserole can range from a cream sauce to a stock. Using parcooked (partially cooked) potatoes or a warmed sauce will decrease the required baking time.

Procedure for Preparing Potato Casseroles

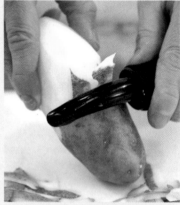

1. Wash and peel the potatoes. Place them in cold water to avoid browning.

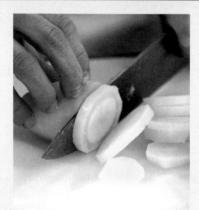

2. Slice potatoes to a uniform thickness. If precooked potatoes are desired, steam or simmer them until they are half done. Drain the cooking liquid from the potatoes and allow them to cool slightly.

3. Lay a single layer of the potatoes in a casserole dish.

4. Cover the potatoes with a warm, well-seasoned sauce.

5. Add another layer of potatoes and cover them with sauce. Continue this process until the potatoes and sauce are used.

6. Top with cheese or bread crumbs and cover the pan with foil.

7. Bake at 350°F until potatoes are almost tender. Uncover and allow the potatoes to brown.

Figure 14-14. *Potatoes used in casseroles may be sliced raw or partially cooked prior to slicing.*

Sautéing Potatoes

Waxy potatoes are best for sautéing. Large waxy potatoes are sliced for sautéing and small potatoes are usually cut in half. Sautéed potatoes are typically steamed or boiled first and then finished in a sauté pan to give them a golden-brown exterior. Potatoes that are steamed or boiled first should be removed from the cooking liquid and allowed to steam dry before they are sautéed. Never rinse steamed potatoes to cool them because they absorb even more water and will not brown properly.

Frying Potatoes

Deep-fried potatoes, such as French fries and potato chips, are a popular side. Fried potatoes can be prepared in a variety of shapes including waffle-cut fries, shoestring fries, steak fries, cottage fries, and hash browns. *See Figure 14-15.* Low-moisture, mealy potatoes, such as russets, are the best choice for deep-frying. They crisp nicely on the outside and remain light and fluffy on the inside. Low-moisture potatoes also spatter less and stay crisp longer.

Fried Potatoes

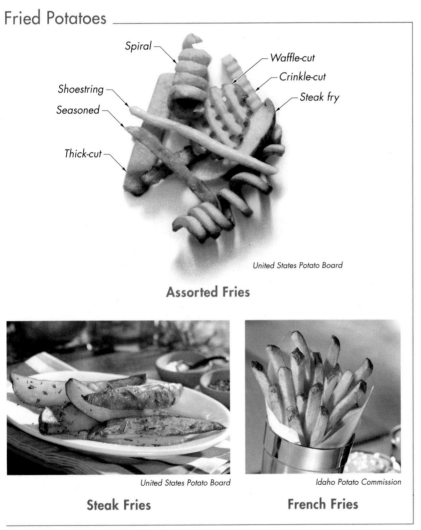

Spiral
Waffle-cut
Crinkle-cut
Shoestring
Seasoned
Steak fry
Thick-cut

United States Potato Board

Assorted Fries

United States Potato Board

Steak Fries

Idaho Potato Commission

French Fries

Figure 14-15. *Fried potatoes can be prepared in a variety of shapes including waffle-cut fries, shoestring fries, steak fries, cottage fries, and hash browns.*

Thinly cut potatoes, such as potato chips and shoestring fries, can be deep-fried in a single step, usually in hot oil between 350°F and 375°F. French fries need to be blanched and cooled before being deep-fried. *See Figure 14-16.* Blanching potatoes in 275°F oil removes some of the moisture, causing them to cook more evenly when they are deep-fried. Potatoes that are cooked in 375°F oil without being blanched first instantly sear, preventing their interior moisture from escaping. The trapped internal moisture causes the potatoes to become limp and soggy soon after frying.

Procedure for Deep-Frying Potatoes

Henny Penny Corporation

1. Wash, peel (if desired), and cut potatoes to the desired size and place them in cold water to avoid browning. When ready for blanching, drain the precut potatoes and carefully pat dry.

2. Blanch the potatoes in 275°F oil until slightly undercooked.

3. Remove the potatoes from the hot oil and drain well.

4. Place the parcooked potatoes on sheet pans and refrigerate until cool. *Note:* Potatoes can be held in the refrigerator for a few hours or can be frozen for approximately one month.

5. When the potatoes are ready to be fried, place the parcooked potatoes in 375°F oil and fry until golden-brown and crispy. *Note:* Do not overfill the frying basket.

Figure 14-16. *French fries need to be blanched and cooled before being deep-fried.*

Refer to DVD for
Deep-Frying Potatoes
Media Clip

Simmering Potatoes

Any type of potato can be simmered with good results. If the simmered potatoes are to be mashed, mealy potatoes are the best choice. Waxy potatoes are used when preparing whole, quartered, or sliced potatoes or potato salad because they retain their shape and do not fall apart as easily. *See Figure 14-17.*

Simmered Potatoes

United States Potato Board
Mashed

Idaho Potato Commission
Sliced

Figure 14-17. *If simmered potatoes are to be mashed, mealy potatoes are best. Waxy potatoes are used for whole, quartered, or sliced potatoes and potato salad because they do not fall apart as easily.*

Herbs and aromatics can add subtle flavors to simmered potatoes. To add even more flavor when simmering potatoes, stocks or juices can be used as the cooking liquid instead of water. *See Figure 14-18.*

Procedure for Simmering Potatoes

1. Cut large potatoes into consistently sized pieces so that they will cook evenly. *Note:* If the potatoes are to be cooked whole, make sure that all the potatoes are similar in size.

2. Place potatoes in a cold, salted liquid. Bring the liquid to a boil and then lower the heat to a simmer.

3. Pierce the potatoes with a fork to test for doneness. Drain the potatoes in a colander to remove all of the cooking liquid.

Figure 14-18. *To add flavor when simmering potatoes, stocks or juices can be used as the cooking liquid instead of water.*

If simmered potatoes are to be served whole, they should be placed on a sheet pan in a single layer to dry. Simmered potatoes have sufficiently dried and are ready to garnish and serve when steam no longer rises from the potatoes. Simmered potatoes are typically seasoned with butter and chopped fresh parsley. Simmered potatoes and baked potatoes can be whipped or puréed before service. *See Figure 14-19.*

Procedure for Whipping Potatoes

1. Return drained, cooked potatoes to the pot they were cooked in and cover.

2. In a separate pot, heat flavoring ingredients, such as milk, cream, stock, or butter, to a temperature equal to that of the potatoes.

3. Place the warm potatoes in a stand mixer. Use the paddle attachment to mash the potatoes until they are fairly smooth.

4. Pour the heated flavoring ingredients in the mixing bowl. Use the whip attachment to thoroughly mix all of the ingredients together until light and fluffy. Season to taste.

Figure 14-19. *Simmered potatoes and baked potatoes can be whipped or puréed before service.*

To purée simmered potatoes, the drained potatoes are passed through a ricer or food mill until a fine, smooth texture is achieved. Milk, butter, and salt are often blended into puréed potatoes. Ingredients such as puréed roasted garlic, fresh herbs, sun-dried tomato purée, grated cheeses, and pesto can also be added to simmered potatoes.

GRAINS

A *grain* is the edible fruit, in the form of a kernel or seed, of a grass. The name of the grass is often the same name given to the grain from that plant. For instance, wheat grass produces the grain called wheat. Common grains include rice, wheat, corn, and barley, but there are many other types of edible grains used in the professional kitchen.

Grain Composition

A kernel of grain is composed of a husk, bran, endosperm, and germ. *See Figure 14-20.* The *husk,* also known as the hull, is the inedible, protective outer covering of grain. The *bran* is the tough outer layer of grain that covers the endosperm. While it is often removed during processing, bran provides necessary fiber, complex carbohydrates, vitamins, and minerals. Studies indicate that bran can help lower cholesterol.

Grain Composition

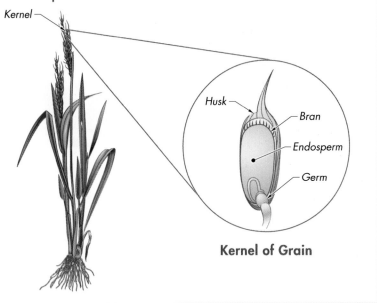

Kernel of Grain

Figure 14-20. *A kernel of grain is composed of a husk, bran, endosperm, and germ.*

The *endosperm* is the largest component of a grain kernel and consists of carbohydrates and a small amount of protein. It is milled to produce flours and other grain products. Endosperm consists of carbohydrates (in the form of starches) and a small amount of protein. The *germ* is the smallest part of a grain kernel and contains a small amount of natural oils as well as vitamins and minerals.

Grains come in many shapes and sizes. Most grains are hard and all have a husk that is indigestible in its natural form. It is necessary to process grains to some extent to make them easier to digest. However, the amount of processing directly affects the nutritional values of the grain.

Whole Grains and Refined Grains

A *whole grain* is a grain that only has the husk removed. *See Figure 14-21.* Because the rest of the grain is intact, whole grains require more time to cook than processed grains. Examples of whole grains include brown rice, wild rice, wheat berries, bulgur, oats, barley, rye, quinoa, millet, and spelt. A *cracked grain* is a whole kernel of grain that has been cracked by being placed between rollers. Bulgur wheat is a cracked grain. Some whole grains are parcooked to reduce the required amount of cooking time.

> **Nutrition Note**
>
> Grains are rich in complex carbohydrates, low in fat, and a good source of fiber, vitamins, and minerals. Eating whole grains can reduce the risk of certain diseases and help individuals maintain a healthy weight.

Whole Grains

Indian Harvest Specialtifood, Inc./Rob Yuretich

Figure 14-21. *Whole grains, such as brown rice, only have the husk removed.*

Refined Grains

United States Department of Agriculture

Figure 14-22. *Refined grains have been processed to remove the germ, bran, or both.*

A *refined grain* is a grain that has been processed to remove the germ, bran, or both. *See Figure 14-22.* Refined grains are easier to digest, have a longer shelf life, can have their color changed, and take less time to cook. Processing often goes beyond removing the husk and can include removing the nutrient-rich bran and germ. The bran is removed from common all-purpose flour to make the product white in color instead of brown. Removing the germ can help preserve the product so that it does not spoil quickly. However, grains lose vitamins, minerals, and fiber during processing. Grains are milled, tumbled, or rolled to produce meal, flour, pearled grains, or flaked grains.

There are three common types of refined grains. A *milled grain* is a refined grain that has been ground into a fine meal or powder. Meal, such as cornmeal, and all varieties of flour are milled grains. A *pearled grain* is a refined grain with a pearl-like appearance that results from having been scrubbed and tumbled to remove the bran. A *flaked grain,* also known as a rolled grain, is a refined grain that has been rolled to produce a flake. Oatmeal is a flaked grain.

TYPES OF GRAINS

The three most abundant grains used around the world are rice, wheat, and corn. Rice is the most used grain in everyday diets. However, corn and wheat are the most cultivated grains, with rice being third.

Rice

Rice is a staple food source for two-thirds of the world's population. There are more than 40,000 varieties of rice. Rice can be cooked whole or made into flour, noodles, paper, milk, wine, vinegar, and a wide variety of processed foods. Rice is a complex carbohydrate that contains only a trace amount of fat. It is cholesterol, sodium, and gluten free and a rich source of thiamine, niacin, iron, phosphorus, and potassium. Its high starch content allows rice to absorb flavor from the ingredients it is cooked with. It is an ideal flavor carrier for Indian, Chinese, Japanese, Vietnamese, Spanish, Brazilian, Salvadoran, and Turkish dishes. The volume of rice triples when it is cooked.

All varieties of rice are milled whole and then rolled to remove the husk. Many varieties are sold as either brown or white rice, depending upon how they were milled. *Brown rice* is rice that has had only the husk removed. It is chewier and nuttier in flavor than white rice and contains more vitamins and fiber. White rice has had the husk, bran, and germ removed and is more tender and delicate than brown rice. Rice can also be red, green, or black. Regardless of its color, rice is classified by the size of the grain. Grain sizes include short-grain, medium-grain, or long-grain. **See Figure 14-23.**

Nutrition Note

Brown, red, and black unmilled rice are whole grains. They contain the nutrient-dense bran and germ and are high in dietary fiber, folic acid, vitamin E, and minerals.

Rice Classifications

**Short-Grain
(Arborio)**

**Medium-Grain
(Forbidden Black)**

**Long-Grain
(Himalayan Red)**

Indian Harvest Specialtifoods, Inc./Rob Yuretich

Figure 14-23. *Rice is often classified by the size of the grain. Grain sizes include short-grain, medium-grain, and long-grain.*

Short-Grain Rice. *Short-grain rice* is rice that is almost round in shape and has moist grains that stick together when cooked. It is high in starch and commonly used to make risotto and rice pudding. Arborio rice is a short-grain rice used to make risotto, pilaf, and paella. Sweet rice, also known as sticky rice, is a short-grain rice used to make desserts. It is not used in sushi. Sweet rice is soaked for several hours and then steamed. It is sometimes ground into flour to make dumplings, pastries, or rice puddings.

Medium-Grain Rice. *Medium-grain rice* is a rice that contains slightly less starch than short-grain rice but is still glossy and slightly sticky when cooked. It has a similar appearance to short-grain varieties, with the grains being only slightly longer and plumper. Medium-grain rice is widely used to make sushi. Forbidden rice, also known as emperor's rice, is a black medium-grain rice that turns indigo when cooked. Thai sticky rice is a medium-grain rice that turns a deep-purple color when cooked. Each grain is purple and brown interspersed with flecks of white.

Long-Grain Rice. *Long-grain rice* is a rice that is long and slender and remains light and fluffy after cooking. *Basmati rice* is a long-grain rice that only expands lengthwise when it is cooked. It has a sweet, nutty flavor and is high in fiber. The dry nature of basmati makes it pair well with sauces. Jasmine rice, also known as Thai basmati rice, is a long-grain rice that releases a floral aroma when cooked. Himalayan red rice is a long-grain rice with a rose-colored bran, firm texture, and nutty flavor. Wild rice is not true rice, but its long, black grains are high in nutrients and have a rich, nutty flavor. Wild rice comes from an aquatic grass and is prepared in the same manner as rice.

Corn

Corn, also known as maize, is a cereal grain cultivated from an annual grass that bears kernels on large woody cobs called ears. ***See Figure 14-24.*** It is prepared as a grain and as a vegetable. Dried kernels of corn are called popcorn. Tiny ears of immature corn, also known as baby corn, are often added to stir-fries and salads or eaten whole.

Corn

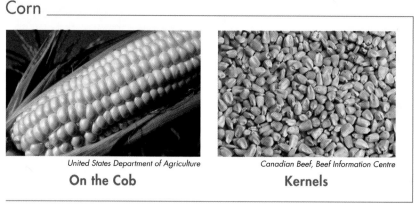

United States Department of Agriculture
On the Cob

Canadian Beef, Beef Information Centre
Kernels

Figure 14-24. *Corn, also known as maize, is cultivated from an annual grass that bears kernels on large woody cobs called ears.*

Corn is used to make cornstarch, corn oil, corn syrup, and whiskey. Hominy, grits, and cornmeal are also made from corn. ***See Figure 14-25.***

Hominy. *Hominy* is the hulled kernels of corn that have been stripped of their bran and germ and then dried. White hominy is made from white corn kernels, and yellow hominy is made from yellow corn kernels. Hominy can be ground into flour and used to make tortillas or tamales. Hominy is boiled whole or ground into grits. Hominy is also used to make posole.

Corn By-Products

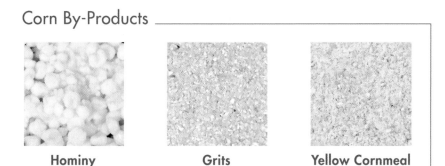

| Hominy | Grits | Yellow Cornmeal |

Figure 14-25. *Hominy, grits, and cornmeal are made from corn.*

Cuisine Note

Polenta is a thick paste often made with medium- or coarsely-ground, yellow cornmeal. It can be served soft or cooled until it stiffens and then sliced and fried. The word polenta stems from the Latin word "pollen," which means "fine flour." Before the 1600s, polenta was made using spelt, buckwheat, chickpea, or barley flour. In Northern Italy, polenta is made using semolina wheat.

Grits. *Grits* are a coarse type of meal made from ground corn or hominy. Grits are traditionally served with butter, salt, and pepper. Cheese, bacon, or hot sauce may also be added.

Cornmeal. Cornmeal is coarsely ground corn. It is commonly used to make cornbread and as a coating for fried foods.

Wheat

The most common form of wheat used in cooking is milled flour. Durum wheat, bulgur wheat, and wheat berries are commonly used in the professional kitchen. *See Figure 14-26.*

Wheat By-Products

Canadian Beef, Beef Information Centre
Whole Wheat

Indian Harvest Specialtifoods, Inc./Rob Yuretich
Wheat Berries

Indian Harvest Specialtifoods, Inc./Rob Yuretich
Moroccan Couscous

Indian Harvest Specialtifoods, Inc./Rob Yuretich
Israeli Couscous

Figure 14-26. *Whole wheat, wheat berries, Moroccan couscous, and Israeli couscous are commonly used in the professional kitchen.*

Durum Wheat. Durum is the hardest type of wheat. Its high protein content and gluten strength make durum the wheat of choice for making pasta dough. Durum kernels are amber colored and larger than other wheat kernels. Its yellow endosperm gives pasta its golden hue.

Semolina is the granular product that results from milling the endosperm of durum wheat. When semolina is mixed with water it forms a stiff dough that can be forced through metal dies to create different pasta shapes.

Couscous is a tiny, round pellet made from durum wheat that has had both the bran and germ removed. It is very fine in texture and similar in size to cornmeal. Israeli couscous is a larger variety. Couscous is usually steamed or simmered like pasta.

Bulgur Wheat. Bulgur wheat is golden-brown, nutty-tasting wheat kernels. The husks and bran are removed and it is steamed, dried, and ground. Bulgur comes whole or is cracked into fine, medium, or coarse grains. Bulgur wheat is commonly simmered and seasoned with herbs, spices, and vegetables and cooks in less time than wheat berries. Tabbouleh is made from cooked, chilled bulgur wheat mixed with mint, lemon, olive oil, and parsley.

Wheat Berries. A wheat berry is a chewy wheat kernel with only the husk removed. It contains both the bran and germ. Wheat berries are simmered using a procedure similar to the procedure used for rice. After simmering for an extended period, they can be finished by sautéing them in a pan with various herbs and spices.

Oats

Oats are derived from the berry of oat grass and can be purchased in many different forms. *See Figure 14-27.* An *oat groat* is an oat grain that only has the husk removed. *Steel-cut oats* are oat groats that have been toasted and cut into small pieces. *Rolled oats,* also known as old-fashioned oats, are oats that have been steamed and flattened into small flakes. Rolled oats require less cooking time than steel-cut oats. Oats are packed with cholesterol-fighting soluble fiber and often served as a hot cereal or used to make breads and desserts.

Oats _____

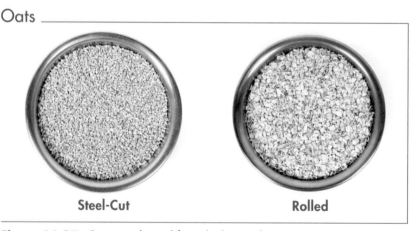

Steel-Cut   **Rolled**

Figure 14-27. *Oats are derived from the berry of oat grass.*

Barley

Barley contains high levels of soluble fiber, takes longer to cook than rice, and has a chewy texture. *See Figure 14-28.* Pearled barley is polished barley with the bran removed. It is often used in salads and pilafs. Barley is often added to soups and stews for its earthy flavor and because it is a natural thickener.

Quinoa

Quinoa is a small, round, gluten-free grain that is classified as a complete protein. *See Figure 14-29.* Quinoa is one of the oldest known grains and is native to the South American Andes. It is available in ivory, red, pink, brown, and black varieties. Quinoa contains fiber, protein, vitamins, and minerals. It is a rich source of the amino acid lysine, which promotes tissue growth and repair and supports the immune system.

Quinoa must be rinsed before cooking to remove its bitter coating. It cooks quickly, has a mild flavor, and has a slightly crunchy texture. Quinoa is often used in salads, stuffing, quick breads, and as a side dish.

Barley

Canadian Beef, Beef Informaton Centre

Figure 14-28. *Barley contains high levels of soluble fiber, takes longer to cook than rice, and has a chewy texture.*

Quinoa

Ivory

Red

Black

Indian Harvest Spectialtifoods, Inc./Rob Yuretich

Figure 14-29. *Quinoa is a small, round, gluten-free grain that is classified as a complete protein.*

Rye

Figure 14-30. Rye is a hearty grain that is used to make flour, breads, crackers, and whiskey.

Farro

Indian Harvest Specialtifoods, Inc./Rob Yuretich

Figure 14-32. Farro is a hearty grain that tastes similar to wheat, yet resembles light brown rice.

Rye

Rye is a hearty grain with dark-brown kernels that are longer and thinner than wheat. Rye has a distinctive flavor. Rye is used to make flour, breads, crackers, and whiskey. *See Figure 14-30.* Rye flour is heavier and darker in color than wheat flours. Soaked and cooked rye berries are sometimes added to breads for extra texture or used to make pilafs or hot breakfast cereals. Triticale is a hybrid grain made by crossing rye and wheat. It has a sweet, nutty flavor and contains more protein and less gluten than wheat. Like rye, triticale makes heavy, hearty loaves of bread. Triticale is sold as whole berries, flaked, and flour.

Buckwheat

Buckwheat is a dark, three-cornered seed of a plant unrelated to wheat that has a nutty, earthy flavor. Buckwheat is commonly ground into a gritty flour and used to make everything from pancakes to soba noodles. *See Figure 14-31.* Buckwheat flour lacks gluten and is loaded with nutrients. A *buckwheat groat* is a crushed, coarse piece of whole-grain buckwheat that can be prepared like rice. *Kasha* is roasted buckwheat.

Buckwheat

Figure 14-31. Buckwheat is commonly ground into a gritty flour and used to make everything from pancakes to soba noodles.

Farro

Farro is an ancient grain that is native to Italy. This hearty grain tastes similar to wheat, yet resembles light brown rice. *See Figure 14-32.* Farro is low in gluten and high in fiber and vitamins. It can be purchased as a whole grain or flour. Like barley, farro can be used to make risottos and is often pearled.

Millet

Millet is a small, round, butter-colored grain that is gluten-free. It is high in iron and B vitamins. *See Figure 14-33.* A native of Asia, millet resembles couscous, yet is prepared like rice. Millet is a quick-cooking grain with a mild flavor that can be toasted to bring forth a nuttier flavor and deep-yellow and light-brown coloring. It pairs nicely with chives, green onions, and garlic. Millet can be used in salads, casseroles, and stuffing.

Spelt

Spelt is an ancient grain with a nutty flavor and high protein content that is also a good source of riboflavin, zinc, and dietary fiber. Spelt is commonly mistaken for farro because of their similar appearance. *See Figure 14-34.* Spelt can be used as a hot breakfast cereal or as a substitute for wheat in many dishes. Though it contains gluten, spelt is often tolerated by people with wheat allergies.

Millet

Figure 14-33. *Millet is a small, round, butter-colored grain that is gluten-free.*

Spelt
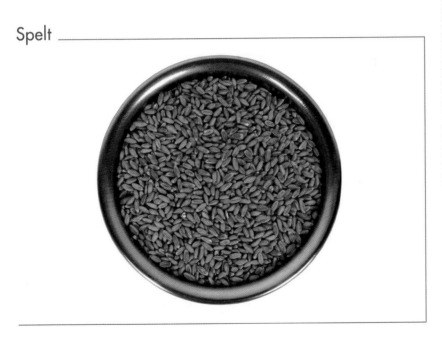

Figure 14-34. *Spelt is commonly mistaken for farro because of their similar appearance.*

STORING GRAINS

Grains should be stored in a cool, dry place in an airtight container to keep out moisture and prevent insects from getting in the product. *See Figure 14-35.* Some grains can absorb strong aromas, so they should be stored away from foods such as garlic and onions. Grains that still contain germ should be used quickly or kept in a refrigerator or freezer to prevent the germ from becoming rancid. Milled grains will keep indefinitely in sealed containers that are stored in a cool, dry place. Brown rice will last for several months.

Storing Grains

Carlisle FoodService Products

Figure 14-35. *Grains should be stored in a cool, dry place in an airtight container to keep out moisture and prevent insects from getting into the product.*

Chef's Tip

When using juices as a cooking liquid to flavor grains, mix one part vegetable or fruit juice and one part water for the best results.

COOKING GRAINS

Grains are most commonly simmered in a hot liquid until all of the liquid has been absorbed by the grain. With the exception of flaked grains and hominy, grains expand in volume when they are cooked. *See Figure 14-36.* The flavors of grains are often enhanced by the liquid that is use to cook them. Stocks, bouillons, consommés, or juices may be used instead of water to add flavor. The addition of aromatic herbs and vegetables heightens the flavor of grains. For example, a pinch of saffron can add flavor and color to grains.

Determining Doneness

Grains are done when they are tender enough to eat or all of the cooking liquid has been absorbed. Cooked grains that are being held for service should be kept at 140°F or above. Hot grains should be cooled to 70°F within 2 hours and then covered, dated, and refrigerated for no more than seven days. When reheating grains, they should reach a core temperature of 165°F before being served.

Simmering Grains

Simmering is the most common method of cooking grains. The grain is simply added to a measured amount of boiling cooking liquid. The appropriate amount of cooking liquid depends on the type and amount of grain being cooked. When the water returns to a boil, it is lowered to a simmer and the pot is covered until the grain is fully cooked and the liquid has been absorbed.

Cooking Grains					
Dry Grain (1 cup)	**Liquid** (in cups)	**Yield*** (in cups)	**Dry Grain** (1 cup)	**Liquid** (in cups)	**Yield*** (in cups)
Arborio rice	2½	2½	Hominy	5	3
Barley, pearled	3	3½–4	Jasmine rice	1½	2
Basmati rice, brown	2	3½	Millet	3	5
Basmati rice, white	1¾	3½	Oats, steel-cut	4	2
Brown rice, long-grain	2	3½	Quinoa	2	4
Brown rice, short-grain	2	3¾	Rye, berries	3	3
Buckwheat groats, unroasted	2	3½	Rye, flakes	3	2½
Bulgur wheat	2	2½–3	Spelt, soaked overnight	3½	2½
Cornmeal polenta	2½	3½	Sweet rice	2	2
Couscous	1¼	2¼	Wheat berries	2½	3
Forbidden rice	1¾	2¾	Wheat flakes	4	2
Grits	3	3½–4	White rice	2	2½

* Yields are approximate

Figure 14-36. *With the exception of flaked grains and hominy, grains expand in volume when they are cooked.*

Risotto and pilaf are two grain preparations that are sautéed and then simmered. *See Figure 14-37.* Risotto preparation begins with lightly sautéing a grain, such as rice. A liquid is then incorporated gradually in small amounts. In a pilaf preparation, a grain is sautéed prior to adding a simmering liquid. However, all of the liquid is added at once. A pilaf can be finished on the stovetop or in an oven.

Simmered Grains

Tanimura & Antle®

Risotto

National Cancer Institute
Daniel Sone (photographer)

Pilaf

Figure 14-37. *Risotto and pilaf are two grain preparations that are sautéed and then simmered.*

Risotto Method. Risotto is a classic Italian dish traditionally made with Arborio rice. Rice cooked using the risotto method has a pudding-like texture. Risotto is cooked slowly to release the starches from the grain, resulting in a creamy finished product. *See Figure 14-38.* Garnishes such as vegetables, meat, or shrimp can be added near the end of the cooking process.

Procedure for Preparing Risottos

1. Melt fat in a hot saucepan and sweat onions or shallots.

2. Add grain and stir to coat with fat. Sauté until the grain appears translucent.

3. Add white wine and cook until the wine is almost completely reduced.

4. Pour a small amount of stock into the saucepan and continue to stir the grain until all the liquid has been absorbed.

5. Repeat step 4 until the grain is cooked al dente.

Figure 14-38. *Risotto is cooked slowly to release the starches from the grain, resulting in a creamy finished product.*

Pilaf Method. When using the pilaf method, the flavoring ingredients and grains are sautéed in fat before adding the liquid to prevent clumping. *See Figure 14-39.* All of the hot liquid or stock is then added along with any seasonings, and the grain is covered and left to simmer until the liquid has been absorbed. A classic pilaf is finished on the stove, although it can also be finished in the oven.

1. Melt fat in a hot saucepan and, if desired, sweat onions and garlic.

2. Add the grain and stir to coat until slightly toasted.

3. Add hot liquid to the saucepan. Bring the liquid to a boil and then reduce to a simmer.

4. Cover and allow to simmer on the stovetop or in an oven until all of the liquid has been absorbed.

5. Remove from the heat and fluff the pilaf with a fork.

Figure 14-39. *When using the pilaf method, the flavoring ingredients and grains are sautéed in fat before adding the liquid to prevent clumping.*

Refer to DVD for
Preparing Pilafs
Media Clip

PASTAS

Pasta is a term for rolled or extruded products made from a dough composed of flour, water, salt, oil, and sometimes eggs. Pasta is one of the most versatile food products used in the professional kitchen. It can be made fresh or purchased in dried or frozen form. Pasta has a mild flavor that does not compete with the flavors of other ingredients or with the sauces that are often added to it. Many pasta dishes, such as macaroni, fettuccine, spaghetti, rigatoni, lasagna, and tortellini, are named for the type of pasta used in the dish. *See Figure 14-40.*

Most pasta is made from wheat flours that contain a very high percentage of gluten, which provides the strength to hold the shape, form, and texture of the dough when cooked. Pasta can be formed into many different shapes and sizes. Ingredients are sometimes added to pasta dough to produce colored pastas. For example, tomato paste produces red pasta, spinach purée produces green pasta, and squid ink produces black pasta.

National Pasta Association

Pasta-Named Dishes

Macaroni

Fettuccine

Spaghetti

Rigatoni

Lasagna

Tortellini

Barilla America, Inc.

Figure 14-40. *Some pasta dishes, such as macaroni, fettuccine, spaghetti, rigatoni, lasagna, and tortellini, are named for the type of pasta used in the dish.*

Pasta Machine

Browne-Halco (NJ)

Figure 14-41. *Dough can be fed through a pasta machine to create various shapes of pasta.*

When pasta dough is soft, it can be shaped by rolling it flat and cutting it to the desired size or by forcing it through the metal dies of a pasta machine to create various shapes. ***See Figure 14-41.*** The dough is then allowed to dry, resulting in a hard, mold-resistant product. The shape of the pasta chosen for a given dish is often determined by the sauce and how it will cling to the particular pasta shape. In addition, the pasta shape should complement the appearance of the finished dish. Pasta is classified by size and shape into four general categories. These categories are shaped, tube, ribbon, and stuffed.

Shaped Pastas

A *shaped pasta* is a pasta that has been extruded into a complex shape such as a corkscrew, bowtie, shell, flower, or star. ***See Figure 14-42.*** Shaped pastas are used in soups, salads, casseroles, and stir-fries. They add visual interest and texture. Common types of shaped pastas include campanelle, conchiglie, farfalle, fiori, fusilli, gemelli, jumbo shells, orecchiete, orzo, radiatori, rotini, and stelline.

Shaped Pastas			
Pasta	**Description**	**Pasta**	**Description**
Campanelle 	Campanelle, also known as gigli, is a fluted sheet of pasta that has been rolled into a cone shape	**Jumbo Shells** 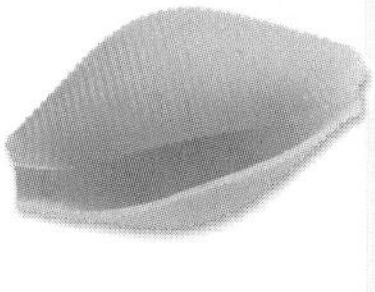	Jumbo shells are large, shell-shaped pasta
Conchiglie (Large Shells) 	Conchiglie, also known as large shells, are shell-shaped pasta	**Orecchiette** 	Orecchiette are small, ridged, bowl-shaped pasta
Farfalle 	Farfalle, also known as bow tie pasta, are flat squares of pasta that are pinched in the center in the shape of bow ties	**Orzo** 	Orzo are small oval pasta with an appearance similar to that of a grain
Fiori 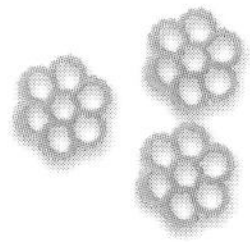	Fiori are flower-shaped pasta	**Radiatori** 	Radiatori are deeply-ridged, curled sheets of pasta
Fusilli 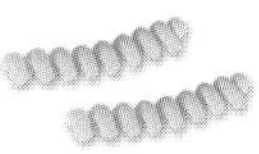	Fusilli are pasta shaped like a corkscrew	**Rotini** 	Rotini are 2 inch long, twisted pasta
Gemelli 	Gemelli is formed by two strands of pasta twisted together	**Stelline** 	Stelline are tiny, star-shaped pasta

Barilla America, Inc.

Figure 14-42. *Shaped pastas are made into complex shapes, such as corkscrews, shells, and stars, using an extruder.*

Barilla America, Inc.

Tube Pastas

A *tube pasta* is a pasta that has been pushed through an extruder and then fed through a cutter that cuts the tubes to desired length. *See Figure 14-43.* Tube pastas are often stuffed with cheese or meat and are often used in casseroles. Short tubes are often used in soups. Cellentani, ditalini, elbows, manicotti, penne, pipettes, rigatoni, and ziti are common types of tube pasta.

Tube Pastas			
Pasta	**Description**	**Pasta**	**Description**
Cellentani	Cellentani, also known as cavatappi, are twisted, hollow tubes of pasta with a ribbed surface	Penne	Penne are hollow, diagonally cut tubes of pasta approximately 1½–2 inches in length with a smooth or ribbed surface
Ditalini	Ditalini are short, hollow tubes of pasta with a smooth or ridged surface	Pipettes	Pipettes are curved, hollow, ridged tubes of pasta with one open end and one pinched end
Elbows	Elbows are relatively short, slightly curved, hollow tubes of pasta	Rigatoni	Rigatoni are wide, hollow, ridged tubes of pasta
Manicotti	Manicotti, also known as cannelloni, are large, round tubes of pasta approximately 4 inches long and 1 inch in diameter; they can be straight cut or diagonal cut	Ziti	Ziti are straight, round tubes of pasta approximately ¼ inch in diameter of various lengths with a smooth or ribbed surface

Barilla America, Inc.

Figure 14-43. *Tube pastas are pushed through a tube-shaped pasta extruder and then fed through a tube cutter that cuts the tubes to desired length.*

Ribbon Pastas

A *ribbon pasta* is a thin, round strand or flat, ribbonlike strand of pasta. *See Figure 14-44.* To form ribbon pasta, the dough is rolled out and cut to the desired width by hand or using a pasta machine. Round strands of pasta are formed by passing the dough through small, round dies. Ribbon pastas are often dressed in sauces. Common types of ribbon pasta include capellini, egg noodles, fettuccine, lasagna, linguine, and spaghetti.

Ribbon Pastas			
Pasta	**Description**	**Pasta**	**Description**
Capellini	Capellini, also known as angel hair, are very fine, round, strandlike pasta approximately 1/64 inch in diameter	Lasagna	Lasagna are flat, ripple-edged pasta, approximately 2 inches wide
Egg noodles	Egg noodles are flat, ribbon-shaped pasta that can be cut long or short and thin, medium, or wide; to be labeled egg noodles, the pasta must contain at least 5.5% egg solids	Linguine	Linguine are long, thin, flat strips of pasta, about 1/8 inch wide; linguine is a good shape for all sauces
Fettuccine	Fettuccine are long, thin, flat strips of pasta approximately 1/4 inch wide	Spaghetti	Spaghetti are long, round rods of pasta approximately 3/32 inch in diameter; very thin strands of spaghetti are known as spaghettini

Barilla America, Inc.

Figure 14-44. *Ribbon pasta is a thin, round strand or a flat, ribbonlike strand of pasta.*

Barilla America, Inc.

Daniel NYC

Stuffed Pastas

A *stuffed pasta* is a pasta that has been formed by hand or machine to hold fillings. *See Figure 14-45.* Individual portions of filling are added to a sheet of pasta and then a wash is applied around each filling. Another sheet of pasta is laid over the first sheet and the mounds of filling, the two sheets are sealed with the mounds of filling inside, and then the pasta is cut into individual portions. Stuffed pastas can be filled with savory and sweet cheeses, puréed meats, poultry, seafood, or vegetables or a combination of ingredients made into a paste. Common types of stuffed pasta include ravioli, tortellini, and tortelloni.

Stuffed Pastas	
Pasta	**Description**
Ravioli	Ravioli are formed from equal size squares, or other shapes, of flat pasta and are filled
Tortellini	Tortellini are formed by wrapping filled half circles of dough around a finger and pressing the ends together
Tortelloni	Tortelloni resemble pot stickers, are larger than tortellini, and will hold more filling

Barilla America, Inc.

Figure 14-45. *Ravioli, tortellini, and tortelloni are stuffed pastas.*

PREPARING PASTA DOUGHS

Pasta is often purchased ready to use. Purchased pasta should be placed in an airtight container and stored in a cool, dry place to keep moisture out and prevent insects from getting in the product.

Some chefs prefer to make their own pasta. Fresh pasta is more tender and allows the chef to flavor and color it as desired. If pasta is made fresh, the dough can be processed in a mixer or kneaded by hand. *See Figure 14-46.*

Using a Mixer

1. In a mixer with a paddle attachment, combine the eggs, salt, and oil. *Note:* The eggs can be whisked before being added.

2. Add approximately one-third of the flour and mix until the mixture forms a smooth, soft dough.

3. Replace the paddle attachment with the dough hook. Add the rest of the flour and knead the mixture thoroughly.

4. Remove the dough, cover it, and allow it to rest for 20 minutes.

5. Roll the dough to the appropriate thickness and cut into desired shapes.

By Hand

1. On a clean work surface, place flour in a mound and form a well in the center. Add eggs, oil, and salt to the well.

2. Slowly work the flour into the well with the egg mixture until all ingredients are mixed.

3. Knead the mixture until a smooth, dry ball of dough has formed. Cover the dough and allow it to rest for 20 minutes.

4. Roll the dough to the appropriate thickness and cut into desired shapes.

Figure 14-46. *If pasta is to be made from scratch, the dough can be processed in a mixer or kneaded by hand.*

Refer to DVD for
Preparing Pasta Dough
Media Clip

Chapter 14 — Potatoes, Grains, and Pastas **601**

Preparing Ravioli

To make ravioli, one sheet of pasta is topped with a small amount of filling, and a second sheet of pasta is laid on top. *See Figure 14-47.* The top sheet is pressed down around the filling, and individual ravioli are formed by cutting around the mounds of filling with a pastry wheel, ravioli cutter, or knife.

Procedure for Preparing Ravioli

1. Roll out a 12 inch, square sheet of pasta dough as thin as possible.

2. Use a pastry bag to deposit ¼ oz portions of filling on the dough, approximately 2 inches apart.

3. Use a pastry brush to egg wash the pasta around each portion of filling.

4. Prepare a second sheet of dough the same size as the first and place it on top of the first sheet.

5. Press down around each portion of filling to seal all edges of the two sheets of dough together.

6. Use a pastry wheel or knife to cut between the filled mounds.

7. Separate the ravioli and place on a sheet pan. Cook or cover and refrigerate or freeze for future use.

Figure 14-47. *To make ravioli, one sheet of pasta is topped with a filling, and a second sheet of pasta is laid on top.*

Preparing Tortellini

When making tortellini, thin pasta dough is cut into circles and a filling is placed in the center of each circle. *See Figure 14-48.* The circle is then folded in half and the edges are pressed together to hold the filling in place. The half circle is then formed into a ring by wrapping it around a finger and pressing the two ends together.

Procedure for Preparing Tortellini

1. Roll out pasta dough as thin as possible.

2. Use a 2 inch cutter to cut the dough into rounds.

3. Place a small portion of filling in the center of each round.

4. Moisten the edge of each round with water or egg wash.

5. Fold each circle in half and then press the edges together until they are tightly sealed.

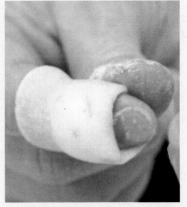

6. Slightly stretch the tips of each half circle to form a ring of dough around a finger. Press the tips together until they are tightly sealed. Cook or cover and refrigerate or freeze.

Figure 14-48. *When making tortellini, thin pasta dough is cut into circles and a filling is placed in the center of each circle.*

COOKING PASTA

A pound of fresh pasta yields approximately 2–2½ lb of cooked pasta. Pasta at least doubles in volume when cooked. Cooking time varies depending on the shape, size, and quality of the pasta. *See Figure 14-49.* Fresh pasta requires less cooking time than dried pasta. Pasta should always be cooked uncovered. A large head of froth should not boil over the pot. After the minimum cooking time is reached, the pasta should be tested frequently to ensure the proper doneness.

Approximate Cooking Times for Dried Pasta							
Pasta	Minutes	Pasta	Minutes	Pasta	Minutes	Pasta	Minutes
Spaghetti	10–12	Fettucini	10–12	Penne and Mostaccioli	9–11	Fusilli	12–14
Vermicelli	5–7	Lasagna	11–13	Manicotti and Connelloni	10–12	Orzo	5–7
Capellini	3–5	Egg Noodles	8–14	Conchiglie	9–12	Farfalle	9–12
Linguine	9–12	Elbow Macaroni	9–12	Jumbo shells	20–25	Tortellini	10–12

Figure 14-49. *Cooking time varies depending on the shape, size, and quality of the pasta.*

Dried pasta cooks in approximately 8–12 minutes, while fresh pasta cooks in approximately 3–5 minutes. All pastas are cooked using the same basic procedure. *See Figure 14-50.*

Procedure for Cooking Pastas

1. Place 1 gal. of water per pound of pasta in a stockpot or steam-jacketed kettle.
2. Add salt and bring the water to a rolling boil.
3. Add the pasta and stir gently. *Note:* If the pasta is long, it is best not to break it. Instead, spread it out around the inner wall of the pot and then stir and lift it gently until the pasta becomes submerged.
4. Return the water to a boil. Stir and lift the pasta occasionally as it cooks.
5. When the pasta is cooked al dente, use a colander to drain and rinse the pasta in cold water to stop the cooking process and to remove any starch residue from the exterior of the cooked pasta.

Figure 14-50. *All pastas are cooked using the same basic procedure.*

A variety of flavors can be added to pasta dough. *See Figure 14-51.* For example, squid ink can provide a salty, metallic flavor that will also color the pasta. Herbs, spices, and vegetables may also be added to enhance the flavor and appearance of pasta.

Flavored Pastas

Figure 14-51. *A variety of flavors can be added to pasta doughs.*

Determining Doneness

Pasta should be cooked al dente, meaning "to the tooth." When pasta is cooked al dente, there should be a slight resistance in the center of the pasta when it is chewed.

Reheating Prepared Pasta

In the professional kitchen, pasta is often cooked in advance. When cooked in advance for later use, pasta is rinsed in cold water to stop the cooking process, tossed with a small amount of oil to prevent sticking, and stored. Parcooked pasta can be rinsed in warm water or dropped in simmering water to bring it to the temperature needed for service. *See Figure 14-52.*

ASIAN NOODLES

Noodles combine the best qualities of grains and pastas. The term "Asian noodles" is used to describe noodles that may not contain flour, water, and eggs. Asian noodles are unique both in composition and in preparation method. Popular types of Asian noodles are made from rice, wheat, buckwheat, vegetables, and eggs. It is also important to note that Asian noodles are soaked briefly in hot water rather than cooked.

Reheating Parcooked Pastas

Barilla America, Inc.

Figure 14-52. *Parcooked pasta can be rinsed in warm water or dropped in simmering water to bring it to the temperature needed for service.*

Rice Noodles

Rice noodles are made from ground rice flour and are available dried in various shapes, including sheets, flat sticks of various widths, or vermicelli. *See Figure 14-53.* They should be soaked in hot water before they are cooked and thoroughly rinsed after cooking. If they are not rinsed after cooking, excess starch remains on the noodles, causing them to turn into a sticky clump. Thin rice noodles can be deep-fried to produce a crispy, delicate noodle that can be used as a base for an entrée, in salads, or in desserts.

Rice Noodles

Figure 14-53. *Rice noodles are made from ground rice flour.*

Wheat Noodles

Several popular Asian noodles are made from wheat flour. Udon, somen, and la mian noodles are all made from wheat. *See Figure 14-54.*

Wheat Noodles

Udon   Somen   La Mian

Figure 14-54. *Udon, somen, and la mian are all made from wheat.*

Udon Noodles. Udon noodles are thick, white wheat noodles. They are available fresh or dried and as round or square strands. Udon noodles have a chewy, slippery texture and are served in soups, stews, or stir-fries, with meat dishes, or cold.

Somen Noodles. Somen noodles are long, thin wheat noodles that are white in color and are sold as rods, similar to vermicelli. Somen noodles are often sold bundled. Varieties of somen noodles include egg yolk and green tea. Somen noodles are used in stir-fries and soups or served cold in noodle dishes.

La Mian Noodles. La mian noodles are Chinese wheat noodles that are sold in round strands or ribbons of various widths. They are available fresh or dried. La mian noodles are popular in a wide range of dishes including soups, stir-fries, and chow mein. La mian noodles that have been deep-fried and then dehydrated are sold as ramen noodles.

Buckwheat Noodles

Buckwheat noodles, also known as soba noodles, are brown-gray noodles made from buckwheat flour. *See Figure 14-55.* They can be used hot or cold, in soups, in main dishes, and as side dishes. Buckwheat noodles can be substituted for spaghetti if desired.

Vegetable Noodles

Some Asian noodles are made from vegetables. Cellophane noodles and shirataki noodles are two common types of vegetable noodles. *See Figure 14-56.*

Buckwheat Noodles

Figure 14-55. *Buckwheat noodles, also known as soba noodles, are brown-gray noodles made from buckwheat flour.*

Vegetable Noodles

Cellophane   Shirataki

Figure 14-56. *Cellophane noodles and shirataki noodles are two common types of vegetable noodles.*

Cellophane Noodles. Cellophane noodles, also known as glass noodles, are made from mung bean starch. They are fairly transparent and are commonly rehydrated in hot water. Cellophane noodles can be deep-fried to produce a crispy and crunchy delicate noodle. If deep-frying, they do not need to be presoaked. Cellophane noodles are used hot or cold, in soups or stir-fries, or as fillings, especially in vegetarian dishes. Similar noodles are available made from sweet potato starch.

Shirataki Noodles. Shirataki noodles are long, slender noodles made from Japanese yams. Shirataki noodles have no calories, carbohydrates, gluten, or fat. They have a distinct earthy odor before they are cooked.

Egg Noodles

Asian noodles made from eggs include hokkien and lo mein. Asian egg noodles are often used in stir-fries and soups. *See Figure 14-57.*

Egg Noodles

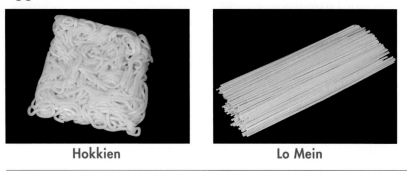

Hokkien   Lo Mein

Figure 14-57. *Egg noodles, such as hokkien and lo mein, are often used in stir-fries and soups.*

Hokkien Noodles. Hokkien noodles are yellow egg noodles that resemble spaghetti. They are sold fresh, dried, or vacuum-sealed. Hokkien noodles work well in soups or stir-fries or served with a sauce.

Lo Mein Noodles. Lo mein noodles, also known as Cantonese noodles, are egg noodles in varying widths that can be purchased fresh, dried, or frozen. Lo mein is a popular stir-fry dish named for the noodle itself.

SUMMARY

Potatoes, grains, and pastas are prepared daily in the professional kitchen. Mealy, waxy, new, and sweet potatoes are used to prepare breakfast, lunch, and dinner items. Mealy potatoes are best for baking, frying, and casseroles. Waxy potatoes are best prepared using any cooking method except baking or frying.

Grains include rice, corn, wheat, oats, barley, quinoa, rye, buckwheat, farro, millet, and spelt. Whole grains are composed of a husk, bran, endosperm, and germ. Due to their hard husks and high spoilage rate, grains are often processed to remove one or more parts of the natural grain. The bran and the germ are the most nutritious parts of grains. Refined grains include meal, flour, pearled grains, and flaked grains. Grains are typically simmered using the risotto method or the pilaf method.

Pastas are made from a flour-based dough made by hand or machine. They can be purchased in dried or frozen form. Tube, ribbon, shaped, and formed pastas are used in the professional kitchen. Pastas are simmered and should be cooked al dente. Pastas can be parcooked and reheated for service. Asian noodles are unique in composition and in preparation method. Popular Asian noodles are made from rice, wheat, buckwheat, vegetables, and eggs. Asian noodles are soaked briefly in hot water rather than cooked.

Refer to DVD for
Quick Quiz® questions

1. Describe the four major classifications of potatoes.
2. Identify five market forms of potatoes.
3. Describe the guidelines for receiving and storing potatoes.
4. Explain how to determine the doneness of potatoes.
5. Identify six different methods of cooking potatoes.
6. Identify the four parts of a whole grain.
7. Differentiate between whole grains and refined grains.
8. Explain how whole grains are cracked.
9. Name three types of refined grains.
10. Explain how grains are milled.
11. Explain how grains are pearled.
12. Explain how grains are flaked.
13. Describe the three major classifications of rice.
14. Identify forms of corn used in the professional kitchen.
15. Identify forms of wheat used in the professional kitchen.
16. Identify forms of oats used in the professional kitchen.
17. Describe barley and how it can be used in the professional kitchen.
18. Describe quinoa and how it can be used in the professional kitchen.
19. Describe rye and how it can be used in the professional kitchen.
20. Describe buckwheat and how it can be used in the professional kitchen.
21. Describe farro and how it can be used in the professional kitchen.
22. Describe millet and how it can be used in the professional kitchen.
23. Describe spelt and how it can be used in the professional kitchen.
24. Explain the importance of storing grains in an airtight container and in a cool, dry place.
25. Demonstrate the risotto method and the pilaf method of preparing grains.
26. Identify three forms of pasta used in the professional kitchen.
27. Describe shaped, tube, ribbon, and stuffed pastas.
28. Explain how pasta dough is formed.
29. Explain how to prepare ravioli.
30. Explain how to prepare tortellini.
31. Identify the cooking method commonly used to prepare pastas.
32. Explain how to determine if pasta is cooked al dente.
33. Explain how to reheat prepared pasta.
34. Explain the term "Asian noodles."
35. Describe nine types of Asian noodles.

Refer to DVD for
Review Questions

Refer to DVD for
Application Scoresheets

1. Grill potatoes and evaluate the results.
2. Roast potatoes and evaluate the results.
3. Bake potatoes and evaluate the results.
4. Sauté potatoes and evaluate the results.
5. Fry potatoes and evaluate the results.
6. Simmer potatoes and evaluate the results.
7. Prepare a rice dish and evaluate the results.
8. Prepare hominy, grits, or polenta and evaluate the results.
9. Prepare a dish using a wheat grain and evaluate the results.
10. Prepare steel-cut and rolled oats. Compare the two dishes.
11. Prepare a dish using barley and evaluate the results.
12. Prepare a dish using quinoa and evaluate the results.
13. Prepare a dish using rye or buckwheat and evaluate the results.
14. Prepare a dish using farro, millet, or spelt and evaluate the results.
15. Prepare fresh pasta dough and evaluate the results.
16. Prepare a tubed pasta dish and evaluate the results.
17. Prepare a ribbon pasta dish and evaluate the results.
18. Prepare a shaped pasta dish and evaluate the results.
19. Prepare a formed pasta dish and evaluate the results.
20. Reheat parcooked pasta and evaluate the results.
21. Prepare a dish using rice noodles and evaluate the results.
22. Prepare a dish using wheat noodles and evaluate the results.
23. Prepare a dish using buckwheat noodles and evaluate the results.
24. Prepare a dish using vegetable noodles and evaluate the results.
25. Prepare a dish using egg noodles and evaluate the results.

Potato Pancakes

Yield: *8 servings, 2 pancakes each*

Ingredients

russet potatoes, peeled and grated finely	2 lb
medium onion, grated	1 ea
large eggs, beaten	2 ea
milk	4 fl oz
baking powder	1 tsp
kosher salt	1 tsp
flour	1 c
canola oil	8 fl oz
salt	TT

Preparation

1. Place potatoes, onions, beaten eggs, and milk in a large mixing bowl.
2. In a separate bowl, sift baking powder and salt into flour.
3. Add the flour mixture to the potato mixture and stir until incorporated.
4. Pour canola oil ½ inch deep in a skillet and heat. *Note:* Test a small drop of potato mixture in the oil to check the temperature before pan-frying. The mixture should bubble when added to the hot oil.
5. Using a No. 16 portion control scoop, drop mixture into the oil. Flatten each pancake with the back of an oiled spatula.
6. When browned, turn the pancakes over and brown on other side.
7. Transfer pancakes to paper towels to drain off excess oil.
8. If desired, sprinkle pancakes with salt. Serve immediately.

NUTRITION FACTS
Per serving: 233 calories, 47 calories from fat, 5.3 g total fat, 48 mg cholesterol, 364.6 mg sodium, 848.9 mg potassium, 40.6 g carbohydrates, 2.4 g fiber, 1.2 g sugar, 6.5 g protein

Rissole (Oven-Browned) Potatoes

Yield: *8 servings, 4 oz each*

Ingredients

canola oil	1 tbsp
red potatoes, cut into quarters	2 lb
paprika	½ tsp
salt and pepper	TT

Preparation

1. Place oil, potatoes, paprika, salt, and pepper in a mixing bowl and toss to coat the potatoes.
2. Place potatoes in a roasting pan and roast in a 375°F oven, turning the potatoes frequently until they become golden brown and tender.

NUTRITION FACTS
Per serving: 105 calories, 17 calories from fat, 1.9 g total fat, 0 mg cholesterol, 43.2 mg sodium, 619.1 mg potassium, 20.5 g carbohydrates, 1.9 g fiber, <1 g sugar, 2.4 g protein

Variations:

Minute (Cabaret) Potatoes: After potatoes are cooked, sprinkle with garlic powder and 1 tbsp melted butter. Toss to coat evenly.

O'Brien Potatoes: Prepare minute potatoes, but omit the garlic and add ½ oz each of sautéed, fine-diced onions, green peppers, and red peppers.

Vesuvio Potatoes: Peel a raw potato and cut it into six equally sized wedges. Roast the potato wedges in a 375°F oven, turning the potatoes frequently until they become golden brown and tender. When almost cooked, toss the potatoes with a mixture of 1 tbsp lemon juice and 1 tbsp olive oil. Season with oregano, salt, and black pepper and roast until fully cooked.

Château Potatoes: Sauté a blanched, tournéed potato in clarified butter seasoned with salt and white pepper until it is golden on all sides. Alternatively, peel a whole russet potato and cut it into a tourné shape. Then cut it in half lengthwise from end to end. Brush the potatoes with clarified butter, season them, and roast in a 350°F oven until tender. If desired, sprinkle with finely chopped parsley for service.

Duchess Potatoes

Yield: *14 servings, 4 oz each*

Ingredients

baking potatoes	3 lb
unsalted butter, softened	4 tbsp
nutmeg, grated	⅛ tsp
salt	TT
white pepper	¼ tsp
egg yolks	3 ea
large egg	1 ea
half and half, heated	2 fl oz
unsalted butter, melted	2 tsp

Preparation

1. Peel potatoes and cut into quarters. Place peeled and cut potatoes in cold water to prevent browning.
2. Simmer the potatoes in salted water until tender.
3. Drain potatoes and allow them to steam dry on a sheet pan.
4. After potatoes have stopped emitting steam and while still warm, put them through a ricer or mash them well.
5. Put potatoes in a stand mixer and use the paddle attachment to beat in softened butter, nutmeg, salt, and white pepper until incorporated.
6. Add the egg yolks and the whole egg, mixing each time to fully incorporate.
7. Add the half and half and beat the mixture until it is smooth.
8. Transfer the potato mixture to a pastry bag fitted with a large star tip.
9. Pipe 4 oz pinecone-shaped mounds onto a sheet pan.
10. Drizzle potatoes with melted butter and bake in 375°F oven until ridges are golden brown (approximately 10 minutes).

> **NUTRITION FACTS**
> **Per serving:** 131 calories, 50 calories from fat, 5.7 g total fat, 63.6 mg cholesterol, 35.7 mg sodium, 424.8 mg potassium, 17.4 g carbohydrates, 2.2 g fiber, <1 g sugar, 3.2 g protein

Variation:

Croquette Potatoes: Pipe duchess mixture from a pastry bag (without a tip) onto a sheet pan. Score each croquette with a knife about every 2 inches. Place the pan in the freezer to firm up the croquettes. Once firm, remove from freezer and snap apart at the scores. Bread the cork-shaped croquettes using the standard breading procedure and bake them in a 375°F oven until ridges are golden brown.

Scalloped Potatoes

Yield: *6 servings, 6 oz each*

Ingredients

nonstick cooking spray	1 tsp
light béchamel sauce	1 pt
nutmeg	pinch
kosher salt	TT
white pepper	TT
russet potatoes, peeled	2¼ lb

Preparation

1. Spray the inside and bottom of a 2 inch deep half pan with nonstick cooking spray and set aside.
2. Heat béchamel sauce in a separate pan until hot. Add nutmeg, salt, and pepper to taste.
3. Slice the potatoes to a ¼ inch thickness. Place slices in cold water to prevent browning.
4. Remove potato slices from water and drain well.
5. Place a layer of potatoes on the bottom of the oiled half pan.
6. Pour enough hot béchamel over the potatoes to cover and then add another layer of potatoes. Repeat adding layers of sauce and potatoes until finished.
7. Gently shake the pan so the potatoes are evenly distributed.
8. Cover the potatoes and bake in a 350°F oven for 30 minutes.
9. Uncover the potatoes and bake an additional 30 minutes or until golden brown and cooked through.
10. Remove potatoes and let cool for 10–15 minutes before serving.

Note: Sliced potatoes can be steamed until al dente and then placed in a hotel pan, topped with béchamel, and finished in the oven.

> **NUTRITION FACTS**
> **Per serving:** 303 calories, 86 calories from fat, 9.5 g total fat, 19.1 mg cholesterol, 117.8 mg sodium, 1295.7 mg potassium, 48.7 g carbohydrates, 3 g fiber, 4.4 g sugar, 7.2 g protein

Variation:

Potatoes au Gratin: Add 3 cloves of mashed or creamed garlic and ½ c Gruyère or Parmesan cheese to the hot béchamel before pouring it over the sliced potatoes.

Twice-Baked Potatoes

Yield: *10 potatoes, 1 potato per serving*

Ingredients

russet potatoes	10 ea
vegetable oil	3 tbsp
kosher salt	4 tsp
unsalted butter, softened	4 oz
white pepper	¼ tsp
fresh nutmeg, grated	¼ tsp
egg yolks	6 ea
large egg	1 ea
half and half or milk, heated	2 fl oz
Parmesan cheese, finely grated	4 oz
parsley, minced and rinsed	1 tbsp
unsalted butter, melted	1 tbsp

Preparation

1. Scrub and rinse potatoes and pierce with a fork.
2. Rub potatoes with vegetable oil and season with half of the kosher salt.
3. Bake potatoes at 400°F until tender and cooked through (approximately 1 hour).
4. Remove potatoes from oven and slice the tops of the potatoes. Scoop out the potato flesh and place it in the bowl of a stand mixer, using caution to not break the potato shells.
5. Add butter, salt, white pepper, and nutmeg and beat the potatoes until incorporated.
6. Add egg yolks and then whole egg, mixing each time to fully incorporate.
7. Add the hot half and half, Parmesan, and parsley and beat the mixture until it is smooth.
8. Transfer the potato mixture to a pastry bag fitted with a large star tip. Pipe the potato mixture back into the potato shells in a decorative manner until filled to the top.
9. Brush or drizzle the tops of the potatoes with melted butter and bake in 375°F oven until the ridges are golden brown (approximately 10 minutes).

Note: Potatoes can be baked a day ahead and baked the second time prior to service.

> **NUTRITION FACTS**
> **Per serving:** 517 calories, 194 calories from fat, 21.9 g total fat, 166.3 mg cholesterol, 959.8 mg sodium, 1583.9 mg potassium, 67.9 g carbohydrates, 4.8 g fiber, 2.5 g sugar, 14.8 g protein

American Fries (Home Fries)

Yield: *8 servings, 4 oz each*

Ingredients

new potatoes, peeled	2 lb
canola oil	2 fl oz
salt and pepper	TT
parsley, chopped	1 tbsp

Preparation

1. Simmer or steam potatoes until they are tender, but still al dente.
2. Allow the potatoes to cool overnight in the refrigerator.
3. Slice the chilled potatoes to a ¼ inch thickness.
4. Heat oil in a skillet and sauté potatoes until golden brown on both sides.
5. Season fries to taste and garnish with chopped parsley.

> **NUTRITION FACTS**
> **Per serving:** 177 calories, 63 calories from fat, 7.1 g total fat, 0 mg cholesterol, 44 mg sodium, 786.2 mg potassium, 27 g carbohydrates, 1.9 g fiber, 0 g sugar, 2.8 g protein

Variations:

Lyonnaise Potatoes: Cook, cool, and slice potatoes as for American fries. Heat skillet and add 1 tbsp clarified butter and 6 oz julienned onion. Cook onions until softened but not browned. Remove onions and set aside. Add 2 oz clarified butter to a hot skillet. Add potatoes and cook until golden brown on both sides. Return onions to the skillet and cook until the onions are slightly brown. Season to taste.

Hash Brown Potatoes: Peel and cook whole potatoes as for American fries. Place in refrigerator until cooled. Grate cooled potatoes. Portion into 3 oz patties and pan-fry in 2 fl oz of canola oil until golden brown on both sides. Remove cooked patties and drain on paper towels or screen. If desired, season with salt, pepper, and parsley.

French-Fried Potatoes

Yield: *8 servings, 4 oz each*

Ingredients

russet potatoes, peeled and batonnet cut	2 lb
canola oil	(for deep-frying)
salt	TT

Preparation

1. Place peeled and cut potatoes in cold water to prevent browning.
2. Strain the potatoes and place in a deep-fry basket.
3. Blanch the potatoes for 5 minutes in 250°F fat that is deep enough to completely cover potatoes. *Note:* Do not allow the potatoes to brown.
4. Drain the potatoes and place on sheet pans to completely cool.
5. Before service, fry the potatoes in 350°F fat until golden brown and crisp.
6. Season potatoes to taste. Serve immediately.

> **NUTRITION FACTS**
> **Per serving:** 144 calories, 30 calories from fat, 3.4 g total fat, 0 mg cholesterol, 43.8 mg sodium, 783.6 mg potassium, 27 g carbohydrates, 1.9 g fiber, 0 g sugar, 2.8 g protein

Candied Sweet Potatoes

Yield: *14 servings, 4 oz each*

Ingredients

sweet potatoes	3 lb
butter	6 tbsp
brown sugar	4 oz
water	2 fl oz
orange juice	2 fl oz
salt	1 tsp
cinnamon	1/8 tsp
vanilla extract	2 tsp

Preparation

1. Wash sweet potatoes and roast in a 350°F oven until almost cooked but still firm in the center (approximately 30–40 minutes).
2. Remove potatoes from oven and let stand to cool slightly.
3. Place butter, brown sugar, water, orange juice, salt, and cinnamon in a saucepot and bring to a boil to dissolve the sugar.
4. Add vanilla and remove the saucepot from the heat.
5. Peel and slice the sweet potatoes into 1/4 inch thick slices.
6. Place a single layer of potato slices in a baking dish or hotel pan.
7. Top the potatoes with the hot sugar syrup mixture.
8. Bake the potatoes at 350°F, basting with syrup until the syrup is bubbly and the potatoes are completely tender (approximately 20–30 minutes).

> **NUTRITION FACTS**
> **Per serving:** 162 calories, 44 calories from fat, 5 g total fat, 13.1 mg cholesterol, 222.7 mg sodium, 349.7 mg potassium, 28.1 g carbohydrates, 2.9 g fiber, 12.4 g sugar, 1.6 g protein

Basmati Rice with Pineapple and Coconut

Yield: *4 servings, 4–5 oz each*

Ingredients

celery, minced	1 oz
onion, minced	2 oz
sesame oil	1 tsp
vegetable oil (or peanut oil)	2 tsp
fresh garlic, minced	1 tsp
hot chile pepper, seeded and minced	½ tbsp
basmati rice	4 oz
water	7 fl oz
tamari soy sauce	½ fl oz
lime juice	½ fl oz
grated coconut	1 tbsp
salt and pepper	TT
pineapple, small diced	2 oz
scallion, small diced	1 ea
cilantro, chopped coarse	½ tsp

Preparation

1. In a medium saucepot, sauté celery and onion in sesame oil and vegetable oil until translucent.
2. Add garlic and chile pepper and cook for 1 minute.
3. Add rice and sauté until grains are coated with oil and translucent.
4. Add water, soy, lime juice, and coconut.
5. Season to taste and bring mixture to a boil.
6. Cover and lower to a simmer. Cook for 20 minutes.
7. Turn off heat and allow the mixture to rest for 5 minutes covered.
8. Uncover and fold in the pineapple, scallion, and cilantro. Serve immediately.

> **NUTRITION FACTS**
> **Per serving:** 159 calories, 36 calories from fat, 4.1 g total fat, 0 mg cholesterol, 281.7 mg sodium, 136.9 mg potassium, 27.8 g carbohydrates, 1.3 g fiber, 2.8 g sugar, 3 g protein

Risotto

Yield: *4 servings, 5 oz each*

Ingredients

onions, minced	2 oz
garlic, minced	1 tsp
butter	1 oz
Arborio rice	8 oz
white wine	2 fl oz
chicken stock, hot	20 fl oz
cream	1 fl oz
Parmesan cheese	2 oz
salt and pepper	TT

Preparation

1. In a saucepot, sweat onions and garlic in butter until onions are translucent but not browned.
2. Add rice and stir to coat all of the grains with butter.
3. Cook until a lightly toasted aroma can be detected. Do not allow the rice to brown.
4. Add white wine and stir continuously until almost dry.
5. Add one-third of the hot stock and stir continuously until all of the moisture has been absorbed.
6. Continue adding the stock in thirds while stirring until it has all been absorbed.
7. Add cream and Parmesan cheese and stir to incorporate.
8. Season to taste. Serve immediately.

> **NUTRITION FACTS**
> **Per serving:** 408 calories, 126 calories from fat, 14.3 g total fat, 42.2 mg cholesterol, 507.2 mg sodium, 207.8 mg potassium, 51.7 g carbohydrates, <1 g fiber, 3.1 g sugar, 13.5 g protein

Variations:

Risotto Milanese: Add ¼ tsp of saffron to the hot stock. Finish with another 2 oz grated Parmesan cheese and 1 tbsp butter.

Mushroom Risotto: Sweat 2 oz diced mushrooms with the onions and garlic.

Asparagus Risotto: Blanch 4 oz of asparagus in salted water until al dente. Remove and shock in cold water and drain. Purée asparagus. Add puréed asparagus to the risotto and stir well.

Shrimp Fried Rice

Yield: *4 servings, 8 oz each*

Ingredients

vegetable oil	1½ tbsp
sesame oil	1 tsp
onions, small diced	2 oz
carrots, blanched and cut	
medium dice	2 oz
napa cabbage, medium diced	2 oz
shrimp, medium diced	4 oz
ginger, grated fine	1 tsp
garlic, minced	1 tsp
long-grain rice, cooked and cooled	1 lb
snow peas, medium dice	2 oz
egg, beaten	1 ea
tamari soy sauce	1 tbsp
salt and black pepper	TT

Preparation

1. Heat two-thirds of the vegetable oil and the sesame oil in a wok or very hot sauté pan.
2. Add onions and stir-fry until onions begin to turn brown.
3. Add carrots and stir-fry until they brown slightly.
4. Add cabbage and stir-fry for 1 minute.
5. Add shrimp and stir-fry until almost cooked through.
6. Add ginger and garlic and stir-fry for 10 seconds.
7. Add cooked rice and stir-fry until rice begins to brown.
8. Add snow peas and stir-fry for 1 minute.
9. Push rice to the sides of wok or sauté pan and add remaining vegetable oil to the center.
10. Add beaten egg directly on top of the oil and let stand for 10 seconds so the egg begins cooking.
11. Stir the egg into the rice to incorporate.
12. Add the soy sauce and adjust seasonings as needed. Serve immediately.

> **NUTRITION FACTS**
> **Per serving:** 265 calories, 73 calories from fat, 8.3 g total fat, 82.2 mg cholesterol, 516 mg sodium, 210.2 mg potassium, 36.9 g carbohydrates, 1.5 g fiber, 2 g sugar, 9.9 g protein

Sesame and Ginger Wild Rice with Almonds

Yield: *4 servings, 4 oz each*

Ingredients

wild rice	4 oz
water	13 fl oz
kosher salt	1 tsp
vegetable oil	½ tsp
sesame oil	½ tsp
fresh ginger, grated	1 tsp
fresh garlic, minced	1 tsp
sliced almonds	2 tsp
scallions, diced small	2 ea
fresh cilantro, chiffonade	1 tbsp
salt and black pepper	TT

Preparation

1. Combine rice, water, and salt in a medium saucepan and bring to a boil.
2. Cover rice and lower the heat to a simmer. Cook for 1 hour.
3. Add vegetable oil and sesame oil to a hot sauté pan. Add ginger and garlic and sauté for 1 minute.
4. Add cooked wild rice and sauté for a few seconds.
5. Add almonds and scallions and stir.
6. Toss with cilantro and season to taste.

> **NUTRITION FACTS**
> **Per serving:** 124 calories, 19 calories from fat, 2.2 g total fat, 0 mg cholesterol, 549.7 mg sodium, 164.8 mg potassium, 22.5 g carbohydrates, 2.2 g fiber, <1 g sugar, 4.7 g protein

Curry and Dried Apricot Couscous

Yield: *6 servings, 4 oz each*

Ingredients

curry powder	2 tsp
chicken stock, hot	12 fl oz
couscous	10 oz
dried apricots, small diced	5 oz
scallions, small diced	1½ oz
mint, rough chopped	2 tbsp
lemon juice	1 tsp
olive oil	2 tbsp
salt and pepper	TT

Preparation

1. Add curry powder to chicken stock and bring to a simmer.
2. Place the couscous in a bowl and pour the hot chicken stock over it. Stir to mix well.
3. Cover the bowl and let stand for 10 minutes until all of the stock has been absorbed.
4. Fluff the couscous with a fork and gently stir in the remaining ingredients.
5. Season to taste. Serve immediately.

> **NUTRITION FACTS**
> **Per serving:** 300 calories, 51 calories from fat, 5.7 g total fat, 1.7 mg cholesterol, 138.8 mg sodium, 452 mg potassium, 54.5 g carbohydrates, 4.6 g fiber, 13.7 g sugar, 8.5 g protein

Variation:

Chilled Curry and Dried Apricot Couscous: Add 1 tbsp sherry vinegar and toss to combine. Cool under refrigeration.

Wheat Berries with Walnuts and Dried Cherries

Yield: *6 servings, 4 oz each*

Ingredients

wheat berries	8 oz
chicken stock (or water)	16 fl oz
vegetable oil	2 tsp
shallot, minced	2 tsp
fresh garlic, minced	1 tsp
walnuts	1 tbsp
dried cherries	2 oz
thyme	½ tsp
kosher salt and black pepper	TT

Preparation

1. In a small saucepot, bring wheat berries and chicken stock to a boil.
2. Cover and lower heat to a simmer. Cook 15–18 minutes until al dente.
3. Add vegetable oil to a hot sauté pan. Add shallots and sauté until shallots are translucent.
4. Add garlic, walnuts, and cherries and sauté for 1 minute.
5. Add cooked wheat berries and sauté until heated through.
6. Toss with thyme and season to taste.

> **NUTRITION FACTS**
> **Per serving:** 212 calories, 36 calories from fat, 4.1 g total fat, 2.4 mg cholesterol, 163.7 mg sodium, 259.3 mg potassium, 39.8 g carbohydrates, 5.3 g fiber, 1.5 g sugar, 6.8 g protein

Grilled Sun-Dried Tomato Polenta

Yield: *8 servings, 4 oz each*

Ingredients

whole milk	16 fl oz
water	16 fl oz
salt	1 tsp
white pepper	¼ tsp
yellow corn meal	8 oz
sun-dried tomato, small diced	2 tbsp
Parmesan cheese, grated	2 tbsp
nonstick cooking spray	1 tsp
olive oil	2 tbsp

Preparation

1. Bring milk and water to a boil in a saucepan. Add salt and pepper and stir well.
2. Add the cornmeal in a fine stream while stirring constantly. *Note:* Adding the cornmeal too quickly or not stirring constantly will result in a lumpy polenta.
3. Add sun-dried tomatoes and reduce heat to a simmer. Cook for 35–40 minutes or until the polenta is smooth and tender.
4. Add Parmesan cheese and stir well to incorporate.
5. Pour polenta onto a parchment-lined sheet pan and spread out until about ½ inch thick.
6. Spray the surface of the polenta with a nonstick cooking spray and lay a piece of parchment paper on top of the polenta to prevent a skin from forming. Refrigerate until needed.
7. For service, cut the cooled polenta to the desired shape and size.
8. Brush the polenta with olive oil and grill on each side until golden and heated through.

NUTRITION FACTS
Per serving: 196 calories, 78 calories from fat, 8.6 g total fat, 7.2 mg cholesterol, 365.4 mg sodium, 192.8 mg potassium, 25.3 g carbohydrates, 2.2 g fiber, 3.6 g sugar, 4.8 g protein

Creamy Polenta

Yield: *8 servings, 5 oz each*

Ingredients

whole milk	22 fl oz
water	20 fl oz
salt	1 tsp
white pepper	¼ tsp
fine cornmeal	8 oz
Parmesan cheese, grated	2 tbsp

Preparation

1. Bring milk and water to a boil in a saucepan. Add salt and pepper and stir well.
2. Add the cornmeal in a fine stream while stirring constantly. *Note:* Adding the cornmeal too quickly or not stirring constantly will result in a lumpy polenta.
3. Reduce heat to a simmer. Cook for 35–40 minutes or until the polenta is smooth and tender.
4. Add Parmesan cheese and stir well to incorporate. Serve immediately.

NUTRITION FACTS
Per serving: 159 calories, 36 calories from fat, 4.1 g total fat, 9.5 mg cholesterol, 357.9 mg sodium, 194.1 mg potassium, 25.9 g carbohydrates, 2.1 g fiber, 4.4 g sugar, 5.4 g protein

Variations:

Creamy Gorgonzola Polenta: Add 2 fl oz of heavy cream and 2 oz of Gorgonzola at the end of the cooking process. Stir well to incorporate.

Creamy Mushroom Polenta: Add 3 oz of sliced, sautéed mushrooms after adding the cornmeal. Stir well to incorporate.

Spicy Quinoa and Cilantro Salad

Yield: *8 servings, 4 oz each*

Ingredients _____

Salad

quinoa, rinsed well	8 oz
vegetable stock	12 fl oz
scallions, bias cut	2 ea
tomato, peeled, seeded, and medium diced	4 oz
red bell pepper, roasted, peeled, seeded, and medium diced	1 ea
pumpkin seeds, toasted	2 tbsp

Dressing

cilantro leaves	1 oz
parsley leaves	1 oz
garlic, minced	1 clove
jalapeño, seeded	1 ea
white vinegar	2 fl oz
olive oil	4 fl oz
water	1 tsp
salt and black pepper	TT

Preparation _____

1. In a saucepan, bring rinsed and well-drained quinoa and vegetable stock to a boil.
2. Cover and lower the heat to a simmer. Cook until quinoa is tender and liquid is absorbed. Let cool.
3. Place all of the dressing ingredients in a blender and blend until smooth.
4. Toss remaining salad ingredients with the cooled quinoa.
5. Add the dressing and toss lightly.

> **NUTRITION FACTS**
> **Per serving:** 273 calories, 148 calories from fat, 16.7 g total fat, <1 mg cholesterol, 348.7 mg sodium, 350.6 mg potassium, 26.1 g carbohydrates, 3.5 g fiber, <1 g sugar, 5.8 g protein

Pasta Dough

Yield: *1 lb*

Ingredients _____

eggs	4 ea
olive oil	2 tbsp
salt	1 tsp
water	1 tbsp
bread flour	½ lb
semolina flour	½ lb

Preparation _____

1. In a mixer with a paddle attachment, combine the eggs, salt, and oil.
2. Add approximately one-third of the flour and mix until the mixture forms a smooth, soft dough.
3. Replace the paddle attachment with the dough hook. Add the rest of the flour and knead the mixture thoroughly.
4. Remove the dough, cover it, and allow it to rest for 20 minutes.
5. Roll the dough to the appropriate thickness and cut into desired shapes.

> **NUTRITION FACTS**
> **Entire recipe:** 2160 calories, 461 calories from fat, 52.2 g total fat, 744 mg cholesterol, 2617.3 mg sodium, 925.6 mg potassium, 331.1 g carbohydrates, 14.3 g fiber, 1.4 g sugar, 81.1 g protein

Variations:

Green (Spinach) Pasta Dough: Blanch 2 oz of fresh spinach per pound of dough in heavily salted water until limp (5 seconds). Refresh immediately in cold water to retain the color. Squeeze out all water from the spinach and purée in a blender. Add the puréed spinach to the pasta dough instead of water and knead well to incorporate.

Red (Tomato) Pasta Dough: Add 3 oz of tomato paste to the dough in place of water and knead well to incorporate.

Black (Squid Ink) Pasta Dough: Add ½ tsp of squid ink to pasta dough and knead well to incorporate.

Macaroni and Cheese

Yield: *24 servings, 6 oz each*

Ingredients

butter	4 oz
all-purpose flour	4 oz
water	2 gal.
salt	2 tsp
elbow macaroni, dry	2 lb
milk	2 qt
Tabasco® sauce	¼ tsp
Worcestershire sauce	½ tsp
dry mustard	½ tsp
nutmeg	pinch
Cheddar cheese, grated	2 lb
nonstick cooking spray	1 tsp
whole butter, melted	2 fl oz
panko bread crumbs	5 oz

Preparation

1. Melt butter in a saucepan.
2. Add flour to make a blond roux, but do not allow to brown. When cooked, allow roux to rest for 10 minutes away from the heat.
3. Bring water and salt to a boil in a saucepot. Add pasta and cook per package directions. When almost fully cooked, strain, refresh, and strain again.
4. Heat milk to a scald in heavy-bottomed saucepan.
5. Add the roux to the scalded milk while whisking continuously.
6. Add Tabasco®, Worcestershire, mustard, and nutmeg and simmer for 3–5 minutes to completely cook the roux.
7. Add Cheddar cheese slowly and whisk continuously to incorporate.
8. Remove from heat and pour sauce over cooked pasta and stir well.
9. Pour macaroni and cheese into a sprayed, 2 inch full hotel pan.
10. In a bowl, mix melted butter with panko.
11. Sprinkle the panko on top of the macaroni and cheese.
12. Bake at 350°F uncovered for approximately 30 minutes until golden and bubbly.
13. Allow macaroni and cheese to rest 10–15 minutes before serving.

NUTRITION FACTS
Per serving: 440 calories, 199 calories from fat, 22.5 g total fat, 63.1 mg cholesterol, 520.7 mg sodium, 251.5 mg potassium, 40.5 g carbohydrates, 1.6 g fiber, 5.7 g sugar, 18.3 g protein

Penne Arrabbiata

Yield: *6 servings, 10 oz each*

Ingredients

Arrabbiata Sauce

olive oil	2 tbsp
butter	1 tbsp
onion, diced small	4 oz
garlic cloves, minced	3 ea
crushed red pepper flakes	2 tsp
white stock	4 fl oz
canned diced and peeled tomatoes	2 lb
basil, chiffonade	1 tbsp
parsley, chopped fine and rinsed	½ tbsp
anchovy fillets, crushed	2 ea
kosher salt	2 tsp
black pepper	½ tsp

Pasta

water	1 gal.
salt	2 tsp
penne pasta, dry	1 lb
Parmesan cheese, grated	2 oz

Preparation

1. Heat oil and butter in a medium saucepot over medium heat.
2. Add onions and sweat without browning until translucent.
3. Add garlic and pepper flakes and cook for 1 minute.
4. Add white stock and reduce by half.
5. Add the rest of the sauce ingredients and simmer for 15–20 minutes. Adjust seasonings if needed.
6. Bring water and salt to a boil in a saucepot. Add penne and cook per package directions until almost al dente.
7. Drain penne.
8. Add cooked penne to the sauce and heat through until sauce is thick enough to coat the pasta.
9. Garnish with grated Parmesan and serve immediately.

NUTRITION FACTS
Per serving: 433 calories, 94 calories from fat, 10.6 g total fat, 22 mg cholesterol, 2158.5 mg sodium, 426.7 mg potassium, 65.2 g carbohydrates, 4.8 g fiber, 4.9 g sugar, 17.9 g protein

Fettuccine Alfredo

Yield: *4 servings, 8 oz each*

Ingredients

water	1 gal.
salt	2 tsp
whole butter	3 oz
garlic cloves, creamed	2 ea
white wine	4 fl oz
heavy cream	16 fl oz
Parmesan cheese	3 oz
fettuccine, dry	1 lb
salt and white pepper	TT
fresh parsley, chopped	2 tsp

Preparation

1. Bring water to a boil and season with salt.
2. Add whole butter to a large saucepan over medium heat. Then add the garlic and sweat for 1–2 minutes without browning.
3. Deglaze the saucepan with white wine and reduce by half.
4. Add heavy cream and Parmesan cheese and reduce by one-third. *Note:* A light béchamel can be substituted for heavy cream.
5. Add pasta to boiling water and cook according to package directions until almost al dente.
6. Drain the pasta.
7. Add drained pasta to the sauce and heat through until the sauce is thick enough to coat the pasta.
8. Adjust seasonings if needed and add fresh parsley. Toss and serve immediately.

NUTRITION FACTS
Per serving: 1102 calories, 609 calories from fat, 69.2 g total fat, 227.5 mg cholesterol, 1646.5 mg sodium, 344.8 mg potassium, 90.2 g carbohydrates, 2.8 g fiber, <1 g sugar, 25.4 g protein

Pasta Carbonara

Yield: *6 servings, 9 oz each*

Ingredients

water	1 gal.
salt	2 tsp
fettuccine, dry	1 lb
bacon, diced small	6 oz
garlic, minced	3 cloves
white wine	4 fl oz
heavy cream	18 fl oz
whole butter	1½ oz
Parmesan cheese, grated	6 oz
salt and pepper	TT
egg yolks	3 ea

Preparation

1. Bring water and salt to a boil. Cook fettuccine until al dente. When done, drain and reserve.
2. Render diced bacon in a large sauté pan until crisp and browned.
3. Add garlic to the pan and sauté briefly without browning.
4. Deglaze the pan with white wine and reduce by half.
5. Add cream and butter and bring to a boil.
6. Add the pasta, Parmesan cheese, salt, and pepper.
7. Add the yolks and mix thoroughly. Remove from the heat.
8. Stir well to thicken the sauce. Serve immediately.

NUTRITION FACTS
Per serving: 960 calories, 550 calories from fat, 62.1 g total fat, 283.7 mg cholesterol, 1976.1 mg sodium, 422.2 mg potassium, 61.8 g carbohydrates, 1.9 g fiber, <1 g sugar, 34.4 g protein

Linguine with Scallops, Capers, and Sun-Dried Tomatoes

Yield: *4 servings, 8 oz each*

Ingredients

linguine pasta	8 oz
salt	2 tsp
water	1 gal.
olive oil	2 tbsp
20–30 count sea scallops	1 lb
kosher salt and pepper	TT
garlic, minced	2 tbsp
capers	2 tbsp
sun-dried tomatoes, oil packed, small diced	4 tbsp
white wine	4 fl oz
fresh lemon juice	2 tbsp
lemon zest	2 tsp
fresh thyme, minced	¾ tsp
fresh basil, chiffonade	2 tbsp
unsalted butter	2 tbsp

Preparation

1. Cook pasta in salted boiling water until al dente. When done, drain completely, but reserve pasta water.
2. Rinse pasta in cold water, drain again, and reserve.
3. Heat half of the olive oil in a large sauté pan until it is almost smoking.
4. Season scallops with kosher salt and pepper and sear on both sides in the hot oil until golden brown. Remove scallops from the pan immediately.
5. Add garlic, capers, and remaining olive oil to the hot pan and cook for 1 minute.
6. Add sun-dried tomatoes and cook for 1 minute more.
7. Add white wine and lemon juice and reduce by half.
8. Add lemon zest, thyme, basil, and butter. Whisk well and season with salt and pepper.
9. Add seared scallops to the sauce and toss to coat well.
10. Add pasta to the sauce and scallops and add a small amount of pasta water to refine the sauce.
11. Toss the pasta and serve immediately.

NUTRITION FACTS
Per serving: 240 calories, 124 calories from fat, 14.1 g total fat, 42.5 mg cholesterol, 1855.6 mg sodium, 410.7 mg potassium, 8.8 g carbohydrates, <1 g fiber, <1 g sugar, 14.7 g protein

German Spaetzle

Yield: *4 servings, 4 oz each*

Ingredients

eggs, beaten	3 ea
whole milk	3 fl oz
water	2 fl oz
salt	½ tsp
white pepper	⅛ tsp
nutmeg	pinch
fresh parsley, chopped fine and rinsed	1 tbsp
all-purpose flour	½ lb
water	1 gal.
salt	2 tsp
butter, whole	2 oz

Preparation

1. In a stand mixer, combine the eggs, milk, water, salt, pepper, nutmeg, and parsley. Mix well.
2. Add flour in small batches. Beat slightly until smooth.
3. Allow the batter to rest for at least 1 hour.
4. Bring a pot of salted water to a boil.
5. Use a spaetzle maker to add the dough to the boiling water.
6. When the spaetzle floats to the surface it is ready to be removed. Use a spider, skimmer, or slotted spoon to remove the spaetzle.
7. Immediately shock the spaetzle in ice water to stop the cooking process.
8. Heat whole butter in a sauté pan.
9. Sauté the cooked spaetzle until it is heated through and the butter begins to turn slightly brown.
10. Season spaetzle and garnish with parsley. Serve immediately.

NUTRITION FACTS
Per serving: 377 calories, 145 calories from fat, 16.4 g total fat, 172.3 mg cholesterol, 1548.6 mg sodium, 161.4 mg potassium, 44.8 g carbohydrates, 1.6 g fiber, 1.5 g sugar, 11.5 g protein

Lasagna Bolognese

Yield: *18 servings, 13 oz each*

Ingredients

ground beef	1 lb
ground Italian sausage	1 lb
tomato sauce recipe	3 qt
ricotta cheese	4 lb
mozzarella cheese, grated	3 lb
Parmesan cheese, grated	8 oz
whole eggs, slightly beaten	6 ea
garlic powder	2 tbsp
salt	1 tbsp
fresh parsley, coarsely chopped	2 tbsp
nonstick cooking spray	1 tsp
lasagna noodles, uncooked	3 lb

Preparation

1. Brown ground beef and sausage in a sauté pan until cooked through. Drain off the fat.
2. Add tomato sauce to ground meats and stir well.
3. In a large bowl, add the ricotta, two-thirds of the mozzarella cheese, half of the Parmesan cheese, the eggs, garlic, salt, and parsley. Mix well.
4. Spray a 2 inch full hotel pan with nonstick cooking spray and place enough meat sauce to coat the bottom of the pan.
5. Place a layer of uncooked lasagna noodles on the bottom of the pan with the edges overlapping.
6. Place a 1 inch thick layer of cheese filling over the noodles and then top with a layer of meat sauce.
7. Place another layer of uncooked noodles on top of the meat sauce and repeat with another cheese layer, another sauce layer, and another layer of pasta.
8. Top the last layer of pasta with meat sauce and sprinkle with the remaining mozzarella cheese and remaining Parmesan cheese.
9. Bake the lasagna in a 350°F oven until it reaches an internal temperature of 165°F (approximately 1 hour and 10 minutes).
10. Remove the lasagna from the oven and let rest for 20 minutes before serving.

NUTRITION FACTS
Per serving: 945 calories, 361 calories from fat, 40.7 g total fat, 200.8 mg cholesterol, 1811.7 mg sodium, 1021.4 mg potassium, 83 g carbohydrates, 4.8 g fiber, 8.8 g sugar, 60.7 g protein

Variations:

Cheese Lasagna: Omit the meat.

Stuffed Manicotti: Use the same cheese filling on flat pasta dough and form small rolls.

Stuffed Shells: Use the same filling and stuff into precooked shells. Top with sauce and bake until the filling is hot.

Chapter 15

Garde Manger Fundamentals

Garde manger is a French term used to describe the kitchen station and the chef responsible for preparing cold pantry items such as salads, cheeses, and charcuterie. The garde manger may also be responsible for preparing garnishes, hors d'oeuvres, and appetizers. In many professional kitchens, sandwiches, cold soups, and carved centerpieces are also prepared in this station. This chapter addresses salad, salad dressing, cheese, hors d'oeuvre, appetizer, and charcuterie fundamentals.

Refer to DVD for **Flash Cards**

Chapter Objectives

1. Identify five types of salad presentations.
2. Identity common varieties of salad greens.
3. Store, trim, and wash salad greens.
4. Identify four types of salad ingredients other than salad greens.
5. Prepare a vinaigrette and a mayonnaise.
6. Describe six types of salads.
7. Identify four factors that determine the flavor and texture of a cheese.
8. Contrast fresh and soft cheeses.
9. Identify three ways semisoft cheeses are ripened.
10. Explain why a blue vein runs through blue-veined cheeses.
11. Contrast hard cheeses and grating cheeses.
12. Identify three types of cheese products.
13. Describe how to store cheese for maximum freshness.
14. Differentiate between hors d'oeuvres and appetizers.
15. Prepare canapés using toasted and untoasted bread.
16. Identify four types of small plates.
17. Identify three types of cold starters.
18. Contrast stuffed and filled starters with wrapped starters.
19. Contrast battered and breaded starters with skewered starters.
20. Contrast raw starters with cured and smoked starters.
21. Explain the role of forcemeats in charcuterie.
22. Prepare charcuterie items.

Key Terms

- **salad green**
- **emulsion**
- **tossed salad**
- **composed salad**
- **bound salad**
- **gelatin salad**
- **fresh cheese**
- **soft cheese**
- **semisoft cheese**
- **dry-rind cheese**
- **washed-rind cheese**
- **waxed-rind cheese**
- **blue-veined cheese**
- **hard cheese**
- **hors d'oeuvre**
- **appetizer**
- **amuse bouche**
- **canapé**
- **crudité**
- **brochette**
- **charcuterie**
- **forcemeat**

SALADS

Salads are popular menu items that offer limitless opportunities when combining ingredients and flavors. Salads may contain raw or partially cooked ingredients and can be served cold or hot. Although many salads contain lettuce, salads do not require lettuce as an ingredient. Salads are usually served with a dressing, which may be sweet, savory, or a combination of both. Salads can be presented as an appetizer, a main course, a separate course, a side salad, or a dessert, depending on the ingredients used. *See Figure 15-1.*

Salad Presentations

Daniel NYC
Appetizer Salad

Courtesy of The National Pork Board
Main Course Salad

Daniel NYC
Separate Course Salad

Side Salad

Frieda's Specialty Produce
Dessert Salad

Figure 15-1. *Salads can be presented as an appetizer, a main course, a separate course, a side salad, or a dessert, depending on the ingredients used.*

- An *appetizer salad* is a salad that is served as a starter to a meal. Appetizer salads are the most basic type of salad and often feature common varieties of lettuce, tomato wedges, cucumber slices, croutons, and a dressing.

- A *main course salad* is a large salad containing an assortment of vegetables and a protein such as poultry, seafood, meats, eggs, or legumes and is served as an entrée. A Caesar salad with grilled chicken is a main course salad.

- A *separate course salad* is a salad served as a course of its own, typically following the main course. The purpose of a separate course salad is to refresh and cleanse the palate prior to dessert. These salads are usually light and are often served with a vinaigrette dressing.

- A *side salad* is a small salad served to accompany a main course. If a main course is hearty or heavy, a side salad is typically light and refreshing. If a main course is light, a heavier side salad such as pasta, grain, or potato salad is commonly served. Sometimes a small amount of salad is served alongside an entrée.

- A *dessert salad* is a sweet salad usually consisting of nuts, fruits, and sweeter vegetables such as carrots. Dessert salads are usually served with yogurt, whipped cream, or a citrus-flavored dressing.

Tanimura & Antle®

SALAD GREENS

A *salad green* is an edible leaf used in raw salads or as a garnish. Salad greens can be varying shades of green as well as red, yellow, white, and rusty brown. Salad greens are low-calorie, low-fat ingredients that are high in vitamin A, vitamin C, and iron. However, when items such as meat, cheese, croutons, and salad dressing are added to salad greens, a salad can quickly become a high-calorie, high-fat dish.

Salad greens include different types of lettuce, spinach, chard, arugula, cabbage, chicory, dandelion greens, mesclun greens, and garnishing greens. Lettuces are used in many salads. Lettuces are categorized as crisphead, butterhead, romaine, or looseleaf lettuces. *Artisan lettuces* are petite-sized lettuces that offer the full flavor, volume, and texture of mature lettuces.

Crisphead Lettuces

A *crisphead lettuce* is a lettuce with a large, round, tightly packed head of pale-green leaves. This lettuce is crisp in texture, very mild in flavor, and wilt-resistant. Iceberg is the most common type of crisphead lettuce. *See Figure 15-2.* Other varieties of crisphead lettuce include Great Lakes, Imperial Valley, vanguard, and western lettuce. Crisphead lettuce remains crisp even after being processed. It only requires a light washing because the compact head shields the leaves from dirt. The outer leaves and the core are removed before use.

Butterhead Lettuces

Butterhead lettuce has pale-green leaves with a sweet, buttery flavor, tender texture, and delicate structure. Before use, the leaves must be thoroughly washed, and blemished leaves or spots must be removed. Butterhead lettuce may be mixed with other greens or served alone. The natural shape of the leaf forms a shallow bowl that can be used as an edible serving dish for portion-size salads or single-portion foods. Butterhead lettuce can also be used as a wrap in place of a tortilla.

Boston lettuce and Bibb lettuce are two varieties of butterhead lettuce that are similar in appearance and flavor, but they have differently sized leaves. *See Figure 15-3.*

Crisphead Lettuce

Tanimura & Antle®

Figure 15-2. *A crisphead lettuce, such as iceberg, is a large, round, tightly packed head of pale-green leaves.*

Tanimura & Antle®

Boston   **Bibb**

Figure 15-3. *Boston lettuce and Bibb lettuce are two varieties of butterhead lettuce that are similar in appearance and flavor, but they have differently sized leaves.*

Boston Lettuce. Boston lettuce, also known as buttercup lettuce, grows in a loosely packed, round head. The outer leaves of Boston lettuce are light green, and the inner leaves are light yellow. Care must be taken when cleaning and cutting Boston lettuce because the leaves are fragile and bruise easily. For this reason, it does not ship well and must be grown close to market.

Bibb Lettuce. Bibb lettuce, also known as limestone lettuce, is similar to Boston lettuce, but the leaves are darker green, crisper, and smaller than those of Boston lettuce. Hydroponically grown Bibb lettuce is typically sold with the roots still attached, which extends shelf life.

Romaine Lettuces

Romaine lettuce has long, green leaves that grow in a loosely packed, elongated head. *See Figure 15-4.* The outer edges of the leaves are darker in color and lighten to a pale celadon near the thick, crisp center rib. Romaine lettuce has a mild, sweet flavor and blends well with other greens. It is high in vitamin A, vitamin C, vitamin K, and folate. Romaine lettuce does not bruise easily when cut. Because of its loose head, dirt collects in the ridges of the leaves during growth, so it must be washed thoroughly before use.

Romaine Lettuces

Romaine   **Hearts**   **Red and Green Petite Gems**

Tanimura & Antle®

Figure 15-4. *Romaine lettuce has long, green leaves that grow in a loosely packed, elongated head.*

Romaine Hearts. Romaine hearts are the very center leaves of a head of romaine lettuce. Harvested as a full head and trimmed down, these lettuce leaves are sweet and crisp.

Petite Gems. Gem lettuce is a petite variety of lettuce that is similar to romaine. The head is dense and compact with a crunchy texture and a mild, sweet flavor. Gem lettuce is often used for lettuce cups.

Looseleaf Lettuces

Looseleaf lettuces have rich-colored, soft leaves and a very mild flavor. These lettuces are a cascade of leaves held loosely together at the root. Some varieties have thick leaves, and others have thin leaves. Some leaves are flat, while others are frilled or curled. Red and green looseleaf lettuces are commonly used in salads and on sandwiches. Red and green Royal Oak lettuces have thicker leaves and are extremely tender. *See Figure 15-5.*

Looseleaf Lettuces

Tanimura & Antle®

Red Leaf **Green Leaf** **Red and Green Royal Oak**

Tanimura & Antle®

Figure 15-5. *Red and green Royal Oak lettuces have thicker leaves and are extremely tender.*

Spinach

Tanimura & Antle®

Figure 15-6. *Spinach is a dark-green, edible leaf with a slightly bitter flavor and may have flat or curly leaves.*

Spinach

Spinach is a dark-green, edible leaf with a slightly sweet flavor and may have flat or curly leaves, depending on the variety. *See Figure 15-6.* Fresh spinach is rich in vitamin A, vitamin C, folate, potassium, iron, and magnesium. In preparation, the long tough stem at the base of each leaf must be removed. Each leaf must be thoroughly washed to remove dirt and grit that can collect in the ridge of the stem. Spinach leaves should be firm and crisp. Yellowed or limp leaves should be discarded. Spinach salad is often topped with hot bacon dressing, diced bacon, sliced mushrooms, and chopped hard-cooked eggs. Spinach may also be steamed or sautéed.

Chard

Figure 15-7. *Chard is a large, dark-green, edible leaf with white or reddish stalks.*

Arugula

Tanimura & Antle®

Figure 15-8. *Arugula has flat, oval leaves with frilled edges.*

Chard

Chard is a large, dark-green, edible leaf with white or reddish stalks. *See Figure 15-7.* It is rich in vitamin A, vitamin C, vitamin K, iron, and potassium. Swiss chard is grown for its silvery stalks and crinkly leaves. It is sometimes called rhubarb chard and has a strong flavor similar to spinach. Ruby chard has deep-red leaves tinged with green and bright-red stalks. The flavor of ruby chard is milder than Swiss chard. Chard can be used in the same manner as spinach.

Arugula

Arugula has flat, oval leaves with frilled edges. *See Figure 15-8.* Arugula has a strong peppery flavor and is seldom served by itself. Larger leaves are not used because they have an overly strong flavor. However, when used in a mixed greens salad or puréed into a cream sauce, arugula can add zesty flavor. Arugula also contains vitamin A, vitamin C, and calcium. It is most often eaten raw in salads or added to stir-fries.

Cabbages

Head cabbage, Napa cabbage, and Savoy cabbage are often used to make salads. *See Figure 15-9.* Coleslaw is a salad made from cabbages.

Cabbages

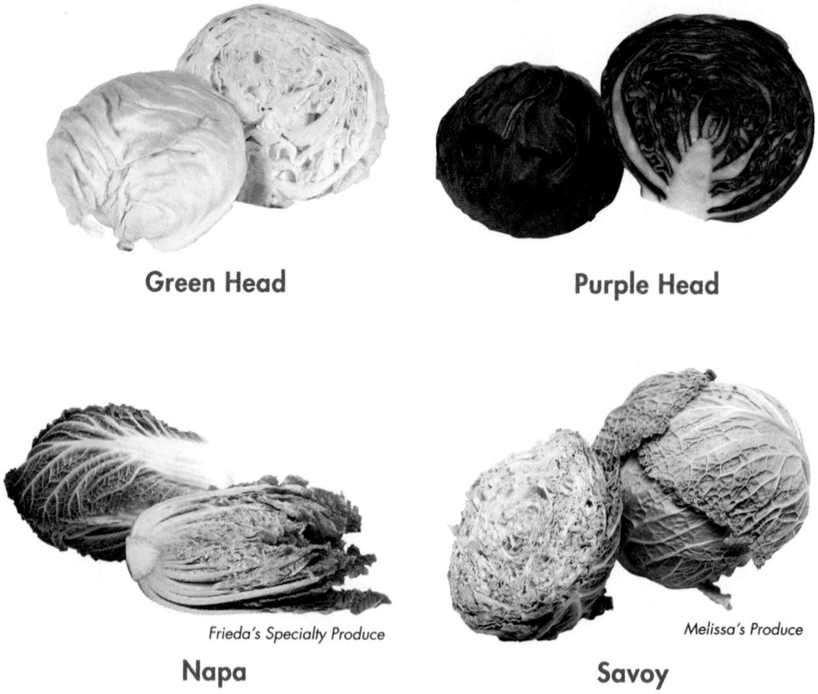

Green Head

Purple Head

Frieda's Specialty Produce

Napa

Melissa's Produce

Savoy

Figure 15-9. *Head cabbage, Napa cabbage, and Savoy cabbage are often used to make salads.*

Head Cabbage. *Head cabbage* is a tightly packed, round head of overlapping edible leaves that can be green, purple, red, or white in color. The inner leaves are usually lighter in color than the outer leaves because they have been exposed to less sunlight. The base of the head where the leaves attach to the stalk is known as the heart. The inedible heart is removed during preparation. The best heads are heavy and compact, with shiny, unblemished leaves. Head cabbage can be eaten raw, steamed, braised, roasted, or stir-fried.

Napa Cabbage. *Napa cabbage,* also known as celery cabbage, is an elongated head of crinkly and overlapping edible leaves that are a pale yellow-green color with a white vein. The leaves are more tender than those of head cabbage and have a very delicate flavor. Napa cabbage is most often used raw in salads or stir-fries.

Savoy Cabbage. *Savoy cabbage* is a conical-shaped head of tender, crinkly, edible leaves that are blue-green on the exterior and pale green on the interior. The leaves have a distinct, sweet flavor. Savoy cabbage can be used raw, stir-fried, or stuffed.

Chicory

Chicory, also known as escarole, is a curly, edible leaf with a slightly bitter-tasting flavor. Raw chicory is often blended with sweeter, milder greens. Chicory comes in various shades including green, yellow, and red. Escarole, frisée, Belgian endive, and radicchio are varieties of chicory. *See Figure 15-10.* Belgian endive and radicchio differ in color and shape from escarole and frisée.

Chicory

Escarole   Frisée   Belgian Endive   Radicchio

Figure 15-10. *Escarole, frisée, Belgian endive, and radicchio are all varieties of chicory.*

Daniel NYC

Belgian Endive. Belgian endive has a slender, tightly packed, elongated head that forms a point. It is the sprouting head of a chicory root and is usually about 4–6 inches in length. The leaves are creamy white in the center and yellow, slightly green, or purple along the edges. Often the leaves are removed one by one to keep their shape intact. Whole leaves can be filled with a spread or dip and served as hors d'oeuvres. A half-head portion set on another contrasting green as a base makes an attractive salad. The head can also be split in half lengthwise before it is cleaned and then braised or grilled.

Frisée. Frisée, also known as curly endive, has twisted, thin leaves that grow in a loose bunch. The leaves vary in color from dark green on the outer leaves to pale green or white in the center and base. Frisée has a bitter flavor and pairs well with strongly flavored cheeses and acidic vinaigrettes. It can also be used as a garnish.

Radicchio. Radicchio is a small, compact head of red leaves, similar to a small head of red cabbage. Radicchio is usually mixed with other salad greens to add flavor and color. The leaves form a bowl shape and can hold individually sized salads. Radicchio is also commonly sautéed or braised and served as a side.

Dandelion Greens

A *dandelion green* is the dark-green, edible leaf of the dandelion plant. *See Figure 15-11.* Dandelion greens can taste quite bitter. Cultivated dandelion greens are milder in taste than those grown in the wild. The greens have a slightly rough, irregular edge. Young dandelion greens can be blended with other salad greens or served alone. They also can be steamed or sautéed.

Dandelion Greens

Frieda's Specialty Produce

Figure 15-11. *A dandelion green is the dark-green, edible leaf of the dandelion plant.*

Mesclun Greens

Mesclun greens are a mix of young greens that range in color, texture, and flavor. *See Figure 15-12.* Mesclun greens can be tender and sweet or crisp and peppery. Mesclun is most often a mix of 10–12 varieties of greens, often including romaine, radicchio, endive, baby spinach, red oak, leaf lettuce, arugula, frisée, and tatsoi. Some mesclun mixes contain as many as 30 different plants, including flowers and herbs. These flavorful salads are often drizzled with olive oil or tossed in a vinaigrette. With its varying shades of green and bits of red and white, mesclun is also an eye-appealing presentation base for appetizers.

Common Mesclun Greens		
Green	**Shape and Color**	**Flavor**
Spinach	Oval-shaped, delicate green leaves	Slightly sweet flavor
Romaine	Green and red leaves	Mild and sweet flavor
Oak leaf	Notched, green and red leaves	Subtle flavor
Leaf	Green, red, or two-tone	Light hazelnut taste
Arugula	Oval leaves with frilled edges	Peppery flavor
Frisée	Tapered green leaves	Hint of bitterness
Radicchio	Red leaves with ivory-white veins	Strong taste and crunch
Tatsoi	Spoon-shaped, dark-green leaves	Cabbage-like flavor

Vita-Mix® Corporation

Figure 15-12. *Mesclun greens are a mix of young greens that range in color, texture, and flavor.*

Garnishing Greens

Greens are often used to garnish sandwiches, stir-fries, and appetizers. Watercress, mâche, microgreens, kale, herb leaves, flowers, sprouts, and seeds are often used to enhance the flavor and appearance of salads.

Watercress. *Watercress* is a small, crisp, dark-green, edible leaf that is a member of the mustard family. Watercress has a pungent, yet slightly peppery flavor. It is typically sold in bouquets. *See Figure 15-13.* The leaves are tender and fragile and should be carefully removed from the stem for use. While watercress can be added to other mixed greens, it is most commonly used as a garnish for salads, sandwiches, stir-fries, and soups.

Mâche. Mâche, also called lamb's lettuce, is a small, leafy green with a velvety texture. *See Figure 15-14.* Mâche can have dark-green, scoop-shaped leaves or elongated, pale-green leaves. It is tender and mild in flavor and pairs well with mildly flavored vinaigrettes.

Watercress

Tanimura & Antle®

Figure 15-13. *Watercress has a pungent, yet slightly peppery flavor. It is typically sold in bouquets.*

Mâche

Figure 15-14. *Mâche, also called lamb's lettuce, is a small, leafy green with a velvety texture.*

Figure 15-15. *Microgreens are the first sprouting leaves of an edible plant.*

Microgreens. *Microgreens* are the first sprouting leaves of an edible plant. *See Figure 15-15.* In addition to being used in salads, microgreens are commonly used to garnish hors d'oeuvres. Common varieties of microgreens include beet greens, turnip greens, spinach, and kale.

Kale. *Kale* is a large, frilly, edible leaf that varies in color from green and white to shades of purple. *See Figure 15-16.* Although all varieties of kale are edible, the green varieties are better for cooking, and the other varieties are used primarily as garnishes.

Kale

Melissa's Produce

Figure 15-16. *Kale is a large, frilly, edible leaf that varies in color from green and white to shades of purple.*

Herb Leaves. The leaves of fresh herbs, such as basil, cilantro, rosemary, chive, dill, mint, oregano, parsley, sage, and savory, can be included in salads. *See Figure 15-17.* It is important not to add too many herb leaves as their flavor can overpower other salad ingredients. Small herb leaves can be added whole, but larger leaves should be torn or cut chiffonade. The stems should be discarded.

Flowers. Flowers and flowering herbs can add color, flavor, and aroma to salads. *See Figure 15-18.* Only edible flowers grown without the use of pesticides should be used. The most common varieties of edible flowers used in salads include roses, nasturtiums, pansies, primroses, and violets. Flowering herbs such as chive blossoms, oregano flowers, and thyme flowers may also be used.

Herb Leaves

— Oregano

— Sage

— Rosemary

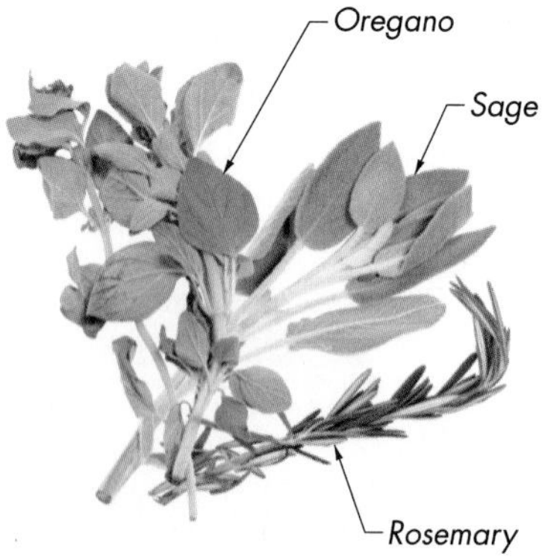

Figure 15-17. *The leaves of fresh herbs, such as oregano, sage, and rosemary, can be included in salads.*

Flowers

Figure 15-18. *Flowers and flowering herbs can add color, flavor, and aroma to salads.*

Sprouts. A *sprout* is an edible strand with an attached bud that comes from a germinated bean or seed. *See Figure 15-19.* Varieties of sprouts include mung bean, soybean, alfalfa, and radish sprouts. Depending on the plant of origin, sprouts range in taste from mild to spicy. Common types of sprouts include chickpea, daikon, green lentil, and mustard. Sprouts are most often used to garnish salads and sandwiches. They also may be sautéed or stir-fried for less than 30 seconds.

Seeds. A *seed* is the fruit or unripened ovule of a nonwoody plant. Ground flax seeds, roasted pumpkin seeds, and roasted sunflower seeds are often added to salads. *See Figure 15-20.* A *flax seed* is a dark-amber colored seed from the flowering Mediterranean flax plant. Ground flax seeds add a mild nutty flavor to salads. A *pumpkin seed,* also known as a pepita, is a green seed that comes from a pumpkin plant. A *sunflower seed* is a tan seed that comes from the sunflower plant.

Sprouts

Bean   Alfalfa

Figure 15-19. *A sprout is an edible strand with an attached bud that comes from a germinated bean or seed.*

Seeds

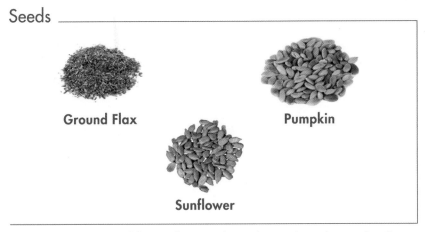

Ground Flax   Pumpkin

Sunflower

Figure 15-20. *Ground flax seeds, roasted pumpkin seeds, and roasted sunflower seeds are often added to salads.*

SALAD GREEN PREPARATION

When guests order a salad, they expect it to be fresh and full of texture. If salad greens are handled incorrectly, they can bruise, brown, and lose crispness. To retain the integrity of salad greens, they must be stored, prepared, and washed with care.

Storing Salad Greens

All salad greens should be stored between 35°F and 38°F, which is colder than the storage temperature of other produce. If any part of a green is frozen, it should be discarded. Delicate greens, such as butterhead lettuces, deteriorate faster than firmer varieties such as romaine.

Salad greens must be stored away from tomatoes, apples, pears, and other fruits that emit ethylene gas. Storing salad greens near these items will cause the greens to wilt and deteriorate rapidly. Washed salad greens should be stored in a perforated stainless steel pan with a second solid pan as an under-liner to hold the water drippings. The salad greens should be covered with a damp paper towel or plastic wrap to retain crispness.

Trimming Salad Greens

Solidly packed head lettuces, such as iceberg, must have the core removed and be cut to an appropriate size before washing the leaves. *See Figure 15-21.* If the leaves are not cut to an appropriate size, it is nearly impossible for water to reach the center of the tightly packed heads to effectively remove dirt, pests, and chemicals.

Procedure for Removing the Core from Head Lettuce

1. Hold the head of lettuce in the palm of one hand with the core facing down.
2. Carefully hit the core against a clean work surface to free the core from the rest of the head.
3. Grab the core and pull it to remove it from the head.
4. Cut the lettuce and then wash, dry, and store it.

Figure 15-21. *Solidly packed head lettuces, such as iceberg, must have the core removed and be cut to an appropriate size before washing the leaves.*

Chef's Tip

Hands and utensils must be sanitized when preparing items that are not cooked or heated to a safe temperature prior to serving. Salads that contain mayonnaise, poultry, or meat are potentially hazardous foods and extra care must be taken to prepare and serve them safely.

Romaine lettuce needs to be cut or trimmed prior to being washed to make sure that all dirt, pests, and chemicals have been washed away. *See Figure 15-22.* The flavorful ribs are not removed.

Procedure for Preparing Romaine Lettuce

1. Trim and remove any damaged, discolored, or bruised outer leaves.

2. Use a chef's knife to make 2–3 lengthwise cuts, leaving the base intact.

3. Cut the leaves perpendicular to the ribs at 1–1½ inch intervals to produce bite-size pieces.

4. Wash, dry, and store leaves.

Figure 15-22. *Romaine lettuce needs to be cut prior to being washed to make sure that all dirt, pests, and chemicals have been washed away.*

Loose greens, such as chicory and spinach, often collect dirt in the center rib. The ribs of loose greens must be removed prior to preparation. *See Figure 15-23.*

Procedure for Removing Ribs from Loose Greens

1. Fold each leaf in half so that the two sides of the top surface meet, exposing the rib.

2. With the opposite hand, pinch the rib and carefully pull it apart from the leaf. Discard the rib.

3. Wash, dry, and store leaves.

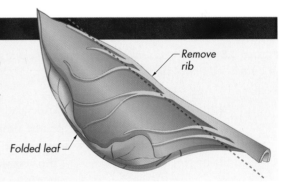

Figure 15-23. *The ribs of loose greens must be removed prior to preparation.*

Washing Salad Greens

Salad greens grow close to the ground and are usually dirty. Even if salad greens look clean, insects, dust, dirt, pesticides, and fertilizers may be hidden between the leaves. It is essential to properly wash salad greens before use. If prepared incorrectly, greens can bruise, brown, and lose crispness.

To wash salad greens, they should first be cut or torn to the desired size and then completely submerged in a sanitized sink of clean, cold water. *See Figure 15-24.* The greens are gently stirred to rinse away dirt and then removed from the dirty water. The sink is rinsed out completely and refilled with clean, cold water. The rinsing process is repeated until the water in the sink remains clean. If the leaves are to be kept whole, each leaf should be washed. This task must be done gently because tender greens can bruise easily. It may also be necessary to cut elongated heads in half lengthwise in order to remove all of the dirt, grit, and sand.

Procedure for Washing Salad Greens

1. Cut or tear greens to the appropriate size.

2. Submerge greens completely in a sink of cold water and gently stir greens to rinse.

3. Remove greens from the dirty water and drain the sink. Rinse the sink and refill it with cold water.

4. Repeat the process until there is no dirt on the bottom of the sink.

5. Remove greens from the water and spin dry.

6. Store greens in a stainless steel pan with a perforated pan insert and cover with damp paper towels or plastic wrap until needed.

Figure 15-24. *To wash salad greens, they should first be cut or torn to the desired size and then completely submerged in a sanitized sink of clean, cold water.*

All salad greens must be dried completely after washing. Wet greens become limp in a short amount of time. Also, oil-based vinaigrettes or mayonnaise-based dressings do not stick to wet leaves. Using a salad spinner is the best way to dry lettuce, as the spinning of the internal basket throws water from the leaves without damaging them.

SALAD INGREDIENTS

A variety of ingredients can be used to create endless combinations of salads that delight the palate. It is important to remember that a salad should complement other courses. The most common ingredients used in salads are vegetables, starches, fruits, and proteins.

Vegetables

A *vegetable* is an edible root, bulb, tuber, stem, leaf, flower, or seed of a non-woody plant. Vegetables, fruit-vegetables, sea vegetables, and mushrooms make nice additions to salad greens. *See Figure 15-25.* Vegetables add color and flavor to salad greens and are also low in fat and calories.

Vegetable Ingredients	
Type	**Ingredients**
Roots	Carrots, parsnips, salsify, radishes, turnips, rutabagas, beets, celeriac, jicamas, lotus roots, and bamboo shoots
Bulbs	Garlic, shallots, onions, scallions, and leeks
Tubers	Potatoes, sweet potatoes, yams, ocas, sunchokes, and water chestnuts
Stems	Asparagus, celery, fennel, rhubarb, kohlrabi, and hearts of palm
Leaves	Cabbages, bok choy, Brussels sprouts, lettuces, chicory, watercress, spinach, sorrel, chard, kale, collards, mustard greens, turnip greens, beet greens, dandelion greens, nopales, tatsoi, and fiddlehead ferns
Flowers	Squash blossoms, broccoli, cauliflower, and artichokes
Seeds	Legumes (beans, peas, and lentils) and sprouts
Fruit-vegetables	Tomatoes, cucumbers, eggplants, sweet peppers, hot peppers, okra, sweet corn, summer squashes, winter squashes, and pumpkins
Sea vegetables	Nori, kombu, arame, wakame, and dulse
Mushrooms*	Button, portobello, enokitake, wood ear, shiitake, oyster, chanterelle, morel, and porcini

* Mushrooms are fungi, not vegetables, but can be added to salads in the same manner as vegetables.

Figure 15-25. *Vegetables, fruit-vegetables, sea vegetables, and mushrooms make nice additions to salad greens.*

Vegetables used in salads can be raw or cooked. Hard vegetables, such as cauliflower, should be blanched prior to use in a salad. Soft vegetables, such as tomatoes, can be added raw. An entire salad can be made from a single type of vegetable, such as coleslaw made from cabbage or pickled beet salad made from blanched or roasted beets. Examples of cooked vegetable salads include potato salad and roasted pepper and artichoke salad. A cucumber and onion salad is a raw vegetable salad.

Starches

Starches such as grains, pastas, and breads can add variety, texture, and nutritional value to many salads. Wheat berries, barley, and quinoa are used in salads. Grains used in a salad should be cooked slightly less than al dente so they retain their shape and are firm. Overcooked grains become mushy as they absorb dressing.

Pasta salads may be traditional, such as rotini pasta tossed with a vinaigrette, or contemporary, such as couscous salad with pomegranate, mint, and curry. Pasta used in salads should be slightly undercooked so that it does not become soggy or break apart when tossed. *See Figure 15-26.*

Starch Ingredients

Figure 15-26. *Pasta used in salads should be slightly undercooked so that it does not become soggy or break apart when tossed.*

Breads, such as toasted croutons, are often used as a salad garnish. Breads can also be an ingredient in a tossed salad. In panzanella, also known as bread salad, crusty bread is toasted or grilled and then torn into bite-sized pieces and tossed with tomatoes, basil, other ingredients, and a red wine vinaigrette.

Fruits

When used in a salad, fruits can be fresh, dried, or roasted. Fruits add texture, flavor, and aroma to a salad. *See Figure 15-27.* Fruits can impart acidic qualities, sweetness, chewiness, or crispness. For example, a microgreen salad could be topped with dried cherries, fresh pears, and a balsamic vinaigrette.

Proteins

Common proteins used in salads include cooked poultry, seafood, meats, legumes, nuts, and cheese. Any protein can be used as the main ingredient of a salad or to make a salad more hearty and flavorful. *See Figure 15-28.*

Fruit Ingredients

Figure 15-27. *Fruits add texture, flavor, and aroma to a salad.*

Protein Ingredients

Figure 15-28. *Any protein can be used as the main ingredient of a salad or to make a salad more hearty and flavorful.*

Legumes are used in salads because they are rich in protein and fiber and contain little or no fat. A *legume* is the edible seed of a nonwoody plant and grows in multiples within a pod. Beans, peas, pulses, and lentils are legumes.

Likewise, nuts enhance salads by adding a crunchy texture and protein and provide a sweet, salty, or spicy flavor. For example, maple-glazed walnuts and blue cheese add protein to a salad.

SALAD DRESSINGS

A *salad dressing* is a sauce for a salad. A low-quality dressing can ruin a well-prepared salad. This makes it essential to prepare salad dressings using the finest ingredients in the correct proportions. Most dressings are prepared by making an emulsion that forms a vinaigrette or mayonnaise base. An *emulsion* is a combination of two unlike liquids that have been forced to bond with each other. *See Figure 15-29.* The result is a creamy, smooth product with a uniform appearance. Emulsions can be either temporary or permanent.

Rapidly whisking oil and vinegar together creates a temporary emulsion. As the oil is broken into tiny droplets, the droplets are surrounded with a coating of vinegar. If the mixture is allowed to rest, it will separate into a pool of oil floating on a pool of vinegar. A temporary emulsion that separates can be emulsified again by shaking or stirring it thoroughly.

Emulsions

Oil

Vinegar and egg
yolk mixture

Figure 15-29. *An emulsion is a combination of two unlike liquids that have been forced to bond with each other.*

Mayonnaise is an example of a permanent emulsion. A permanent emulsion is formed when an emulsifier, such as egg yolk, is added to stabilize the combined liquids.

Oils and Vinegars

Regardless of the principal ingredients, flavorings, or seasonings, it is important to match the oil to the appropriate vinegar. Some oils, such as olive oil, have a stronger flavor and are better paired with more intense vinegars, such as balsamic vinegar. Other oils, such as canola or safflower, are lighter or more neutral in flavor and are better paired with a more delicate vinegar, such as an herbal or fruit vinegar. Nut oils work best with sherry or champagne vinegar because the crispness of the vinegar brings out the earthiness of the nut oil.

Vinaigrettes

A basic vinaigrette is commonly referred to as "basic French" in the food-service industry and should not be confused with orange-red French or Russian dressing. A *basic French vinaigrette* is a temporary emulsion of oil and vinegar that may also include additional flavorings and seasonings. *See Figure 15-30.* The acid in a vinaigrette can liven up the overall flavor of a dish. In addition to salads, vinaigrettes are often served over broiled or grilled vegetables, poultry, and seafood.

> **Chef's Tip**
>
> When preparing an emulsion, ingredients should sit at room temperature for 1 hour prior to preparation. Ingredients at room temperature are much easier to emulsify than cold ingredients.

Procedure for Preparing a Basic French Vinaigrette

1. Select a vinegar (or another acid) and an oil that pair well and that complement the item they are intended to dress.
2. Place seasonings, flavoring ingredients, and vinegar in a mixing bowl and whisk to incorporate.
3. Slowly add the oil in a fine stream while whisking continuously to form a temporary emulsion.
4. Let the vinaigrette rest 1–2 hours to marry the flavors.
5. Whisk as needed to incorporate oil and vinegar prior to use.

Figure 15-30. *A basic French vinaigrette is a temporary emulsion of oil and vinegar that may also include additional flavorings and seasonings.*

In a basic French vinaigrette, the ratio of oil to vinegar is 3:1, or 3 parts oil to 1 part vinegar. When using intensely flavored vinegar, more than 3 parts oil is recommended to balance the flavors. If a mildly flavored vinegar is used, less than 3 parts oil is needed. When preparing citrus vinaigrettes where half or more of the vinegar is replaced with a citrus juice, less than 3 parts oil is needed.

An emulsified vinaigrette is more stable than a basic French vinaigrette. It takes more time to prepare an emulsified vinaigrette and the ingredients must be at room temperature before starting the preparation. *See Figure 15-31.*

1. Collect all ingredients and allow them to sit at room temperature for 1 hour prior to preparation.

2. Place pasteurized yolks or pasteurized whole eggs in a mixing bowl and whisk until foamy and light.

3. Add all dry ingredients and any flavoring ingredients to the eggs and whisk to incorporate.

4. Combine any liquid ingredients, except the oil. Add one-fourth of the liquid mixture and whisk well to incorporate.

5. While whisking rapidly, carefully begin adding the oil very slowly in a fine stream until a smooth emulsion forms.

6. Continue to add the oil very slowly in a fine stream until the mixture begins to thicken.

7. Add a little more of the liquid mixture, then add a little more oil while whisking rapidly to keep the oil and vinegar from separating. Repeat until all ingredients are incorporated.

8. Check the consistency and season to taste. *Note:* If the vinaigrette is too thick, a little water can be whisked in to thin it.

Figure 15-31. *The ingredients used to prepare an emulsified vinaigrette must be at room temperature before starting the preparation.*

Flavors, seasonings, and other ingredients can be added to basic French vinaigrette or emulsified vinaigrette to produce flavored dressings. Minced garlic, minced shallots, Dijon mustard, and sugar are common additions to vinaigrettes. Garlic and shallots are used to add a savory flavor, mustard adds a touch of zest, and sugar mellows the sharpness of vinegar. Ingredients such as mustard and paprika help stabilize the emulsion. Fresh herbs, spices, or acidic fruit juices can be added to further intensify flavor.

Creamy Dressings

Mayonnaise is the foundation for many popular dressings, such as ranch, thousand island, blue cheese, and green goddess. Mayonnaise is a permanent emulsion formed by dripping oil into a small amount of vinegar and egg yolks while whisking rapidly and continuously. *See Figure 15-32.* The oil naturally repels the water in the vinegar, but the egg yolks act as an emulsifier, preventing the separation of the two opposing liquids.

Chef's Tip
Fresh herbs can intensify the flavor of a dressing quickly. Dried herbs should be warmed slightly to reactivate their essential oils before being used in a dressing.

1. Allow all ingredients to sit at room temperature for 1 hour prior to preparation.

2. In a nonreactive mixing bowl, whisk the seasonings and half of the vinegar into the yolks until the mixture is lightly colored and foamy.

3. Add oil very slowly (one drop at a time), until the emulsion begins to form and the mixture thickens slightly.

4. Once one-fourth of the oil has been added, add the oil in a fine stream while whisking rapidly to incorporate.

5. As the mixture thickens and becomes harder to whisk, add a small amount of vinegar to thin it slightly.

6. Continue alternating the oil and vinegar while whisking constantly until all the oil and vinegar have been added.

7. Transfer the mayonnaise into a chilled storage container and refrigerate until needed.

Figure 15-32. *Mayonnaise is a permanent emulsion formed by dripping oil into a small amount of vinegar and egg yolks while whisking rapidly and continuously.*

Each yolk can emulsify up to 7 fl oz of oil. If more oil is added, the excess oil separates from the vinegar, breaking the emulsion. The emulsion can also break if the mixture is not whisked rapidly and continuously or if the oil is not added slowly enough to the vinegar.

The quality of a mayonnaise depends on the quality of oil used. It is usually best to prepare mayonnaise with oil that is fairly neutral in flavor, such as corn, canola, or cottonseed oil. Olive oil is not recommended because it has a stronger flavor than many other oils and might limit the use of the mayonnaise as a base for preparing other dressings. If olive oil is used, it is best to blend it with another oil to reduce the strong flavor.

SALAD TYPES

The term "leafy green salad" can refer to any salad made with a base of greens. Leafy green salads can be as basic as iceberg lettuce, a wedge of tomato, and grated carrots topped with a choice of salad dressing. Leafy green salads can also be as complex as mixed baby field greens with diced pears, Gorgonzola cheese, glazed pecans, and a balsamic vinaigrette.

All salads fall into one of six types based on how the salad is to be plated or served. The six types of salads are tossed, composed, bound, vegetable, fruit, and gelatin.

Tossed Salads

A *tossed salad* is a mixture of leafy greens, such as lettuce, spinach, chicory, or fresh herbs, and other ingredients, such as fruits, vegetables, nuts, cheese, meats, and croutons, served with a dressing. There are endless combinations of ingredients that can be used to construct tossed salads, but care should be taken to choose complementary flavors and textures. The dressing should not overpower or conflict with the ingredients. To prepare a tossed salad, five simple steps can be followed. *See Figure 15-33.*

Vita-Mix® Corporation

Procedure for Preparing Tossed Salads

1. Select a variety of leafy greens so the salad will be colorful and have a variety of textures and flavors.

2. Cut, wash, dry, and store the greens until ready for use.

3. Prepare and store any additional ingredients and garnishes that will be added to the salad.

4. Prepare the appropriate salad dressing. *Note:* Hearty greens pair well with any dressing, while delicate greens require lighter dressings.

5. Carefully combine the greens, additional ingredients, garnishes, and dressing in a mixing bowl and toss gently to distribute all components evenly.

Figure 15-33. *To prepare a tossed salad, five simple steps can be followed.*

Composed Salads

A *composed salad* is a salad that consists of a base, body, garnish, and dressing attractively arranged on a plate. *See Figure 15-34.*

Daniel NYC

The Base. The base of a composed salad serves as the foundation on which a salad is built. A base typically is a bed of salad greens, fruits, or vegetables. The base is not only an edible component of the salad, but a colorful liner between the rest of the salad and the plate. For example, a base may be a cup-shaped leaf in which the rest of the ingredients can be placed.

The Body. The body of a composed salad consists of the main ingredients. Tossed salad greens, a fruit medley, vegetables, cooked grains, cooked pastas, or precooked proteins such as chicken salad can be used as the body of a composed salad.

The Garnish. The garnish for a composed salad should add color, flavor, and texture to a dish. It may be as simple as a tomato wedge or a fanned strawberry or as substantial as a sliced hard-cooked egg.

The Dressing. A dressing should bring all the flavors, textures, and components of a salad together. It should complement the ingredients and not mask the other flavors in the salad. Dressing can be applied to a composed salad by tossing the salad in the dressing, by ladling it on top of the salad, or by being sprayed on the salad. In some banquet operations, salads are composed on sheet pans and then sprayed with a light dressing before being lifted onto individual plates.

Composed Salads

Charlie Trotter's

Figure 15-34. *A composed salad is a salad that consists of a base, body, garnish, and dressing attractively arranged on a plate.*

Bound Salads

A *bound salad* is a salad made by combining a main ingredient, often a protein, with a binding agent such as mayonnaise or yogurt and other flavoring ingredients. *See Figure 15-35.* Popular bound salads include egg salad, chicken salad, tuna salad, potato salad, and kidney bean salad. Many ingredients can be used for the main item in a bound salad including hard-cooked eggs, cooked poultry, seafood, meats, fruits, vegetables, legumes, potatoes, grains, or pastas. Minced chives, onions, and celery are common flavoring ingredients, and mayonnaise is often used as a binding agent. Other binding agents include vinaigrettes, yogurt, and creamy dressings.

Bound Salads

American Egg Board

Figure 15-35. *A bound salad is a salad made by combining a main ingredient, often a protein, with a binding agent such as mayonnaise or yogurt and other flavoring ingredients.*

Some bound salads are served as a side dish, while others are primarily served as a filling. For example, potato salad is often served as side dish. In contrast, chicken salad is often served as a filling but is not typically served as a side dish.

Tips for Preparing Bound Salads

- Chilled ingredients should be used so that a salad is not in the temperature danger zone during or after assembly.
- Cooked items should be prepared in advance and chilled completely before use (unless the item is served hot, such as hot German potato salad).
- Ingredients should be cut into consistent sizes for a pleasing appearance.
- Ingredients should be evenly combined with dressing to bind the salad together and to distribute flavors throughout.
- Ingredients should be mixed as close to the time of service as possible to ensure safe food handling and freshness.

Courtesy of The National Pork Board

Vegetable Salads

A *vegetable salad* is a salad that is primarily made of vegetables. ***See Figure 15-36.*** Raw or precooked vegetables can be used. Raw vegetables must be washed and trimmed to the desired size. Examples of vegetable salads include coleslaw, cucumber and onion, carrot and raisin, green bean salad, and roasted beet and goat cheese salad.

Vegetable Salads _____

Artichoke and Frisée   **Beet and Radicchio**

Daniel NYC

Figure 15-36. *A vegetable salad is a salad that is primarily made of vegetables.*

When making vegetable salads, knowing how the vegetables react to acidity is important. For example, asparagus and broccoli can turn yellow and unappealing if left in an acidic solution, such as a vinaigrette, too long. For this reason, when making a mushroom, zucchini, red pepper, and asparagus salad tossed with a vinaigrette, it is best to leave the asparagus out of the salad until immediately prior to service.

Fruit Salads

A *fruit salad* is a salad that is primarily made of fruits. Fruit salads can be prepared from fresh, frozen, or canned fruits. ***See Figure 15-37.*** Fresh fruits are superior to frozen and canned fruits in taste and texture. However, professional kitchens often rely on the convenience of frozen or canned products in order to meet production demand and reduce labor costs.

Some fruit salads can be mixed or tossed, but most require the fruit to be arranged in an attractive manner, like a composed salad. Fruit salads are fragile, can discolor rapidly, and become soft when cut fruit is exposed to air too long. They should be prepared as close to serving time as possible and served chilled.

Some fresh fruits, such as apples, avocados, bananas, and pears, discolor when cut and exposed to air. To prevent rapid discoloration, these fruits should be dipped or lightly tossed in liquids that contain citric acid, such as lemon or lime juice. Using a nonreactive stainless steel knife also helps prevent discoloration.

Fruit Salads

Figure 15-37. *Fruit salads can be prepared from fresh, frozen, or canned fruits.*

In place of a dressing, fruit salads can be tossed with a small amount of fruit juice or fruit purée, a chiffonade of fresh mint, or a splash of a sweet liqueur. If a dressing is used with a fruit salad, it is usually a lighter dressing made with yogurt, honey, or whipped cream mixed into a fruit purée.

Gelatin Salads

A *gelatin salad* is a salad made from flavored gelatin. Gelatin salads can be presented in many different forms, colors, and flavors, and are easy to prepare. *See Figure 15-38.*

In any gelatin preparation, the correct ratio of gelatin powder to liquid must be used. Fruit-flavored gelatin powder packaged for commercial use typically contains 1 lb 8 oz of gelatin powder. A total of 1 gal. of water or fruit juice should be added. Half of the water or juice should be boiling liquid, which is used to dissolve the gelatin. The other half should be cold liquid, which is used to cool and set the mixture.

Gelatin Salads

Figure 15-38. *Gelatin salads can be presented in many different forms, colors, and flavors.*

- Gelatin should be mixed in a stainless steel container.
- All of the gelatin must be thoroughly dissolved in the hot liquid before the cold liquid is added.
- Gelatin sets more quickly in the coldest part of the refrigerator. Adding ice, frozen fruit, or frozen juice to the cold liquid can accelerate the action.
- Never place gelatin in the freezer. Freezing will cause gelatin to crystallize and the gelatin will become liquid upon defrosting.

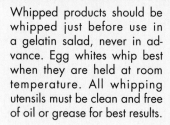

Chef's Tip

Whipped products should be whipped just before use in a gelatin salad, never in advance. Egg whites whip best when they are held at room temperature. All whipping utensils must be clean and free of oil or grease for best results.

Many gelatin recipes call for whipped cream or beaten pasteurized egg whites to be folded in. This procedure is used to give a spongy texture to the gelatin. The cream or egg whites should be whipped until soft peaks form and are stiff enough to hold their shape. If whipped cream, beaten egg whites, mayonnaise, sour cream, grated cheese, fruits, or vegetables are to be added to the gelatin, the gelatin mixture must be chilled until slightly thickened before the additions are made. This ensures even distribution of the added ingredients.

When layering two or more gelatin mixtures in the same pan or mold, each layer poured into the pan or mold must be chilled until slightly firm before adding the next layer. If one layer is set too firm, the layer placed on top may slip off or separate from the completed mold when the entire salad is unmolded. Before adding gelatin to a layer that has already set, be sure the gelatin being added is completely cool. If the gelatin being added is too warm, it can melt the set layer and the two mixtures will run together.

When preparing a gelatin mold for display on a buffet, the selected mold is filled with water, and then the water is poured into a liquid measure to determine the volume of the mold. If the amount of water held by the mold is equal to or less than the recipe yield, the mold will hold the amount of gelatin stated on the recipe.

CHEESES

Most cheeses used in the professional kitchen are made from the milk of cows, sheep, or goats. Cheese is most often made from milk that has been coagulated or curdled using rennet. *Rennet* is a substance that contains acid-producing enzymes or an acid. After milk curdles, it is separated into the cheese curds and whey. The *curd* is the thick, casein-rich part of coagulated milk. The *whey* is the watery part of milk. The flavor and texture of each cheese differs depending on what type of animal produced the milk, the diet of the animal, the percentage of butterfat in the cheese, and how long the cheese was aged. Cheeses are generally classified as fresh, soft, semisoft, blue-veined, hard, or grating cheese. Cheese products are processed foods made of natural cheeses that also include emulsifiers.

Fresh Cheeses

A *fresh cheese* is a cheese that is not aged or allowed to ripen. Fresh cheeses spoil easily. Mozzarella, cottage cheese, ricotta, cream cheese, Neufchâtel, mascarpone, feta, chèvre, and baker's cheese are fresh cheeses. These cheeses should be used soon after they are purchased.

Mozzarella. *Mozzarella* is a very tender fresh cheese with a soft, elastic-like curd. *See Figure 15-39.* Mozzarella is primarily made from cow milk. In the process of making mozzarella, the whey is drained from the curd and reserved for making ricotta cheese. When mozzarella is melted it has an elastic or rubbery consistency. It is commonly used in pizza and lasagna. Mozzarella di bufala is a fresh cheese made from the milk of water buffalo or a combination of cow and water buffalo milk. It is prized for its rich, slightly sour flavor.

Cabot Creamery Cooperative

Mozzarella

Wisconsin Milk Marketing Board, Inc.

Figure 15-39. *Mozzarella is a very tender fresh cheese with a soft, elastic-like curd.*

Cottage Cheese. *Cottage cheese* is a pebble-shaped fresh cheese with a mildly sour taste. *See Figure 15-40.* It is marketed in different varieties including small curd, large curd, flake curd, home-style, and whipped. Cottage cheese is often called Dutch cheese, pot cheese, smearcase, and popcorn cheese because of its curds. Cottage cheese is highly perishable and should always be refrigerated.

Ricotta. *Ricotta* is a creamy fresh cheese that looks similar to cottage cheese but is made from the whey of other cheeses instead of primarily from milk. *See Figure 15-41.* It has a bland, yet slightly sweet flavor. Ricotta made in Europe is made from the whey of other cheeses, while ricotta made in North America is made using a mixture of whey and whole or skim milk. Ricotta blends well with the flavors and textures of other foods. It is an important ingredient in pasta dishes such as lasagna and manicotti.

Cottage Cheese

Wisconsin Milk Marketing Board, Inc.

Figure 15-40. *Cottage cheese is a pebble-shaped fresh cheese with a mildly sour taste.*

Ricotta

Figure 15-41. *Ricotta is a creamy fresh cheese that looks similar to cottage cheese but is made from the whey of other cheeses instead of primarily from milk.*

Cream Cheese. *Cream cheese* is a soft, fresh cheese with a rich, mild flavor and smooth consistency. ***See Figure 15-42.*** It is made from cream or a mixture of cream and milk. Gum arabic is often used as a stabilizer to extend the product life of cream cheese. Cream cheese that does not contain gum arabic has a lighter, more natural texture but does not keep as well. Cream cheese is used in spreads, salad dressings, sandwiches, and cheesecakes.

Cream Cheese

Figure 15-42. *Cream cheese is a soft, fresh cheese with a rich, mild flavor and smooth consistency.*

Neufchâtel. *Neufchâtel* is a soft, fresh cheese made from whole or skim milk or a mixture of milk and cream. *See Figure 15-43.* Neufchâtel is similar to cream cheese, but has more moisture and less fat. When young, Neufchâtel has a mild, slightly salty flavor that becomes more pungent as the cheese ripens. Because of the smooth texture of this cheese, it is used in spreads, salad dressings, and desserts. Neufchâtel that is made in the United States has a smoother texture than the French product.

Mascarpone. *Mascarpone* is a cream cheese of Italian origin that has a smooth texture, is white or pale yellow in color, and has a buttery, somewhat sweet flavor. *See Figure 15-44.* It can be used in savory dishes but is most often seen in desserts, such as tiramisu. Mascarpone works well as a spread or can be served on its own with fruit, a liqueur, or a dusting of cocoa.

Feta. *Feta* is a fresh cheese of Greek origin made from sheep milk or goat milk. *See Figure 15-45.* Feta is slightly cured for a period that can range from a few days to four weeks. It has a salty taste and, when aged for a long period, becomes very salty and dry. Because of this, it is wise to taste feta before making a purchase. When aged properly, feta has a creamy texture, pleasant saltiness, and a soft to semisoft consistency. The smell is similar to cider vinegar and the taste has a faint trace of olives. Feta can be used in snacks, salads, sandwiches, and cooked dishes such as lasagna and omelets.

Neufchâtel

Figure 15-43. *Neufchâtel is a soft, fresh cheese made from whole or skim milk or a mixture of milk and cream.*

Mascarpone

Figure 15-44. *Mascarpone is white or pale yellow in color and has a buttery, somewhat sweet flavor.*

Feta

Figure 15-45. *Feta is a fresh cheese of Greek origin made from sheep milk or goat milk.*

Chèvre

Figure 15-46. *Chèvre is a fresh cheese made from goat milk.*

Chèvre. *Chèvre* is a fresh cheese made from goat milk. ***See Figure 15-46.*** While chèvre is available in textures ranging from soft to firm, the soft, fresh varieties are most popular. Chèvre has a pure white color and a soft, spreadable texture that is slightly dry. It has a mild, slightly peppery flavor and is often blended with herbs or spices. It can be used in cooking and is also used as a spread. Montrachet is among the most popular varieties.

Baker's Cheese. *Baker's cheese* is a fresh cheese made from skim milk that is like cottage cheese but softer and finer grained. ***See Figure 15-47.*** Baker's cheese can be kept moist by placing it in a container, smoothing the surface until even, and covering it with a thin layer of fine sugar. Baker's cheese is used in making cheesecakes, pies, and some pastries.

Baker's Cheese

Figure 15-47. *Baker's cheese is a fresh cheese made from skim milk that is like cottage cheese but softer and finer grained.*

Soft Cheeses

A *soft cheese,* also known as a rind-ripened cheese, is a cheese that has been sprayed with a harmless live mold to produce a thin skin or rind. A *rind* is an exterior layer of a food. The mold ripens the soft cheese by reacting with the rind. The result is a soft, suede-like outer coating and a soft interior. A soft cheese will become extremely soft and somewhat runny once it is fully ripened. Brie and Camembert are soft cheeses. ***See Figure 15-48.***

Brie. *Brie* is a creamy, white soft cheese with a strong odor and a sharp taste. It originated in France but is now made in many countries. Brie is usually produced in small disks and is best known as a buffet or dessert cheese. It is also used in hors d'oeuvres, salads, sandwiches, and hot entrées.

Soft Cheeses

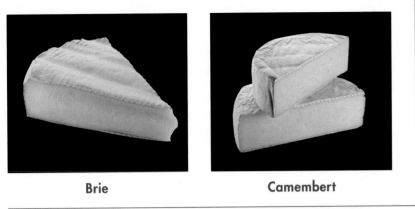

Brie   Camembert

Wisconsin Milk Marketing Board, Inc.

Figure 15-48. *Brie and Camembert are soft cheeses.*

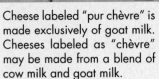
Camembert. *Camembert* is a soft cheese made from cow milk and has a yellow color and a waxy, creamy consistency. Like Brie, Camembert is made in many countries, including the United States. It is similar to Brie, but there are differences in flavor and aroma due to variations in manufacturing and ripening. The thin rind has the appearance of felt. It is served most often as a dessert accompanied by crackers and fruit. Camembert should be left at room temperature prior to serving.

Semisoft Cheeses

A *semisoft cheese* is a cheese that is firmer than a soft cheese but not as hard as a hard cheese. Semisoft cheeses are produced using one of three different ripening processes resulting in a dry-rind, washed-rind, or waxed-rind cheese.

Dry-Rind Cheeses. A *dry-rind cheese* is a semisoft cheese that is allowed to ripen through exposure to air. The air dries out the exterior, producing a dry, almost woody rind. Although the rind becomes hard and dry, the interior of the cheese remains tender and smooth. Common examples of dry-rind semisoft cheeses are Monterey Jack, Havarti, and bel paese. *See Figure 15-49.*

Dry-Rind Semisoft Cheeses

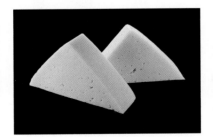

Monterey Jack   Havarti   Bel Paese

Figure 15-49. *Monterey Jack, Havarti, and bel paese are dry-rind semisoft cheeses.*

Monterey Jack. *Monterey Jack* is a semisoft dry-rind cheese that has a smooth texture, a creamy white color, and a mild taste. Monterey Jack that is aged for a long period of time becomes harder in texture and zestier in flavor.

Havarti. *Havarti* is a Danish, semisoft dry-rind cheese made from cow milk and has a buttery, somewhat sharp flavor. It is aged for approximately three months. During the aging process, it develops very small holes, similar to those of Swiss cheese, but smaller.

Bel Paese. *Bel paese* is a lightly colored, semisoft dry-rind cheese with a buttery flavor that melts easily. It originated in Italy but is now produced in both Italy and the United States. It is allowed to mature for nearly eight weeks. Bel paese can be substituted for mozzarella.

Washed-Rind Cheeses. A *washed-rind cheese* is a semisoft cheese with an exterior rind that is washed with a brine, wine, olive oil, nut oil, or fruit juice. Washing the rind generates the growth of harmless bacteria that penetrate the cheese to ripen the interior. Common examples of washed-rind semisoft cheeses are brick, Limburger, Muenster, and Port Salut. *See Figure 15-50.*

Washed-Rind Semisoft Cheeses

Wisconsin Milk Marketing Board, Inc.
Brick

Wisconsin Milk Marketing Board, Inc.
Limburger

Wisconsin Milk Marketing Board, Inc.
Muenster

Port Salut

Figure 15-50. *Brick, Limburger, Muenster, and Port Salut are washed-rind semi-soft cheeses.*

Brick. *Brick cheese* is a washed-rind semisoft cheese made from cow milk. It has a mild, sweet flavor and a texture that is firm, yet elastic, with many small holes. Brick cheese is sold in a brick shape. It slices and melts well and is often used for cheese platters and sandwiches.

Limburger. *Limburger* is a washed-rind semisoft cheese with a strong aroma and flavor. Limburger cheese was first marketed in Limburg, Belgium. This cheese is also made in Germany and the United States. Limburger cheese is made from whole or skim cow milk. It has a creamy texture that is developed through ripening in a damp atmosphere for a period of two months. Limburger cheese is often served with crackers and onions as a dessert or buffet cheese.

Wisconsin Milk Marketing Board

Muenster. *Muenster* is a washed-rind semisoft cheese with a flavor between that of brick cheese and Limburger. It was first produced in the vicinity of Munster, Germany. European Muenster ranges from mild to sharp in taste and may have a strong aroma due to aging. A version of this cheese is produced in North America and is comparatively mild. Muenster is marketed in cylindrical form and is used as a buffet or sandwich cheese.

Port Salut. *Port Salut* is a washed-rind semisoft cheese that has a soft, smooth, orange-colored rind and a glossy, ivory-colored interior. Its flavor can range from mellow to robust, depending on the age of the cheese. The aroma is often compared to a mild Limburger, and the flavor is compared to Gouda. Port Salut is used for appetizers, used as a dessert cheese, and may be served with apple pie.

Waxed-Rind Cheeses. A *waxed-rind cheese* is a semisoft cheese produced by dipping a wheel of freshly made cheese into a liquid wax and allowing the wax to harden. The cheese ripens while encased in the wax. Although the outer coating is hard, the cheese inside the wax stays soft. Common waxed-rind semisoft cheeses include Edam, Gouda, and fontina. *See Figure 15-51.*

Waxed-Rind Semisoft Cheeses

| Edam | Gouda | Fontina |

Wisconsin Milk Marketing Board, Inc.

Figure 15-51. *Edam, Gouda, and fontina are waxed-rind semisoft cheeses.*

Cabot Creamery Cooperative

Edam. *Edam* is a waxed-rind semisoft cheese made from cow milk and has a firm, crumbly texture. It is usually shaped like a ball with a slightly flattened top and bottom and has a distinctive red wax coating. Edam is named after its place of origin in the Netherlands. Edam cheese made in the Netherlands is rubbed with oil but is not colored. However, the cheese that is exported is coated in red wax before it is shipped. Edam that is made in the United States is coated in red wax. It is used most often as a dessert cheese and is used on platters and buffets.

Gouda. *Gouda* is a waxed-rind semisoft cheese that is similar to Edam but contains more fat. It also originated in the Netherlands in the province of Gouda. Gouda is usually shaped like a flattened ball or formed into a loaf. It is generally served as a dessert or buffet cheese.

Fontina. *Fontina* is a waxed-rind semisoft cheese made from cow milk. It has been produced in the Alps since the 12th century. Young fontina cheese is somewhat soft and becomes harder as it ages. It has a nutty, herbal flavor and is perfect for use in fondues and on platters and buffets.

Blue-Veined Cheeses

A *blue-veined cheese* is a cheese produced by inserting harmless live mold spores into the center of ripening cheese with a needle. The blue vein that runs through these cheeses indicates where the needle was inserted and the mold spores were released. The mold decreases the time required for the cheese to ripen and is safe to eat. It also adds a distinctive flavor to the cheese. After the mold spores are injected, the exterior of the cheese wheel is salted to help keep the surface dry and prevent the mold from overtaking the exterior. Common blue-veined cheeses include blue cheese, Gorgonzola, Roquefort, and Stilton. *See Figure 15-52.*

Blue Cheese. *Blue cheese,* also known as bleu cheese, is a blue-veined cheese made from cow milk and is characterized by the presence of a blue-green mold. Blue cheese is generally produced in wheels weighing approximately 7 lb. Blue cheese is used in blue cheese dressing, on sandwiches, and on cheese platters.

Gorgonzola. *Gorgonzola* is a blue-veined cheese that is mottled with blue-green veins. It originated in Gorgonzola, Italy but it is now made in the Lombardy and Piedmont regions of Italy as well as the United States. Gorgonzola is generally cured for a period of 6–12 months. Gorgonzola is used in salads and salad dressings and as a dessert and buffet cheese.

Blue-Veined Cheeses

Blue Cheese

Gorgonzola

Roquefort

Stilton

Figure 15-52. *Blue cheese, Gorgonzola, Roquefort, and Stilton are blue-veined cheeses.*

Roquefort. *Roquefort* is a blue-veined cheese made from sheep milk and is characterized by a sharp, tangy flavor. The blue-green veins are created by spreading a powdered bread mold over the curd before ripening. The cheese is aged for a period of two to five months, depending on the sharpness desired. It was first made in the village of Roquefort, France. Roquefort is used primarily as a dessert cheese but is also used in salads.

Stilton. *Stilton* is a blue-veined cheese made from cow milk with a flavor that is milder than Roquefort or Gorgonzola. It was first made in the village of Stilton, England. Stilton cheese has a crumbly texture and veins of blue-green mold running through the curd and wrinkled rind. Although some Stilton is available in North America, it is not imported in great quantities. It is primarily used as a dessert and buffet cheese.

Hard Cheeses

A *hard cheese* is a firm, somewhat pliable and supple cheese with a slightly dry texture and buttery flavor. Because hard cheeses have a firmer overall texture than softer varieties, they slice well, making them well suited for sandwiches. All hard cheese varieties grate well. Cheddar, Swiss, provolone, Gruyère, Cheshire, and Manchego are hard cheeses.

Cheddar. *Cheddar* is a hard, aged cheese that is yellow or white and ranges in taste from mild to sharp. *See Figure 15-53.* It originated in Cheddar, England, but accounts for a significant portion of all cheeses made in the United States. The sharpness of Cheddar cheese varies based on the length of the aging period. Longer aging produces a sharper taste. Cheddar is used in the preparation of fondue, soufflés, spreads, and sandwiches. It is also eaten on crackers, used to top chili, and served with apple pie.

Cheddar

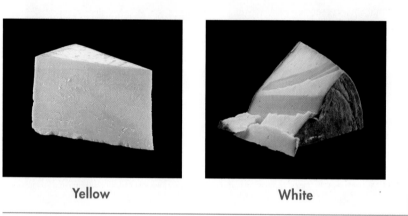

Yellow   White

Wisconsin Milk Marketing Board, Inc.

Figure 15-53. *Cheddar is a hard, aged cheese that is yellow or white and ranges in taste from mild to sharp.*

Swiss

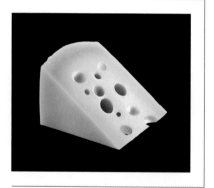

Wisconsin Milk Marketing Board, Inc.

Figure 15-54. *Swiss cheese is a term for varieties of hard cheese with large holes.*

Swiss Cheese. *Swiss cheese* is the name used to describe varieties of hard cheeses with large holes, also known as eyes. *See Figure 15-54.* Swiss cheese has an elastic body and a mild, sweet flavor. The holes in Swiss cheese are developed by gas-producing bacteria that release carbon dioxide during the ripening period, forming bubbles that result in the holes.

Emmentaler is a Swiss-made cheese. Other varieties of Swiss cheese are produced elsewhere. In the United States, Swiss cheese is typically cured three to four months, while in Switzerland it is cured for six to ten months. The longer curing period that is used in Switzerland produces cheese with larger holes and a more pronounced flavor. Swiss cheese is a traditional cheese for fondue and is used in many preparations, from sandwiches to stuffed veal chops.

Provolone. *Provolone* is a hard cheese with an elastic texture and a mild to sharp taste depending on the age. *See Figure 15-55.* Provolone originated in southern Italy but is now produced throughout Italy and North America. It is light in color and can be cut without crumbling. A distinguishing characteristic of provolone is that it is often formed into the shape of a pear, sausage, or cone and corded for hanging. Provolone is used in the preparation of many Italian dishes, especially pizza.

Provolone

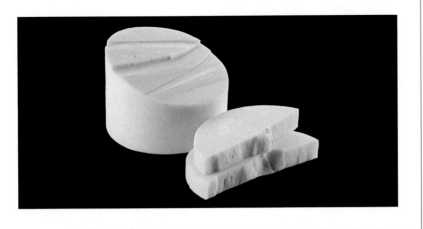

Figure 15-55. *Provolone is a hard cheese with an elastic texture and a mild to sharp taste depending on age.*

Gruyère. *Gruyère* is a hard cheese that is similar in texture to Emmentaler but has a sharper flavor. *See Figure 15-56.* Gruyère was originally made in Switzerland and is an excellent cheese for cooking. Gruyère is often used in fondue, veal cordon bleu, and sautéed veal chops.

Cheshire. *Cheshire* is a hard cheese and is the oldest of the English-named cheeses. Cheshire is often sold as a Cheshire-Cheddar blend. *See Figure 15-57.* Cheshire derives its unique, mildly salty taste from the soil in the Cheshire area of England. The cows that graze on the grass grown there produce salty milk, which is used in making the cheese. It is similar to cheddar but has a more crumbly and less dense texture. Cheshire is available in a natural white or a deep-yellow color. The yellow is the result of adding vegetable dye to the curd. Cheshire cheese is used in the same manner as cheddar and can be used for fondue.

Manchego. *Manchego* is a hard cheese with a tangy, slightly salty flavor and is produced from sheep milk in the La Mancha region of Spain. *See Figure 15-58.* It is aged for at least three months and has a rich golden color. Manchego is the most well-known cheese from Spain. The criss-crossed textured rind is the result of the cheese being aged in grass molds.

Gruyère

Figure 15-56. *Gruyère is a hard cheese that is similar in texture to Emmentaler but has a sharper flavor.*

Cheshire-Cheddar Blends

Figure 15-57. *Cheshire is often sold as a Cheshire-Cheddar blend.*

Manchego

Figure 15-58. *Manchego is a hard cheese produced from sheep milk in the La Mancha region of Spain.*

Browne-Halco (NJ)

Grating Cheeses

A *grating cheese* is a hard, crumbly, dry cheese grated or shaved onto food prior to service. The crumbly texture makes it difficult to slice. Grating cheeses are usually produced as large wheels, many weighing close to 100 lb. The most common grating cheeses are Parmesan, Romano, and Asiago. *See Figure 15-59.*

Grating Cheeses

Parmesan

Romano

Asiago

Wisconsin Milk Marketing Board, Inc.

Figure 15-59. *Parmesan, Romano, and Asiago are grating cheeses.*

Parmesan. *Parmesan* is a grating cheese that originated in Parma, Italy. It is extremely hard, has a granular texture, and can keep indefinitely when produced properly. Parmesan cheese is widely produced in the United States and Argentina, but Parmesan from Italy is considered superior. It is available in grated or whole form and is considered a seasoning cheese. Parmesan is used to season items such as soups and pasta dishes.

Romano. *Romano* is a grating cheese that is similar to Parmesan but softer in texture. It was first made in the vicinity of Rome from sheep milk. Today it is made in other parts of Italy and other countries from cow or goat milk. Romano has a granular texture, sharp flavor, and brittle, black rind. It can be aged from 5–12 months. A longer aging period sharpens the flavor. In the professional kitchen, Romano is used in various ways, such as a seasoning in pasta dishes or as an au gratin topping.

Asiago. *Asiago* is a grating cheese with a nutty, toastlike flavor. High-quality Asiago often contains a crunchy material that results from an amino acid in milk that crystallizes during the aging process. Italian Asiago is produced from the milk of cows that graze in fields where herbs and flowers grow, which gives it a distinct flavor. For this reason, Asiago produced outside of Italy has a slightly different taste.

Cheese Products

A *cheese product* is a processed food made of natural cheeses that may include additional ingredients such as emulsifiers. Cheese products can be made exclusively from natural cheeses, but they often contain other ingredients in order to increase shelf life and to produce a product that spreads and melts easily. Cheese products include cold-pack cheese, processed cheese, and processed cheese food.

Cold-Pack Cheeses. A *cold-pack cheese* is a creamy cheese product made by blending natural cheeses without the addition of heat. *See Figure 15-60.* Cold-pack cheeses usually consist of two or more varieties of mild and sharp natural cheeses that are ground together. Cold-pack cheeses range from white to orange in color. They are available in many different flavors and are often mixed with a variety of spices and seasonings.

Cold-Pack Cheeses

Wisconsin Milk Marketing Board, Inc.

Figure 15-60. *A cold-pack cheese is a creamy cheese product made by blending natural cheeses without the addition of heat.*

Processed Cheeses. A *processed cheese* is a blend of fresh and aged natural cheeses that are heated and melted together. The result is a cheese product that can be packaged in just about any shape or size and is uniform in flavor and texture. A common type of processed cheese is American cheese. *See Figure 15-61.*

Processed Cheese Food. A *processed cheese food* is a cheese-based product that may contain as little as 51% cheese. The remaining 49% is made of dairy or nondairy products, including emulsifiers. Emulsifiers allow the processed cheese food to melt into a smooth, silky texture when heated.

Processed Cheeses

Wisconsin Milk Marketing Board, Inc.

Figure 15-61. *A common type of processed cheese is American cheese.*

Cabot Creamery Cooperative

CHEESE STORAGE AND PREPARATION

Cheese is a dairy product that contains a considerable amount of butterfat, absorbs odors easily, spoils relatively quickly, and can dry out if left exposed to air. In order to serve cheese at its best, it is important to understand proper storage and serving requirements.

Storing Cheese

Cheese is perishable and should be kept tightly wrapped in plastic wrap or plastic bags and stored in the refrigerator. Wrapping cheese helps it retain moisture and keeps out odors from other foods. When refrigerated properly, cheese should retain freshness as follows:

- Fresh cheeses keep for seven to ten days.
- Soft cheeses keep for about two weeks.
- Semisoft cheeses keep for two to three weeks.
- Hard cheeses keep for about one month.
- Grating cheeses keep for several months.

Serving Cheese

In North America, cheese is often served as a snack food, as a salad ingredient, on sandwiches and vegetables, or in entrées. In Europe, cheese is traditionally served with a continental breakfast or as a separate course at the end of a meal.

Cheese should ideally be served at room temperature. If cheese has been refrigerated, it should be removed approximately 1 hour prior to service. Room-temperature cheese exhibits the best flavor and texture. However, cheeses left at room temperature for more than a few hours may become oily or dry out. When preparing a cheese platter, it is important to include an assortment of cheeses with various textures, flavors, and degrees of ripeness. *See Figure 15-62.*

Serving Cheeses

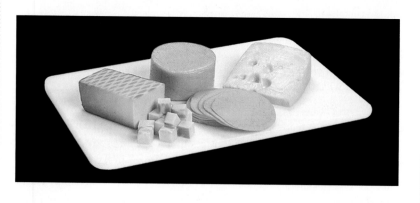

Carlisle FoodService Products

Figure 15-62. *A cheese platter should include an assortment of cheeses with various textures, flavors, and degrees of ripeness.*

Cooking with Cheese

Cheeses are generally cooked at the lowest temperature possible to prevent proteins in the cheese from hardening. This allows the cheese to maintain an even texture and smooth consistency. Cheese becomes tough and stringy when overheated. For this reason, cheese should be added at the end of the cooking process so it melts smoothly and incorporates evenly. For example, when making cheese sauce for nachos, cheese should be added after the béchamel sauce has thickened. If cheese is added directly to the milk during the béchamel preparation, it will separate and settle on the bottom of the pot.

HORS D'OEUVRES AND APPETIZERS

The terms hors d'oeuvre and appetizer are often used interchangeably. However, there is a difference. An *hors d'oeuvre* is an elegant, bite-size portion of food that is creatively presented and served apart from a meal. An *appetizer* is food that is larger than a single bite and is typically served as the first course of a meal. Most appetizers are meant to be shared. Because hors d'oeuvres and appetizers commonly precede a meal, their flavor should complement the courses to come. They should also be appropriately seasoned so that no further seasoning is necessary.

Hors d'oeuvres are often served butler style, which consists of servers carrying plates or small platters of individual hors d'oeuvres to guests. A wide variety of hors d'oeuvres are usually offered. Although they are very small portions, hors d'oeuvres may become a substitute for an entire meal when many varieties are served.

A guest should never need a knife to eat an hors d'oeuvre. However, some hors d'oeuvres may require a cocktail fork or toothpick because they are coated in a sauce, making them difficult to eat with the fingers. Hors d'oeuvres can be made from a countless number of fillings, spreads, and garnishes and may be served hot or cold. *See Figure 15-63.* They can be served in many forms and made with a wide variety of ingredients.

Hors d'oeuvres may also be served buffet style. Buffet presentations usually combine hot and cold varieties of hors d'oeuvres. If a food was prepared hot, it should be served hot or held in a manner that would enable it to be eaten hot. The same is true for cold items. Serving items at the appropriate temperature signifies a commitment to providing guests with high-quality food and enhances the dining experience.

Most appetizers are served with a dipping sauce and may require the use of one or more utensils. *See Figure 15-64.* Appetizers are often simple foods but can also be elegant. Common appetizers include fried mozzarella sticks, battered onion rings, nachos, and Buffalo wings. More elegant appetizers include beef carpaccio, spring rolls, or bruschetta.

Serving Hors d'Oeuvres

Figure 15-63. *Hors d'oeuvres may be served hot or cold.*

Photo Courtesy of Perdue Foodservice,
Perdue Farms Incorporated

Shared

Daniel NYC

Individual

Figure 15-64. *Most appetizers are served with a dipping sauce and may require the use of one or more utensils.*

Amuse Bouche

Tru, Chicago

Figure 15-65. *Many upscale restaurants offer a petite hors d'oeuvre as a complimentary amuse bouche.*

In the past, the differences between hors d'oeuvres and appetizers were very distinct. Today, the differences can be as simple as the manner in which the item is served. For example, a single tortilla chip may be topped with a spicy avocado purée, a slice of seared ahi tuna, a dollop of wasabi sour cream, and a cilantro leaf. The individual nachos may be served at a reception as a passed hors d'oeuvre or at an upscale restaurant as a complimentary hors d'oeuvre that is not included as a menu item. In contrast, a casual restaurant may serve a platter of the tuna nachos as an appetizer, enabling guests to share.

Amuse bouche, canapés, small plates, and starters are terms used by foodservice operations to describe various types of hors d'oeuvres and appetizers. Regardless of the term used, small bites and starters are popular menu items.

Amuse Bouche

Many upscale restaurants offer a complimentary petite hors d'oeuvre called an amuse bouche. *See Figure 15-65.* An *amuse bouche* is a single, bite-sized portion of food selected by the chef and not ordered by the guest. Amuse bouche is a French term meaning "entertain the mouth." An amuse bouche is meant to awaken the taste buds and excite guests for the dining experience to follow. Chefs commonly offer an amuse bouche in a simple yet elegant presentation to highlight the flavor, color, and texture of the food.

Canapés

A *canapé* is an hors d'oeuvre that looks like a miniature open-faced sandwich. *See Figure 15-66.* Canapés are often served at receptions, teas, and other events where a lighter fare is desired over a full meal. Canapés are composed of a base, a spread, and an edible garnish.

Bases. Canapés can be made on slices of fresh, firm vegetables. In this type of preparation, firm vegetables, such as cucumbers, yellow or green squash, or carrots, are thinly sliced into rounds and topped with a spread. However, bread is the most common base for canapés. Toasted bread can handle more moisture and holds its shape better than fresh bread. If freshly sliced bread is used, it must be firm enough to hold the topping without becoming soggy.

Canapés

Figure 15-66. *A canapé is an hors d'oeuvre that looks like a miniature open-faced sandwich.*

Spreads. Canapé spreads can be smooth and spreadable or thick with texture. Spreads start with a main ingredient, such as butter, cream cheese, or mayonnaise, or a puréed food, such as meats, seafood, vegetables, or legumes. Flavoring ingredients, such as lemon juice, herbs, Dijon mustard, or roasted garlic, are then blended with the main ingredient to create a spread. Smooth spreads include compound butters, pâtés, and hummus. A pastry bag is often used to apply spreads to a base. To prevent the pastry tip from clogging and to keep a smooth texture, solid ingredients should be puréed.

Textured spreads include bound salads made from vegetables, legumes, poultry, seafood, or meats. The salad ingredients are finely chopped or slightly puréed and mixed with a binding agent such as mayonnaise, vinaigrette, or yogurt. Examples of bound salads include chicken salad, tuna salad, and grilled vegetable salad.

Garnishes. The garnish for a canapé should be delicate and provide an elegant finished appearance. Examples of garnishes include the leaf of a delicate herb, a thin slice of smoked salmon, or a dab of caviar or fried capers.

When making canapés using bread as a base, a decision has to be made regarding the use of toasted or fresh bread. Slightly different procedures are used for making canapés with toasted and untoasted bread. *See Figure 15-67.*

Using Toasted Bread

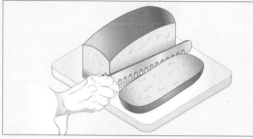

1. Remove the crusts and reserve for making croutons or bread crumbs.
2. Slice the bread lengthwise into evenly sliced pieces.

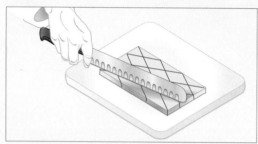

3. Cut the bread into desired shapes.
4. Brush both sides of the cutouts with melted butter and place in a 450°F oven for 8–10 minutes.
5. When evenly golden, remove slices from oven and cool completely.

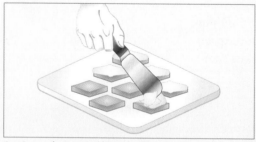

6. Cover the toasted bread with a thin layer of spread.

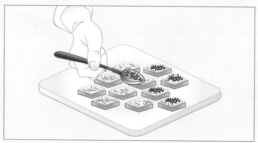

7. Garnish as desired.

Using Untoasted Bread

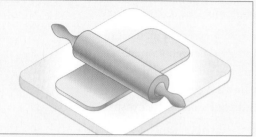

1. Remove the crusts and reserve for making croutons or bread crumbs.
2. Slice the bread lengthwise into evenly sliced pieces.
3. Using a rolling pin, firmly roll over the bread slices to flatten them.

4. Cover the flattened bread with a thin layer of spread.

5. Cut the bread into desired shapes.

6. Garnish as desired.

Figure 15-67. *Slightly different procedures are used for making canapés with toasted and untoasted bread.*

Small Plates

The practice of serving small plates of food as either a starter course or a complete meal is common all over the world. Spain calls their small portions tapas, China has dim sum, Italy has antipasti, and Greece has mezes.

Tapas. The word "tapas" is Spanish for lids. Restaurateurs were trying to think of a way to keep insects from getting into people's wine glasses as they dined outdoors. The solution was to provide bite-sized portions of food on round slices of bread. The bread was used both as a snack and as a lid to cover the top of the wine glasses. Tapas include a wide variety of ingredients and presentations. *See Figure 15-68.*

Dim Sum. Dim sum, translated as "touch the heart," consists of items that can be eaten in one or two bites. Dim sum is traditionally served as a mid-afternoon meal, but may be served as a starter course. In general, a server wheels a cart that is stacked with round steamer baskets made of bamboo or stainless steel to the table. The server removes one lid from each stack to display the food inside. Dim sum offers a wide variety of one- or two-bite items. *See Figure 15-69.*

Tapas

Tapas Small Plates
• Assorted miniature grilled sausage
• Marinated white anchovies
• Fried calamari
• Thinly sliced Serano ham
• Spicy tomato sauce and goat cheese spread
• Spiced hot or cold shrimp
• Shellfish or other seafood
• Bacon wrapped figs
• Assorted Spanish olives

Irinox USA

Figure 15-68. *Tapas include a wide variety of ingredients and presentations.*

Dim Sum

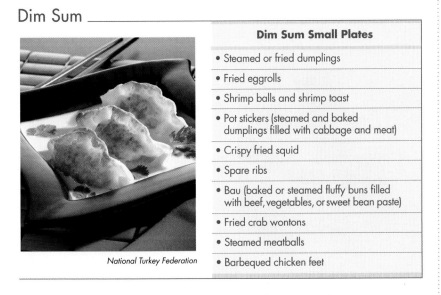

Dim Sum Small Plates
• Steamed or fried dumplings
• Fried eggrolls
• Shrimp balls and shrimp toast
• Pot stickers (steamed and baked dumplings filled with cabbage and meat)
• Crispy fried squid
• Spare ribs
• Bau (baked or steamed fluffy buns filled with beef, vegetables, or sweet bean paste)
• Fried crab wontons
• Steamed meatballs
• Barbequed chicken feet

National Turkey Federation

Figure 15-69. *Dim sum offers a wide variety of one- or two-bite items.*

Antipasti. Italian antipasti consist of a variety of foods arranged on platters rather than individually served portions. Antipasti presentations often include colorful selections of meats, cheeses, and marinated, pickled, or grilled vegetables. Antipasti plates are designed to whet the appetite without being too filling. *See Figure 15-70.*

Antipasti —————————————

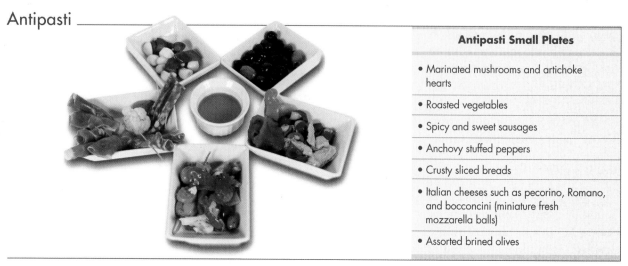

Antipasti Small Plates
• Marinated mushrooms and artichoke hearts
• Roasted vegetables
• Spicy and sweet sausages
• Anchovy stuffed peppers
• Crusty sliced breads
• Italian cheeses such as pecorino, Romano, and bocconcini (miniature fresh mozzarella balls)
• Assorted brined olives

Figure 15-70. *Antipasti plates are designed to whet the appetite without being too filling.*

Mezes. Mezes is the Greek word for "snacks." These bite-sized portions of food are commonly served as a starter course, but may also be eaten as a full meal. Mezes vary according to the ingredients found in different regions of Greece. *See Figure 15-71.*

Mezes —————————————

Courtesy of The National Pork Board

Mezes Small Plates
• Dolmades (grape leaves stuffed with currents, pine nuts, and rice)
• Revythosalata (chickpeas puréed with coriander and garlic)
• Spanikopita (spinach and feta cheese filled phyllo pastry)
• Taramosalata (purée of salted fish roe, bread crumbs, and potato)
• Souvlakia (skewered and grilled cubes of pork seasoned with lemon)
• Melitzanes (miniature eggplant filled with stewed onions, tomatoes, and herbs)
• Assorted Greek olives

Figure 15-71. *Mezes vary according to the ingredients found in different regions of Greece.*

Cold Starters

A variety of cold starters are some of the most recognizable to guests. Some well-known cold starters include cocktails, crudités, and relishes. These starters are served in a range of foodservice operations from casual restaurants to banquets and buffets.

Cocktails. A *cocktail* is a chilled appetizer served in a stemmed glass. Poached and chilled shrimp placed on the edge of a glass surrounding cocktail sauce is a well-known cocktail. *See Figure 15-72.* A cocktail can be as simple as cubed fruit or as complex as seafood tossed with avocado, jalapeño, cilantro, lime juice, and chile sauce. Cocktails should always be served as fresh and as cold as possible.

Crudités. Crudité is a French term meaning "raw things." A *crudité* is a group of raw vegetables arranged on a platter and served with a dipping sauce. Any vegetable that is acceptable to eat raw can be cut into bite-size pieces and arranged beautifully on a serving platter as crudités. A crudité platter may consist of carrots, celery, broccoli, bell peppers, mushrooms, and snow peas. *See Figure 15-73.* Cherry tomatoes, cauliflower, radishes, zucchini, yellow squash, cucumbers, and asparagus are also popular crudités.

Cocktails

Harbor Seafood, Inc.

Figure 15-72. *Poached and chilled shrimp placed on the edge of a glass is a well-known cocktail.*

Crudités

Mushrooms — Broccoli

Snow peas — Carrots

Bell peppers — Celery

Carlisle FoodService Products

Figure 15-73. *A crudité platter may consist of carrots, celery, broccoli, bell peppers, mushrooms, and snow peas.*

Relishes. A *relish* is an assortment of uncooked vegetables that are served raw, marinated, or pickled. *See Figure 15-74.* A variety of vegetables are suitable for marinating or pickling. Examples of these vegetables include cucumbers, olives, beets, peppers, artichoke hearts, baby corn, and mushrooms. Marinated or pickled vegetables can range in flavor from sour and tangy to sweet and spicy. Relishes should be served cold with a small amount of the marinade or pickling liquid.

Relishes

Carlisle FoodService Products

Figure 15-74. *A relish is an assortment of raw, marinated, or pickled vegetables.*

Grilled and Roasted Vegetable Starters

Grilled and roasted vegetables are often marinated briefly and then grilled over an open flame or roasted in an oven at a high temperature. Grilling and roasting vegetables causes the natural sugars to caramelize and intensify in flavor. *See Figure 15-75.* Hard vegetables, such as carrots, should be blanched before grilling to make them tender.

Grilled Vegetables

Tanimura & Antle®

Figure 15-75. *Grilling and roasting vegetables causes the natural sugars to caramelize and intensify in flavor.*

Stuffed and Filled Starters

Stuffed and filled starters are similar to canapés. However, they have side walls which support more filling than a canapé. Savory or sweet fillings can be placed in bases that range from pastries, such as barquettes, tartlets, profiteroles, bouchées, and phyllo shells, to vegetable and protein food bases.

Barquettes and Tartlets. A *barquette* is a miniature, boat-shaped pastry shell that contains a savory or sweet filling. A *tartlet* is a miniature, round pastry shell that contains a savory or sweet filling. *See Figure 15-76.* Barquettes and tartlets are made from dough that resembles thin pie dough. After they are baked and cooled, barquettes and tartlets can be used to hold hot or cold fillings. Common fillings include savory or sweet mousses, braised meats or vegetables, purées, and custards.

Tartlets

Courtesy of The National Pork Board

Figure 15-76. *A tartlet is a miniature round pastry shell that contains a savory or sweet filling.*

Profiteroles and Bouchées. A *profiterole* is a miniature pastry made from choux paste that is filled with a sweet or savory filling. Choux paste is a very soft pastry dough that is piped onto sheet pans and baked. The steam created during the baking process causes this dough to rise. If choux pastry is even slightly underbaked, the profiteroles will deflate once they cool.

A *bouchée* is a puff pastry that is filled with a savory filling. *See Figure 15-77.* Puff pastry dough is made by repeatedly folding and refolding thin layers of butter and dough together. As it bakes, the puff pastry dough rises as a result of the steam created between the layers of dough. The term "bouchée" is French for mouthful. Although bouchées are commonly round or square in shape, they can be made into shapes such as fish or crescents.

Phyllo Shells. A *phyllo shell* is a shell made by layering buttered sheets of phyllo dough in miniature muffin tins and baking them until golden brown. The delicate cups can be filled with an assortment of fillings from brie and pesto to roasted minced pears with caramelized onions and blue cheese. *See Figure 15-78.*

Bouchées

Figure 15-77. *A bouchée is a puff pastry that is filled with a savory filling.*

Phyllo Shells

Grated cheese and crème fraîche garnish

Phyllo formed into cups and baked

Hot filling

Figure 15-78. *Phyllo shells can be filled with an assortment of fillings.*

Vegetables. Many petite vegetables can be filled and served as a starter. *See Figure 15-79.* For example, Belgium endive, celery stalks, and tomatoes can be filled with savory cold fillings and served as casual or elegant starters depending on the filling used. Mushroom caps can be stuffed with cooked or cured meats and cheeses. Cherry tomatoes can be filled with a savory cheese spread. Scooping out some of the flesh from roasted new potatoes and filling them with sour cream and caviar, braised pork and truffle, or a seafood salad makes an elegant starter.

Stuffed Vegetables

Figure 15-79. *Petite tomatoes can be filled and served as a starter.*

Stuffed Proteins

National Turkey Federation

Figure 15-80. *Proteins, such as meats, can be stuffed and presented as starters.*

Proteins. Proteins such as eggs, seafood, and meats can be stuffed and presented as starters. *See Figure 15-80.* For example, the yolk of a hard cooked egg can be removed and seasoned with flavorful ingredients, such as curry, wasabi, or mustard, and then piped back into the egg white. Clams and oysters served in their shells can be topped with a variety of savory ingredients. Shrimp can be butterflied to create a pocket and stuffed with a savory bread dressing. Chicken drummettes may also be stuffed by making a small slice just above the joint and pulling back the flesh to reveal a small pocket, which can be filled with a flavorful sausage or gorgonzola cheese.

Wrapped Starters

Small portions of food can be wrapped with another food to make an attractive starter. The ingredient wrapped around the main food item imparts color, texture, and flavor and enhances the presentation. For example, bacon wrapped around scallops and prosciutto wrapped around asparagus make flavorful, eye-appealing starters. Phyllo and thinly sliced vegetables are often used as wrappings.

Pastry Wrappings. In many cultures, pastry dough is used to wrap seafood, meat, and vegetable fillings to make bite-sized starters. For example, turmeric-seasoned pastry dough can be used to wrap a potato and lamb filling into turnovers called samosas. Empanadas consist of pastry dough wrapped around meat, vegetables, chicken, or caramelized onions and are baked or fried. Empanadas are shaped like turnovers. Phyllo dough is often wrapped around spinach and feta to form triangles known as spanakopita. Thin sheets of pastry dough can be wrapped around julienned vegetables to form spring rolls or wontons. *See Figure 15-81.* Puff pastry sheets are often used to prepare miniature pastries such as morel mushroom and thyme turnovers.

Wrapped Starters

National Turkey Federation   Courtesy of The National Pork Board

Figure 15-81. *Thin sheets of pastry dough can be wrapped around julienned vegetables to form spring rolls or wontons.*

Vegetable Wrappings. Vegetables such as zucchini, yellow squash, seedless cucumbers, sweet potatoes, or russet potatoes can be sliced into very thin strips and wrapped around a main food ingredient. Vegetable wrapped items are commonly steamed, sautéed, or baked.

Battered and Breaded Starters

Virtually any food can be battered or breaded and then fried or baked. Common battered or breaded items include onion rings, zucchini, mushrooms, hot peppers, cheese sticks, shrimp, calamari, and chicken wings. *See Figure 15-82.* Battered and breaded starters are commonly served in casual dining restaurants, but they may also be served in upscale restaurants and at receptions. The batter and breading used in casual dining restaurants is often heavier than the type used in upscale dining environments. This is because upscale operations are more concerned with presenting a quality item cooked to perfection, rather than serving hearty starters.

Battered and Breaded Starters

Onion Rings

Hot Peppers

Photo Courtesy of McCain Foods USA

Figure 15-82. *Common battered or breaded starters include battered onion rings and breaded hot peppers.*

Skewered Starters

Almost any solid food can be cooked and served on skewers. A *brochette* is a food that is speared onto a wooden, metal, or natural skewer and then grilled or broiled. *See Figure 15-83.* Brochettes are also called shish kebabs, or kebabs for short. While brochette is a French term, the words "shish kebab" reflect the Middle Eastern origins of this skewered presentation. For example, chicken and pork satay are kebabs commonly offered on menus. Natural skewers include twigs of rosemary, shoots of lemon grass, and slivers of sugar cane.

Skewered Starters

National Honey Board
Bacon-Wrapped Scallops

National Turkey Federation
Turkey Satay

Figure 15-83. *A brochette is a food that is speared onto a wooden, metal, or natural skewer and then grilled or broiled.*

Classical examples of brochettes include a brochette of fresh fruit, a seafood brochette of shrimp and scallops, and a vegetarian brochette of mushrooms, zucchini, cherry tomatoes, peppers, and onions. Brochettes are often marinated to intensify flavors. When the portion size is enlarged, brochettes may be served as entrées. Upscale versions of brochettes include the following:

- olive, feta, mint, and orange segments
- ginger and lime basted scallops
- coconut and curried shrimp

Tips for Preparing Skewered Starters

- Soak wooden skewers in water for at least 30 minutes prior to use. Soaking allows the wooden skewer to absorb water so that it will not burn or catch fire.
- Spray metal skewers with a nonstick cooking spray prior to use.
- Leave a portion of the skewer exposed at each end for the guest to use as a holder.

Dips

A *dip* is a sauce or condiment that is served with foods such as vegetables or chips. *See Figure 15-84.* Dips can be cold or hot, sweet or savory, and smooth or chunky. A dip must be thick enough to adhere to the food item and be complementary in flavor. From chips served with salsa to pitas served with hummus, a wide variety of dips can be prepared.

Dips

Vita-Mix® Corporation

Figure 15-84. *A dip is a sauce or condiment that is served with foods such as vegetables or chips.*

Cold creamy dips often have sour cream, yogurt, mayonnaise, or cream cheese as the base ingredient. Other cold dips, such as guacamole, hummus, or curried lentil dip, have a purée as the main ingredient. Almost any ingredient, such as crabmeat, roasted garlic, cooked spinach, or Parmesan cheese, can be added to a base. Cold dips can be served in a small dish, bread bowl, or a hollowed vegetable.

Hot dips often have a tomato, béchamel, or Mornay sauce as the base ingredient. A variety of items can be added as a flavoring ingredient. For example, jalapeños are a flavorful accompaniment to a cheddar-based Mornay sauce. Pesto, roasted garlic, or goat cheese is an appetizing addition in a fresh tomato sauce. Hot dips should be served from a bain marie or chafing dish and held at the appropriate temperature.

Carlisle FoodService Products

Raw Meat and Seafood Starters

Raw meats and seafood are often served as starters. Only fresh meats and seafood that have been handled with extreme care can be served raw. Raw meats and seafood are popular dishes that include items such as carpaccio, tartare, caviar, and sushi. Some foodservice operations have raw bars that feature a variety of raw seafood presentations.

Carpaccio. *Carpaccio* is thin slices of meats or seafood that are served raw. *See Figure 15-85.* Typically, the meat or seafood is partially frozen in order to make slicing it easier. The thinly sliced meat is then attractively arranged on a serving plate, seasoned with salt and pepper, and often drizzled with extra virgin olive oil or lemon or lime juice.

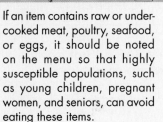

Chef's Tip

If an item contains raw or under-cooked meat, poultry, seafood, or eggs, it should be noted on the menu so that highly susceptible populations, such as young children, pregnant women, and seniors, can avoid eating these items.

Carpaccio

Courtesy of Chef Gui Alinat

Figure 15-85. *Carpaccio is thin slices of meats or seafood that are served raw.*

Tartare

Irinox USA

Figure 15-86. *Steak tartare is often served as a starter.*

Tartare. *Tartare* is freshly ground or chopped raw meat or seafood. Tartare is usually seasoned and served with toast points or another cracker-like accompaniment. Steak tartare and tuna tartare are often served as starters. **See Figure 15-86.**

Caviar. *Caviar* is the harvested roe of sturgeon. Although roe, or eggs, are harvested from other varieties of fish, such as salmon and whitefish, only sturgeon roe can be labeled as caviar. Other fish roe must be labeled by the type of fish it comes from, such as salmon caviar or whitefish caviar. The highest quality caviar is harvested from beluga, osetra, and sevruga. **See Figure 15-87.**

Caviar

Tru, Chicago

Figure 15-87. *The highest quality caviar is harvested from beluga, osetra, and sevruga.*

Caviar is often served on toast points, small savory pancakes called blinis, or chilled, cooked fingerling potatoes with a dollop of sour cream. Standard accompaniments for caviar include finely minced hard-cooked eggs and red onions. Caviar should only be handled with a mother of pearl, china, or wooden spoon. Metal spoons cause a chemical reaction which gives the caviar a metallic flavor.

Sushi. *Sushi* is a vinegar-seasoned rice dish garnished with raw fish, cooked seafood, eggs, or vegetables. *See Figure 15-88.* The term sushi is derived from "zushi," the Japanese term for cold, vinegar-seasoned, cooked rice. Sushi rice is made by steaming or simmering a sticky, short-grain rice until it is slightly firm and chewy. The cooked rice is gently mixed with rice wine vinegar and sugar syrup. Once the sushi rice cools, it is formed by hand into various shapes. The most common forms of sushi include nigirizushi, makizushi, inarizushi, and chirashizushi.

- *Nigirizushi* is a type of sushi made with small, hand-formed mounds of vinegar-seasoned cooked rice that is garnished with a topping.

- *Makizushi*, also known as a maki roll, is a type of sushi made with vinegar-seasoned cooked rice layered with other ingredients and rolled in a dried seaweed paper called nori. Maki rolls form a cylinder shape that is then sliced into bite-size portions.

- *Inarizushi* is a type of sushi made with vinegar-seasoned rice and toppings stuffed into a small purse of fried tofu.

- *Chirashizushi,* also known as scattered sushi, is a type of sushi made by scattering toppings over or mixing toppings into vinegar-seasoned rice.

Sushi

Nigirizushi

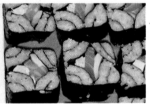

Maki Rolls

Figure 15-88. *Sushi is a vinegar-seasoned rice dish garnished with raw fish, cooked seafood, eggs, or vegetables.*

Raw Bars. A *raw bar* is a presentation of a variety of raw and steamed seafood presented and served on a bed of ice. *See Figure 15-89.* Common items found on raw bars include shrimp, crab legs, clams, mussels, and oysters. Shellfish shooters are also popular. Shellfish shooters consist of raw or cooked seafood presented in a shot glass or on the half shell with a flavorful sauce or liquid. Common shellfish shooters include oysters in horseradish cocktail sauce, mussels on the half shell with a creamy vinaigrette, and shrimp in a roasted pepper and spicy tomato juice cocktail.

Raw Bars

Figure 15-89. *A raw bar is a presentation of a variety of raw and steamed seafood served on a bed of ice.*

Cured and Smoked Starters

Cured and smoked items such as bacon, prosciutto, gravlax, and kippered herring are often served as starters. Cured items are often combined with other foods to create complex flavor profiles such as bacon-wrapped figs topped with goat cheese. Smoked items, such as salmon, may be served alone. *See Figure 15-90.* Likewise, prosciutto can be wrapped around shrimp or melon. Cured and smoked meats and cheeses may also be placed on rosemary skewers to create a fragrant and colorful starter.

Cured and Smoked Starters

Courtesy of The National Pork Board

Cured

Smoked

Figure 15-90. *Cured, bacon-wrapped figs topped with goat cheese offer a complex flavor profile as does smoked salmon.*

Foods were traditionally cured or smoked to preserve them from spoiling. Today, items such as cheese, vegetables, poultry, seafood, and meats are cured or smoked to develop exceptional flavors.

Curing. The methods used to cure food items include dry curing and wet curing, or brining. *Dry curing* is a curing method that involves the use of salt to dehydrate the protein in food. Proteins are dry-cured by completely covering them with salt or a salt-cure mixture, which includes salt, sugar, seasonings, and pink curing salt.

Wet curing is a curing method in which foods are submerged in a brine. A *brine* is a salt solution that usually consists of 1 cup of salt per 1 gal. of water. Chefs may choose to add sugar, spices, or herbs to a brine.

Smoking. A smoker is used to cold smoke or hot smoke foods. A small compartment inside the smoker holds wet wood chips that, when ignited, smolder and emit smoke into the smoker. Food can be cold smoked or hot smoked. Cold smoking exposes food to smoke at temperatures below 100°F. Smoke is released into a container holding the food. The container is then immediately covered and stored in the refrigerator. Cold smoking does not actually cook the food. In contrast, hot smoking food cooks food while it is being smoked. The longer the food is left in the smoker, the stronger the flavor and aroma.

Indian Harvest Specialtifoods, Inc./Rob Yuretich

CHARCUTERIE

Charcuterie is the art of making sausages and other preserved items such as pâtés, terrines, galantines, and ballotines. Charcuterie began in the 15th century as a way of preserving pork products prior to refrigeration. Items were smoked, cured, or processed with a high amount of fat to preserve and protect them from being exposed to the air and spoiling. Today, charcuterie refers to cured, smoked, or preserved items made from a variety of meats, poultry, game, seafood, and vegetables. *See Figure 15-91.*

Charcuterie _____

Daniel NYC

Figure 15-91. *Charcuterie refers to cured, smoked, or preserved items made from a variety of meats, poultry, game, seafood, and vegetables.*

Common forms of preserving charcuterie items include curing and smoking. Cured charcuterie items are often served as hors d'oeuvres or appetizers. Prosciutto, gravlax, and kippered herring are examples of cured meats. Smoked charcuterie items can be prepared by either hot or cold smoking. While poultry, seafood, and meats are commonly smoked, other foods such as cheeses and vegetables can also be smoked. The length of time an item is exposed to smoke determines the intensity of the smoke flavor.

Forcemeats

Forcemeat is the main ingredient of sausages, pâtés, terrines, galantines, and ballotines. A *forcemeat* is a mixture of raw or cooked meat, poultry, seafood, or vegetables, fat, seasonings, and sometimes other ingredients. Additional ingredients may include cubed bread, heavy cream, eggs, and edible garnishes that are strategically placed inside the forcemeat for both flavor and appearance. Depending on the intended use, a forcemeat can be ground smooth like a purée, coarsely, or somewhere in between. When uncooked, properly prepared forcemeat should have a shiny and slick appearance. When cooked, it should appear moist and have a smooth texture.

A *grinder* is a tool that consists of an auger, which pushes the forcemeat forward, and a cutting blade, which cuts the forcemeat as it is pushed through perforated stainless steel dies. ***See Figure 15-92.*** Dies come in several sizes, including small, medium, and large. Forcemeat is usually ground in a progressive style, using a larger die for the first grind and then replacing it with a smaller die to make the next grind and so forth. When grinding forcemeats, it is important to refrigerate all parts of the grinder after it has been thoroughly cleaned and sanitized.

Grinders

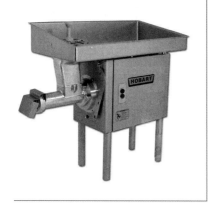

Hobart

Figure 15-92. *A grinder cuts the forcemeat as it is pushed through perforated stainless steel dies.*

Chef's Tip

All of the connective tissue must be trimmed from all meat used in forcemeats prior to grinding.

Meats, Poultry, Game, or Seafood. The main protein used in most charcuterie preparations is meat, poultry, game, or seafood. Almost all forcemeats incorporate some amount of pork, because pork adds a smooth texture to forcemeats that would otherwise be dry or rubbery. Pork is usually added to forcemeats without a pork base in a ratio of 1 part pork to 2 parts other protein. Veal, game meats, or vegetables may also be used.

Liver. Liver is often added to forcemeats as a binding agent and as a flavoring. The most common types of liver used in forcemeats are from poultry. Poultry livers are small, tender, and milder in flavor as compared to pork, lamb, or beef liver. It is important to process livers appropriately before grinding. ***See Figure 15-93.***

Procedure for Preparing Liver for Use in a Forcemeat

1. Trim liver of all connective tissue and cut liver into 1 inch pieces.
2. Wash liver thoroughly under cold running water and drain well.
3. Place liver in a bowl, cover completely with milk, and soak for at least 24 hours.
4. Drain the liver from the milk. Wash the liver well under cold running water and drain again. *Note:* At this stage the liver can be added to pâtés or terrines where larger pieces of liver may be desirable.
5. Purée the liver in a blender.
6. Strain the liver purée through a chinois or cheesecloth-lined china cap.
7. Add the liver purée to a forcemeat.

Figure 15-93. *It is important to process livers appropriately before grinding.*

Fats. Fat adds moisture and flavor to a forcemeat. In classical forcemeat recipes, the percentage of fat to lean meat is typically 50% meat and 50% fat. However, the protein may be as high as 65%. Higher percentages of protein result in a dry product.

A hard fat, such as pork fatback, is the most desirable fat to use in a forcemeat. The dense structure of fatback prevents it from melting as easily as a softer fat, allowing it to easily bind with the meat. When forcemeats are not processed properly, the fat and any additional moisture will separate from the meat when it is cooked and yield a dry and grainy forcemeat. A forcemeat with a perfect emulsification will be smooth, moist, and have a pleasant texture.

Binding Agents. Forcemeats that include liquids, such as wine, brandy, or heavy cream, may need an additional binding agent other than the ground protein from which it is made. The addition of a liquid may overwhelm the protein's binding and emulsification properties, causing the fat and liquids to leak out of the forcemeat during the cooking process. Common binding agents include whole eggs, egg whites, dried bread, grains, and flour. Binding agents help bind protein to liquids, resulting in a better quality forcemeat. *See Figure 15-94.*

Seasonings and Flavorings. Forcemeats must be seasoned prior to cooking or baking. Almost any herb or spice may be incorporated in a forcemeat, depending on the desired flavor profile. However, salt is the most important seasoning to add to a forcemeat. Salt will intensify the flavor of the other ingredients in the forcemeat and facilitate the emulsification of the protein and the fat. Salt also acts as an agitator during the grinding process.

Curing salt, also known as TCM, is a combination of table salt and sodium nitrite. *See Figure 15-95.* Curing salts are used to reduce the spoilage caused by harmful bacteria and enhance the flavor of the meat being cured. Forcemeats that do not include curing salt can have a grayish appearance, whereas those produced with curing salt look fresher.

Marinades add flavor and moisture to a forcemeat. Meat absorbs the aromas, flavors, and moisture from a marinade. Oils, alcohol, such as wine or brandy, herbs, spices, citrus zest, and aromatic vegetables are common marinade ingredients.

Binding Agents

Figure 15-94. *Binding agents, such as eggs, help bind protein to liquids, resulting in a better quality forcemeat.*

Curing Salts

Figure 15-95. *Curing salt, also known as pink salt, is a combination of iodized salt and sodium nitrite.*

Garnishes. Garnishes, such as batonnet of blanched carrots, bell pepper julienne, pistachios, hard-cooked eggs, whole eggs, peeled eggs, slices of lean cooked tenderloin, cooked seafood, diced ham, or truffles, can be placed in the middle of a forcemeat in a uniform pattern prior to cooking. When a garnished forcemeat is sliced, the colorful garnishes make a beautiful presentation. *See Figure 15-96.* Garnishes must be cooked prior to being added to a forcemeat because shrinkage occurs during the cooking process and can create unattractive air gaps in the finished product.

Forcemeat Garnishes

Figure 15-96. *When a garnished forcemeat is sliced, the colorful garnishes make a beautiful presentation.*

Forcemeat Varieties

Varieties of forcemeats include straight, country-style, mousseline-style, emulsified, and gratin. Straight, country-style, and mousseline forcemeats are commonly used for the production of pâtés, terrines, and galantines. Emulsified and gratin style forcemeats are not commonly used in foodservice operations.

Straight Forcemeats. A *straight forcemeat* is a forcemeat that consists of seasoned ground meat emulsified with ground fat. Some straight forcemeats may incorporate liver or eggs. Pâtés and terrines are typically prepared using a straight forcemeat.

Country-Style Forcemeats. A *country-style forcemeat,* also known as pâté de campagne, is a coarsely ground forcemeat with a strong flavor. *See Figure 15-97.* Country-style forcemeats are typically marinated with a substantial amount of garlic, onion, bay leaves, juniper berries, freshly ground pepper, and wine or brandy. They also often include pork or chicken liver. Most varieties of sausage are country-style forcemeats.

Country-Style Forcemeats

Figure 15-97. *A country-style forcemeat, also known as pâté de campagne, is a coarsely ground forcemeat with a strong flavor.*

Mousseline-Style Forcemeats. A *mousseline-style forcemeat* is a forcemeat made from less-fibrous proteins such as poultry, seafood, veal, or pork. The protein is commonly ground first and then puréed. It is finished by folding in heavy cream and eggs or egg whites. A mousseline-style forcemeat is commonly used to make seafood sausages, hot or cold timbales, and terrines. When properly prepared, mousseline-style forcemeats have a light and tender consistency and a more finished appearance than the other styles of forcemeats. Adding too much cream results in too fragile a forcemeat, and too many eggs or egg whites results in a rubbery forcemeat.

Emulsified Forcemeats. An *emulsified forcemeat,* also known as the 5-4-3 style, is a forcemeat made from a ratio of five parts protein, to four parts fat, to three parts crushed ice. Hot dogs, knockwurst, and mortadella are examples of emulsified forcemeats.

Gratin Forcemeats. A *gratin forcemeat* is a forcemeat made from meat that is partially seared prior to grinding and a starchy binding agent. Dried white bread or a cooked grain that has been soaked in milk, known as a panada, is typically ground with the meat to help emulsification occur.

Forcemeat Preparation

When preparing forcemeats all of the ingredients and equipment must be ice cold. Having everything cold will help the meat grind easier and prevent it from becoming tough during processing. Even the part of the grinder used to process the forcemeat needs to be refrigerated prior to grinding to ensure it does not warm the forcemeat as it is ground. Because charcuterie products have a high ratio of fat added while grinding, keeping items cold helps prevent the fat from getting soft or melting and results in a better emulsification. ***See Figure 15-98.***

Procedure for Preparing Forcemeats

1. Completely chill all ingredients and equipment at or below 41°F.
2. Trim meat of all connective tissue and cut into approximately 2 inch cubes.
3. Mix meat and fat in a marinade and refrigerate until needed. *Note:* Meat can be partially frozen as the cold will help the emulsification process.
4. Cut fatback into 2 inch cubes and place the cubes in the freezer until needed.
5. If the specific recipe calls for liver, trim the liver of all veins and connective tissue. Grind the liver to a fine texture and force it through a fine-meshed strainer or tamis. Place the ground liver in the freezer until needed.
6. Place a medium die in the cold grinder and grind all of the meat, including the liver (if applicable).
7. Change to a small die and grind the meat again.
8. Grind the partially frozen fatback and then mix well with the ground meat. *Note:* When using a food processor, add a small amount of crushed ice to offset the heat generated by the friction of the blades.
9. Add other ingredients such as eggs, panada, and seasonings and mix well to incorporate.
10. Check forcemeat for consistency and overall flavor by making and tasting a quenelle.
11. Refrigerate forcemeat until needed.

Figure 15-98. *Because charcuterie products have a high ratio of fat added while grinding, keeping items cold helps prevent the fat from getting soft or melting and results in a better emulsification.*

All forcemeats should be checked to make sure the flavor, seasonings, and texture are correct prior to cooking the entire recipe. This can be done by preparing the forcemeat, making a quenelle from the forcemeat, and tasting it. A *quenelle* is a small dumpling made from a forcemeat by using two spoons. *See Figure 15-99.* First, a small portion of forcemeat is placed on one of the spoons. The other spoon is used to scrape the forcemeat from the first spoon. Next, the original spoon is used to scrape the quenelle from the second spoon. This procedure is repeated until the resulting dumpling is oval in shape. The shaped quenelle is then poached.

Quenelles

Figure 15-99. *A quenelle is a small dumpling made from a forcemeat by using two spoons.*

As the poached quenelle rises to the surface of the poaching liquid, it is removed and cut in half to reveal the center texture. The quenelle is then tasted for flavor and texture and the forcemeat recipe can be adjusted if needed.

FORCEMEAT APPLICATIONS

Forcemeats are often the primary component in sausages, pâtés, terrines, galantines, and ballotines. Forcemeats may be molded into various shapes such as rectangles, ovals, and cylinders.

Sausages

A *sausage* is a forcemeat that is typically stuffed into a casing or shaped into a patty. The common ratio of fat to lean meat in most sausages is 1 to 3 or about 33% fat. Fat percentages lower than 25% will result in a dry and mealy sausage. Adding a starch, such as rice or oats, and crushed ice to a lean sausage forcemeat helps lower the fat percentage without drying out the sausage because the starch absorbs the moisture.

Sausage seasonings vary depending on the variety of sausage being prepared and the flavor profile desired. Curing salts can be either added to the diced meat during the marinating process or can be added during the grinding procedure. Some sausages are simply composed of ground meat, fat, and seasonings, while others may include garnishes such as a chiffonade of fresh oregano, roasted red peppers, feta cheese crumbles, or sliced black truffles. Garnishes are always added after the grinding process to keep the identity of the garnish intact.

Sausages can be produced from pork, veal, seafood, poultry, beef, game meats, or vegetables. The three basic categories of sausages are fresh, cured, and smoked.

Fresh Sausages. *Fresh sausage* is a type of sausage that is freshly made and has not been cured, smoked, dried, or further processed in any way. According to the USDA, a fresh sausage contains no nitrites, which are added to other categories of sausages to help preserve them and aid in the processing. Fresh sausages are made solely of seasoned ground forcemeat. Game meats such as rabbit and venison are commonly used to make fresh sausages.

Cured Sausages. *Cured sausage* is a type of raw sausage that contains sodium nitrite mixed into the forcemeat. Salt and sodium nitrite help to greatly extend the shelf life of cured sausages as they work to prevent the development of foodborne pathogens. Cured sausages can be sold in many forms from raw and soft, such as with Italian sausage, to hard, dry, and firm, such as with salami. ***See Figure 15-100.***

Daniel NYC

Cured Sausages

Photo Courtesy of D'Artagnan, Photography by Doug Adams Studio

Figure 15-100. *Cured sausages can be sold in many forms from raw and soft, such as with Italian sausage, to hard, dry, and firm, such as with salami.*

Smoked Sausages. *Smoked sausage* is a type of cured sausage that is smoked during the curing process. Smoked sausages can be either hot smoked or cold smoked. Hot-smoked sausages are placed in a heated smoker, which cooks the sausages as they are being smoked. Cold-smoked sausages are placed in a smoker set below 100°F so the sausages are not cooked during the smoking process. Cold-smoked sausages are then cured or cooked in another manner before being served to guests.

Sausage Casings

Sausage casings are used for almost every type of sausage produced. A *natural casing* is a casing produced from the intestines of sheep, hogs, or cattle. Natural casings are cleaned thoroughly and then salted heavily before being stuffed. They are sold in salt-packed bundles called hanks. Sheep casings are the most tender natural casings and are no more than 1 inch in diameter. Hog casings are a little tougher than sheep casings and about 1½ inches in diameter. Cattle casings range from between 1¾–4 inches in diameter and are used for specialty style meats. Prior to using a natural casing, the casing must be cleaned properly. *See Figure 15-101.*

Procedure for Cleaning Natural Casings

1. Carefully remove desired amount of casing from a salt pack to avoid creating knots that cannot be easily undone. *Note:* It is always a good idea to prepare more casing than might be needed to allow for holes or unusable sections.
2. Place a large bowl or bain-marie under the faucet in a clean, sanitized sink.
3. Place the untangled casing in the bowl with one end near the faucet opening. Turn on the faucet and flush the casing to rinse out the salt and impurities.
4. Search for any large holes in the casing. *Note:* Small holes the size of a pin are natural. If large holes are discovered, cut the casing and tie it off at the hole to make that the end of the link.

Figure 15-101. *Prior to using a natural casing, the casing must be cleaned properly.*

Smoked Sausages

Figure 15-102. *Sausages that are hung for smoking require durable casings.*

To store unused rinsed casings, fill a container about 1 inch from the top with cold water and add the casings one at a time while holding on to one end of each casing. Drape one end of each casing over the top edge of the container and then secure the lid. This allows one end of each casing to be found without searching through the whole tub and helps prevent the casings from becoming tangled.

A *collagen casing* is a casing produced from collagen. It is uniform in size and more durable than natural casings. The consistent size of collagen casings makes them ideal for portion control. Collagen casings are produced from collagen that has been removed from the hides of cattle. Collagen casings come in different strengths to accommodate fresh sausages, which use more delicate casings, and sausages that are hung for smoking, which require durable casings. *See Figure 15-102.* Collagen casings are often dipped in water before use to make them more pliable.

A *synthetic fibrous casing* is a casing produced from a plastic-like synthetic material. Synthetic fibrous casings are not edible and must be peeled off prior to eating. Since they are a plastic composite, they do not need to be refrigerated. However, they need to be submerged in water for a few minutes before use to make them slightly more pliable.

Stuffing Sausages

Foodservice operations can prepare an assortment of sausages using a meat grinder and a sausage stuffer. A *sausage stuffer* is a manual or electric piece of equipment that uses a piston to pump forcemeat through a nozzle into casings. Casings are fed onto the nozzle, and the forcemeat is pumped from the stuffer reservoir to the casing to create a sausage link. *See Figure 15-103.*

Procedure for Stuffing Sausage

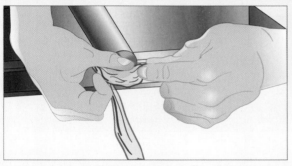

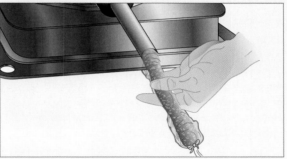

1. Completely chill the sausage stuffer to prevent the forcemeat from heating up during processing.
2. Add the forcemeat to the sausage stuffer.
3. Inspect the cleaned casings before use. Then, slide the sausage casing over the stuffing nozzle.

4. Tie a knot in the end of the casing hanging off the nozzle and then poke a small hole in it with a toothpick to allow air to escape while the casing is being filled.

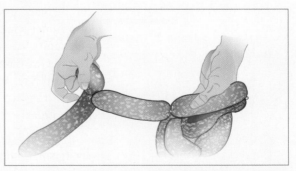

5. Hold the end of the casing on the stuffer nozzle to guide the casing as it fills with forcemeat.
6. Fill all of the casing with forcemeat.

7. Twist each long link into individually-sized sausage links.

Figure 15-103. *Casings are fed onto the nozzle, and the forcemeat is pumped from the stuffer reservoir to the casing to create a sausage link.*

Pâtés

A *pâté* is a forcemeat that is typically layered with garnishes and baked in a mold. Pâté is often encased in a pastry crust and baked in a hinged mold. Pâtés prepared with a crust are commonly referred to as pâté en croûte, which translates to pâté in a crust. *See Figure 15-104.* The most common shapes of pâté molds are oval and rectangular.

Pâte en Croûtes

Figure 15-104. *Pâtés prepared with a crust are commonly referred to as pâté en croûte, which translates to pâté in a crust.*

Terrines

A *terrine* is a forcemeat that is baked in a mold without a crust. A terrine can be made using any style of forcemeat. Some terrines are a combination of products, such as game birds or vegetables. *See Figure 15-105.* Rectangular and oval molds have always been the most popular shapes for terrines. Terrines are baked in earthenware, glass, or enamel-coated cast-iron molds. Classically, terrine molds were lined with thin sheets of fatback and then the forcemeat was placed inside and encased in fatback on all sides. Today many chefs omit the fatback or replace it with a much thinner piece of caul fat.

Galantines and Ballotines

Galantines and ballotines are often confused because they are made in a similar manner. A *galantine* is forcemeat that is wrapped in the skin of the animal the meat was taken from, shaped into a cylinder, trussed, and poached in a flavorful stock. For example, a duck galantine would be a duck forcemeat wrapped in duck skin. Galantines may be roasted instead of poached. A cooked galantine is cooled, sliced, and glazed with gelatin.

Terrines

Photo Courtesy of D'Artagnan, Photography by Doug Adams Studio

Pheasant Terrine

Vegetable Terrine

Figure 15-105. *Some terrines are a combination of products, such as game birds or vegetables.*

A traditional galantine is made from boned-out poultry stuffed with poultry forcemeat. It is chilled, glazed with aspic, and garnished with items included in the filling, such as olives or truffles. Galantines are always served cold, whereas ballotines may be served hot or cold. *See Figure 15-106.* A *ballotine* is a boned poultry leg that has been stuffed with forcemeat. The stuffed leg is then trussed to maintain its shape. A ballotine is braised or roasted and is normally served hot but can be served cold.

Galantines and Ballotines

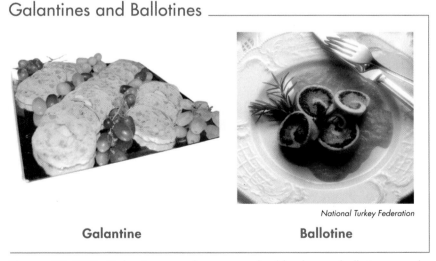

National Turkey Federation

Galantine

Ballotine

Figure 15-106. *Galantines are always served cold, whereas ballotines may be served hot or cold.*

GLAZES

Many charcuterie items are presented on platters for elaborate buffets. Pâtés, terrines, roasted meats, hors d'oeuvres, and other items can be sliced and arranged on platters to entice the guest and illustrate the skill and time needed to prepare them. Since the exposed surface of the terrine, pâté, or other item is exposed to air as it sits on a platter, there is a possibility that it may dry out, darken in color, and lose quality. Glazing items with aspic or chaud froid will help to preserve them, keep them from losing quality, and give them a more beautiful finished appearance. *See Figure 15-107.*

Glazes _____

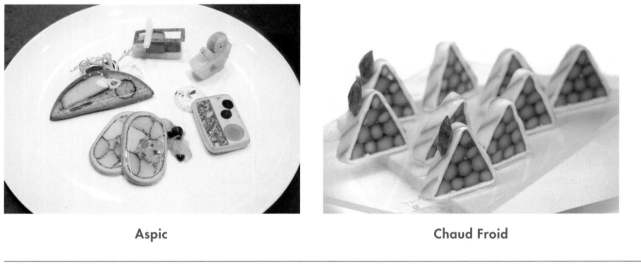

| Aspic | Chaud Froid |

Figure 15-107. *Glazing items with aspic or chaud froid will help to preserve them, keep them from losing quality, and give them a more beautiful finished appearance.*

Aspic

Aspic, also known as aspic gelée, is a savory jelly made from a consommé or clarified meat, fish, or vegetable stock that produces a clear finish when coating foods. It can be made from the natural gelatin extracted from bones, a packaged gelatin, or a combination of both. Aspic made from stock must be clarified in the same manner as a consommé to ensure that it is very clear. *See Figure 15-108.*

Aspic is used as a protective coating and as a binding agent. As a protective coating, aspic can be used to preserve the color of foods as well as prevent them from drying out due to exposure to air. Food with an aspic glaze has an artificial appearance because it has such a shiny, clear finish. It is important to always lift aspic-coated foods with an offset spatula because aspic retains fingerprints.

1. Prepare a rich white or brown stock from poultry, meat, or fish bones and mirepoix.

2. Test the stock for its gelatin properties by ladling a small amount onto a cold plate and then refrigerating the plate until the stock is completely chilled.

3. Remove the plate from the refrigerator and tilt it from side to side to see if the aspic has set completely. *Note:* If the stock is too watery, dilute instant gelatin in a small amount of cold water until completely softened. Add the softened gelatin to the stock and clarify the stock in the same way as when preparing a consommé. Then, strain the stock to remove all traces of fat from the surface. Retest the stock to ensure that it will set properly.

Figure 15-108. *Aspic made from stock must be clarified in the same manner as a consommé to ensure that it is very clear.*

Chaud Froid

Chaud froid is a French term meaning "hot-cold." Chaud froid is similar to clear aspic except that it is bright white in color. Like aspic, chaud froid protects foods from drying out when exposed to air. An additional benefit to coating foods with chaud froid is that it serves as a white canvas for colorful decorations or garnishes. For example, a cooked veal roast could be completely cooled, glazed with chaud froid, and then decorated with colorful garnishes.

Preparing a chaud froid can be as simple as adding gelatin to heavy cream and a white stock. It also can be prepared in a more traditional manner using a roux or an egg yolk liaison. Although chaud froid can be eaten, it is typically removed from food prior to eating. Chaud froid may also be used as a gelatin-based binding in terrines. Any charcuterie item can be glazed with aspic or chaud froid. *See Figure 15-109.*

Chef's Tip

It is important that the correct amount of aspic or chaud froid be carefully poured over the surface of an item to ensure the entire item is coated and that any drips and unevenness in the glaze are prevented.

1. Lay items on a glazing screen presentation-side up. Tightly cover the items and refrigerate them until needed.

2. Heat the aspic or chaud froid until it has completely melted.

3. Place the hot glaze in a nonreactive bowl over an ice bath and stir constantly while allowing it to cool until it becomes slightly thicker in consistency but not gelatinous. *Note:* If the glaze cools too much and becomes gelatinous it will not yield a smooth coating. If the glaze is too thin or warm, the glaze will not be thick enough to coat the items.

4. Remove items from the refrigerator. Using a spoon or ladle, completely cover the item with aspic or chaud froid.

5. Place glazed items in the refrigerator to allow the aspic or chaud froid to set.

6. Repeat steps 4 and 5 at least three times to completely glaze each item.

Figure 15-109. *Any charcuterie item can be glazed.*

CHARCUTERIE PRESENTATIONS

Charcuterie items are commonly presented on platters. *See Figure 15-110.* Aspic or chaud froid glazed items and smoked or cured products are neatly arranged to draw the eye from one end of the platter to the other end. Typically, a decorative piece relating to the types of food being served on the platter is also included. The decorative item may be a vegetable carving or a large uncut portion of a terrine. Often, the piece is a whole cooked meat item that has been coated with chaud froid and decorated with vegetable garnishes.

Charcuterie Presentations

Daniel NYC

Figure 15-110. *Charcuterie items are commonly presented on platters.*

Once items are placed on a charcuterie platter, it is best not to move them. Adjusting or moving items after they are placed can result in the platter being smudged and can potentially damage the shiny coatings. Charcuterie platters also showcase smoked and cured items as well as complementary accompaniments. For example, cured salmon may be served with capers, diced red onions, and chopped hard-cooked eggs. Smoked meats, pâtés, and terrines are often served with gherkins and assorted mustards.

SUMMARY

Salads are versatile dishes that can be presented as appetizers, main courses, separate courses, side salads, and dessert salads. Care must be taken to properly prepare, wash, and store salad greens. In addition to salad greens, vegetables, starches, fruits, nuts, and proteins are common salad ingredients. Salads are often served with a dressing. Temporary emulsions, such as vinaigrettes, and permanent emulsions, such as mayonnaise, form the foundation for countless varieties of dressings. There are many different types of salads including tossed, composed, bound, vegetable, fruit, and gelatin salads.

Cheese varieties include fresh, soft, semisoft, blue-veined, hard, and grating cheese. Cheese can be served as a separate course or used as an ingredient in a savory dish or a dessert. Cheeses selected for a cheese platter should vary in texture, degree of ripeness, color, and flavor. Cheese products are often used for their melting qualities when making au gratin dishes.

Hors d'oeuvres are elegant, bite-size portions of food that are creatively presented and served apart from a meal. Most appetizers are larger than a single bite, meant to be shared, and served as the first course of a meal. Amuse bouche, canapés, small plates, and a wide variety of starters are terms used by foodservice operations to describe various types of hors d'oeuvres and appetizers. Regardless of the name, chefs create starters that please the eye and satisfy the palate when served at the proper temperature.

Charcuterie is the art of making sausages and other preserved items such as pâtés, terrines, galantines, and ballotines. Forcemeats are made by adding fat and seasoning to meats, game, poultry, seafood, or vegetables. Charcuterie platters often showcase smoked items, cured items, and complementary accompaniments. Glazing items with aspic or chaud froid keeps items from losing quality and gives them a beautiful finished appearance.

Refer to DVD for
Quick Quiz® questions

Review

1. Identify five types of salad presentations.
2. Define salad greens.
3. Describe four types of lettuces.
4. Contrast spinach, chard, and arugula.
5. Describe three types of cabbages used as salad greens.
6. Distinguish between the different varieties of chicory.
7. Contrast dandelion greens and mesclun greens.
8. Describe eight types of garnishing greens.
9. Describe how to properly store salad greens.
10. Explain how to remove the core from head lettuce.
11. Explain how to trim romaine lettuce.
12. Explain how to remove the rib from loose greens.
13. Explain how to wash salad greens.
14. Identify four types of salad ingredients other than salad greens.
15. Define emulsion.
16. Identify a temporary emulsion and a permanent emulsion.
17. Explain why it is important to match an oil with an appropriate vinegar.
18. Explain how to prepare a basic French vinaigrette.
19. Explain how to prepare an emulsified vinaigrette.
20. Explain how to prepare a mayonnaise.
21. Name six types of salads.
22. Explain how to prepare a tossed salad.
23. Identify the four parts of composed salads.

Refer to DVD for
Review Questions

24. List five tips for preparing bound salads.
25. Explain why it is important to know how vegetables will react to acidity when making vegetable salads.
26. Explain how to prevent fruits from discoloring while making fruit salads.
27. Describe the ratio of gelatin powder to liquid used to make gelatin salads.
28. List four tips for preparing gelatin salads.
29. Explain why it is important to chill gelatin layers until slightly firm before combining the layers.
30. Explain the purpose of rennet in making cheese.
31. Differentiate between curds and whey.
32. Identify four factors that determine the flavor and texture of a cheese.
33. Describe nine types of fresh cheese.
34. Contrast fresh and soft cheeses.
35. Describe two types of soft cheese.
36. Identify three ways semisoft cheeses are ripened.
37. Describe three types of dry-rind cheese.
38. Describe four types of washed-rind cheese.
39. Describe three types of waxed-rind cheese.
40. Explain why a blue vein runs through blue-veined cheeses.
41. Describe four types of blue-vein cheese.
42. Describe six types of hard cheese.
43. Contrast hard cheeses and grating cheeses.
44. Describe three types of grating cheese.
45. Identify three types of cheese products.
46. Describe how to store cheese for maximum freshness.
47. Explain why cheeses are generally cooked at low temperatures.
48. Differentiate between hors d'oeuvres and appetizers.
49. Define amuse bouche.
50. Describe canapés.
51. Explain how to prepare canapés using toasted and untoasted bread.
52. Describe four types of small plates.
53. Describe three types of cold starters.
54. Explain what happens when vegetables are grilled or roasted.
55. Describe seven types of stuffed and filled starters.
56. Contrast stuffed and filled starters with wrapped starters.
57. Describe two types of wrapped starters.
58. Identify common battered and breaded starters.
59. Contrast battered and breaded starters with skewered starters.
60. List three tips for preparing skewered starters.
61. Define dip.
62. Describe five types of raw meat and seafood starters.
63. Contrast raw meats and seafood starters with cured and smoked starters.
64. Explain the purpose of curing foods.
65. Explain the purpose of smoking foods.
66. Identify the main ingredient of sausages, pâtés, terrines, galantines, and ballotines.

67. Explain why forcemeat is ground using a progressive style of grinding.
68. Explain the role of fat as a forcemeat ingredient.
69. Explain the role of a binding agent as a forcemeat ingredient.
70. Identify the two main ingredients in a curing salt.
71. Explain why garnishes have to be cooked prior to being added to a forcemeat.
72. Name three styles of forcemeats used in foodservice operations.
73. Describe how forcemeat is prepared.
74. Explain the purpose of a quenelle and how to make one.
75. Identify the common ratio of fat to lean meat in most sausages.
76. Identify the three categories of sausages.
77. Explain how sausages are stuffed into clean casings and made into links.
78. Differentiate between a pâté and a terrine.
79. Differentiate between a galantine and a ballotine.
80. Explain the purpose of glazes.
81. Differentiate between aspic and chaud froid.
82. Explain why glazed items should not be moved after being placed on a presentation platter.

Applications

1. Remove the core from a head of lettuce.
2. Trim romaine lettuce.
3. Remove the rib from loose greens.
4. Wash and store salad greens.
5. Prepare a basic French vinaigrette. Evaluate the quality of the vinaigrette.
6. Prepare an emulsified vinaigrette. Evaluate the quality of the vinaigrette.
7. Prepare a mayonnaise. Evaluate the quality of the mayonnaise.
8. Prepare a tossed salad. Evaluate the quality of the prepared dish.
9. Prepare a composed salad. Evaluate the quality of the prepared dish.
10. Prepare a bound salad. Evaluate the quality of the prepared dish.
11. Prepare a vegetable salad. Evaluate the quality of the prepared dish.
12. Prepare a fruit salad. Evaluate the quality of the prepared dish.
13. Prepare a gelatin salad. Evaluate the quality of the prepared dish.
14. Prepare canapés using toasted and untoasted bread. Evaluate the quality of each type of canapé.
15. Prepare tapas, dim sum, antipasti, and mezes. Evaluate the quality of each prepared dish.
16. Prepare a cocktail, crudité tray, and a relish tray. Evaluate the quality of each prepared dish.
17. Prepare a forcemeat and then poach a quenelle to test the texture and flavor of the forcemeat. Evaluate the quality of the forcemeat.
18. Stuff and then smoke sausages. Evaluate the quality of the finished items.
19. Prepare a pâté, a terrine, a galantine, or a ballotine. Evaluate the quality of the finished item.
20. Prepare and glaze charcuterie items with aspic or chaud froid and then plate them for presentation. Evaluate the quality of the glazed items and the way in which they are presented.

Refer to DVD for
Application Scoresheets

French Vinaigrette

Yield: *16 servings, 1 fl oz each*

Ingredients

red wine vinegar	4 fl oz
kosher salt	1½ tsp
black pepper, ground	¼ tsp
granulated sugar	pinch
canola oil	12 fl oz

Preparation

1. In a stainless steel mixing bowl, combine vinegar, salt, pepper, and sugar and whisk to incorporate.
2. Add oil in a steady stream while whisking continuously to form a temporary emulsion.
3. Whisk again prior to service.

NUTRITION FACTS
Per serving: 190 calories, 188 calories from fat, 21.3 g total fat, 0 mg cholesterol, 176.9 mg sodium, 3.4 mg potassium, <1 g carbohydrates, <1 g fiber, <1 g sugar, <1 g protein

Sweet Balsamic Vinaigrette

Yield: *16 servings, 1 fl oz each*

Ingredients

balsamic vinegar	5 fl oz
granulated sugar	2 tbsp
olive oil	10 fl oz
salt and pepper	TT

Preparation

1. Place vinegar and sugar in mixing bowl and whisk until sugar is almost dissolved.
2. Add olive oil while whisking continuously.
3. Season to taste.

NUTRITION FACTS
Per serving: 170 calories, 155 calories from fat, 17.5 g total fat, 0 mg cholesterol, 20.9 mg sodium, 11.9 mg potassium, 3.3 g carbohydrates, 0 g fiber, 3.1 g sugar, <1 g protein

Red Wine Oregano Vinaigrette

Yield: *16 servings, 1 fl oz each*

Ingredients

red wine vinegar	4 fl oz
kosher salt	1½ tsp
black pepper, ground	¼ tsp
granulated sugar	pinch
oregano, dried	2 tsp
scallions, diced	1 tbsp
canola oil	12 fl oz

Preparation

1. In a stainless steel mixing bowl, combine vinegar, salt, pepper, sugar, oregano, and scallions and whisk to incorporate.
2. Add oil in a steady stream while whisking continuously to form a temporary emulsion.
3. Whisk again prior to service.

NUTRITION FACTS
Per serving: 190 calories, 188 calories from fat, 21.3 g total fat, 0 mg cholesterol, 177 mg sodium, 6.8 mg potassium, <1 g carbohydrates, <1 g fiber, <1 g sugar, <1 g protein

Greek Vinaigrette

Yield: *16 servings, 1 fl oz each*

Ingredients

red wine vinegar	½ c
sugar	1 tsp
olive oil	1½ c
lemon juice	2 tsp
oregano	2 tsp
scallions, minced	2 tbsp
salt and pepper	TT

Preparation

1. Place vinegar and sugar in a mixing bowl and whisk until most of the sugar has dissolved.
2. Slowly add the olive oil while whisking.
3. Add lemon juice, oregano, and scallions and mix well.
4. Season to taste.

NUTRITION FACTS
Per serving: 185 calories, 179 calories from fat, 20.3 g total fat, 0 mg cholesterol, 19.4 mg sodium, 9.5 mg potassium, 1.1 g carbohydrates, <1 g fiber, <1 g sugar, <1 g protein

Mayonnaise

Yield: *32 servings, 1 tbsp each*

Ingredients

pasteurized egg yolks	2 ea
kosher salt	½ tsp
white pepper	pinch
dry mustard powder	½ tsp
white wine vinegar	1½ tbsp
canola oil	14 fl oz
lemon juice	½ tsp
Tabasco® sauce	dash
Worcestershire sauce	dash

Preparation

1. Allow all ingredients to sit at room temperature for about 1 hour.
2. In a nonreactive mixing bowl, whisk the yolks with the salt, pepper, mustard, and half of the vinegar until light colored and foamy.
3. Add about one-fourth of the oil very slowly, a drop at a time, while whisking rapidly and continuously, until the emulsion begins to form and the mixture thickens slightly.
4. Continue to add the oil in a fine stream while whisking rapidly and continuously to incorporate.
5. As the mixture becomes difficult to whisk, add a small amount of vinegar to thin the mixture slightly.
6. Continue to add oil and then vinegar, alternating them while whisking, until all the oil and vinegar have been added.
7. Add lemon juice, Tabasco®, and Worcestershire. Adjust seasonings if necessary.
8. Place in a chilled storage container and refrigerate until needed.

> **NUTRITION FACTS**
> **Per serving:** 113 calories, 112 calories from fat, 12.7 g total fat, 11.3 mg cholesterol, 30.1 mg sodium, 2.3 mg potassium, <1 g carbohydrates, <1 g fiber, <1 g sugar, <1 g protein

Thousand Island Dressing

Yield: *16 servings, 1 fl oz each*

Ingredients

mayonnaise	1 pt
ketchup	2¼ tbsp
chili sauce	2½ fl oz
hard-cooked egg, chopped	1 ea
small shallot, minced	1 ea
garlic, crushed	½ tsp
pimentos, chopped	2 tsp
gherkins, chopped	1 tbsp
capers, chopped	2 tsp
green peppers, minced	2 tsp
chives, minced	1 tsp
parsley, minced	½ tsp
sugar	1 tsp
Worcestershire sauce	dash
Tabasco® sauce	dash
lemon juice	1 tsp
salt and pepper	TT

Preparation

1. Place all ingredients in a nonreactive bowl and mix well.
2. Season to taste.

> **NUTRITION FACTS**
> **Per serving:** 149 calories, 91 calories from fat, 10.2 g total fat, 20.9 mg cholesterol, 276.4 mg sodium, 118 mg potassium, 14.1 g carbohydrates, <1 g fiber, 3.6 g sugar, 1.5 g protein

Creamy Garlic Dressing

Yield: *16 servings, 1 fl oz each*

Ingredients

sour cream	6 oz
mayonnaise	6 oz
garlic, minced	2 cloves
garlic powder	½ tsp
Worcestershire sauce	dash
Tabasco® sauce	dash
French vinaigrette	3 fl oz
salt and pepper	TT

Preparation

1. Place all ingredients except the vinaigrette in a nonreactive mixing bowl and mix well.
2. Add vinaigrette and mix well.
3. Season to taste.

NUTRITION FACTS
Per serving: 98 calories, 85 calories from fat, 9.6 g total fat, 8.3 mg cholesterol, 136 mg sodium, 19.5 mg potassium, 3.1 g carbohydrates, <1 g fiber, 1.1 g sugar, <1 g protein

Blue Cheese Dressing

Yield: *16 servings, 1 fl oz each*

Ingredients

sour cream	6 oz
mayonnaise	6 oz
garlic, minced	½ tsp
blue cheese, crumbled	3 oz
Worcestershire sauce	dash
Tabasco® sauce	dash
French vinaigrette	3 fl oz
salt and pepper	TT

Preparation

1. Place all ingredients except the vinaigrette in a nonreactive mixing bowl and mix well.
2. Add vinaigrette and mix well.
3. Season to taste.

NUTRITION FACTS
Per serving: 116 calories, 98 calories from fat, 11.2 g total fat, 12.3 mg cholesterol, 210.1 mg sodium, 30.8 mg potassium, 3 g carbohydrates, 0 g fiber, 1.1 g sugar, 1.5 g protein

Green Goddess Dressing

Yield: *16 servings, 1 fl oz each*

Ingredients

chives, minced	1 tbsp
parsley, chopped	2 tbsp
garlic, minced	1 clove
fresh lemon juice	2 tbsp
tarragon vinegar	½ fl oz
mayonnaise	1¼ c
sour cream	½ c
Worcestershire sauce	dash
Tabasco® sauce	dash
salt and pepper	TT

Preparation

1. Place herbs in a blender with lemon juice and vinegar and pulse until finely chopped.
2. Add mayonnaise, sour cream, Worcestershire, and Tabasco® and blend well.
3. Season to taste.

NUTRITION FACTS
Per serving: 86 calories, 67 calories from fat, 7.6 g total fat, 8.5 mg cholesterol, 155.3 mg sodium, 18.5 mg potassium, 4.9 g carbohydrates, <1 g fiber, 1.5 g sugar, <1 g protein

Buttermilk Ranch Dressing

Yield: *16 servings, 1 fl oz each*

Ingredients

kosher salt	½ tsp
garlic, minced	3 cloves
buttermilk	6 fl oz
mayonnaise	9 oz
white vinegar	2 tsp
parsley, chopped	3 tbsp
chives, minced	3 tbsp
salt and black pepper	TT

Preparation

1. Place salt on the minced garlic and use the blade of a chef's knife to cream the garlic.
2. Whisk together the creamed garlic and buttermilk.
3. Add mayonnaise, vinegar, parsley, and chives and mix well.
4. Season to taste.

NUTRITION FACTS
Per serving: 68 calories, 48 calories from fat, 5.4 g total fat, 4.6 mg cholesterol, 202.9 mg sodium, 27.3 mg potassium, 4.7 g carbohydrates, <1 g fiber, 1.6 g sugar, <1 g protein

Caesar Dressing

Yield: *16 servings, 1 fl oz each*

Ingredients

buttermilk ranch dressing	13 fl oz
Parmesan cheese, grated	3 oz
red wine vinegar	1 tbsp
lemon juice	½ tsp
garlic powder	1 tsp
black peppercorns, cracked	1 tsp
salt	TT

Preparation

1. Whisk all ingredients together.
2. Season to taste.

NUTRITION FACTS
Per serving: 80 calories, 52 calories from fat, 6 g total fat, 8.4 mg cholesterol, 264.5 mg sodium, 34.4 mg potassium, 4.3 g carbohydrates, <1 g fiber, 1.4 g sugar, 2.6 g protein

Croutons

Yield: *20 servings, 1 oz each*

Ingredients

white bread, crusts removed and diced	10 slices
butter, melted	2½ fl oz
salt	½ tsp
pepper	¼ tsp

Preparation

1. Toss all ingredients together to coat the bread.
2. Place bread on sheet pan and bake in a 350°F oven until lightly browned on all sides (approximately 15 minutes).

NUTRITION FACTS
Per serving: 41 calories, 27 calories from fat, 3.1 g total fat, 7.6 mg cholesterol, 89.2 mg sodium, 7.2 mg potassium, 3.1 g carbohydrates, <1 g fiber, <1 g sugar, <1 g protein

Parmesan Croutons

Yield: *20 servings, 1 oz each*

Ingredients

white bread, crusts removed and diced	10 slices
butter, melted	2½ fl oz
salt	½ tsp
pepper	¼ tsp
Parmesan cheese	1 tbsp

Preparation

1. Toss bread, melted butter, salt, and pepper together to coat the bread.
2. Place buttered bread on sheet pan and bake in a 350°F oven until lightly browned on all sides (approximately 15 minutes).
3. Sprinkle croutons with Parmesan cheese immediately after removing from the oven.

NUTRITION FACTS
Per serving: 43 calories, 28 calories from fat, 3.2 g total fat, 7.9 mg cholesterol, 93 mg sodium, 7.5 mg potassium, 3.1 g carbohydrates, <1 g fiber, <1 g sugar, <1 g protein

Grilled Chicken Caesar Salad

Yield: *4 servings, 9 oz each*

Ingredients

romaine lettuce	1 lb
Caesar dressing	6 fl oz
grilled chicken breasts, sliced into thin strips and chilled	½ lb
seasoned croutons	4 oz
Parmesan cheese, grated	2 oz

Preparation

1. Cut, wash, and pat the lettuce dry.
2. Toss lettuce with Caesar dressing and portion on plates.
3. Top salad with chicken breast, croutons, and Parmesan cheese.

NUTRITION FACTS
Per serving: 326 calories, 136 calories from fat, 15.4 g total fat, 30.6 mg cholesterol, 825.6 mg sodium, 401.3 mg potassium, 31.6 g carbohydrates, 4 g fiber, 3.5 g sugar, 16.1 g protein

Insalata Kalamata

Yield: *4 servings, 5 oz each*

Ingredients

mesclun greens	4 oz
kalamata olives	2 oz
crumbled feta	2 oz
grape tomatoes, sliced in half	4 oz
sun-dried tomatoes	2 oz
cucumbers, medium dice	2 oz
red wine oregano vinaigrette	4 fl oz

Preparation

1. Gently toss all ingredients together except for the vinaigrette.
2. Lightly toss the salad with the vinaigrette.

NUTRITION FACTS
Per serving: 312 calories, 253 calories from fat, 28.6 g total fat, 12.6 mg cholesterol, 861.6 mg sodium, 634.2 mg potassium, 11.8 g carbohydrates, 2.6 g fiber, 6.9 g sugar, 4.7 g protein

Praline Pecan and Pear Salad with Sweet Balsamic Vinaigrette and Crumbled Blue Cheese

Yield: *4 servings, 4 oz each*

Ingredients

Praline Pecans

shelled pecans	½ c
brown sugar	½ tbsp
water	2 tbsp
salt and pepper	TT
butter, unsalted	2 tbsp

Salad

mesclun greens	4 oz
ripe pear, medium dice	½ c
blue cheese, crumbled	1 oz
sweet balsamic vinaigrette	2 fl oz

Preparation

1. Place shelled pecans in hot sauté pan and dry pan-roast until edges of pecans are brown. *Note:* Pecans may also be browned in oven.
2. When pecans are slightly browned, add brown sugar and let the sugar melt slightly.
3. Add water and season to taste.
4. Stir continuously until all of the sugar has melted and all the water has completely evaporated.
5. Pour nuts onto buttered aluminum foil to cool.
6. Toss all salad ingredients gently to coat thoroughly.
7. Top the salad with the praline pecans.

NUTRITION FACTS
Per serving: 277 calories, 228 calories from fat, 26.4 g total fat, 20.6 mg cholesterol, 186.2 mg sodium, 152.8 mg potassium, 9.1 g carbohydrates, 2.3 g fiber, 5.8 g sugar, 3.2 g protein

Thai Noodle Salad

Yield: *8 servings, 4 oz each*

Ingredients _____

Dressing

garlic, minced	2 cloves
scallions, minced	3 ea
chili garlic sauce	2 tsp
sugar	2 tbsp
lime juice	1 tbsp
rice wine vinegar	2 fl oz
fish sauce	1 tbsp
salad oil	6 fl oz
sesame oil	1 tbsp
cilantro	2 tbsp

Salad

rice noodles (vermicelli), cooked and cooled	1 lb
bean sprouts	3 oz
green onions, bias cut in 1 inch lengths	4 ea
carrot, julienned	4 oz
red pepper, julienned	1 ea
Napa cabbage, julienned	4 leaves

Preparation _____

1. Combine all dressing ingredients and mix well to incorporate.
2. Place salad ingredients and dressing in a mixing bowl and toss gently.

> **NUTRITION FACTS**
> **Per serving:** 299 calories, 203 calories from fat, 23 g total fat, 0 mg cholesterol, 230.5 mg sodium, 232.8 mg potassium, 25.5 g carbohydrates, 2.1 g fiber, 5.7 g sugar, 1.7 g protein

Tuna Niçoise Salad

Yield: *8 servings, 8 oz each*

Ingredients _____

Dressing

red wine vinegar	2 fl oz
Dijon mustard	2 tsp
shallots, finely minced	2 tsp
vegetable oil	6 fl oz
sugar	½ tsp
mixed herbs (basil, parsley, tarragon)	2 tbsp
salt and black pepper	TT

Salad

Bibb lettuce leaves	8 ea
mixed greens	4 oz
red potatoes, cooked	4 ea
green beans, blanched	¾ lb
tomatoes, cut into wedges	1 ea
roasted red peppers	4 oz
Niçoise olives, pitted	12 ea
eggs, hard cooked	2 ea
croutons	4 oz
tuna steaks, grilled and chilled	4 ea

Preparation _____

1. Combine vinegar, Dijon mustard, and shallots.
2. Slowly whisk in oil.
3. Season with sugar, herbs, salt, and pepper.
4. Place Bibb lettuce on plate as a liner.
5. Add a small amount of dressing to the mixed greens and toss well.
6. Place on top of Bibb lettuce.
7. Carefully arrange red potatoes, green beans, tomatoes, roasted red peppers, olives, eggs, and croutons on the mixed greens.
8. Top with tuna steaks.

> **NUTRITION FACTS**
> **Per serving:** 559 calories, 275 calories from fat, 30.9 g total fat, 94.7 mg cholesterol, 314.5 mg sodium, 1013.2 mg potassium, 37.7 g carbohydrates, 5 g fiber, 2.8 g sugar, 32.8 g protein

Zesty Potato Salad with Dill

Yield: *4 servings, 4 oz each*

Ingredients

new red potatoes	12 oz
mayonnaise	2 tbsp
sour cream	1½ tbsp
garlic, minced	1 tsp
dill, minced	2 tsp
whole-grain mustard	2 tsp
red pepper, small dice	1 tbsp
green pepper, small dice	1 tbsp
celery, small dice	1 tbsp
red onion, small dice	2 tbsp
celery seed	½ tsp
fresh lemon juice	1 tsp
salt and black pepper	TT

Preparation

1. Cook potatoes until al dente.
2. Refrigerate potatoes to cool completely.
3. Add remaining ingredients and whisk together.
4. When potatoes are cool, cut into quarters.
5. Add potatoes to dressing and toss gently to coat.

NUTRITION FACTS
Per serving: 334 calories, 35 calories from fat, 4 g total fat, 4.3 mg cholesterol, 186.9 mg sodium, 2019.9 mg potassium, 69 g carbohydrates, 6.1 g fiber, <1 g sugar, 8.1 g protein

Waldorf Salad

Yield: *10 servings, 5 oz each*

Ingredients

red apples	3 ea
green apples	2 ea
lemon juice	½ tsp
mayonnaise	6 oz
whipped cream, whipped stiff	½ c
vanilla extract	1 tsp
sugar	2 tsp
salt and pepper	TT
walnuts, chopped	4 oz
leaf lettuce, washed and separated	10 ea
maraschino cherries, halved	5 ea
parsley	10 sprigs

Preparation

1. Peel and cut apples into small dice. Toss with lemon juice to prevent browning.
2. Mix mayonnaise, whipped cream, vanilla, sugar, salt, and pepper to make dressing.
3. Add apples and walnuts to the dressing and mix well.
4. Serve a scoop of salad on a lettuce leaf. Garnish with a half cherry and parsley.

NUTRITION FACTS
Per serving: 208 calories, 133 calories from fat, 15.5 g total fat, 12.6 mg cholesterol, 158.6 mg sodium, 161.1 mg potassium, 17.8 g carbohydrates, 2.6 g fiber, 9.6 g sugar, 2.4 g protein

Spinach and Roasted Potato Salad with Caraway Bacon Vinaigrette

Yield: *6 servings, 7 oz each*

Ingredients

bacon, diced	4 oz
shallots, minced	1 ea
garlic, minced	1 clove
caraway seeds, pan-toasted	1 tsp
brown sugar	1 oz
cider vinegar	3 fl oz
vegetable oil	6 fl oz
whole-grain mustard	1 tbsp
fresh thyme, minced	1½ tsp
salt and black pepper	TT
fresh leaf spinach, ribs removed	6 oz
roasted potatoes, cubed	1 lb

Preparation

1. Render the bacon. Remove bacon from heat and reserve.
2. Add shallots, garlic, and caraway seeds to the bacon fat and sweat until tender.
3. Blend in brown sugar until it melts. Do not burn.
4. Remove pan from heat and whisk in vinegar, oil, mustard, and thyme until emulsified.
5. Season to taste.
6. Place spinach, potatoes, bacon, and vinaigrette in a large bowl and toss to mix well.

NUTRITION FACTS
Per serving: 504 calories, 326 calories from fat, 36.8 g total fat, 20.8 mg cholesterol, 559.4 mg sodium, 920.1 mg potassium, 33.9 g carbohydrates, 2.6 g fiber, 5.7 g sugar, 11.7 g protein

Green Bean, Tomato, and Prosciutto Salad with Warm Bacon Vinaigrette

Yield: *10 servings, 4 oz each*

Ingredients

Dressing

bacon, diced	5 oz
shallot, minced	2 tbsp
garlic, minced	1 clove
sugar	1 tbsp
cider vinegar	4 fl oz
Dijon mustard	1 tsp
vegetable oil	1½ tbsp
salt and black pepper	TT

Salad

green beans, trimmed, blanched, and shocked	2 lb
tomatoes, peeled, cut into wedges, and blanched	2 ea
red onion, julienned	4 oz
prosciutto, sliced thin and julienned	4 oz
croutons	4 oz

Preparation

1. Render diced bacon until crisp and brown. Remove from pan and drain on paper towels.
2. Reserve one-fourth of the rendered bacon fat in the pan. Discard the remainder.
3. Add shallots and garlic to the bacon fat and sweat over medium heat for 1–2 minutes.
4. Add sugar and vinegar and heat to dissolve the sugar.
5. Remove from heat and whisk in mustard and oil.
6. Season to taste.
7. Combine all salad ingredients in a mixing bowl.
8. Add dressing for salad at time of service.

NUTRITION FACTS
Per serving: 203 calories, 84 calories from fat, 9.4 g total fat, 23.5 mg cholesterol, 751.1 mg sodium, 365.8 mg potassium, 18.3 g carbohydrates, 4 g fiber, 2.2 g sugar, 11.9 g protein

Asparagus and Tomato Salad with Shaved Fennel and Gorgonzola

Yield: *4 servings, 6 oz each*

Ingredients

Dressing

olive oil	3 tbsp
red wine vinegar	½ tbsp
balsamic vinegar	½ tbsp
garlic, minced	½ tsp
shallots, minced	2 tsp
thyme, minced	¼ tsp
tarragon, minced	¼ tsp

Salad

fennel, shaved	¼
red onion, julienned	2 oz
tomatoes, peeled and cut into wedges	2 ea
asparagus, cut into 1½ inch sections, blanched and cooled	½ lb
Gorgonzola cheese	2 oz

Preparation

1. Combine all dressing ingredients in a bowl and mix well.
2. Toss salad ingredients in dressing. *Note:* Asparagus can be left whole for a composed salad presentation, with the other ingredients placed on top of the dressed asparagus.

> **NUTRITION FACTS**
> **Per serving:** 177 calories, 129 calories from fat, 15 g total fat, 12.7 mg cholesterol, 211.1 mg sodium, 349 mg potassium, 8.3 g carbohydrates, 2.7 g fiber, 3.2 g sugar, 5.3 g protein

Roasted Corn and Black Bean Salad

Yield: *8 servings, 4 oz each*

Ingredients

Salad

corn, shucked	2 ears
vegetable oil	2 tsp
red onion, brunoise, rinsed	3 tbsp
red pepper, brunoise	3 tbsp
green pepper, brunoise	3 tbsp
black beans, cooked and rinsed	¾ c

Dressing

garlic, creamed	1 clove
shallot, minced	1 ea
honey	2 tsp
lime juice	3 tbsp
white vinegar	1 tsp
vegetable oil	3 fl oz
cumin	¼ tsp
chipotle pepper, roasted and minced	½ ea
cilantro, chopped	3 tbsp
salt and pepper	TT

Preparation

1. Lightly oil the corn.
2. Roast corn until charred slightly. Cool completely.
3. Remove kernels of corn from the cobs.
4. Mix all dressing ingredients well.
5. Add salad ingredients to dressing and toss gently.

> **NUTRITION FACTS**
> **Per serving:** 196 calories, 108 calories from fat, 12.3 g total fat, 0 mg cholesterol, 86.3 mg sodium, 316.9 mg potassium, 20.2 g carbohydrates, 2.3 g fiber, 3.2 g sugar, 3.8 g protein

Cucumber Salad

Yield: *6 servings, 5 oz each*

Ingredients

white wine vinegar	1½ fl oz
garlic, minced	1 clove
dill	1 tbsp
vegetable oil	3 fl oz
sugar	1 tsp
kosher salt and white pepper	TT
pickled ginger, julienned	1 tbsp
cucumber, peeled, seeded, cut in half moons	1 ea
red onion, julienned	½ ea

Preparation

1. Mix all ingredients except cucumber and onion.
2. Add cucumber and onion and mix well.

> **NUTRITION FACTS**
> **Per serving:** 140 calories, 126 calories from fat, 14.3 g total fat, 0 mg cholesterol, 48.6 mg sodium, 77.4 mg potassium, 3.3 g carbohydrates, <1 g fiber, 1.4 g sugar, <1 g protein

Spicy Garlic and Citrus Shrimp

Yield: *4 servings, 4 oz each*

Ingredients

olive oil	1 tbsp
medium shrimp, peeled and deveined	1 lb
shallots, minced	2 ea
red pepper flakes	1 tsp
garlic, minced	4 cloves
sun-dried tomatoes, julienned	2 tbsp
piquillo peppers, julienned	2 tbsp
green onions, sliced in ¼ inch slices	2 tbsp
lemon juice, fresh squeezed	2 tbsp
lime juice, fresh squeezed	2 tbsp
dry sherry	2 tbsp
paprika	1 tsp
parsley, chopped	2 tbsp
kosher salt and white pepper	TT

Preparation

1. In a sauté pan over medium-high heat, heat the olive oil.
2. Add shrimp and sauté until the shrimp begin to turn pink on one side.
3. Turn shrimp over and add shallots and red pepper flakes to the pan. Sauté for 1 minute until shallots are cooked.
4. Add garlic, tomatoes, peppers, and green onions. Sauté the mixture for 1 minute.
5. Add lemon juice, lime juice, sherry, and paprika to shrimp and stir well.
6. Cook shrimp mixture for 1 minute more until shrimp are cooked through.
7. Toss in parsley and season to taste.

> **NUTRITION FACTS**
> **Per serving:** 280 calories, 43 calories from fat, 4.9 g total fat, 142.9 mg cholesterol, 774.7 mg sodium, 921.6 mg potassium, 38.8 g carbohydrates, <1 g fiber, 1.3 g sugar, 21.1 g protein

Mini Serrano Ham and Arugula Wraps

Yield: *4 servings, 2 pieces each*

Ingredients

Vinaigrette

sherry vinegar	1½ tsp
granulated sugar	pinch
walnut oil	2 tsp
olive oil	2 tsp
kosher salt and pepper	TT

Serrano ham slices, sliced very thin	8 ea
arugula, washed and spun dry	1 c
mixed baby salad greens, washed and spun dry	½ c
Manchego cheese, cut into 1 × 3 inch slices	2 oz

Preparation

1. Place vinegar, sugar, walnut oil, and olive oil in mixing bowl and whisk to incorporate. Season to taste.
2. Lay the ham slices side by side on a flat work surface.
3. Place arugula and greens in a mixing bowl. Add vinaigrette and toss gently to coat.
4. Top each slice of ham with small portion of greens, divided evenly among the slices.
5. Top each with 1 slice of cheese.
6. Roll the ham, leaving a tuft of greens protruding from one end.

NUTRITION FACTS
Per serving: 153 calories, 96 calories from fat, 10.7 g total fat, 34.6 mg cholesterol, 1006.9 mg sodium, 188.2 mg potassium, 1.2 g carbohydrates, <1 g fiber, <1 g sugar, 12.6 g protein

Potstickers

Yield: *6 servings, 2 each*

Ingredients

Filling

Napa cabbage, shredded	2 oz
ground pork	4 oz
green onions, chopped	2 ea
white wine	1 tsp
cornstarch	½ tsp
sesame oil	¼ tsp
salt and pepper	TT

Wontons

all-purpose flour	½ c
water	½ c
vegetable oil	2 tbsp

Sauce

soy sauce	2 tbsp
sesame oil	2 tsp
garlic, minced	1 clove

Preparation

1. Combine all filling ingredients in a stainless steel bowl and set aside.
2. Bring half the water to a boil.
3. Pour the boiling water into a separate bowl and combine it with the flour and half of the vegetable oil. Knead the mixture until smooth.
4. Roll dough into a 6 inch long roll and cut into equally sized pieces.
5. Roll each piece into a 3 inch circle.
6. Place 1 tbsp of the filling in the center of each circle.
7. Bring up the edges of each circle to create 5 pleats. Pinch the edges together at the top to make a pouch.
8. Heat remaining vegetable oil in a large skillet. Cook wontons in hot oil until the bottoms are lightly browned.
9. Add the remaining water to the skillet, cover, and let steam for approximately 7 minutes or until all the water is absorbed by the wontons.
10. Combine the soy sauce, sesame oil, and garlic in a serving bowl. Serve immediately.

NUTRITION FACTS
Per serving: 162 calories, 97 calories from fat, 10.9 g total fat, 13.6 mg cholesterol, 397.9 mg sodium, 114.3 mg potassium, 10.1 g carbohydrates, <1 g fiber, <1 g sugar, 5.2 g protein

Ahi Tuna Nacho

Yield: *16 servings, 1 tortilla each*

Ingredients

Wasabi Sour Cream

sour cream	2 tbsp
wasabi powder	⅛ tsp

Avocado Purée

avocado	½ ea
lemon juice	⅛ tsp
cilantro, finely minced	1 tsp
scallion, finely minced	¼ tsp
garlic, minced	⅛ tsp
Tabasco® sauce	dash

ahi tuna, cut into a log	6 oz
coriander, ground	pinch
cumin, ground	pinch
kosher salt and ground pepper	TT
canola oil	1 tsp

Tortilla Chips

6 inch white corn tortillas, cut into quarters	4 ea
vegetable oil	(for deep frying)

Preparation

1. Mix wasabi powder and sour cream until completely incorporated.
2. Place the mixture in a small pastry bag without a tip.
3. Place all of the avocado purée ingredients in a food processor and purée until smooth.
4. Season the tuna on all sides with coriander, cumin, salt, and pepper.
5. Place sauté pan over medium-high heat and add oil.
6. Place the seasoned tuna in a sauté pan and sear quickly on all sides, but do not cook through.
7. Remove tuna from pan and let rest a minute before slicing.
8. Slice tuna into equal slices about ¼ inch thick.
9. Using the swimming method, deep-fry the quartered tortillas in oil heated to 350°F until golden and crisp (approximately 1 minute).
10. Drain on paper towels.
11. Apply about 1½ tsp of the avocado mixture to the top of each chip.
12. Lay a slice of seared tuna on top of the avocado.
13. Add dollops of wasabi sour cream an equal distance from each other where the chips will eventually be placed. *Note:* This will anchor the chips to the plate.
14. Place a chip on top of each dollop with the narrowest end facing toward the center of the plate.
15. Top each slice of tuna with a dollop of wasabi sour cream and serve.

NUTRITION FACTS
Per serving: 43 calories, 17 calories from fat, 1.9 g total fat, 4.9 mg cholesterol, 27.3 mg sodium, 89.9 mg potassium, 3.5 g carbohydrates, <1 g fiber, <1 g sugar, 3.1 g protein

Seafood Canapés

Yield: *4 serving, 4 each*

Ingredients

white bread, crust removed	4 slices
crab meat, flaked	⅓ c
salad shrimp	⅓ c
celery, finely chopped	2 tbsp
green onion, finely chopped	1 tbsp
mayonnaise	¼ c
sea salt	1 tsp
fresh dill	2 sprigs

Preparation

1. Arrange the bread on a cutting board and flatten it slightly with a rolling pin.
2. In a small bowl, combine the crab meat, shrimp, celery, onion, and mayonnaise.
3. Divide the seafood spread evenly among the bread slices, coating each slice evenly.
4. Cut each bread slice into quarters.
5. Season each canapé with sea salt and garnish with a bit of fresh dill.
6. Serve canapés chilled.

> **NUTRITION FACTS**
> **Per serving:** 123 calories, 50 calories from fat, 5.7 g total fat, 39.5 mg cholesterol, 937 mg sodium, 95.8 mg potassium, 10 g carbohydrates, <1 g fiber, 1.6 g sugar, 7.9 g protein

Tomato and Mozzarella Canapés with Basil Pesto

Yield: *8 servings, 2 each*

Ingredients

French baguette	½ loaf
olive oil	¼ c
fresh basil	2 c
Parmesan cheese, grated	2 oz
pine nuts, toasted	2 c
garlic, minced	1 clove
fresh mozzarella	½ lb
grape tomatoes	16 ea
salt and pepper	TT

Preparation

1. Slice the baguette on the diagonal into 16 slices about ½ inch thick each.
2. Lightly brush each slice with olive oil and place on a baking sheet.
3. Bake in a 350°F oven for 8–10 minutes or until lightly golden.
4. Blanch basil leaves in boiling water for 2 seconds. Immediately place the blanched basil leaves in ice water and then remove them.
5. In a food processor, purée the basil, Parmesan cheese, pine nuts, and garlic.
6. Add the remaining olive oil and mix well.
7. Spread the pesto on toasted bread slices.
8. Thinly slice the mozzarella and place one slice on each canapé.
9. Thinly slice the grape tomatoes and arrange the slices, one tomato on each canapé.
10. Season to taste.

> **NUTRITION FACTS**
> **Per serving:** 476 calories, 319 calories from fat, 37.4 g total fat, 24.4 mg cholesterol, 495.8 mg sodium, 379.6 mg potassium, 21.9 g carbohydrates, 2.7 g fiber, 2.5 g sugar, 17.4 g protein

Cucumber and Smoked Salmon Canapés

Yield: *8 servings, 4 each*

Ingredients

medium cucumber	1 ea
cream cheese, softened	8 oz
sour cream	½ c
chives, finely chopped	1 tbsp
grape tomatoes	16 ea
smoked salmon, sliced thin into ½ oz pieces	1 lb
parsley	1 sprig

Preparation

1. Peel the skin off the cucumber. Slice the cucumber into ¼ inch thick slices.
2. In a mixing bowl, combine cream cheese, sour cream, and chives until well blended.
3. Spread 2 tsp of cream cheese spread on each cucumber slice.
4. Slice grape tomatoes in half.
5. Garnish each cucumber slice with half of a grape tomato, a piece of smoked salmon, and a bit of fresh parsley.
6. Serve canapés cold.

NUTRITION FACTS
Per serving: 203 calories, 134 calories from fat, 15.2 g total fat, 51.7 mg cholesterol, 1240.4 mg sodium, 283.5 mg potassium, 3.9 g carbohydrates, <1 g fiber, 1.9 g sugar, 12.9 g protein

Shrimp Cocktail with Cocktail Sauce

Yield: *4 servings, 3 each*

Ingredients

jumbo shrimp, fully cooked	12 ea
lemon	1 ea

Cocktail Sauce

ketchup	¼ cup
chili sauce	1 tbsp
tomato purée	1 tbsp
lemon juice	1 tsp
horseradish, prepared	½ tbsp
salt	TT

Preparation

1. Arrange shrimp around the rim of a cocktail glass filled with crushed ice.
2. In a small bowl, combine all cocktail sauce ingredients until well blended.
3. Place a tiny serving dish in the center of the cocktail glass and fill with cocktail sauce.
4. Garnish with lemon wedges and serve.

NUTRITION FACTS
Per serving: 43 calories, 4 calories from fat, <1 g total fat, 34.8 mg cholesterol, 404.8 mg sodium, 168.7 mg potassium, 7.7 g carbohydrates, 1.5 g fiber, 3.9 g sugar, 4.5 g protein

Variation:

Shrimp Cocktail with Mignonette: Replace the cocktail sauce ingredients with 1 tsp freshly cracked black pepper, ¼ cup of red wine vinegar, 1 minced shallot, and salt to taste.

Bloody Mary Shrimp Shooters

Yield: *6 servings, 1 shot each*

Ingredients

Bloody Mary

tomato juice	8 fl oz
horseradish	1 tsp
freshly squeezed lemon juice	1 tbsp
Worcestershire sauce	dash
Tabasco® sauce	¼ tsp
salt and black pepper	TT

Garnish

lemon zest, grated	1 tbsp
Italian parsley, finely chopped	1 tbsp
garlic, finely chopped	2 cloves
large shrimp, cooked, medium dice	6 ea

Preparation

1. Combine the Bloody Mary ingredients together in a nonreactive bowl and chill for 1 hour.
2. In a small bowl, combine the lemon zest, parsley, and garlic.
3. Place 1 oz Bloody Mary mixture in each shot glass.
4. Add approximately ½ oz shrimp to each shot glass.
5. Sprinkle with the lemon-parsley mixture and serve.

NUTRITION FACTS
Per serving: 17 calories, 1 calories from fat, <1 g total fat, 11.6 mg cholesterol, 109.9 mg sodium, 116.9 mg potassium, 2.6 g carbohydrates, <1 g fiber, 1.6 g sugar, 1.7 g protein

Steak and Cheese Quesadillas

Yield: *4 servings, ½ quesadilla each*

Ingredients

vegetable oil	2 tbsp
flank steak, thinly sliced	1 lb
cumin, ground	½ tsp
salsa, chunky	¼ c
flour tortillas, 7 inches in diameter	4 ea
queso Chihuahua (Mexican melting cheese), crumbled	½ lb
fresh cilantro	1 tbsp
sour cream	4 tbsp

Preparation

1. In a sauté pan, heat half the oil. Add flank steak and cumin. Cook until browned.
2. Add salsa and cook until heated through.
3. Arrange tortillas on a cutting board.
4. Divide cheese evenly and sprinkle on the bottom tortillas.
5. Cover the cheese evenly with the meat mixture and cilantro.
6. Top each tortilla with a second tortilla.
7. Heat the remaining oil in a sauté pan.
8. Cook filled tortillas in hot oil until cheese is melted and the tortilla is golden brown.
9. Turn the quesadillas over and brown the other side.
10. Serve quesadillas with sour cream.

NUTRITION FACTS
Per serving: 606 calories, 318 calories from fat, 36.1 g total fat, 136.1 mg cholesterol, 812.6 mg sodium, 562.2 mg potassium, 28.3 g carbohydrates, 1.8 g fiber, 5 g sugar, 41.1 g protein

Mushroom Barquettes

Yield: *4 servings, 3 each*

Ingredients

pâte dough	12 oz

Filling

fresh mushrooms, sliced	¼ c
sour cream	¼ c
cream cheese, softened	¼ c
mozzarella cheese, shredded	½ c
fresh rosemary, finely chopped	1 sprig

Preparation

1. On a lightly floured surface, roll out chilled pâte dough in a 13 inch square.
2. Arrange 12 barquette molds close together. Drape the dough over the molds.
3. With a rolling pin, roll over the molds to cut the dough.
4. Press the dough into each mold.
5. Prick the dough with a fork and chill the molds for 20 minutes.
6. Arrange the molds on a baking sheet and cover with parchment paper.
7. Bake at 375°F for approximately 8 minutes.
8. Cool the barquette shells for 10 minutes.
9. Remove the molds and let the shells cool completely on a wire rack.
10. Sauté the mushrooms and let them cool.
11. Blend the mushrooms with the remaining filling ingredients.
12. Fill each barquette shell with a spoonful of mushroom filling.
13. Place filled barquette shells on a baking sheet and broil for 3 minutes.
14. Remove the golden barquettes and serve immediately.

NUTRITION FACTS
Per serving: 368 calories, 104 calories from fat, 11.7 g total fat, 41.6 mg cholesterol, 646.9 mg sodium, 139 mg potassium, 50.9 g carbohydrates, 1.6 g fiber, 1.2 g sugar, 13.6 g protein

Tomato Cream Cheese Bouchées

Yield: *4 servings, 4 each*

Ingredients

olive oil	2 tbsp
garlic, minced	1 clove
medium tomato, diced	1 ea
fresh basil	½ tbsp
salt and pepper	TT
cream cheese, softened	8 oz
puff pastry dough	½ lb

Preparation

1. In a small skillet, heat half of the olive oil. Sauté garlic and diced tomato for approximately 3 minutes.
2. Add basil, salt, and pepper. Cook for 1 minute and remove from heat.
3. Purée the tomato mixture and then cool it to room temperature.
4. In a mixing bowl, combine the cream cheese and the tomato purée until well blended.
5. On a lightly floured surface, roll out the puff pastry dough. Cut into 3 inch squares.
6. Place a small spoonful of the mixture on each square.
7. Fold each square diagonally and pinch the edges to contain the filling.
8. Brush both sides of each bouchée lightly with olive oil and place the bouchées on a baking sheet.
9. Bake in 375°F oven for 10–12 minutes, or until the bouchées are golden brown.
10. Allow bouchées to cool slightly on wire racks. Serve immediately.

NUTRITION FACTS
Per serving: 577 calories, 427 calories from fat, 48.1 g total fat, 62.4 mg cholesterol, 399.9 mg sodium, 190.3 mg potassium, 29.7 g carbohydrates, 1.2 g fiber, 3.1 g sugar, 7.9 g protein

Spinach and Bacon Tartlets

Yield: *4 servings, 3 each*

Ingredients

pâte dough	12 oz

Spinach and Bacon Filling

bacon	4 oz
small onion, diced	1 ea
fresh spinach	½ c
Monterey Jack cheese, grated	2 oz
feta cheese, crumbled	2 oz
eggs	3 ea
cottage cheese	⅓ c
salt and pepper	TT

Preparation

1. On a lightly floured surface, roll out the chilled dough in a 13 inch square.
2. Arrange the tartlet molds closely together. Drape the dough over the molds.
3. With a rolling pin, roll over the molds to cut the dough.
4. Press the dough into each mold.
5. Prick the dough with a fork and chill the molds for 20 minutes.
6. Arrange the molds on a baking sheet and cover with parchment paper. Bake at 375°F for approximately 8 minutes.
7. Cool the tartlet shells for 10 minutes.
8. Remove the molds and let the shells cool completely on a wire rack.
9. In a skillet, cook the bacon over medium-high heat until crispy. Place the bacon on paper towels to drain excess fat. Reserve the rendered fat.
10. Finely chop the cooked bacon.
11. In the reserved bacon fat, sauté diced onion until translucent.
12. Add fresh spinach and sauté the mixture for 3 minutes, stirring frequently.
13. Add the chopped bacon and mix well.
14. Arrange the tartlet shells on a baking sheet and evenly divide the spinach mixture among the shells.
15. Top with grated Monterey Jack and feta.
16. In a bowl, beat eggs with a wire whisk. Add the cottage cheese, salt, and pepper and whisk again.
17. Pour egg mixture over the spinach and cheese.
18. Bake the tartlets at 375°F for approximately 25 minutes, or until the filling is cooked through.
19. Serve the tartlets slightly cooled.

NUTRITION FACTS
Per serving: 320 calories, 208 calories from fat, 23.2 g total fat, 197.8 mg cholesterol, 1080.6 mg sodium, 291.3 mg potassium, 3.6 g carbohydrates, <1 g fiber, 2.1 g sugar, 23.2 g protein

Scallop Rumaki

Yield: *4 servings, 2 each*

Ingredients

bacon	4 slices
sea scallops	8 ea

Preparation

1. Stretch the bacon strips to make them thinner. Cut each strip into two 4 inch pieces.
2. Wrap one piece of bacon around each scallop and secure with a toothpick.
3. Place scallops in a 350°F oven and cook until the bacon is done and the scallops are opaque.

NUTRITION FACTS
Per serving: 64 calories, 31 calories from fat, 3.5 g total fat, 16 mg cholesterol, 302.4 mg sodium, 106.7 mg potassium, 1.1 g carbohydrates, 0 g fiber, 0 g sugar, 6.6 g protein

Prosciutto-Wrapped Baked Brie en Croûte

Yield: *4 servings, 2 ½ oz each*

Ingredients

prosciutto	¼ lb
Brie cheese (1 wheel)	8 oz
egg, beaten	1 ea
milk	1 oz
puff pastry	1 sheet

Preparation

1. Lay out sheets of thinly sliced prosciutto and wrap around the wheel of Brie.
2. Mix beaten egg and milk to make an egg wash.
3. Lay prosciutto-wrapped Brie on top of a sheet of puff pastry. Apply egg wash to the edges of the pastry.
4. Wrap edges of pastry around Brie to completely encase it.
5. Place Brie in freezer for 30 minutes to firm.
6. Apply egg wash to the top and sides of the puff pastry.
7. Bake at 350°F for 25–30 minutes or until golden brown in color. Serve warm.

NUTRITION FACTS
Per serving: 332 calories, 213 calories from fat, 24 g total fat, 123.8 mg cholesterol, 1171.1 mg sodium, 265.3 mg potassium, 6.1 g carbohydrates, <1 g fiber, <1 g sugar, 22.3 g protein

Bacon-Wrapped Blue Cheese Stuffed Dates with Balsamic Glaze

Yield: *12 servings, 2 pieces each*

Ingredients

bacon slices	12 ea
Medjool dates, pitted	4 ea
blue cheese	3 oz
cream cheese	1 oz
balsamic vinegar	¼ c
honey	1 tsp

Preparation

1. Cut each piece of bacon in half horizontally.
2. In small bowl, mix blue cheese and cream cheese thoroughly.
3. Slice one side of each date open to expose the hollow center.
4. Fill each date with approximately ½ tbsp of cheese mixture.
5. Wrap each stuffed date with a half slice of bacon, pulling the bacon tight, and place on a lightly sprayed sheet pan. *Note:* Make sure the ends of the bacon strips are on the bottom of the dates to prevent them from coming undone during broiling.
6. In a small saucepan, simmer balsamic vinegar and honey until reduced by one-third. Remove from heat and reserve.
7. Broil the stuffed dates for 2–5 minutes on each side, turning them over after a few minutes to prevent burning the bacon. *Note:* The bacon should be crispy when done.
8. Arrange broiled dates on a serving platter and drizzle with balsamic reduction.

NUTRITION FACTS
Per serving: 216 calories, 56 calories from fat, 6.3 g total fat, 16.7 mg cholesterol, 293 mg sodium, 406.9 mg potassium, 37.8 g carbohydrates, 3.2 g fiber, 33.3 g sugar, 5.5 g protein

Crab Rangoons

Yield: *12 servings, 2 each*

Ingredients

vegetable oil   (for deep-frying)

Wontons

egg	1 ea
water	¼ c
salt	1 tsp
all-purpose flour	2 c

Filling

cream cheese, softened	4 oz
crab meat, cooked and flaked	4 oz
garlic, minced	1 clove
shallot, finely chopped	1 ea
black pepper	TT

Preparation

1. To prepare wontons, beat the egg with a wire whisk. Add water and salt to the egg and whisk again.
2. Sift flour into a medium bowl.
3. Form a well in the center of the flour and pour the egg mixture into the well.
4. Mix wet and dry ingredients together. Add more water as necessary to form a dough.
5. Knead the dough for 5 minutes.
6. Allow dough to rest for 30 minutes.
7. Roll dough out paper thin on a floured surface.
8. Cut into 3 inch squares to form wonton wrappers.
9. Layer the wonton wrappers with wax paper and refrigerate until needed.
10. Place all of the filling ingredients in a stainless steel bowl and use a fork to blend them together.
11. Lay each wonton wrapper on a flat working surface and wet all four edges of the wrapper with water.
12. Place 1 tsp of filling in the center of each wrapper.
13. Fold each wrapper in half, carefully matching and sealing the wet edges.
14. Fold each closed wrapper in half lengthwise.
15. Grab two opposite corners and pinch together.
16. Pinch the other two opposite corners together to form a purse. *Note:* Cover prepared wontons with a wet paper towel to prevent them from drying out.
17. Use the swimming method to deep-fry rangoons in vegetable oil heated to a temperature between 350°F and 375°F until golden brown.
18. Drain rangoons on paper towels.

NUTRITION FACTS
Per serving: 158 calories, 45 calories from fat, 5.2 g total fat, 30.9 mg cholesterol, 335.5 mg sodium, 177.9 mg potassium, 22 g carbohydrates, <1 g fiber, <1 g sugar, 5.9 g protein

Onion Rings

Yield: *8 servings, 4 oz each*

Ingredients

medium onions, sliced and separated into rings	4 ea
eggs	2 ea
milk	1 c
cake flour	8 oz
baking powder	1 tsp
salt	¼ tsp
vegetable oil	(for deep-frying)

Preparation

1. Submerge the raw onion rings in ice-cold water and reserve.
2. Break the eggs into a stainless steel bowl and beat slightly with a wire whisk.
3. Whisk the milk into the eggs.
4. In a separate bowl, sift the flour, baking powder, and salt together.
5. Slowly add the dry ingredients to the liquid mixture while whisking briskly. Whip until a smooth batter forms.
6. Remove the onion rings from the ice water and pat dry.
7. Coat the onions with batter.
8. Using the swimming method, deep-fry the onions in vegetable oil that is heated to between 350°F and 375°F until golden brown.
9. Remove the onion rings and place on paper towels to drain.

NUTRITION FACTS
Per serving: 182 calories, 52 calories from fat, 5.9 g total fat, 49.6 mg cholesterol, 166.3 mg sodium, 132 mg potassium, 26.7 g carbohydrates, 1 g fiber, 3 g sugar, 5.2 g protein

Buffalo Wings

Yield: *4 servings, 6 pieces each*

Ingredients

chicken wings	12 ea
all-purpose flour	1¾ c
salt	2 tsp
pepper	1 tsp
eggs	2 ea
milk	½ c
bread crumbs	2 c
vegetable oil	(for deep-frying)
Tabasco® sauce	4 tbsp
vinegar	1 tsp
cayenne pepper	1 tsp
butter, melted	2 tbsp

Preparation

1. Cut the chicken wings in half at the joint. Trim the tips of the wings.
2. Mix the flour with the salt and pepper.
3. Coat chicken pieces in seasoned flour.
4. Beat eggs with a wire whisk in a separate bowl.
5. Add milk and mix well.
6. Dip floured chicken pieces in egg mixture.
7. Coat chicken with dried bread crumbs.
8. Deep-fry chicken in oil heated to a temperature between 350°F and 375°F until golden brown.
9. To make the sauce, combine Tabasco® sauce, vinegar, cayenne pepper, and butter in a small bowl.
10. Drizzle sauce over cooked chicken and toss to coat. Serve immediately.

NUTRITION FACTS
Per serving: 894 calories, 369 calories from fat, 41.3 g total fat, 224.5 mg cholesterol, 1805.3 mg sodium, 505.7 mg potassium, 83 g carbohydrates, 4.3 g fiber, 5.2 g sugar, 44.3 g protein

Chicken Satay with Teriyaki Sauce

Yield: *8 servings, 4 oz each*

Ingredients

peanut butter	¼ c
fresh cilantro, chopped	¼ c
lime juice	3 tbsp
water	3 tbsp
fresh ginger, peeled and grated	1 tbsp
hot chile sauce	1 tsp
salt	TT
chicken breast, cut into 2 inch pieces	2 lb

Teriyaki Sauce

soy sauce	⅔ c
mirin (sweet rice wine)	¼ c
cider vinegar	⅓ c
garlic, minced	2 cloves
light brown sugar, packed	3 tbsp
fresh ginger, peeled and grated	3 tbsp

Preparation

1. In a large mixing bowl, combine peanut butter, cilantro, lime juice, water, ginger, chile sauce, and salt.
2. Add chicken pieces and toss to coat.
3. Soak wooden skewers in water for 10 minutes. Place chicken on wooden skewers and grill for 15–20 minutes or until cooked through.
4. In a small saucepan, bring soy sauce, mirin, and cider vinegar to a simmer.
5. Add garlic, brown sugar, and ginger. Stir until sugar is completely dissolved.
6. Transfer sauce to a small bowl. Place the bowl in an ice bath and cool to room temperature.
7. Serve chicken hot with dipping sauce at room temperature.

NUTRITION FACTS
Per serving: 113 calories, 39 calories from fat, 4.6 g total fat, 11.4 mg cholesterol, 1439.9 mg sodium, 222.8 mg potassium, 9.5 g carbohydrates, <1 g fiber, 6.5 g sugar, 8.5 g protein

Tropical Fruit Kebabs

Yield: *8 servings, 8 oz each*

Ingredients

fresh pineapple, large dice	2 c
red seedless grapes	1 c
pears, large dice	2 c
mangoes, large dice	2 c
kiwifruit, peeled and quartered	4 ea
lemon juice	2 tbsp

Preparation

1. Place all fruit in a large mixing bowl.
2. Drizzle lemon juice over fruit and toss to coat.
3. Arrange layers of pineapple, grape, pear, mango, and kiwifruit on wooden skewers.
4. Serve chilled.

NUTRITION FACTS
Per serving: 107 calories, 4 calories from fat, <1 g total fat, 0 mg cholesterol, 3.9 mg sodium, 346.4 mg potassium, 27.3 g carbohydrates, 4 g fiber, 16.1 g sugar, 1.3 g protein

Curried Summer Squash Hummus

Yield: *8 servings, 2 oz each*

Ingredients

medium zucchini (4 oz)	1 ea
medium yellow squash (4 oz)	1 ea
garbanzo beans	4 oz
tahini	3 oz
lemon juice	2 tbsp
garlic	2 cloves
yellow curry powder	1 tsp
paprika	¼ tsp
olive oil	¼ c
kosher salt	½ tsp

Preparation

1. Cut the zucchini and yellow squash into large cubes.
2. Place the remaining ingredients in a food processor and purée until smooth.
3. Add squash to the food processor and purée again until smooth.
4. Serve chilled.

> **NUTRITION FACTS**
> **Per serving:** 147 calories, 105 calories from fat, 12.2 g total fat, 0 mg cholesterol, 170.6 mg sodium, 179.7 mg potassium, 8 g carbohydrates, 2.2 g fiber, 1.3 g sugar, 3.2 g protein

Artichoke and Goat Cheese Spread Crostini

Yield: *10 servings, 5 pieces each*

Ingredients

Spread

artichoke hearts, drained	14 oz
goat cheese	1 lb
olive oil	2 tbsp
lemon juice	2 tsp
lemon zest, finely grated	1 tsp
garlic, minced	1 clove
roasted red bell pepper, diced	¼ c
parsley, chopped	2 tbsp
chives, chopped	2 tbsp
basil, chiffonade	1 tbsp
red pepper flakes, crushed	¼ tsp
salt and black pepper	TT

Crostini

French bread	½ baguette
extra virgin olive oil	½ c
kosher salt	½ tsp
black pepper, ground	½ tsp

Preparation

1. In a food processor, purée all of the spread ingredients until smooth.
2. Taste and adjust seasoning, if necessary.
3. Store the spread in a nonreactive container and refrigerate until ready to serve.
4. Cut bread into ¼ inch slices and place on baking sheets.
5. Oil and season one side of each slice of bread.
6. Bake in a 350°F oven until toasted.
7. Cool before serving.

> **NUTRITION FACTS**
> **Per serving:** 369 calories, 245 calories from fat, 27.8 g total fat, 35.8 mg cholesterol, 533.4 mg sodium, 253.1 mg potassium, 17.9 g carbohydrates, 3 g fiber, 1.2 g sugar, 13.3 g protein

Swiss Cheese and Roasted Garlic Fondue

Yield: *4 servings, 4 oz each*

Ingredients

garlic	2 cloves
olive oil	1 tsp
Swiss cheese, grated	6 oz
béchamel sauce, prepared	12 fl oz

Preparation

1. Place garlic in aluminum foil with olive oil.
2. Place in a 300°F oven and roast until soft and lightly browned (approximately 20 minutes).
3. Heat béchamel sauce.
4. Using a whisk, slowly add grated Swiss cheese to the béchamel. Continue to whisk until the cheese is incorporated and the fondue is smooth.
5. Add roasted garlic and whisk well to incorporate.

NUTRITION FACTS
Per serving: 294 calories, 183 calories from fat, 20.8 g total fat, 60.6 mg cholesterol, 149 mg sodium, 173.9 mg potassium, 11.9 g carbohydrates, <1 g fiber, 5.5 g sugar, 15 g protein

Blue Cheese Dip

Yield: *4 servings, 2 oz each*

Ingredients

onion juice	1 tsp
cream cheese	6 oz
blue cheese, crumbled	2 oz
sour cream	3 tbsp
salt	TT

Preparation

1. Place a peeled onion in a food processor and grind to a fine consistency.
2. Place ground onion in cheesecloth and squeeze the juice from the puréed onion.
3. Place all ingredients in a food processor and blend thoroughly at low speed. *Note:* Use extra cream cheese or sour cream as necessary to reach desired consistency.
4. Chill dip until ready to serve.

NUTRITION FACTS
Per serving: 213 calories, 179 calories from fat, 20.4 g total fat, 62.1 mg cholesterol, 414.2 mg sodium, 108.9 mg potassium, 2.4 g carbohydrates, <1 g fiber, 1.8 g sugar, 5.8 g protein

Nacho Cheese Sauce

Yield: *8 servings, 3.5 oz each*

Ingredients

béchamel sauce	1 pt
sharp cheddar cheese, grated	½ lb
canned jalapeño slices	2 tbsp
canned jalapeño juice	2 tbsp
garlic powder	½ tsp
ground cumin	½ tsp
salt and pepper	TT

Preparation

1. Heat béchamel until hot.
2. Slowly add grated cheese to the hot béchamel, whisking constantly.
3. Once all of the cheese has been incorporated, add the remaining ingredients and mix well.

NUTRITION FACTS
Per serving: 209 calories, 129 calories from fat, 14.7 g total fat, 44.1 mg cholesterol, 294.6 mg sodium, 129.7 mg potassium, 10.1 g carbohydrates, <1 g fiber, 3.5 g sugar, 9.4 g protein

Spicy Spinach-Artichoke Dip

Yield: *8 servings, 2½ oz of dip each*

Ingredients

baby spinach	10 oz
fresh basil	1 c
Great Northern beans, drained	¾ c
cream cheese	6 oz
garlic, minced	1 clove
chicken stock	½ c
artichoke heart quarters, drained	14 oz
Parmesan cheese, grated	¼ c
mozzarella cheese, shredded	¾ c
cayenne pepper	pinch
Worcestershire sauce	2 dashes
salt and black pepper	TT

Preparation

1. Blanch the spinach and basil and then shock in ice water. Remove the spinach and basil from the ice water and drain thoroughly.
2. Chop the blanched spinach and basil.
3. Purée the beans, cream cheese, garlic, and chicken stock in a food processor until smooth.
4. Transfer the purée to a mixing bowl and fold in the spinach, basil, artichokes, Parmesan, and half of the mozzarella.
5. Add the cayenne, Worcestershire sauce, and season to taste.
6. Place mixture in a crock and top with the remaining mozzarella.
7. Bake until bubbling and golden, approximately 6–8 minutes. Serve immediately.

NUTRITION FACTS
Per serving: 186 calories, 94 calories from fat, 10.7 g total fat, 34.5 mg cholesterol, 328 mg sodium, 537.4 mg potassium, 14.2 g carbohydrates, 4.8 g fiber, 1.3 g sugar, 10.6 g protein

California Rolls

Yield: *4 servings, 4 each*

Ingredients

rice vinegar	¼ c
sugar	1 tbsp
salt	1 tsp
sushi rice, cooked	1 c
nori, 8 × 7 inch sheets	2 ea
avocado, julienned	1 tbsp
cucumber, peeled, julienned	1 tbsp
Alaskan king crab meat	½ c
wasabi paste	1 tbsp
pickled ginger	1 tbsp

Preparation

1. Combine rice vinegar, sugar, and salt over medium-low heat. Heat until sugar and salt are dissolved.
2. Slowly add the rice vinegar mixture to the cooked sushi rice. Let stand until cool.
3. Refrigerate 2 hours or longer.
4. Dry-roast each sheet of nori by moving it quickly over direct heat for 30 seconds or until it turns bright green.
5. Place a piece of nori on a sudare (bamboo sushi mat). With wet hands, spread half of the sushi rice evenly on the piece of nori, leaving a 1 inch border on the top edge.
6. Arrange avocado and cucumber in a horizontal line across the center of the rice. Top with half of the crab meat.
7. From the bottom edge, lift the nori and the sudare slightly and roll tightly and evenly while pressing down.
8. Remove the roll from the sudare.
9. Seal the nori by placing a bit of water on the top 1 inch border. Press the seam closed.
10. Repeat the procedure with the second sheet of nori.
11. Trim the two ends of the rolls with a sharp knife.
12. Cut each roll into six 1 inch pieces.
13. Garnish plates with wasabi paste and pickled ginger.
14. Serve rolls cold with soy sauce.

NUTRITION FACTS
Per serving: 98 calories, 7 calories from fat, <1 g total fat, 11.9 mg cholesterol, 832.7 mg sodium, 216 mg potassium, 21.5 g carbohydrates, <1 g fiber, 3.5 g sugar, 6.5 g protein

Straight Forcemeat

Yield: *38 servings, 1 oz each*

Ingredients

fatback, medium dice	16 oz
shallots, minced	1 oz
butter	½ oz
white wine	1 fl oz
lean pork, medium dice	16 oz
brandy	1 fl oz
pâté spice	1 tsp
salt	1½ tsp
white pepper	¼ tsp
bay leaves	2 ea
whole eggs	2 ea

Preparation

1. Dice fatback and place in freezer until needed.
2. Sweat shallots in butter until translucent.
3. Add wine and reduce by half. Refrigerate to cool completely.
4. Combine the pork with the shallot and wine reduction. *Note:* Any protein can be substituted for all or a portion of the lean pork.
5. Add the brandy, pâté spice, salt, white pepper, and bay leaves and mix well. Refrigerate overnight or for a minimum of 4 hours.
6. Remove the bay leaves from the refrigerated mixture. Grind the mixture using the medium die on the grinder and place the ground mixture in a pan over ice.
7. Change to a small die and grind the mixture again.
8. Add the frozen fatback to the grinder and grind directly into the mixture.
9. Transfer the ground mixture to a food processor and mix until smooth and emulsified.
10. Add eggs to mixture and process until fully incorporated.
11. Make a poached quenelle of the forcemeat to test for seasoning and texture.
12. Refrigerate the forcemeat until needed.

NUTRITION FACTS
Per serving: 120 calories, 103 calories from fat, 11.4 g total fat, 25.2 mg cholesterol, 103.4 mg sodium, 62.5 mg potassium, <1 g carbohydrates, <1 g fiber, <1 g sugar, 3.2 g protein

Country-Style Forcemeat

Yield: *44 servings, 1 oz each*

Ingredients

fatback, medium dice	16 oz
shallots, minced	1 oz
butter	½ oz
white wine	1 fl oz
lean pork, medium dice	16 oz
brandy	1 fl oz
pâté spice	1 tsp
salt	1½ tsp
white pepper	¼ tsp
bay leaves	2 ea
liver	4 oz
garlic	1½ tsp
fresh parsley, minced	2 tbsp
whole eggs	3 ea

Preparation

1. Dice fatback and place in freezer until needed.
2. Sweat shallots in butter until translucent.
3. Add wine and reduce by half. Refrigerate to cool completely.
4. Combine the pork with the shallot and wine reduction. *Note:* Any protein can be substituted for all or a portion of the lean pork.
5. Add the brandy, pâté spice, salt, white pepper, and bay leaves and mix well. Refrigerate overnight or for a minimum of 4 hours.
6. Trim the liver of any connective tissue.
7. Purée the liver until smooth.
8. Force the purée through a tamis or sieve. Refrigerate the purée until needed.
9. Remove the bay leaves from the refrigerated mixture. Grind the fatback and the meat mixture using the large die on the grinder and place the ground mixture in a pan over ice.
10. Change to a medium grinding die. Grind half of the mixture along with the fresh garlic and parsley into a separate pan placed over ice.
11. Beat the eggs slightly and then add to the puréed liver.
12. Combine the liver mixture with the meat mixture. Mix well to incorporate.
13. Make a poached quenelle of the forcemeat to test for seasoning and texture.
14. Refrigerate the forcemeat until needed.

NUTRITION FACTS
Per serving: 109 calories, 91 calories from fat, 10.1 g total fat, 33 mg cholesterol, 92.8 mg sodium, 65 mg potassium, <1 g carbohydrates, <1 g fiber, <1 g sugar, 3.5 g protein

Kielbasa

Yield: *14 servings, 3.5 oz each*

Ingredients

boneless beef top round, diced	12 oz
pork butt, diced	14 oz
fatback, diced	10 oz
kosher salt	3 tbsp
pink curing salt	½ tsp
dextrose	¾ tbsp
white pepper	1 tsp
dry mustard	1 tsp
garlic powder	1 tsp
coriander	½ tsp
crushed ice	12 oz
nonfat dry milk	1½ oz

Preparation

1. Dice the beef, pork, and fatback. Place the diced beef, pork, and fatback on separate trays and place the trays in the freezer.
2. Toss the pork with one-third of the kosher salt, curing salt, and dextrose and grind the mixture using a large die.
3. Toss beef with remaining two-thirds of the kosher salt, curing salt, and dextrose. Then add all of the spices and mix well.
4. Chill the beef mixture until semifrozen.
5. Grind the semifrozen beef mixture using a small die on the grinder and place the ground mixture in a pan over ice.
6. Grind the fatback using a small die. Refrigerate until needed.
7. Transfer the beef mixture to a buffalo chopper and add crushed ice. Process until the temperature reaches 40°F on a probe thermometer.
8. Add the fatback to the buffalo chopper and process to 45°F.
9. Add the dry milk to the buffalo chopper and process to 58°F.
10. Make a poached quenelle to test the forcemeat for seasoning and texture.
11. Remove the forcemeat from the buffalo chopper. Fold the pork into the forcemeat over an ice bath.
12. Stuff casings with forcemeat.
13. Twist sausages into appropriately sized links and refrigerate overnight.
14. Hot smoke sausages at 160°F for 1½–2 hours.
15. Poach the smoked sausages in 165°F water until an internal temperature of 150°F is reached on a probe thermometer.

NUTRITION FACTS
Per serving: 258 calories, 197 calories from fat, 21.8 g total fat, 43.6 mg cholesterol, 1352.4 mg sodium, 276.4 mg potassium, 2.9 g carbohydrates, <1 g fiber, 1.9 g sugar, 12.2 g protein

Pork Terrine

Yield: *22 servings, 2.5 oz each*

Ingredients

Gratin Forcemeat

shallots, minced	1 ea
garlic clove, minced	2 ea
vegetable oil	2 tbsp
brandy	3 fl oz
pork, trimmed and diced	18 oz
fatback	6 oz
pink curing salt	¼ tsp
kosher salt	2 tsp
sage, chopped	1 tbsp
thyme, chopped	1 tbsp
nutmeg	pinch
ginger powder	½ tsp
black pepper, ground	½ tsp
white bread panada slices	3 ea
egg	1 ea
juniper berries, crushed	6 ea
flat leaf parsley, chiffonade	2 tbsp
ham, thinly sliced	10 oz

Garnishes

dried apricots, diced	2 oz
water	2 fl oz
brandy	2 fl oz
ham, ¼ inch dice	6 oz
pistachios, blanched and peeled	2 oz

Preparation

1. In a hot sauté pan sweat shallots and garlic in half of the oil. Deglaze with brandy and reduce by three-fourths. Let cool.
2. Bring apricots, water, and brandy to a simmer. Set aside until the apricots rehydrate.
3. In a hot sauté pan, lightly brown half of the pork in the unused oil. Remove from heat and chill immediately.
4. Combine the uncooked pork with the fatback, curing salt, kosher salt, shallots, garlic, sage, thyme, nutmeg, ginger, and pepper. Let marinate overnight or for a minimum of 1 hour.
5. Grind the marinated pork using a large die and then grind again using a small die.
6. Transfer the ground pork into a chilled food processor. Add the panada, egg, juniper berries, and parsley. Purée until smooth, frequently scraping the bowl.
7. Drain the apricots. Fold the apricots, diced ham, and pistachios into the puréed forcemeat over an ice bath.
8. Line a terrine mold with plastic wrap and then with ham slices, overlapping the slices slightly.
9. Pack the forcemeat into the terrine mold, ensuring there are no air pockets.
10. Fold the ham slices over the top of the forcemeat. Then fold over the plastic wrap.
11. Cover the terrine mold and poach in 150°F water until an internal temperature of 150°F is reached on a probe thermometer.
12. Refrigerate overnight.

NUTRITION FACTS
Per serving: 186 calories, 96 calories from fat, 10.7 g total fat, 37.2 mg cholesterol, 451.2 mg sodium, 362.1 mg potassium, 8 g carbohydrates, <1 g fiber, 1.8 g sugar, 10.7 g protein

Aspic for Coating Food for Presentation Only

Yield: *1 gal.*

Ingredients

water, cool	1 gal.
powdered gelatin, 225 bloom	16 oz

Preparation

1. Place the cold water in a bain-marie and sprinkle the gelatin over the surface.
2. Cover the bain-marie with plastic wrap and place in a warm water bath until the gelatin dissolves.
3. Remove from the water bath and cool to room temperature for 1 hour.
4. Test the consistency of the aspic before glazing items.
5. Place the bain-marie back in a 140°F water bath until the aspic melts and is clear. Do not over heat.
6. Cool the aspic to approximately 90°F before using.

Aspic for Glazing Edible Foods

Yield: *256 servings, 1 tbsp each*

Ingredients

consommé, cold	1 gal.
powdered gelatin, 225 bloom	4–6 oz

Preparation

1. Place the cold consommé in a bain-marie and sprinkle the gelatin over the surface. *Note:* The amount of gelatin needed will vary due to the gelatin content of the consommé. If the consommé solidifies when chilled, add a smaller amount of gelatin.
2. Cover the bain-marie with plastic wrap and place in a warm water bath until the gelatin dissolves.
3. Remove from the water bath and cool to room temperature for 1 hour.
4. Test the consistency of the aspic before glazing items.
5. Place the bain-marie back in a 140°F water bath until the aspic melts and is clear. Do not over heat.
6. Cool the aspic to approximately 90°F before using.

> **NUTRITION FACTS**
> **Per serving:** 7 calories, 1 calorie from fat, <1 g total fat, 1.7 mg cholesterol, 23.2 mg sodium, 30.9 mg potassium, <1 g carbohydrates, <1 g fiber, <1 g sugar, 1.3 g protein

Basic Chaud Froid

Yield: *32 oz*

Ingredients

unflavored gelatin	1 oz
heavy cream	16 fl oz
white stock	16 fl oz

Preparation

1. Soften the gelatin in one-fourth of the heavy cream. Set aside.
2. Bring the white stock to a boil and then lower to a simmer.
3. Add a small amount of heated stock to the remaining cream to temper the cream.
4. Add the tempered cream to the heated stock and stir well to incorporate.
5. Add the gelatin-cream mixture and stir well to complete the chaud froid.
6. Strain the chaud froid through a chinois.

Poultry, Ratites, and Related Game

*P*oultry is the term used to describe chickens, turkeys, ducks, geese, guinea fowls, and pigeons. Other edible birds include ratites, such as ostriches, emus, and rheas, and farm-raised game birds, such as pheasants, quails, grouses, partridges, and wild turkeys. Some edible birds have both light and dark flesh, while others have only dark flesh. The size or the cut of the bird determines the cooking method used. Most edible birds are low in fat, calories, and cholesterol if the skin is removed before cooking.

Refer to DVD for **Flash Cards**

Chapter Objectives

1. Describe six kinds of poultry recognized by the USDA and how each kind is further classified.
2. Explain the advantage of purchasing whole poultry.
3. Identify common fabricated cuts of poultry.
4. Explain the meaning of the USDA inspection stamp.
5. Describe the qualities of Grade A poultry.
6. Identify precautions to take when receiving and storing poultry.
7. Truss whole poultry.
8. Cut poultry into halves, quarters, and eighths.
9. Cut poultry into boneless and airline breasts.
10. Bone a leg and a thigh.
11. Partially bone a leg and thigh.
12. Bone whole poultry.
13. Use marinades, barding, and stuffing to enhance the flavor of poultry.
14. Explain the four methods used to determine the doneness of poultry.
15. Prepare poultry using 10 different cooking methods.
16. Describe three kinds of ratites.
17. Describe five kinds of farm-raised game birds.

Key Terms

- **poultry**
- **poussin**
- **capon**
- **Cornish hen**
- **confit**
- **foie gras**
- **squab**
- **tender**
- **wing tip**
- **wing paddle**
- **drummette**
- **giblets**
- **caul fat**
- **poêléing**
- **ratite**
- **farm-raised game bird**

POULTRY

Poultry is the collective term for various kinds of birds that are raised for human consumption. The six different kinds of poultry recognized by the USDA include chickens, turkeys, ducks, geese, guinea fowls, and pigeons. Each kind of poultry is also subdivided into classes based on the age, gender, and tenderness of the bird.

Chickens

Chicken is the most common kind of poultry consumed. Chickens are classified by age and sometimes gender. Common chicken classifications include poussins, Cornish hens, broilers/fryers, roasters, capons, and stewers. **See Figure 16-1.**

Common Chicken Classifications

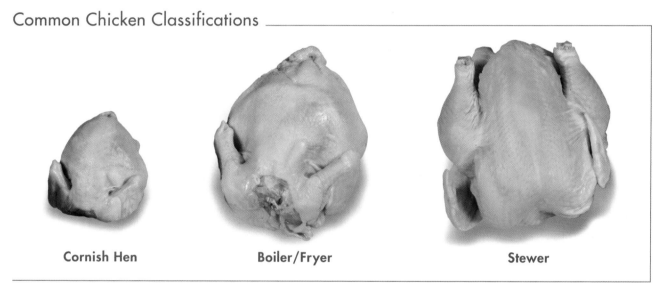

Cornish Hen   Boiler/Fryer   Stewer

Figure 16-1. *Chicken classifications include Cornish hens, broilers/fryers, and stewers. Poussins, roasters, and capons are also available.*

Poussins. A *poussin* is a very young female or male chicken that weighs 1 lb or less. The flesh is very tender and the flavor is very light. Any cooking method can be used to prepare poussins.

Cornish Hens. A *Cornish hen* is either a female or male chicken that is five to six weeks old and weighs 1½ lb or less. The term "hen" is commonly reserved for female birds, but the Cornish hen is the exception. The flesh of a Cornish hen is very tender and mildly flavored. Cornish hens are most often stuffed and roasted whole.

Broilers/Fryers. A *broiler/fryer* is a young male or female chicken under 13 weeks old that weighs 1½–3½ lb. Broilers/fryers have very tender flesh and smooth skin. They contain a slightly higher percentage of fat than Cornish hens. Any cooking method can be used to prepare broilers/fryers.

Roasters. A *roaster* is a young male or female chicken from 3–5 months old that weighs 3½–5 lb. Roasters have tender flesh and smooth skin. Roasters can be prepared using any cooking method.

Capons. A *capon* is a surgically castrated male chicken that is less than eight months old and weighs approximately 4–7 lb. Capons are castrated to produce large, well-formed breasts with flesh that is more tender than that of broilers/fryers. Capons are most often roasted.

Stewers. A *stewer,* also known as a stewing hen, is a female chicken that has laid eggs for one or more seasons, is usually more than 10 months old, and weighs from 3–8 lb. The flesh and skin of a stewer is tough, but flavorful due to the age of the bird. The tough flesh and skin require slow moist-heat cooking methods, such as stewing or braising.

Turkeys

Like chicken, turkeys are also classified by age. Classifications of turkeys include fryer/roaster turkeys, young turkeys, yearling turkeys, and mature turkeys.

Fryer/Roaster Turkeys. A *fryer/roaster turkey* is a male or female turkey that is less than 16 weeks old and weighs approximately 4–9 lb. Fryer/roaster turkeys are very tender with soft, flexible skin. Male fryer/roaster turkeys are commonly called "toms." Female turkeys are commonly called "hens." Common preparation methods include roasting, sautéing, and pan-frying.

Young Turkeys. A *young turkey* is a male or female turkey that is less than eight months old and is 8–22 lb. Young turkeys have tender flesh, smooth skin, and a firm breastbone. Young turkeys are commonly roasted or stewed.

Yearling Turkeys. A *yearling turkey* is a mature turkey that is less than 15 months old and weighs 10–30 lb. The flesh of a yearling is still tender. Yearlings are commonly roasted or stewed.

Mature Turkeys. A *mature turkey* is a turkey that is more than 15 months old. Mature turkeys have toughened flesh and coarse skin. Like yearlings, mature turkeys range in weight from 10–30 lb. Mature turkeys are often roasted or stewed. *See Figure 16-2.*

Daniel NYC

Roasted Turkeys

Photo Courtesy of Perdue Foodservice, Perdue Farms Incorporated

Figure 16-2. *Mature turkeys are often roasted or stewed.*

Ducks

Like other forms of poultry, ducks are classified by age. Classifications of duck include broiler/fryer ducklings, roaster ducklings, and mature ducks. Because ducks can fly long distances, they do not have white breast flesh. Ducks also have less flesh in proportion to bone and fat than most other kinds of poultry. For example, a duck yields half as much flesh as a chicken of the same size.

Broiler/Fryer Ducklings. A *broiler/fryer duckling* is a duck less than eight weeks old and weighs between 3–4 lb. Broiler/fryer ducklings have very tender flesh, a soft bill, and a soft windpipe. They are commonly roasted or stewed.

Roaster Ducklings. A *roaster duckling* is a duck that is less than 16 weeks old and weighs between 4–6 lb. The flesh of a roaster duckling is tender, and the bill and windpipe are just starting to harden. Roaster ducklings are most often roasted.

Mature Ducks. A *mature duck* is a duck more than six months of age and weighs 4–6 lb. The flesh of a mature duck is fairly tough and the bill and windpipe are hardened. Mature ducks are typically braised.

The majority of fat in a duck is located in and just beneath the skin. With the fatty skin removed, duck is lean in comparison to other kinds of poultry. When cooking duck, the fatty skin must be rendered slowly or it becomes nearly inedible.

Duck breasts are often seared in a sauté pan, finished in an oven, and served medium-rare. *See Figure 16-3.* Duck legs are most often braised or roasted until well-done and tender. Whole ducks are usually roasted on a rack to allow the fat to render, such as when making duck à l'orange or Peking duck.

Duck Breasts

Raw

Cooked

Photo Courtesy of D'Artagnan, Photography by Doug Adams Studio

Figure 16-3. *Duck breasts are often seared in a sauté pan, finished in the oven, and served medium-rare.*

Duck legs and thighs may be prepared as confit. *Confit* is a French term for meat that has been cooked and preserved in its own fat. To make a duck confit, pieces of duck leg and thigh are salted and then simmered in rendered duck fat until tender. The confit can be served whole, or the flesh can be pulled from the bone and used as a flavorful ingredient in salads or other dishes. A *cassoulet* is a dish that consists of white beans stewed with duck fat, fresh sausage, and duck confit.

Foie gras is the fattened liver of a duck or goose. ***See Figure 16-4.*** Foie gras is a delicacy that is considered a staple of classic fine dining. A duck or goose that is used to make foie gras is force-fed a rich diet until the liver becomes almost solid fat. The fattened liver is removed and sold as the most expensive part of the duck or goose. Because foie gras is almost all fat, it is typically seared quickly in a very hot sauté pan and served immediately. Foie gras may also be poached, cooled, and puréed to make a pâté de foie gras.

Geese

Geese have dark flesh and a large amount of fat in both the skin and the flesh. A goose is classified as either young or mature. Only young geese are used in foodservice operations.

A *young goose* is a goose that is usually less than six months of age and weighs approximately 4–10 lb. ***See Figure 16-5.*** Young geese have tender flesh and a windpipe that is easily dented. Their flesh has a rich taste due to a high fat content. There are large flaps of thick, fatty skin around the neck and tail that should be removed with a knife or kitchen shears prior to cooking. A young goose is commonly roasted at very-high temperatures to aid in rendering some of the fat from the skin and the flesh. Roasted goose is frequently served with an acidic fruit sauce to cut some of the fatty richness.

Photo Courtesy of D'Artagnan, Photography by Doug Adams Studio

Raw

Daniel NYC

Cooked

Figure 16-4. *Foie gras is the fattened liver of a duck or goose.*

Goose

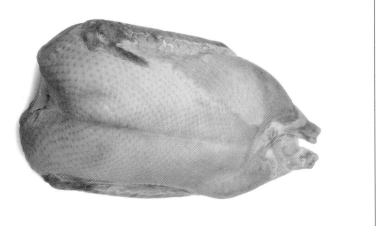

Czimer's Game & Seafoods, Inc.

Figure 16-5. *A young goose weighs approximately 4–10 lb.*

Pigeons

Pigeons that are raised for human consumption are slaughtered at a very young age. Only young pigeons are used in foodservice operations. A *squab* is a young pigeon that is less than four weeks old and weighs approximately 1 lb or less. Squabs have dark flesh and are commonly roasted. However, they can also be sautéed or broiled. ***See Figure 16-6.***

Squabs

Photo Courtesy of D'Artagnan, Photography by Doug Adams Studio

Raw

Daniel NYC

Cooked

Figure 16-6. *Squabs have dark flesh and are commonly roasted. However, they can also be sautéed or broiled.*

Guinea Fowls

A *guinea fowl,* also known as an African pheasant, is a farm-raised bird that has both light and dark flesh that is lean and tender. ***See Figure 16-7.*** A guinea fowl has a 50/50 ratio of flesh to carcass. Guinea fowls are substantially leaner than chickens. Chefs prefer guinea fowls to pheasants because the flavor is similar but less gamey.

A *young guinea* is a guinea fowl that has tender flesh, a flexible breastbone, and weighs between 1–1½ lb. Young guineas are most commonly roasted whole. A *mature guinea* is a guinea fowl that has tough flesh, a hardened breastbone, and weighs between 1–2 lb. Mature guineas must be braised or stewed to make the flesh tender enough to eat.

MARKET FORMS OF POULTRY

Poultry is sold after the head, neck, feet, and feathers have been removed from the bird. Poultry can be sold fresh or frozen, whole or cut-up, boneless or bone in, ground, or processed into a prepared form such as chicken nuggets. Foodservice operations purchase all market forms of poultry.

Guinea Fowls

Photo Courtesy of D'Artagnan, Photography by Doug Adams Studio

Figure 16-7. *Guinea fowl is a farm-raised bird that has both light and dark flesh that is lean and tender.*

Some poultry has both light and dark flesh. Other poultry has only dark flesh. Leg and thigh flesh is always dark, but breasts and wings can be either light or dark. For example, chickens spend most of their time on their feet and therefore have dark flesh in their legs and thighs. Ducks have dark flesh throughout their entire bodies, because all of their muscle groups are used for slow, sustained movement as they fly. Because the breast and wing muscles get more exercise, more blood flows through them. Blood contains the protein myoglobin, which causes the darkening of muscle tissue. The more a muscle is used, the darker and more flavorful the muscle becomes.

The fat in poultry is found just beneath the skin, around the tail, and on the abdomen. *See Figure 16-8.* Because poultry does not have much intramuscular fat, it can become very dry if it is overcooked even slightly. As a bird ages, its breastbone becomes less flexible and the flesh and skin toughen, intensifying the overall flavor. Therefore, younger birds are often desired for their tenderness and mild flavor.

Poultry Fat

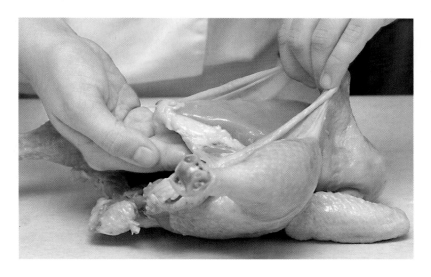

Figure 16-8. *The fat in poultry is found just beneath the skin, around the tail, and on the abdomen.*

Whole Poultry and Fabricated Cuts

Poultry is available in a variety of market forms, including whole birds and fabricated cuts. Knowledge of these market forms is necessary for accurate ordering. Purchasing whole poultry allows a chef to be more creative with the menu. It is also less expensive to purchase whole poultry and debone it in-house.

Common fabricated cuts of poultry include breasts, tenders, tenderloins, wings, legs, leg quarters, breast quarters, halves, and ground. *See Figure 16-9.*

Each wing can be cut into wing tips, wing paddles, and drummettes, and each leg can be divided into drumsticks and thighs. Cutlets and sausages are also available.

- A *breast* is the top front portion of the flesh above the rib cage. The breast consists of white flesh in birds that only fly in quick, short bursts and dark flesh in birds that have the ability to fly long distances.
- A *tender* is a small strip of a breast.
- A *tenderloin* is the inner pectoral muscle that runs alongside the breastbone.
- A *wing* consists of a tip, paddle, and drummette. A *wing tip* is the outermost section of a wing. It is often used to make stock as it contains little flesh. A *wing paddle*, also known as a wing flat, is the second section of a wing located between the two wing joints. A *drummette* is the innermost section of a wing located between the first wing joint and the shoulder.

Whole and Fabricated Poultry Cuts

Whole Turkey

Ground Turkey

Turkey Wings

Turkey Breast

Turkey Thighs

Turkey Legs

National Turkey Federation

Figure 16-9. *Poultry is sold whole and in common fabricated cuts, including ground, wings, breasts, thighs, and legs.*

Nutrition Note

Ground turkey has become a popular substitute for ground beef or ground pork. Ground turkey is lower in cholesterol and saturated fat.

- A *leg* consists of a drumstick and thigh. A *drumstick* is the lower portion of the leg located below the hip and above the knee joint. The *thigh* is the upper section of the leg located below the hip and above the knee joint.
- A *leg quarter* is a thigh, a drumstick, and a portion of the back.
- A *breast quarter* is half of a breast, a wing, and a portion of the back.
- A *half* is a full half-length of a bird split down the breast and spine.
- *Ground poultry* is ground fabricated cuts of poultry.

Giblets

When purchasing whole poultry, it is common to find a small bag containing the giblets inside the cavity of the bird. *Giblets* is the name for the grouping of the neck, heart, gizzard, and liver of a bird. **See Figure 16-10.** When ordering whole poultry it is common to specify whether the giblets are desired. When purchasing poultry without giblets, the acronym WOG is used by suppliers.

The neck, heart, and gizzard are often used to make giblet gravy. The neck contains an abundant amount of gelatin, which makes a rich and flavorful stock. Turkey hearts are served with a béchamel sauce in a classic preparation called creamed hearts. Gizzards are often trimmed of connective tissue, breaded or battered, and then fried. Livers are often breaded then fried or sautéed and served with caramelized onions. Livers are also used to make liver pâtés.

INSPECTION AND GRADES OF POULTRY

The USDA inspection stamp indicates that the poultry was processed at a USDA-inspected plant. **See Figure 16-11.** The round USDA inspection stamp is found either on a tag attached to the wing of the bird or on a label on the packaging. While inspection is mandatory, it does not indicate the overall quality of the bird. It is imperative that the handling and processing of poultry is monitored from the moment it enters a foodservice operation to ensure that it is safe to prepare and eat.

In addition to the round inspection stamp, a USDA grading stamp indicating the quality may also appear on a tag clipped to the bird or be stamped on the packaging material. Although grading is not mandatory, most poultry is graded. Poultry is graded based on overall quality with respect to shape, distribution of fat, condition of the skin, and general appearance of the bird. USDA grades for poultry include Grade A, Grade B, and Grade C. For the most part, the only poultry sold to consumers is Grade A.

To be classified as Grade A, poultry must be free of deformities and have plump, meaty flesh and a thin layer of fat under the skin. **See Figure 16-12.** Grade A poultry cannot contain any pinfeathers, discoloration, bruising, cuts or tears on the flesh, or broken bones. Poultry not meeting these criteria is graded as either Grade B or Grade C, depending on the extent of the flaws. Grade B and Grade C poultry is often used to make chicken nuggets, potpies, and canned soups.

Giblets

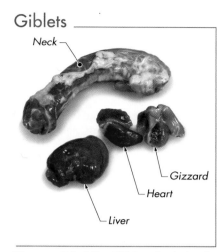

Figure 16-10. *Giblets include the neck, gizzard, heart, and liver of a bird.*

USDA Inspection Stamps

Figure 16-11. *The USDA inspection stamp indicates that the poultry was processed at a USDA-inspected plant.*

USDA Grade Stamps

Figure 16-12. *Grade A poultry is free of deformities, has plump, meaty flesh, and a thin layer of fat under the skin.*

RECEIVING AND STORING POULTRY

Poultry spoils rapidly. Although it develops an odor as it spoils, it may be unsafe for consumption prior to developing offensive odors. Fresh, whole birds should always be packed in crushed ice and in self-draining containers. The ice must be changed and the containers sanitized on a regular basis. Poultry must be stored at an internal temperature of 41°F or below. It should always be stored beneath other foods to prevent the juices from contaminating other foods.

If fresh poultry is not going to be used for two to three days, it should be frozen immediately to prevent loss of quality and potential spoilage. Frozen poultry should be frozen in its original packaging at 0°F or below. Frozen poultry should be moved to a refrigeration unit a day prior to use in order to thaw safely. *See Figure 16-13.* Whole turkeys may need an additional one to two days to thaw because of their size. Poultry should never be refrozen once it has thawed.

Thawed Poultry

Photo Courtesy of Perdue Foodservice, Perdue Farms Incorporated

Figure 16-13. *Frozen poultry should be moved to a refrigeration unit a day prior to use in order to thaw safely.*

FABRICATING POULTRY

There are many ways that poultry can be cut into portions. Typical fabrication methods for poultry include cutting them into halves, quarters, and eighths. Poultry may also be fabricated to produce boneless breasts and airline breasts. Fabricating techniques can be applied to almost any type of poultry because all types have similar bodies and bone structures. Chicken is the most economical bird to fabricate. Understanding the skeletal structure of a chicken can aid the fabrication process. *See Figure 16-14.*

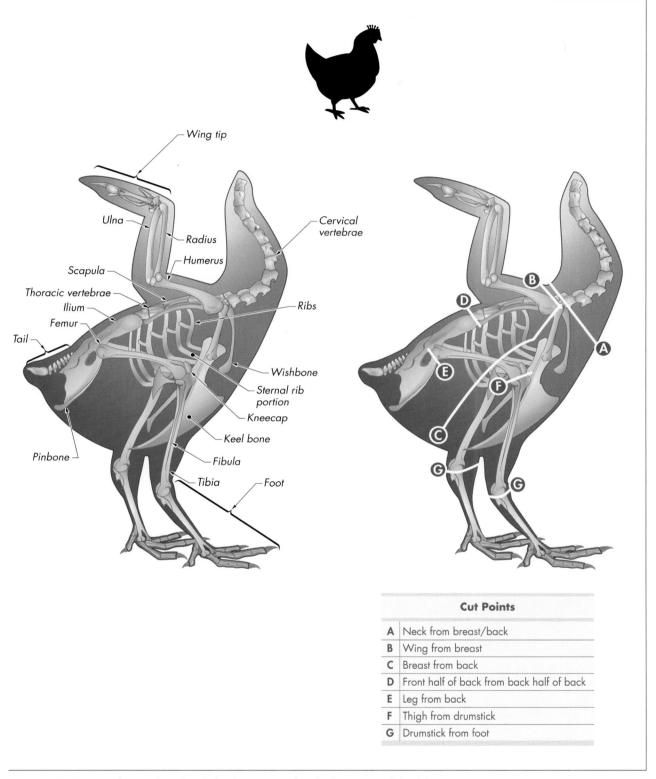

Cut Points

A	Neck from breast/back
B	Wing from breast
C	Breast from back
D	Front half of back from back half of back
E	Leg from back
F	Thigh from drumstick
G	Drumstick from foot

Figure 16-14. *Understanding the skeletal structure of a chicken can aid the fabrication process.*

Trussing Whole Poultry

Refer to DVD for
Trussing Whole Poultry
Media Clip

When roasting a whole bird, the bird is usually trussed. *Trussing* is the process of tying the legs and wings of a bird tightly to the body to keep a compact shape. *See Figure 16-15.* Trussing helps the bird cook evenly and retain moisture. Trussing also gives the bird a pleasing finished appearance. Prior to trussing, any excess fat should be trimmed from around the neck area and tail portion of the bird. The skin is then pulled tightly and evenly across the breast to cover any exposed flesh and prevent the breast from drying out during cooking. If desired, the wing tips can be removed, as they have a tendency to burn during roasting. If the wings are left intact, the first joint is tucked behind the second joint for a neat appearance.

Procedure for Trussing Whole Poultry

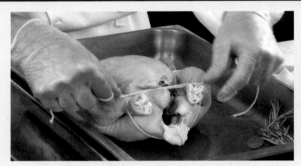

1. Cut a length of butcher's twine approximately three times the length of the bird to be trussed.

2. With the breast up, place the center of the twine beneath the bird, about 1 inch under the tail.

3. Bring the twine up around the legs and cross the ends, creating an "X" between the legs.

4. Pass the ends of the twine under the legs and pull tight.

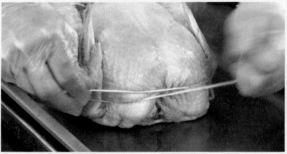

5. Turn the bird around, and pull the twine across the wings and cross at the neck.

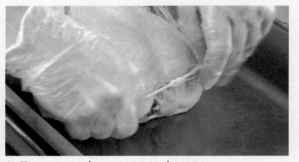

6. Tie a square knot to secure the truss.

Figure 16-15. *Trussing is the process of tying the legs and wings of a bird tightly to the body to keep a compact shape.*

Cutting Poultry into Halves

Poultry is commonly cut into halves. The bird is split from top to bottom between the breasts and along the backbone to the tail. This results in two equal portions. *See Figure 16-16.*

1. Square the bird by firmly squeezing the legs and wings toward the body.

2. With the breast side down, use a stiff (curved) boning knife to split the bird along both sides of the backbone from the neck to the tail.

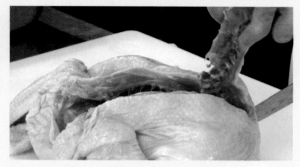

3. Remove the backbone from the carcass.

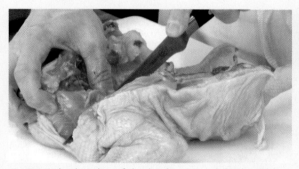

4. Open both sides of the bird to reveal the keel bone (breastbone). Cut through the keel bone and wishbone lengthwise from neck to tail. If necessary, hit the spine of the blade with the heel of the hand.

5. Cut through the flesh and skin behind the breastbone to separate the bird into halves.

Poultry cut into halves.

Figure 16-16. *Poultry is commonly cut into halves by splitting the bird from top to bottom between the breasts and along the backbone to the tail.*

Cutting Poultry into Quarters and Eighths

Poultry is often cut into quarters for grilling, broiling, or roasting. First, the bird is divided into leg and thigh sections and wing and breast sections. There are two in each section, yielding four quarters.

Poultry is commonly cut into eighths for pan-frying, deep-frying, grilling, broiling, and roasting. To cut a bird into eighths, quarters are cut into two breasts, two wings, two thighs, and two legs. *See Figure 16-17.*

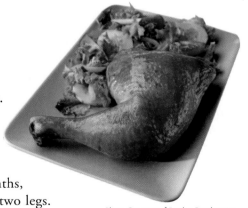

Photo Courtesy of Perdue Foodservice, Perdue Farms Incorporated

1. Cut the bird into two halves.

2. Cut through the flap of skin between the breast and thigh and pull the thigh away from the breast to expose the joint. Then, cut the joint to separate the thigh from the breast.

Poultry cut into quarters.

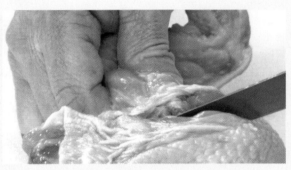

3. Cut the joint between the breast and the thigh to separate the breast and thigh from one half.

4. Hold the wing away from the breast, and cut the wing from the breast at the wing joint.

5. Hold the leg so that the inside thighbone is visible and locate the thin line of fat that separates the leg and thigh muscles. Cut along this line to separate the leg and thigh joint.

Poultry cut into eighths.

Figure 16-17. *Poultry is commonly cut into quarters or eighths by separating the bird into breast and wing quarters and leg and thigh quarters that can be cut into two breasts, two wings, two thighs, and two legs.*

Refer to DVD for
Cutting Poultry into Eighths
Media Clip

Cutting Boneless Breasts

Boneless breasts are the most popular poultry cut because of their versatility. They can be grilled, broiled, roasted, sautéed, pan-fried, poached, or stuffed. Larger breasts, such as those from a turkey, can be roasted whole or sliced into medallions and sautéed. A boneless breast is often flattened before being sautéed or stuffed. The breast is pounded flat with a meat mallet until the desired thickness is achieved. Boneless breasts are often fabricated in-house. *See Figure 16-18.*

Procedure for Cutting Boneless Breasts

1. With the breast side down, use a stiff (curved) boning knife to split the bird along both sides of the backbone from the neck to the tail. Remove the backbone from the carcass.
2. Pull the leg and thigh away from the breast. Locate the thigh joint and cut the flesh down to the joint.
3. Twist the leg to break the thigh joint. Cut through the joint and separate the leg from the carcass.
4. Open the bird to reveal the keel bone. Cut through the keel bone and wishbone lengthwise from the neck to the tail. If necessary, hit the spine of the blade with the heel of the hand.
5. Cut through the flesh and skin behind the keel bone to separate the bird into halves.
6. Lay the bird skin-side down and cut the breast away from the rib bones. *Note:* Use care when cutting around the rib cage to keep the tenderloin attached to the breast.

Figure 16-18. *Boneless breasts are often fabricated in-house.*

Cutting Airline Breasts

Airline breasts, also known as suprêmes, are similar to boneless breasts but contain one drummette bone. *See Figure 16-19.* An airline breast makes an elegant presentation due to the exposed drummette bone.

Procedure for Cutting Airline Breasts

1. With the breast side down, use a stiff (curved) boning knife to split the bird along both sides of the backbone from the neck to the tail. Remove the backbone from the carcass.
2. Open the bird to reveal the keel bone. Cut through the keel bone and wishbone lengthwise from the neck to the tail. If necessary, hit the spine of the blade with the heel of the hand.
3. Cut through the flesh and skin behind the keel bone to separate the bird into halves.
4. Cut through the skin between the breast and the thigh. Pull the thigh away from the breast to expose the joint.
5. Cut the joint to separate the breast from the thigh.
6. Cut along one side of the breastbone, following the curve of the ribs, to separate the flesh from the bone.
7. Separate the wing from the rib cage by cutting the joint. Keep the wing attached to the breast.
8. Cut the breast meat free from the carcass.
9. Make a cut on the back of the joint between the drummette and the paddle bones.
10. Break the joint and pull back the flesh and skin to expose the drummette bone. Trim the end of the drummette bone to remove the cartilage. *Note:* If desired, the skin can be removed.

Figure 16-19. *Airline breasts are similar to boneless breasts, but contain one wing bone.*

Boning Legs and Thighs

Legs and thighs have more flavor than breasts because they have dark flesh and additional fat. Boneless legs and thighs are often stuffed and roasted. When stuffing and roasting legs and thighs, the bones are removed but the flesh is left intact. *See Figure 16-20.*

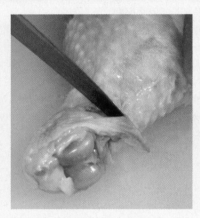

1. Place a leg quarter on a cutting board with the inside of the thigh facing upward.

2. Use a stiff (curved) boning knife to cut down the length of the thighbone and leg bone on each side and around the cartilage of the joint to free the bones from the flesh.

3. Pull the flesh away from the leg bone and cut the flesh off where it connects to the end joint.

4. With smooth, even strokes cut around the joint at the end of the leg until the L-shaped leg, thighbones, and cartilage are free from the flesh. Reserve the bones for stock.

5. Repeat with the other leg quarter. If desired, remove the skin.

Figure 16-20. *When stuffing and roasting legs and thighs, the bones are removed but the flesh is left intact.*

Boneless legs and thighs may be cut into smaller pieces for use in stir-fries and soups. They can also be flattened prior to being sautéed or stuffed. To flatten a boneless leg or thigh, the boneless leg or thigh is wrapped in plastic and then pounded flat with a meat mallet until the desired thickness or diameter is achieved. Partially boned legs and thighs are often desired to create a more elegant presentation. *See Figure 16-21.*

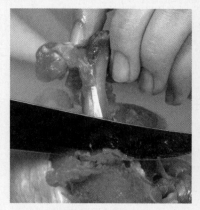

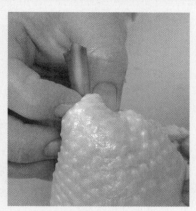

1. Use a stiff (curved) boning knife to cut down the length of the thighbone of a leg quarter.
2. Scrape the thigh flesh off the thighbone down to the joint.
3. Cut between the leg and thigh-bone at the joint to remove the thighbone.
4. Use a chef's knife to chop off the joint.
5. Push the flesh away from the joint for a finished presentation.

Figure 16-21. *Partially boned legs and thighs are often used to create a more elegant presentation.*

Fully boned and partially boned legs and thighs may be stuffed. They are trussed or wrapped in caul fat before cooking to help to maintain the shape of the flesh. *Caul fat* is a meshlike fatty membrane that surrounds sheep or pig intestines. It can be wrapped around items to be roasted to add additional moisture and maintain a consistent shape.

Caul fat melts almost completely away during the cooking process, but not before it helps set the shape of the item being cooked. Stuffed legs and thighs can be seared in a sauté pan and finished in the oven, or they can be roasted or braised. Before roasting, each stuffed leg is brushed with oil to provide moisture and aid in browning.

Boning Whole Poultry

Boning a whole bird is one of the more difficult fabrication techniques performed in the professional kitchen. *See Figure 16-22.* In this technique the bones of the bird are removed while keeping the flesh intact. Small birds, such as quail, are often boned prior to being stuffed. Larger birds, such as chicken, are boned for charcuterie preparations.

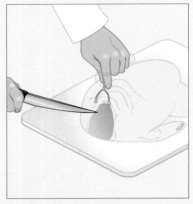

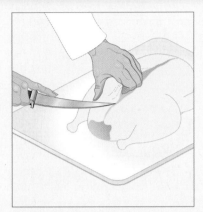

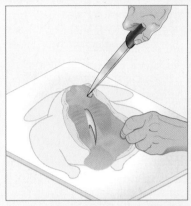

1. Place the bird breast-side up and stretch each wing flat against the cutting board by pulling on the tip. Use a stiff (curved) boning knife to cut off the wing tip and the next joint, leaving the drummette still attached.

2. Pull the skin of the neck area out of the way, and slide the knife along the underside of the wishbone. Cut around and under the wishbone until it is free and can be pulled out by hand.

3. Turn the bird breast-side down. Use short strokes to cut along the backbone from the neck to the tail, keeping the knife close to the bones. Carefully pull the flesh away from the carcass as the flesh is cut.

4. Cut through the ball-and-socket joints connecting each wing and thighbone to the carcass to separate them from the carcass while leaving them attached to the skin. *Note:* When this step is complete on both sides, the flesh will only be attached along the ridge of the breastbone.

5. Gently pull to separate the breastbone and carcass from the flesh.

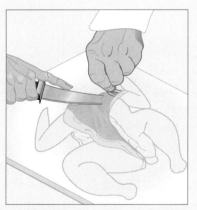

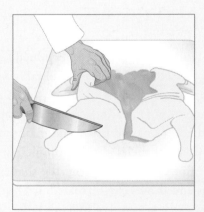

6. Cut the flesh from the curved bone near each wing to remove the bone.

7. Cut down the length of the thighbone and scrape the flesh down to the joint.

8. Cut between the leg and thighbones at the joint to remove the thighbone from each side.

9. While holding the exposed portion of the wing bone, cut through the tendons and scrape the meat from the bones.

10. Pull out the wing bone and repeat on the other side.

11. Use a chef's knife to cut off the end of the leg bone.

Figure 16-22. *Boning a whole bird is one of the more difficult fabrication techniques performed in the professional kitchen.*

FLAVOR ENHANCERS FOR POULTRY

After fabrication, poultry is often prepared further prior to cooking. Most poultry has a mild flavor and may be enhanced by marinades, barding, or stuffings.

Marinades

Marinades are an effective way to add flavor to poultry. *See Figure 16-23.* A *marinade* is a flavorful liquid used to soak uncooked foods such as meat, poultry, and fish to impart flavor and sometimes to tenderize. Poultry is typically marinated 30–60 minutes prior to cooking. Common poultry marinades contain wine, citrus juice, or buttermilk and oil, herbs, and spices. Teriyaki and mojo are popular poultry marinades. Leftover marinade that came in contact with raw poultry is contaminated and must be discarded. Marinades are never used as a sauce.

Barding

Barding is the process of laying a piece of fatback across the surface of a lean cut of meat to add moisture and flavor. This additional layer of fat helps keep the thin skin of poultry from burning and adds moisture and flavor to the flesh. Small birds benefit from barding, particularly those cooked at high temperatures. The fatback is trussed to the bird to prevent it from falling off during the cooking process. As poultry cooks, the fat is rendered from the fatback and absorbed into the flesh. The fatback prevents the flesh from drying out and browning. It is often removed for the last 10 minutes of the roasting process to allow browning.

Stuffings

In the professional kitchen, it is common to stuff small birds, such as Cornish hens and squabs, but not larger birds. *See Figure 16-24.* If stuffing is to be served with a large bird, such as a turkey, it is prepared and baked separately. Stuffing that is prepared and served separately is called "dressing." Once poultry has been stuffed, it should not be marinated.

Marinated Duck Breasts

Photo Courtesy of D'Artagnan, Photography by Doug Adams Studio

Figure 16-23. *Marinades are an effective way to add flavor to poultry.*

Chef's Tip

Strips of raw bacon can be used to bard poultry, but bacon adds a strong flavor.

Stuffed Birds

National Honey Board

Figure 16-24. *Small birds, such as Cornish hens and squabs, are often stuffed.*

COOKING POULTRY

Poultry can be cooked whole or in pieces. Poultry can also be cooked using a variety of cooking methods. *See Figure 16-25.* Smaller cuts cook more quickly than larger cuts. Legs and thighs have more connective tissue than breasts or wings, lending them to moist-heat cooking methods and longer combination cooking methods. Breasts must be cooked quickly but thoroughly, so as not to overcook.

Cooking Poultry, Ratites, and Game Birds		
Bird	**Class**	**Cooking Methods**
Chicken	Poussin	Any
	Cornish hen	Broiling, grilling, or roasting
	Broiler/fryer	Any
	Roaster	Any
	Capon	Roasting
	Stewer	Braising or stewing
Turkey	Fryer/roaster turkey	Roasting, sautéing, or pan-frying
	Young turkey	Roasting or stewing
	Yearling turkey	Roasting or stewing
	Mature turkey	Roasting or stewing
Duck	Broiler/fryer duckling	Roasting or sautéing
	Roaster duckling	Roasting
	Mature duck	Braising
Goose	Young goose	Roasting
Pigeon	Squab	Broiling, roasting, or sautéing
Guinea fowl	Young guinea	Roasting
	Mature guinea	Braising or stewing
Ratite	Ostrich	Grilling, broiling, sautéing, braising, or stewing
	Emu	Grilling, broiling, sautéing, braising, or stewing
	Rhea	Grilling, broiling, sautéing, braising, or stewing
Game birds	Pheasant	Roasting or braising
	Quail	Grilling, broiling, roasting, or sautéing
	Grouse	Sautéing or roasting
	Partridge	Broiling, roasting, or sautéing
	Wild turkey	Roasting, smoking, or braising

Figure 16-25. *Poultry can be cooked whole or in pieces using a variety of cooking methods.*

Regardless of the cooking method used, poultry should be cooked to an internal temperature of 165°F, except in the case of duck. Duck breast is commonly served medium-rare. When cooked, it has the same appearance as a lean cut of beef. Overcooked duck breast is very dry and tough. Poultry continues to cook after it is removed from the heat due to carryover cooking.

Larger birds should be cooked slowly to reduce shrinkage and retain moisture. Smaller birds should be cooked at temperatures between 375°F and 400°F to avoid drying out while cooking.

National Honey Board

Determining Doneness

Poultry is always cooked well-done, with the exception of duck and squab breasts. Well-done cuts of poultry should still be moist and juicy. The four methods that chefs use to determine the doneness of poultry are temperature, touch, joints, and juices (TTJJ). *See Figure 16-26.*

Temperature. Poultry should always be cooked to an internal temperature of 165°F. To determine doneness, the temperature is taken by inserting an instant-read thermometer in the innermost part of the thigh, wing, and/or breast. The thermometer should be inserted close to, but not touching, any large bones. The difficulty with using temperature to indicate doneness is that smaller cuts, such as boneless chicken breasts, may be too thin for a thermometer to provide accurate readings.

Touch. Touch is not as accurate as temperature. However, experienced chefs know that the firmness of the poultry increases in proportion to its doneness. When poultry is thoroughly cooked, the flesh is firm but springs back when touched.

Joints. Testing joints is not as accurate as temperature. The joints of raw or undercooked poultry are firmly connected. In contrast, the joints become soft and tender as the cartilage that holds them together cooks. For example, when the ends of a thoroughly cooked leg and thigh portion are twisted, the bones will separate easily at the knee joint.

Juices. Testing the color of juices from poultry is not as accurate as taking the temperature. However, juices from raw poultry are red, and juices from fully cooked birds are clear. As poultry cooks, the color of the juices changes, which explains why chicken stock (made from chicken flesh and bones) is clear. The color of the internal juices, not the external juices, is used to determine doneness.

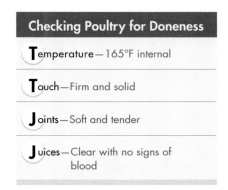

Checking Poultry for Doneness

Temperature—165°F internal

Touch—Firm and solid

Joints—Soft and tender

Juices—Clear with no signs of blood

Figure 16-26. *The doneness of poultry can be determined by temperature, touch, joints, and juices (TTJJ).*

Poaching

Poaching is a moist-heat cooking method in which food is cooked in a liquid that is held between 160°F and 180°F. Poaching is often used for tender poultry cuts, such as chicken breasts that can be cooked slowly. When poaching, the cooking liquid flavors the cooked poultry. The bird is most often submerged in a stock. If any portion of the bird is not completely submerged, it will cook improperly and become tough. A mirepoix, a bouquet garni, fresh herbs, wine, or lemon juice can be added to a poaching liquid to help flavor the product. A poaching temperature between 160°F and 180°F, with almost no steam or bubbles forming in the pot, produces the most desirable results.

Poêléing

Poêléing is a combination cooking method that is often referred to as butter roasting. ***See Figure 16-27.*** Whole chickens are most often poêléd. First, a whole bird is placed on top of a matignon in a covered pot. If desired, the bird may be seared prior to being placed in the pot. Then the pot is placed in an oven. While the poultry steams in its own juices, it can be occasionally basted with pan drippings. The bird can be uncovered near the end of the cooking process if it was not seared prior to being poêléd. A pan sauce is typically made from the cooking juices and the matignon.

Poêléing Poultry

Figure 16-27. *Poêléing is a combination cooking method that is often referred to as butter roasting.*

Grilling and Broiling

Most poultry can be grilled or broiled. Small birds, such as quails, are typically skewered prior to being cooked on a grill or in a broiler so that they will lie flat on the grates. Cornish hens should be split in half or butterflied before being grilled or broiled. Birds larger than Cornish hens should be cut into quarters or eighths before grilling or broiling. A whole turkey is too thick to be grilled or broiled because the exterior will dry out and burn before the interior can be fully cooked. The heat from a grill or a broiler is intense, so care must be taken to not overcook or dry out poultry.

Poultry cuts are typically seasoned prior to grilling and broiling so that the flavors of the seasoning can penetrate the flesh during the cooking process. It is important to remember that most poultry cuts contain very little fat and are prone to drying out. Grilled or broiled poultry should be moist and cooked throughout. ***See Figure 16-28.*** Duck and squab breasts are an exception. They are cooked medium-rare or medium.

Chef's Tip

Some chefs prefer to remove poultry skin prior to grilling or broiling. This prevents burning the skin and reduces the amount of fat in the final dish.

Grilled and Broiled Poultry

Turkey Burgers

Chicken Fajitas

Chicken Breasts

Chicken Quarter

Photo Courtesy of Perdue Foodservice, Perdue Farms Incorporated

Figure 16-28. *Grilled or broiled poultry should be moist and cooked throughout.*

Barbequing

Some forms of barbeque are similar to smoking, while other forms are similar to grilling. ***See Figure 16-29.*** Smoking flavors and cooks food using the heat and smoke from burning wood. Ducks and turkeys are the only types of poultry that are commonly smoked. Turkey is most often smoked with wood. Wood or tea can be used to smoke duck. Chicken is barbequed more often than other kinds of poultry. In Cantonese cuisine, barbequed duck is a popular dish.

Smoked and Barbequed Poultry

National Turkey Federation

Smoked Jerk Turkey

National Honey Board

Barbequed Chicken

Figure 16-29. *Turkey is most often smoked with wood, and chicken is most often barbequed.*

Roasting

Roasting is one of the most common methods of cooking whole poultry. Cornish hens and chickens can be roasted at higher temperatures for shorter cooking times. Larger birds, such as geese and turkeys, need to cook longer and at lower temperatures to allow the inner flesh to cook thoroughly without the outer flesh becoming dry and overcooked. Some chefs prefer to start larger birds at a temperature greater than 400°F for a few minutes to allow the skin to begin to brown. The key to roasting poultry is to determine how long the bird should be roasted and at what temperature. *See Figure 16-30*

Roasted Poultry

National Chicken Council

Figure 16-30. *The key to roasting poultry is to determine how long the bird should be roasted and at what temperature.*

High-temperature roasting is used for birds that are 2 lb or less, such as squabs, pheasants, and Cornish hens. The birds are roasted between 375°F and 400°F to produce a crisp, golden-brown skin quickly without drying out the flesh.

Very-high-temperature roasting is used for thick-skinned or fatty-skinned birds, such as ducks and geese. Roasting at a temperature between 400°F and 425°F allows the fat to melt out and away from the skin. The fat also insulates the flesh, preventing it from drying out at higher temperatures. As the hot fat seeps out of the skin, it fries the skin slightly, turning the skin crispy. It is important to prick the skin of fatty-skinned or thick-skinned birds prior to roasting and again during the roasting process to ensure that the skin is not seared in the high heat.

All poultry should be seasoned prior to roasting, but the size of the bird determines where the seasoning should be placed. Whole herbs and mirepoix are added to the internal cavity of a large bird, such as a turkey or capon. These aromatics infuse the flesh with their aromas and flavors as a bird slowly roasts. A smaller bird can be seasoned directly on the skin with salt and pepper. Herbs should not be placed on the skin of poultry that is being roasted at a high temperature because the herbs will burn.

Basting poultry while it is roasting adds moisture to the bird as it cooks. Poultry should only be basted with fat that escapes from the bird during cooking or with fat that is added to the roasting pan prior to cooking. Fatty birds, such as ducks and geese, should not be basted. Basting can make fatty birds greasy and unpleasant to eat.

Photo Courtesy of Perdue Foodservice, Perdue Farms Incorporated

Cuisine Note

Poultry does not have much internal fat. Basting poultry with stock or wine will wash away any fat that would otherwise be absorbed by the flesh, causing the texture of the bird to be dry.

Large birds, such as turkeys, are often carved into portions. *Carving* is a process of slicing a large piece of cooked poultry or meat into service-sized portions. *See Figure 16-31.*

Procedure for Carving Large Poultry

1. Cut off both leg and thigh pieces.
2. To portion the thigh and leg flesh, hold the leg with one hand and cut the thigh into thin slices parallel to the bone. When the thigh flesh is completely sliced, continue to the leg flesh, cutting parallel to the bone.
3. Repeat on the second leg and thigh.
4. With the tip of the knife, trim along the wishbone to remove it completely.
5. Begin to cut along the breastbone all the way down the breast, following the natural curve of the rib bones to remove the breast from the carcass completely. *Note:* If desired, the breast can be carved directly on the bird by making a horizontal slice just above the wing joint where it joins the breast.
6. Starting at the neck and working down one side, cut the breast on the bias. *Note:* Cutting on a bias makes the slices appear larger.
7. Repeat on other half of the breast.

Figure 16-31. *Carving is the process of slicing a large piece of cooked poultry or meat into service-sized portions.*

Sautéing

Boneless, uniformly sized cuts of poultry are often sautéed to ensure even cooking. The best cuts for sautéing are boneless breasts and medallions. Chicken Marsala, chicken saltimbocca, and turkey piccata are sautéed dishes. Although it is most common to sauté boneless cuts of poultry, small birds such as quail and squab can be sautéed with the bone in. Stir-frying is a light sauté method that is often used on boneless cuts of poultry. *See Figure 16-32.*

Prior to being sautéed, poultry is commonly dredged in seasoned flour to help seal in moisture, promote even browning, and allow the seasonings to stick to the flesh. Items to be sautéed should be cooked over medium-high heat. If the pan is too hot, the food or particles of flour can burn. If the sauté pan is not hot enough, the food absorbs some of the cooking fat and sweats in its own juices, preventing it from turning the appropriate color.

Sautéed poultry is often served with a sauce that was made in the same pan used to cook the food. Once the poultry is cooked through, it is removed from the pan. While the pan is still hot, flavoring ingredients, such as garlic, shallots, a tomato product, and spices, are added to the pan and sautéed in the remaining fat. The pan is then deglazed with stock, wine, or another flavorful liquid and reduced to make a flavorful pan sauce. Cream or fresh herbs may also be added to the pan and reduced to make a richer sauce. The cooked poultry is often added back to the pan, glazed in the sauce, and reheated for service.

Stir-Fried Chicken

U.S. Apple Association

Figure 16-32. *Stir-frying is a light sauté method that is often used on boneless cuts of poultry.*

Frying

Frying poultry produces a moist product because portioned cuts are typically breaded or battered to seal in the moisture before they are fried. *See Figure 16-33.* Chicken that has been cut into eighths, boneless chicken breasts, and turkey cutlets are often fried. The common seasoning method for fried poultry is to add salt and white pepper to the flour, bread crumbs, or batter. Prior to breading, the poultry can be placed in a marinade containing various herbs, seasonings, or flavorings. For example, fried chicken is often marinated in buttermilk.

Fried Poultry

| Buffalo Tenders | Sliced Tenders | Whole Tenders |

Photo Courtesy of Perdue Foodservice, Perdue Farms Incorporated

Figure 16-33. *Frying poultry produces a moist product because portioned cuts are typically breaded or battered to seal in the moisture before they are fried.*

When deep-frying poultry, it is important to get the fat hot enough, typically between 300°F and 375°F, to crisp the exterior of the poultry. However, the fat should not be too hot or it will burn the poultry in the time required for the interior to cook fully.

Pan-frying requires a lower temperature and a slightly longer cooking time than deep-frying. Poultry is turned over when the side being fried turns golden brown. Larger pieces of poultry need to be fried at a lower heat. Smaller items, especially those without bones, require higher heat. For the best presentation, the side facing up on the plate should always be pan-fried first.

The best method for determining doneness of fried poultry is to insert an instant-read thermometer into the thickest part of the flesh. Testing for doneness using the touch, joints, or juices method is not recommended for fried poultry. The crunchy exterior of fried poultry prevents an accurate measure of firmness. Joint twisting breaks the golden coating, and cooking oil can easily be misinterpreted as clear juices.

Braising and Stewing

Braising and stewing are used to prepare tough or less flavorful birds that will benefit from cooking in a flavorful liquid for a long period. A large part of the flavor of a braised or stewed dish comes from the cooking liquid. *See Figure 16-34.* Almost all whole or bone-in cuts can be braised or stewed. For example, whole bone-in duck legs are often braised.

An instant-read thermometer can be inserted in the thickest portion of a bird to make sure it has reached 165°F. This is the best method of determining the doneness of braised and stewed poultry. When cooked properly, braised and stewed poultry should be tender but not fall off the bone. If the flesh falls apart, the poultry has been overcooked. After the item has been properly cooked, it is usually removed from the cooking liquid. Cooked poultry is generally served in or with the braising or stewing liquid.

RATITES

Like poultry, ratites are raised for human consumption. A *ratite* is any of a large variety of flightless birds that have flat breastbones and small wings in relation to their body size. Types of ratites include ostriches, emus, and rheas. Unlike poultry, ratites have almost no breast flesh. Most of the flesh used is taken from the back, thigh, and forequarter of the bird. The *fan,* taken from a muscle in the thigh, is the most tender cut of a ratite. The top loin is also a very tender cut. *See Figure 16-35.* Ratite flesh is dark red and has a similar appearance to beef after it is cooked.

Ratite Cuts

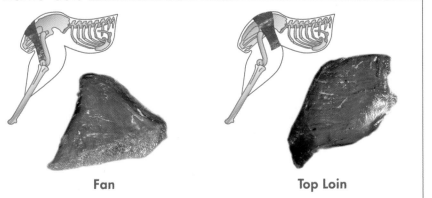

Fan

Top Loin

Emu Today and Tomorrow

Figure 16-35. *Ratite cuts that are most tender include the fan and the top loin.*

The flavor of ratite flesh is similar to beef, but sweeter. It is sold as steaks, fillets, medallions, whole roasts, or ground product. Ratites are prepared similarly to lean cuts of beef. Most cuts are best prepared medium-rare or medium by grilling, broiling, or sautéing. If ratite flesh is to be cooked well-done, it is best to braise or stew it. Mushrooms, garlic, red wine, and compotes complement the flavor of ratites.

National Turkey Federation

Figure 16-34. *A large part of the flavor of a stewed dish comes from the cooking liquid.*

United States Department of Agriculture

Ostrich Meat

MacFarlane Pheasants, Inc.

Figure 16-36. *Ostrich is best prepared with fast, dry-heat cooking methods.*

Nutrition Note

Emu meat has more than twice the amount of iron as beef. It is almost fat free and very low in cholesterol.

Ostriches

An ostrich is the largest ratite and can weight 300–400 lb. Ostriches can reach 8 feet in height. Ostriches are native to Africa. The most tender cuts of ostrich are found in the fan or thigh. Ostrich is best prepared with fast, dry-heat cooking methods such as grilling or broiling. *See Figure 16-36.* Tougher cuts from the leg are often marinated and then roasted or braised. Ostrich can also be sautéed or stewed.

Emus

An emu is the second largest ratite and can weigh between 125–140 lb. Emus can reach 6 feet in height. Emus are native to Australia. Emu meat is prepared in the same manner as ostrich meat. Tender cuts are cooked quickly using dry-heat methods such as grilling or broiling. Tougher cuts are slow-cooked using a combination cooking method such as braising or stewing. The flavor of emu is similar to, yet slightly different from, that of ostrich.

Rheas

A rhea is a small ratite that can weigh 60–100 lb. Rheas can reach 5 feet in height. Rheas are native to South America. Their meat is very lean and is similar in flavor to ostrich and emu. Rhea meat can be prepared in the same manner as ostrich and emu meat.

GAME BIRDS

In addition to poultry and ratites, game birds can be found on many foodservice menus. A *game bird* is a wild bird that is hunted for human consumption. In the United States, birds caught or hunted in the wild cannot be sold, so game birds are raised on farms. A *farm-raised game bird* is a game bird that is raised for legal sale under state regulations. Farm-raised game birds are processed at USDA-inspected facilities, but are not graded by the USDA.

Farm-raised game birds are generally leaner than poultry, so care must be taken not to overcook these birds. Farm-raised game birds include pheasants, quails, grouses, partridges, and wild turkeys.

Pheasants

A pheasant is a fairly large, long-tailed game bird. Pheasants range in size from 1½–2½ lb. The flesh is light colored and has a mild flavor. Pheasant is often roasted or braised. *See Figure 16-37.* The female pheasant is preferred over the male pheasant because the female pheasant is more tender and moist. A consommé made from a pheasant carcass is extremely flavorful.

Pheasants

Photo Courtesy of D'Artagnan,
Photography by Doug Adams Studio
Raw

MacFarlane Pheasants, Inc.
Cooked

Figure 16-37. *A pheasant is a long-tailed game bird with light-colored flesh and a mild flavor. Pheasants are often roasted.*

Quails

A quail is the smallest game bird. The breasts of quails weigh only about 1–2 oz. *See Figure 16-38.* The amount of flesh on quails is so minimal that quails are commonly boned out, filled with stuffing, rice, or forcemeat, roasted, and served whole. Quails are most often skewered and grilled, broiled, or sautéed. A common preparation is to season boneless quail, sauté it over high heat, and then deglaze the pan with balsamic vinegar and sherry. Quail eggs are considered a delicacy.

Grouses

A grouse is a game bird that resembles a chicken in appearance but has thicker, stronger legs and dark flesh. Common types of grouses include the ruffled grouse, the sage grouse, and the blue or dusty grouse. Small grouses are most often sautéed, while larger grouse are usually roasted.

Partridges

A partridge is a game bird that weighs about 1 lb, has white flesh, and yields an edible portion for two people. *See Figure 16-39.* The flavor of partridge is gamier and the flesh is somewhat chewier than that of pheasant. Roasting is the most popular method of cooking partridges. They are also broiled.

Wild Turkeys

The flesh of farm-raised wild turkeys is all dark. Wild turkey flesh is tougher, leaner, less moist, and less meaty than domesticated turkey flesh. Wild turkey can have a strong gamey flavor. Most cuts are braised but can also be slow roasted or smoked.

Quails

Photo Courtesy of D'Artagnan,
Photography by Doug Adams Studio

Figure 16-38. *Quail is the smallest variety of game bird.*

Partridges

Photo Courtesy of D'Artagnan,
Photography by Doug Adams Studio

Figure 16-39. *A partridge yields an edible portion for two people.*

SUMMARY

Poultry, ratites, and farm-raised game birds are offered on foodservice menus. The kinds of poultry recognized by the USDA include chickens, turkeys, ducks, geese, guinea fowls, and pigeons. Some birds have both light and dark flesh, while others have only dark flesh. Legs and thighs are always dark. Breasts and wings can be either light or dark meat, depending on the muscles exercised the most. Fabricating poultry is a skill aspiring chefs need to learn.

Poultry used in foodservice operations must be procured from USDA-inspected plants. Refrigerated poultry should maintain an internal temperature of 41°F or below. Frozen poultry is stored at temperatures of 0°F or below. Marinades, barding, and stuffing may be used to enhance the flavor of poultry. Poultry can be poached, poêléd, grilled, broiled, roasted, barbequed, sautéed, fried, braised, or stewed. The size or cut of the poultry determines the cooking method used.

Ostriches, emus, and rheas are ratites. They have flesh that is more like beef in color and flavor, but can be prepared in the same manner as poultry. Pheasants, quails, grouses, partridges, and wild turkeys are farm-raised game birds that are also prepared like poultry.

Refer to DVD for
Quick Quiz® questions

Review

1. Describe six kinds of poultry recognized by the USDA and how each kind is further classified.
2. Explain the advantage of purchasing whole poultry.
3. Explain why some poultry has both light and dark flesh and some poultry has only dark flesh.
4. Identify common fabricated cuts of poultry.
5. Name the items in the bag located inside whole poultry.
6. Explain the meaning of the USDA inspection stamp.
7. Describe the qualities of Grade A poultry.
8. Identify special precautions to follow when receiving and storing poultry.
9. Explain how frozen poultry should be thawed.
10. Describe how to truss whole poultry.
11. Describe how to cut poultry into halves.
12. Describe how to cut poultry into quarters.
13. Describe how to cut poultry into eighths.
14. Describe how to cut a boneless breast.
15. Describe how to cut an airline breast.
16. Describe how to bone legs and thighs.
17. Describe how to partially bone legs and thighs.
18. Describe how to bone whole poultry.

Refer to DVD for
Review Questions

19. Explain the process of barding.
20. Explain the four methods used to determine the doneness of poultry.
21. Name the safest method of determining the doneness of poultry.
22. Identify the internal temperature required of cooked poultry (with the exception of duck).
23. List 10 different methods of cooking poultry.
24. Explain how to poêlé poultry.
25. Describe how to carve large poultry.
26. Describe three kinds of ratites.
27. Describe the color and flavor of ratite flesh.
28. Define farm-raised game bird.
29. Name the category of birds that are processed at USDA-inspected facilities but are not graded by the USDA.
30. Identify five kinds of farm-raised game birds.

Applications

1. Truss whole poultry.
2. Cut poultry into halves.
3. Cut poultry into quarters.
4. Cut poultry into eighths.
5. Cut a boneless breast and an airline breast.
6. Bone a leg and a thigh.
7. Partially bone a leg and a thigh.
8. Bone whole poultry.
9. Use a marinade to enhance the flavor of a poultry dish. Evaluate the result.
10. Bard and cook a cut of poultry. Cook a similar cut of poultry using the same cooking method, but do not bard the flesh. Compare the results.
11. Stuff a Cornish hen to enhance the flavor. Evaluate the result.
12. Poach a chicken and poêlé a chicken. Compare the results.
13. Grill a piece of poultry. Broil an identical piece of poultry. Compare the results.
14. Roast one chicken and barbeque another chicken. Compare the results.
15. Sauté, pan-fry, and deep-fry identical cuts of poultry. Contrast the results.
16. Braise one piece of poultry and stew a similar piece of poultry. Compare the results.
17. Prepare a ratite dish.
18. Prepare a dish using a farm-raised game bird.

Refer to DVD for
Application Scoresheets

Grilled Chicken with Garlic-Herb Marinade

Yield: *4 servings, 1 breast each*

Ingredients

fresh garlic, minced	2 tbsp
fresh thyme leaves, stems removed	1 tbsp
fresh rosemary leaves, stems removed	1 tbsp
fresh parsley	1 tbsp
fresh lemon juice	6 fl oz
extra virgin olive oil	1 c
salt and pepper	TT
boneless chicken breasts	4 ea

Preparation

1. Place garlic, thyme, rosemary, parsley, lemon juice, and olive oil in a blender or food processor and purée until smooth. Season to taste.
2. Place chicken in cold marinade for 30 minutes to 1 hour.
3. Remove breasts from marinade. Grill or broil chicken until cooked through.

NUTRITION FACTS
Per serving: 368 calories, 136 calories from fat, 15.3 g total fat, 151 mg cholesterol, 348.7 mg sodium, 953.2 mg potassium, 5.2 g carbohydrates, <1 g fiber, 1.2 g sugar, 50.7 g protein

Grilled Chicken Breast with Sweet Teriyaki Marinade

Yield: *4 servings, 2 breasts each*

Ingredients

soy sauce	4 fl oz
mirin or sweet sherry	½ c
rice wine vinegar or cider vinegar	2 tbsp
vegetable oil	2 tbsp
granulated sugar	2 oz
fresh ginger, grated or minced	2 tbsp
lager-style beer (optional)	2 fl oz
boneless chicken breasts	8 ea

Preparation

1. Mix soy sauce, mirin or sherry, vinegar, vegetable oil, sugar, ginger, and beer in a saucepan and bring to a boil to dissolve the sugar.
2. Remove the marinade from the heat and cool over an ice bath.
3. Place chicken in cold marinade and let sit for 1 hour.
4. Remove breasts from the marinade and grill or broil the chicken until cooked through.

NUTRITION FACTS
Per serving: 723 calories, 173 calories from fat, 19.3 g total fat, 302.1 mg cholesterol, 2134.6 mg sodium, 1903.1 mg potassium, 22.3 g carbohydrates, <1 g fiber, 14.7 g sugar, 103.4 g protein

Grilled Chicken with Ginger-Lemongrass Marinade

Yield: *4 servings, 1 breast each*

Ingredients _____

fresh ginger, grated	2 tbsp
fresh lemongrass, minced	
(white part only)	1 tbsp
garlic, minced	1 tbsp
rice wine vinegar	2 fl oz
brown sugar	4 tbsp
hot sauce	½ tsp
lime juice, unsweetened	2 fl oz
cilantro, minced	1 tbsp
canola oil	2 fl oz
salt and pepper	TT
boneless chicken breasts	4 ea

Preparation _____

1. Mix ginger, lemongrass, garlic, vinegar, brown sugar, and hot sauce in a saucepan and bring to boil to dissolve the sugar.
2. Remove from heat and cool over an ice bath.
3. Whisk in lime juice, cilantro, and oil. Season to taste.
4. Place chicken in cold marinade for 1 hour.
5. Remove breasts from marinade. Grill or broil chicken until cooked through.

> **NUTRITION FACTS**
> **Per serving:** 460 calories, 181 calories from fat, 20.4 g total fat, 151 mg cholesterol, 370.9 mg sodium, 1048.8 mg potassium, 22.6 g carbohydrates, <1 g fiber, 13.6 g sugar, 50.4 g protein

Roast Chicken

Yield: *4 servings, 1 quarter each*

Ingredients _____

mirepoix	
onions, small dice	1 c
celery, small dice	½ c
carrots, small dice	½ c
roaster chicken, trussed	1 ea
butter, melted	1 tbsp
salt and pepper	TT
chicken stock	2 fl oz

Preparation _____

1. Place the mirepoix in the bottom of the roasting pan.
2. Rub or brush the surface of the chicken with butter. Place in roasting pan, breast-side up. Season chicken to taste.
3. Roast chicken at 375°F for approximately 1 hour or until an internal temperature of 165°F is achieved.
4. Strain drippings through a china cap. Reserve drippings for making gravy.
5. Deglaze the roasting pan with chicken stock and reserve the liquid for making gravy.
6. Cut chicken into quarters.

> **NUTRITION FACTS**
> **Per serving:** 777 calories, 505 calories from fat, 56.1 g total fat, 251.9 mg cholesterol, 345.1 mg sodium, 819 mg potassium, 6.2 g carbohydrates, 1.4 g fiber, 3 g sugar, 58.3 g protein

Chicken Cacciatore

Yield: *4 servings, 2 pieces each*

Ingredients

flour	1 c
salt and pepper	TT
frying or roasting chicken, cut in eighths	1 ea
cooking oil	2 tbsp
mushrooms, medium dice	1 c
onions, minced	1 c
garlic, minced	1 clove
Marsala wine	½ c
crushed tomatoes in juice	16 oz
tomato purée	1 c
fresh basil, crushed	1 tsp
fresh oregano, crushed	1 tsp
fresh chives, minced	2 tsp

Preparation

1. In a medium bowl, mix flour, salt, and pepper.
2. Coat each piece of chicken in the seasoned flour.
3. Heat oil in a large skillet and sauté chicken until golden brown.
4. Transfer chicken to paper towels or a wire rack to drain excess oil.
5. Using the same oil, sauté the mushrooms, onions, and garlic until tender, but do not brown.
6. Add Marsala and simmer for 5 minutes.
7. Add the crushed tomatoes in juice, tomato purée, basil, oregano, and chives. Stir well.
8. Cover the bottom of a roasting pan with the sautéed chicken. Pour the sauce over the chicken.
9. Bake chicken at 325°F for 45 minutes or until it is tender.
10. To serve, place two pieces of chicken on each plate and cover with sauce.

> **NUTRITION FACTS**
> **Per serving:** 368 calories, 160 calories from fat, 18 g total fat, 51.8 mg cholesterol, 389 mg sodium, 874.5 mg potassium, 30.4 g carbohydrates, 4.3 g fiber, 9.5 g sugar, 17.5 g protein

Roasted Chicken Roulades with Prosciutto

Yield: *2 servings, 4 roulades each*

Ingredients

boneless, skinless chicken breasts	2 ea
salt and black pepper, cracked	TT
olive oil	2 tbsp
thyme, chopped	½ tsp
sage, chopped	½ tsp
rosemary, chopped	½ tsp
garlic, creamed	2 cloves
prosciutto	2 slices

Preparation

1. Spread a 1 foot sheet of plastic wrap on a clean surface. Arrange the chicken breasts on half of the plastic wrap and fold the other half over the breasts to cover.
2. With a mallet, pound the breasts to approximately a ¼ inch thickness. Uncover the top of each chicken breast and season to taste.
3. Drizzle half of the olive oil over the chicken breasts.
4. Combine herbs and garlic and rub on each oiled chicken breast.
5. Lay a slice of prosciutto on top of each chicken breast.
6. Tightly roll up each chicken breast into a roulade (an even log).
7. Discard the plastic and tie each roulade with butcher's twine to hold its shape.
8. Heat the remaining olive oil in a sauté pan and gently sear the chicken on all sides.
9. Roast the roulades in a 400°F oven until an internal temperature of 165°F is reached.
10. Remove roulades from the oven, cover, and allow to rest for 10 minutes.
11. Slice each roulade crosswise into four pieces and plate.

> **NUTRITION FACTS**
> **Per serving:** 408 calories, 180 calories from fat, 20.3 g total fat, 156 mg cholesterol, 611.2 mg sodium, 926 mg potassium, 1.3 g carbohydrates, <1 g fiber, <1 g sugar, 52.3 g protein

Herbed Pan-Fried Chicken

Yield: *2 servings, 1 quarter each*

Ingredients

flour	2 c
dried oregano	1 tbsp
dried basil	1 tbsp
dried thyme	1 tbsp
dried sage	1 tbsp
dried mustard	2 tsp
garlic powder	2 tsp
onion powder	2 tsp
chicken, leg and thigh portion	2 quarters
salt and pepper	TT
buttermilk	1 c
vegetable oil	12 fl oz

Preparation

1. Combine 1 c of flour with oregano, basil, thyme, sage, dried mustard, garlic powder, and onion powder.
2. Season chicken with salt and pepper.
3. Dredge chicken in plain flour, dip in buttermilk, and then dredge in seasoned flour, pressing firmly to adhere the flour to the chicken. Place chicken in refrigerator for 15 minutes.
4. Heat 1 inch of vegetable oil in sauté pan over low to medium heat.
5. Place chicken in pan, skin-side down. Pan-fry chicken until light golden brown.
6. Turn chicken over and cook until second side is light golden brown.
7. Transfer chicken to a sheet pan and finish cooking in 350°F oven.
8. Check the thigh with a fork or knife. When juices run clear or meat has reached an internal temperature of 165°F, remove from oven.
9. Drain on paper towels to remove excess fat.

> **NUTRITION FACTS**
> **Per serving:** 616 calories, 326 calories from fat, 36.5 g total fat, 143.5 mg cholesterol, 413.6 mg sodium, 731.2 mg potassium, 32.8 g carbohydrates, 4.3 g fiber, 6.4 g sugar, 39 g protein

Chicken Kiev

Yield: *4 servings, 1 breast each*

Ingredients

boneless, skinless chicken breasts	4 ea
butter, chilled	8 tbsp
garlic, minced	1 clove
fresh chives, minced	1 tsp
salt and pepper	TT
fresh marjoram	pinch
eggs, beaten	2 ea
milk	½ c
flour	1 c
breadcrumbs	1½ c
vegetable oil	12 fl oz

Preparation

1. Spread a 2 foot sheet of plastic wrap on a clean surface. Arrange the chicken breasts on half of the plastic wrap and fold the other half over the breasts to cover.
2. With a mallet, pound the breasts to approximately a ¼ inch thickness. Remove plastic wrap.
3. Mix chilled butter in mixer at slow speed until a smooth consistency is achieved.
4. Add minced garlic, chives, salt, and pepper to butter.
5. Crush the marjoram in the palm of the hand and add to mixture. Blend mixture at low speed until evenly seasoned. *Note:* If needed, chill the mixture for a short time in a refrigerator.
6. Roll chilled butter into four pieces, each about the size and shape of a finger. Place a piece of the butter in the center of each breast.
7. Roll the chicken around the butter and fold in the ends to enclose the butter completely. Wrap chicken tightly in plastic wrap and place in freezer until butter is very firm.
8. Combine the eggs and milk to create an egg wash.
9. With one hand, coat a chicken breast with flour.
10. With the opposite hand, dip the floured breast into the egg wash.
11. With the dry hand, coat the breast in breadcrumbs.
12. Repeat the process for each chicken breast.
13. Heat vegetable oil in a large skillet. Fry chicken in hot oil until golden brown and cooked through. Turn chicken as needed to cook evenly.
14. Transfer chicken to paper towels or a wire rack to drain excess oil.

> **NUTRITION FACTS**
> **Per serving:** 864 calories, 371 calories from fat, 42 g total fat, 308.2 mg cholesterol, 695.4 mg sodium, 1071.6 mg potassium, 54.9 g carbohydrates, 2.7 g fiber, 4.3 g sugar, 63.1 g protein

Stuffed Breast of Chicken Provençale with a Tomato-Pesto Cream Reduction

Yield: *4 servings, 1 breast with 2 fl oz of sauce each*

Ingredients

boneless, skinless chicken breasts	4 ea
ground turkey meat	8 oz
frozen chopped spinach, thawed and drained	6 oz
garlic	2 tsp
salt and white pepper	TT
cream	2 tsp
seasoned flour	¼ c
olive oil	1 tbsp

Tomato-Pesto Cream Sauce

tomato paste or tomato purée	1 oz
olive oil	2 tsp
garlic, minced	2 cloves
white wine	½ c
light cream	½ c
pesto sauce or fresh basil purée	2 fl oz
salt and pepper	TT

Preparation

1. Spread a 2 foot sheet of plastic wrap on a clean surface. Arrange the chicken breasts on half of the plastic wrap and fold the other half over the breasts to cover.
2. With a mallet, pound the breasts to approximately a ¼ inch thickness. Refrigerate until needed.
3. In a food processor, combine cold turkey meat and spinach. Purée until smooth.
4. Add garlic, salt, and white pepper and mix well.
5. Add cream and pulse to incorporate.
6. Divide the turkey mixture evenly. Place half of the mixture in the center of each breast.
7. Tightly roll up each stuffed breast and secure with frill picks or butcher's twine.
8. Roll each stuffed breast in seasoned flour.
9. Sauté stuffed breasts in olive oil to brown the outside slightly.
10. Place stuffed breasts in oven to finish cooking at 325°F for 10–15 minutes or until the internal temperature reaches 165°F.
11. While the chicken is cooking, sauté the tomato paste in olive oil until slightly browned.
12. Add garlic and sauté for 1 minute.
13. Deglaze the pan with white wine.
14. Add cream and reduce slightly.
15. Whisk in pesto sauce or basil purée until thoroughly incorporated. Season to taste.

NUTRITION FACTS
Per serving: 660 calories, 317 calories from fat, 35.8 g total fat, 231.3 mg cholesterol, 758 mg sodium, 1187 mg potassium, 10.5 g carbohydrates, <1 g fiber, <1 g sugar, 66.7 g protein

Chinese Five Spice Duck Breast with Cherry and Ginger Lacquer

Yield: *4 servings, 1 breast each*

Ingredients

boneless duck breasts, skin intact	4 ea
Chinese five spice powder	2 tsp
ginger powder	½ tsp
Cherry and Ginger Lacquer	
red wine	2 fl oz
fresh ginger, minced	1 tsp
soy sauce	1 fl oz
teriyaki sauce	1 fl oz
cherry jelly or preserves	3 oz
water	1 fl oz

Preparation

1. Score the skin of the duck breasts without cutting into the meat.
2. Rub duck with five spice powder and ginger powder until well coated.
3. In a sauté pan over medium heat, render fat from duck breasts, skin-side down.
4. When slightly cooked, turn each breast over and sear the bottom.
5. Cook until the internal temperature of the breast reaches 135°F.
6. Remove duck from the pan and set aside to rest.
7. Place red wine and fresh ginger in a separate saucepan and reduce by half.
8. Add soy sauce, teriyaki sauce, cherry jelly or preserves, and water. Cook until mixture is slightly thickened and smooth.
9. Slice duck breasts on the bias.
10. Plate and top duck with cherry and ginger lacquer and serve.

> **NUTRITION FACTS**
> **Per serving:** 314 calories, 64 calories from fat, 7.1 g total fat, 127.8 mg cholesterol, 846 mg sodium, 527.1 mg potassium, 23.1 g carbohydrates, <1 g fiber, 16.1 g sugar, 34.4 g protein

Chicken Marsala

Yield: *4 servings, 1 breast each*

Ingredients

boneless, skinless chicken breasts	4 ea
extra virgin olive oil	2 tbsp
all-purpose flour, seasoned	1 c
porcini or button mushrooms, sliced thick	½ lb
prosciutto, sliced thin	¼ lb
salt and pepper	TT
Marsala wine	4 fl oz
fresh chicken stock	4 fl oz
butter	2 tbsp
fresh parsley, chopped	1 tbsp

Preparation

1. Spread a 2 foot sheet of plastic wrap on a clean surface. Arrange the chicken breasts on half of the plastic wrap and fold the other half over the breasts to cover.
2. With a mallet, pound the breasts to approximately a ¼ inch thickness. Remove plastic wrap.
3. In large sauté pan, heat olive oil over medium-high heat.
4. Dredge the breasts in the seasoned flour. Shake off any excess flour.
5. Place breasts carefully in the hot oil and sauté until golden brown. Turn breasts and sauté on other side. When golden brown on both sides, remove chicken from pan and reserve.
6. Add mushrooms to the pan and sauté in remaining oil until slightly browned. Remove mushrooms.
7. Lower the heat to medium and add the prosciutto to the sauté pan. Sauté prosciutto for 1 minute to render some of the fat.
8. Return the mushrooms to the pan and sauté until the moisture has evaporated, about 5 minutes. Season with salt and pepper.
9. Add Marsala and reduce by half.
10. Add chicken stock and simmer for 1 minute to reduce the sauce slightly.
11. Stir in butter, return the chicken to the pan, and simmer gently until chicken is cooked through.
12. Season to taste and garnish with chopped parsley before serving.

> **NUTRITION FACTS**
> **Per serving:** 751 calories, 197 calories from fat, 22.1 g total fat, 187.1 mg cholesterol, 1164 mg sodium, 1991 mg potassium, 68.2 g carbohydrates, 7.3 g fiber, <1 g sugar, 67.4 g protein

Sautéed Chicken Provençale

Yield: *4 servings, 1 breast each*

Ingredients

olive oil	4 fl oz
partially boneless chicken breast, skin on	4 ea
salt and pepper	TT
onion, diced	1 c
garlic, minced or sliced thin	4 cloves
fresh tomatoes, concassé	2 c
white wine	1 c
chicken stock	1 c
capers, optional	4 tsp
kalamata olives, sliced	4 tsp
fresh basil, chiffonade	2 oz
parsley, chopped	2 oz

Preparation

1. Heat a sauté pan over medium heat. Add half of the olive oil.
2. Season chicken with salt and pepper.
3. Place skin-side down in oil and cook until skin is crisp, golden brown in color, and the fat is completely rendered.
4. Turn breasts over and place sauté pan in the oven. Continue cooking until chicken is slightly underdone.
5. Remove chicken from the pan and discard the oil.
6. Add the other half of the fresh olive oil to the pan along with the diced onions. Sweat onions until translucent.
7. Add garlic and cook until fragrant.
8. Add tomatoes and deglaze the pan with white wine. Reduce by half.
9. Add stock, capers, olives, basil, and parsley. Bring stock to a boil.
10. Reduce heat to bring the sauce to a simmer and adjust seasoning to taste.
11. Add the chicken to the sauce, and cook until the chicken is completely cooked through.

NUTRITION FACTS
Per serving: 525 calories, 342 calories from fat, 38.5 g total fat, 57.5 mg cholesterol, 349.5 mg sodium, 701.2 mg potassium, 13.5 g carbohydrates, 2.6 g fiber, 5.8 g sugar, 22 g protein

Turkey Tetrazzini

Yield: *6 servings, 8 oz each*

Ingredients

boneless turkey	1 lb
thin spaghetti	16 oz
mushrooms, sliced	½ lb
butter	3 tbsp
flour	1 oz
chicken stock	2 c
cream	½ c
dry sherry	1 tbsp
salt and pepper	TT
Parmesan cheese	¼ c

Preparation

1. Simmer the turkey in lightly salted water until thoroughly cooked. Dice into 1 inch cubes. Cover and set aside.
2. Boil spaghetti in lightly salted water. Drain and then rinse in cold water. Set aside.
3. Sauté the mushrooms in one-third of the butter until tender. Remove from heat and reserve.
4. In a saucepot, melt the remaining butter. Add flour and continue cooking to make a blonde roux.
5. Add chicken stock and whisk until thickened and smooth.
6. In a separate saucepot, heat the cream, being careful not to scald it.
7. Add the heated cream and the sherry to the thickened turkey stock. Whisk until the sauce is smooth.
8. Add the diced turkey and sautéed mushrooms to the sauce. Mix well and season to taste.
9. Reheat the spaghetti in warm water.
10. Arrange a fourth of the spaghetti in the bottom of each serving dish.
11. Pour the turkey mixture over the spaghetti.
12. Sprinkle with Parmesan cheese and brown slightly in the broiler. Serve immediately.

NUTRITION FACTS
Per serving: 488 calories, 152 calories from fat, 17.3 g total fat, 52.4 mg cholesterol, 244.6 mg sodium, 415.9 mg potassium, 65.3 g carbohydrates, 2.9 g fiber, 4.1 g sugar, 16.8 g protein

Chicken à la King

Yield: *8 servings, 7 fl oz each*

Ingredients

boneless chicken	1 lb
green peppers, diced	½ c
pimientos, diced	1 tbsp
mushrooms, diced	½ c
butter	2 tbsp + 2 oz
dry sherry	2 tbsp
flour	2 oz
chicken stock	1 c
milk	1 c
light cream	½ c
salt	TT

Preparation

1. Dice chicken into 1 inch cubes and sauté until cooked through. Cover and set aside.
2. Sauté peppers, pimientos, and mushrooms in one-third of the butter and deglaze with sherry. Reserve the sauce.
3. In a saucepot, melt the remaining butter. Add flour and continue cooking to make a blonde roux.
4. Add chicken stock and whip with a wire whisk until thickened and smooth.
5. In a separate saucepot, heat the milk and cream, being careful not to scald it.
6. Add the milk and cream mixture to the thickened chicken stock. Whisk until smooth.
7. Season to taste.
8. Add the cooked chicken and sautéed vegetables and mix well.
9. Serve over a slice of toast or in a prepared pastry shell.

> **NUTRITION FACTS**
> **Per serving:** 248 calories, 143 calories from fat, 16.2 g total fat, 79.7 mg cholesterol, 165.3 mg sodium, 340.2 mg potassium, 9.3 g carbohydrates, <1 g fiber, 2.4 g sugar, 15.1 g protein

Lemongrass Poached Chicken Breasts

Yield: *4 servings, 1 breast and 4 fl oz of sauce each*

Ingredients

salt and white pepper	TT
boneless, skinless chicken breasts	4 ea
butter	2 tbsp
shallot, minced	2 tbsp
white wine	¾ c
white wine vinegar	¼ c
chicken stock	3 c
ginger, chopped	2 inch piece
star anise	10 pods
peppercorns	16 ea
lemongrass, chopped	2 tbsp
heavy cream	1 c
fresh lemon juice	1 fl oz
basil	4 tsp
cilantro	4 tsp

Preparation

1. Season chicken breasts to taste and set aside.
2. Melt butter in a medium sauté pan. Add shallots, white wine, vinegar, stock, ginger, star anise, peppercorns, and lemongrass and slowly heat to 160°F.
3. Add the chicken breasts and poach until the meat is no longer pink, but still moist.
4. Remove the chicken breasts. Reduce liquid to one-fourth the original volume.
5. Add heavy cream and reduce to a nappe consistency.
6. Strain the sauce. Add lemon juice, basil, and cilantro to the sauce.
7. Return chicken breasts and sauce to the pan. Reheat to 165°F.
8. Serve with sauce poured over breast.

> **NUTRITION FACTS**
> **Per serving:** 673 calories, 330 calories from fat, 37.5 g total fat, 253.2 mg cholesterol, 633.5 mg sodium, 1372.8 mg potassium, 18.2 g carbohydrates, 1.9 g fiber, 3.6 g sugar, 58 g protein

Fish, Shellfish, and Related Game

*F*ish, shellfish, and related game, such as turtles, frogs, snails, snakes, and alligators, are often grouped together on restaurant menus and referred to as seafood. Seafood refers to any edible animal that lives in freshwater or saltwater. There are thousands of fish species and hundreds of shellfish species. Seafood can be purchased live, whole, portion-controlled, frozen, and processed. The cooking methods used to prepare seafood vary depending on the fat content of the fish, shellfish, or related game.

Refer to DVD for **Flash Cards**

Chapter Objectives

1. Differentiate between lean and fatty fish.
2. Describe three classifications of fish based on external shape and structure.
3. Identify five types of freshwater fish.
4. Identify two types of anadromous fish.
5. Identify 17 types of saltwater fish.
6. Identify three common cartilaginous fish.
7. Describe the various market forms of fish.
8. Name the government organization in charge of voluntary fish inspections.
9. Explain how fresh and frozen fish are received and stored.
10. Fabricate fish into steaks and fillets.
11. Cook fish using eight different cooking methods.
12. Identify four types of crustaceans.
13. Identify 11 types of mollusks.
14. Describe the various market forms of shellfish.
15. Explain how live and frozen shellfish are received and stored.
16. Fabricate eight types of shellfish.
17. Cook crustaceans, mollusks, and cephalopods.
18. Describe five types of related game.

Key Terms

- **fish**
- **roundfish**
- **flatfish**
- **cartilaginous fish**
- **anadromous fish**
- **drawn fish**
- **dressed fish**
- **individually quick-frozen (IQF)**
- **shellfish**
- **crustacean**
- **mollusk**
- **univalve**
- **bivalve**
- **siphon**
- **adductor muscle**
- **cephalopod**
- **glazing**
- **en papillote**

Daniel NYC

SEAFOOD

Menus often list fish, shellfish, and related game as seafood. There are thousands of types of seafood and hundreds of them are offered on restaurant menus. Different types of seafood live in freshwater, saltwater, and sometimes both. Many varieties of fish have bones, while other varieties of fish, shellfish, and related game do not have a skeleton. Understanding the composition and structure of fish, shellfish, and related game helps a foodservice professional prepare a wider array of dishes.

FISH

A *fish* is any of a classification of animal that has fins for moving through the water, gills for breathing, and an internal skeleton made of bones or cartilage. Fish have edible flesh that contains protein, carbohydrates, fat, water, and trace amounts of vitamins and minerals. Unlike land animals, fish have delicate connective tissue. Because the connective tissue is very thin, fish flesh naturally flakes and comes apart easily. Fish are considered lean or fatty based on the amount of fat they contain. *See Figure 17-1.*

Approximate Fat Content of Common Fish			
Lean Fish		**Fatty Fish**	
Species	Fat Content*	Species	Fat Content*
Cod	1 g	Farmed catfish	6 g
Flounder	1 g	Rainbow trout	4 g
Halibut	2 g	Coho salmon	6 g
Striped bass	2 g	Shad	10 g

* per 3 oz serving

Figure 17-1. *Fish are considered lean or fatty based on the amount of fat they contain.*

The fat content of a fish can affect both flavor and the cooking method required. Lean fish contain very little fat and are often prepared using moist-heat cooking methods. Pacific cod is an example of a lean fish. Fatty fish contain more fat and are rich in omega-3 fatty acids, vitamin A, and vitamin D. Fatty fish, such as salmon, are best prepared using dry-heat cooking methods. The fat content of a particular type of fish can vary slightly by season.

Fish are most often grouped by freshwater or saltwater habitat and by their external shape and structure. Fish are classified as roundfish, flatfish, or cartilaginous fish, based on their shape and structure.

Roundfish. A *roundfish* is any fish with a cylindrical body, an eye located on each side of the head, and a backbone that runs from head to tail in the center of the body. *See Figure 17-2.* Roundfish are the most common type of fish and are found in freshwater lakes and streams as well as in saltwater. Trout and salmon are examples of roundfish.

Roundfish Shape and Structure

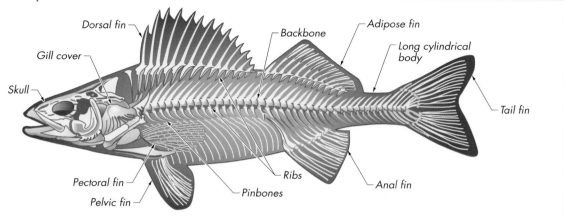

Figure 17-2. *A roundfish has a cylindrical body, an eye located on each side of the head, and a backbone that runs from head to tail in the center of the body.*

Flatfish. A *flatfish* is any thin, wide fish with both eyes located on one side of the head and a backbone that runs from head to tail through the midline of the body. *See Figure 17-3.* Flatfish swim parallel to the surface of the water with one side facing down and the side having both eyes facing toward the surface. The skin on the top side of a flatfish is typically dark greenish-brown and may change color to blend in with its environment. The bottom side of a flatfish is light in color. Flounder is an example of a flatfish.

Flatfish Shape and Structure

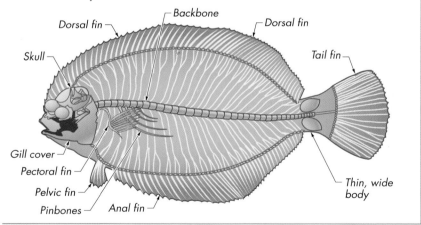

Figure 17-3. *A flatfish has a thin, wide body with both eyes located on one side of the head and a backbone that runs from head to tail down the center of the body.*

Cartilaginous Fish. A *cartilaginous fish* is any fish that has a skeleton composed of cartilage instead of bones. *See Figure 17-4.* Cartilaginous fish often have a smooth, tough outer skin without scales. Sharks, skates, and stingrays are examples of cartilaginous fish.

Cartilaginous Fish Structure

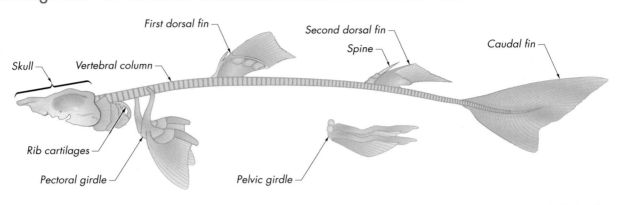

Figure 17-4. *Cartilaginous fish have a skeleton composed of cartilage instead of bones.*

FRESHWATER FISH

Freshwater fish are roundfish that live in freshwater lakes, rivers, and streams. Some varieties of freshwater fish, as well as saltwater fish and shellfish, are raised on aquafarms. *Aquafarming* is a process of raising fish, shellfish, or related game in a controlled inland environment. A wide variety of freshwater fish appear on foodservice menus. Common examples of freshwater fish include trout, perch, catfish, tilapia, and smelt.

Trout

A *trout* is a fatty, freshwater roundfish with tender flesh. Trout are rich yet delicate in flavor. Trout vary in weight from ½–10 lb. Much of the trout sold commercially come from aquafarms. Common varieties of trout include lake trout and rainbow trout. *See Figure 17-5.* Lake trout is the largest trout and has dark-gray to pale-gray skin covered with white spots and a fairly large head. Rainbow trout are named for the colorful band that extends along their sides from head to tail. Trout flesh may be red, pink, or white depending on the lake from which the fish was taken. Trout is most often broiled or sautéed.

Trout Varieties

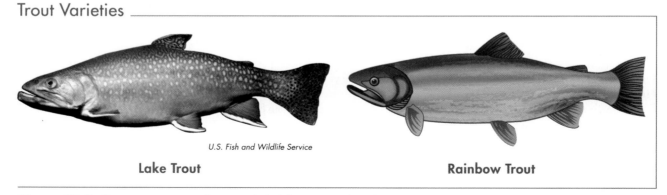

U.S. Fish and Wildlife Service

Lake Trout   **Rainbow Trout**

Figure 17-5. *Common varieties of trout include lake trout and rainbow trout.*

Perch

A *perch* is a lean, freshwater roundfish native to the Great Lakes and northern Canada. Common varieties of perch include yellow perch and walleye. *See Figure 17-6.* Yellow perch, also known as lake perch, average 12 inches in length and weigh about 1 lb. They have a mild-tasting white flesh. Yellow perch have skin that is dark olive-green on the back, blends into a golden yellow on the sides, and get lighter in color on the belly. The green portion is marked with six to eight dark, broad vertical bands that run from the back to just above the belly. Yellow perch can be broiled, sautéed, or fried.

Perch Varieties

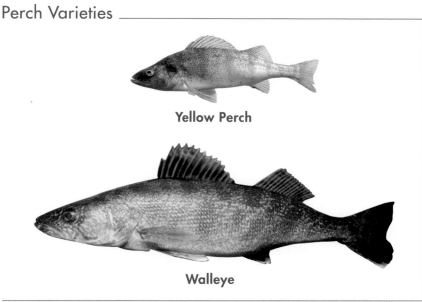

Yellow Perch

Walleye

U.S. Fish and Wildlife Service

Figure 17-6. *Common varieties of perch include yellow perch and walleye.*

A *walleye* is a lean fish and a variety of perch that resembles a pike. Walleye have exceptionally large, shiny eyes. Walleye vary in size, with an average weight between 2–5 lb. Walleye do not have the distinctive vertical bands found on the yellow perch. The flesh of a walleye has a relatively fine grain with a mild flavor and contains many small bones. Walleye are most often sautéed or fried.

Catfish

A *catfish* is a fatty, round freshwater fish named for the whiskers that protrude off the sides of its face. Instead of scales, catfish have tough skin that adheres tightly to the flesh, making it difficult to skin them. *See Figure 17-7.* The skin is always removed prior to cooking. The flesh is firm and flaky. Most aquafarmed catfish have a more mild taste than wild-caught catfish. The most common size of catfish used in the professional kitchen is ¾–1 lb. Catfish are usually sautéed or breaded and fried whole, in fillets, or strips called fingers.

Catfish

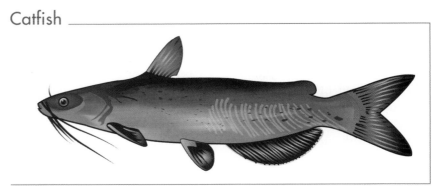

Figure 17-7. *Catfish have whiskers and tough skin that adheres tightly to the flesh.*

Tilapia

A *tilapia* is a lean, freshwater roundfish with firm, white flesh. ***See Figure 17-8.*** Tilapia is primarily aquafarmed and averages between 2–3 lb. Tilapia have a very mild flavor and are most often grilled, sautéed, or fried. Tilapia fillets average 4–8 oz.

Tilapia

Figure 17-8. *A tilapia is a lean, freshwater roundfish with firm, white flesh.*

Smelt

A *smelt* is a lean roundfish with a slender body, pointed head, and deeply forked tail. ***See Figure 17-9.*** Smelt are olive to dark green in color along the top with a silver cast along the sides. Their bellies are silver, and their fins are speckled with tiny spots. Based on the variety of smelt, they can be found in either freshwater or saltwater. The largest catch of freshwater smelt comes from Lake Michigan. Smelt are typically 4–7 inches in length and 6–8 per lb. Smelt are marketed whole and, once dressed, are sautéed or fried whole.

Smelt

U.S. Fish and Wildlife Service

Figure 17-9. *Smelt is a lean roundfish with a slender body, pointed head, and deeply forked tail.*

SALTWATER FISH

Saltwater fish can be roundfish, flatfish, or cartilaginous fish. Saltwater flat-fish do not leave their natural habitat. However, some saltwater roundfish are anadromous. An *anadromous fish* is a fish that begins life in freshwater, spends most of its life in saltwater, and returns to freshwater to spawn. There are many varieties of saltwater fish.

Salmon

A *salmon* is a fatty, anadromous saltwater fish found in both the northern Atlantic Ocean and Pacific Ocean. Salmon instinctively swim many miles, sometimes hundreds of miles, against the current to return to the freshwater where they were spawned. Pacific salmon die after they lay their eggs. This is not true of Atlantic salmon. Salmon can be grilled, broiled, baked, smoked, sautéed, and poached.

Alaska produces more than 90% of the salmon consumed in North America. However, most aquafarmed salmon are Atlantic salmon. The flesh of an Atlantic salmon is medium pink. Atlantic salmon are typically between 6–12 lb but can weigh as much as 18 lb. Pacific salmon varieties include king, sockeye, coho, and pink. *See Figure 17-10.* King salmon, also known as chinook salmon, are the largest of the Pacific salmon and have a reddish-orange flesh and a rich flavor. They range in size from 4–30 lb and average 20 lb.

> **Nutrition Note**
>
> Salmon contains omega-3 fatty acids, which are polyunsaturated and may help prevent heart disease, inflammation, and some cancers.

Salmon Varieties

King Salmon
U.S. Geological Survey

Red Salmon
U.S. Fish and Wildlife Service

Pink Salmon
U.S. Fish and Wildlife Service

Figure 17-10. *Pacific salmon varieties include king, sockeye (red), and pink salmon.*

Sockeye salmon, also known as red salmon, have deep-red flesh and weigh between 2–9 lb. Coho salmon, also known as silver salmon, have deep-pink flesh and are typically between 8–10 lb. Pink salmon are the smallest of the Pacific salmon, weighing 3–5 lb on average. They have soft, pink flesh and a lower fat content than other salmon. Pink salmon are sometimes called humpback salmon because of the noticeable hump that develops in front of the dorsal fin of males during the spawning season. Care should be taken not to overcook pink salmon because they dry out more quickly than other varieties. Pink salmon are often canned.

Arctic Char

Arctic char is a fatty, anadromous saltwater fish with a pale belly and dark sides that are speckled with light spots. *See Figure 17-11.* They are native to the north Atlantic and the Arctic Ocean. Most of the Arctic char sold today are aquafarmed in Canada, Iceland, and Norway. Arctic char range from 2–6 lb dressed. Although they have a fine flake, they stay firm when cooked. They have a similar flavor to both salmon and trout.

Sole

Sole is a lean, saltwater flatfish with pale-brown skin on the top side. *See Figure 17-12.* True Dover sole averages 10 inches in length. Dover sole has a pearly white, mild flesh that pairs well with many classical sauces. Sole fillets are available from 3–16 oz. Sole is often sautéed whole and then filleted tableside in classic preparations such as sole à la meunière (with a lemon-butter sauce) or sole amandine (with almonds).

Flounder

A *flounder* is a lean, saltwater flatfish with dark-brown skin on the top side. *See Figure 17-13.* Atlantic flounder varieties include summer flounder, winter flounder, yellowtail flounder, and sand dab. California flounder, which is also known as California halibut, is a common variety of Pacific flounder. Flounder fillets are similar in size to sole fillets and are usually broiled, sautéed, or fried.

Halibut

A *halibut* is a very large, lean, saltwater flatfish that resembles a giant flounder. *See Figure 17-14.* Only Atlantic halibut and Pacific halibut are recognized by the FDA as true halibut. Mature halibut range between 10–100 lb. Halibut have a thick, white, sweet-flavored flesh and are often sold as steaks. Halibut cheeks are considered a delicacy and can be purchased fresh or frozen. Halibut may be grilled, broiled, sautéed, fried, baked, poached, or steamed. Because halibut is fairly dry, it is often served with a sauce.

Arctic Char

Fortune Fish Company

Figure 17-11. *Arctic char has a pale belly and dark sides that are speckled.*

Sole

Plitt Seafood

Figure 17-12. *Sole is a lean, saltwater flatfish with pale-brown skin on the top side.*

Flounder

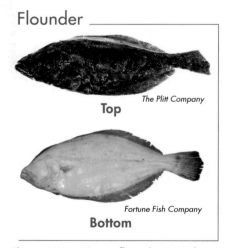

The Plitt Company

Top

Fortune Fish Company

Bottom

Figure 17-13. *A flounder is a lean, saltwater flatfish with dark-brown skin on the top side.*

Halibut

Fortune Fish Company

Figure 17-14. *A halibut is a very large, lean, saltwater flatfish that resembles a giant flounder.*

Tuna

A *tuna* is a very large, fatty, saltwater roundfish that is a member of the mackerel family. There are many different varieties of tuna, some weighing more than 300 lb. Common varieties of tuna include albacore, bigeye, yellowfin, skipjack, and bluefin. Tuna is typically sold as steaks. Tuna steaks are brushed with oil and seasoned or marinated just before grilling or broiling.

Albacore tuna is the variety most often used for canned white tuna. Bigeye and yellowfin tuna are both marketed as ahi tuna, which is most often seared and served rare or medium-rare. Yellowfin and skipjack varieties of tuna are most often used in canned products that are not specified as white or albacore tuna. Bluefin tuna is used to make sushi and sashimi. *See Figure 17-15.*

Bluefin Tuna

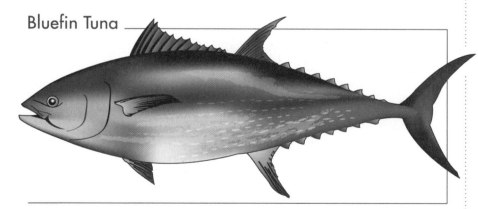

Figure 17-15. *Bluefin tuna is used to make sushi and sashimi.*

Cod

Cod is a lean, saltwater roundfish that can range from 1½–100 lb. *See Figure 17-16.* Although cod can reach over 200 lb, they commonly weigh 10 lb. Scrod is the term for young cod between 1½–2½ lb in size. Varieties of cod include Pacific cod, Atlantic cod, haddock, and hake. Cod is commonly purchased as skinless, boneless fillets. Cod flesh is white with a flaky grain and often broiled, baked, steamed, or poached. Whole cod is stuffed before baking. Cod may also be smoked, salted, or dried.

Atlantic Cod

National Oceanic and Atmospheric Administration

Figure 17-16. *Cod is a lean, saltwater roundfish that can range from 1½–100 lb.*

Pacific cod, also known as grey cod, is most often marketed as "true cod." Atlantic cod is sold as fillets or steaks and in frozen fish products. Pollock, commonly referred to as blue cod, is a cod variety with pinkish flesh that turns white when it is cooked. Pollock is the main ingredient in processed items such as surimi, crab sticks, and some fish sticks.

Haddock is a cod variety similar to Atlantic cod. The typical weight of haddock is 4 lb, and they are often sold skinless. Haddock is available year-round, but is in peak season in the spring. Broiling, baking, and steaming are the best methods of cooking haddock. Haddock is often served with a sauce.

Hake is a cod variety that has a slender body with two sets of dorsal fins and a pointed snout. Hake can be purchased whole or as fresh or smoked fillets. Hake produces lean, delicate fillets that break apart very easily. It can be baked or poached and served with a Creole or Dugléré sauce. Hake is sometimes substituted for haddock.

Herring

A *herring* is a long, thin, and somewhat fatty saltwater fish with shiny, silvery-blue skin. Herring are generally found in the northern Atlantic and in parts of the Pacific. A typical herring weighs 1–1½ lb and is commonly smoked or brined because it spoils rapidly. Herring can also be grilled or broiled. A sardine is a very small young herring. *See Figure 17-17.* Sardines are typically sold packed in oil and used in salads and on sandwiches. However, sardines can be grilled, broiled, or smoked.

Bass

A *bass* is a lean, spiny-finned roundfish with white-colored flesh that produces sweet-tasting, delicate fillets. There are many unrelated species of bass in both saltwater and freshwater. Common varieties of commercially available bass include black sea bass and striped bass. *See Figure 17-18.*

Sardines

Fortune Fish Company

Figure 17-17. *A sardine is a very small young herring.*

Bass Varieties

Black Sea Bass

Striped Bass

Figure 17-18. *Common varieties of commercially available bass include black sea bass and striped bass.*

A black sea bass is a saltwater fish with mottled black skin. It has lean, white, flavorful flesh and is typically 1–3 lb in weight. Black sea bass are usually sautéed, broiled, or baked whole. Striped bass begin life as a freshwater fish and migrate to saltwater at maturity. They are native to the Atlantic coast and marked with seven to eight well-defined dark stripes running from the head to the tail. The white-fleshed fish are sweet and rich tasting and often broiled, baked, or sautéed.

Grouper

A *grouper* is a lean, white-fleshed, saltwater roundfish with a tough, strong-flavored skin that is always removed before cooking. Although some varieties can weigh as much as 700 lb, grouper typically weigh 5–15 lb. Common varieties of grouper include black and red grouper. *See Figure 17-19.* Grouper is a moist fish that is best grilled or sautéed and is often served blackened.

Grouper

Black **Red**

Figure 17-19. *Common varieties of grouper include black and red grouper.*

Red Snapper

A *red snapper* is a lean, saltwater roundfish with pink flesh that becomes pearly white and flakes easily when cooked. Red snapper are named for their deep-red fins and red skin, which is lighter around the throat and on the belly. *See Figure 17-20.* A distinguishing factor of red snapper is that its fillets do not curl when cooked. Red snapper are typically between 4–7 lb, but can reach 25 lb. The bones of the red snapper make it slightly difficult to fillet. However, the head and bones are prized for the flavor they give to stocks and soups. Red snapper can be broiled, baked, steamed, or poached.

Red Snapper

Figure 17-20. *Red snapper are named for their deep-red fins and red skin, which is lighter around the throat and on the belly.*

Barramundi

Barramundi are large, fatty, saltwater roundfish that are silver in color. They are native to Australia but are aquafarmed sustainably in the United States. *See Figure 17-21.* Barramundi can weigh up to 88 lb and grow to be nearly 60 inches long. They have mild tasting, large-flake white meat with a delicate texture. Barramundi are high in omega-3 fatty acids and can be substituted in recipes that include red snapper, Chilean sea bass, and grouper.

Barramundi

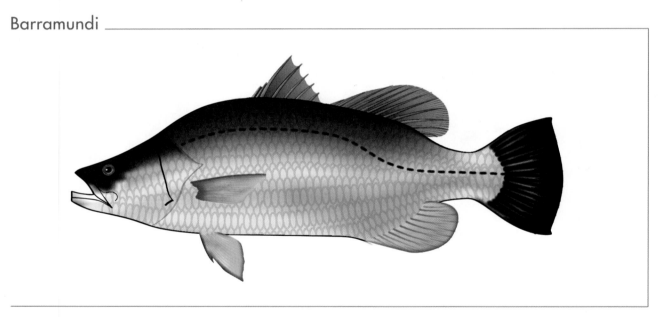

Figure 17-21. *Barramundi are a rather large, fatty, saltwater roundfish native to Australia that are aquafarmed sustainably in the United States.*

Mahi-Mahi

A *mahi-mahi* is a lean, saltwater roundfish that has colorful skin and firm, pink flesh. *See Figure 17-22.* The flesh of a mahi-mahi has a sweet flavor. Mahi-mahi is also marketed as dorado. Mahi-mahi are typically between 8–25 lb, but can weigh up to 50 lb. Mahi-mahi is typically grilled, broiled, baked, or poached, and served with a flavorful sauce.

Mahi-Mahi

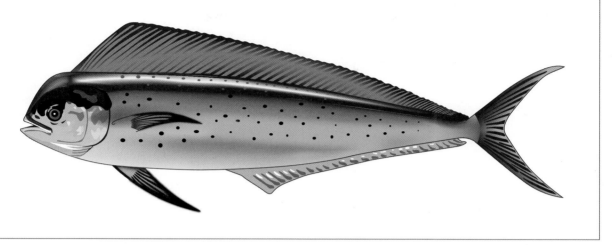

Figure 17-22. *A mahi-mahi is a lean, saltwater roundfish that has colorful skin and firm, pink flesh.*

Orange Roughy

An *orange roughy* is a lean, round, saltwater fish with orange to light-gray skin that has a rough appearance. ***See Figure 17-23.*** The firm, large-flaked flesh is pearly white and has a sweet, delicate flavor. Most orange roughy is from New Zealand. It is available year-round and most often sold as 8–12 oz skinless, boneless fillets. Orange roughy is commonly baked or stuffed and then baked. It also can be broiled, sautéed, fried, or steamed.

Orange Roughy

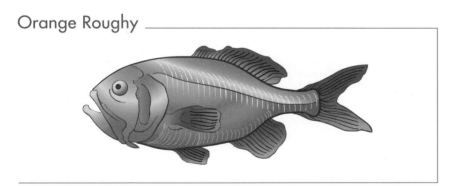

Figure 17-23. *Orange roughy is a lean, round, saltwater fish with orange to light-gray skin that has a rough appearance.*

Ocean Perch

An *ocean perch,* also known as redfish, is a round, saltwater fish native to the north Atlantic that has red skin and pink flesh. ***See Figure 17-24.*** The typical weight of an ocean perch is 1–1½ lb, but some reach 5 lb. The flesh has a medium-firm texture and a sweet, mild taste. Although it shares a name with yellow perch and lake perch, ocean perch is not related to those species. Ocean perch can be broiled, sautéed, or fried.

Ocean Perch

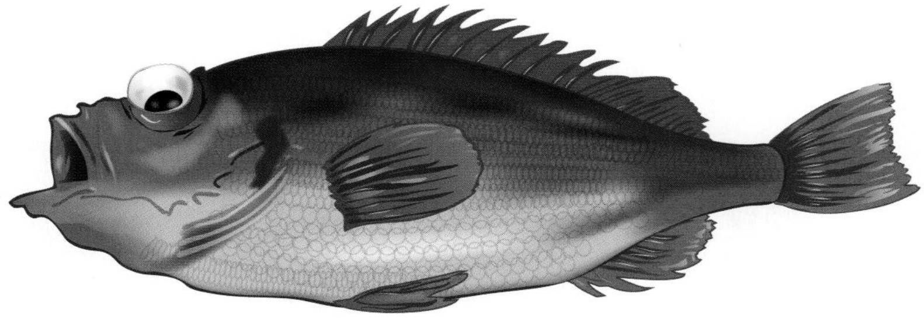

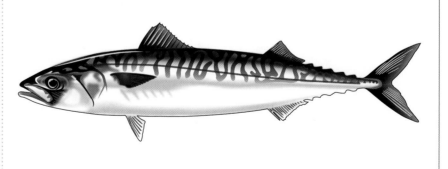

Figure 17-24. *Ocean perch, also known as redfish, are round, saltwater fish native to the north Atlantic and have a red skin and pink flesh.*

Mackerel

A *mackerel* is a fatty, saltwater roundfish with a streamlined shape and dark bluish-brown or black skin. ***See Figure 17-25.*** The firm, grayish-pink flesh has a strong flavor. Boston and Spanish mackerel average between 14–18 inches in length and can weigh up to 5 lb. King mackerel is a larger variety that is commonly sold cut into steaks. Wahoo is an even larger variety of mackerel that has a flavor similar to that of tuna. Most wahoo are 20–40 lb. Mackerel is typically grilled, broiled, baked, or smoked.

Nutrition Note

Methylmercury is a toxin that results from industrial pollution in open bodies of water. Many types of fish eat plant matter that contains methylmercury. Levels of this harmful toxin are highest in fish such as mackerel, swordfish, and shark.

Mackerel

Figure 17-25. *A mackerel is a fatty, saltwater roundfish with a streamlined shape and dark bluish-brown or black skin.*

Swordfish

A *swordfish* is a very large, fatty roundfish with an upper jaw that extends to form a long and flat, sharp-edged sword. ***See Figure 17-26.*** Swordfish can weigh over 300 lb and be nearly 15 feet long. Their slightly pink flesh is quite dense and turns white when cooked. Due to their size, swordfish are commonly sold as steaks. They have a sweet, mild flavor and are commonly grilled or broiled.

Swordfish

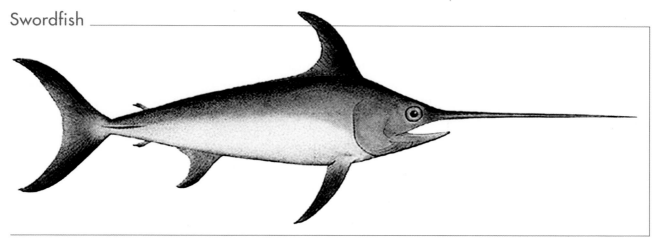

Figure 17-26. *A swordfish is a very large, fatty roundfish with an upper jaw that extends to form a long and flat, sharp-edged sword.*

Sturgeon

A *sturgeon* is a fatty roundfish with a wedge-shaped snout and a body style similar to that of a shark. Sturgeon do not have teeth and are bottom feeders. They can be found in freshwater and saltwater, and some varieties of sturgeon are anadromous. Although prized for their flesh, sturgeon are primarily caught to obtain their roe (eggs), better known as caviar. ***See Figure 17-27.*** Varieties of sturgeon include beluga, osetra, sevruga, and lake.

- Beluga sturgeon are the most prized variety of sturgeon because of their caviar and firm, oily flesh. Some beluga sturgeon weigh close to 2000 lb, although most do not weigh over 1000 lb.

- Osetra sturgeon provide one of the highest-quality caviars available and have firm, flavorful fillets.

- Sevruga sturgeon from the Black Sea and the Caspian Sea offer high-quality caviar and fillets.

- Lake sturgeon from North America are commonly smoked or brined. Caviar can be produced from some varieties of lake sturgeon.

Sturgeon

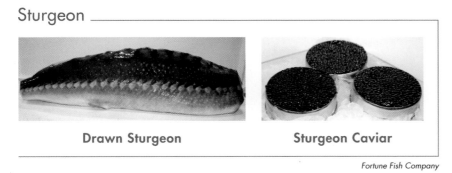

Drawn Sturgeon   **Sturgeon Caviar**

Fortune Fish Company

Figure 17-27. *Although prized for their flesh, sturgeon are primarily caught to obtain their roe (eggs), better known as caviar.*

Monkfish

Figure 17-28. *Monkfish is a very large, lean, saltwater roundfish.*

Cuisine Note

Shark fin soup is a soup for special occasions in Chinese cuisine. Shark fins are fairly bland but add to the texture of the soup.

Monkfish

A *monkfish* is a very large, lean, saltwater roundfish that can weigh up to 50 lb. *See Figure 17-28.* Only the tail section and liver of a monkfish are edible. The firm-textured flesh is usually sold as whole tails and has similar qualities to lobster flesh. The liver is used in Japanese cuisine. Any cooking method can be used to prepare monkfish. When cooked by dry-heat methods, monkfish is served with a sauce. It is also used to make soups and stews.

CARTILAGINOUS FISH

Cartilaginous fish have a smooth outer skin and cartilage instead of bones. Sharks, skates, and eels are cartilaginous fish. *See Figure 17-29.*

Sharks

A *shark* is a lean, cartilaginous fish that is generally found in saltwater. There are approximately 250 species of shark, but only a few provide quality flesh. Sharks used for human consumption usually range in weight between 30–200 lb. Sharks are most commonly grilled or broiled, but can also be baked, fried, or poached.

Cartilaginous Fish

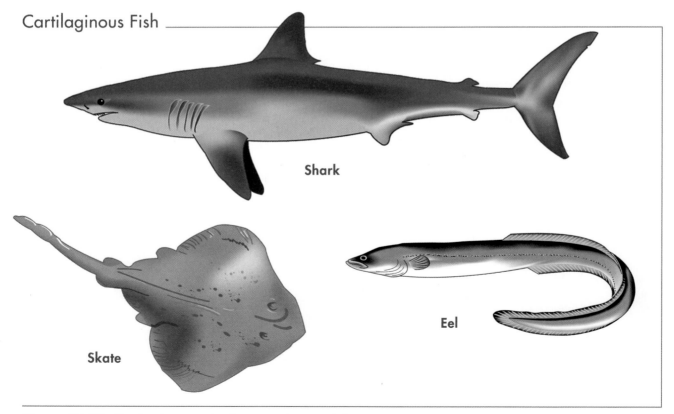

Shark

Eel

Skate

Figure 17-29. *Sharks, skates, and eels are cartilaginous fish.*

Skates

A *skate,* also known as a ray, is a lean, cartilaginous fish with two winglike sides found in saltwater. Skates are grayish brown on top and white on the bottom. Typically only the wings of the skate are removed and sold. The edible flesh is divided by a layer of cartilage that runs through the center of each wing. Each wing produces two separate fillets. Skate wings are commonly grilled or sautéed.

Eels

An *eel* is a fatty cartilaginous fish with a long, slender body that resembles a snake. The skin of an eel is similar to that of a catfish, in that it is smooth, tough, tight to the flesh, and does not have scales. Eels have a strong, slightly sweet flavor and can be prepared using any cooking method.

MARKET FORMS OF FISH

Fish may be purchased fresh, frozen, or processed. The more that is done to a fish prior to delivery, the higher the cost per pound. Fish are commonly sold whole, drawn, dressed or pan-dressed, as steaks, and as fillets. *See Figure 17-30.* Fish may also be sold as loins, roasts, or portions. A fish loin is cut lengthwise from either side of the backbone of large roundfish, such as tuna. A fish roast is a crosswise section cut from behind the head and just short of the tail of large fish. A fish portion is cut from fillets into smaller pieces that are sold individually.

Market Forms of Fish

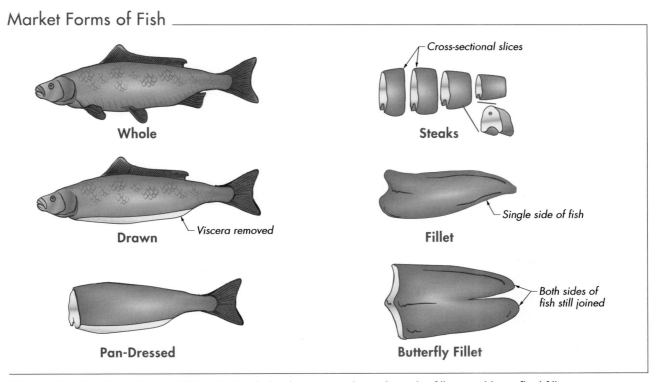

Figure 17-30. *Market forms of fish include whole, drawn, pan-dressed, steaks, fillets, and butterflied fillets.*

Florida Department of Agriculture and Consumer Services, Bureau of Seafood and Aquaculture Marketing

Whole Fish

A *whole fish* is the market form of a fish that is taken from the water and sold as is. Nothing has been done to process it. Whole fish have the shortest shelf life of any market form because all the internal organs (viscera) are still present. Fish purchased whole cost less per pound, require more preparation, and yield more waste than any other market form.

Drawn and Dressed Fish

A *drawn fish* is a fish that has had only the viscera removed. Drawn fish can be prepared whole. A *dressed fish* is a fish that has been scaled and has had the viscera, gills, and fins removed. A *pan-dressed fish,* also known as a headed and gutted fish, is a dressed fish that has had its head removed.

Fish Steaks

A *fish steak* is a cross section of a dressed fish. Steaks are ready to cook when purchased. Generally, the only bone present in a steak is a small section of the backbone. Steaks from very large fish, such as swordfish, are boneless. Tuna, salmon, swordfish, and sharks are commonly sold as steaks.

Fish Fillets

A *fish fillet* is the lengthwise piece of flesh cut away from the backbone. Roundfish have two fillets, one on each side. Flatfish have four fillets, two on each side. Fillets can be purchased with or without bones and with or without skin. Fillets with the skin left on are sold scaled. A *butterflied fillet* is two single fillets from a dressed fish that are held together by the uncut back or belly of the fish.

Frozen Fish

Frozen fish are served in restaurants more often than fresh fish. Using frozen fish allows a foodservice operation to serve a wider variety of fish year-round. Frozen fish cannot be purchased whole. For example, a large percentage of salmon is sold frozen. Frozen salmon is held at 32°F until it is flash-frozen. It is protected from dehydration by glazing. *Glazing* is the process of covering an item with water to form a protective coating of ice before the item is frozen. Proper glazing of frozen salmon results in a product that tastes like fresh salmon.

Processed Fish

Processed fish are available canned, smoked, salted, or pickled. Sardines, anchovies, and tuna are the most common varieties of canned fish. Canned fish should be checked for signs of damage or bulging before they are used. Damaged cans should be discarded. Salmon is often smoked. *See Figure 17-31.* Cod is sold salted. Herring is available pickled.

Processed Fish _____

Figure 17-31. *Salmon is often smoked.*

Popular smoked and cured fish include cod, haddock, salmon, sturgeon, and herring. Cod is often salted, whereas salmon and herring are often pickled. *Gravlax* is a traditional Scandinavian dish where salmon is cured for 24–48 hours with salt, sugar, and dill. *Lox* is salmon that has been brine cured and then cold smoked. *Nova* is a brine-cured, cold-smoked salmon that is less salty than lox.

Edward Don & Company

INSPECTION AND GRADES OF FISH

Fresh fish are not subject to a mandatory federal inspection. Instead, optional inspections are carried out by the National Marine Fisheries Service, which is part of the United States Department of Commerce (USDC). There are three types of optional inspections. These types of inspections include Type 1, Type 2, and Type 3.

A Type 1 inspection guarantees the fish or shellfish product is safe and wholesome for human consumption, is accurately labeled, has a good odor, and was processed in a sanitary, inspected facility. After being processed under a Type 1 inspection, a PUFI (processed under federal inspection) mark is affixed to the packing carton. *See Figure 17-32.* Type 1 inspection involves continual inspection of the fresh product from the time it arrives at the processing plant to the moment it is packaged for sale.

A Type 2 inspection takes place in a warehouse or cold storage facility. The product is randomly inspected to ensure it meets the product specifications listed on a specification sheet.

A Type 3 inspection involves the examination of the fishing boats and processing plants to ensure that they are adhering to sanitation guidelines when handling and processing the product.

Only fish that is processed under a voluntary Type 1 inspection is eligible for grading. Fish may be graded A, B, or C. Because there are so many varieties of fish, the USDC only sets grade standards for the most common varieties. Grade A fish are of the best quality and do not have any visible defects. Grade B and Grade C are typically used for processed products.

PUFI Marks

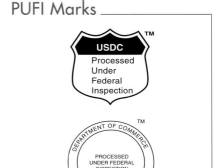

Figure 17-32. *After being processed under a Type 1 inspection, a PUFI (processed under federal inspection) mark is affixed to the packing carton.*

RECEIVING AND STORING FRESH FISH

Fresh fish spoils rapidly so it is essential to check the smell, appearance, and flesh upon receipt. *See Figure 17-33.* The smell of a fish is one of the easiest ways to determine freshness, but it can also be misleading. Fresh fish have a light, fishy smell. Strong fishy smells or ammonia odors are signs of deterioration. Fish that give off such smells should not be accepted.

The eyes and gills of fresh fish indicate age. The eyes should be round, slightly bulging, and clear. Cloudy or sunken eyes are signs of age and the fish should be rejected. The gills of a fresh fish should be bright red or reddish pink. If the gills are brown or missing, the fish should be rejected.

L. Isaacson and Stein Fish Company

Consideration for Purchasing Fresh Fish		
	Acceptable	**Unacceptable**
Smell	• Slight smell of seaweed or the ocean	• Strong fishy smell
	• Slight smell of fish	• Ammonia odor
Appearance	• Round, clear, bulging eyes	• Cloudy or sunken eyes
	• Bright-red or reddish-pink gills	• Brown gills or missing gills
	• Properly covered flesh	• Bruised, discolored, or damaged flesh
	• Moist and solid fillets	• Flesh of fillets separates when slightly bent
Touch	• Wet, slightly slippery exterior surface	• Slimy internal cavity
	• Smooth scales lying flat against body	• Rough scales
	• Moist, intact fins	• Dry fins
	• Firm flesh	• Mushy flesh

Figure 17-33. *Fresh fish spoils rapidly so it is essential to check the smell, appearance, and flesh upon receipt.*

If fish has been cut into fillets, they should be moist and solid. Gently pressing on the fillet should not leave an indentation. The flesh should be firm and spring back when touched. A fillet should not be slimy, and the flakes of flesh should not separate when the fillet is bent slightly. The flesh should be of the proper color for the particular variety of fish. There also should not be any bruises or other signs of visible damage. If a fish shows any of these signs, it should be rejected.

The external surface of a fresh fish should be wet and a little slippery. Scales should be smooth and lie flat against the body of the fish. Scales sticking out are signs of age and these fish should be rejected. Fins should be moist and intact. If the fins are dried out or removed, the fish may be older and the flesh may be starting to dry out. Fish with dry fins should be rejected.

Even if the fish being delivered is from an inspected facility, it may not have been held at proper temperatures since leaving the plant. Fresh fish should have an internal temperature of 45°F or lower to be accepted at delivery. Fish above 45°F should be rejected. Once accepted, fish must be stored at 41°F or below and used as quickly as possible to maintain quality and freshness.

Fresh fish can be stored a maximum of one to two days. If a fresh fish is not cooked upon receipt, it should be removed from the ice it was packaged in, placed in a self-draining container, covered with plastic wrap, and then covered with ice. *See Figure 17-34.* Then the container should be placed in the coldest part of the refrigerator. The plastic wrap helps prevent the flesh of the fish from developing freezer burn and from absorbing water as the ice melts. The ice helps hold the proper temperature and reduces deterioration. The ice should be replaced daily. Fish must be stored away from other food in the refrigerator to prevent the odors from affecting the smell or taste of other foods.

Storing Fresh Fish

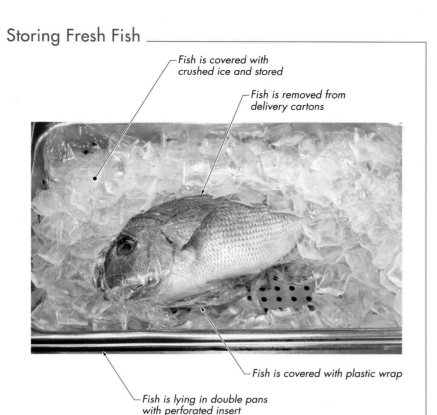

Fish is covered with crushed ice and stored

Fish is removed from delivery cartons

Fish is covered with plastic wrap

Fish is lying in double pans with perforated insert

Figure 17-34. *If a fresh fish is not cooked upon receipt, it should be removed from the ice it was packaged in, placed in a self-draining container, covered with plastic wrap, and then covered with ice.*

RECEIVING AND STORING FROZEN FISH

When receiving frozen fish, it should be received frozen solid. If the fish is even slightly thawed on the edges, it should be rejected. Fish that has been refrozen will have a poor texture. If there are dry-looking spots on the fish, it has slightly thawed and been refrozen and should be rejected. Signs of moisture on the outside of the box may also indicate that the product has thawed and been refrozen. Ice crystals are another sign that fish thawed or was not stored at or below 0°F. Frozen fish quickly deteriorate at temperatures above 0°F, so fish displaying ice crystals should be rejected.

Fish stored in the freezer must be wrapped properly with moistureproof wrapping to prevent freezer burn. Fatty fish should not be stored in the freezer for more than two months because they deteriorate quickly. Lean fish can be stored frozen for up to six months. Stock should be rotated so that the oldest fish is used first. Frozen fish must be thawed in the refrigerator as close as possible to the time the fish is needed. Once frozen fish has thawed, it must be used or discarded.

Frozen fish can be individually quick-frozen (IQF), layer packed, cello-packed, or block frozen. *Individually quick-frozen (IQF)* is a designation for products preserved using a method in which each item is glazed with a thin layer of water and frozen individually. *See Figure 17-35.* IQF portions can be packaged together without sticking to other portions. This allows as much of an item to be used as needed without having to thaw an entire package. IQF products are packaged according to an average size, such as 2–3 oz fillets. IQF packaging makes it easy to remove the exact number of items needed.

Individually Quick-Frozen (IQF) Fish

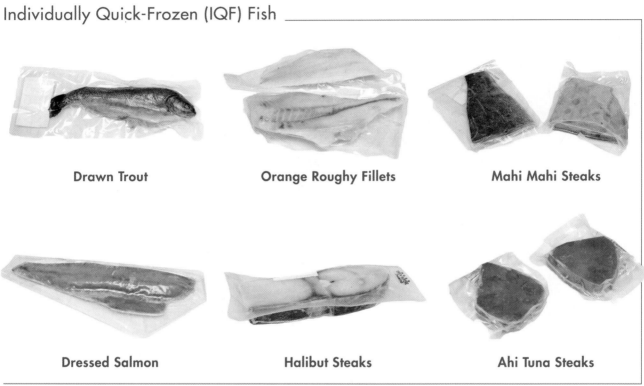

Drawn Trout

Orange Roughy Fillets

Mahi Mahi Steaks

Dressed Salmon

Halibut Steaks

Ahi Tuna Steaks

Czimer's Game & Seafoods, Inc.

Figure 17-35. *Individually quick-frozen (IQF) products have been glazed with a thin layer of water and frozen individually.*

Layer packs, also known as shatter packs, consist of high-quality, graded fish fillets layered on polyethylene sheets. The edges slightly overlap so that entire layers can be removed when desired.

Cello-packs contain ungraded fish fillets that are frozen in packets, typically one to three fillets per packet, wrapped in cellophane, frozen, and packaged six packets per box. Fillets packaged in this manner may be inconsistent in size and are relatively inexpensive.

Block-frozen fish is placed in a block-shaped form between two hollow stainless steel plates that have refrigerant flowing through them. The plates freeze the fish into a solid block within 2–4 hours. Ungraded fish are often packaged block frozen. Block-frozen items are often breaded or battered and then fried.

FABRICATION OF FISH

Fish may be purchased whole and fabricated in-house or may be purchased in a portion-controlled or processed form. Knowing and using proper fabrication techniques for fish is an important skill to have in the professional kitchen.

Scaling Fish

Scales must always be removed from fish prior to preparation. *See Figure 17-36.* However, care should be taken to ensure the flesh is not damaged as the scales are removed.

Procedure for Scaling Fish

1. In a sink, firmly hold the tail of the fish with the guiding hand. Use a fish scaler to scrape against the scales down toward the head. *Note:* Use caution not to damage the flesh of the fish by applying too much pressure.
2. When finished scaling one side, turn the fish over and repeat the process.
3. Rinse the fish under cold running water to remove any loose scales.

Figure 17-36. *Scales must always be removed from a fish prior to preparation.*

A fish steak is a popular cut for many large roundfish varieties. *See Figure 17-37.* For example, tuna is often purchased in loins. The loins are commonly cut into steaks. The smaller end portions of the loin are cut into small cubes. Belly loins commonly have a higher fat content than back loins and are therefore more desirable for sashimi. A dark-red, almost black, muscle runs along the lateral line of the fish and should be removed prior to preparation of the fish. There may also be a vein running along the loin that should be removed, as it has a strong flavor and is not desirable.

Procedure for Cutting Roundfish into Steaks

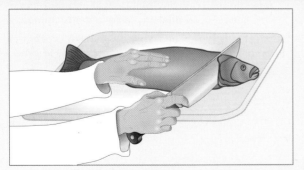 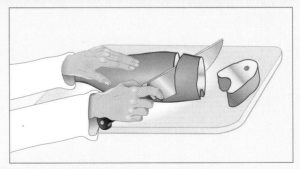

1. Use a boning knife to remove the scales, fins, viscera, head, and tail from the fish.

2. Slice the fish horizontally into equal sized pieces.

Figure 17-37. *A fish steak is a popular cut for many large roundfish varieties.*

Refer to DVD for
Filleting Roundfish
Media Clip

Filleting Roundfish

A fillet is the most common serving portion from a fish. *See Figure 17-38.* It is important to remove all bones from a fillet before cooking it.

Procedure for Filleting Roundfish

1. Use a boning knife to make a cut about ½ inch behind the gills and down to, but not through, the backbone.

2. Make a second cut along the backbone from just behind the head all the way to the tail. Do not cut through the backbone.

3. Starting at the tail, carefully slice toward the head to cut the flesh away from the backbone.

4. Carefully lift the fillet and cut away any rib bones that are still attached to the fillet. Trim any belly fat from the fillet.

5. Run fingers gently along the surface of the fillet to raise the ends of any pinbones that may remain. Use needle-nose pliers to remove the pinbones.

6. Turn the fish over and repeat the entire process on the other side.

Figure 17-38. *A fillet is the most common serving portion from a fish.*

Skinning Fillets

A fillet can be prepared with or without the skin intact. For a particular presentation or dish, it may be required to remove the skin from a fillet. *See Figure 17-39.*

Skinned fish fillet.

1. Place the fillet skin-side down. Starting at the tail, use a boning knife to carefully cut down through the flesh to the skin while using the guiding hand to firmly hold the tail skin.

2. Angle the edge of the blade slightly downward and cut between the skin and the flesh while moving the blade toward the head end.

Figure 17-39. *For a particular presentation or dish, it may be required to remove the skin from a fillet.*

Filleting Flatfish

The backbone of a flatfish runs through the midline of the fish. As a result, flatfish yield four fillets, two on each side of the fish. ***See Figure 17-40.***

Refer to DVD for **Filleting Flatfish** Media Clip

1. Use a boning knife to cut along the backbone from the head to the tail.

2. Insert the tip of the blade near the head of the fish and carefully slice the flesh away from the bones on one side of the backbone.

3. Remove the second fillet using the same process.

4. Turn the fish over and repeat steps 1–3 to remove the other two fillets.

Figure 17-40. *A flatfish yields four fillets, two on each side of the fish.*

Cuisine Note

Cooked tuna should have a pink center similar to that of a medium-rare filet mignon. Using Grade A tuna reduces the risk of foodborne illness when serving seared or medium-rare tuna.

COOKING FISH

Most fish are naturally tender and require very little cooking time. The flesh of fish is fragile, flakes apart easily, and overcooks quickly. The fat content of a fish affects the cooking methods used. Both lean and fatty fish can be sautéed or fried with excellent results. Lean fish, such as red snapper, haddock, halibut, and perch, are best steamed or poached. Fatty fish, such as mackerel, tuna, trout, and salmon, are best grilled, broiled, or baked because their natural fat prevents them from drying out during cooking. A lean fish may also be broiled or baked if it is basted during the cooking process.

Determining Doneness

Determining the doneness of fish takes practice because it involves sight and touch more than temperature. A fish that is done will have a caramelized skin, firm and dense interior edges, and a moist, opaque center. *See Figure 17-41.* An instant-read thermometer that registers 145°F when inserted into the thickest part of a thick-fleshed fish indicates the fish is cooked through. When a fork is used to lightly prod the thickest part of a thin fish, done flesh will be opaque and the juices will be milky white. Undercooked fish is translucent and the juices will be clear and watery. Overcooked fish is dry and falls apart easily.

Determining Doneness

Firm and dense interior edges

Caramelized skin

Moist and opaque center

Figure 17-41. A fish that is done will have a caramelized skin, firm and dense interior edges, and a moist, opaque center.

Grilling and Broiling

Many thick cuts of fish are grilled or broiled. For example, salmon is perfect for grilling and broiling because the thick flesh can withstand the intense heat of open flames. *See Figure 17-42.* When cooking fish fillets with the skin left on, the skin should be scored with a sharp knife first so that the fillet does not curl during cooking.

Broiled Salmon

Figure 17-42. *Salmon is perfect for grilling and broiling because the thick flesh can withstand the intense heat of open flames.*

Sautéing and Frying

Fish is often sautéed or fried. Sautéing uses a small amount of hot fat to sear and cook the fish. Only uniformly sized cuts should be sautéed to ensure even cooking. Care should be taken when sautéing not to burn the oil or the surface of the food by using too high a temperature. A pan sauce can be made in the same pan after the cooked fish has been removed.

Fish is typically cut into fillets or strips for frying. These cuts are commonly breaded or battered then deep-fried. Breading protects the fish from the hot oil and helps it retain moisture. *See Figure 17-43.*

Fried Fish

Florida Department of Agriculture and Consumer Services, Bureau of Seafood and Aquaculture Marketing

Figure 17-43. *Breading protects the fish from the hot oil and helps it retain moisture.*

Chef's Tip

Shark flesh has a slight ammonia smell. Soaking shark in milk or cream before cooking it will neutralize this odor. If the shark has a really strong odor it should be discarded.

Steaming and Poaching

Steaming adds no additional calories or fat to a fish. Steaming also helps fish retain nutritional value by gently cooking the fish without immersing it in a liquid. Fish is often steamed on fresh herbs or tea leaves so that as the steam penetrates and cooks the fish, it also imparts a delicate aroma.

En papillote is a a technique in which food is steamed in a parchment paper envelope as it bakes in an oven. *See Figure 17-44.* When baking fish en papillote, the fish is often topped with julienned vegetables, butter, white wine, and lemon. The paper is then folded over and the edges are sealed. The parchment envelope is then placed in a 450°F oven and baked for approximately 8–10 minutes, until the paper is browned slightly and puffed. The envelope is removed from the oven, plated, and opened tableside to allow the aromas to escape in front of the guest.

En Papillote

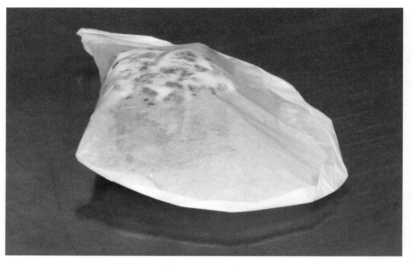

Figure 17-44. *Fish may be cooked en papillote, or in a parchment paper envelope.*

Fish is commonly poached in a court bouillon or a fumet. These flavorful poaching liquids add moisture and a delicate flavor to the fish. When poaching, it is essential to ensure that the poaching liquid never reaches a simmer, because the intense heat would cause the fish to toughen and dry out. Common poaching liquids used with fish include wine and lemon juice. The small amount of acid they add to the liquid makes items taste fresher and retain more of their natural color. Fish, such as salmon and trout, are sometimes poached whole and then chilled for a cold presentation or cold buffet. *See Figure 17-45.*

1. Place a whole fish on a lightly oiled rack or glazing screen. Secure the fish with butcher's twine to prevent it from tipping over.

2. Lower the rack or screen into a well-seasoned court bouillon and poach the fish at a temperature between 170°F and 180°F for 30 minutes.

3. Turn off the heat and let the fish rest in the court bouillon for 20–30 minutes until completely cooked through.

4. Carefully remove the rack or glazing screen and allow the fish to drain completely.

5. If the fish is to be served warm, serve immediately. If the fish is to be served cold, immediately chill the fish on the rack or glazing screen until cold.

Figure 17-45. *Fish, such as salmon and trout, are sometimes poached whole and then chilled for a cold presentation or cold buffet.*

SHELLFISH

Shellfish is the classification of aquatic invertebrates that may or may not have a hard, external shell. An external shell functions as a skeleton and is called an exoskeleton. Shellfish are commonly categorized as crustaceans or mollusks. *See Figure 17-46.* A *crustacean* is a shellfish that has a hard, segmented shell that protects soft flesh and does not have an internal bone structure. A *mollusk* is a shellfish with a soft, nonsegmented body. Some mollusks, such as clams, have a hard, external shell. Other types of mollusks, such as squid, do not have an external shell. Shellfish are popular menu items and are a large part of Creole and Cajun cuisine.

Shellfish

Agricultural Research Service, USDA

Crustaceans

Florida Department of Agriculture and Consumer Services, Bureau of Seafood and Aquaculture Marketing

Mollusks with Shells

National Oceanic and Atmospheric Administration

Mollusks without Shells

Figure 17-46. *Shellfish are commonly categorized as crustaceans or mollusks.*

CRUSTACEANS

Crustaceans live in both freshwater and saltwater. Unlike fish, crustaceans can live out of water for a few days if they are kept moist. Crustaceans include shrimp, prawns, lobsters, crayfish, and crabs.

Shrimp and Prawns

Shrimp and prawns are crustaceans with a tender white flesh and a distinctive flavor. In the United Kingdom and Australia, the term prawn is used almost exclusively for both shrimp and prawns. In North America, larger shrimp are commonly marketed as prawns. *See Figure 17-47.* In the culinary industry, the terms shrimp and prawns are used interchangeably.

Shrimp and Prawns _____

Shrimp   **Prawns**

Figure 17-47. *In North America, larger shrimp are commonly marketed as prawns. In the culinary industry, shrimp and prawns are used interchangeably.*

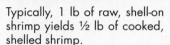

Chef's Tip

Typically, 1 lb of raw, shell-on shrimp yields ½ lb of cooked, shelled shrimp.

Four common types of shrimp used in the professional kitchen are white, brown, pink, and black tiger. Although these four types of shrimp vary in color when caught, they differ very little in appearance when cooked. They also have similar flavor and nutritional value. Grilling, broiling, sautéing, and frying are popular methods of preparing shrimp. Cooked shrimp also can be chilled and served cold as appetizers, salads, or entrées.

Lobsters

A *lobster* is a saltwater crustacean with a brown to bluish-black external shell and two large claws. *See Figure 17-48.* Lobsters are sold live and must be kept alive until cooking. A *sleeper* is a lobster that is dying. Sleepers are sold at a reduced price and must be cooked immediately to retain the firm flesh. Cooking and eating dead lobsters can be harmful.

Anatomy of Lobsters _____

Tail meat   Coral   Stomach

Intestinal vein

Tomalley   Claw meat

Figure 17-48. *A lobster is a saltwater crustacean with a brown to bluish-black external shell and two large claws.*

Almost all lobster flesh is edible except for a small section near the eye area. The soft green substance found in the body cavity of a lobster is the liver and pancreas, which are collectively called the tomalley. The tomalley is considered a delicacy. It may be eaten alone or used to thicken sauces. Lobster paste, also known as lobster pâté, is a mixture of tomalley and lobster roe (coral).

Most of the lobster flesh is found in the tail and the claws. Lobsters are most often steamed or poached. Common varieties of lobster include Maine lobsters, spiny lobsters, and langoustines.

Maine Lobsters. A *Maine lobster* is a large lobster with a dark bluish-green shell, two large, heavy claws, and four slender legs on each side of its body. Maine lobsters are harvested from cold north Atlantic waters.

Spiny Lobsters. A *spiny lobster,* also known as a rock lobster, is a lobster that has spines covering its body and five slender legs on each side. Spiny lobsters have smaller claws than Maine lobsters. The tail section of a spiny lobster is typically eaten. The flesh is less delicate in flavor than that of a Maine lobster.

Langoustines. A *langoustine,* also known as a Norway lobster, is a small lobster that resembles a very large shrimp, except it has very long front arms with long, thin claws. Langoustines are harvested from the northeastern Atlantic Ocean, the North Sea, and the Mediterranean Sea. Only the tail section of a langoustine is eaten.

Crayfish

A *crayfish,* also known as crawfish or crawdad, is a freshwater crustacean that resembles a tiny lobster. *See Figure 17-49.* Crayfish are dark brown to black in color and can range from 3–7 inches in length. Most of the crayfish harvested come from Louisiana and the Pacific Northwest. Crayfish have a flavor similar to that of shrimp, but a slightly tougher texture. They are commonly used in Creole, Cajun, and French cuisine.

Crabs

Crabs are a popular variety of shellfish with tender, sweet-tasting flesh that can be used in many menu items. Crabmeat is available fresh, frozen, and canned. When using fresh crabs, only live crabs should be used. Fresh crabs that have died should be discarded. Hard-shell crabs are commonly simmered in a flavorful poaching liquid until the shell turns bright red and the flesh is cooked through. Soft-shell crabs are often sautéed or fried.

Most edible crabs can be distinguished from inedible varieties by counting the pairs of legs. Most edible crabs have five pairs of legs, four for walking and one pair that serve as arms. Most inedible crabs have only four pairs of legs. King crabs are an exception, as they have only four pairs of legs but are still edible. King crabs, blue crabs, Dungeness crabs, snow crabs, and stone crabs are frequently used in the professional kitchen. *See Figure 17-50.*

Crayfish

Figure 17-49. *A crayfish, also known as a crawfish or crawdad, is a freshwater crustacean that resembles a tiny lobster.*

Crab Varieties

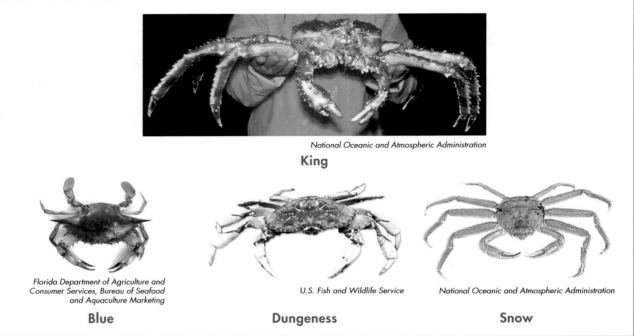

National Oceanic and Atmospheric Administration

King

Florida Department of Agriculture and Consumer Services, Bureau of Seafood and Aquaculture Marketing

Blue

U.S. Fish and Wildlife Service

Dungeness

National Oceanic and Atmospheric Administration

Snow

Figure 17-50. *Crab varieties include king crab, blue crab, Dungeness crab, and snow crab.*

King Crabs. *King crab* is the largest-sized variety of crab, typically weighing between 6–20 lb and measuring as much as 10 feet from the tip of one leg to the tip of the opposite leg. King crabs are one of the few edible crabs that have four pairs of legs. The raw flesh has a pinkish tinge that becomes snowy white when cooked. King crab is available in the shell as cooked frozen legs, as legs and claws, as leg and body flesh, or as shredded flesh. Canned or frozen king crabs are also available shucked or pulled (removed from the shell). King crab is not typically sold live because the legs are removed at catch and flash-frozen to preserve their quality.

Blue Crabs. *Blue crab* is a North American crab with blue claws and a dark blue-green, oval shell. Blue crabs measure approximately 5 inches across and weigh approximately 5 oz. Hard-shell blue crab is sold live, cooked and frozen, or cooked, pickled, and canned. A *soft-shell crab* is a blue crab that has been harvested within 6 hours of molting, or shedding its shell in order to grow a larger shell. Molting season is from mid-May to the beginning of September. Soft-shell crab is only sold live during molting season. Soft-shell crabs must be handled with special care to ensure that they arrive at their destination alive. They must be kept alive until they are cooked or cleaned and immediately quick-frozen.

Dungeness Crabs. *Dungeness crab* is a Pacific crab with a sweet-tasting flesh. Approximately 25% of the body weight of a Dungeness crab is flesh, which is the highest percentage found in any variety of crab. Most of the edible flesh in a Dungeness crab is body flesh, not leg flesh. Dungeness crabs are larger than blue crabs, weighing approximately 1¾–4 lb each. Dungeness crab is commonly available frozen or canned but can be purchased live.

Snow Crabs. A *snow crab* is similar to a king crab but is smaller and available in greater supply. Also known as a spider crab, snow crab is typically sold as cooked, frozen leg clusters. A snow crab has only a fraction of the flesh that a king crab has. Frozen leg clusters are commonly steamed or poached just enough to warm the flesh because they are already fully cooked. Snow crab can be steamed and served whole or served cold in a crab cocktail.

Stone Crabs. A *stone crab* is an Atlantic crab with a brownish-red shell and large claws of unequal size. Unlike other varieties of crab, when a stone crab is caught one claw is removed and the crab is placed back in the water. The claws are the only parts of the stone crab that are harvested. In approximately 12–24 months, the stone crab generates a new claw in place of the one that was removed. Each time a claw is removed it grows back slightly larger than before. Stone crab claws range from 2–5 oz and are typically served cold with most of the shell removed. They can also be served hot. Stone crabs have a similar taste and texture to lobster.

Florida Department of Agriculture and Consumer Services, Bureau of Seafood and Aquaculture Marketing

MOLLUSKS

Like crustaceans, mollusks do not have an internal skeleton. The soft body of a mollusk is often protected by a hard external shell. Mollusks should always be alive prior to cooking, unless they are frozen or canned. If a mollusk dies prior to cooking, it should be discarded. It should not be eaten under any circumstance or illness could result. The three classifications that edible mollusks are divided into are univalves, bivalves, and cephalopods.

Univalves

A *univalve* is a mollusk that has a single solid shell and a single foot. A univalve uses its foot to move along the surface of underwater structures or on land. Univalve varieties include abalone and conch. *See Figure 17-51.*

Abalones. An *abalone* is a univalve contained in a brown, bowl-shaped shell with an iridescent, multicolored interior. The California abalone variety is considered the highest in quality, but all varieties of abalone have a sweet, slightly salty flesh. The texture of an abalone is tender like lobster, but has a creamier taste. It is very tough when overcooked. Aquafarmed abalones are raised in saltwater pens. Abalones are also imported from the coastal waters of Japan, New Zealand, and Mexico and are sold in canned and frozen form.

Conchs. A *conch* is a univalve that has a pinkish-orange shell with an interior that resembles a large snail. Conchs are found in the warm waters off the Florida Keys and throughout the Caribbean. Conch flesh is rubbery and is typically sliced thin and tenderized before it is cooked. It has a sweet flavor similar to that of a clam. Fresh conch is often eaten raw with lime juice and hot sauce.

Univalves

National Oceanic and Atmospheric Administration
Abalone

U.S. Fish and Wildlife Service
Conch

Figure 17-51. *Abalones and conchs are types of univalves.*

Florida Department of Agriculture and Consumer Services, Bureau of Seafood and Aquaculture Marketing

Bivalves

A *bivalve* is a mollusk with a top shell and a bottom shell connected by a central hinge that can close for protection. Live bivalves should close tightly when gently tapped. Bivalves that do not close when tapped are dead and should be discarded. A bivalve that is noticeably heavier than others is probably dead and should be discarded. Common types of bivalves include clams, cockles, mussels, oysters, and scallops.

Clams. A *clam* is a bivalve found in both freshwater and saltwater. Clams can be fried or steamed with excellent results and are the key ingredient in clam chowders. Different varieties of clams are found in the Atlantic Ocean and the Pacific Ocean. *See Figure 17-52.* The Atlantic coast produces soft-shell, hard-shell, and surf clams.

Clams

United States Department of Agriculture
Littlenecks

Plitt Seafood
Manila

Figure 17-52. *Different varieties of clams are found in the Atlantic Ocean and the Pacific Ocean. Littleneck clams are found in the Atlantic and Manila clams are found in the Pacific.*

A *soft-shell clam,* also known as a long-neck or steamer clam, is an Atlantic clam with a thin, brittle shell that breaks easily. Soft-shell clams have a protruding siphon that prevents the shell from closing completely, which causes them to dry out more quickly than hard-shell clams once removed from the water. A *siphon* is a tubular organ that is used to draw in or eject fluids. Soft-shell clams also have a tendency to be gritty due to excess sand settling inside the shell. They need to be soaked in a solution of salted water and cornmeal to remove the sand. Soft-shell clams have a tender, sweet-tasting flesh and are commonly steamed or fried.

A *hard-shell clam,* also known as a quahog, is an Atlantic clam with a blue-grey shell that contains a chewy flesh. These clams are rarely sold by the name hard-shell clam or quahog. Instead, they are sold by classifications that indicate their size. These classifications include littleneck, cherrystone, topneck, and chowder. Littlenecks are 1–2 inches in size. These bite-size clams are served raw on the half shell or steamed. Cherrystones are 2–3 inches in size. They are commonly steamed or served raw on the half shell. Topnecks are 3 inches in size and are often stuffed and baked. Chowder clams are larger and typically cut into strips or minced for making soups and chowders.

A *surf clam* is a large species of Atlantic clam that can grow to 8 inches in size. Half of a shucked surf clam is the siphon, which is typically cut into strips and then breaded and fried. The other half of the shucked surf clam is the adductor muscle. The *adductor muscle* is a muscle that opens and closes the shell of a bivalve. It is commonly chopped or ground for use in chowders. Surf clams have a sweet, mild flavor.

Clams found on the Pacific coast have a tougher texture than Atlantic varieties. Pacific clams include Manila clams, butter clams, razor clams, and geoducks. Manila clams are the most common variety of Pacific clam. The shell is covered with slight ridges from the lip to the hinge. Manila clams can be steamed or served raw and have a sweet and salty flavor. Butter clams have a mild, buttery flavor. Razor clams are named for their narrow, oval-shaped shell. They have a sweet flavor.

A *geoduck* is a very large Pacific clam with a meaty siphon that protrudes from its shell. Geoducks can weigh up to 3 lb. The siphon of a geoduck is typically split in half lengthwise and then cut into very thin slices and served carpaccio-style. It can also be quickly sautéed, steamed, or poached. Geoducks become very tough if overcooked.

Cockles. A *cockle* is a 1 inch wide bivalve with a shell that has deep, straight ridges. *See Figure 17-53.* The small, fleshy cockles are often served shucked with a squeeze of lemon juice. Cockles are often used to make paella.

Mussels. A *mussel* is a freshwater or saltwater bivalve with whisker-like threads that extend outside the shell to allow the animal to attach to items for protection. These threads are referred to as a "beard." In the wild, mussels are commonly found attached by their beards to rocks. Aquafarmed mussels attach their beards to ropes and hang there until they are large enough to be harvested. Common types of mussels include blue mussels and greenlip mussels.

Blue mussels are the most common variety of edible mussel. *See Figure 17-54.* Aquafarmed blue mussels have a thinner, blue-black shell, while wild-caught blue mussels have a thicker, silver-blue shell. Blue mussels have tender, sweet flesh that is bright orange in color. They vary in size, but typically there are 10–20 mussels per pound. Blue mussels can be steamed and served either hot or cold or shucked and then sautéed or simmered in a sauce.

Greenlip mussels have a distinctive green-edged shell and are larger than blue mussels. They average 8–12 mussels per pound. They have a sweet, plump, and tender flesh. Greenlip mussels are often steamed with white wine, lemon, and herbs or cooked and served cold with cocktail sauce and lemon wedges.

Oysters. An *oyster* is a saltwater bivalve with a very rough shell that is coated with calcium deposits. Although oysters are not found in freshwater, they can be found in brackish water (a mixture of freshwater and saltwater). Most oysters are cultivated from oyster beds that require care and attention if they are to continue to produce. Common varieties include Atlantic oysters, Pacific oysters, and European oysters. The variety of oyster and where it comes from impacts the flavor profile.

Cockles _____

Fortune Fish Company

Figure 17-53. *A cockle is a 1 inch wide bivalve with a shell that has deep, straight ridges.*

Blue Mussels _____

New Zealand Greenshell™ Mussels

Figure 17-54. *Blue mussels are the most common variety of edible mussel.*

Atlantic Oysters

Florida Department of Agriculture and Consumer Services, Bureau of Seafood and Aquaculture Marketing

Figure 17-55. *Atlantic oysters have fairly flat shells and distinctive, salty-flavored, plump, and tender flesh.*

Scallop Varieties

Bay

Sea

Figure 17-56. *Bay scallops and sea scallops are the two primary varieties of scallops.*

An *Atlantic oyster,* also known as an Eastern oyster, is a variety of oyster that has a fairly flat shell and a distinctive, salty-flavored, plump, and tender flesh. *See Figure 17-55.* Atlantic oysters account for roughly 70% of all oyster production. Common varieties of Atlantic oysters include Blue Point oysters, Chesapeake Bay oysters, and Long Island oysters.

A *Pacific oyster,* also known as a Japanese oyster, is a variety of large oyster that has fragile, curvy shells and a briny, sweet, and mild-tasting flesh. The plump, moist flesh is silver, gold, or white in color. Common varieties include Olympia, Penn Cove, and Kumamoto oysters.

A *European oyster* is a variety of oyster with a relatively flat, cup-shaped shell and salty-sweet flavored, creamy-textured flesh. European oysters are commonly served on the half shell. Belon oysters are a rare variety of European oysters.

Oysters are available year-round, but most are at peak quality from September to April. Oysters may be eaten raw, roasted, baked, breaded and fried, or poached. A common preparation is to top an oyster with a savory filling. The key to preparing oysters is to apply just enough heat to heat them through, leaving them plump and tender.

Scallops. A *scallop* is a bivalve with a fan-shaped shell and a cream-colored adductor muscle with a sweet, delicate flavor. The well-developed adductor muscle is the lean and juicy edible portion of the scallop. The rest of the scallop body is made up of white or red roe called coral, which is often removed prior to sale. Coral is considered a delicacy and is sometimes served with scallops in upscale restaurants. Some chefs use scallop shells as serving dishes when featuring seafood appetizers or entrées. Bay scallops and sea scallops are the two primary varieties of scallops. *See Figure 17-56.*

A *bay scallop* is a fairly small scallop harvested from shallow saltwater. Bay scallops average 100 scallops per pound. They are typically cleaned prior to sale and packed wet (soaked in a preservative that whitens the scallop and helps prevent spoiling), dry (untreated, without any preservatives), or individually quick-frozen. Bay scallops are commonly sautéed in butter, battered and fried, or marinated in lemon juice and served seviche-style.

A *sea scallop* is a large scallop with a coarse texture that is harvested from deep saltwater. Sea scallops are typically 2–5 times larger than bay scallops and average 30 scallops per pound. Sea scallops have a sweet, somewhat briny taste.

Color is the best way to judge the quality of scallops. High-quality scallops have a creamy, almost translucent color. If scallops are white, they have been packed wet and the flavor and texture have been impaired. If they are a brownish color, they are old and should not be used. Scallops are typically broiled, sautéed, fried, or poached.

Cephalopods

A *cephalopod* is any of a variety of mollusks that do not have an external shell. Some cephalopods have an internal bone called a cuttlebone. Cephalopods have keen vision, a highly-developed nervous system, and a birdlike beak that is used to crack the shells of shrimp, crabs, lobsters, and other prey. Squid, octopuses, and cuttlefish are cephalopods. *See Figure 17-57.*

Cephalopods

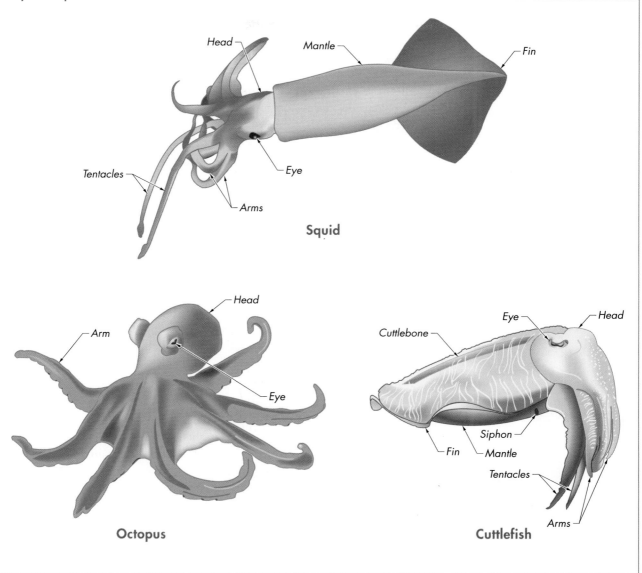

Figure 17-57. *Squid, octopuses, and cuttlefish are cephalopods.*

Squid. A *squid* is a translucent, head-footed cephalopod that has two tentacles, eight sucker-equipped arms, two lateral fins, and a flat, internal cuttlebone. The arms are attached near the eyes at the bottom of the head. The two tentacles are longer than the arms. Squid can change their skin color at will and expel a dark cloud of ink to confuse their prey. The ink is often used to color and flavor grain and pasta dishes. Most whole squid are packed 8–10 per pound. Larger squid are cut into steaks and sold frozen. Squid is often called by its Italian name, calamari. Squid is commonly sautéed, breaded and fried, or used to make stews. The flesh becomes tough if overcooked.

Octopuses. An *octopus* is a gray cephalopod with eight sucker-equipped arms, a birdlike beak, well-developed vision, and no internal or external shell. Octopuses can be small or quite large. However, they do not grow as large as giant squid. Octopuses are usually sold whole by the pound and are available fresh and frozen. When cooked, the gray skin turns deep purple or reddish-purple. Octopus flesh is white, firm, and sweet. The flesh becomes tough if overcooked. Cold, cooked octopus is used to make pulpo salad.

Cuttlefish. A *cuttlefish* is a translucent cephalopod with two tentacles, eight sucker-equipped arms, a hard internal cuttlebone, and large eyes at the base of its head. They can change their color at will. Cuttlefish expel a dark-brown cloud of ink to confuse their prey. The tentacles, arms, and ink are the only parts of a cuttlefish that are eaten. The ink is commonly used to make cuttlefish risotto, which is often dark in color. The tentacles and arms are typically cut into rings and sautéed, breaded and fried, or cooked in soups or stews. The flesh becomes tough if overcooked.

MARKET FORMS OF SHELLFISH

Each market form of shellfish has advantages and disadvantages in terms of cost, convenience, and labor. Shellfish are available live, shucked, or frozen. Shucked and cooked shellfish are also available canned.

National Oceanic and Atmospheric Administration

Live Shellfish

Lobsters are usually purchased live and come in various sizes. On the coast both hard-shell and soft-shell crabs are sold live. Away from the coast only soft-shell crabs are available live.

Live clams, oysters, and scallops can be purchased by the gallon or the pound. They are usually packed by count. A higher count indicates a smaller size. If a container of oysters is marked 350 per gallon and another container is marked 225 per gallon, the 350 count container holds more oysters that are smaller. Live oysters and clams have tightly closed shells. If a shell is open or does not close fully when lightly tapped, the animal is dead and must be discarded. Oyster and clam shells that are unusually heavy probably contain mud and should also be discarded.

Shucked Shellfish

Clams, oysters, and scallops are often sold with the shell removed and are available both fresh and frozen. Shrimp may also be purchased cooked and canned. Cooked shrimp may be purchased by the pound either in the shell or peeled. Crabmeat can be cooked (fresh or frozen) and canned. Crabmeat also may be purchased fresh or frozen in the shell in a variety of cooked forms. *See Figure 17-58.* Cooked crabmeat should be kept packed in ice and refrigerated until it is used.

Forms of Cooked Crab	
Forms	**Descriptions**
Jumbo lump	Large pieces of white body flesh attached to the swimming legs
Back fin, lump, or special	Small pieces of white flesh from the body
Claws and fingers	Brownish flesh from the claws and legs
Cocktail claws	Claw flesh with a protion of the shell remaining as a handle for eating

Figure 17-58. *Cooked crabmeat may be purchased fresh or frozen in the shell in a variety of cooked forms.*

Imitation crabmeat, known as surimi, is also available. *Surimi* is a fish product made from a mixture of fish and/or shellfish and other ingredients. Surimi looks, cooks, and tastes like crabmeat. ***See Figure 17-59.*** It is high in protein and low in calories, sodium, fat, and cholesterol. Surimi is precooked and frozen. It can be purchased as legs, chunk meat, or flake meat.

Frozen Shellfish

Prepared lobster tails may be purchased frozen. The head and thorax (body) of shrimp and prawns are typically removed, and the tails are frozen while still at sea to ensure maximum freshness. Frozen shellfish is commonly packaged by count. ***See Figure 17-60.*** For example, shrimp packed and labeled 21/25 count indicates that there is an average of 21–25 shrimp per 1 lb package. Frozen shrimp can be purchased green (uncooked), peeled or unpeeled, cooked and peeled, or peeled, deveined, and breaded. Frozen scallops are usually sold in 5 lb blocks.

Surimi

Harbor Seafood, Inc.

Figure 17-59. *Surimi is a fish product that looks, cooks, and tastes like crabmeat.*

Packaging by Count

Sizing Shrimp	
Count	**Size**
25 or fewer per lb	Jumbo
25 to 30 per lb	Large
30 to 42 per lb	Medium
42 or more per lb	Small

Frozen Shrimp

Fortune Fish Company

King Crab Legs

Figure 17-60. *Frozen shellfish is commonly packaged by count.*

RECEIVING AND STORING SHELLFISH

To be acceptable upon delivery, live shellfish should be delivered with an air temperature of 45°F and shucked shellfish must have an internal temperature of 45°F or below. Once accepted, it is imperative to store shellfish between 30°F and 34°F or below and to use it as quickly as possible to maintain quality. If the shellfish will not be used within two days of purchase, it should be wrapped in moistureproof freezer paper and foil, to protect it from air and moisture, and then frozen. Shrimp, scallops, and crabmeat are often block frozen.

New Zealand Greenshell™ Mussels

Live Shellfish

Lobsters, crabs, clams, oysters, and mussels are often purchased live in the shell. Live crustaceans should be covered with wet seaweed or damp newspaper to keep them from drying out. Crustaceans may also be stored in a saltwater tank. In the absence of a tank, live crustaceans live two to four days. When received, the lobsters must be carefully inspected. Live lobsters have a tightly curled tail. Live lobsters and crabs should show leg movement when received and must be kept alive until they are cooked. If a shellfish is dead, it should be rejected.

Fresh shellfish are not subject to a mandatory federal inspection. However, all mollusks must be delivered with a shellstock tag. Shellstock tags are waterproof, tear-resistant tags attached to containers of mollusks. *See Figure 17-61.* Shellstock tags list the dealer's contact information and identification number, the original harvester (if different than the dealer), the harvest date and general area, the type and quantity of shellfish, a 90 day retention notice, and a consumer advisory. Mollusks delivered without shellstock tags should be rejected. Shellstock tags are kept on file for 90 days after the mollusks are harvested in case of foodborne illness.

Live bivalves should be refrigerated in the original box or netted bag and placed in a pan to prevent drips from contaminating other foods. Live mollusks should not be stored in a sealed container or plastic bag because they will die from a lack of oxygen. Under ideal conditions, fresh mollusks can live in refrigerated storage for up to a week. *See Figure 17-62.* Live oysters and clams have tightly closed shells. If a shell is open and does not close when handled, the oyster or clam is dead and should be discarded.

Shucked oysters are packed in metal containers and must be kept refrigerated and packed in ice at all times to prevent spoilage. If handled in the proper manner, oysters remain fresh for up to a week.

Chef's Tip

Live bivalves should never be covered with or placed on top of ice because the intense cold from the ice can cause them to die.

Shellstock Tags

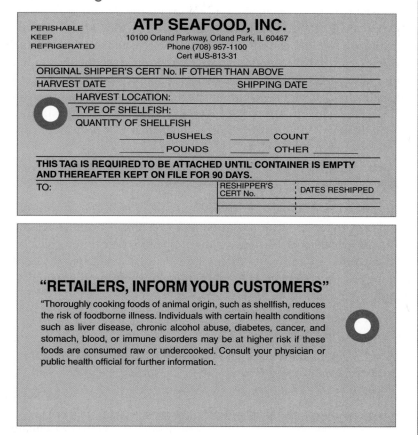

Figure 17-61. *Shellstock tags are waterproof, tear-resistant tags attached to containers of mollusks.*

Storing Live Mollusks

Figure 17-62. *Under ideal conditions, fresh mollusks can live in refrigerated storage for up to a week.*

Frozen Shellfish

Fortune Fish Company

Figure 17-63. *When receiving frozen shellfish, verify the product is frozen and has not been thawed and refrozen.*

Frozen Shellfish

When receiving frozen shellfish, it is important to verify that the product is frozen and has not been thawed and refrozen. ***See Figure 17-63.*** If shellfish is thawed on the edges, it should be rejected. Shellfish that has been refrozen will have a poor texture. Dry spots on the shellfish indicate that it has thawed slightly and been refrozen. Ice crystals are another sign that shellfish has been thawed or not stored at appropriate temperatures. Shellfish displaying dry spots or ice crystals should be rejected. Maximum shelf life is obtained by storing frozen shellfish at 0°F or below.

FABRICATION OF SHELLFISH

Shellfish are often purchased whole and fabricated in-house. Knowing how to use proper fabrication techniques for shellfish is a valuable skill to have in the professional kitchen. Common shellfish fabrication techniques include deveining shrimp, splitting lobster tails, cleaning soft-shell crabs, shucking oysters, shucking clams, debearding mussels, cleaning squid, and cleaning octopuses.

Deveining Shrimp

Shrimp contain a sand vein, or intestinal tract, that must be removed prior to further preparation. ***See Figure 17-64.*** The shells are also removed.

Procedure for Deveining Shrimp

1. Hold the tail fins in one hand with the underside of the shrimp facing up. Pinch the legs between the thumb and index finger of the other hand and gently pull to remove.

2. Grasp the tail between the thumb and index finger of the guiding hand. Use the other hand to gently grasp the shell and twist it slightly to separate the shell from the body.

3. Use a paring knife to make a shallow slice along the back of the shrimp to expose the sand vein. *Note:* If the shrimp is to be butterflied, make a deeper slice to open the flesh so that it lays flat.

4. Gently pull the sand vein to remove it. Discard the vein. *Note:* Holding the shrimp under cold running water may facilitate the removal of the sand vein.

Figure 17-64. *Shrimp contain a sand vein, or intestinal tract, that must be removed prior to further preparation.*

Refer to DVD for
Deveining Shrimp
Media Clip

Splitting Lobster Tails

Broiled lobster tail is a preparation that requires the shell of the tail to be split open. The tail meat is pulled up from, but left inside, the shell for service. *See Figure 17-65.*

Procedure for Splitting Lobster Tails

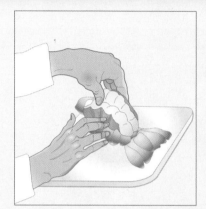
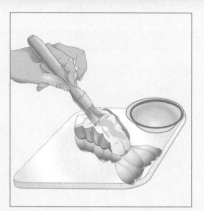

1. With the underside of the lobster facing down, insert the tip of a chef's knife into the center of the tail just above the bottom fins.
2. Use the guiding hand to push the handle of the knife down to slice through the top portion of the shell lengthwise.

3. Spread the cut shell all the way open to reveal the tail flesh.
4. Gently pull the tail flesh up through the cut shell.
5. Make a few shallow slices along the underside of the tail flesh.

6. Brush the exposed flesh with clarified butter and broil the tail.

Figure 17-65. *Broiled lobster tail is a preparation that requires the shell of the tail to be split open and the tail meat pulled up from, but left inside, the shell for service.*

Cleaning Soft-Shell Crabs

Cleaning soft-shell crabs involves removing the inedible portions, such as the eyes and the apron. *See Figure 17-66.* The apron is tucked under the body and used to carry and conceal eggs.

Procedure for Cleaning Soft-Shell Crabs

1. Peel back the pointed top shell of the crab to reveal the stringlike gills. Use a boning knife to carefully scrape away the gills.
2. Locate the eyes on the top of the head. Use kitchen shears to remove the mouth and head by cutting just behind the eyes.
3. Gently squeeze the body just behind where the head was removed to release a green bubble of fluid. Rinse the green fluid away.
4. Turn the crab over and locate the apron. Firmly twist and then pull the apron to remove it. Discard the apron. *Note:* The intestinal tract will be attached to the apron.

Figure 17-66. *Cleaning soft-shell crabs involves removing the inedible portions, such as the eyes and the apron.*

Refer to DVD for
Shucking Oysters
Media Clip

Shucking Oysters

Shucking is the process of opening a bivalve. Bivalves such as oysters and clams must be shucked to access the edible flesh inside the shell. An oyster knife is used to shuck oysters. *See Figure 17-67.*

Procedure for Shucking Oysters

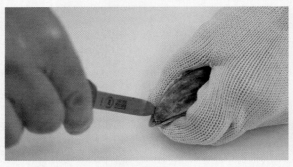

1. While wearing a mesh glove on the guiding hand, firmly hold the oyster in the palm, hinge side up.
2. Insert the tip of an oyster knife near the hinge of the shell.

3. Carefully twist the blade to pop the hinge open. *Note:* Care should be taken not to cut the flesh of the oyster.

4. Slide the oyster knife under the flesh of the top shell to cut it away from the shell.

5. Slide the oyster knife under the adductor muscle to separate the flesh from the bottom shell.
6. Remove any shell fragments that may have mixed with the oyster flesh.

Figure 17-67. *An oyster knife is used to shuck oysters.*

Shucking Clams

Clams are shucked with a clam knife. *See Figure 17-68.* Clams also may be steamed open to access the flesh.

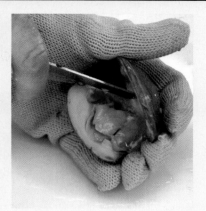

1. While wearing a mesh glove on the guiding hand, firmly hold the clam with the hinge resting against the base of the thumb and the lip facing the fingertips.

2. Insert the blade of a clam knife between the top shell and bottom shell until halfway inserted.

3. Carefully twist the blade to pop the shells open.

4. Use the tip of the clam knife to separate the flesh from the top shell. Discard the top shell.

5. Use the knife blade to separate the adductor muscle from the bottom shell.

6. Remove any shell fragments that may have mixed with the clam flesh.

Figure 17-68. *Clams are shucked with a clam knife.*

Debearding Mussels

Mussels have hairlike threads called beards that need to be removed prior to preparation. *See Figure 17-69.* If the beards are removed too soon, the mussels will die.

1. Brush the mussel under running water to remove excess mud, sand, and debris.

2. Use a pair of needle-nose pliers to grab the beard and gently pull it free from the mussel.

Figure 17-69. *Mussels have hairlike threads called beards that need to be removed prior to further preparation.*

Cleaning Squid

Squid is often shipped cleaned and ready for cooking. If a squid is white and the eyes have been removed, it has already been cleaned. If a squid is dark in color and the eyes are intact, it must be cleaned before being prepared. ***See Figure 17-70.***

Procedure for Cleaning Squid

1. Gently pull the arms and tentacles of the squid away from the body (mantle).
2. Using a boning knife, cut just below each of the eyes to remove the eight arms and two tentacles from the head.
3. Remove the ink sac from the head and reserve.
4. Pull back the tentacles to expose the beak. Carefully pull out the beak and discard. *Note:* The tentacles may be left whole or cut into smaller sections.
5. Hold the body (mantle) in guiding hand and carefully pull out the transparent quill with the other hand. Discard the quill.
6. Carefully pull as much skin as possible from the body (mantle) to reveal the lighter-colored flesh below. Discard the skin. *Note:* The mantle can be cut into rings or sections or left whole.
7. Thoroughly rinse all of the squid flesh before use.

Figure 17-70. *If a squid is dark in color and the eyes are intact, it must be cleaned before being prepared.*

Cleaning Octopuses

Octopuses typically are shipped cleaned. Occasionally octopuses are shipped as caught and additional cleaning is necessary. ***See Figure 17-71.***

Procedure for Cleaning Octopuses

1. Locate the beak of the octopus and use a boning knife or a paring knife to carefully remove it from the body. Discard the beak.
2. Locate the eyes of the octopus and cut just beneath each eye to remove the head. Discard the head.
3. Firmly, but carefully, pull the dark-colored skin off of the body. Discard the skin.
4. Carefully and firmly pull the suction-cup covered skin from each of the eight arms. Discard the skin.

Figure 17-71. *Occasionally octopuses are shipped as caught and additional cleaning is necessary.*

COOKING SHELLFISH

Shellfish require short cooking times and overcook quickly. Some shellfish need to be tenderized prior to cooking or they will be too tough. Color is a good indicator of the doneness of shellfish. For example, a lobster shell turns red when it is done, and shrimp curl and turn slightly pink when cooked. ***See Figure 17-72.*** Crayfish turn a reddish-brown when cooked, and the center of a scallop turns opaque when it is done.

Cooked Shellfish

Harbor Seafood, Inc.
Cooked Lobster Tail

Daniel NYC
Cooked Shrimp

Figure 17-72. *A lobster shell turns red when it is done, and shrimp curl and turn slightly pink when cooked.*

Chef's Tip

Cooked shellfish cannot be placed on plates that previously held raw seafood because the bacteria from the raw food can contaminate the cooked seafood.

Cooking Crustaceans

Crustaceans can be cooked using several different cooking methods. Shrimp, prawns, and crayfish are often sautéed, fried, or steamed. Lobsters are commonly poached whole and then cracked tableside or the tail flesh is removed from the shell, placed on top of the shell, and then broiled. Lobster flesh is not as delicate as fish flesh and can handle the high heat of a broiler. Lobsters are commonly poached in a court bouillon. ***See Figure 17-73.***

Procedure for Poaching Whole Lobsters

1. Place the live lobster on its belly and use the tip of a chef's knife to pierce the center of its head.

2. Place the lobster in a stockpot filled with 180°F court bouillon and poach 5 minutes per pound.

3. Turn off the heat and allow the lobster to rest in the court bouillon for 10–15 minutes.

4. Remove the lobster and drain the excess liquid. Serve immediately.

Figure 17-73. *Lobsters are commonly poached in a court bouillon.*

Cooked lobster that is to be used in other preparations must be removed from the shell. *See Figure 17-74.* If removed in one piece, the claw flesh can make a nice presentation.

Procedure for Removing Cooked Lobster from the Shell

1. Remove each claw by firmly pulling and twisting it away from the body.
2. Use the back of a chef's knife to crack each claw just beneath the pinchers all the way around the claw.

3. Carefully remove the flesh from inside the claw. *Note:* Claw flesh that is removed in one piece makes a nice presentation.

4. Twist the tail section away from the rest of the body and pull the thin membrane away from the underside of the shell.
5. Use a chef's knife to split the tail shell lengthwise.
6. Remove the flesh from the tail.

Figure 17-74. *Cooked lobster that is to be used in other preparations must be removed from the shell.*

Soft-shell crabs are usually dusted in seasoned flour and sautéed or lightly breaded and then fried. Because the shell is soft, the entire crab can be eaten. Hard-shell crabs are commonly steamed or poached. Precooked crab legs or clusters of crab legs only need to be heated through prior to serving.

Cooking Mollusks

Mollusks are prepared using a variety of cooking methods. Clams are most commonly shucked and stuffed with a filling or topped with butter and seasoned bread crumbs and then baked. Clams may also be chopped for use in soups and chowders. Mussels are commonly steamed or simmered in their shells until the shells open. Mussels are also sautéed and added to pasta or rice dishes.

Oysters are often served raw on the half shell or baked with a topping, such as with oysters Rockefeller. *See Figure 17-75.* Oysters may also be lightly breaded and fried. An oyster po'boy sandwich includes fried oysters. Scallops are typically sautéed to produce a golden exterior. They may also be grilled or broiled. Scallops are tender and moist when not overcooked.

New Zealand Greenshell™ Mussels

Raw and Cooked Oysters

Raw Oysters on the Half Shell

Oysters Rockefeller

Oyster Po' Boy

*Florida Department of Agriculture and Consumer Services,
Bureau of Seafood and Aquaculture Marketing*

Figure 17-75. *Oysters are often served raw on the half shell, baked with a topping, or deep-fried.*

Cooking Cephalopods

Cephalopods can be prepared using a variety of cooking methods. Squid is commonly grilled as a whole tube or stuffed and baked. Squid is often grilled whole, cut into rings and quickly sautéed, or cut into rings, lightly breaded, and fried to make calamari. *See Figure 17-76.* If squid is overcooked it becomes rubbery and almost inedible. The ink sac is often used as a black food coloring for risotto and pasta dishes.

Cooked Cephalopods

Harbor Seafood, Inc.
Grilled Squid   **Fried Calamari**

Figure 17-76. *Squid is often grilled whole, cut into rings and quickly sautéed, or cut into rings, lightly breaded, and fried to make calamari.*

Like squid, if octopus is overcooked it becomes rubbery and inedible. Cooking octopus quickly yields a chewy texture. It is often used to make sushi. Tenderized octopus can be sautéed in butter and garlic and served with lemon as a side dish or an appetizer. Poached octopus is often chilled, tossed with vinaigrette, and served as a cold salad. Cuttlefish is commonly cut into rings and lightly sautéed. It can also be stir-fried or gently simmered in cooking liquid when making risotto.

Turtles

Czimer's Game & Seafoods, Inc.

Figure 17-77. *The two most common varieties of edible turtles are green turtles and snapping turtles.*

Frog Legs

Figure 17-78. *The hind legs of frogs are commonly sold for human consumption.*

Snails

L. Issacson and Stein Fish Company

Figure 17-79. *Snails have a coiled or spiral shell that grows as the snail grows.*

RELATED GAME

Some edible amphibians and reptiles are often grouped with fish and shell-fish on menus even though they are not technically classified as seafood. These items are often prepared in a manner similar to fish and shellfish. Some of these related game animals include turtles, frogs, snails, snakes, and alligators.

Turtles

The two most common varieties of edible turtles are green turtles and snapping turtles. ***See Figure 17-77.*** Green turtles have a smooth shell that is olive-green in color. The flesh of a green turtle can range from white to green in color. Green flesh has a better flavor and is slightly more tender than the white flesh. The shell of a snapping turtle has bumps and sharp pointed tips. The turtle also has a very sharp beak-like mouth. Snapping turtles have both white and dark flesh. The legs are dark flesh, while the neck and back strap are both white flesh.

Turtles have been part of culinary tradition for many years in most cultures. Turtle flesh is most commonly used to make turtle soup, stews, and gumbos. It may also be fried or pressure cooked and served with turtle gravy. Because turtle flesh tends to be somewhat chewy, it is often pounded thin to tenderize the flesh or simmered before cooking. Turtle flesh has a different flavor depending on which part of the turtle it comes from. Most turtle flesh is sold boneless, but it can also be purchased with the bone in.

Frogs

The hind legs of frogs are commonly sold for human consumption. ***See Figure 17-78.*** The best frog legs come from bullfrogs raised on frog farms. The hind legs of bullfrogs contain a large amount of white flesh. A large percentage of the frog legs used in the professional kitchen come from India or Japan. Frog legs are available all year, but are most plentiful from April to October. Frog legs are sold by the pair, with the most desirable legs averaging 2–3 pairs per pound. Frog legs are typically sautéed with garlic, parsley, wine, and butter or can be breaded and fried.

Snails

Snails have a coiled or spiral shell that grows as the snail grows. ***See Figure 17-79.*** *Escargot* is the French term for snail and refers to any variety of snail fit for human consumption. It is important to purge the stomach of the snail prior to cooking because the stomach contents may be toxic to humans. Purging is accomplished by feeding the snail cornmeal and allowing them to purge. After purging, snails are commonly removed from the shell, eviscerated, and quickly sautéed with garlic, white wine, and butter. Snail flesh may also be stuffed back into the shell, topped with garlic butter and wine, and then baked.

Rattlesnakes

Rattlesnake is the most common variety of snake raised for human consumption. Farm-raised rattlesnake flesh is sold fabricated because care needs to be taken when removing the head to not allow any venom to come in contact with the snake flesh or the hands of the person fabricating the snake. Snake flesh is similar in color to veal, has a flavor similar to chicken, and a texture similar to fish. *See Figure 17-80.* Snake can be chewy, yet delicate and tender when cooked properly. Because the meat is so delicate, care should be taken to not overcook it. Overcooking snake meat will produce a tough, dry, and rubbery product.

Rattlesnakes

Topside **Underside**

Czimer's Game & Seafoods, Inc.

Figure 17-80. *Snake flesh is similar in color to veal, has a flavor similar to chicken, and a texture similar to fish.*

Alligators

Farm-raised alligators are fed a natural diet of fish and frogs. Alligators are raised for two cuts of flesh, the tenderloin and the tail. The tail flesh of a farm-raised alligator is almost white in color and the tenderloin has a color similar to veal. Flesh is also taken from just under the jaw and the legs, but these are much smaller cuts than the tenderloin and tail. Alligator leg and jaw flesh is sold as smaller pieces that can be breaded, fried, and sold as "alligator bites." *See Figure 17-81.* Alligator flesh can be used as a substitute for turtle flesh in soups or ground to make alligator sausage.

Alligator flesh has a texture and flavor somewhere between chicken and lobster. It is very low in cholesterol and extremely high in iron. Alligator flesh is often soaked in milk for up to an hour before cooking in order to soften its strong flavor. It can be sautéed, fried, or braised.

Alligators

Czimer's Game & Seafoods, Inc.

Figure 17-81. *Alligator leg and jaw flesh is sold as smaller pieces.*

SUMMARY

Seafood includes many varieties of freshwater fish, saltwater fish, crustaceans, mollusks, cephalopods, and related game. There are thousands of fish varieties and hundreds of shellfish varieties. Fish are commonly sold whole, drawn, dressed, as steaks, as fillets, frozen, and processed. They may also be sold as loins, roasts, or portions. Processed fish is available canned, smoked, cured, salted, or pickled. Only fish that is processed under a voluntary Type 1 inspection is eligible for grading. Grade A fish are of the best quality, with no visible defects.

Being able to identify high-quality fish and shellfish and store each type properly helps ensure that a foodservice operation will serve high-quality seafood. Shellfish can be purchased live, shucked, frozen, and processed. Knowing and using proper fabrication techniques for fish and shellfish is an important skill to have in the professional kitchen. Most fish and shellfish have naturally tender flesh and do not require much cooking time. The cooking method used depends on the type of fish, shellfish, or related game being prepared. Turtles, frogs, snails, snakes, and alligators are related game animals that are often grouped with seafood on menus.

Refer to DVD for
Quick Quiz® questions

Review

1. Differentiate between lean and fatty fish.
2. Describe three classifications of fish based on external shape and structure.
3. Identify five types of freshwater fish.
4. Identify two types of anadromous fish.
5. Identify 17 types of saltwater fish.
6. Identify three common cartilaginous fish.
7. Describe the various market forms of fish.
8. Name the government organization in charge of voluntary fish inspections.
9. Describe what happens during a Type 1 inspection.
10. Name the grades that can be assigned to fish.
11. Explain how fresh fish are received and stored.
12. Explain how frozen fish are received and stored.
13. Explain how to scale a fish.
14. Explain how to cut a roundfish into steaks.
15. Explain how to fillet a roundfish.
16. Explain how to skin a roundfish fillet.
17. Explain how to fillet a flatfish.
18. Explain how to determine the doneness of fish.
19. Identify eight methods used to cook fish.
20. Identify four types of crustaceans.
21. Identify 11 types of mollusks.
22. Describe the various market forms of shellfish.
23. Explain how live shellfish are received and stored.

Refer to DVD for
Review Questions

24. Describe the purpose of a shellstock tag.
25. Explain how frozen shellfish are received and stored.
26. Explain how to devein shrimp.
27. Explain how to split lobster tails.
28. Explain how to clean soft-shell crabs.
29. Explain how to shuck oysters.
30. Explain how to shuck clams.
31. Explain how to debeard mussels.
32. Explain how to clean squid.
33. Explain how to clean octopuses.
34. Explain how to cook crustaceans.
35. Explain how to cook mollusks.
36. Explain how to cook cephalopods.
37. Describe five types of game often grouped with seafood.

Applications

1. Scale a fish.
2. Cut a roundfish into steaks.
3. Fillet a roundfish.
4. Skin a roundfish fillet.
5. Fillet a flatfish.
6. Demonstrate how to determine the doneness of fish.
7. Grill one fish fillet and broil another fillet identical in size and variety. Compare the quality of the two prepared dishes.
8. Sauté a fish portion and stir-fry another portion identical in size and variety. Compare the quality of the two prepared dishes.
9. Braise a fish portion and stew another portion identical in size and variety. Compare the quality of the two prepared dishes.
10. Poach one fish and steam another fish identical in size and variety. Compare the quality of the two prepared dishes.
11. Devein shrimp.
12. Split a lobster tail.
13. Clean a soft-shell crab.
14. Shuck oysters.
15. Shuck clams.
16. Debeard mussels.
17. Clean a squid.
18. Clean an octopus.
19. Prepare crustaceans using two different cooking methods. Compare the quality of the prepared dishes.
20. Prepare mollusks using two different cooking methods. Compare the quality of the prepared dishes.
21. Prepare a cephalopod. Evaluate the quality of the prepared dish.
22. Prepare a turtle, frog, snail, snake, or alligator dish. Evaluate the quality of the prepared dish.

Refer to DVD for
Application Scoresheets

Grilled Indian-Spiced Mahi-Mahi with Mango and Mint Chutney

Yield: *4 servings, 7 oz fillet with 1 oz of chutney each*

Ingredients

cumin, ground	½ tsp
nutmeg, ground	½ tsp
cardamom, ground	½ tsp
green curry paste	½ tsp
ginger, ground	¼ tsp
garlic, crushed	1 clove
water	1 tbsp
olive oil	1 tbsp
salt and pepper	TT
6–8 oz mahi-mahi fillets	4 ea

Mango and Mint Chutney

mangoes, small diced	4 oz
dried currants	1 tbsp
fresh ginger root, grated	½ tsp
jalapeño pepper, roasted, peeled, and seeded	1 ea
fresh mint, cut chiffonade	2 tsp
lime juice	2 tsp
yellow curry powder	¼ tsp
salt and pepper	TT

Preparation

1. Make a paste by mixing the cumin, nutmeg, cardamom, curry paste, ginger, garlic, water, olive oil, salt, and pepper.
2. Rub the paste over the fish fillets.
3. Let fish marinate for 30 minutes.
4. Combine the chutney ingredients in a bowl and mix well.
5. Place fish on a well-oiled grill for 4–5 minutes per side or until cooked through.
6. Plate each fillet and garnish with chutney.

NUTRITION FACTS
Per serving: 312 calories, 109 calories from fat, 12.1 g total fat, 64.6 mg cholesterol, 215.7 mg sodium, 578.4 mg potassium, 8.7 g carbohydrates, 1.8 g fiber, 5.5 g sugar, 40.6 g protein

Smoked Fish Croquettes with Watercress-Almond Pesto

Yield: *8 servings, 2 croquettes each*

Ingredients

smoked fish	1 lb
peeled potatoes, cooked, riced	1 lb
egg yolks	3 ea
horseradish, shredded	1 oz
dry mustard	2 tsp
salt and white pepper	TT
vegetable oil	(for deep frying)

Watercress-Almond Pesto

fresh watercress, blanched and drained	2 bunches
garlic, minced	4 cloves
extra virgin olive oil	1 c
almonds, sliced and toasted	3 oz
Parmesan, grated	1 c
salt and pepper	TT

Breading

all-purpose flour	4 oz
salt and pepper	TT
whole eggs, slightly beaten	4 ea
panko bread crumbs	1 c

Preparation

1. To make the pesto, combine watercress, garlic, and half of the olive oil in a blender or food processor.
2. Add almonds, Parmesan, and remaining olive oil to the pesto and purée. Season to taste and set aside.
3. To make breading, season flour in a pan and set aside.
4. Place the beaten eggs in a second pan.
5. In a third pan, season the panko bread crumbs.
6. Coarsely flake the fish and then mix it gently with the potatoes, egg yolks, horseradish, and mustard. Season to taste.
7. Form the fish into 2 oz patties.
8. Follow standard breading procedure to bread patties.
9. Deep-fry patties in 350°F oil.
10. Drain patties and serve with pesto.

NUTRITION FACTS
Per serving: 897 calories, 634 calories from fat, 71.9 g total fat, 184.6 mg cholesterol, 1590.3 mg sodium, 704.4 mg potassium, 38.2 g carbohydrates, 3.3 g fiber, 2 g sugar, 26.6 g protein

Sautéed Tilapia with Watercress Cream Sauce

Yield: *4 servings, 6 oz fillet and 2 oz of sauce each*

Ingredients

6 oz tilapia fillets	4 ea
salt and pepper	TT
all-purpose flour	2 oz
vegetable oil	1 tbsp

Watercress Cream Sauce

shallot, minced	1 ea
butter	1 tbsp
garlic, minced	1 clove
white wine	2 fl oz
heavy cream	4 fl oz
fresh watercress	½ bunch
salt and white pepper	TT

Preparation

1. Season tilapia on both sides with salt and pepper.
2. Dredge tilapia in flour and shake off the excess.
3. Heat oil in a sauté pan over medium-high heat. Carefully lay the fillets in the hot pan and cook until golden brown on both sides.
4. In a separate pan, sauté the shallots in the butter until translucent. Add garlic and cook 1 minute more.
5. Add white wine to deglaze and reduce by half.
6. Add cream and simmer until reduced by one-fourth. *Note:* Cream can be replaced with whole milk or chicken stock and then thickened with a cornstarch slurry for a lower-calorie sauce.
7. Transfer cream mixture to a blender. Reserve four pieces of watercress for a garnish.
8. Add the rest of the watercress and purée the mixture until smooth.
9. Season to taste and strain if desired.
10. Plate each fillet on top of cream sauce and garnish with a sprig of watercress.

NUTRITION FACTS
Per serving: 459 calories, 181 calories from fat, 20.5 g total fat, 133.4 mg cholesterol, 261.1 mg sodium, 916.8 mg potassium, 29 g carbohydrates, <1 g fiber, <1 g sugar, 38.9 g protein

Sautéed Sole Meunière

Yield: *4 servings, 1 fillet each*

Ingredients

10 oz sole, pan dressed	4 ea
all-purpose flour	2 oz
kosher salt and white pepper	TT
canola oil	2 tbsp
clarified butter	2 oz
yellow squash, cut brunoise	1 oz
zucchini, cut brunoise	1 oz
carrot, cut brunoise	1 oz
fresh lemon juice	2 fl oz
fresh parsley, chopped	1 tbsp

Preparation

1. Pat sole dry with paper towels. Season the flour to taste. Dredge sole in the seasoned flour.
2. In a hot sauté pan over medium heat, add oil and sauté the sole until golden brown on both sides (approximately 4–5 minutes).
3. Remove fish from pan and keep warm.
4. Wipe remaining oil out of the pan. Lower the heat and add the butter and brunoise vegetables to the pan.
5. Lightly sauté the vegetables. Remove them from the pan and set aside.
6. Continue cooking the remaining butter until it reaches a light-brown color and has a nutty aroma.
7. Add lemon juice to deglaze the pan.
8. Remove the pan from the heat and add chopped parsley.
9. Plate fish and top with vegetables.
10. Spoon sauce over the top and serve immediately.

NUTRITION FACTS
Per serving: 349 calories, 204 calories from fat, 23.1 g total fat, 106.1 mg cholesterol, 559.8 mg sodium, 354.5 mg potassium, 13.1 g carbohydrates, <1 g fiber, 1.2 g sugar, 22 g protein

Wasabi and Peanut-Crusted Ahi Tuna with Soy-Peanut Reduction

Yield: *4 servings, 1 steak with 1 oz of sauce each*

Ingredients

egg white	1 ea
4 oz ahi tuna steaks	4 ea
salt and pepper	TT
wasabi peas, coarsely chopped	3 tbsp
roasted peanuts, chopped fine	1 tbsp
vegetable oil	3 tbsp

Soy-Peanut Reduction

white wine	2 tbsp
soy sauce	1 tbsp
teriyaki sauce	2 tbsp
water	2 tbsp
creamy peanut butter	2 oz

Preparation

1. Beat egg white in a small bowl.
2. Season fish on both sides with salt and pepper.
3. In a separate bowl, mix the wasabi peas and roasted peanuts.
4. Dip the fish in the egg white to coat.
5. Coat fish in the wasabi-peanut mixture to coat completely.
6. To prepare soy-peanut reduction, add white wine to a small saucepan and reduce by half.
7. Add all the other ingredients and bring to a simmer.
8. Heat a sauté pan over medium heat and add vegetable oil. Sauté the tuna steaks until golden brown on both sides and cooked to medium-rare or medium.
9. Plate each tuna steak and drizzle with soy-peanut reduction.

NUTRITION FACTS
Per serving: 341 calories, 171 calories from fat, 19.8 g total fat, 44.2 mg cholesterol, 785.4 mg sodium, 689.3 mg potassium, 5.7 g carbohydrates, 1.4 g fiber, 2.4 g sugar, 33.9 g protein

Sole en Papillote

Yield: *4 servings, 1 fillet each*

Ingredients

red onion, julienne	½ c
yellow squash, julienne	½ c
zucchini, julienne	½ c
garlic, minced	1 clove
extra virgin olive oil	1 tbsp
salt and pepper	TT
parchment paper, 16 × 20 sheets	4 ea
6 oz sole fillets	4 ea
butter	8 tsp
dry white wine	¼ c
thyme, sprigs	4 ea
lemon, thinly sliced	1 ea

Preparation

1. Combine red onion, yellow squash, zucchini, garlic, and olive oil. Season to taste.
2. Separate the sheets of parchment paper and fold each one in half.
3. Place one fillet parallel to the fold and approximately 2 inches away from the fold of each sheet.
4. Evenly divide the vegetables and place on each fillet.
5. Place 2 tsp butter, 1 tbsp wine, and a sprig of thyme on each fillet.
6. Cover the vegetables with lemon slices.
7. Fold the parchment paper over the fillet and then carefully fold over the edges to close the packet.
8. Place the packet on a baking sheet in a preheated 375°F oven. Bake until the fish is cooked through (approximately 12 minutes). *Note:* Adjust the time for larger or thicker fish.

NUTRITION FACTS
Per serving: 255 calories, 130 calories from fat, 14.8 g total fat, 96.9 mg cholesterol, 582.5 mg sodium, 425.6 mg potassium, 7.9 g carbohydrates, 2.9 g fiber, 1.2 g sugar, 22.4 g protein

Poached Pistachio-Crusted Salmon

Yield: *4 servings, 6 oz each*

Ingredients

salmon fillets	2 lb
water, ice cold	2 tsp
egg	1 ea
heavy cream	2 fl oz
salt and white pepper	TT
shelled pistachios, coarsely chopped	½ lb
lemon, sliced	1 ea

Court Bouillon
sachet d'épices	
bay leaf	1 ea
thyme	½ tsp
peppercorns	4 ea
water	1 qt
white wine	2 fl oz
shallot, minced	1 ea

Preparation

1. To begin the court bouillon, make a sachet d'épices of bay leaf, thyme, and peppercorns.
2. Place water, wine, shallots, and the sachet d'épices in a large saucepan and bring to a boil.
3. Lower the temperature to between 175°F and 180°F and hold at a simmer.
4. Cut salmon fillets into 6 oz portions.
5. Place remaining salmon in a food processor with the ice-cold water and purée.
6. Add egg, cream, salt, and white pepper to food processor and pulse to incorporate ingredients evenly.
7. Remove puréed salmon from the food processor and spread on top of the salmon fillets, coating evenly.
8. Dip purée-topped side into chopped pistachios to coat well.
9. Slowly lower fillets into court bouillon and poach for approximately 12–14 minutes or until cooked through.
10. Remove the fillets and plate immediately. Garnish with lemon slices.

NUTRITION FACTS
Per serving: 951 calories, 550 calories from fat, 63.1 g total fat, 191.6 mg cholesterol, 251.3 mg sodium, 1826.8 mg potassium, 36.6 g carbohydrates, 7.4 g fiber, 4.6 g sugar, 62.6 g protein

Poached Dover Sole and Salmon Roulade

Yield: *4 servings, 3 roulades each*

Ingredients

6 oz Dover sole fillets	4 ea
fresh spinach leaves	20 ea
salmon, cut into 1 inch chunks, ice cold	6 oz
cream	2 fl oz
salt	1 tsp
white pepper	½ tsp

Preparation

1. Place the fillets side by side on plastic wrap and lay another sheet of plastic wrap over the top.
2. Gently pound the fillets until they are about half as thick. Set aside.
3. Submerge spinach leaves for 3–5 seconds in a saucepan of boiling water until just limp. Remove immediately and shock in an ice bath.
4. When the spinach is cold, remove from the ice bath and lay on paper towels to drain. Set aside.
5. Place ice-cold salmon in a food processor and purée. When it is fairly smooth, add cream, salt, and white pepper. Purée again until very smooth. *Note:* Do not overpurée or the cream will curdle.
6. Remove the top sheet of plastic from the fillets and spread the salmon mixture on each fillet.
7. Lay spinach leaves on top of each fillet to completely cover the salmon mixture.
8. Using the bottom sheet of plastic, tightly roll each fillet into a log with the salmon and spinach on the inside (to form roulades).
9. When completely rolled, wrap the roulades in silicone-coated parchment paper and poach for 15–20 minutes or until cooked through. *Note:* Use caution to not allow the water to boil.
10. Remove the roulades from the poaching water and let rest for a few minutes.
11. Remove the parchment paper.
12. Slice the roulades into ½ inch thick slices and serve.

NUTRITION FACTS
Per serving: 266 calories, 130 calories from fat, 14.6 g total fat, 117.1 mg cholesterol, 1134.1 mg sodium, 705.7 mg potassium, 2.4 g carbohydrates, 1.2 g fiber, <1 g sugar, 30.7 g protein

Pan-Fried Buttermilk Walleye

Yield: *4 servings, 8 oz each*

Ingredients

buttermilk	2 fl oz
Dijon mustard	1 tsp
6–8 oz walleye fillets	4 ea
all-purpose flour	4 oz
cayenne	¼ tsp
paprika	¼ tsp
Old Bay® seasoning	½ tsp
kosher salt	½ tsp
white pepper	¼ tsp
canola oil	4 fl oz

Preparation

1. Combine buttermilk and Dijon mustard in a large bowl and mix thoroughly.
2. Add fish fillets to mixture and coat well. Let fish marinate for 30 minutes.
3. In another bowl combine flour, cayenne, paprika, Old Bay®, salt, and white pepper. Mix well.
4. Drain the fish from the buttermilk mixture and then dredge in the seasoned flour. Let rest for a few minutes.
5. Heat oil in a saucepan to 375°F.
6. Dredge the fish in the seasoned flour again and then ease the fillets into the hot oil.
7. Pan-fry the fish for 5–7 minutes or until cooked through.
8. Drain on paper towels. Plate and garnish as desired.

NUTRITION FACTS
Per serving: 322 calories, 84 calories from fat, 9.5 g total fat, 137.4 mg cholesterol, 356.5 mg sodium, 679.2 mg potassium, 22.7 g carbohydrates, <1 g fiber, <1 g sugar, 34 g protein

Seafood Seviche Tacos

Yield: *4 servings, 2 tacos, 3.5 oz each*

Ingredients

red bell peppers, small dice	2 oz
green bell pepper, small dice	2 oz
red onion, small dice	2 oz
jalapeño, roasted and small dice	1 ea
garlic, minced	1 tbsp
cilantro, minced	¼ c
coriander, ground	¼ tsp
fresh lime juice	6 tbsp
olive oil	1½ tbsp
salt and pepper	TT
bay scallops	4 oz
shrimp, peeled, deveined, and medium dice	4 oz
fish fillet, firm white flesh	4 oz
crabmeat	4 oz
taco shells	8 ea
lime, cut into wedges	1 ea

Preparation

1. Combine all ingredients except seafood, taco shells, and lime in a large bowl and mix well.
2. Add seafood and gently toss the mixture to coat well.
3. Refrigerate for at least 1 hour.
4. Divide the mixture among the taco shells. Place two tacos on each plate and garnish each with a lime wedge.

NUTRITION FACTS
Per serving: 281 calories, 106 calories from fat, 12 g total fat, 80 mg cholesterol, 496 mg sodium, 445.2 mg potassium, 25 g carbohydrates, 3.6 g fiber, 1.8 g sugar, 19.8 g protein

Crab and Corn Pancakes

Yield: *6 servings, 3 pancakes each*

Ingredients

all-purpose flour	¾ c
baking powder	2 tbsp
water	1 c
cornmeal	1 c
honey	2 tbsp
whole eggs, slightly beaten	2 ea
whole milk	1 c
melted butter	2 tbsp
lump crabmeat	½ lb
leek tops, minced	¼ c
sweet corn, puréed and cooked	¼ c
parsley, chopped fine	2 tsp
salt and pepper	TT

Preparation

1. Sift flour and baking powder together and set aside.
2. Bring water to a boil in a saucepot. Remove from the heat and add the cornmeal and honey and mix well. Allow to cool.
3. In another bowl, mix eggs, milk, and melted butter.
4. Combine with the cornmeal mixture.
5. Add the flour mixture to the cornmeal mixture. Stir just enough to combine, using caution to not overmix.
6. Fold in crabmeat, leeks, corn, parsley, salt, and pepper.
7. Preheat griddle to 400°F and scoop out batter into 3 oz portions.
8. When bubbles appear on the top of each pancake, flip the pancake over and cook until done. Serve immediately.

> **NUTRITION FACTS**
> **Per serving:** 277 calories, 72 calories from fat, 8.1 g total fat, 105.7 mg cholesterol, 699.3 mg sodium, 307.5 mg potassium, 38.4 g carbohydrates, 2.1 g fiber, 8.6 g sugar, 13.8 g protein

Seafood Cioppino with Grilled Fennel Root

Yield: *4 servings, 18 oz each*

Ingredients

olive oil	2 tbsp
onions, small diced	2 oz
celery, small diced	2 oz
red peppers, diced	2 oz
green peppers, diced	2 oz
fennel root, diced	2 oz
minced garlic	1 tbsp
bay leaves	2 ea
dried oregano	2 tsp
dried red pepper flakes	1 tsp
thyme	½ tsp
tomato paste	1 tbsp
canned diced tomatoes, with juice	1 lb
white wine	¾ c
mild fish stock	1 qt
salt and pepper	TT
mussels	8 ea
slipper lobsters, cut in half	4 ea
snapper or salmon, cut into 1 oz cubes	12 oz
medium shrimp, peeled and deveined	12 ea
fresh fennel root, cut into 1½ inch wedges	8 ea
parsley, chopped fine	1 tbsp
bread, crusty	8 slices

Preparation

1. Add olive oil to a hot saucepan. Add the vegetables and sweat until slightly cooked.
2. Add the herbs and tomato products and gently sauté.
3. Deglaze the pan with white wine and reduce by half.
4. Add the fish stock. Season to taste and let the stock simmer until it is flavorful.
5. Add the seafood and stir well. Cover and let simmer for 15 minutes or until cooked through.
6. Blanch the fennel and then grill it.
7. Place the seafood cioppino in a deep plate and garnish with chopped parsley.
8. Serve with grilled fennel and slices of crusty bread.

> **NUTRITION FACTS**
> **Per serving:** 639 calories, 125 calories from fat, 14.1 g total fat, 173.4 mg cholesterol, 1794.4 mg sodium, 3519.1 mg potassium, 70.4 g carbohydrates, 19.3 g fiber, 7.3 g sugar, 54.4 g protein

Baked Clams with Pesto and Pine Nut Crust

Yield: *6 servings, 8 oz each*

Ingredients

cornmeal	½ c
salt	2 tbsp
cold water	½ gal.
littleneck clams	3 lb
butter	2 tbsp
shallots, minced	2 ea
garlic, crushed	3 tbsp
dry white wine	4 fl oz
lemon juice, fresh	2 fl oz
fresh parsley, chopped fine	2 tbsp
pesto	4 tbsp
whole pine nuts	2 tbsp
coarse bread crumbs	3½ c
Parmesan, freshly grated	4 oz
olive oil	4 fl oz
melted butter	4 fl oz
kosher salt	1 tsp
black pepper	½ tsp

Preparation

1. Mix cornmeal and salt in a large bowl with cold water.
2. Add clams to bowl and, if needed, add more water to completely submerge clams. *Note:* This will purge the clams of sand and grit as the clams will open just enough to eat the cornmeal, rinsing the sand and grit from inside the shells.
3. Add butter to a hot sauté pan. Then, add the shallots and garlic and sweat for 2 minutes.
4. Deglaze pan with wine and lemon juice. Reduce by half.
5. Remove the pan from the heat and add parsley, pesto, and pine nuts. Stir to incorporate.
6. In a separate bowl, combine bread crumbs, Parmesan, olive oil, melted butter, salt, and pepper. Mix well.
7. Add the pesto mixture to the bread crumb mixture. Stir to incorporate and reserve until needed.
8. Shuck clams. Discard the top shells and carefully free clams from the bottom shells.
9. Top each clam with a dollop of pesto and pine nut mixture.
10. Bake clams in a 350°F oven until golden brown (approximately 12–15 minutes).

NUTRITION FACTS
Per serving: 445 calories, 237 calories from fat, 26.9 g total fat, 40.3 mg cholesterol, 785.6 mg sodium, 361.9 mg potassium, 37.2 g carbohydrates, 1.7 g fiber, 2.4 g sugar, 13.4 g protein

Crab Cakes

Yield: *12 servings, 4 oz each*

Ingredients

mayonnaise	4 oz
onion, minced	2 oz
green onion, minced	1 oz
fresh dill, minced	1 tsp
eggs, lightly beaten	4 ea
Worcestershire sauce	1 tsp
dry mustard	½ tsp
kosher salt	½ tsp
cayenne pepper	½ tsp
Old Bay® seasoning	½ tsp
panko bread crumbs	2 c
lump crabmeat	2 lb
canola oil	4 fl oz
unsalted butter	2 tbsp
fresh lemon, cut into 6 wedges	2 ea

Preparation

1. Add mayonnaise, onions, green onions, dill, egg, Worcestershire, mustard, salt, cayenne, and Old Bay® to mixing bowl and mix well to incorporate.
2. Gently fold in one-fourth of the panko and the crabmeat into the mayonnaise mixture, using caution to not break apart the chunks of crab.
3. Gently form the crab mixture into 4 oz cakes, approximately 1 inch thick.
4. Dip the crab cakes into the remaining crumbs and coat well on all sides.
5. Place crab cakes in refrigerator for at least 30 minutes to firm. *Note:* Resting will allow the crumbs to absorb moisture and bind the ingredients.
6. Add the oil to a large skillet over medium heat. Add the butter and let it melt in the oil.
7. When the foam from the butter disappears, add the crab cakes and pan-fry until golden brown in color on both sides.
8. Place crab cakes on paper towels to drain the excess oil.
9. Plate each crab cake with a lemon wedge and serve immediately.

NUTRITION FACTS
Per serving: 305 calories, 159 calories from fat, 18 g total fat, 128.5 mg cholesterol, 530.8 mg sodium, 356.1 mg potassium, 18.1 g carbohydrates, 1.8 g fiber, 2.1 g sugar, 18.6 g protein

Oysters Rockefeller

Yield: *10 servings, 2 oysters each*

Ingredients

fresh oysters	20 ea
bacon, cooked and cut small dice	4 oz
unsalted butter	1 tbsp
shallots, minced	2 ea
fennel, minced	2 oz
celery, minced	2 oz
garlic, minced	2 cloves
anisette liquor	1 tsp
white wine	2 fl oz
cayenne pepper	TT
Italian parsley, finely chopped	1 tbsp
baby spinach, finely chopped	2 oz
watercress, finely chopped	2 oz
béchamel sauce, heated	4 fl oz
bread crumbs, coarsely chopped	4 tbsp
salt and pepper	TT

Preparation

1. Shuck fresh oysters and refrigerate until needed.
2. Scrub shells to clean thoroughly and reserve.
3. Add butter, shallots, fennel, and celery to a sauté pan and sweat until translucent.
4. Add garlic and cook for 1 minute more.
5. Add anisette liquor, white wine, and cayenne pepper. Reduce by half.
6. Add parsley, spinach, and watercress and cook until wilted and bright green (approximately 1 minute).
7. Add hot béchamel, bread crumbs, and diced bacon.
8. Stir well and season to taste.
9. Remove from heat and place on a sheet pan to cool in the refrigerator until needed.
10. Place each oyster back into a clean shell and top with 2 tsp of the spinach mixture, using care to spread mixture over the entire surface of each oyster.
11. Bake at 450°F until bubbly (approximately 5–7 minutes).
12. Serve hot oysters on a bed of rock salt.

NUTRITION FACTS
Per serving: 248 calories, 85 calories from fat, 9.5 g total fat, 68.4 mg cholesterol, 450.3 mg sodium, 617.9 mg potassium, 23 g carbohydrates, <1 g fiber, 1 g sugar, 16.9 g protein

Steamed Mussels in Tomato Kalamata Olive Sauce

Yield: *4 servings, 1 lb each*

Ingredients

olive oil	2 tbsp
butter	1 tbsp
shallots, minced	2 ea
celery, minced	3 oz
garlic, minced	3 tbsp
canned diced tomatoes, in juice	1 qt
kalamata olives, pitted and cut in half	4 oz
crushed red pepper	¼ tsp
thyme, fresh	2 tsp
basil, cut chiffonade	3 tbsp
mussels, scrubbed and debearded	4 lb
dry white wine	3 fl oz
fish fumet	4 fl oz
fresh parsley, chopped fine	2 tbsp

Preparation

1. Heat olive oil and butter in a saucepot.
2. Add shallots and celery and sweat for 1–2 minutes.
3. Add garlic and cook for 1 minute more.
4. Add tomatoes, olives, crushed red pepper, thyme, basil, and mussels.
5. Deglaze with white wine and fish fumet.
6. Stir well and cover. Reduce heat to medium, and steam until mussels have opened (approximately 5–7 minutes). *Note:* Discard any mussels that do not open.
7. Place mussels in serving bowls and spoon sauce over the mussels.
8. Garnish with chopped fresh parsley.

NUTRITION FACTS
Per serving: 978 calories, 429 calories from fat, 48.4 g total fat, 188.3 mg cholesterol, 2406 mg sodium, 2936 mg potassium, 72.4 g carbohydrates, 4.3 g fiber, 10.4 g sugar, 62.7 g protein

Battered Fish

Yield: *4 servings, 2 pieces each*

Ingredients

all-purpose flour	2 c
baking powder	2 tsp
salt	1 tsp
egg, beaten just enough to emulsify	1 ea
soda water	12 fl oz
white vinegar	½ tsp
cod	24 oz
vegetable oil	(for deep-frying)

Preparation

1. Mix three-fourths of the flour, baking powder, salt, and beaten egg in a mixing bowl.
2. Add soda water and vinegar and mix just enough to moisten.
3. Let mixture rest for 15 minutes.
4. Slice cod in 3 oz pieces (approximately 1–2 inches wide and 3 inches long).
5. Dip cod pieces into the remaining fourth of the flour. Then dip the cod pieces in the batter mixture. Allow the excess batter to run off.
6. Use the swimming method to deep-fry the cod in hot oil. When a portion floats it is done.
7. Plate two pieces each and serve immediately.

NUTRITION FACTS
Per serving: 424 calories, 82 calories from fat, 9.3 g total fat, 126.5 mg cholesterol, 1378.2 mg sodium, 486.8 mg potassium, 48.5 g carbohydrates, 1.7 g fiber, <1 g sugar, 34 g protein

---

Fried Calamari with Lemon-Garlic Butter

Yield: *4 servings, 4.5 oz each*

Ingredients

calamari rings and tentacles	1 lb
buttermilk	2 c
all-purpose flour	2 c
chopped parsley	2 tsp
cayenne pepper	⅛ tsp
garlic powder	½ tsp
salt and pepper	TT
vegetable oil	(for deep-frying)
butter	2 tbsp
garlic, minced	1 clove
fresh lemon juice	1 tbsp

Procedure

1. Soak calamari in buttermilk for 20–30 minutes.
2. Mix flour, parsley, cayenne pepper, garlic powder, salt, and pepper together in a separate bowl.
3. Remove calamari from the buttermilk marinade and shake off excess buttermilk.
4. Dip calamari in the flour mixture and shake off excess flour.
5. Deep-fry calamari in oil that is between 350°F and 375°F until slightly golden (approximately 1–2 minutes). *Note:* Calamari overcooks quickly and becomes tough and rubbery.
6. Mix butter, garlic, salt, and pepper in a small saucepan and cook for 2 minutes, using caution to not brown the garlic.
7. Add lemon juice to the butter mixture and remove it from the heat.
8. Remove calamari from the fryer and drain well.
9. Season with salt and pepper and toss with butter mixture. Serve immediately.

NUTRITION FACTS
Per serving: 498 calories, 142 calories from fat, 16.1 g total fat, 284.4 mg cholesterol, 254 mg sodium, 548.7 mg potassium, 57.9 g carbohydrates, 1.8 g fiber, 6.2 g sugar, 28.4 g protein

---

Coconut-Breaded Shrimp with Minted Orange Dipping Sauce

Yield: *4 servings, 5 shrimp and 1 oz of sauce each*

Ingredients

all-purpose flour	1 c
curry powder	¾ tsp
cayenne pepper	¼ tsp
baking powder	1 tsp
salt	½ tsp
eggs, slightly beaten	2 ea
club soda	12 fl oz
vegetable oil	(for deep-frying)
shredded unsweetened coconut	2 c
shrimp, tail-on, peeled and deveined	20 ea

Minted Orange Dipping Sauce

orange marmalade	4 oz
fresh lime juice	1 fl oz
rice wine vinegar	1 tsp
Madras curry powder	¼ tsp
kosher salt	⅛ tsp
fresh mint leaves, cut chiffonade	2 tsp

Preparation

1. In a large bowl, combine flour, curry, cayenne pepper, baking powder, and salt and mix well.
2. Add beaten eggs and club soda. Whisk just to incorporate while still leaving batter somewhat lumpy. Let batter rest for 10–15 minutes.
3. Heat oil to 375°F.
4. Spread coconut evenly in a hotel pan and set aside.
5. Dip each shrimp (21–25 count or larger) into the batter while holding the tail.
6. Remove shrimp from batter, shake off excess and roll in coconut to coat evenly.
7. Deep-fry the shrimp in hot oil for 2–3 minutes or until completely cooked.
8. Remove shrimp and drain on paper towels.
9. To prepare sauce, place orange marmalade, lime juice, vinegar, curry, and salt in small saucepan and heat until the marmalade has melted completely.
10. Add fresh mint and stir to incorporate. Serve with sauce.

> **NUTRITION FACTS**
> **Per serving:** 636 calories, 378 calories from fat, 44.7 g total fat, 113.9 mg cholesterol, 752.9 mg sodium, 443.2 mg potassium, 52.5 g carbohydrates, 10.4 g fiber, 21.5 g sugar, 13.7 g protein

Cornmeal-Crusted Oysters with Spicy Fruit Salsa

Yield: *4 servings, 6 oysters each*

Ingredients

flour	¾ c
salt and white pepper	TT
eggs, beaten	3 ea
cornmeal or polenta	4 oz
shucked oysters	24 ea
vegetable oil	(for deep frying)
oyster shells, scrubbed and reserved	24 ea

Salsa

mango, diced	2 oz
papaya, diced	2 oz
banana, diced	2 oz
honeydew melon, diced	2 oz
lime juice	½ tsp
cilantro, minced	2 tsp
sugar	½ tsp
cider vinegar	½ tsp
Tabasco® sauce	2 shakes
salt and white pepper	TT

Preparation

1. Place half of the flour in a bowl and season with salt and white pepper.
2. Place the beaten eggs in a second bowl.
3. In a third bowl, mix cornmeal with the other half of the flour and season with salt and pepper.
4. Bread the oysters using standard breading procedure.
5. Prepare the salsa by gently tossing the fruit with remaining salsa ingredients.
6. Deep-fry the oysters until crispy and drain.
7. Place each fried oyster back into a cleaned half shell.
8. Plate oysters on a bed of rock salt.
9. Place salsa in a bowl in the center of the plated oysters or top each oyster with a dollop of salsa and serve.

> **NUTRITION FACTS**
> **Per serving:** 493 calories, 135 calories from fat, 15.1 g total fat, 289.5 mg cholesterol, 532.6 mg sodium, 783.8 mg potassium, 51.7 g carbohydrates, 3.3 g fiber, 6.8 g sugar, 36.6 g protein

Culinary Arts
PRINCIPLES AND APPLICATIONS

Photo Courtesy of the **Beef Checkoff**

Beef, Veal, and Bison

Beef, veal, and bison are popular items on most foodservice menus. They are similar in composition and can be prepared using most cooking methods. Veal is commonly sautéed, roasted, or stewed. Beef and bison can be prepared using any cooking method. Prior to cooking, some cuts of beef are aged, barded, larded, cured, or covered with a rub or marinade to develop the flavor further. Bison is lean, high in iron and protein, and contains less cholesterol than skinless chicken.

Refer to DVD
for **Flash Cards**

Chapter Objectives

1. Describe the composition of beef.
2. Differentiate between grain-fed and grass-fed animals.
3. Identify the eight primal cuts of beef.
4. Identify the cuts fabricated from each primal cut of beef.
5. Explain how to prepare beef offals.
6. Explain the purpose of Institutional Meat Purchase Specifications.
7. Describe the composition of veal.
8. Identify the five primal cuts of veal.
9. Identify the cuts fabricated from each primal cut of veal.
10. Explain how to prepare veal offals.
11. Describe the USDA inspection and grading of beef and veal.
12. Identify four traits to check upon receiving beef and veal.
13. Trim and cut beef tenderloin.
14. Cut a boneless strip into steaks.
15. French veal chops.
16. Tenderize beef and grind fresh meat.
17. Describe seven ways to enhance the flavor of beef and veal.
18. Explain how to determine the doneness of beef and veal.
19. Cook beef and veal using different cooking methods.
20. Describe how bison is similar to and different from beef.

Key Terms

- **grain-fed beef**
- **grass-fed beef**
- **collagen**
- **silverskin**
- **marbling**
- **fat cap**
- **primal cut**
- **fabricated cut**
- **brisket**
- **offals**
- **tripe**
- **oxtail**
- **cutlet**
- **sweetbreads**
- **wet aging**
- **dry aging**
- **barding**
- **larding**
- **shrinkage**

BEEF

Beef is the flesh of domesticated cattle. Types of domesticated cattle include steers, heifers, cows, bulls, and stags. The age and gender of these animals affects the quality and flavor of the meat.

Most beef comes from steers. A *steer* is a male calf that has been castrated prior to reaching sexual maturity. Steers produce the best-quality beef and are 15–24 months old when they are sold. Most steers are grain-fed, which increases the marbling of the meat. *Grain-fed beef* is meat from cattle that were grain-fed in confined feeding operations for 90 days to 1 year. *Grass-fed beef* is meat from cattle that were raised on grass with little or no special feed. Grass-fed beef contains less saturated fat than grain-fed beef.

It is important to understand the basic composition of cattle before working with the meat. Raw beef is red in color and made up of bundles of muscle fibers held together by two types of connective tissues called collagen and silverskin. *See Figure 18-1. Collagen* is a soft, white, connective tissue that breaks down into gelatin when heated. *Silverskin* is a tough, rubbery, silver-white connective tissue that does not break down when heated. Silverskin is trimmed from the meat prior to cooking because it is inedible. The amount of connective tissue increases as the animal ages.

Composition of Beef

National Cattlemen's Beef Association

Figure 18-1. *Beef is composed of muscle held together by connective tissues called collagen and silverskin. Marbling is found within the muscle, and fat cap surrounds the muscle.*

Collagen and fat give beef a rich taste and help to thicken and flavor sauces made from meat juices. Beef contains both marbling and fat cap. *Marbling* is the fat found within a muscle. The amount of marbling can affect the taste, tenderness, and quality of the meat. However, marbling does not guarantee flavor, tenderness, or juiciness. *Fat cap* is the fat that surrounds a muscle. Because fat cap increases as an animal ages, beef contains more fat cap than veal. Most of the fat can be trimmed prior to or after the cooking process. Fat cap is often removed prior to service.

Cuisine Note

Grain-fed beef contains more marbling than grass-fed beef.

Bones may or may not be removed from cuts of beef prior to cooking. Understanding the skeletal structure of cattle can aid in identifying the market forms of beef. *See Figure 18-2.*

Skeletal Structure of Cattle

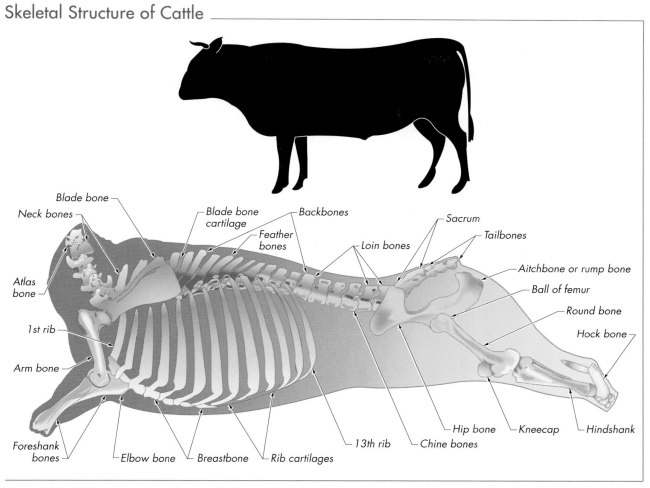

Figure 18-2. *Understanding the skeletal structure of cattle can aid in identifying the market forms of beef.*

MARKET FORMS OF BEEF

Beef is available in a variety of market forms. Knowledge of each of these market forms is necessary for accurate product ordering. Common market forms of beef include partial carcasses, primal cuts, fabricated cuts, and offals.

Partial Carcasses of Beef

Sides are the most common partial carcass of beef. A *side of beef* is a half of a carcass split along the backbone. There are two sides to each carcass. A side can be purchased for less per pound than any primal or fabricated cut. However, in order for purchasing sides of beef to be cost-effective, adequate facilities and equipment for fabrication and storage must be available.

In addition, foodservice staff capable of cutting a side of beef is required. Because of the amount of space, time, and labor required to break down a side of beef, sides are rarely purchased by foodservice operations.

A quarter is one side of a beef carcass that has been divided into two parts between the 12th and 13th ribs. Quarters are not often purchased by foodservice operations.

Primal and Fabricated Cuts of Beef

A *primal cut* is a large cut from a whole or a partial carcass. A single beef carcass has two of each primal cut (a left side and a right side). Primal cuts of beef are very large. The eight primal cuts of beef are the chuck, rib, short loin, sirloin, round, flank, short plate, and brisket and shank. *See Figure 18-3.*

Primal Cuts of Beef

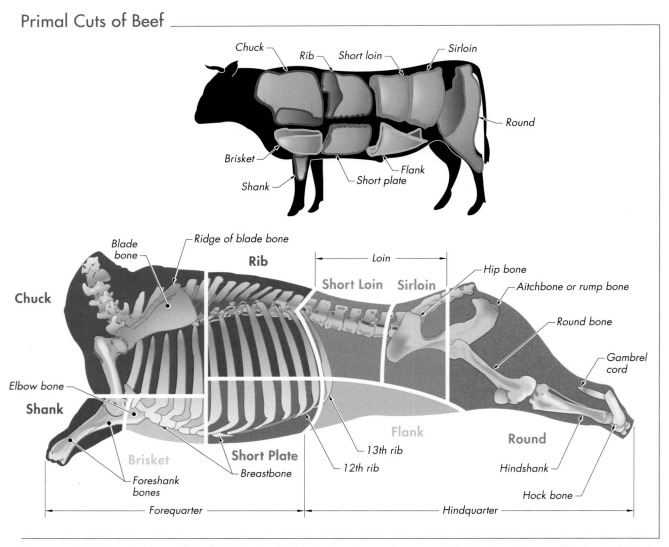

Figure 18-3. *The primal cuts of beef include the chuck, rib, short loin, sirloin, round, flank, short plate, and brisket and shank.*

Each primal cut of beef is further divided into fabricated cuts. A *fabricated cut* is a ready-to-cook cut that is packaged to certain size and weight specifications. Fabricated beef cuts include short ribs and tenderloins, as well as top sirloin, flat-iron, eye-roll, T-bone, porterhouse, Delmonico, butt, and skirt steaks. Fabricated cuts are a convenient way of providing uniform portions while reducing labor costs. The price per pound for fabricated cuts is higher than the price per pound for primal cuts. Foodservice operations often purchase some primal cuts and some fabricated cuts of beef.

Chucks. A *beef chuck* is a shoulder primal cut of beef that contains the first five rib bones, some of the backbone, and a small amount of the arm and blade bones. The chuck is the largest primal cut and its average weight is approximately 26% of the total carcass weight. The shoulder is one of the most-exercised muscles on the animal, so it is a tough cut of meat with a lot of connective tissue. However, chuck is also quite lean and has an excellent flavor. Larger pieces of chuck lend themselves to braising and stewing. Smaller pieces of chuck and trimmings produce very flavorful ground meat.

A small, thin muscle section on top of the chuck yields a fabricated cut known as flat-iron steak. This cut is somewhat tender and lends itself to marinating and then grilling or broiling. Other fabricated cuts from the chuck include shoulder clods, clod tenders, chuck rolls, top blade chucks, flat-iron steaks, and strips. *See Figure 18-4.* Short ribs are fabricated cuts produced from the small rib bone ends that are sawed off of the primal rib as the rib roast is removed. Short ribs have a sizable portion of lean meat on them.

Chuck Cuts of Beef

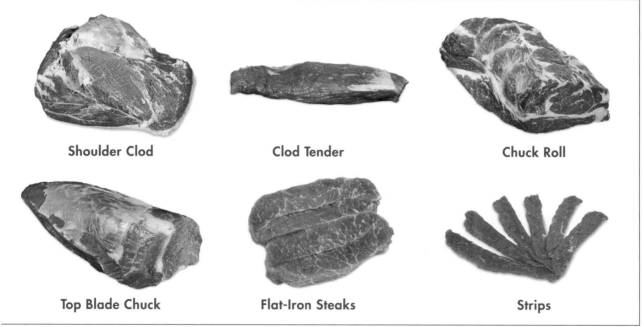

Shoulder Clod   Clod Tender   Chuck Roll

Top Blade Chuck   Flat-Iron Steaks   Strips

Figure 18-4. *A beef chuck is often fabricated into a variety of cuts, including shoulder clods, clod tenders, chuck rolls, top blade chucks, flat-iron steaks, and strips.*

Ribs. A *beef rib* is the primal cut of beef located between the chuck and short loin and contains seven rib bones, from the 6th to the 12th rib. Its average weight is approximately 10% of the total carcass weight. The meat is tender and well marbled. The rib bones (finger bones) make the rib cut a good cut for roasting, because they form a natural rack on which the meat can cook.

A beef rib is often fabricated into a variety of cuts. *See Figure 18-5.* A beef rib contains the prime rib roast. The rib bones can be left on for roasting to produce a moist roast. If the bones are removed, a boneless rib eye roast can be further cut into fabricated bone-in or boneless rib eye steaks. A *rib eye* is a large, eye-shaped muscle within the rib that is a continuation of the sirloin muscle. Meat from eye-shaped muscles, such as the rib eye or tenderloin, is often referred to as eye meat. The 6th to the 12th ribs can be prepared as smoked or barbequed beef ribs. If the rib bones are removed, the meat can be rolled and tied into a rolled-rib roast.

Rib Cuts of Beef

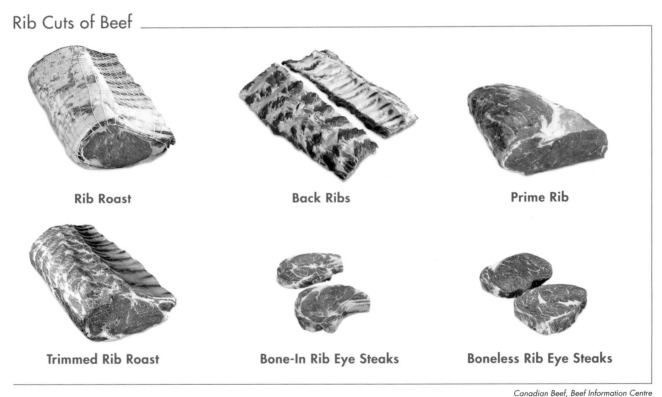

Rib Roast   Back Ribs   Prime Rib

Trimmed Rib Roast   Bone-In Rib Eye Steaks   Boneless Rib Eye Steaks

Canadian Beef, Beef Information Centre

Figure 18-5. *A beef rib is often fabricated into a variety of cuts.*

Short Loins. A *beef short loin* is a primal cut of beef located just to the rear of the primal rib and includes the 13th rib and a small section of the backbone. Its average weight is approximately 8% of the total carcass weight. The short loin can be cut in cross sections to produce some of the most popular fabricated cuts. *See Figure 18-6.* Cuts from the short loin are commonly grilled, broiled, or roasted.

Short Loin Cuts of Beef

Canadian Beef, Beef Information Centre

Strip Loin

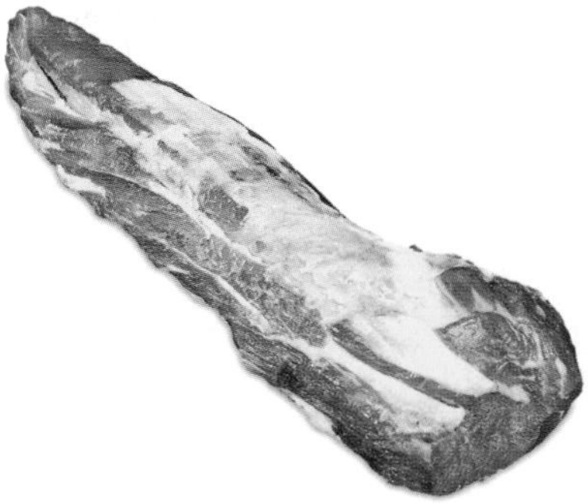

Canadian Beef, Beef Information Centre

Tenderloin

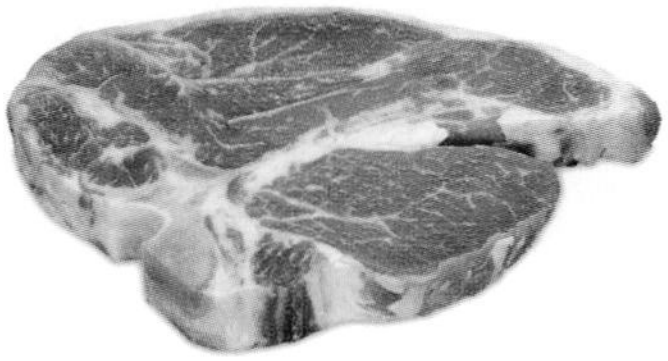

Canadian Beef, Beef Information Centre

Porterhouse Steak

Canadian Beef, Beef Information Centre

Strip Steaks

Canadian Beef, Beef Information Centre

Tenderloin Steaks

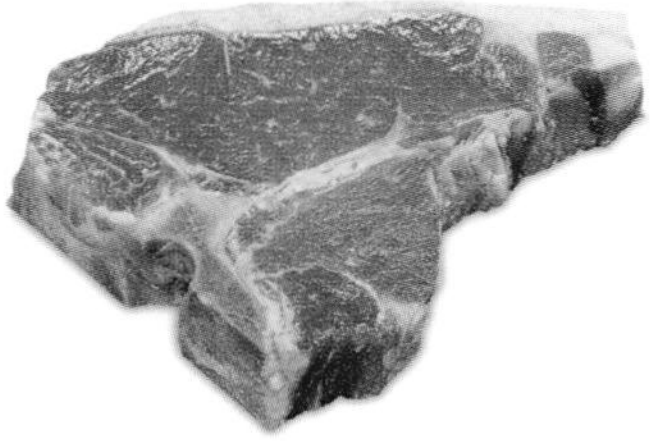

Canadian Beef, Beef Information Centre

T-Bone Steak

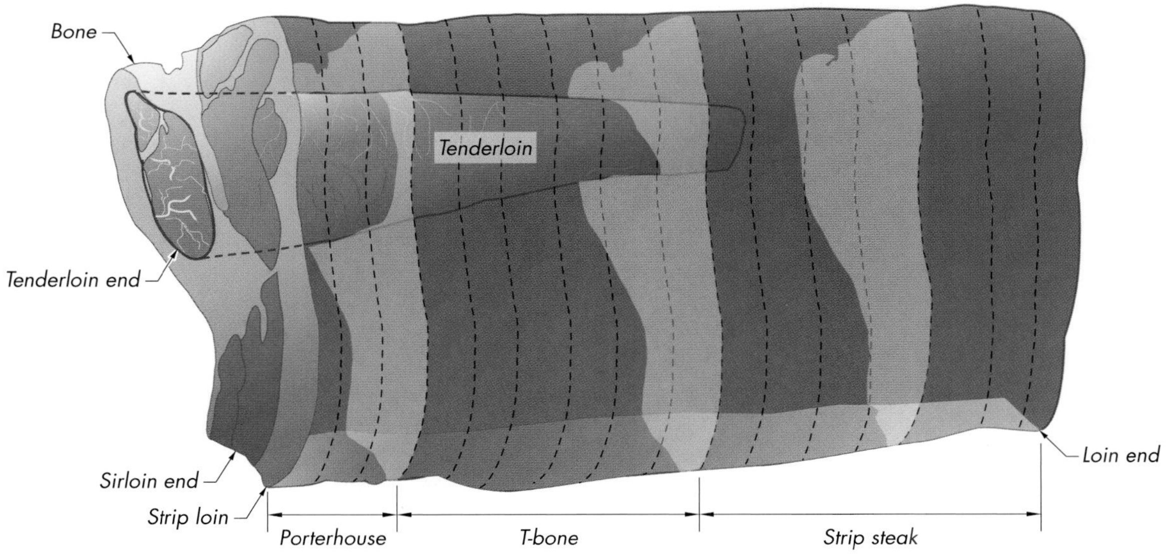

Figure 18-6. *The short loin can be cut in cross sections to produce some of the most popular fabricated cuts.*

When the short loin and sirloin are split apart, the smaller portion of tenderloin is part of the short loin. A *beef tenderloin* is an eye-shaped muscle running from the primal rib cut into the primal leg. The tenderloin is located just beneath the strip loin and is the most tender piece of beef. Sometimes the entire tenderloin is removed prior to dividing the short loin and sirloin. The whole tenderloin can be roasted whole or divided into chateaubriand, filets mignons, and tournedos.

A *beef strip loin* is a short loin without a tenderloin. A strip loin can be cut into boneless strip steaks or roasted whole. Fabricated cuts from the short loin are often aged, as they have ample fat covering and marbling and are very tender. Aging beef loin intensifies the flavor and tenderness of the meat.

The short loin produces many popular fabricated cuts. Starting from the end nearest the primal rib, cross-section cuts produce Delmonico steaks, T-bone steaks, and porterhouse steaks. Delmonico steaks do not include any tenderloin, T-bone steaks include only a small section of tenderloin, and porterhouse steaks include a large section of tenderloin.

Sirloins. A *beef sirloin* is a primal cut of beef situated just behind the short loin and contains some of the backbone and hip bone. Its average weight is approximately 9% of the total carcass weight. With the exception of the butt tenderloin muscle, meat from the sirloin is not quite as tender as meat from the short loin. However, the sirloin can be cut into butt steaks that can be marinated, skewered, and then grilled or broiled. Ball-tip steaks, sirloin flaps, sirloin tri-tips, sirloin tri-tip steaks, top sirloin roasts, top sirloin steaks, top sirloin caps, and top sirloin cap steaks can be fabricated from the sirloin primal cut. *See Figure 18-7.*

Sirloin Cuts of Beef

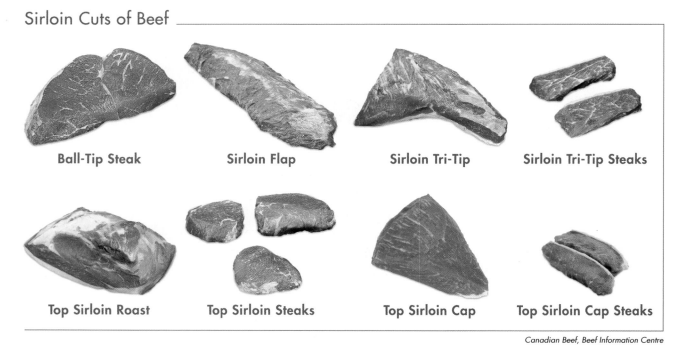

Ball-Tip Steak   Sirloin Flap   Sirloin Tri-Tip   Sirloin Tri-Tip Steaks

Top Sirloin Roast   Top Sirloin Steaks   Top Sirloin Cap   Top Sirloin Cap Steaks

Figure 18-7. *Ball-tip steaks, sirloin flaps, sirloin tri-tips, sirloin tri-tip steaks, top sirloin roasts, top sirloin steaks, top sirloin caps, and top sirloin cap steaks can be fabricated from the sirloin primal cut.*

Rounds. A *beef round* is a primal beef cut that includes a large grouping of muscles that represent the hind hip and thigh of the carcass. The average weight of the round is approximately 27% of the total carcass weight. A round contains large bones including the round bone (leg), aitchbone (pelvis), shank, and tailbone.

A round can be slow-roasted whole. However, due to its size, the round is commonly broken down and sold as separate subprimal cuts. Subprimal and fabricated cuts from the round include the top round, bottom round, knuckle, and shank. The bottom round can be cut into the outside round and the eye of round. *See Figure 18-8.*

Round Cuts of Beef

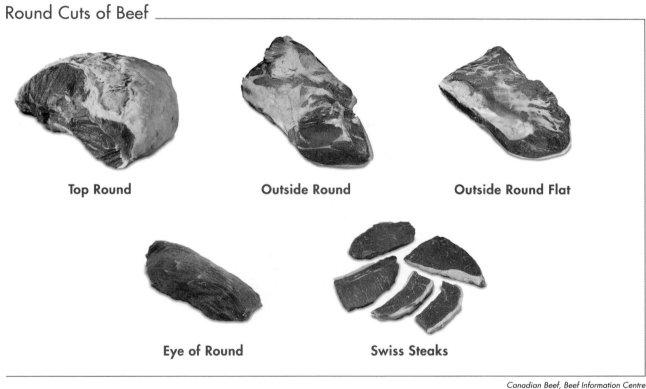

Top Round

Outside Round

Outside Round Flat

Eye of Round

Swiss Steaks

Canadian Beef, Beef Information Centre

Figure 18-8. *The round can be broken down into a variety of subprimal and fabricated cuts.*

When the round is trimmed for cooking it yields a large amount of usable meat. A *beef rump roast* is a roast cut from the primal round, above the back end of the hip bone. If the bone is left in, it is called a standing rump roast. A boneless rump roast is rolled and tied prior to being sold. The rump roast is a very flavorful cut that is often braised. A *steamship round roast* is the beef round with the shank and rump removed. The top round and knuckle portions can be roasted. The outside round and eye of round are often braised.

Flank Cuts of Beef

Figure 18-9. *A beef flank is fabricated into flank steaks.*

Flanks. The *beef flank* is a primal cut of beef that includes the thin, flat section of the hindquarters located beneath the loin. Its average weight is approximately 4% of the total carcass weight. The flank has more fat than lean meat and contains one thin, oval-shaped, boneless flank steak. The flank steak can be scored or cubed before it is cooked. Prior to cooking, the fat covering should be removed and the meat can be marinated to produce a more tender and flavorful piece of meat. If cooked whole, flank steak should be cut across the grain or it will be tough. *See Figure 18-9.*

Short Plates. A *beef short plate* is a primal cut of beef that includes a thin portion of the beef forequarter located just beneath the rib cut. The average weight of the short plate is approximately 5% of the total carcass weight. The bones attached to this cut are the remaining sections of the rib bones. The small bones from the short plate are called short ribs. Short ribs are not quite as meaty as those from the rib cut. In addition to short ribs, the short plate yields the flavorful skirt steak. *See Figure 18-10.*

Short Plate Cuts of Beef

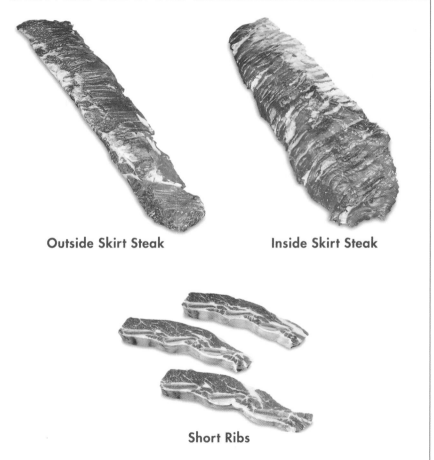

Outside Skirt Steak   **Inside Skirt Steak**

Short Ribs

Figure 18-10. *A beef short plate is fabricated into skirt steak and short ribs.*

Briskets and Shanks. The brisket and shank are two separate muscle groups that make up one primal cut of beef that is located just below the chuck. *See Figure 18-11.* The weight of the brisket and shank is approximately 9% of the total carcass weight. The *beef brisket* is a thin section of beef that contains some of the ribs, the breastbone, and layers of lean muscle, fat, and connective tissue. The ribs and breastbone are always removed prior to cooking the brisket.

Brisket and Shank Cuts of Beef

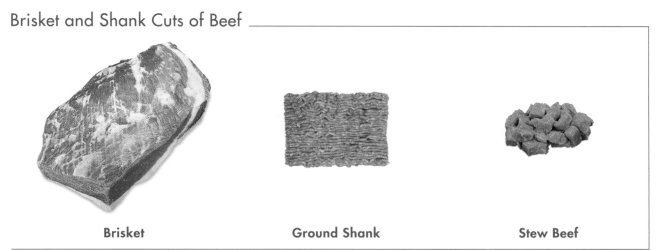

| Brisket | Ground Shank | Stew Beef |

Canadian Beef, Beef Information Centre

Figure 18-11. *A beef brisket and shank are two separate muscle groups that make up one primal cut of beef that is located just below the chuck.*

Brisket is a tough cut of beef with excellent flavor that can become tender when cooked properly. It has long muscle fibers that run in several directions, making it difficult to slice. Brisket is often braised or simmered. Brisket can be cured, peppered to make pastrami, braised as sauerbraten, simmered as New England-style brisket, and corned or pickled as corned beef.

The *beef shank* is a bony section of beef that is surrounded by a small amount of very tough but flavorful meat. Shanks are used for making stocks and rich reduction sauces. Shank meat is usually ground to flavor and clarify consommés, because the meat has a high concentration of collagen that converts to gelatin when cooked. Shanks are generally cut perpendicular across the bone and then braised.

Beef Offals

An *offal* is an edible part of an animal that is not part of a primal cut. Commonly used beef offals include liver, tongue, tripe, oxtail, and kidney. *See Figure 18-12.* With the exception of liver, beef offals are prepared using moist-heat cooking methods. Beef liver is covered with a thin membrane that should be removed before slicing. Beef liver is commonly broiled, sautéed, or pan-fried.

Chef's Tip

For best results, liver should be partially frozen when it is sliced. Larger slices are cut at a 45° angle.

Beef Offals

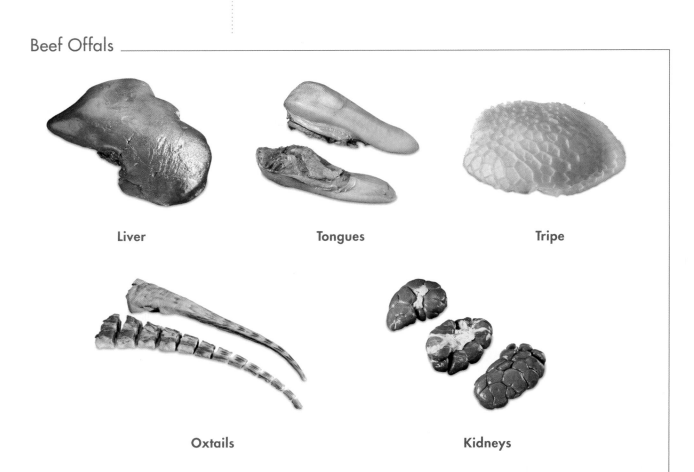

Liver

Tongues

Tripe

Oxtails

Kidneys

Canadian Beef, Beef Information Centre

Figure 18-12. *Beef offals include liver, tongue, tripe, oxtail, and kidney.*

Beef Tongues

U.S. Wellness Meats

Figure 18-13. *Beef tongue can be served cold or can be reheated and served hot.*

Beef tongue is available fresh, smoked, pickled, and corned, but smoked is the most popular form. Tongue is always prepared using moist-heat cooking methods. To determine doneness, the tip of the tongue is felt to test tenderness. When the tip is soft, the tongue is done. Cooked tongue is cooled in cold water and then skinned. Beef tongue can be served cold or can be reheated and served hot. *See Figure 18-13.*

Tripe is the muscular inner lining of a stomach of an animal, such as cattle or sheep. *Honeycomb tripe* is the lining of the second stomach found in cattle. Honeycomb tripe may be purchased fresh, pickled, or canned. It is often fried, creamed, used in soups, or served cold with a vinaigrette dressing.

An *oxtail* is the tail from a cattle carcass. Oxtail contains a considerable amount of bone and a good portion of richly flavored flesh. Oxtail is most often used in stews or braised in rich cooking liquids. *See Figure 18-14.* The thin end of the oxtail can be used to make oxtail soup.

Oxtail Stews

Figure 18-14. *Oxtail is most often used in stews or braised in rich cooking liquids.*

Beef kidneys have irregularly shaped lobes divided by deep cracks. Before cooking, the kidney is split lengthwise, and the suet (fat) and urinary canals are carefully removed. Kidneys are commonly braised.

Institutional Meat Purchase Specifications for Beef

To help ensure that foodservice operations and suppliers communicate efficiently, the USDA publishes the Institutional Meat Purchase Specifications (IMPS) for commonly purchased meats and meat products. Cuts of beef are numbered by category. *See Figure 18-15.*

USDA Institutional Meat Purchase Specifications (IMPS) for Beef				
Series	**Item No.**	**Product**	**Item No.**	**Product**
100 (fresh beef)	100	Carcass	135	Diced beef
	101	Side	136, 137	Ground beef
	102	Forequarter	138, 139	Beef trimmings
	103–112	Rib	139	Special trim, boneless
	113–116	Chuck	140	Hanging tender
	117	Foreshank	155	Hindquarter
	118–120	Brisket	157	Hindshank
	121	Plate, short plate	158–171	Round
	122	Plate, full	172	Loin, full loin, trimmed
	123	Short ribs	173	Loin, short loin
	124	Rib, back ribs	175	Loin, strip loin
	125, 126	Chuck, arm bone	176	Loin, steak tails
	127, 128	Chuck, cross-cut	180	Loin, strip loin, boneless
	130	Chuck, short ribs	181–186	Loin, sirloin
	132, 133	Triangle	188–192	Loin, tenderloin
	134	Beef bones	193	Flank, flank steak

Figure 18-15. *(continued on next page)*

(continued from previous page)

USDA Institutional Meat Purchase Specifications (IMPS) for Beef				
Series	**Item No.**	**Product**	**Item No.**	**Product**
600 (cured, dried, and smoked beef)	600, 601	Beef brisket	622	Beef, sliced, cooked, cured, chunked, and formed
	602, 603	Beef knuckle		
	604, 605	Beef top (inside) round	623	Beef top (inside) round, cooked
	606, 607	Beef bottom, (gooseneck) round	624	Beef outside round, corned, cooked
	608	Beef outside round, corned	625, 626	Brisket, boneless, deckle off, corned, cooked
	609	Beef rump butt, corned		
	611	Beef pastrami	627	Beef knuckle, peeled, cooked
	612	Beef fajita strips	628	Beef loin, top sirloin butt, center cut, boneless, cooked
	613, 614	Beef tongue, cured, trimmed		
	617	Processed dried beef	629	Bottom (gooseneck) round, heel out, cooked
	618	Sliced processed dried beef		
	619, 620	Sliced dried beef	630	Beef rib eye roll, boneless, cooked
	621	Beef, cooked, cured, chunked, and formed	631	Charbroiled beef patties
700 (variety meats)	701–703	Beef liver	724	Head meat
	716	Beef tongue, short cut	725	Beef, lamb, or pork brains
	717	Tongue, Swiss cut	726	Beef tripe, scalded, bleached (denuded)
	720	Beef heart, trimmed		
	721	Beef oxtail, trimmed	727	Beef tripe, honeycomb, bleached
	722	Beef, lamb, or pork kidney	731	Edible tallow
	723	Cheek meat	732	Lard (edible)
800 (sausage)	800	Frankfurters	813	Polish sausage
	801	Bologna	814	Meat loaves
	804	Cooked salami	815	Meat food product loaves
	805	Minced luncheon meat	816	Knockwurst
	806	Lebanon bologna	817	Breakfast sausage, cooked
	807	Thuringer	818	Italian sausage
	808	Dry salami	820	Head cheese
	809	Cervelat	821	Pepperoni
	810	Breakfast sausage	822	Bratwurst
	811	Smoked sausage	825	Canned luncheon meat
	812	New England brand sausage	826	Scrapple
1100 (fresh beef portion cuts)	1100, 1101	Cubed steak	1150	Top side steak, boneless
	1102	Braising steak, Swiss	1167	Round, steak
	1103	Rib, rib steak	1169	Round, top (inside) round steak
	1112	Rib, rib eye roll steak	1170	Round, bottom (gooseneck) round steak
	1114	Chuck, shoulder steak		
	1116	Chuck, chuck eye roll steak	1173	Loin, porterhouse steak
	1121	Plate, skirt steak	1174	Loin, T-bone steak
	1123	Short ribs, flanken style	1179, 1180	Loin, strip loin steak
	1136, 1137	Ground beef patties	1184	Loin, top sirloin butt steak, boneless
	1138	Beef steaks, flaked and formed, frozen	1185	Loin, bottom sirloin butt steak
	1139	Beef slices	1189, 1190	Loin, tenderloin steak
	1140	Hanger steaks		

Figure 18-15. *Cuts of beef are numbered by category in the Institutional Meat Purchase Specifications (IMPS).*

VEAL

Veal is the flesh of calves, which are young cattle. Nearly all of the calves slaughtered are male. Veal is often grouped into four categories. The categories of veal include bob veal, crate-raised veal, group-raised veal, and pasture-raised veal. Crate-raised veal and group-raised veal are further divided into special-fed and formula-fed veal.

Daniel NYC

- Bob veal is the flesh of a male dairy calf that is sold at 8 days of age and slaughtered at no more than 10 days of age. Bob veal is an inconsistent product that is generally sold as cutlets.

- Crate-raised veal comes from calves that are special-fed or formula-fed an iron-deficient liquid diet for 22 weeks. Special-fed crate-raised veal is the most common type of veal sold. Veal from these calves often has a bland flavor because the animals do not get any exercise.

- Group-raised veal is managed in two phases. During phase one, calves are raised in individual pens and may or may not be tethered. After eight weeks the calves are moved to phase two, where they are allowed to move around freely.

- Pasture-raised calves are not confined and their diet consists of grass and their mother's milk. Their meat is redder than bob veal and crate-raised veal. Pasture-raised veal is more sustainable than other types of veal.

It is important to understand the basic composition of calves before working with veal. Due to the young age of the calves, veal has less connective tissue and fat and is more tender than beef. *See Figure 18-16.* Raw veal is naturally pink or rosy in color. Extremely pale veal indicates a lack of iron in the diet.

Composition of Veal

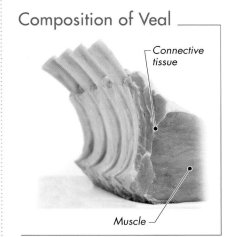

Connective tissue

Muscle

Strauss Free Raised

Figure 18-16. *Veal has less connective tissue and fat and is more tender than beef.*

MARKET FORMS OF VEAL

Veal is available in a variety of market forms, including whole and partial carcasses, primal cuts, fabricated cuts, and offals. Knowledge of these different market forms is necessary for accurate product ordering. Understanding the skeletal structure of calves can aid in identifying the market forms of veal. *See Figure 18-17.*

Partial Carcasses of Veal

In order for the purchasing of partial carcasses of veal to be cost-effective, skilled labor and storage space are required. Veal is typically not split into sides like beef. Veal is split into head and tail sections known as the foresaddle and hindsaddle. The foresaddle and hindsaddle are split between the 11th and 12th ribs, not down the backbone. A *veal foresaddle* is the front half of a carcass consisting of the primal shoulder, rack, breast, and shank cuts. A *veal hindsaddle* is the rear half of a carcass consisting of the loin and leg.

Skeletal Structure of Calves

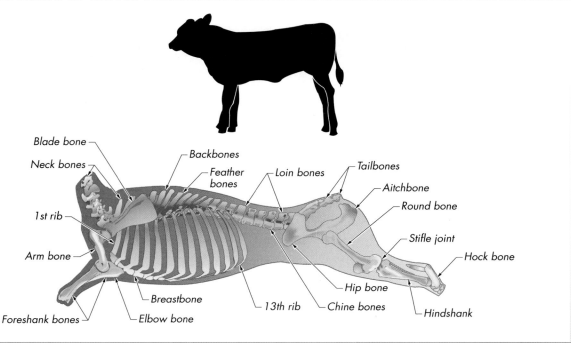

Figure 18-17. *Understanding the skeletal structure of calves can aid in identifying the market forms of veal.*

Primal and Fabricated Cuts of Veal

The left and right primal cuts of veal remain joined together and are sold as a single cut. For example, the leg primal cut has two joined legs. The primal cuts of veal include the shoulder, rack, loin, leg, and foreshank and breast. *See Figure 18-18.*

Each primal cut is further divided into fabricated cuts. Fabricated cuts of veal include frenched chops, baby T-bone steaks, boneless cutlets, and ossobuco-cut shanks. Some foodservice operations may choose to purchase only fabricated cuts of veal. However, the price per pound is higher for fabricated cuts than the price per pound for primal cuts.

Shoulders. A *veal shoulder* is a primal cut that contains the first four rib bones, some of the backbone, and a small amount of the arm and blade bones. The average weight of a veal shoulder is approximately 21% of the total carcass weight. Veal shoulder is tough but flavorful. It can be fabricated into steaks or chops. However, it is most often ground, cut into cubes for stewing, or cooked whole. *See Figure 18-19.*

Racks. The *veal rack* is located between the shoulder and loin and contains seven rib bones. Its average weight is approximately 9% of the total carcass weight. The meat is tender and well marbled. A veal rib is different from a beef rib in that veal is not split into two halves along the backbone (chine bones). An unseparated veal rack is called a hotel rack and consists of two very tender veal rib loins. A veal rack can be split into halves and tied into a circle to form a crown rib roast.

Primal Cuts of Veal

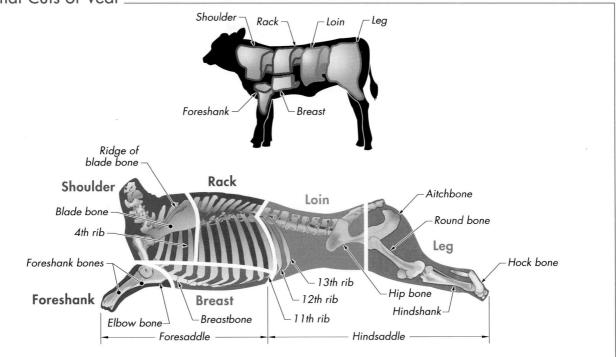

Figure 18-18. *The primal cuts of veal include the shoulder, rack, loin, leg, and foreshank and breast.*

Shoulder Cuts of Veal

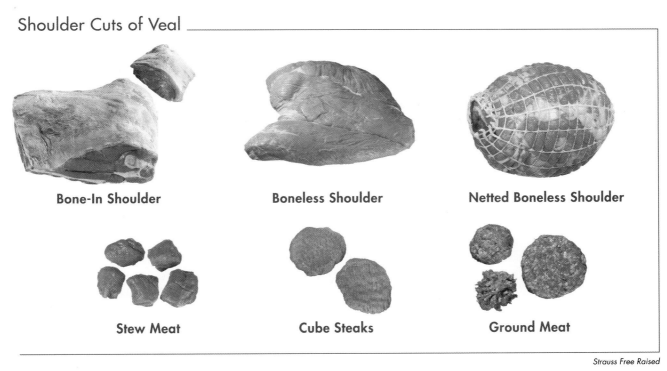

Bone-In Shoulder

Boneless Shoulder

Netted Boneless Shoulder

Stew Meat

Cube Steaks

Ground Meat

Strauss Free Raised

Figure 18-19. *A veal shoulder can be fabricated into steaks or chops. However, it is most often ground, cut into cubes for stewing, or cooked whole.*

Veal racks can be trimmed, frenched, and cut into veal chops. *See Figure 18-20.* Other fabricated cuts from the rack include a small portion of the tenderloin, known as the short tenderloin, and the boneless veal rib eye roast.

Rack Cuts of Veal

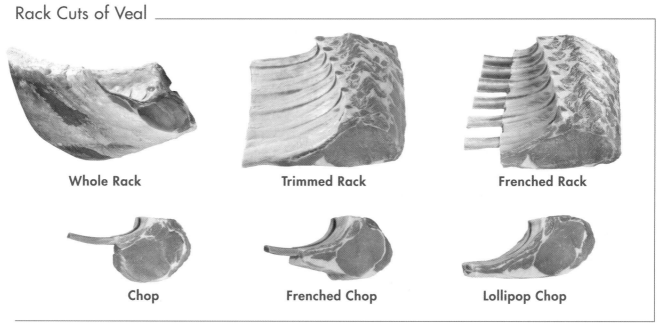

Whole Rack

Trimmed Rack

Frenched Rack

Chop

Frenched Chop

Lollipop Chop

Strauss Free Raised

Figure 18-20. *Veal racks can be trimmed, frenched, and cut into veal chops.*

Loins. A *veal loin* is a primal cut located between the primal rack and leg and includes the 12th and 13th rib, the loin eye muscle, the center section of the tenderloin, the strip loin, and flank meat. The average weight of the loin is approximately 10% of the total carcass weight. A complete, unsplit primal loin from a veal carcass is commonly referred to as a saddle. Veal loins are sometimes roasted whole, however they are often divided into fabricated cuts such as baby T-bones and cutlets. *See Figure 18-21.*

A *baby T-bone* is a 6–8 oz steak cut from the loin of veal. It contains loin meat on one side and tenderloin on the other side. Cuts from the loin are usually grilled, broiled, or roasted. A *veal cutlet* is a thin slice of veal. Wiener schnitzels are prepared by pounding veal cutlets until very thin and then breading and frying them. Wiener schnitzels are commonly served with lemon, topped with a variety of vegetables, or topped with woodland cream sauce.

Legs. A veal leg is a primal cut from the hind leg that contains the leg, sirloin, last portion of the backbone, pelvis (hip bone and aitchbone), round bone, hindshank, and tailbone. Its average weight is approximately 42% of the total carcass weight. The leg is the most versatile cut of veal because it contains solid, lean, fine-textured meat. Tender meat is located near the sirloin end, and tougher meat is located toward the shank. The entire leg is typically boned and cut into scallops or cutlets rather than being roasted whole. The leg is boned by following the muscle structure of the meat so that pieces of equal tenderness are removed.

Loin Cuts of Veal

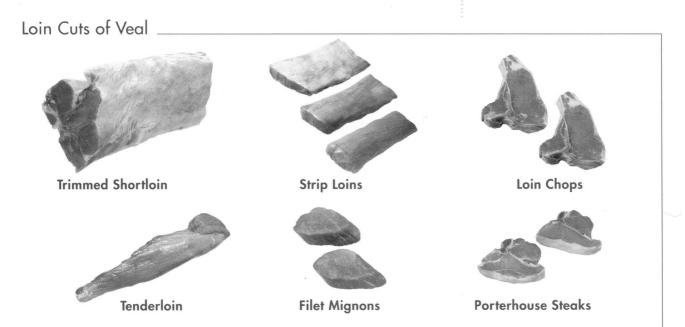

Trimmed Shortloin

Strip Loins

Loin Chops

Tenderloin

Filet Mignons

Porterhouse Steaks

Strauss Free Raised

Figure 18-21. *Veal loins are sometimes roasted whole, however they are often divided into fabricated cuts.*

Veal legs can be divided into leg, hindshank, ossobuco, inside round, eye round, and scallopini cuts. ***See Figure 18-22.*** These cuts are commonly sliced against the grain and pounded until thin to tenderize them. Ossobuco-cut shanks are cut perpendicular to the bone. A *scallopini* is a small, ¼ inch thick slice of veal (generally leg meat) that is 2–3 inches in diameter.

Leg Cuts of Veal

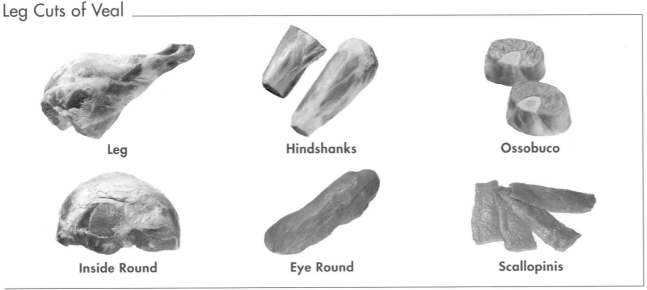

Leg

Hindshanks

Ossobuco

Inside Round

Eye Round

Scallopinis

Strauss Free Raised

Figure 18-22. *Veal legs can be divided into leg, hindshank, ossobuco, inside round, eye round, and scallopini cuts.*

Foreshanks and Breasts. The foreshank and breast form the primal cut of veal from the lower foresaddle. *See Figure 18-23.* Its average weight is approximately 16% of the total carcass weight. A *veal foreshank* is the upper portion of the front leg of a calf. It can be braised whole or sliced in cross sections across the bone. A *veal breast* is a thin, flat cut of meat located under the shoulder and ribs and contains the breastbone, tips of the rib bones, and cartilage. The breastbone is typically still cartilage because the animal is so young. The thin, flat shape makes the breast easy to stuff, roll, and tie into a tender rolled roast that can be braised to break down the connective tissue.

Foreshank and Breast Cuts of Veal

Whole Foreshank   **Ossobuco-Cut Shanks**   **Breast**

Figure 18-23. *The foreshank and breast form the primal cut of veal from the lower foresaddle.*

Veal Sweetbreads

Photo Courtesy of D'Artagnan, Photography by Doug Adams Studio

Figure 18-24. *Sweetbreads are often sautéed in brown butter or pan-fried.*

Veal Offals

Veal offals are more tender than beef offals. Common veal offals include sweetbreads, liver, and kidney. *Sweetbreads* are the thymus glands of a calf, located in the neck. High-quality sweetbreads should be plump and somewhat firm with a thin protective membrane. Sweetbreads are often sautéed in brown butter or pan-fried. *See Figure 18-24.*

Veal liver is commonly broiled, pan-fried, or sautéed and should be cooked medium. Veal kidneys can be broiled with excellent results, unlike beef kidneys, which must be cooked with moist heat. Veal kidneys are typically used to prepare entrées such as kidney stew and kidney pie.

Institutional Meat Purchase Specifications for Veal

To help ensure that foodservice operations and suppliers communicate efficiently, the USDA publishes the Institutional Meat Purchase Specifications (IMPS) for commonly purchased meats and meat products. Cuts of veal are numbered by category. *See Figure 18-25.*

USDA Institutional Meat Purchase Specifications (IMPS) for Veal				
Series	**Item No.**	**Product**	**Item No.**	**Product**
300 (veal and calf)	300	Carcass	337	Hindshank
	303	Side	338, 339	Trimmings
	304	Foresaddle, 11 ribs	341	Back, 9 ribs, trimmed
	304A	Forequarter, 11 ribs	342	Back, strip, boneless
	306	Hotel rack, 7 ribs	344	Loin, strip loin, boneless
	307	Rack, rib eye, 7 ribs	346	Leg, butt tenderloin, defatted
	308–311	Chuck	347	Loin, short tenderloin
	312	Foreshank	349	Leg, top round, cap on
	313, 314	Breast	389	Mixed bones
	323	Veal short ribs	390	Marrow bones
	330	Hindsaddle, 2 ribs	391	Marrow
	331, 332	Loins	395	Veal for stewing
	334–336	Legs	396, 397	Ground veal
700 (variety meats)	704, 705	Calf liver	715	Veal sweetbreads
	707, 708	Veal liver		
800 (sausage)	800	Frankfurters	813	Polish sausage
	801	Bologna	814	Meat loaves
	803	Liver sausage	815	Meat food product loaves
	804	Cooked salami	816	Knockwurst
	805	Minced luncheon meat	817	Breakfast sausage, cooked
	807	Thuringer	818	Italian sausage
	808	Dry salami	821	Pepperoni
	810	Breakfast sausage	822	Bratwurst
	811	Smoked sausage	826	Scrapple
1300 (veal and calf portion cuts)	1300, 1301	Cubed steak	1336	Cutlets
	1302	Veal slices	1337	Ossobuco, hindshank
	1306	Rack, rib chops	1338	Veal steak, flaked and formed, frozen
	1309	Chuck, shoulder arm chops	1349	Leg, top round, cap off, cutlets
	1312	Ossobuco, foreshank	1396, 1397	Ground veal patties
	1332	Loin chops		

Figure 18-25. *Cuts of veal are numbered by category in the Institutional Meat Purchase Specifications (IMPS).*

INSPECTION AND GRADES OF BEEF AND VEAL

All beef and veal used in foodservice operations must be procured from USDA-inspected plants. At the time of slaughter, the carcasses are stamped to indicate that the animal was slaughtered at a USDA-inspected plant. The USDA inspection stamp will be directly on large cuts or on the packaging of fabricated cuts of beef and veal. *See Figure 18-26.* The number on the stamp identifies the plant where the animal was processed. It does not indicate quality.

Unlike inspection, USDA quality and yield grading is optional for beef and veal producers. Beef can be graded for quality, yield, or both. Veal can only be graded for quality, but the grade of veal is less important than beef because calves are young animals that have tender meat. Quality and yield grading stamps are stamped on carcasses in the same manner as inspection stamps.

Cuisine Note

Beef labeled with a trademark or trade name, such as "Angus" or "Wagyu," indicates a packer's grade. These brands are sometimes placed on a product even if the USDA has graded the meat.

**Stamp for
Whole Carcass**

**Stamp for Fabricated
or Processed Meats**

United States Department of Agriculture

Figure 18-26. *The USDA inspection stamp will be directly on large cuts or on the packaging of fabricated cuts of beef and veal.*

Quality Grading

USDA quality grading of beef and veal is voluntary. Although the grading for beef and veal is based on specific details, the overall terms have relatively the same criteria. However, the quality grade is not a guarantee of quality. The grades of beef most commonly used in foodservice operations are USDA Prime, USDA Choice, and USDA Select. *See Figure 18-27.*

USDA Quality Grade Stamps

Figure 18-27. *The grades of beef most commonly used in foodservice operations are USDA Prime, USDA Choice, and USDA Select.*

USDA Prime. USDA Prime beef is well marbled and has a thick, firm fat covering. It is the juiciest and most flavorful of all meats, but it is also the most expensive because of the large amount of fat that must be trimmed from it before cooking. Of all the beef marketed in the United States, only a small percentage is graded USDA Prime.

USDA Choice. USDA Choice has slightly less marbling and fat covering than USDA Prime. Most foodservice operations prefer this grade because there is less waste.

USDA Select. USDA Select beef has minimal marbling and a soft fat covering. This grade of beef is often used in quick service operations.

Yield Grading

Beef and lamb are the only animals graded for yield. Yield grades are numbered 1 to 5 and indicate how much usable meat can be obtained from a carcass. A grade of 1 indicates the highest yield of meat, and a grade of 5 indicates the lowest yield of meat. Most high-quality beef will have a yield grade of 3 or higher. *See Figure 18-28.*

RECEIVING AND STORING BEEF AND VEAL

Beef and veal are potentially hazardous foods that must be checked for color, odor, texture, and temperature upon receipt. *See Figure 18-29.* Beef should be red in color with white fat. Veal should be pink in color with white fat. There should be no odor, and the texture should be firm and not dry or slick. Refrigerated beef and veal should maintain an internal temperature of 41°F or below. Beef or veal that is in the temperature danger zone should be rejected.

Receiving Beef and Veal

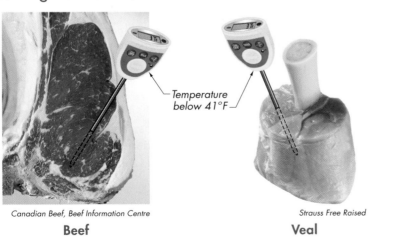

Canadian Beef, Beef Information Centre
Beef

Strauss Free Raised
Veal

— Temperature below 41°F

Figure 18-29. *Beef and veal are potentially hazardous foods that must be checked for color, odor, texture, and temperature upon receipt.*

USDA Yield Grades

Figure 18-28. *Most high-quality beef will have a yield grade of 3 or higher.*

Frozen beef and veal should be kept at temperatures that will allow the meat to remain frozen. When frozen meats need to be thawed, they should be placed in the refrigerator overnight. Larger cuts of meat may take more than one day to thaw under refrigeration.

Vacuum-packed beef and veal should never be opened until needed for service or preparation and can be stored refrigerated for three to four weeks. *See Figure 18-30.* Once the vacuum seal is broken, meat has a shelf life of only two to three days. Cut meats should be rewrapped airtight, refrigerated immediately, and used as soon as possible.

Vacuum-Packed Beef

Edlund Co.

Figure 18-30. *Vacuum-packed beef and veal should never be opened until needed for service or preparation and can be stored refrigerated for three to four weeks.*

FABRICATION OF BEEF AND VEAL

The size of storage facilities and availability of staff with the fabrication skills often determine whether a foodservice operation purchases primal cuts, fabricated cuts, or a combination. Beef and veal fabrication require trimming, cutting, tenderizing, and tying.

Trimming and Cutting Beef Tenderloin

The tenderloin is often trimmed and then cut into portion-controlled cuts. These cuts include tenderloin tips, chateaubriand, filets mignons, and tournedos. *See Figure 18-31.*

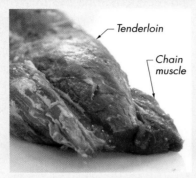

1. With a rigid boning knife, carefully remove the chain muscle from the side of the tenderloin and reserve.

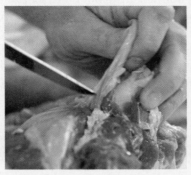

2. Trim and pull the thick fat covering away from the tenderloin.

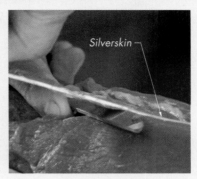

3. Insert the tip of the boning knife just beneath the silverskin at the tail end of the tenderloin. Draw the blade slightly upward along the length of the tenderloin, just beneath the silverskin, toward the head of the tenderloin.

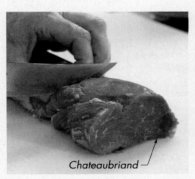

4. Starting at the largest end, cut off the uneven tip of the tenderloin. Cut the tip across the grain into tenderloin tips.

5. Make a cut across the grain just after the large portion ends to remove the chateaubriand.

6. Cut the center of the tenderloin across the grain to desired thickness to produce filet mignons.

7. Cut the smallest third of the tenderloin across the grain to produce tournedos ½–¾ inch thick and approximately 2½ inches in diameter.

Fabricated beef tenderloin cuts.

Figure 18-31. *The tenderloin is often trimmed and then cut into portion-controlled cuts such as tenderloin tips, chateaubriand, filets mignons, and tournedos.*

Refer to DVD for
Trimming Beef Tenderloin
Media Clip

Cutting Boneless Strip Loin into Steaks

A boneless strip loin is often trimmed and cut into portion-controlled cuts. *See Figure 18-32.* The sirloin end of the strip loin contains connective tissue that does not break down during cooking. Steaks cut from this end of the loin are referred to as vein steaks. Although the meat is as tender as the rest of the loin, the amount of connective tissue makes vein steaks less desirable than steaks from the rest of the strip loin.

Refer to DVD for
Cutting Boneless Strip Loin
Media Clip

Procedure for Cutting Boneless Strip Loin into Steaks

1. Trim the surface fat to approximately ¼ inch thickness.
2. Turn the loin over and trim off any additional fat or connective tissue that could affect the quality of the steaks.

3. Cut the loin across the grain into steaks of desired weight or thickness.

Figure 18-32. *A boneless strip loin is often trimmed and cut into portion-controlled cuts.*

Frenching Veal Chops

Veal chops are a popular menu item and are available boneless or with the bone in. Bone-in veal chops are commonly frenched for service. *See Figure 18-33.*

Procedure for Frenching Veal Chops

1. Peel back the fat cap where it still is connected to the rack and remove it completely.
2. Trim the entire rack of any excess fat, leaving a thin covering between ¼–⅛ inch thick.
3. Starting on the top side, score the meat by making a straight cut all the way to the bone, approximately 1 inch above the eye meat.
4. Turn the rack over and use the tip of the knife to score the thin membrane on the back of each rib bone from the location of the cut to the tip of the bone.
5. From the scored mark, scrape the meat away from the top of the ribs to expose the bones.
6. Cut between each rib down to the scored mark to remove the meat.

Figure 18-33. *Bone-in veal chops are commonly frenched for service.*

Tenderizing Beef

Beef often benefits from tenderizing. Pounding or cutting meat tenderizes the muscle and connective tissue. *See Figure 18-34.* The following methods are used to tenderize meat:

- pounding meat with a mallet to break up the protein structure and muscle tissue
- using a hand tenderizer with needlelike knives that pierce and gently cut the connective tissues and muscle fibers
- slicing the meat across the grain to produce thin slices with shorter muscle fibers
- grinding the meat to completely break apart strands of connective tissue and muscle tissue

Tenderizing Meat

Pounding with a Mallet

Using a Hand Tenderizer

Slicing across the Grain

Figure 18-34. *Pounding or cutting meat tenderizes the muscle and connective tissue.*

Ground beef and veal can be used to make meatballs, sausages, meatloafs, and meat fillings. Meat that is to be ground should be very cold before grinding, as should all parts of the grinder that will contact the meat. Room temperature meat and meat that is only somewhat cool has a tendency to be smashed while grinding, which causes it to lose its texture and appear puréed instead of ground. When grinding meat, it should be passed through progressively smaller dies on the grinder in order to evenly distribute the fat and give the mixture a smooth appearance. *See Figure 18-35.*

Grinding Meats

Figure 18-35. *When grinding meat, it should be passed through progressively smaller dies on the grinder in order to distribute the fat evenly and give the mixture a smooth appearance.*

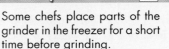

Chef's Tip

Some chefs place parts of the grinder in the freezer for a short time before grinding.

Tied Roasts

National Cattlemen's Beef Association

Figure 18-36. *Some cuts of meat are rolled to form roasts and tied so that they do not unroll during cooking.*

Wet Aging

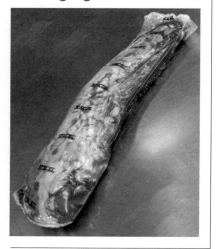

Figure 18-37. *Beef is allowed to wet age from one to six weeks in vacuum packaging under normal refrigeration.*

Tying Beef and Veal

Tying beef and veal helps maintain a consistent shape and ensures that the meat will cook evenly without falling apart. Some cuts of meat are rolled to form roasts and tied so that they do not unroll during cooking. *See Figure 18-36.* Stuffed veal breast is commonly tied to retain its shape and hold the stuffing.

FLAVOR ENHANCERS FOR BEEF AND VEAL

Certain cuts of beef and veal may require or benefit from flavor enhancers prior to cooking. Common flavor enhancers for beef and veal include aging, barding, larding, curing, and the use of rubs and marinades.

Aging

Aging is the period of rest that occurs for a length of time after an animal has been slaughtered. Immediately after slaughter, the carcass is very limp. However, within 5–24 hours the carcass goes through a period known as rigor mortis where natural enzymes in the meat cause the tissues to seize and become stiff. Rigor mortis causes the limpness of the carcass to disappear and the joints and muscles to become stiff. This stiffening stage lasts for two to three days under normal refrigeration.

Meat that is experiencing rigor mortis is commonly called "green meat" because it is freshly slaughtered. Eating green meat is not recommended because the meat is extremely tough and almost flavorless. If meat is frozen immediately after slaughter, the process is paused and rigor mortis sets in once the meat is thawed. Natural enzymes in the tissue of the meat begin to relax and tenderize the muscles two to three days after slaughter. The meat is then ready to be fabricated, frozen, or cooked.

Veal is typically not aged any longer than the initial two to three days for rigor mortis to leave the meat. Beef, however, benefits from longer aging periods as the enzymes in the meat continue to break down and tenderize the tissues, making it more flavorful. Beef can be wet or dry aged for an extended period.

Wet Aging. *Wet aging* is the process of aging meat in vacuum-sealed plastic. Beef carcasses are typically broken down into manageable cuts and placed in vacuum-sealed plastic packaging. Beef is allowed to wet age from one to six weeks in vacuum packaging under normal refrigeration. *See Figure 18-37.* During this period, natural enzymes further break down and tenderize the meat, intensifying its natural flavor.

Dry Aging. *Dry aging* is the process of aging larger cuts of meat that are hung in a well-controlled environment. During dry aging temperature, humidity, and airflow are monitored around the clock for up to six weeks. Over this time, natural enzymes break down the tissue of the meat, making it more flavorful and tender.

858 Culinary Arts · PRINCIPLES AND APPLICATIONS

It is important that the meat is not wrapped and does not touch other pieces of meat during the dry aging process because harmful bacteria could form. The surface of the meat dries out like leather during the dry aging process and usually develops a layer of harmless mold. The dry surface and the mold are trimmed off and disposed of prior to cooking. Dry aging commonly results in 5–20% weight loss through the evaporation of moisture and the eventual trimming of inedible surface material. Dry-aged meats are typically found only at premium butchers, high-end distributors, or elite steak houses.

Barding

Barding is the process of laying a piece of fatback across the surface of a lean cut of meat to add moisture and flavor. ***See Figure 18-38.*** The fatback is tied to the meat to prevent it from falling off during the roasting process. As the roast cooks, the fat is rendered from the fatback and absorbed into the meat. While the fatback prevents the meat from drying out, it also prevents the meat from browning. Occasionally, the fatback is removed for the last 10 minutes of roasting to allow the item to gain color.

Barding

Figure 18-38. *Barding is the process of laying a piece of fatback across the surface of a lean cut of meat to add moisture and flavor.*

Larding

Beef can be larded. *Larding* is the process of inserting thin strips of fatback into lean meat with a larding needle. ***See Figure 18-39.*** The fatback is cut into thin strips and drawn through the beef to increase the juiciness when the meat is roasted. Larded meats are commonly braised to allow ample time for the fat to be rendered and absorbed by the meat.

Larding Needles

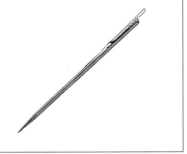

Paderno World Cuisine

Figure 18-39. *Larding is the process of inserting thin strips of fatback into lean meat with a larding needle.*

Curing

Beef is more commonly cured than veal. *Curing* is the process of using salt and sodium nitrite alone or with flavorings or sugar to preserve a food item. Summer sausage, beef jerky, and bresaola are common cured-beef products. *See Figure 18-40. Bresaola,* also known as brisaola, is air-dried cured beef that is aged two to three months until it becomes hard and turns almost purple in color. Bresaola is made from the top round and is lean and tender with a musty smell.

Rubs and Marinades

Rubs and marinades can be used to add intense flavor to beef and veal. Both dry and wet rubs can be used. Dry rubs are made by grinding herbs and spices together into a fine powder and then rubbing the mixture into the meat prior to cooking. A wet rub is made by incorporating wet ingredients, such as Dijon mustard, flavored oils, puréed garlic, or honey, into a dry rub mixture and then rubbing it into the meat prior to cooking.

Cured Beef Products

Summer Sausage   **Beef Jerky**

Figure 18-40. *Summer sausage and beef jerky are common cured-beef products.*

Photo Courtesy of the Beef Checkoff

Marinades add flavor to meat and tenderize it at the same time. Marinades have an acidic liquid base, such as wine or lemon juice. The acidic base tenderizes the meat by breaking down the protein structure of the meat. Any herb, spice, or condiment can be added to the base to create the desired flavor. If wine is used, a red wine should be used for beef and a white wine for veal. The length of time that meat should be left in the marinade is determined by the size of the piece of meat. Once the meat has been removed from the marinade, the remaining liquid must be discarded to prevent cross-contamination.

COOKING BEEF AND VEAL

Tender cuts of beef and veal can be cooked using dry-heat cooking methods. Tougher cuts of beef are cooked slowly using moist-heat cooking methods. The majority of animal muscle is made up of water. As meat loses water, it shrinks. *Shrinkage* is the loss of volume and weight of a piece of food as the food cooks. Shrinkage is the reason why a 20 lb roast may only be 18.5 lb when fully cooked. ***See Figure 18-41.***

> **Chef's Tip**
>
> Large roasts need to be cooked at lower temperatures to reduce shrinkage.

Shrinkage

Raw Rib Roast

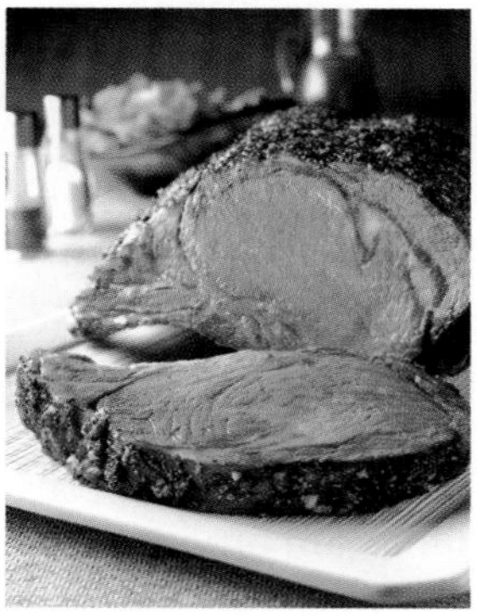

Photo Courtesy of the Beef Checkoff

Cooked Rib Roast

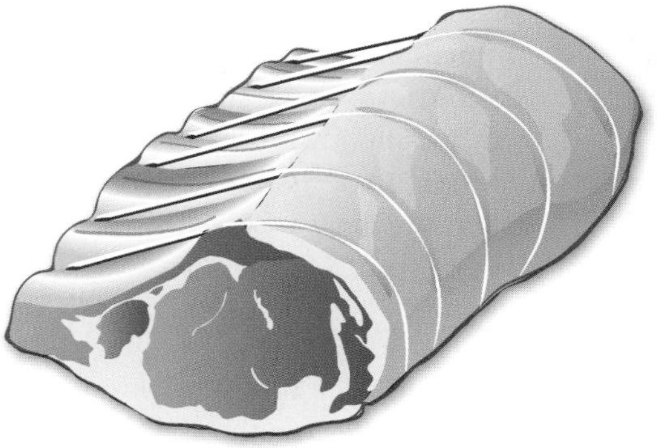

Figure 18-41. *Shrinkage is the reason why a 20 lb roast may only be 18.5 lb when fully cooked.*

Cooking meat at too high a temperature can toughen the protein. However, grilling and frying use very high temperatures for short periods and result in only the exterior of an item receiving high amounts of heat. High heat quickly cooks the exterior of meat to a crispy texture while slowly cooking the interior of the meat. This is the reason that a grilled steak is crispy and somewhat dry on the outside yet remains tender and juicy on the inside.

Determining Doneness

When cooking meat to the desired degree of doneness, the type of meat, the thickness of the meat, the temperature of the meat when it begins to cook, and the intensity of the heat are factors that must be considered. Because of these variables, time is not an accurate way to determine the doneness of meats.

The most accurate way to determine the degree of doneness of meat, with the exception of braised or stewed meats, is by measuring the internal temperature of the meat. To do this, a probe type instant-read thermometer is inserted in the thickest part of the meat. Beef and veal cuts that are done have been cooked to an internal temperature of 145°F for at least 15 seconds and rested for 3 minutes. Ground beef and veal must be cooked to an internal temperature of 160°F.

Beef and veal may be served to varying degrees of doneness as requested. *See Figure 18-42.* For example, steaks may be requested very rare, rare, medium-rare, medium, medium-well, or well-done. Braised or stewed beef is cooked until it is fork tender, or until the meat can be easily separated with a fork.

Determining Degrees of Doneness

Degree of Doneness	Internal Temperature
Very rare	130°F
Rare	140°F
Medium-rare	145°F
Medium	160°F
Medium-well	165°F
Well-done	170°F

Eloma Combi Ovens

Figure 18-42. *Beef and veal may be served to varying degrees of doneness as requested.*

Smaller cuts that are grilled or broiled can be tested for doneness with the touch method, which uses the sense of touch to check the firmness of cooked meat. For example, a steak or veal chop that is cooked well-done is firm and springs back immediately when gently pressed. In contrast, a rare steak or chop is soft and slightly mushy to the touch. The texture feels almost the same as when the meat was raw.

Grilled Beef

National Cattlemen's Beef Association

Figure 18-43. *Beef and veal cuts that are grilled or broiled are commonly taken from the rib and the loin.*

Cuisine Note

Kalbi is the name of Korean barbequed short ribs. These barbequed ribs are served sliced and rolled in romaine lettuce leaves alongside white rice, lightly browned slices of garlic, and Korean red pepper paste.

Grilling and Broiling

Grilling and broiling use a hot flame to sear and cook foods quickly. Only tender cuts of beef or veal should be cooked with these methods. Well-done beef cooked using these methods will be somewhat dry, because most of the moisture cooks out as the meat cooks. Removing the fat after cooking can help the meat retain flavor and moisture. Beef and veal cuts that are grilled or broiled are commonly taken from the rib and the loin. *See Figure 18-43.*

Smoking and Barbequing

Some forms of barbeque are similar to smoking, while other forms are similar to grilling. Smoking flavors and cooks a product using the heat and smoke from burning wood. For example, beef brisket is commonly smoked. *See Figure 18-44.* Beef ribs and veal chops are often barbequed.

Smoked Beef Brisket

Photo Courtesy of the Beef Checkoff

Figure 18-44. *Beef brisket is commonly smoked.*

Roasting

Many cuts of beef and veal from the rib, loin, and leg are roasted. Smaller roasts should be cooked at higher temperatures, between 400°F and 450°F, to allow them to caramelize on the exterior without overcooking the interior. Larger roasts require a longer cooking time and should be roasted at lower temperatures, between 275°F and 325°F, to prevent excessive shrinkage. It is important to allow for carryover cooking when roasting meats. Lean cuts of meat will become dry when roasted unless some form of fat is added. Barding or larding can be done to add additional fat.

Some cuts of beef are roasted whole and then sliced to order. Cuts from the inside round are relatively easy to slice against the grain. Slicing against the grain produces a cut of meat that is more tender than a cut sliced in the direction of the grain. Some cuts of meat contain bones that must be removed prior to slicing or that require the meat to be sliced off the bone. Prime rib and steamship round roast are commonly sliced for service.

A bone-in prime rib is a tender and juicy cut of meat that is often carved before service. *See Figure 18-45.* Slices may be carved with or without the bone.

Photo Courtesy of the Beef Checkoff

Procedure for Carving a Bone-In Prime Rib

1. Use a slicer to remove the end cut.

2. Continue to cut the prime rib into thick slices. The second slice should include a bone. The third slice should not include a bone, and so forth. *Note:* Cutting each slice with a bone offers larger portions.

Figure 18-45. *A bone-in prime rib is a tender and juicy cut of meat that is often carved before service.*

A bone-in steamship round roast can be difficult to carve because large bones are present in the center of the roasted meat. *See Figure 18-45.* A steamship round roast is typically prepared for large buffets and special events.

Procedure for Carving Steamship Round Roasts

1. Place the cooked roast on a cutting board or carving station with the large hip joint end facing downward and the long exposed leg bone facing upward. *Note:* Use a ball of foil to stabilize the bottom of the roast if necessary.
2. Hold the shank bone firmly in place. Trim the excess fat from the exterior of the roast to expose the meat.
3. Using a slicer, make a vertical cut along the bone, approximately 1 inch from the end of the shank meat.
4. Use long, smooth stokes to slice horizontally toward the bone. As slices become larger near the sirloin end of the leg, slightly angle the slicer to create a smaller surface.
5. Rotate the leg to remove slices from the other sides of the roast.

Figure 18-46. *A bone-in steamship round roast can be difficult to carve because large bones are present in the center of the roasted meat.*

Sautéing and Frying

Tender cuts of beef and veal are typically sautéed or fried. Sautéing uses a small amount of hot fat to sear and cook meat. Only uniformly sized cuts should be sautéed to ensure even cooking. Care should be taken when sautéing not to burn the oil or the surface of the food by using too high of a temperature. A pan sauce can be made in the same pan after the meat has been removed.

Tender cuts of beef and veal can be breaded then pan-fried or deep-fried. Meats to be pan-fried should not be so thick as to prevent them from fully cooking. Veal is typically cut into cutlets or scallops and pan-fried. Veal chops are pan-fried and typically finished in the oven. ***See Figure 18-47.*** Breading protects the meat from the hot oil and helps the meat retain moisture.

Pan-Fried Veal Chops

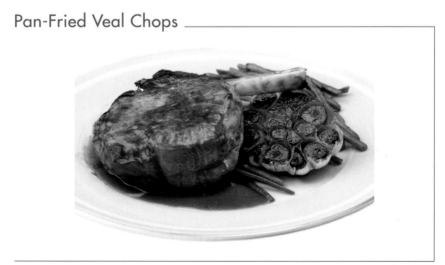

Daniel NYC

Figure 18-47. *Veal chops are pan-fried and typically finished in the oven.*

Braising and Stewing

Braising and stewing are both combination cooking methods that can be used to prepare tough or tender cuts of beef and veal. Braising is typically used for larger cuts of beef, such as the chuck and shank, that have a good amount of fat. Searing the meat adds flavor to the beef and the resulting sauce that can be made by thickening the cooking liquid. The meat is then slow cooked until tender. The braising liquid may be thickened slightly and served as an accompanying sauce. Tender cuts of veal, such as veal chops, may also be braised. The only difference between braising tender cuts and tougher cuts is the amount of time the meat is cooked.

Stewing bite-sized pieces of meat requires that most of the fat be removed prior to cooking. The meat is completely covered with liquid and cooked slowly. Both brown and white stews can be made.

Veal is used to make white stews such as a fricassee and a blanquette. To make a fricassee, the meat is seared in a small amount of hot fat but not allowed to brown. The cooking liquid is then added and the meat is cooked until tender. To make a blanquette, the meat is blanched in simmering water, rinsed to remove any impurities, and then added to the cooking liquid. White stews should have an ivory color when finished. *See Figure 18-48.*

White Stews

Paderno World Cuisine

Figure 18-48. *White stews should have an ivory color when finished.*

Simmering

Cuts of beef and veal that benefit from extended cooking times may be simmered. Simmering makes meat more tender by breaking down connective tissue. Simmering uses longer cooking times and low heat to cook tougher cuts of meat until they are tender. For example, the simmering method is used for beef brisket. Simmered meats should be cooked in liquid that is between 185°F and 205°F until the meat is fork tender.

BISON

Bison is a large animal similar to cattle that is over 6 feet in height and 10 feet in length and can weigh over 2000 lb. Most bison are raised free range and grass fed, but some are fed grain. Grass-fed bison meat contains greater amounts of unsaturated fat and lower amounts of saturated fat than beef. Bison is extremely lean with very little marbling. *See Figure 18-49.* It is higher in iron and protein than beef. Bison is also lower in cholesterol than skinless chicken.

Bison Cuts

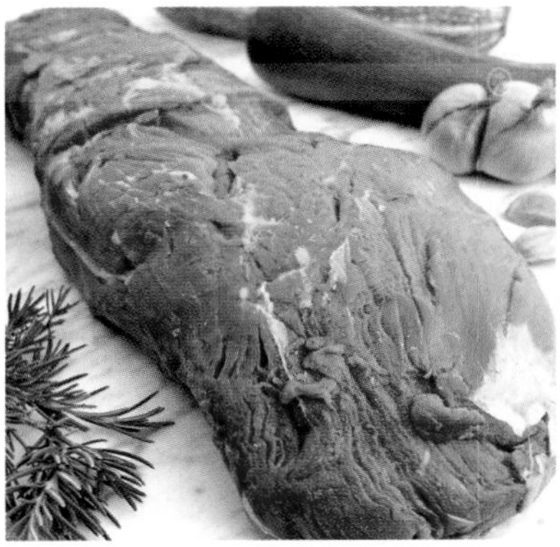

Tenderloin

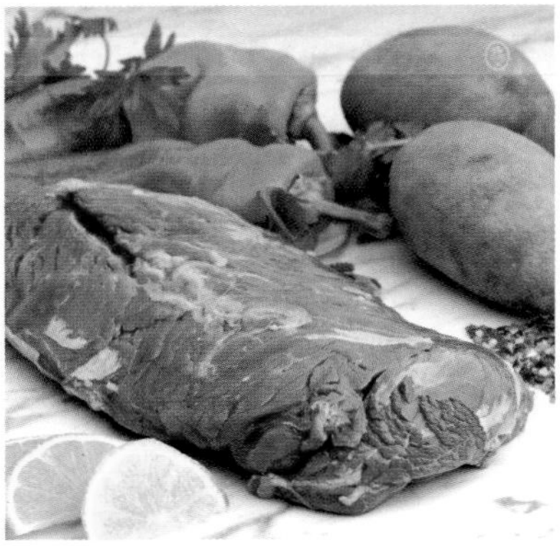

Hanger Steak

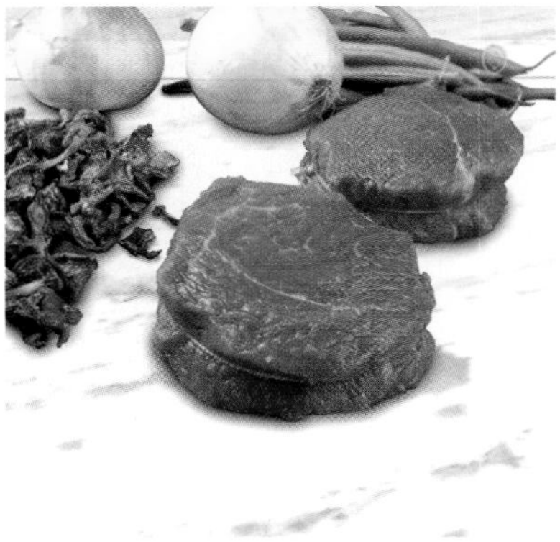

Medallions

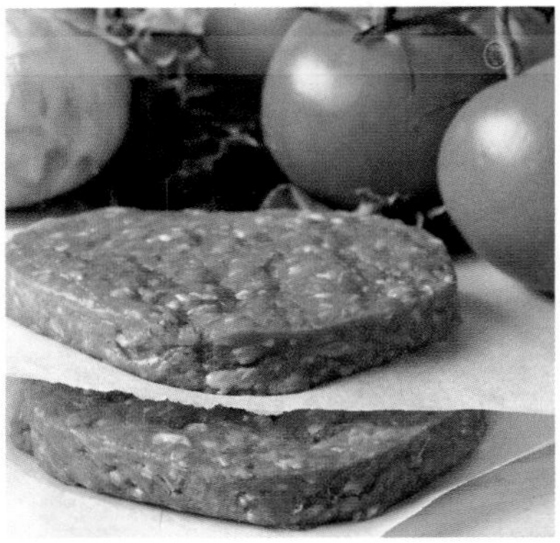

Ground Patties

Photos Courtesy of D'Artagnan, Photography by Doug Adams Studio

Figure 18-49. *Bison is extremely lean with very little marbling.*

Bison is sold in market forms similar to those of beef and must be procured from USDA-inspected plants. However, bison is not graded for quality or yield. Bison is stored and prepared in a similar manner to beef. Bison meat is similar in flavor to beef, but much richer and sweeter. Because it has a similar flavor and texture to beef, bison can be used in any recipe that calls for beef. ***See Figure 18-50.*** However, bison has less marbling than beef and is more likely to be overcooked.

Bison Ribeye Steaks

Venison World

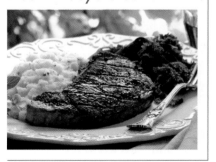

Figure 18-50. *Because bison has a similar flavor and texture to beef, it can be used in any recipe that calls for beef.*

SUMMARY

Beef is the flesh of domesticated cattle, most of which are steers. Most steers are grain fed, which increases the marbling of the meat. Market forms of beef include partial carcasses, primal cuts, fabricated cuts, and offals. Primal cuts of beef include the chuck, rib, short loin, sirloin, round, flank, short plate, and brisket and shank. Beef offals include liver, tongue, tripe, oxtail, and kidney.

Veal is the meat of calves, which are young cattle. Though both beef and veal are graded, beef grades are of greater importance than veal grades.

Market forms of veal include partial carcasses, primal cuts, fabricated cuts, and offals. Primal cuts of veal include the shoulder, rack, loin, leg, and foreshank and breast. Veal offals include sweetbreads, liver, and kidney.

Refrigerated beef and veal should be received and stored at temperatures that maintain the internal temperature of the meat at 41°F or below. Frozen meat should be kept at temperatures that will allow the meat to remain frozen. Prior to cooking, some cuts of beef are aged, barded, larded, cured, or covered with a rub or marinade to develop the flavor further. Beef can be cooked using almost any cooking method. Veal is commonly sautéed, roasted, or stewed.

Bison is extremely lean with very little marbling and is higher in iron and protein than beef. Grass-fed bison contains greater amounts of unsaturated fat and lower amounts of saturated fat than beef. It also contains less cholesterol than skinless chicken. Bison can be used in any recipe that calls for beef, but because bison has less marbling than beef it cooks faster and is more likely to be overcooked.

Refer to DVD for
Quick Quiz® questions

Review

1. Describe the composition of beef.
2. Differentiate between grain-fed and grass-fed animals.
3. Identify the eight primal cuts of beef.
4. Identify the cuts fabricated from each primal cut of beef.
5. Describe the advantages and disadvantages of purchasing fabricated cuts.
6. Explain how to prepare beef offals.
7. Explain the purpose of Institutional Meat Purchase Specifications.
8. Describe the composition of veal.
9. Identify the five primal cuts of veal.
10. Identify the cuts fabricated from each primal cut of veal.
11. Explain how to prepare veal offals.
12. Explain the significance of the USDA inspection stamp.
13. Describe the most common USDA quality grades of beef.
14. Explain why the quality grade of veal is less important than the quality grade of beef.
15. Identify the two types of animals that are yield graded.
16. Identify four traits to check upon receiving beef and veal.
17. Identify the temperature at which refrigerated beef and veal must be kept.

Refer to DVD for
Review Questions

18. Explain why vacuum-sealed packages should only be opened at the time of use.
19. Describe how to trim and cut a beef tenderloin.
20. Describe how to cut a boneless strip into steaks.
21. Describe how to french veal chops.
22. Describe four ways to tenderize beef.
23. Describe how to grind fresh meat.
24. Explain the purpose of tying meat.
25. Describe green meat and why it should not be eaten.
26. Contrast wet aging and dry aging.
27. Explain the process of barding.
28. Explain the process of larding.
29. Identify three commonly cured beef products.
30. Explain how rubs and marinades add flavor to beef and veal.
31. Define shrinkage.
32. Explain the safest way to determine the doneness of beef and veal.
33. Identify the different cooking methods used to cook beef.
34. Identify the different cooking methods used to cook veal.
35. Describe how bison is similar to and different from beef.

Applications

1. Trim and cut a beef tenderloin.
2. Cut a boneless strip into steaks.
3. French veal chops.
4. Tenderize identical cuts of beef four different ways.
5. Grind fresh beef.
6. Compare a piece of wet-aged beef with a piece of dry-aged beef. Evaluate the two pieces for flavor, tenderness, and moistness.
7. Bard a piece of beef or veal. Lard an identical piece of beef or veal. Compare the results.
8. Compare two pieces of cured beef for flavor, tenderness, and moistness.
9. Use a dry rub and a marinade to enhance the flavor of two identical cuts of beef or veal. Cook both cuts and compare the quality of the two prepared dishes.
10. Demonstrate three ways to determine the doneness of beef and veal.
11. Grill and broil a cut of beef or veal. Compare the quality of the two prepared dishes.
12. Roast and barbeque a cut of beef or veal. Compare the quality of the two prepared dishes.
13. Sauté and fry a cut of veal. Compare the quality of the two prepared dishes.
14. Braise and stew a cut of beef. Compare the quality of the two prepared dishes.
15. Prepare bison using two different cooking methods. Compare the quality of the two prepared dishes.

Refer to DVD for
Application Scoresheets

Grilled Asian Flank Steak

Yield: *6 servings, 8 oz steak and ¼ c of grilled peppers each*

Ingredients

dry red wine	1 c
dry sherry	4 tbsp
sesame oil	2 tbsp
soy sauce, reduced sodium	2 tbsp
rice wine vinegar	2 tbsp
ginger root, freshly grated	4 tsp
garlic, crushed	4 cloves
flank steak, well-trimmed	3 lb
bell peppers, cut into 2 inch slices	3 c

Preparation

1. Combine red wine, sherry, sesame oil, soy sauce, vinegar, ginger, and garlic. Reserve one-fourth of the marinade.
2. Place flank steak and the rest of the marinade in a resealable plastic bag and close securely. Turn the bag to coat the steak completely.
3. Place plastic bag in the refrigerator for 6–8 hours, turning occasionally.
4. Remove the steak from the marinade. Discard the marinade.
5. Place the steak and the pepper slices on a grill over medium heat. Grill the steak 12–15 minutes for medium-rare steak. Grill the peppers until they are tender. Brush steak with the reserved marinade before turning it. Turn the peppers once.
6. Cut the steak diagonally across the grain into thin slices.
7. Plate steak with peppers and serve immediately.

> **NUTRITION FACTS**
> **Per serving:** 361 calories, 123 calories from fat, 13.7 g total fat, 140.6 mg cholesterol, 172.3 mg sodium, 941 mg potassium, 5.5 g carbohydrates, 1.4 g fiber, 1.9 g sugar, 49.8 g protein

Roasted Standing Rib of Beef

Yield: *8 servings, 12 oz each*

Ingredients

standing rib of beef, tied	6 lb
salt and pepper	TT
mirepoix	
onions, rough cut	4 oz
celery, rough cut	2 oz
carrots, rough cut	2 oz
Sauce	
beef stock	1 c
Worcestershire sauce	2 tsp

Preparation

1. Place rib in a roasting pan, bone-side up. Season to taste.
2. Roast in 350°F oven for 1 hour or until the roast is evenly browned on the surface. Remove from oven.
3. Turn the roast rib-side down. Add the mirepoix to the roasting pan.
4. Roast at 350°F until vegetables turn light brown.
5. Reduce temperature to 325°F and continue to roast to desired doneness.
6. Remove the roast and allow to rest for 20 minutes before slicing.
7. Pour the drippings and vegetables in a large saucepot. Deglaze the roasting pan with stock and bring to a boil while scraping the fond from the bottom of the pan.
8. Add deglazed juices to the reserved drippings. Add Worcestershire sauce and simmer for 15 minutes.
9. Skim off fat and strain sauce through a china cap.
10. Remove the butcher's twine and feather bones from the roast.
11. Stand the roast up by placing the large end down. Carve the roast as desired.
12. Serve with 1½ fl oz of sauce per serving.

> **NUTRITION FACTS**
> **Per serving:** 832 calories, 622 calories from fat, 68 g total fat, 184.9 mg cholesterol, 256.1 mg sodium, 771.1 mg potassium, 2.8 g carbohydrates, <1 g fiber, 1.4 g sugar, 49.1 g protein

Italian Meatballs

Yield: *4 servings, 2 (3 oz) meatball each*

Ingredients

olive oil	1 tbsp
onions, small dice	4 oz
garlic, minced	1 tsp
bread, cubed	2 slices
milk	2 tbsp
ground beef	1 lb
Parmesan cheese, grated	1 tbsp
oregano	½ tsp
basil	TT
salt and pepper	TT
egg, beaten	1 ea
fresh parsley, chopped	2 tsp

Preparation

1. Heat olive oil in a skillet over medium heat. Sauté onions and garlic until tender. Do not allow the onions to brown.
2. In a mixing bowl, combine cubed bread and milk and mix well.
3. Add sautéed onions and garlic, ground beef, Parmesan cheese, oregano, basil, salt, and pepper to the bread mixture. Mix by hand until evenly combined.
4. Pass the mixture through a food grinder twice while using a medium-sized hole chopper plate. *Note:* If the mixture is too wet add more bread crumbs
5. Add egg and parsley and blend into the mixture thoroughly.
6. Form into balls 1½ inch in diameter and place on a greased sheet pan. *Note:* Rubbing hands with olive oil will prevent the meat from sticking to the hands.
7. Bake meatballs at 350°F until cooked through.

NUTRITION FACTS
Per serving: 348 calories, 204 calories from fat, 22.6 g total fat, 125.5 mg cholesterol, 253.2 mg sodium, 427 mg potassium, 9.9 g carbohydrates, <1 g fiber, 2.2 g sugar, 24.7 g protein

Herbed Veal Roast with Apricot-Thyme Chutney

Yield: *4 servings, 5 oz of veal and 2 oz of chutney each*

Ingredients

fresh sage, chopped	2 tbsp
garlic, crushed	1 clove
black pepper	½ tsp
veal roast	2 lb
Chutney	
vegetable oil	2 tbsp
medium onion, sliced	1 ea
dried apricots, coarsely chopped	3 oz
chicken broth	½ c
sugar	½ tbsp
cider vinegar	½ tsp
dried thyme, crushed	½ tsp

Preparation

1. Combine sage, garlic, and pepper to create a rub. Press the rub into the surface of the veal roast.
2. Place the roast rib-end down in a roasting pan. Insert a meat thermometer in the thickest part of the meat, not touching bones.
3. Roast for 45 minutes at 325°F or until thermometer registers 155°F. Allow the meat to rest for 15 minutes.
4. Heat oil in a large skillet over medium heat. Add onions and cook slowly for 15–20 minutes, stirring occasionally.
5. Add remaining chutney ingredients to the onions. Cover and simmer 20–25 minutes.
6. Carve into 5 oz slices and serve with 2 oz of chutney.

NUTRITION FACTS
Per serving: 387 calories, 123 calories from fat, 13.9 g total fat, 188.2 mg cholesterol, 207.1 mg sodium, 1049.8 mg potassium, 17.3 g carbohydrates, 2 g fiber, 13.6 g sugar, 47.4 g protein

Sautéed Beef Tenderloin Tips in Mushroom Sauce

Yield: *4 servings, 6 oz each*

Ingredients

vegetable oil	4 tbsp
salt and pepper	TT
beef tenderloin tips, sliced 1 inch thick on bias	1½ lb
onion, diced	3 oz
mushrooms, sliced thick	8 oz
Burgundy wine	3 fl oz
espagnole sauce	3 c
parsley	4 sprigs

Preparation

1. Heat half of the oil in a saucepan over medium heat. Season beef tips to taste and add tips to the pan. Cook until browned. Remove meat from heat and set aside.
2. Add the remaining oil to the pan. Sauté onions and mushrooms until onions are translucent and mushrooms are slightly browned.
3. Add beef tips back to the saucepan and deglaze with wine. Reduce the wine by half.
4. Add espagnole sauce and bring to a simmer until heated through, or about 15 minutes.
5. Garnish with parsley.

> **NUTRITION FACTS**
> **Per serving:** 760 calories, 488 calories from fat, 54.6 g total fat, 169.2 mg cholesterol, 717 mg sodium, 1387.9 mg potassium, 20.5 g carbohydrates, 2.4 g fiber, 5.6 g sugar, 42.8 g protein

Beef Stroganoff

Yield: *4 servings, 8 oz each*

Ingredients

wide egg noodles	12 oz
vegetable oil	4 tsp
beef round steak, cut into thin strips	1 lb
garlic, crushed	1 clove
mushrooms, sliced	6 oz
espagnole sauce	2 c
sour cream	¼ c
Worcestershire sauce	1 tsp
salt and pepper	TT

Preparation

1. Cook noodles according to package directions. Keep warm.
2. Heat three-fourths of vegetable oil in a large skillet over medium heat. Stir-fry beef and garlic for 1 minute.
3. Remove beef and garlic from skillet.
4. In the same skillet, heat the rest of the vegetable oil over medium heat. Sauté mushrooms for 2 minutes.
5. Stir in espagnole sauce and bring to a simmer.
6. Add sour cream and Worcestershire sauce and mix thoroughly.
7. Return beef to skillet and heat thoroughly. Season to taste.
8. Divide the noodles into equal portions and add stroganoff.

> **NUTRITION FACTS**
> **Per serving:** 672 calories, 195 calories from fat, 22.3 g total fat, 164.9 mg cholesterol, 556.9 mg sodium, 1222.2 mg potassium, 73.8 g carbohydrates, 4.2 g fiber, 5.5 g sugar, 43.6 g protein

Veal Piccata with a Lemon Caper Sauce

Yield: *4 servings, 2 cutlets and 3 oz of sauce each*

Ingredients

veal cutlets, 3 oz each	8 ea
all-purpose flour	4 tbsp
salt and white pepper	TT
olive oil	4 tbsp

Sauce

dry white wine	1 c
fresh lemon juice	4 tbsp
capers, drained	4 tsp
cream	6 fl oz

Preparation

1. Pound the veal cutlets to ⅛ inch thickness.
2. In a small bowl, combine flour, salt, and white pepper.
3. Lightly coat cutlets with flour mixture.
4. In a large skillet, heat oil over medium heat. Add cutlets and cook 3–4 minutes for medium doneness, turning once.
5. Remove cutlets and keep warm.
6. Drain oil from the pan. Add wine and lemon juice to deglaze the pan. Cook until sauce begins to thicken.
7. Remove sauce from heat. Stir in capers and cream.
8. Return to heat and reduce to nappe consistency. Season to taste.
9. Spoon sauce over cutlets and serve.

> **NUTRITION FACTS**
> **Per serving:** 552 calories, 317 calories from fat, 35.8 g total fat, 197.2 mg cholesterol, 354.2 mg sodium, 652.1 mg potassium, 10 g carbohydrates, <1 g fiber, 1 g sugar, 36.2 g protein

Oxtail Stew

Yield: *4 servings, 20 fl oz each*

Ingredients

shortening	2 oz
oxtails	2 lb 6 oz
salt and pepper	TT
all-purpose flour	1 tbsp
onions, small dice	3 oz
garlic, minced	2 tsp
beef stock	2½ c
tomato purée	½ c
thyme	½ tsp
bay leaf	1 ea
carrots, julienned	4 oz
celery, julienned	4 oz
turnips, julienned	4 oz
peas	4 oz
water	8 fl oz
pearl onions, peeled	4 oz
butter	2 tbsp
salt and pepper	TT

Preparation

1. Thoroughly heat shortening in a braiser in a 400°F oven.
2. Add oxtails and season with salt and pepper. Turn as needed until evenly browned on all sides.
3. Sprinkle flour over the meat and brown evenly.
4. Add diced onions and garlic and cook until tender.
5. Add brown stock, tomato purée, thyme, and bay leaf. Cover and cook until the meat is tender and the sauce is slightly thick. Remove the bay leaf.
6. In a large saucepan, simmer carrots, celery, turnips, and peas in water until tender.
7. Strain and reserve one-third of the peas for garnishing.
8. Add the vegetables to the stew.
9. Sauté pearl onions in butter. Add the onions to the stew.
10. Season to taste and garnish with remaining peas.

> **NUTRITION FACTS**
> **Per serving:** 397 calories, 218 calories from fat, 24.5 g total fat, 26 mg cholesterol, 577.2 mg sodium, 1018.5 mg potassium, 32.5 g carbohydrates, 10.4 g fiber, 8.5 g sugar, 14 g protein

Ossobuco

Yield: *6 servings, 10 oz veal and 4 oz of sauce each*

Ingredients

10 oz veal shanks	6 ea
olive oil	4 tbsp
butter	2 tbsp
red onion, medium dice	1 c
green pepper, diced	½ c
red pepper, diced	½ c
carrots, diced	1½ c
celery stalk, diced	1 c
garlic, crushed	6 cloves
Marsala wine	⅓ c
brandy	⅓ c
balsamic vinegar	1 tbsp
sachet d'épices	
rosemary	1 spring
fresh thyme	4 sprigs
bay leaves	2 ea
tomato paste	2 tbsp
chicken stock	2 c
salt and pepper	TT

Preparation

1. In a rondeau, brown the veal shanks in half of the olive oil. Remove the browned shanks from the pan and reserve.
2. Add remaining oil and butter to the hot pan. Sauté onions, peppers, carrots, celery, and garlic over medium heat for 10 minutes.
3. Add the Marsala, brandy, and balsamic vinegar to the sautéed vegetables. Cover the pan and simmer for 6–8 minutes.
4. Make a sachet of rosemary, thyme, and bay leaves.
5. Add the sachet and tomato paste to the simmering vegetables. Increase the heat and add the veal shanks and chicken stock.
6. When the mixture has returned to a simmer, reduce the heat and cover the pan. Cook for approximately 1–1½ hours until the veal separates from the bone easily.
7. Season to taste. Plate and serve ossobuco with pan sauce.

NUTRITION FACTS
Per serving: 257 calories, 132 calories from fat, 14.9 g total fat, 33.8 mg cholesterol, 234.7 mg sodium, 462.7 mg potassium, 12.8 g carbohydrates, 2.6 g fiber, 5.1 g sugar, 8.9 g protein

Bison Short Ribs Provençal

Yield: *8 servings, 12 oz bison and 5 oz of sauce each*

Ingredients

olive oil	2 tbsp
bison short ribs	6 lb
onion, small dice	1 lb
carrots, small dice	8 oz
celery, small dice	8 oz
garlic, peeled and crushed	12 cloves
all-purpose flour	2 tbsp
herbes de Provence	1 tbsp
dry red wine	2 c
beef stock	2½ c
tomato concassé	16 oz
bay leaf	1 ea
water	½ c
baby carrots, peeled	1 lb
Niçoise olives, pitted	½ c
fresh parsley, finely chopped	3 tbsp

Preparation

1. Preheat the oven to 325°F. Heat the olive oil in a large, heavy pot over medium-high heat.
2. Working in batches, add ribs to pot and brown well (approximately 8 minutes), turning often. Transfer ribs to a large bowl as browned. Reserve.
3. Add diced onions, carrots, and celery to the pot and cook over medium-low heat until vegetables are soft (approximately 10 minutes). Stir frequently.
4. Add garlic, flour, and herbes de Provence. Stir for 1 minute.
5. Add wine and beef stock to deglaze the pot.
6. Add the tomato concassé and bay leaf. Return the ribs and any accumulated juices to the pot. If necessary, add enough water to barely cover the ribs. Bring to a simmer.
7. Cover pot tightly and transfer to the oven. Bake at 325°F for approximately 2 hours, 15 minutes until ribs are very tender.
8. Add peeled baby carrots and olives to the pot. Gently press the carrots to submerge. Cover the pot, return it to oven, and continue cooking at 350°F until carrots are tender.
9. Discard bay leaf. Remove short ribs and carrots from pan. If necessary, reduce the sauce to thicken. Season to taste.
10. Plate the short ribs. Pour sauce over the ribs and sprinkle with parsley.

NUTRITION FACTS
Per serving: 805 calories, 380 calories from fat, 42.3 g total fat, 200.7 mg cholesterol, 813.5 mg sodium, 1955.5 mg potassium, 23.7 g carbohydrates, 4.7 g fiber, 9.8 g sugar, 68.9 g protein

Courtesy of **The National Pork Board**

Pork and Related Game

Pork is unique in that it is the only meat of which all cuts can be cured. The meat of young hogs is generally leaner than beef and lamb and can be cooked using almost any cooking method. Prior to cooking, many cuts of pork are cured. Some cured pork cuts are also smoked. Rubs and marinades may also be used to enhance the flavor of pork. Wild boar is a related game meat that can be used in any recipe that calls for pork.

Refer to DVD
for **Flash Cards**

Chapter Objectives

1. Describe the composition of pork.
2. Explain the advantage of purchasing a whole carcass.
3. Identify the five primal cuts of pork.
4. Identify the cuts fabricated from each primal cut of pork.
5. Describe four offals that are only fabricated from pork.
6. Explain the purpose of Institutional Meat Purchase Specifications.
7. Describe the USDA inspection and grading of pork.
8. Identify four traits that should be checked upon receiving pork.
9. Explain why vacuum-sealed packages should only be opened at the time of use.
10. Remove and trim a tenderloin.
11. Tie a boneless pork roast.
12. Butterfly boneless pork chops.
13. Identify three curing methods used on pork cuts.
14. Explain the purpose of salt in the curing process.
15. Explain why combination curing is used on most cuts of pork.
16. Use rubs and marinades to enhance the flavor of pork.
17. Explain how to determine the doneness of pork.
18. Cook pork using eight different cooking methods.
19. Contrast wild boar meat with pork meat.

Key Terms

- **pork**
- **suckling pig**
- **picnic shoulder**
- **shoulder butt**
- **cottage ham**
- **clear plate**
- **pork tenderloin**
- **baby back ribs**
- **fatback**
- **ham**
- **side pork**
- **spareribs**
- **pork belly**
- **bacon**
- **pancetta**
- **jowl**
- **dry curing**
- **wet curing**
- **combination curing**
- **wild boar**

PORK

Pork is the meat from slaughtered hogs that are less than a year old. It is important to understand the basic composition of pork meat and the bone structure before working with pork. Pork is pink in color when raw and similar in color to white-meat poultry when cooked. Pork meat is made up of bundles of long, string-like muscle fibers held together by two types of connective tissues called collagen and silverskin. *See Figure 19-1.*

National Pork Producers Council

Composition of Pork

Photo Courtesy of D'Artagnan, Photography by Doug Adams Studio

Figure 19-1. *Pork meat is composed of muscle fibers and fat held together by connective tissues called collagen and silverskin.*

Collagen is a soft, white, connective tissue that breaks down into gelatin when heated. *Silverskin* is a tough, rubbery, silver-white connective tissue that does not break down when heated. Silverskin is trimmed from the meat prior to cooking, as it is inedible.

Muscles that receive the most exercise have more connective tissue and are less tender. For example, the legs and the shoulder of the hog have more connective tissue and generally yield tougher, yet more flavorful meat. Pork muscles contain very little fat. However, most cuts of pork have a fat cap, or thick layer of fat that surrounds the muscle. The fat cap may be left on meats, such as pork chops, while they are being cooked. This practice makes the meat juicier and prevents it from drying out. The fat cap may be removed prior to service.

Bones are often not removed from many cuts of pork prior to cooking in order to enhance flavor and presentation. Understanding the bone structure of a hog can aid in identifying the market forms of pork. *See Figure 19-2.*

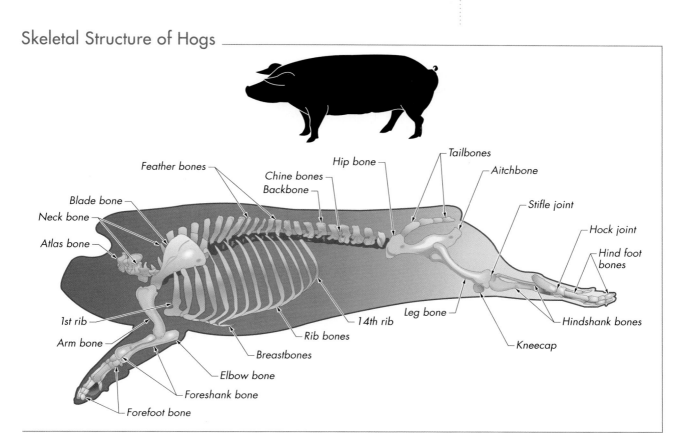

Figure 19-2. *Understanding the bone structure of a hog can aid in identifying the market forms of pork.*

MARKET FORMS OF PORK

Pork is available in a variety of market forms, including whole carcasses, primal cuts, fabricated cuts, and variety meats. Knowledge of these market forms is necessary for accurate product ordering.

Whole Carcasses of Pork

Both whole hog and suckling pig carcasses are available. Purchasing a whole hog carcass allows a chef to be more creative with the menu, rather than being restricted to pre-portioned cuts that can be found in most restaurants. A *suckling pig* is a pig 4–6 weeks old that weighs 20–35 lb dressed. Suckling pig carcasses are sold with the head attached and are priced per pig rather than by the pound.

Primal and Fabricated Cuts of Pork

A *primal cut* is a large cut from a whole or a partial carcass. A hog carcass has two sets of primal cuts. One set is on the left side and one set is on the right side. Each primal cut is divided into fabricated cuts. The five primal cuts of pork are the picnic shoulder, shoulder butt, loin, leg, and belly. *See Figure 19-3.* Some operations have a butcher on staff and order only primal cuts. Other operations order only fabricated cuts.

Primal Cuts of Pork

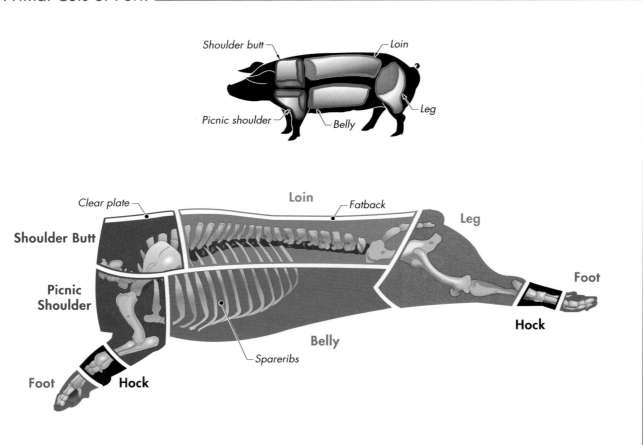

Figure 19-3. *The primal cuts of pork include the picnic shoulder, shoulder butt, loin, leg, and belly*

Picnic Shoulder Cut of Pork

National Pork Producers Council

Figure 19-4. *Picnics are fabricated from the picnic shoulder primal cut.*

A *fabricated cut* is a ready-to-cook cut that is packaged to certain size and weight specifications. Fabricated cuts of pork include hams, roasts, pork chops, pork cutlets, stew meat, and ground pork. Fabricated cuts are a convenient way of providing uniform portions while reducing labor costs. However, the price per pound for fabricated cuts is higher than the price per pound for primal cuts.

Picnic Shoulders. The *picnic shoulder* is a primal cut of pork that is the lower half of the shoulder of a hog. The average weight of a picnic shoulder is approximately 9% of the total carcass weight. Picnic shoulders are fabricated from the picnic shoulder primal cut. *See Figure 19-4.* A *picnic* is fabricated from the upper part of the foreleg that includes a portion of the shoulder. A picnic resembles a ham in shape, but is smaller and contains more bone and less lean meat. Fresh picnic can be used to prepare chop suey, pork patties, or pork sausage. Pulled pork is prepared from smoked picnic meat.

Shoulder Butts. The *pork shoulder butt,* also known as Boston butt, is a square, compact area of the shoulder located just above the front legs of a hog. Its average weight is approximately 8% of the total carcass weight.

The shoulder butt contains the blade bone and a large portion of lean meat. It is usually sold fresh with the bone. The meat is moderately tough due to the amount of connective tissue, so it is typically roasted or braised. A shoulder butt can also be fabricated into blade steaks, cottage ham, or ground meat. *See Figure 19-5.* A *cottage ham* is the smoked, boneless meat extracted from the blade section of the shoulder butt. Boneless shoulder butts are often tied with string because they fall apart easily when cooked.

Shoulder Butt Cuts of Pork

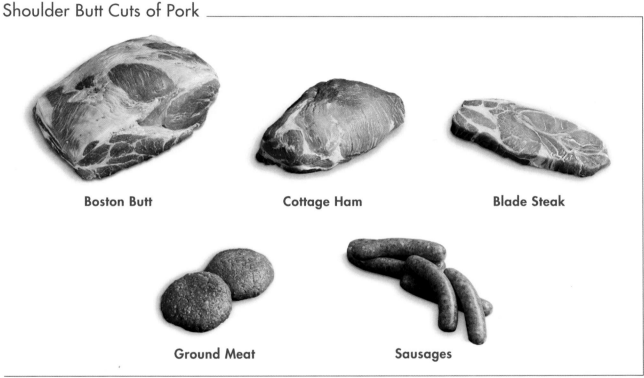

Boston Butt

Cottage Ham

Blade Steak

Ground Meat

Sausages

National Pork Producers Council

Figure 19-5. *A shoulder butt can be fabricated into cottage ham, blade steaks, ground meat, or sausages.*

The *clear plate* is a rectangular slab of fat that contains a few strips of lean meat located just above the shoulder butt. Clear plate that has been cured in salt is called salt pork. It is often used as a flavoring ingredient in dishes such as beans and bitter greens. Salt pork is often blanched to extract excess salt before it is used as a flavoring ingredient.

Loins. The *pork loin* is a primal cut that extends along the greater part of the backbone, from about the second rib, through the rib and loin area of a hog. Its average weight is approximately 18% of the total carcass weight and can be cut into a variety of fabricated cuts. *See Figure 19-6.* The *pork tenderloin* is a fairly long, tapered strip of lean meat taken from the underside of the loin. Tenderloin is the most tender pork cut and can be prepared using any cooking method.

Loin Cuts of Pork

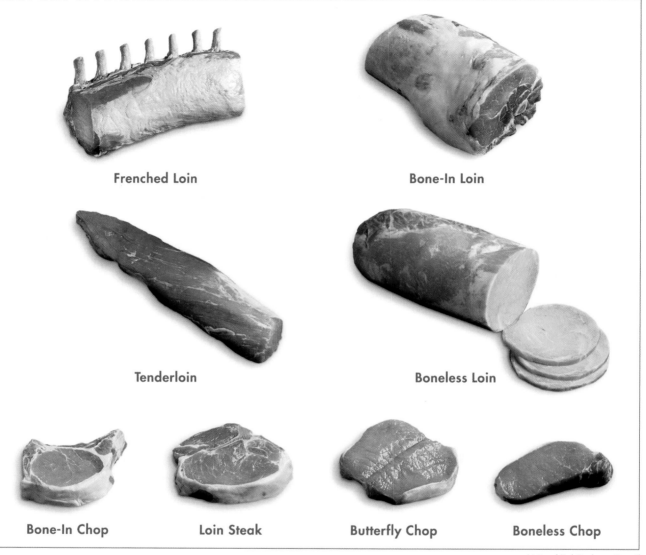

Frenched Loin

Bone-In Loin

Tenderloin

Boneless Loin

Bone-In Chop

Loin Steak

Butterfly Chop

Boneless Chop

Figure 19-6. *A variety of cuts are fabricated from the loin primal cut.*

Baby back ribs are the meaty bones on the rib end of the pork loin. These meaty ribs are only 3–6 inches long and are curved where they meet the backbone. A full slab of baby back ribs has 11–13 ribs. A standing rib roast is the whole pork loin muscle with the baby back ribs attached. A boneless loin and pork cutlets are also fabricated cuts from the loin. Canadian bacon is the trimmed, pressed, and smoked boneless loin of pork. *Fatback* is the layer of fat that runs along the back of the hog. It can be used to flavor dishes such as beans and collard greens or added to sausage or ground pork. Lard is usually rendered from fatback.

Legs. A *pork leg,* also known as a ham, is a primal cut of pork that is composed of the hind thigh and buttock of a hog. It contains a high proportion of lean meat and the average weight is approximately 24% of the total carcass weight. A ham is a fabricated cut from the hind leg of a hog that is typically cut from the middle of the shank bone to the aitchbone, or hip bone. *See Figure 19-7.* Unprocessed ham is called fresh ham. Hams are sold boneless, bone-in, and partially boned. The most popular form of ham is cured in a solution of salt, sodium nitrite, and sugar and then smoked. The skin may be left on a ham or it may be removed.

Leg Cuts of Pork

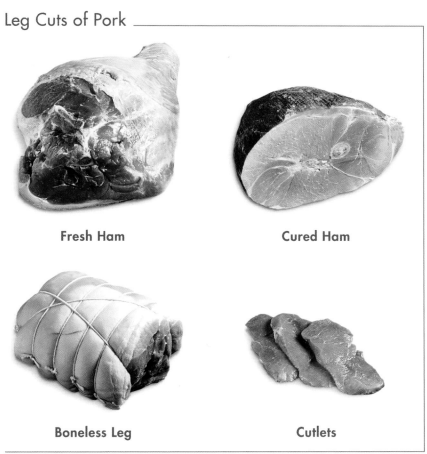

Fresh Ham

Cured Ham

Boneless Leg

Cutlets

Figure 19-7. *A primal leg cut is fabricated into several cuts.*

Ham is often cut into steaks or cutlets and broiled or pan-fried or it is sliced and used to make sandwiches. *Prosciutto* is a type of dry-cured Italian ham that is sliced very thin and used to make hors d'oeuvres or appetizers.

Belly. A *belly* is a primal cut of pork that is the lower portion of the hog between the shoulder and the leg. Its average weight is approximately 19% of the total carcass weight. Spareribs, pork belly, and bacon are fabricated cuts from the belly primal cut. *See Figure 19-8.*

Side Cuts of Pork

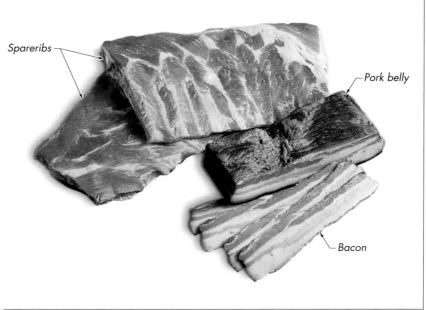

Sspareribs

Pork belly

Bacon

National Pork Producers Council

Figure 19-8. *Spareribs, pork belly, and bacon are fabricated from the belly primal cut.*

Pork spareribs are the long, narrow ribs and breastbone of a hog. They are quite fatty, yet the meat is tender and has an excellent flavor. A full rack contains 14 ribs. Spareribs may be purchased fresh or smoked. Fresh spareribs should be cooked slowly over low heat until the meat is tender and the fat is rendered. Spareribs are typically seasoned prior to cooking with a rub or a marinade. Cooked spareribs can be browned on the grill or under a broiler to caramelize the meat. They are commonly barbequed, broiled, or roasted and served as an appetizer or an entrée.

Pork belly is a fatty slab of meat and skin from the side and belly of a hog. It can be cooked as fresh pork belly or cured and served as bacon or pancetta. *Bacon* is side pork that has been cured and usually smoked. There are three basic cuts of bacon: thin, regular, and thick. Thin bacon is sliced into 22–26 strips per lb. Regular bacon is cut into 16–20 strips per lb, and thick bacon is cut into 12–16 strips per lb. Bacon is used to make entrées, appetizers, and sandwiches. It also is used to garnish soups. Bacon and the fat rendered from bacon are often used to season other foods.

Pancetta, also known as Italian bacon, is unsmoked pork belly that has been cured in salt and spices, such as nutmeg and pepper, and then dried for a few months. Pancetta is typically cut into paper-thin slices before being added to a dish. It adds a distinctive pork flavor, especially to pastas, that is not smoky like bacon.

Pork Offals

Unlike most domesticated meat animals, just about every part of a hog is used for consumption. An *offal* is an edible part of an animal that is not part of a primal cut. Offals are often ground and used to prepare sausage, meatballs, and fillings. Pork that is to be ground should be very cold before grinding, as should all parts of the grinder that come in contact with the pork. Pork that is cool instead of cold has a tendency to lose its texture and can appear puréed instead of ground. Some chefs place pork in the freezer for a short time before grinding it for better results.

In addition to common offals such as the heart, liver, and kidneys, pork variety meats include jowls, hocks, pig feet, and headcheese. A *jowl* is meat from the cheek of a hog. Jowls are cured and smoked in the same manner as bacon. Jowls are also used to flavor items such as baked beans or long beans.

Hocks are cut from the lower part of the front and hind legs of a hog. They have very little meat, but good flavor, with a good amount of fat, bone, and gristle. Hocks are available fresh, cured, or cured and smoked. They are often used to flavor soups or served with bean dishes. *See Figure 19-9.*

Pork Offals

National Pork Producers Council

Figure 19-9. *Pork offals, such as hocks, are often served with foods such as red beans and rice.*

Pig feet, also known as trotters, can be purchased fresh, smoked, or pickled. Fresh pig feet are commonly broiled or served boneless on a cold plate.

Headcheese is the spiced, pressed, and jellied meat from the head of a hog. The tongue, heart, feet, and quality trimmings may also be ground and included in headcheese. It is often served cold like lunchmeat or used as a binding agent in stocks.

Institutional Meat Purchase Specifications

To help ensure that foodservice operations and suppliers communicate efficiently, the USDA publishes the Institutional Meat Purchase Specifications (IMPS) for commonly purchased meats and meat products. Cuts of pork are numbered by category. *See Figure 19-10.*

Cuisine Note

Pork intestines are purged and used as sausage casings. Chitterlings are hog intestines that have been emptied and thoroughly rinsed before being fried.

	USDA Institutional Meat Purchase Specifications (IMPS) for Pork			
Series	**Item No.**	**Product**	**Item No.**	**Product**
400 **(fresh pork)**	400 401, 402 403–407 408, 409 410–414 415 416 417 418	Carcass Leg (fresh ham) Shoulder Belly Loin Tenderloin Spareribs Shoulder hocks Trimmings	419 420 421 422 423 424 435 496	Jowl Pigs feet, front Neck bones Loin, back ribs Loin, country-style ribs Loin, riblet Diced pork Ground pork
500 **(cured, smoked,** **and fully cooked** **pork)**	500–512 514 515–530 531 535, 537 536, 538 539–541 545, 546 547	Ham Pork, diced (cured) Pork shoulder Pork Boston butt Belly Bacon, slab Bacon, sliced Pork loin (cured and smoked) Pork center loin, boneless, (cured and smoked)	548 550 555 556 558, 559 560 561 562 563	Pork center-cut loin, 8 ribs (cured and smoked) Canadian style bacon (cured and smoked), unsliced Jowl butts, cellar trim (cured) Jowl squares (cured and smoked) Spareribs Hocks, ham (cured and smoked) Hocks, shoulder (cured and smoked) Clear fatback (cured) feet, front (cured)
700 **(variety meats)**	710 722 723 724	Pork liver Beef, lamb, or pork kidney Cheek meat Head meat	725 728 729 732	Beef, lamb, or pork brains Pork chitterlings Pork stomach (maws), scalded Lard (edible)
800 **(sausage)**	800 801 802 803 804 805 806 807 808 809 810 811 812	Frankfurters Bologna Pork sausage Liver sausage Cooked salami Minced luncheon meat Lebanon bologna Thuringer Dry salami Cervelat Breakfast sausage Smoked sausage New England brand sausage	813 814 815 816 817 818 819 820 821 822 824 825 826	Polish sausage Meat loaves Meat food product loaves Knockwurst Breakfast sausage, cooked Italian sausage Ham links Head cheese Pepperoni Bratwurst Pork rib shape patty Canned luncheon meat Scrapple
1400 **(fresh pork** **portion cuts)**	1400, 1401 1402 1406 1407	Steak cubed Cutlets Boston butt steaks Shoulder butt steaks, boneless	1410–1413 1438 1495 1496	Loin chops Steaks, flanked and formed, frozen Coarse chopped pork Ground pork patties
1500 **(cured, smoked,** **and fully cooked** **pork portion cuts)**	1513 1531 1545 1548 1596	Ham patties (cured), fully cooked Ham steaks (cured and smoked), boneless Pork loin chops (cured and smoked) Pork loin chops, boneless, center cut (cured and smoked) Pork patty, precooked		

Figure 19-10. *Cuts of pork are numbered by category in the Institutional Meat Purchase Specifications (IMPS).*

INSPECTION AND GRADES OF PORK

All pork used in foodservice operations must be procured from USDA inspected plants. At the time of slaughter, a hog carcass is stamped with the round USDA inspection stamp, indicating that the hog was slaughtered at an inspected plant. This stamp does not indicate anything about the quality of the meat. *See Figure 19-11.* The purple inspection stamp is used for whole carcasses and all fabricated and processed meats. It is found either on the meat itself or on the case in which it is packed. The number on the stamp identifies the plant where the animal was processed.

Pork is not graded like beef, veal, and lamb. It is produced from young hogs that were bred and fed to produce uniformly tender meat. Quality pork has very little fat covering on the surface. The meat is firm and grayish-pink in color.

RECEIVING AND STORING PORK

Like all meats, pork is a potentially hazardous food. The color, odor, texture, and temperature must be checked when received. The meat should be pink in color and the fat should be white. There should be no odor, and the texture should be firm and not dry or slick. Pork that is in the temperature danger zone should be rejected. Refrigerated pork should maintain an internal temperature of 41°F or below.

Frozen pork should be stored at temperatures below 0°F. Frozen pork that needs to be thawed should be placed in the refrigerator overnight. Larger cuts of pork may take more than one day to thaw under refrigeration. If necessary, pork can be thawed under running water if it is cooked immediately after thawing. Some frozen cuts, such as sausage patties or breaded pork tenderloins, can be thawed as part of the cooking process.

Vacuum-packed pork must not be opened until needed for service or preparation. The airtight plastic of vacuum packaging helps to preserve meat for three to four weeks. *See Figure 19-12.* Once the vacuum seal is broken, the meat has a shelf life of only two or three days. Cut pork that is not vacuum-sealed should be wrapped tightly, refrigerated immediately, and used as soon as possible.

In addition to packaging, pork can also be irradiated to reduce the risk of potentially harmful microorganisms. The irradiation process does not cook pork and does not have an adverse effect on its appearance or taste. The USDA requires all irradiated food be labeled with an irradiation symbol.

FABRICATION OF PORK

The size of storage facilities and availability of staff with the fabrication skills often determine whether a foodservice operation purchases primal cuts, fabricated cuts, or a combination. Some operations purchase whole hog carcasses and fabricate all the cuts in-house, while others only fabricate the loin. Pork fabrication requires the removal, trimming, and tying of meat as well as butterfly cuts.

USDA Inspection Stamps

Mark for raw whole carcass meat   Mark for fabricated or processed meats

Figure 19-11. *A round USDA inspection stamp indicates that a hog was slaughtered at a USDA-inspected plant.*

Chef's Tip

Due to a lack of refrigeration or poor hygiene practices by foodservice handlers, pork can become a prime breeding ground for staphylococcal gastrocerteritis.

Vacuum-Packaged Pork

Figure 19-12. *Airtight vacuum packaging helps preserve meat for 3–4 weeks.*

Removing Tenderloins

Some food service operations purchase primal loins and fabricate cuts in-house. If a loin is being fabricated in-house, the tenderloin must first be removed from the loin. *See Figure 19-13.*

Procedure for Removing the Tenderloin from a Loin

1. Trim the surface fat of the loin to approximately ¼ inch thick on both sides.
2. Slide the knife under the exposed rib and slice through to the backbone to separate the tenderloin muscle from the vertebrae.
3. Carefully cut down the center of the backbone between the muscle and the exposed vertebrae.
4. Continue cutting to the end of the vertebrae until the tenderloin is completely removed from the loin.

Figure 19-13. *Some foodservice operations purchase primal loins and fabricate cuts in-house.*

Refer to DVD for
Fabricating Boneless Pork Loin
Media Clip

Once the tenderloin has been removed from the loin, it needs to be trimmed. Trimming a tenderloin involves removing all the fat and silverskin. *See Figure 19-14.*

Procedure for Trimming a Tenderloin

1. With a rigid boning knife, carefully remove the chain muscle from the side of the tenderloin and reserve. *Note:* Some chefs prefer to leave the chain muscle. The chain muscle is a thin strip of tender meat surrounded with fat that is located next to the tenderloin. It can be used in stocks, soups, or ground meat.
2. Use a rigid boning knife to pull back and trim the thick fat covering from the tenderloin.
3. Position the tenderloin so that the head (wide end) is near the cutting hand and the tail (narrow end) is near the guiding hand. Insert the tip of the boning knife just beneath the silverskin at the tail of the tenderloin. Angle the knife blade upward slightly and draw it along the length of the tenderloin, just beneath the silverskin, toward the head of the tenderloin.
4. Use the guiding hand to hold the silverskin firmly as the cutting process is repeated until all the silverskin has been removed. *Note:* If silverskin tapers into the tenderloin muscle, scrape the muscle away in order to remove the silverskin.

Figure 19-14. *Trimming a tenderloin involves removing all the fat and silverskin.*

Tying Boneless Pork Roasts

A boneless pork roast falls apart when cooked. To prevent this, the roast is often rolled and then tied with butcher's twine prior to cooking. *See Figure 19-15.*

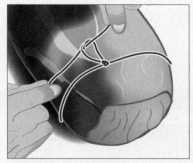

1. Place the outside of the meat face-down and roll the sides tightly toward the center.

2. Wrap butcher's twine around one end of the roast using a slipknot to tighten firmly. Then tie a square knot to secure the tie.

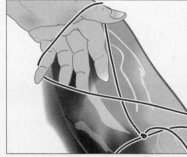

3. Wrap the twine around one hand to form a loop. Then twist the hand over the meat until the palm is facing down.

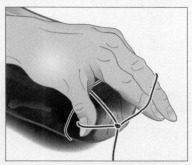

4. Move the palm to the front of the roast and slide the loop over the meat.

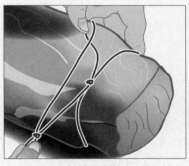

5. Remove the hand from the loop and secure the twine around the meat.

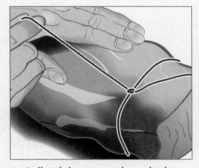

6. Pull tightly to straighten the loop.

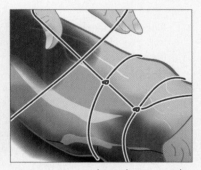

7. Continue making loops in the same fashion 1½ inches apart down the length of the roast. *Note:* Check each loop to make sure it is an even distance from the last loop.

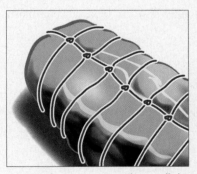

8. Turn the roast over. Then pull the twine over the first loop and tuck it under and around the loop, pulling it toward the front of the roast.

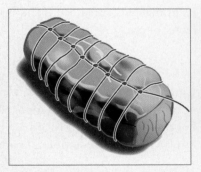

9. Continue pulling over the loops until the underside of the roast is fully tied.

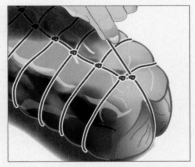

10. Turn the roast over and bring the twine up to tie a square knot to secure the roast. Cut the twine close to the knot.

Figure 19-15. *Tying a boneless pork roast with butcher's twine prevents it from falling apart during the cooking process.*

Refer to DVD for
Tying a Pork Loin
Media Clip

Making Butterfly Cuts

Thicker cuts such as pork chops can be butterfly cut to make a pocket for stuffing or to a create a thinner product that will cook more quickly. *See Figure 19-16.* To make a butterfly cut, the chop is cut almost completely in half horizontally.

Procedure for Making a Butterfly Cut

1. Place the pork chop near the edge of a cutting board to prevent the knuckles of the knife hand from hitting the board.

2. Place the guiding hand flat on top of the chop to hold it in place. While holding the knife blade parallel to the cutting board, place the blade edge at the midpoint of the chop.

3. Slice the chop almost all the way through, leaving ¼–½ inch connected at the back side.

Figure 19-16. *A butterfly cut creates a pocket for stuffing or a thinner product that will cook more quickly.*

FLAVOR ENHANCERS FOR PORK

Several methods are used to enhance the mild flavor of pork after fabrication and prior to cooking. For example, pork is the only meat that can have all its primal cuts cured. Rubs and marinades also can be used to enhance the flavor of pork.

Curing

Curing is the process of using salt and sodium nitrite alone or with flavorings or sugar to preserve a food item. More than two-thirds of all pork is cured. For example, ham and bacon are cured-pork products. Pork can be cured using a dry, wet, or combination method. *See Figure 19-17.*

Dry Curing. *Dry curing* is a curing method that involves the use of salt to dehydrate the protein in food. Meats are dry cured by completely covering the meat with salt or a salt-cure mixture, which includes salt, sugar, seasonings, and pink curing salt. Dry curing is done in stages to maximize the absorption of the curing ingredients. Ham, bacon, and sausages that will be air-dried are dry cured.

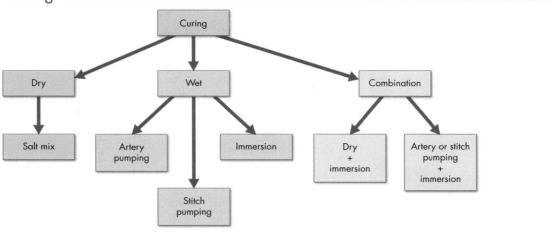

Figure 19-17. *Pork can be cured using a dry, wet, or combination method.*

Adding salt to meat adds flavor and prevents microbial growth. As the salt penetrates the meat, water rises to the outside surface and runs off. This process occurs very fast during the first week and slows as the amount of salt exceeds the amount of water inside the meat. The length of curing time depends on the size of the meat and its composition. Fat and skin will create a barrier for the cure and slow down the process. The curing time takes approximately three days per pound for large cuts of pork and two days per pound for small cuts.

After the dry cure is complete, the meat is rinsed thoroughly to remove any crystallized salt that has accumulated on the surface. If the meat is going to be smoked, the salt crystals will prevent smoke penetration. Dry-cured pork is refrigerated at 40°F or below.

Wet Curing. *Wet curing,* also known as immersion curing, is a curing method in which foods are submerged in a brine. Wet curing is sometimes referred to as brining or sweet pickling. A *brine* is a salt solution that usually consists of 1 cup of salt per 1 gal. of water. A sweet pickle brine is composed of salt, water, and sugar. Sugar is only added when curing is done at refrigerator temperatures because fermentation will start in warmer temperature and spoil the meat. Smaller cuts of pork can be wet-cured in a shorter amount of time than larger cuts.

Wet curing yields a less salty product with a shorter shelf life than dry curing. Butts and hams were traditionally wet cured, but wet curing is a time-consuming process. The meat has to be turned over repeatedly as well as kept from floating to the surface. Since salt tends to sink and sodium nitrite floats, the curing solution also has to be agitated to cure the meat evenly. Pork absorbs different amounts of sodium in relation to curing time and the amount of surface fat. *See Figure 19-18.*

Chef's Tip

It is important to be precise when measuring curing ingredients, especially when nitrites are being used. The proper curing steps also need to be carefully followed to ensure food safety.

Sodium Absorption During Curing

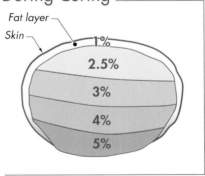

Figure 19-18. *Pork absorbs different amounts of sodium in relation to curing time and the amount of surface fat.*

*Photo Courtesy of D'Artagnan,
Photography by Doug Adams Studio*

Two types of pumps are used to wet cure meats: artery pumps and stitch pumps. An artery pump, also known as a spray pump, uses a long needle connected to a pump to inject brine directly into the artery of a large piece of meat. In the case of a ham, this is fairly simple to do. However, a professional butcher would have to make sure the artery of a pork leg is still intact before an artery pump could be used. Fast curing takes 7 to 14 days and requires around a 10% brine solution in relation to the original weight of the meat. Slow curing takes 30 to 50 days and requires approximately a 5% brine solution in relation to the original weight of the meat.

Stitch pumps apply pressure to the surface of large pieces of meat such as a butt or ham and use a bank of needle injectors connected to a pump to inject the brine directly into the meat. A stronger brine solution is often used to speed up the curing process. The pressure must be strictly controlled to prevent textural damage. Injections are only effective if the meat cures faster without too much damage to its texture.

Combination Curing. *Combination curing* is the process of combining either dry curing or wet curing with artery or stitch pumping to reduce processing time. One option is to rub a dry cure mix into the pieces of meat and then use an artery or stitch pump to inject a brine solution. This option is faster than the second option because no water is added to the meat because the salt solution will draw water out of the meat.

The second option is to use an artery or stitch pump to inject a brine solution into large pieces of meat and then stack them on top of each other inside a container. The container is then filled with a wet curing solution until all of the meat is immersed in brine. The meat must be weighed down to keep it from surfacing and the pieces need to be turned over each day for the duration of the curing process.

Rubs and Marinades

Rubs and marinades can also be used to add intense flavor to pork. Both dry and wet rubs are used. Dry rubs are made by grinding herbs and spices together into a fine powder and then rubbing the mixture into the pork prior to cooking. *See Figure 19-19.* A wet rub is made by incorporating wet ingredients, such as Dijon mustard, flavored oils, puréed garlic, or honey, into a dry rub mixture and then rubbing it into the pork prior to cooking.

Marinades add flavor to pork and tenderize it at the same time. Marinades have an acidic liquid base, such as wine or lemon juice. The acidic base tenderizes the meat by breaking down the protein structure of the pork. Any herb, spice, or condiment can be added to the base to create the desired flavor. The length of time that pork is left in the marinade is determined by the size of the piece of meat. Care should be taken not to use an overly strong marinade on pork, because the marinade can overpower the pork. Once the pork has been removed from the marinade, the remaining liquid must be discarded to prevent cross-contamination.

Dry Rubs

Figure 19-19. *A dry rub can be used to add intense flavor to a cut of pork.*

COOKING PORK

Tougher cuts of pork benefit from being slowly cooked using moist-heat cooking methods, while more tender cuts can be cooked using dry-heat cooking methods. The majority of animal muscle is made up of water. As meat loses water, it shrinks. *Shrinkage* is the loss of volume and weight of a piece of food as the food cooks. Shrinkage is why a 20 lb roast may only be 18.5 lb when fully cooked.

Cooking pork at too high a temperature can toughen the protein in meat. However, grilling and frying use very high temperatures for short periods and result in only the exterior of an item receiving high amounts of heat. High heat quickly cooks the exterior of meat to a crispy texture while slowly cooking the interior of the meat. This is the reason that a grilled pork chop is crispy and somewhat dry on the outside yet remains tender and juicy on the inside.

Determining Doneness

Cooking meat to the desired degree of doneness is determined by the type and thickness of the meat, the temperature of the meat when it begins to cook, and the intensity of the heat. Because of these variables, time is not an accurate way to determine the doneness of meats.

Pork that is done has been cooked to an internal temperature of 145°F for at least 15 seconds and rested for 3 minutes. ***See Figure 19-20.*** Ground pork must be cooked to an internal temperature of 160°F. An instant-read thermometer is inserted into the thickest part of large cuts of pork to take the internal temperature. Braised and stewed pork is cooked until it is fork tender, or until the meat can easily be separated with a fork.

Smaller cuts of pork that are grilled or broiled can be tested for doneness with the touch method, which uses the sense of touch to check the firmness of cooked meat. For example, a pork chop that is cooked well done is firm and springs back immediately when gently pressed with the fingertip. In contrast, a rare pork chop is soft and slightly mushy to the touch. The texture will feel almost the same as when the meat was raw.

Internal Temperature

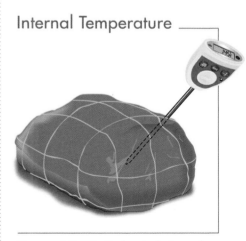

Figure 19-20. *Pork cuts that are done have been cooked to an internal temperature of 145°F for at least 15 seconds and rested for 3 minutes.*

Grilled Pork

National Pork Producers Council

Figure 19-21. *Fabricated loin cuts are commonly grilled or broiled.*

Grilling and Broiling

Grilling and broiling use a hot flame to quickly sear and cook foods. Only tender cuts or ground pork should be used with these methods. Fabricated loin cuts are commonly grilled or broiled. *See Figure 19-21.* Well-done pork cooked with these methods will probably be somewhat dry, because most of the moisture cooks out. Pork to be grilled or broiled should be properly trimmed of fat, while too little fat may cause the meat to dry out.

Smoking and Barbequing

Pork is smoked and barbequed more often than any other meat. The smoking process flavors the pork as it cooks over burning wood. Pork is typically basted in barbeque sauce during the grilling or smoking process. *See Figure 19-22.* Different types of barbeque sauce are served in various regions of the United States.

Barbequed Pork

Barbequed Rib Appetizer

Rack of Barbequed Ribs

Barbequed Pulled Pork

National Pork Producers Council

Figure 19-22. *Some forms of barbeque are similar to smoking while other forms are similar to grilling.*

Roasting

Many cuts of pork are roasted or barbequed. Cuts from the pork loin and leg are commonly roasted. *See Figure 19-23.* Smaller roasts should be cooked at higher temperatures, between 400°F and 450°F, to allow them to caramelize nicely on the exterior without overcooking the interior. Larger roasts require a longer cooking time and should be roasted at lower temperatures, between 275°F and 325°F, to prevent excessive shrinkage. It is important to allow for carryover cooking when roasting meats. Lean cuts of pork will become dry when roasted unless some form of fat is added.

Roasted Rack of Pork

Figure 19-23. *Cuts from the pork loin, such as a frenched rack, are commonly roasted.*

Sautéing and Frying

Tender cuts of pork are typically sautéed or fried. Sautéing uses a small amount of hot oil to sear and cook the pork. Only uniformly sized cuts should be sautéed to ensure even cooking. Care should be taken when sautéing not to burn the oil or the surface of the food by using too high of a temperature. A pan sauce made in the same pan after the cooked pork has been removed can be served with the cooked meat.

Pork is typically cut into strips for stir-frying. ***See Figure 19-24.*** Cutlets and chops can be pan-fried or deep-fried. These cuts should not be too thick to prevent them from fully cooking. Tender cuts of pork are commonly breaded and deep-fried. The breading protects the pork from the hot oil and helps the pork retain moisture.

Braising and Stewing

Large or tougher cuts of pork are typically braised or stewed. Braising is used for larger cuts of pork that have a good amount of fat. The pork is seared, and then a cooking liquid is added about halfway up the side of the meat. Searing the meat adds flavor to the pork and the resulting sauce created by the liquid. The pork is then slow-cooked until tender. The braising liquid may be thickened slightly and served as an accompanying sauce. ***See Figure 19-25.***

Tender cuts, such as pork chops, can also be braised. The only difference between braising tender cuts and tougher cuts is the amount of time the pork is cooked.

Stewing pork requires that most of the fat be removed prior to the cooking process. The meat is completely covered with liquid and cooked slowly.

Stir-Fried Pork

Figure 19-24. *Pork is typically cut into strips for stir-frying.*

Braised Pork

Pork Roast

Carnitas

National Pork Producers Council

Figure 19-25. *Braised pork is slow cooked until tender and may be accompanied by a sauce.*

WILD BOAR

Wild boar is a game animal that is similar in bone structure and muscle composition to domesticated hogs. All wild boar served must be purchased from ranches or farms and processed in USDA-inspected plants. Wild boar are fed a diet of nuts, fruits, roots, tubers, and grasses similar to what they might eat in the wild. The meat is similar in appearance to cuts of pork. However, wild boar meat has a stronger flavor and contains less fat.

Although specific cuts may be the same shape as pork cuts, wild boar meat is a much deeper red color. *See Figure 19-26.* Wild boar meat also has a bolder and slightly nutty flavor that is sweeter than pork. While boar meat is much leaner than pork, it can be used in any recipe that calls for pork. A popular cut of wild boar is the tenderloin, which runs the length of the back from the hip to the shoulder. Boar tenderloin is often grilled or smoked.

Wild Boar Meat Cuts

Shoulder

Tenderloin

Rib Rack

Photos Courtesy of D'Artagnan, Photography by Doug Adams Studio

Figure 19-26. *Wild boar cuts are similar in shape to pork cuts, but a much deeper red color.*

SUMMARY

Pork is the meat of young hogs. Market forms of pork include whole carcasses, primal cuts, fabricated cuts, and offals. The primal cuts of pork include the shoulder, butt, loin, leg, and belly. Various cuts can be fabricated from each primal cut. Pork offals include jowls, hocks, feet, and headcheese.

All of the pork used in foodservice operations must be procured from USDA-inspected plants. Refrigerated pork should be received and stored at temperatures that maintain an internal temperature of 41°F or below. Frozen pork should be kept at temperatures of 10°F or below.

Prior to cooking, many cuts of pork are cured. Pork is unique in that all of its cuts can be cured. Some cured pork cuts are also smoked. Rubs and marinades may also be used to enhance the flavor of pork. Pork can be grilled, broiled, roasted, barbequed, sautéed, fried, braised, and stewed.

Wild boar is a game meat that is leaner and a much deeper red color than pork. It can be used in any recipe that calls for pork.

Refer to DVD for
Quick Quiz® questions

Review

Refer to DVD for
Review Questions

1. Describe the composition of pork.
2. Identify common market forms of pork.
3. Explain the advantage of purchasing a whole carcass.
4. Identify the five primal cuts of pork.
5. Identify the cut fabricated from a picnic shoulder.
6. Identify the cuts fabricated from a shoulder butt.
7. Identify the cuts fabricated from a pork loin.
8. Identify the cuts fabricated from a leg of pork.
9. Identify the cuts fabricated from a side of pork.
10. Describe four offals that are only fabricated from pork.
11. Identify the source of natural sausage casings.
12. Explain the purpose of Institutional Meat Purchase Specifications.
13. Describe the USDA inspection and grading of pork.
14. Identify four traits that should be checked upon receiving pork.
15. Identify the required storage temperature for refrigerated pork and for frozen pork.
16. Explain why vacuum-sealed packages of pork should only be opened at the time of use.
17. Describe the effects of irradiation on pork.
18. Describe how to remove the tenderloin from a loin.
19. Describe how to trim a tenderloin.
20. Describe how to tie a boneless pork roast.
21. Describe how to butterfly boneless pork chops.
22. Identify three curing methods used on pork cuts.
23. Explain the purpose of salt in the curing process.
24. Explain why salt crystals need to be rinsed off of the surface of dry-cured pork.
25. Describe the process of wet (immersion) curing.

26. Explain why combination curing is used on most cuts of pork.
27. Contrast a dry rub and a wet rub.
28. Identify what a marinade does in addition to enhancing flavor.
29. Explain how to determine the doneness of pork.
30. Identify eight cooking methods used to cook pork.
31. Contrast wild boar meat with pork meat.

Applications

Refer to DVD for
Application Scoresheets

1. Remove the tenderloin from a loin.
2. Trim a tenderloin.
3. Tie a boneless pork roast.
4. Butterfly a boneless pork chop.
5. Use a dry rub and a marinade to enhance the flavor of two identical cuts of pork. Cook both cuts and compare the quality of the two prepared dishes.
6. Demonstrate three ways to determine the doneness of pork.
7. Grill and broil a cut of pork. Compare the quality of the two prepared dishes.
8. Roast and barbeque a cut of pork. Compare the quality of the two prepared dishes.
9. Sauté and fry a cut of pork. Compare the quality of the two prepared dishes.
10. Braise and stew a cut of pork. Compare the quality of the two prepared dishes.

Five Spice Pork Tenderloin with Dark Cherry and Soy Demi-Glace

Yield: *6 servings, 5 oz each*

Ingredients

pork tenderloin, trimmed	2 lb

Marinade

orange juice	2 fl oz
vegetable oil	1 tbsp
sesame oil	2 tsp
orange zest	2 tsp
lemon grass, minced, white part only	2 tsp
ginger root, grated	2 tsp
Chinese five spice powder	2 tsp
garlic cloves, minced	2 ea
black pepper, fresh ground	½ tsp

Dark Cherry and Soy Demi-Glace

butter	1 tbsp
dark pitted cherries, rough chopped	6 oz
shallot, minced	1 ea
ginger root, grated fresh	1 tsp
brown sugar	½ tbsp
red wine	1 fl oz
orange juice	4 fl oz
tamari soy sauce	1 tbsp
teriyaki sauce	½ tbsp
demi-glace	6 fl oz

Preparation

1. Mix all marinade ingredients.
2. Place pork tenderloin in marinade and let sit for at least 4 hours or overnight.
3. Place butter in a small saucepan over medium heat.
4. Add cherries, shallot, and ginger and cook until slightly caramelized.
5. Add brown sugar and heat until sugar is melted and begins to smoke.
6. Deglaze pan with wine and reduce by half.
7. Add orange juice and simmer for 2 minutes.
8. Add soy sauce and teriyaki sauce and simmer 2 minutes more.
9. Add demi-glace and heat to incorporate. Simmer for 5 minutes on very low heat.
10. Remove pork tenderloins from the marinade and cook on preheated grill until caramelized on the outside and until an internal temperature of 145°F is reached.
11. Remove from grill and allow to rest 10 minutes.
12. Slice tenderloin on the bias into ⅜ inch slices.
13. Place 5 oz of tenderloin and 2 oz of sauce on each plate.

NUTRITION FACTS
Per serving: 383 calories, 128 calories from fat, 14.4 g total fat, 111.1 mg cholesterol, 617.6 mg sodium, 1272.5 mg potassium, 25.2 g carbohydrates, <1 g fiber, 7.2 g sugar, 37.4 g protein

Apple and Walnut Stuffed Pork Chops

Yield: *6 servings, 1 pork chop each*

Ingredients

8 oz bone in center cut pork chops	6 ea
mirepoix	
onion, small diced	6 oz
carrot, small diced	3 oz
celery, small diced	3 oz
whole butter	½ oz
whole wheat bread, toasted	
and cut into 1 inch cubes	¾ lb
granny smith apple,	
cut medium dice	1 ea
walnuts, chopped medium	3 oz
egg, beaten	1 ea
sage	¼ tsp
cinnamon	⅛ tsp
chicken stock	4 fl oz
salt and pepper	TT

Preparation

1. Butterfly the pork chop to create a pocket that extends all the way to the bone.
2. Sweat mirepoix in butter until tender, then set aside to cool.
3. Mix mirepoix and all remaining ingredients gently in a large mixing bowl until well incorporated, but do not overmix.
4. Divide bread mixture into 6 equal portions and stuff each chop with a portion.
5. Season the chops to taste.
6. Place stuffed chops on a sheet pan and bake at 350°F until an internal temperature of 160°F is reached.

NUTRITION FACTS
Per serving: 652 calories, 229 calories from fat, 26.1 g total fat, 160.6 mg cholesterol, 201.9 mg sodium, 991.1 mg potassium, 51.5 g carbohydrates, 4 g fiber, 5.4 g sugar, 51 g protein

Pork Roast with Mushroom Sauce

Yield: *6 servings, 6 oz of pork and 2 fl oz of sauce each*

Ingredients

pork loin	2¼ lb
salt and black pepper	TT
butter	6 oz
garlic, minced	4 cloves
button mushrooms, sliced	8 oz
balsamic vinegar	1 tbsp

Preparation

1. Season the pork loin with salt and black pepper.
2. In a sauté pan, melt the butter and sauté the garlic and mushrooms until tender.
3. Add vinegar to mushroom mixture.
4. Add pork loin to the pan and spoon the mixture over the roast.
5. Place in a 350°F oven and roast uncovered until the internal temperature reaches 145°F.
6. Let the pork rest for 3 minutes before slicing.

NUTRITION FACTS
Per serving: 433 calories, 256 calories from fat, 29 g total fat, 173.3 mg cholesterol, 137.8 mg sodium, 796.3 mg potassium, 2.4 g carbohydrates, <1 g fiber, 1.2 g sugar, 39.7 g protein

Thai Red Curry Pork with Basil

Yield: *6 servings, 7 oz each*

Ingredients

vegetable oil	1 fl oz + 1 fl oz
blade end pork loin, trimmed and thinly sliced	1½ lb
green beans, fresh cut into 2 inch sections	12 oz
garlic cloves, crushed	3 ea
ginger, freshly grated	1 tbsp
red curry paste	2 tbsp
red potatoes, par boiled until al dente, then cooled and cut into large dice	9 oz
coconut milk	28 fl oz
fresh basil, fine chiffonade	3 tbsp

Preparation

1. In large sauté pan, add 1 fl oz vegetable oil and sauté pork on each side. Remove from pan and set aside.
2. Add remaining 1 fl oz of oil to pan and sauté green beans for 1 minute.
3. Add garlic and ginger and sauté a few seconds until aromatic.
4. Add curry paste and sauté for a few seconds until aromatic.
5. Add potatoes and coconut milk to pan and simmer until coconut milk has reduced by ⅓.
6. Add pork back to pan with basil. Stir to incorporate.
7. Serve over jasmine rice.

NUTRITION FACTS
Per serving: 641 calories, 463 calories from fat, 53.7 g total fat, 71.4 mg cholesterol, 103.9 mg sodium, 1031.6 mg potassium, 17.5 g carbohydrates, 3 g fiber, 1.9 g sugar, 27.4 g protein

Thai Red Curry Paste

Yield: *12 servings, 2 tbsp each*

Ingredients

fresh Thai red chilies, cleaned of stems and seeds	4 oz
garlic cloves, peeled	4 ea
shallot, peeled	2 oz
cilantro, stems only	2 oz
lemon grass stalk, white tender portion only	2 ea
kaffir lime zest	1 oz
galangal, peeled	1 oz
cayenne pepper	1 tsp
sugar	1 tsp
vegetable oil	1 fl oz
coriander seed, toasted and ground	½ tsp
cumin, ground	¼ tsp
cardamom, ground	⅛ tsp

Preparation

1. Place all ingredients in food processor and purée to make a smooth paste.
2. Refrigerate or freeze until needed.

NUTRITION FACTS
Per serving: 56 calories, 43 calories from fat, 4.9 g total fat, 0 mg cholesterol, 4.4 mg sodium, 86.2 mg potassium, 3.2 g carbohydrates, <1 g fiber, <1 g sugar, <1 g protein

Dijon Lemon Pepper Medallions

Yield: *4 servings, 6 oz of pork and 1 fl oz of sauce each*

Ingredients

pork tenderloin, medallions	1½ lb
lemon zest	1 tsp
black pepper, ground	½ tsp
butter	2 tbsp
lemon juice	2 tbsp
Worcestershire sauce	1 tbsp
Dijon mustard	1 tsp
parsley, minced	1 tbsp

Preparation

1. Flatten each medallion between two pieces of plastic wrap.
2. Sprinkle surfaces of pork with lemon zest and black pepper.
3. In sauté pan, melt the butter and brown the pork evenly, about 4 minutes on each side.
4. Remove from pan and tent to keep warm.
5. Deglaze the pan with the lemon juice.
6. Add Worcestershire sauce and mustard to pan.
7. Stir the mixture until heated through.
8. Serve medallions with sauce and sprinkle with parsley.

NUTRITION FACTS
Per serving: 267 calories, 106 calories from fat, 11.9 g total fat, 125.8 mg cholesterol, 147.4 mg sodium, 745.8 mg potassium, 3.3 g carbohydrates, <1 g fiber, 1.2 g sugar, 35.4 g protein

Pork Teriyaki Lettuce Wraps

Yield: *4 servings, 1 wrap each*

Ingredients

Marinade and Sauce

mirin soy sauce	¼ c
ginger, ground	½ tsp
garlic powder	¼ tsp
dark brown sugar	4 tbsp
honey	1 tbsp
cornstarch	2 tbsp
water, cold	1½ c
pork tenderloin, cut into ¼ inch thick strips	1 lb
vegetable oil	1 tbsp
large iceberg lettuce leaves	4 ea
large carrot, peeled and shredded	1 ea
water chestnuts, sliced	8 oz
bean sprouts, rinsed	14 oz

Preparation

1. Mix all sauce ingredients except the cornstarch and half of the water in a saucepan and begin heating.
2. Mix cornstarch and the remaining cold water in a separate container and dissolve. Add to sauce in pan.
3. Heat until sauce thickens to desired thickness. Reserve one-third of the prepared sauce.
4. Place two-thirds of the prepared sauce and pork in a plastic bag and seal bag.
5. Marinate for 2 hours under refrigeration.
6. Heat oil in large skillet or wok over medium-high heat.
7. Remove pork. Discard bag and marinade.
8. Sauté pork over medium-high heat for about 4 minutes or until slightly pink in center and browned on both sides.
9. Fill lettuce leaves with equal amounts of carrot, water chestnuts, bean sprouts, and pork.
10. Drizzle 1 tablespoon of reserved sauce over each wrap. Roll tightly and serve.

NUTRITION FACTS
Per serving: 309 calories, 69 calories from fat, 7.7 g total fat, 73.7 mg cholesterol, 1334.4 mg sodium, 736.6 mg potassium, 32.9 g carbohydrates, 2.6 g fiber, 22.9 g sugar, 28 g protein

Tex-Mex Pork Steak

Yield: *6 servings, 6 oz of pork and 6 oz of rice each*

Ingredients

canola oil	2 tbsp
6 oz pork blade steaks, trimmed	6 ea
garlic, minced	4 cloves
small yellow onion, chopped	1 ea
long-grain brown rice, uncooked	2 c
plum tomatoes, chopped	4 ea
jalapeño, minced	2 ea
chicken stock	3 c
salt and pepper	TT
cilantro, chopped	1 tbsp

Preparation

1. Heat oil in large sauté pan. Sear pork on both sides over medium-high heat just until brown on each side. Remove pork from pan and tent with foil.
2. Add garlic and onion and sauté until tender, scraping the fond from the pan.
3. Add rice. Stir constantly until rice is toasted.
4. Add tomatoes, jalapeños, and stock. Bring to boil. Cover and reduce heat to medium-low and simmer 10 minutes.
5. Place pork on top of rice and cover. Simmer until internal temperature reaches 145°F. Let pork rest for about 3 minutes.
6. Season with salt and pepper, sprinkle with chopped cilantro, and serve.

NUTRITION FACTS
Per serving: 643 calories, 262 calories from fat, 29.3 g total fat, 109.1 mg cholesterol, 327.2 mg sodium, 803 mg potassium, 61.1 g carbohydrates, 4.9 g fiber, 3.6 g sugar, 37.3 g protein

Chipotle Braised Pork with Spicy Roja Sauce

Yield: *15 servings, 2 tacos, 5 oz each*

Ingredients

Spice Rub

cumin, ground	2 tbsp
chipotle pepper chili powder	3 tbsp
white pepper, ground	1 tsp
oregano, dried and ground	1 tbsp
pork shoulder or pork butt, trimmed	5 lb
poblano peppers, roasted, seedless	4 ea
Anaheim peppers, roasted, seedless	2 ea
red bell pepper, roasted, seedless	1 ea
jalapeño pepper, roasted, seedless	4 ea
chipotle peppers in adobo sauce	2 ea
vegetable oil	2 fl oz
onion, cut small dice	½ lb
garlic clove, minced	4 ea
tomato paste	3 tbsp
adobo sauce	3 tbsp
crushed, canned tomatoes	28 oz
fresh cilantro, chopped	1 c

Preparation

1. Mix all spice rub spices and rub over pork one day ahead to coat well. Cover and reserve in refrigerator until needed.
2. Place all peppers in food processor and purée until smooth.
3. Add oil to preheated brazier over medium-high heat and sear well on all sides. Remove and set aside.
4. Lower heat to medium and add onions to brazier. Cook until onions become translucent.
5. Add garlic and cook for 1 minute until aromatic.
6. Add tomato paste and cook slightly to caramelize.
7. Add puréed chilies, adobo sauce, and crushed tomatoes. Stir well and cook for 1 minute to incorporate flavors.
8. Add meat back in with sauce. Reduce heat and simmer covered until meat is tender (approximately 4 hours) or cook in a 300°F oven until tender.
9. After pork is tender, shred and serve on warm tortillas garnished with chopped fresh cilantro.

NUTRITION FACTS
Per serving: 370 calories, 215 calories from fat, 23.9 g total fat, 93.7 mg cholesterol, 836.3 mg sodium, 827 mg potassium, 10.7 g carbohydrates, 3.2 g fiber, 1.4 g sugar, 28.4 g protein

Culinary Arts
PRINCIPLES AND APPLICATIONS

American Lamb Board

Lamb and Specialty Game

Lamb is the meat of young sheep. Because the animal is so young at the time of processing, lamb is very tender and delicate in flavor. The tender meat can be grilled, broiled, roasted, barbequed, sautéed, fried, braised, and stewed. Prior to cooking, some cuts of lamb are covered with a rub, marinated, or barded to develop the flavor further.

Specialty game such as venison, rabbit, goat, kangaroo, and bear are appearing on more restaurant menus. Rabbit can be used as a substitute in almost any chicken recipe, and goat meat is prepared in the same manner as lamb meat.

Refer to DVD
for **Flash Cards**

Chapter Objectives

1. Describe the composition of lamb.
2. Explain how lamb carcasses differ from other meat carcasses.
3. Identify the five primal cuts of lamb.
4. Identify the cuts fabricated from each primal cut of lamb.
5. Describe the types of lamb offals used in some cuisines.
6. Explain the purpose of Institutional Meat Purchase Specifications.
7. Describe the USDA inspection and grading of lamb.
8. Identify four traits that should be checked upon receiving lamb.
9. Explain why vacuum-sealed packages should only be opened at the time of use.
10. Separate a hotel rack.
11. French a rack of lamb.
12. Bone and tie a lamb loin.
13. Bone and tie a leg of lamb.
14. Cut a tenderloin into noisettes.
15. Explain how rubs, marinades, and barding can enhance the flavor of lamb.
16. Explain how to determine the doneness of lamb.
17. Cook lamb using eight different cooking methods.
18. Describe five types of specialty game meats.

Key Terms

- **lamb**
- **foresaddle**
- **hindsaddle**
- **back**
- **bracelet**
- **hotel rack**
- **lamb rack**
- **crown roast**
- **frenching**
- **lamb loin**
- **noisette**
- **lamb breast**
- **riblet**
- **blanquette**

LAMB

Lamb is the meat from slaughtered sheep that are less than a year old. It is important to understand the basic composition and bone structure of a lamb before working with the meat. ***See Figure 20-1.*** Lamb meat has smooth grain and is similar in color to beef. It is also very tender and has a mild flavor.

Composition of Lamb

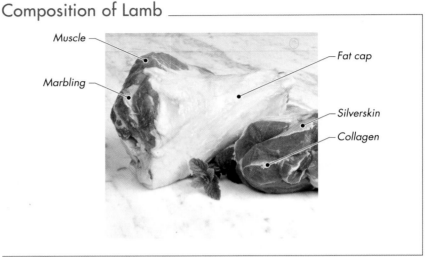

Muscle

Marbling

Fat cap

Silverskin

Collagen

Photo Courtesy of D'Artagnan, Photography by Doug Adams Studio

Figure 20-1. *Lamb meat is composed of muscle held together by connective tissues called collagen and silverskin. Marbling is found within the muscle and fat cap that surrounds the muscle.*

Lamb meat is made up of bundles of muscle fibers held together by two types of connective tissues called collagen and silverskin. *Collagen* is a soft, white, connective tissue that breaks down into gelatin when heated. *Silverskin* is a tough, rubbery, silver-white connective tissue that does not break down when heated. Silverskin is trimmed from the meat prior to cooking because it is inedible.

Muscles that receive the most exercise have more connective tissue and are less tender. For example, the legs and the shoulder areas of lamb have more connective tissue and generally yield tougher, yet more flavorful meat.

Lamb meat contains both marbling and fat cap. *Marbling* is the fat found within a muscle. *Fat cap* is the fat that surrounds a muscle. Most of the fat can be trimmed prior to or after the cooking process. Marbling does not guarantee flavor, tenderness, or juiciness. However, fat cap does affect those qualities if left on the meat while it is being cooked. Fat cap is often removed prior to service.

Bones are often not removed from many cuts of lamb prior to cooking because they enhance both the flavor and presentation of the meat. Lamb bones are porous and add flavor to the meat during the cooking process. They also add a regal look to a finished plate. Examples of this include lamb chops and leg of lamb. Understanding the skeletal structure of lamb can aid in identifying the market forms of lamb. ***See Figure 20-2.***

Skeletal Structure of Lamb

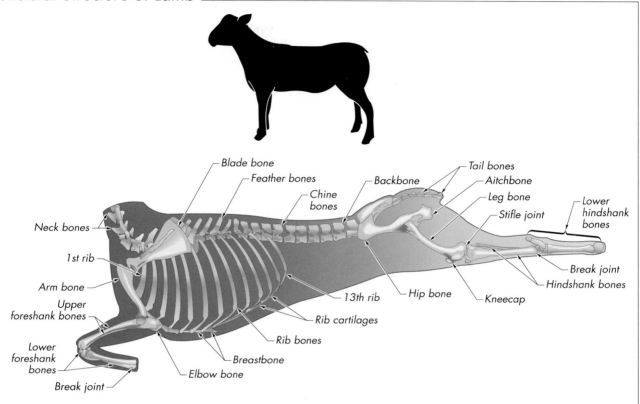

Figure 20-2. *Understanding the bone structure of a lamb can aid in identifying the market forms of lamb.*

MARKET FORMS OF LAMB

Lamb is available in a variety of market forms, including whole and partial carcasses, primal cuts, fabricated cuts, and offals. Knowledge of these different market forms is necessary for accurate product ordering.

Whole and Partial Lamb Carcasses

Both whole and partial lamb carcasses are available, but partial carcasses are not commonly purchased by foodservice operations. A *partial carcass* is a primary division of a whole carcass. Partial carcasses of lamb include foresaddles, hindsaddles, backs, and bracelets. Like whole carcasses, partial carcasses of lamb are often not purchased due to the skilled labor and storage space required to process them. Whole and partial carcasses also yield cuts of lamb that may not be used by the foodservice operation and are therefore wasted.

Foresaddle and Hindsaddle. Lamb is typically split into head and tail sections known as the foresaddle and the hindsaddle. The foresaddle and hindsaddle are split between the 12th and 13th ribs, not down the backbone. The *lamb foresaddle* is the front half of the carcass consisting of the primal shoulder, rack, and breast cuts. The *lamb hindsaddle* is the rear half of the carcass consisting of the loin and leg.

Back and Bracelet. Lamb carcasses are also divided into backs and bracelets. A *back* consists of a rack and a loin that are still joined. A back is often purchased by foodservice operations that sell a lot of lamb chops because chops can be cut from the end of the rack to the end of the loin.

A *bracelet* is a hotel rack with the breast still attached. ***See Figure 20-3.*** A *hotel rack* is a back that remains joined along the backbone. A bracelet is typically fabricated into a hotel rack and a breast. Leaving the breast attached to the rack yields a larger amount of meat and fat.

Lamb Bracelet

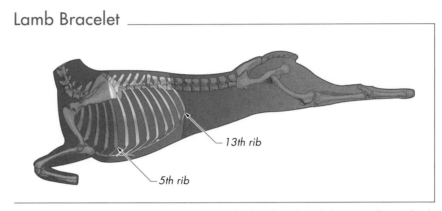

13th rib

5th rib

Figure 20-3. *A lamb bracelet consists of a hotel rack with breast still attached.*

Primal and Fabricated Cuts of Lamb

Unlike beef, lamb is not split into sides before being divided into primal cuts. A *primal cut* is a large cut from a whole or a partial carcass. The left and right primal cuts remain joined together and are purchased as a single cut. For example, the leg primal cut has two joined legs. The five primal cuts of lamb are the shoulder, rack, loin, leg, and breast/shank. ***See Figure 20-4.***

Each primal cut is further divided into fabricated cuts. A *fabricated cut* is a ready-to-cook cut that is packaged to certain size and weight specifications. Some fabricated cuts of lamb include stew meat, ground meat, hotel racks, roasts, chops, and leg of lamb.

Fabricated cuts are a convenient way of providing uniform portions while reducing labor costs. The price per pound for fabricated cuts is higher than the price per pound for primal cuts. Some operations order only primal cuts. Other operations order only fabricated cuts. Most foodservice operations primarily purchase some primal cuts and some fabricated cuts of lamb.

Shoulder. A *lamb shoulder* includes the first four rib bones of each side and the arm and neck bones. Its average weight is about 36% of the total carcass weight. Shoulder meat is quite lean and has excellent flavor. It is seldom cooked whole because of its many small bones and connective tissues. Instead, it is fabricated into a variety of cuts, including roasts, chops, stew meat, and ground meat *See Figure 20-5.* Ground lamb meat can be used in the preparation of sausages, meatballs, and meat fillings.

Primal Cuts of Lamb

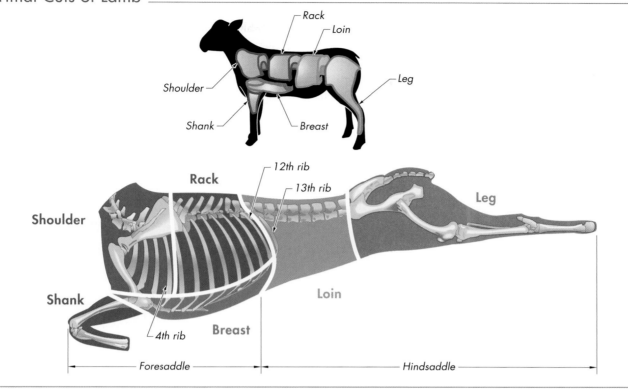

Figure 20-4. *The primal cuts of lamb include the shoulder, rack, loin, leg, and breast/shank.*

Shoulder Cuts of Lamb

Square Cut Whole Shoulder

Saratoga Roast

Boneless Shoulder Roast

Arm Chop

Blade Chop

Stew Meat

Ground Meat

Figure 20-5. *The shoulder is often fabricated into a variety of cuts, including roasts, chops, stew meat, and ground meat.*

Rack. A *lamb rack* is eight rib bones located between the shoulder and loin of a lamb. Its average weight is about 8% of the total carcass weight. The meat is tender and well-marbled because it comes from an area of the back where the muscles are not worked much. A rack is often split along the backbone into two racks. A *lamb crown roast* is a frenched rack with the ribs formed into a circle to resemble a crown. *Frenching* is a method of removing the meat and fat from the end of a bone and is generally applied to chops. *See Figure 20-6.*

Rack Cuts of Lamb

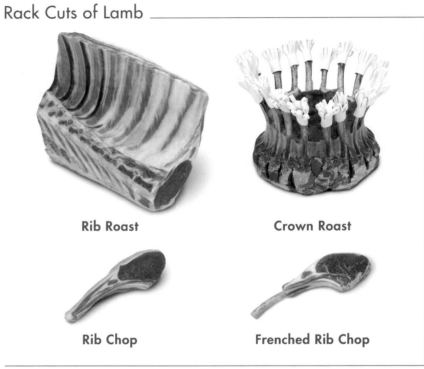

Rib Roast

Crown Roast

Rib Chop

Frenched Rib Chop

Figure 20-6. *A rack is fabricated into rib roasts, crown roasts, and rib chops.*

A rack of lamb is sometimes coated with herbs, roasted whole, and then sliced to order. However, in most cases it is cut into single or double chops and grilled or broiled. A *double rib lamb chop* is a rib chop cut to a thickness equal to two standard rib chops. An *English lamb chop* is a 2 inch thick fabricated cut taken along the entire length of the unsplit loin. Chops may also be frenched.

Loin. A *lamb loin* is a primal cut located between the rack and leg that includes the 13th rib, the loin eye muscle, the center section of the tenderloin, the strip loin, and some flank meat. *See Figure 20-7.* An unsplit primal lamb loin is commonly known as a saddle and has an average weight of about 13% of the total carcass weight.

Loin Cuts of Lamb

Loin Roast   Boneless Loin Strip   Tenderloin

Loin Chop   Double Loin Chop

American Lamb Board

Figure 20-7. *A primal loin can be fabricated into a loin roast, boneless loin strip, tenderlion, loin chop, and double loin chop.*

Fabricated cuts from the loin are best prepared using dry-heat cooking methods such as grilling, broiling, or roasting. Lamb loins are typically cut into boneless or bone-in chops that can be grilled or broiled. The tenderloin can be removed and cut into noisettes or can be roasted whole. A *noisette* is a small, round, boneless medallion of meat.

Leg. Lamb legs are not split into two separate legs. Lamb legs remain joined at the hip. A *leg of lamb* is a primal cut of lamb that contains the last portion of the backbone, hip bone, aitchbone, round bone, hindshank, and tailbone. The aitchbone is the buttock or rump bone and is located at the top of the leg. The leg accounts for approximately 34% of the total carcass weight. Leg of lamb includes part of the sirloin, the top round, bottom round, and knuckle meat. ***See Figure 20-8.***

The leg contains lean, fine-textured meat that is more tender near the sirloin end and tougher toward the shank end. Lamb leg is commonly split in two and partially boned, stuffed, and roasted. It can also be split in two and completely boned, rolled, tied, and roasted. Single lamb legs can be purchased boned, rolled, and tied (BRT). Meat from the sirloin end can also be cut into lamb steaks. Meat from the shank end is most commonly used for stew meat or ground for patties. Shank meat is commonly cut in cross-sections, braised, and served in a rich, flavorful sauce.

Leg Cuts of Lamb

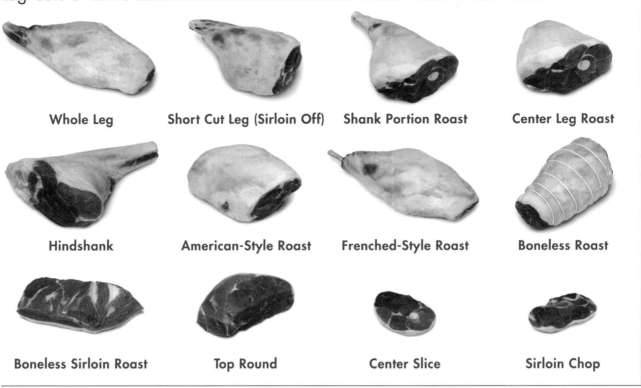

Whole Leg

Short Cut Leg (Sirloin Off)

Shank Portion Roast

Center Leg Roast

Hindshank

American-Style Roast

Frenched-Style Roast

Boneless Roast

Boneless Sirloin Roast

Top Round

Center Slice

Sirloin Chop

American Lamb Board

Figure 20-8. *A primal leg of lamb is fabricated into many cuts.*

Breast/Shank. A *lamb breast* is a thin, flat, primal cut of lamb that contains the breastbone, the tips of the rib bones, and cartilage that is located under the shoulder and ribs. The breastbone is actually cartilage because the animal is so young. The breast also includes the shank and weighs about 17% of the total carcass weight. *See Figure 20-9.* A *lamb shank* is a cut of lamb that contains the upper foreshank bones. The small section of seven or more rib tips can also be braised and is often referred to as Denver ribs or lamb riblets. A *lamb riblet* is a rectangular strip of meat cut from the lamb breast that contains part of a rib bone.

Breast/Shank Cuts of Lamb

Foreshank

Spareribs

Riblets

American Lamb Board

Figure 20-9. *A primal breast/shank is fabricated into a foreshank, spareribs, and riblets.*

The breast is not a popular cut of meat, but the shape and ample connective tissue make the breast a good cut to stuff, roll, tie, and then braise. Braising breaks down the connective tissue and yields a very tender rolled roast.

Lamb Offals

Lamb offals are not commonly used in most foodservice operations. However, lamb tongue, kidneys, liver, heart, and sweetbreads are featured dishes in Mediterranean and Middle Eastern cuisine. Lamb intestines may also be used as casings for small sausages.

Nutrition Note

Pan-fried lamb liver contains more saturated fat than beef liver prepared in the same manner.

Institutional Meat Purchase Specifications

To help ensure that foodservice operations and suppliers communicate efficiently, the USDA publishes the Institutional Meat Purchase Specifications (IMPS) for commonly purchased meats and meat products. Cuts of lamb are numbered by category. ***See Figure 20-10.***

USDA Institutional Meat Purchase Specifications (IMPS) for Lamb

Series	Item No.	Product	Item No.	Product
200 (lamb)	200	Carcass	231, 232	Loins
	202	Foresaddle	233, 234	Legs
	203	Bracelet	235, 236	Back
	204	Rack	238, 239	Trimmings
	206	Shoulders	242, 243	Loins, full
	207, 208	Shoulders, square-cut	244	Loin, boneless
	209	Breast	245	Sirloin
	210	Foreshank	246	Tenderloin
	229	Hindsaddle, long-cut	295	Lamb for stewing
	230	Hindsaddle	296	Ground lamb
700 (variety meats)	713	Lamb liver	722	Beef, lamb, or pork kidney
	717	Tongue, Swiss-cut	725	Beef, lamb, or pork brains
800 (sausage)	805	Minced luncheon meat	826	Scrapple
	811	Smoked sausage		
1200 (lamb portion cuts)	1200, 1201	Cubed steak	1232	Loin chops
	1202	Braising steak, Swiss	1233, 1234	Leg chops, boneless
	1204	Rib chops	1296	Ground lamb patties
	1207	Shoulder chops	1297	Lamb steaks, flaked and formed, frozen

Figure 20-10. *Cuts of lamb are numbered by category in the Institutional Meat Purchase Specifications (IMPS).*

INSPECTION AND GRADES OF LAMB

All lamb used in foodservice operations must be purchased from USDA-inspected plants. At the time of slaughter, the lamb carcass or the inspection tag is stamped with the round USDA inspection stamp, indicating the lamb was slaughtered at an inspected plant. This stamp does not indicate anything about the quality of the meat. The inspection stamp is used for whole and partial carcasses as well as all fabricated and processed meats. It is found either on the meat itself, on the tag attached to the meat, or on the case in which it is packed. *See Figure 20-11.* The number on the stamp identifies the plant where the animal was processed.

USDA Inspection Stamps

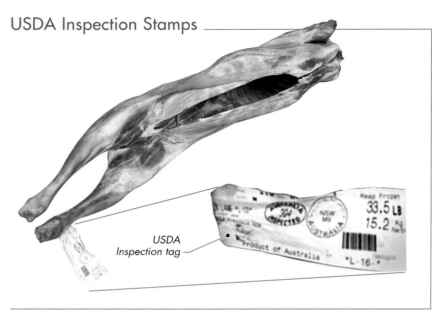

USDA Inspection tag

Figure 20-11. *A USDA inspection stamp is sometimes attached to a lamb carcass in the form of a tag.*

Quality Grades

Unlike inspection, USDA quality and yield grading is optional for lamb producers. Quality and yield grading stamps are stamped onto carcasses in the same manner as inspection stamps. Quality grading is based on the overall tenderness, juiciness, and flavor of the meat. However, quality grades do not guarantee these characteristics.

The grades of lamb commonly used in foodservice operations are USDA Prime and USDA Choice. *See Figure 20-12.* USDA Prime lamb is well marbled. USDA Choice lamb has slightly less marbling, but is still the most popular grade of lamb used in foodservice operations.

Yield Grades

Lamb can also be yield graded for the percentage of edible meat to fat and bone. Yield grade shields are numbered 1 to 5 and indicate how much usable meat can be obtained from a carcass. A grade of 1 indicates the highest yield of meat, and a grade of 5 indicates the lowest yield.

USDA Quality Grade Stamps

Figure 20-12. *USDA Prime and USDA Choice lamb is used in foodservice operations.*

RECEIVING AND STORING LAMB

Like all meats, lamb is a potentially hazardous food and the color, odor, texture, and temperature must be checked upon receipt. The meat should be light red in color and the fat should be white. *See Figure 20-13.* There should be no odor, and the texture should be firm and not dry or slick. Refrigerated lamb should maintain an internal temperature of 41°F or below. Lamb that is in the temperature danger zone should be rejected.

Receiving Standards

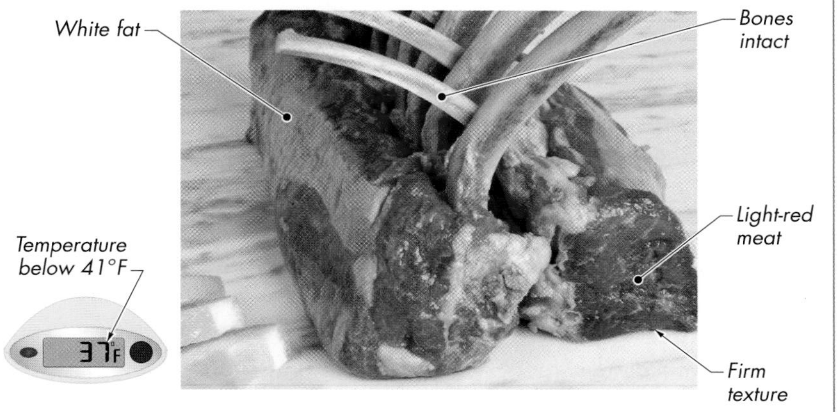

Photo Courtesy of D'Artagnan, Photography by Doug Adams Studio

Figure 20-13. *Upon receiving, lamb cuts should be checked for color, odor, texture, and temperature. The bones should also be intact.*

Frozen lamb should be stored at temperatures below 0°F. Frozen lamb that needs to be thawed should be placed in the refrigerator overnight. Larger cuts of lamb, such as legs, may take more than one day to thaw under refrigeration. If necessary, lamb can be thawed under running water if it is cooked immediately after thawing.

Vacuum-packed lamb must not be opened until needed for service or preparation. Once the vacuum seal is broken, the meat has a shelf life of only 2–3 days. Cut lamb that is not vacuum-sealed should be wrapped tightly, refrigerated immediately, and used as soon as possible.

FABRICATION OF LAMB

The size of storage facilities and availability of staff with the fabrication skills often determine whether a foodservice operation purchases primal cuts, fabricated cuts, or a combination. Lamb fabrication involves separating and frenching racks, boning and tying loins and legs, and cutting noisettes.

Separating Hotel Racks

The first step in fabricating a rack is to separate the hotel rack. *See Figure 20-14.* This yields two separate racks.

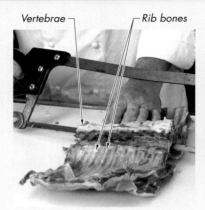

Vertebrae — Rib bones

1. Turn the rack upside-down, with the ribs pointing upward.

2. Use a meat saw to cut at a 45° angle between the base of the rib bones and the vertebrae.

3. Cut completely through the bones, using caution not to cut the meat.

4. Use a knife to finish separating the meat from one side of the vertebrae.

5. Repeat on the other rack.

6. Remove the back strap and fat cap. *Note:* The back strap is a tough tendon that runs parallel to the vertebrae.

Figure 20-14. *A hotel rack is often separated into two individual racks.*

Frenching Racks

A rack of lamb is often frenched for service. *See Figure 20-15.* When frenching a rack, the meat is removed from the end of each bone.

1. Starting on the top side of the rack, score the meat all the way to the bone, 1 inch above the eye meat.

2. Turn the rack over and score the thin membrane on the back of each rib bone from the scored cut to the tip of the bone.

3. From each scored mark, scrape the meat away from the top of the rib to expose the bone.

4. Cut between each rib bone down to the scored mark to remove the meat.

Figure 20-15. *Racks of lamb are often frenched.*

Refer to DVD for
Frenching Racks of Lamb
Media Clip

Boning and Tying Loins

A whole lamb loin is often boned and then tied to form a rolled roast. ***See Figure 20-16.*** It is important to remove all connective tissue from the skin side of the loin.

Procedure for Boning a Loin to Form a Rolled Roast

1. Trim the layer of connective tissue from the skin side of the loin.
2. Place the skin side down and remove the fat from around the tenderloin.
3. Cut under the vertebrae to separate the bones from the tenderloin. Be careful not to remove the tenderloin. Repeat this step on the other side.
4. Use a knife to cut under the rib bones to the backbone. This cut will free the rib eye from the bones.
5. Remove the backbone by pulling the meat away from the bones and making small cuts as needed.
6. Check the trimmed piece of loin for bone or connective tissue.
7. With the skin side facing up, closely trim the surface so there is about ¼ inch of fat remaining.
8. Place the outside of the meat face down and roll the sides tightly toward the center.

Figure 20-16. *A lamb loin is often boned and then formed into a rolled roast.*

Rolled loin roasts are commonly tied before roasting. A rolled roast is tied to maintain a consistent shape and ensure even cooking. ***See Figure 20-17.***

Procedure for Tying a Rolled Roast

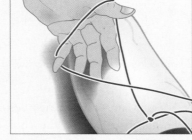

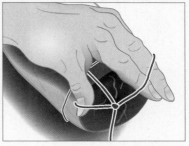

1. Using butcher's twine, tie a slip-knot around one end of the roast and secure it with a square knot.

2. Wrap the twine around one hand to form a loop. Then twist the hand over the meat until the palm is facing down.

3. Move the palm to the front of the roast and slide the loop over the meat.

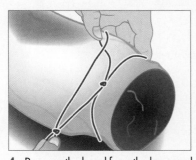

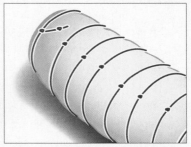

4. Remove the hand from the loop and secure the twine around the meat.
5. Pull tightly to straighten the loop.

6. Continue making loops in the same fashion 1½ inches apart down the length of the roast.

7. Tie a square knot to secure the rolled roast. Cut the twine close to the knot. Then tie another knot at the end of the twine.

Figure 20-17. *A rolled roast is tied to maintain a consistent shape and ensure even cooking.*

Refer to DVD for
Boning a Leg of Lamb
Media Clip

Boning and Tying Legs

The bone is commonly removed from a leg of lamb before cooking. *See Figure 20-18.* A boned leg of lamb makes a nice presentation.

Procedure for Boning a Leg of Lamb

Pelvic (hip) bone

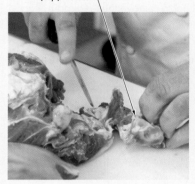

Top round

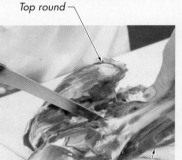

Shank   Knuckle

1. Hold the pelvic bone while cutting around the bone to free the pelvic bone from the leg.
2. Carefully remove the top round by cutting lengthwise along the leg bone.

3. Continue trimming around the knee joint to separate the knuckle meat from the bone.

4. Working near the shank end of the leg, continue to cut lengthwise to free the remaining meat from the bone.

Figure 20-18. *The bone is often removed from a leg of lamb prior to cooking.*

A leg of lamb may be tied before roasting. Tying helps the meat maintain a consistent shape and cook evenly during the cooking process. *See Figure 20-19.*

Procedure for Tying a Leg of Lamb

1. Roll the leg into a consistent shape.
2. Tie a loop of twine around the leg with the knot at the top center of the nearest end of the leg.
3. Wrap the twine around one opened hand to form a loop.
4. Twist the loop backward so that it twists around itself twice.
5. Spread the fingers wider to open the loop wider.
6. Place the loop around the leg and position it 1½ inches from the previous loop.
7. Pull the twine to tighten the loop around the leg.
8. Continue down the entire leg, keeping the loops the same distance apart.
9. Secure the last loop by tying a square knot and cutting off the excess twine.

Figure 20-19. *A boneless leg of lamb is tied to maintain a consistent shape and ensure even cooking.*

Cutting Noisettes

Noisettes are small, round, boneless medallions of meat. A lamb tenderloin may be cut into noisettes. *See Figure 20-20.*

Refer to DVD for
Cutting Noisettes
Media Clip

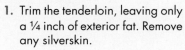

Procedure for Cutting Noisettes

1. Trim the tenderloin, leaving only a ¼ inch of exterior fat. Remove any silverskin.

2. Cut the loin horizontally to produce noisettes of desired thickness.

Figure 20-20. *Noisettes are often cut from the tenderloin prior to cooking.*

FLAVOR ENHANCERS FOR LAMB

Certain cuts of lamb may require or benefit from additional procedures prior to cooking. In addition to slow, moist-heat cooking methods, certain ingredients and methods can help break down collagen and make the lamb meat more tender. Rubs, marinades, and barding are often used to enhance the flavor of lamb.

Rubs and Marinades

Rubs and marinades can be used to enhance the flavor of lamb. Both dry and wet rubs are used. Dry rubs are made by grinding herbs and spices together into a fine powder and then rubbing the mixture into the meat prior to cooking. A wet rub is made by incorporating wet ingredients, such as lemon juice, flavored oils, puréed garlic, or honey, into a dry rub mixture and then rubbing it on the meat prior to cooking. *See Figure 20-21.*

Marinades add flavor to lamb and tenderize it at the same time. Marinades have an acidic liquid base that tenderizes the meat by breaking down the protein structure of the lamb. Any herb, spice, or condiment can be added to the base to create the desired flavor. Common marinades for lamb include red wine, lemon juice, and yogurt. The length of time that lamb is left in the marinade depends on the size and cut of meat. Once the meat has been removed from the marinade, the remaining liquid must be discarded to prevent cross-contamination.

Wet Rubs

Figure 20-21. *A wet rub can be made with oil, lemon juice, and herbs.*

Barding

Very lean cuts of lamb can be barded. *Barding* is the process of laying a piece of fatback across the surface of a lean cut of meat to add moisture and flavor. The fatback is tied to the lamb meat to prevent it from falling off during the roasting process. After the meat has finished cooking the fatback is removed.

COOKING LAMB

Tougher cuts of lamb can be slowly cooked using moist-heat cooking methods, while more tender cuts can be cooked by dry-heat cooking methods. The majority of animal muscle is made up of water. As meat loses water, it shrinks. Shrinkage is the loss of volume and weight of a piece of food as the food cooks.

Cooking lamb at too high a temperature can toughen the protein in meat. However, grilling and frying use very high temperatures for short periods and result in only the exterior of an item receiving high amounts of heat. High heat quickly cooks the exterior of meat to a crispy texture while slowly cooking the interior of the meat. This is the reason that a grilled lamb chop is crispy on the outside yet remains tender and juicy on the inside.

Determining Doneness

Cooking meat to the desired degree of doneness is determined by the type and thickness of the meat, the temperature of the meat when it begins to cook, and the intensity of the heat. Because of these variables, time is not an accurate way to determine the doneness of meats.

To determine the doneness of lamb, an instant-read thermometer is inserted into the thickest part of the meat to take the internal temperature. The temperature reading indicates the doneness of the meat. Lamb is often served at varying degrees of doneness. *See Figure 20-22.* Braised and stewed lamb is cooked until it is fork tender, or until the meat can easily be separated with a fork. Ground lamb must be cooked to an internal temperature of 160°F.

Smaller cuts of lamb that are grilled or broiled can be tested for doneness with the touch method, which involves using the sense of touch to check the firmness of cooked meat. For example, a lamb chop that is cooked well-done is firm and springs back immediately when gently pressed. In contrast, a rare lamb chop is soft and slightly mushy to the touch. The texture will feel almost the same as when the meat was raw.

Grilling and Broiling

Grilling and broiling use a hot flame to quickly sear and cook foods. Only tender cuts or ground lamb should be used with these methods. Fabricated loin cuts are commonly grilled or broiled. *See Figure 20-23.* Well-done lamb cooked with these methods will probably be somewhat dry because most of the moisture is lost as the lamb cooks. Lamb to be grilled or broiled should be properly trimmed of fat, while too little fat may cause the meat to dry out.

Determining Doneness	
Degree of Doneness	**Internal Temperature**
Medium-rare	145°F
Medium	160°F
Medium-well	165°F
Well-done	170°F

Figure 20-22. *The internal temperature of lamb should reach at least 145°F for 15 seconds and the meat should rest for 3 minutes to be considered done.*

Grilled Lamb

Figure 20-23. *Fabricated lamb loin cuts are commonly grilled.*

Roasting

Cuts from the lamb rack, loin, and leg are commonly roasted. *See Figure 20-24.* Smaller loin roasts should be cooked at higher temperatures, between 400°F and 450°F, to allow them to caramelize on the exterior without overcooking the interior. Larger leg roasts require a longer cooking time and should be roasted at lower temperatures, between 275°F and 325°F, to prevent excessive shrinkage. It is important to allow for carryover cooking when roasting meats.

Roasted Lamb

Figure 20-24. *Fabricated cuts from the lamb rack, loin, and leg are commonly roasted.*

Sautéed Lamb

Figure 20-25. *Sautéing uniformly sized cuts of lamb ensures even cooking.*

Sautéing and Frying

Tender cuts of lamb are typically sautéed or fried. Sautéing uses a small amount of hot fat to sear and cook the lamb. *See Figure 20-25.* Only uniformly sized cuts should be sautéed to ensure even cooking. Care should be taken when sautéing not to burn the oil or the surface of the food by using too high of a temperature. Lamb cutlets and chops are typically pan-fried. These cuts should not be too thick to prevent them from fully cooking.

Braising and Stewing

Braising and stewing are combination cooking methods often used to cook tougher cuts of lamb. The meat is first seared using a dry-heat method and then cooked for a long time with liquid. Searing the meat helps to add flavor to the lamb and the resulting sauce.

Both tender and tougher cuts of lamb can be braised. The difference between braising tender cuts or tougher cuts is the amount of time the lamb meat is cooked. The item is first browned in a small amount of hot fat, and then a cooking liquid is added to about halfway up the side of the meat and slowly cooked until fork tender. The resulting liquid is then slightly thickened if necessary and served as an accompanying sauce.

Stewing lamb meat is similar to braising, except the lamb is cut into small pieces, seared, and then completely covered in liquid. This process produces a brown stew. *See Figure 20-26.* However, a white stew called a blanquette can contain lamb. A *blanquette of lamb* consists of pieces of lamb meat that are blanched and then rinsed to remove any impurities before being added to a cooking liquid.

Stewed Lamb

Figure 20-26. *Lamb stew is prepared by searing small pieces of meat, covering the pieces with liquid, and then cooking the stew slowly.*

SPECIALTY GAME

Venison, rabbit, goat, kangaroo, and bear are considered specialty game. These wild animals are being raised on farms and ranches for their meat. Specialty game meats are more commonly found on menus in some regions than others, but they are continuing to grow in popularity. Specialty game must be purchased from insured vendors and stored with care to ensure food safety.

Cooking game meat to the proper temperature is the only way to destroy bacteria and parasites, such as trichinella, that can cause foodborne illness. Game steaks and chops should reach an internal temperature of 145°F for 15 seconds. Game roasts should reach 145°F for 4 minutes, and stuffed game meat should reach 165°F for 15 seconds. Ground game meat should reach 155°F for 15 seconds.

Venison

Venison refers to the meat from deer, elk, antelope, moose, or pronghorn. All venison packaging displays the name of the animal. Venison is a dark, lean meat with little to no marbling. *See Figure 20-27.* Most farm-raised venison is tender. It is often marinated overnight before cooking to lessen its gamey flavor. Dry heat methods are generally used to cook the loin, while combination cooking methods are used for the other cuts. Ground venison is used for sausage and as a pâté ingredient.

> **Nutrition Note**
>
> Venison is high in protein and low in fat. It is a much leaner meat than beef, veal, pork, and lamb.

Venison

Photo Courtesy of D'Artagnan, Photography by Doug Adams Studio
Uncooked

Charlie Trotter's
Cooked

Figure 20-27. *Uncooked venison is a dark, lean meat with little or no marbling. Cooked venison can be prepared using either dry heat or combination cooking methods.*

Rabbit

Rabbit meat is very flavorful. Uncooked rabbit meat looks similar to skinless chicken meat. Rabbit is also similar to chicken in texture and can be used as a substitute in almost any chicken recipe. *See Figure 20-28.* Because rabbit meat is very lean, care should be taken not to overcook it. Rabbit meat is very tender and lends itself to sautéing, grilling, roasting, braising, and stewing.

Rabbit

Photo Courtesy of D'Artagnan, Photography by Doug Adams Studio

Uncooked

Daniel NYC

Cooked

Figure 20-28. *Uncooked rabbit looks similar to uncooked chicken. Rabbit is often cooked in the same manner as chicken.*

Kangaroo

Czimer's Game & Seafoods, Inc.

Figure 20-29. *Hind leg meat is the most commonly prepared cut of kangaroo.*

Bear

Czimer's Game & Seafoods, Inc.

Figure 20-30. *Ranch-raised black bear is the most common type of bear meat*

Goat

Goats were one of the first domesticated animals and are still farmed for both their meat and milk. Many quality cheeses are made from goat milk. The flavor of goat tastes like a combination of lamb and beef. Goat meat is prepared in a similar manner to lamb. It contains more protein and iron and less fat than beef. Young goats, known as kids, have very sweet, tender meat. The most common market forms of goat meat are loin chops, fillets, stew meat, and ground meat.

Kangaroo

Indigenous to Australia, kangaroos are hunted for their meat and hide. Kangaroo meat has a flavor that is a cross between beef and pheasant. It is high in protein and low in fat and cholesterol. Hind leg meat is the most commonly prepared cut of kangaroo. *See Figure 20-29.* The leg meat is cooked at a very high temperature by sautéing or grilling to a medium-rare stage of doneness. Meat from other parts of the kangaroo is used to make sausage, ground meat, and jerky.

Bear

Ranch-raised black bear is the most common type of bear meat served in restaurants. *See Figure 20-30.* The most desirable cuts of bear meat come from the loin and rear legs. The loin is typically roasted or braised, while the leg meat is braised or stewed. It is always best to remove as much fat as possible from bear meat because the fat has a very strong, unpleasant taste and odor. Meat from the neck, shoulders, and front legs is typically used for stew meat or ground meat. Bear meat should be thoroughly cooked to the well-done stage.

SUMMARY

Lamb is the meat of young sheep. Market forms of lamb include foresaddles and hindsaddles, primal cuts, and fabricated cuts. Unlike other meats, lamb is not split into sides before being divided into primal cuts. The left and right primal cuts remain joined together and are purchased as a single cut. The five primal cuts of lamb are the shoulder, rack, loin, leg, and breast/shank. Each primal cut is further divided into fabricated cuts. Some fabricated cuts of lamb include stew meat, ground meat, hotel racks, roasts, chops, and leg of lamb.

USDA Prime and USDA Choice are the grades of lamb most commonly used in foodservice operations. Refrigerated lamb should be received and stored at temperatures that maintain an internal temperature of 41°F or below. Frozen lamb should be kept at temperatures of 10°F or below. Prior to cooking, some cuts of lamb are covered with a rub, marinated, or barded to develop the flavor further. Lamb can be grilled, broiled, roasted, barbequed, sautéed, fried, braised, and stewed.

Specialty game meats used in the professional kitchen include venison, rabbit, goat, kangaroo, and bear. Farm-raised venison is tender and mild in flavor. Rabbit meat is similar in texture to chicken and can be used as a substitute in almost any chicken recipe. Goat meat is prepared in the same manner as lamb meat. Kangaroo meat is high in protein and low in fat and cholesterol. Loin and leg meats are the most desirable cuts of bear.

Refer to DVD for
Quick Quiz® questions

Review

1. Describe the composition of lamb.
2. Identify common market forms of lamb.
3. Explain the difference between a foresaddle and a hindsaddle.
4. Explain the difference between a back and a bracelet.
5. Identify the five primal cuts of lamb.
6. Identify the cuts fabricated from a lamb shoulder.
7. Identify the cuts fabricated from a lamb rack.
8. Identify the cuts fabricated from a lamb loin.
9. Identify the cuts fabricated from a leg of lamb.
10. Identify the cuts fabricated from a lamb breast.
11. Identify the lamb offals that are used in some cuisines.
12. Explain the purpose of Institutional Meat Purchase Specifications.
13. Describe the USDA inspection and grading of lamb.
14. Identify four traits that should be checked upon receiving lamb.
15. Identify the required storage temperature for refrigerated lamb and frozen lamb.
16. Describe how to separate a hotel rack.
17. Describe how to french a rack.

Refer to DVD for
Review Questions

18. Describe how to bone and tie a lamb loin.
19. Describe how to bone and tie a leg of lamb.
20. Describe how to cut noisettes.
21. Explain what a marinade does in addition to enhancing flavor.
22. Explain how to determine the doneness of lamb.
23. Identify eight cooking methods used to cook lamb.
24. Describe five types of specialty game.

Applications

Refer to DVD for
Application Scoresheets

1. Separate a hotel rack.
2. French a rack of lamb.
3. Bone and tie a lamb loin.
4. Bone and tie a leg of lamb.
5. Cut a tenderloin into noisettes.
6. Marinate and cook a lean cut of lamb. Bard and cook an identical cut of lamb. Compare the quality of the prepared dishes.
7. Demonstrate three ways to determine the doneness of lamb.
8. Grill a cut of lamb and broil an identical cut of lamb. Compare the quality of the two prepared dishes.
9. Roast a cut of lamb and barbeque an identical cut of lamb. Compare the quality of the two prepared dishes.
10. Sauté a cut of lamb and stir-fry an identical cut of lamb. Compare the quality of the two prepared dishes.
11. Braise a cut of lamb and stew an identical cut of lamb. Compare the quality of the two prepared dishes.
12. Prepare a specialty game meat. Evaluate the quality of the prepared dish.

Shish Kebabs

Yield: *4 servings, 1 skewer each*

Ingredients

Marinade

vegetable oil	¼ c
olive oil	¼ c
red wine vinegar	2 tbsp
fresh lemon juice	1 tbsp
garlic, minced	1 tsp
thyme	½ tsp
marjoram	½ tsp
basil	½ tsp
oregano	½ tsp
salt and pepper	TT

Kebabs

lamb leg meat, cut in 1 inch cubes	20 oz
small tomatoes, quartered	2 ea
medium mushroom caps	8 ea
small onions, quartered	2 ea

Preparation

1. Mix marinade ingredients well and set aside.
2. Place lamb, tomatoes, mushrooms, and onions in marinade. Cover and refrigerate for 2 hours.
3. Remove ingredients from the marinade and set aside. Reserve the marinade.
4. Skewer lamb cubes, tomato, mushrooms, and onions, alternating meat and vegetables. *Note:* Each skewer should contain 5 cubes of lamb, 2 wedges of tomato, 2 mushroom caps, and 2 wedges of onion.
5. Place the kebabs on a broiler for approximately 15 minutes on low heat. Brush the kebabs frequently with the reserved marinade. Turn kebabs as needed.
6. Serve kebabs on a bed of cooked rice or pilaf.

> **NUTRITION FACTS**
> **Per serving:** 451 calories, 309 calories from fat, 34.8 g total fat, 92.1 mg cholesterol, 169.6 mg sodium, 589.5 mg potassium, 4.4 g carbohydrates, 1.2 g fiber, 2.2 g sugar, 29.7 g protein

Broiled Lamb Chops

Yield: *4 servings, 2 chops each*

Ingredients

vegetable oil	4 tbsp
fresh rosemary (leaves only), minced	2 sprigs
garlic, minced	2 cloves
salt and pepper	TT
5 oz lamb chops	8 ea

Preparation

1. Mix oil, rosemary, garlic, salt, and pepper to create a marinade.
2. Coat lamb chops with marinade. Let sit for 30 minutes.
3. Place chops in the broiler. Brown chops on each side, turning once, cooking to desired degree of doneness.

> **NUTRITION FACTS**
> **Per serving:** 716 calories, 492 calories from fat, 54.8 g total fat, 187.1 mg cholesterol, 240.3 mg sodium, 881.6 mg potassium, <1 g carbohydrates, <1 g fiber, <1 g sugar, 52.1 g protein

Roasted Rack of Lamb

Yield: *8 servings, 2 ribs each*

Ingredients

rack of lamb, frenched	2 ea
marjoram	TT
salt and pepper	TT
veal stock	1 pt
garlic, minced	1 clove

Preparation

1. Season a frenched rack of lamb with marjoram, salt, and pepper to taste.
2. Place the seasoned rack of lamb in a roasting pan with a roasting rack, fat-side up.
3. Roast in 400°F oven for 30–45 minutes or until desired doneness is achieved.
4. Transfer the lamb to a baking pan. Hold in a warm place.
5. To make an au jus, deglaze the roasting pan with veal stock. Pour drippings and stock into a saucepan. Add garlic and simmer over low heat.
6. Strain sauce through a chinois and season to taste.
7. Cut the rack of lamb between the ribs. Serve with au jus.

> **NUTRITION FACTS**
> **Per serving:** 249 calories, 112 calories from fat, 12.4 g total fat, 95.4 mg cholesterol, 237 mg sodium, 497 mg potassium, 2.3 g carbohydrates, <1 g fiber, <1 g sugar, 30.1 g protein

Thyme, Sage, and Parmesan Crusted Rack of Lamb

Yield: *4 servings, 2 chops each*

Ingredients

8 bone rack of lamb	1 ea
salt and pepper	TT
olive oil	1 tbsp
egg white	1 ea
Dijon mustard	2 tbsp
thyme	½ tbsp
sage	½ tbsp
parmesan, grated	½ tbsp
panko bread crumbs	½ c

Preparation

1. Season lamb with salt and pepper.
2. In medium sauté pan over medium-high heat, add the olive oil and cook lamb until browned well on all sides.
3. Remove lamb from pan and set aside for 2 minutes to cool slightly.
4. In small bowl, whisk together egg and mustard.
5. Spread mustard mixture over lamb to coat well.
6. Mix herbs, cheese, and bread crumbs and coat lamb rack well.
7. Place rack in 350°F oven and roast for 20–25 minutes or until internal temperature reaches between 128°F and 130°F.
8. Remove from oven and let rest for 15 minutes before slicing and serving.

> **NUTRITION FACTS**
> **Per serving:** 324 calories, 146 calories from fat, 16.3 g total fat, 94.1 mg cholesterol, 404 mg sodium, 487.9 mg potassium, 10.7 g carbohydrates, 1 g fiber, <1 g sugar, 31.9 g protein

Sautéed Noisettes of Lamb with Roasted Red Pepper Coulis

Yield: *2 servings, 8 oz of lamb and 4 fl oz of sauce each*

Ingredients

Coulis	
canola oil	1 tsp
onion, diced	⅓ c
garlic	1 clove
roasted red peppers, cleaned and diced	6 oz
white wine	¼ c
chicken stock	¼ c
kosher salt and pepper	TT
balsamic vinegar	½ tsp
butter	1 tbsp
olive oil	½ tbsp
kosher salt and black pepper	TT
2 oz lamb noisettes	8 ea

Preparation

1. Heat canola oil in a saucepan over low-medium heat.
2. Sweat onions and garlic until soft.
3. Add roasted red pepper and cook for 1–2 minutes.
4. Add white wine and let reduce by half.
5. Add chicken stock and bring to a simmer. Purée until smooth.
6. Return to saucepan. Adjust consistency and season.
7. Add vinegar and mount in butter.
8. Heat olive oil in a sauté pan over medium-high heat.
9. Season lamb with salt and pepper and sauté until medium-rare.
10. Let rest 1 minute before serving.

> **NUTRITION FACTS**
> **Per serving:** 514 calories, 254 calories from fat, 28.6 g total fat, 197.6 mg cholesterol, 431.8 mg sodium, 523.1 mg potassium, 7.5 g carbohydrates, 1.3 g fiber, 2.1 g sugar, 49.6 g protein

Braised Lamb Shanks with Rosemary Gremolata

Yield: *4 servings, 1 lamb shank and 1 tsp of gremolata each*

Ingredients

Gremolata

rosemary, finely minced	1 tsp
parsley, finely chopped	2 tsp
lemon zest, finely minced	1 tsp
small garlic clove, minced	1 ea
lamb shanks	4 ea
canola oil	2 tsp
onions, diced	½ c
garlic cloves, minced	2 ea
tomato purée	¼ c
white wine	⅔ c
brown lamb stock	2¾ c
lemon zest	1 tsp
sachet d'épices	
bay leaf	1 ea
thyme, dried	½ tsp
unsalted butter	1 tbsp
salt and black pepper	TT

Preparation

1. Combine all gremolata ingredients.
2. Brown lamb shanks over medium-high heat in oil and remove from pan when browned on all sides.
3. Caramelized the onions in same pan as the lamb shanks. When browned, add garlic and tomato purée and cook until lightly browned.
4. Add white wine and reduce until almost dry. Add stock, lemon zest, and sachet d'épices.
5. Return shanks back to pan, bring to a boil, cover, and place in 325°F oven.
6. When tender, remove shanks and degrease liquid.
7. Reduce to a nappe consistency. Mount in butter and adjust seasonings.
8. Sprinkle gremolata over lamb shanks before service.

> **NUTRITION FACTS**
> **Per serving:** 426 calories, 143 calories from fat, 16 g total fat, 173.2 mg cholesterol, 587.9 mg sodium, 909.3 mg potassium, 7.6 g carbohydrates, 1 g fiber, 3 g sugar, 53.5 g protein

Sautéed Venison with a Black Pepper Cherry Glaze

Yield: *4 servings, 6 oz of venison and 3 fl oz of sauce each*

Ingredients

Glaze

honey	1 c
sherry vinegar	½ c
dried cherries	1 c
coriander seeds	1 tsp
cracked black peppercorns	1 tbsp
thyme	2 sprigs
whole unsalted butter	2 tbsp
kosher salt and pepper	TT
canola oil	2 tsp
kosher salt and black pepper	TT
3 oz venison loin medallions, pounded slightly	8 ea
whole unsalted butter	2 tbsp
garlic, crushed	1 clove

Preparation

1. In a saucepan, combine all black pepper glaze ingredients except the thyme, butter, salt, and pepper and let reduce to syrup-like consistency over medium-low heat.
2. Mount in butter and season with salt and black pepper.
3. Heat oil in sauté pan over medium-high heat.
4. Season venison and sauté until rare.
5. Add butter, garlic, and fresh thyme and baste venison medallions while cooking until medium-rare.
6. Remove and let rest 1–2 minutes.

> **NUTRITION FACTS**
> **Per serving:** 704 calories, 169 calories from fat, 19 g total fat, 61.2 mg cholesterol, 152.7 mg sodium, 126.1 mg potassium, 104.4 g carbohydrates, 3.4 g fiber, 69.7 g sugar, 38.4 g protein

All-Clad Metalcrafters

Baking and Pastry Fundamentals

Yeast breads, quick breads, cakes, cookies, pies, pastries, and frozen desserts can be challenging dishes to execute. Baked products and pastries require exact formulas that are based on how ingredients chemically interact with one another. The proper combinations and accurate measuring of flour, fat, liquid, leavening agents, eggs, and other ingredients and the appropriate use of proofing, baking, and cooling techniques are required in order to produce successful baked products and pastries.

Refer to DVD
for **Flash Cards**

Chapter Objectives

1. Identify how ingredients are measured in a bakeshop.
2. Explain how to calculate the baker's percentage of an ingredient.
3. Describe eight types of ingredients used to create baked products.
4. Define terms describing methods of combining ingredients.
5. Identify common equipment, bakeware, and tools used in a bakeshop.
6. Contrast three types of yeast doughs.
7. Describe the steps to follow to prepare yeast doughs.
8. Describe three methods of mixing quick breads.
9. Demonstrate four methods of mixing cake batters.
10. Describe nine types of icings.
11. Form, fill, and use a pastry bag to pipe icing.
12. Describe six methods of preparing cookies.
13. Contrast mealy and flaky piecrusts.
14. Blind bake a piecrust.
15. Prepare a fruit, cream, soft, and chiffon pie filling.
16. Prepare three types of meringues.
17. Explain how laminated dough is prepared.
18. Identify eight types of custard and creams.
19. Contrast six types of frozen desserts.

Key Terms

- **formula**
- **baker's percentage**
- **gluten**
- **yeast**
- **blend**
- **cream**
- **cut-in**
- **fold**
- **scaling**
- **kneading**
- **punching**
- **rounding**
- **panning**
- **proofing**
- **scoring**
- **docking**
- **icing**
- **ratio**
- **blind baking**
- **overrun**

Irinox USA

BAKESHOP MEASUREMENTS

In a bakeshop, ingredients are measured by weight or by volume. Although weight is more accurate than volume, it is important to understand both. Weight is measured in grams (g), pounds (lb), and ounces (oz). Volume is measured in liters (l), pints (pt), cups (c), tablespoons (tbsp), and teaspoons (tsp).

The weight of liquid is measured in ounces (oz) and the volume of liquid is measured in fluid ounces (fl oz). However, 1 oz does not equal 1 fl oz. Because different ingredients have different densities, two items of the same volume can have different weights. For example, 1 cup of honey weighs more than 1 cup of water.

A baker's scale or a digital scale can be used to measure the weight of ingredients in the bakeshop. *See Figure 21-1.* A *baker's scale,* also known as a balance scale, is a scale with two platforms that use a counterbalance system to measure weight. The food to be weighed is placed on one platform, and a weight is placed on the other platform. Weight is added to, or removed from, the second platform until the scale balances. The beam between the two platforms has a smaller weight that is used for fine adjustment of the scale. A baker's scale is used to measure most baking ingredients. Digital scales are often used when weighing small amounts, such as ⅛ oz, of an ingredient.

Refer to DVD for
Measuring Weight
Culinary Math Tutorial

Bakeshop Scales

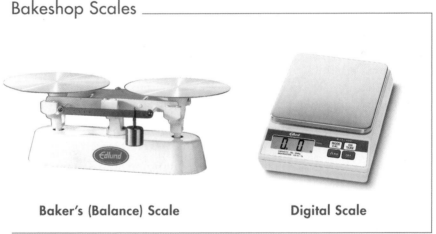

Baker's (Balance) Scale   **Digital Scale**

Edlund Co.

Figure 21-1. *A baker's scale and a digital scale can both be used to measure the weight of ingredients in the bakeshop.*

FORMULAS AND BAKER'S PERCENTAGES

Baking is much different from other cooking techniques because once a product is prepared and placed in the oven, there is no opportunity to make any changes. Because of this, formulas are used in the bakeshop. A *formula* is a recipe format in which all ingredient quantities are provided as baker's percentages. *See Figure 21-2.* A *baker's percentage* is the weight of a particular ingredient expressed as a percentage based on the weight of the main ingredient in a formula. Formulas help to ensure that measurements are performed consistently and accurately.

Bakeshop Formulas		
Ingredient	Ingredient Weight	Baker's Percentage
Bread flour	7 lb 8 oz	100%
Salt	3 oz	3%
Sugar	3½ oz	3%
Shortening	3 oz	3%
Egg whites	3 oz	3%
Water	4 lb 8 oz	60%
Compressed yeast	4½ oz	4%

Figure 21-2. *A formula is a recipe format in which all ingredient quantities are provided as baker's percentages.*

If flour is the main ingredient in a formula, its baker's percentage is set to 100% and all other ingredients in the formula are expressed as a percentage based on the weight of flour. For example, if the formula also uses sugar and the baker's percentage of the sugar in the formula is 50%, no matter how much flour is used, the weight of the sugar will always be equal to 50% of the weight of flour. It does not mean that 50% of the total formula is made up of sugar. In fact, the sum of the baker's percentages in a formula will always be greater than 100%.

If a formula uses two types of flour, the two types of flour combined are considered the main ingredient and will have a total baker's percentage of 100%. For example, a formula for rye bread may contain white flour with a baker's percentage of 75% and rye flour with a baker's percentage of 25%. In this case, the two types of flour will have a total baker's percentage of 100% (75% + 25% = 100%).

Baker's Percentages

Two important rules must be followed when using baker's percentages. First, all of the items must be weighed (not measured by volume). Second, all of the measurements must have the same unit of measure.

The first step when calculating the weight of each ingredient in a formula is to determine the weight of the main ingredient. Sometimes, the weight of the main ingredient is stated directly. For example, a baker may decide to make a batch of French bread using 20 lb of bread flour. In other cases, the weight of the main ingredient may need to be calculated based on the desired yield of the formula. When a specific total yield is desired, the weight of the main ingredient can be calculated using the following formula:

WM = DY ÷ TBP

where

WM = weight of main ingredient

DY = desired yield

TBP = total baker's percentage

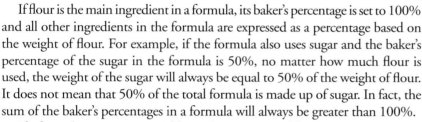

$$\text{Weight of Main Ingredient} = \frac{\text{Desired Yield}}{\text{Total Baker's Percentage}}$$

For example, if the total of the baker's percentages in a hard-roll dough recipe is 182%, and the desired yield is 30 lb of dough, the weight of the main ingredient (bread flour) is calculated as follows:

WM = DY ÷ TBP

WM = 30 lb ÷ 182%

WM = 30 lb ÷ 1.82

WM = **16.5 lb**

Once the weight of the main ingredient is determined, the remaining ingredient weights are calculated. This is done by multiplying the baker's percentage for each of the remaining ingredients in the formula by the weight of the main ingredient. *See Figure 21-3.* The weights of the remaining ingredients are calculated using the following formula:

WI = WM × BP

where

WI = weight of ingredient

WM = weight of main ingredient

BP = baker's percentage

Weight of Ingredient = Weight of Main Ingredient × Baker's Percentage

Using a Baker's Percentage		
Ingredients	**Baker's Percentage**	**Ingredient Weight* (WI = WM × BP)**
Bread flour (main ingredient)	100%	16.5 × 100% = 16.5 lb × 1.00 = **16.5 lb**
Salt	3%	16.5 × 3% = 16.5 lb × 0.03 = **0.5 lb**
Sugar	5%	16.5 × 5% = 16.5 lb × 0.05 = **0.8 lb**
Shortening	6%	16.5 × 6% = 16.5 lb × 0.06 = **1.0 lb**
Egg whites	3%	16.5 × 3% = 16.5 lb × 0.03 = **0.5 lb**
Water	60%	16.5 × 60% = 16.5 lb × 0.60 = **9.9 lb**
Yeast	5%	16.5 × 5% = 16.5 lb × 0.05 = **0.8 lb**
Total	**182%**	**30 lb**

* Results rounded to the tenths place

Figure 21-3. *When using a formula, ingredient weights are calculated by multiplying the baker's percentages for each ingredient by the amount of the main ingredient to be used.*

For example, in the hard-roll dough formula, bread flour is the main ingredient. If 16.5 lb of bread flour is used to make the dough, and the baker's percentage for salt in this formula is 3%, the weight of salt required is calculated as follows:

WI = WM × BP

WI = 16.5 lb × 3%

WI = 16.5 lb × 0.03

WI = **0.5 lb**

Flour is not always the main ingredient in a formula. In fact, some formulas will not contain any flour. For example, dried apricots are the main ingredient in a recipe for an apricot filling. If the baker's percentage for water is 20% and the filling is to be made using 10 lb of dried apricots, the weight of water required is calculated as follows:

WI = WM × BP
WI = 10 lb × 20%
WI = 10 lb × 0.20
WI = **2.0 lb**

Converting Recipes into Formulas

A recipe can be converted into a formula by converting the ingredient quantities to a common weight unit of measure and then dividing that weight by the weight of the main ingredient. *See Figure 21-4.* The first step is to identify the main ingredient. The next step is to convert all of the ingredient amounts in the recipe to the same weight unit of measure. Then, the weight of each ingredient is divided by the weight of the main ingredient. The result, in decimal form, is changed to a percentage. A baker's percentage is calculated by using the following formula:

BP = WI ÷ WM
where
BP = baker's percentage
WI = weight of ingredient
WM = weight of main ingredient

$$\text{Baker's Percentage} = \frac{\text{Weight of Ingredient}}{\text{Weight of Main Ingredient}}$$

Converting Recipes into Formulas			
Ingredient	Ingredient Weight (*W*)	Ingredient Weight Converted to Ounces	Baker's Percentage (*BP = WI ÷ WM*)
Cake flour (main ingredient)	**5 lb (WM)**	5 lb × 16 oz/lb = **80 oz**	80 oz ÷ 80 oz = 1.0 = **100%**
Shortening	3.5 lb	3.5 lb × 16 oz/lb = **56 oz**	56 oz ÷ 80 oz = 0.7 = **70%**
Sugar	6.25 lb	6.25 lb × 16 oz/lb = **100 oz**	100 oz ÷ 80 oz = 1.25 = **125%**
Salt	3 oz	**3 oz**	3 oz ÷ 80 oz = 0.0375 = **3.75%**
Baking powder	5 oz	**5 oz**	5 oz ÷ 80 oz = 0.0625 = **6.25%**
Water	1.75 lb	1.75 lb × 16 oz/lb = **28 oz**	28 oz ÷ 80 oz = 0.25 = **25%**
Nonfat dry milk	5 oz	**5 oz**	5 oz ÷ 80 oz = 0.0625 = **6.25%**
Eggs	1.25 lb	1.25 lb × 16 oz/lb = **20 oz**	20 oz ÷ 80 oz = 0.25 = **25%**
Egg whites	2 lb	2 lb × 16 oz/lb = **32 oz**	32 oz ÷ 80 oz = 0.4 = **40%**
Vanilla extract	2 oz	**2 oz**	2 oz ÷ 80 oz = 0.025 = **2.5%**

Figure 21-4. *A recipe can be converted into a formula by converting the ingredient quantities to a common weight unit of measure and then dividing that weight by the weight of the main ingredient.*

For example, a white cake recipe may call for 5 lb of cake flour, 3.5 lb of shortening, 5 oz of baking powder, and a number of other ingredients. To create a formula from this recipe, all of the original ingredient amounts need to be converted to a common unit of measure. First, the amount of the main ingredient, cake flour, is converted from 5 lb to 80 oz (5 lb × 16 = 80 oz).

To calculate the baker's percentage of shortening, the weight of shortening called for in the recipe is first converted to ounces (3.5 lb × 16 = 56 oz). Then the baker's percentage is calculated as follows:

$$BP = WI \div WM$$
BP = 56 oz ÷ 80 oz
BP = 0.70
BP = **70%**

The same process is repeated for every ingredient in the recipe. Conversion calculations are not necessary for the ingredient measurements provided in ounces.

BAKESHOP INGREDIENTS

Different types of flours, sugars and sweeteners, fats, eggs, and milk produce different results when used in baked products. Thickeners and leavening agents are used to alter the texture and lightness of breads. The flavoring ingredient chosen often determines the flavor of the end product.

Ingredients cannot be easily substituted in baked products. Baking is based upon chemical reactions among ingredients. Bakeshop recipes are known as formulas because they require precise measurements and limited substitutions in order to ensure successful end products. For example, substituting bread flour for cake flour will result in a product with an entirely different look, texture, and flavor.

Flours

Flour is a fine powder that is created by grinding grains. Flour supplies structure and nutritional value to baked products. It also acts as an absorbing agent. Flour is characterized as being hard or soft, depending on the protein content of the grain from which it is ground. *Gluten* is a type of protein found in grains such as wheat, barley, and some varieties of oats. Gluten is the protein in flour that, when combined with water, gives a baked product its structure. *See Figure 21-5.* Common flours used in the professional kitchen include bread, cake, pastry, wheat, rye, and specialty flours.

Bread Flour. Bread flour, also known as hard wheat flour, has low amounts of starch and high amounts of protein that produce a large amount of gluten when water is added. Bread flour is typically used to make bread doughs.

Cake Flour. Cake flour, also known as soft wheat flour, contains low amounts of protein and high amounts of starch. Cake flour does not form gluten as readily as bread flour, which results in a more delicate, tender crumb. Cake flour is typically used to make cakes, cookies, pies, and pastries. Most cake flour is bleached during processing, which results in flour that is more acidic. The acidity of cake flour causes starches to gelatinize and, in turn, reduces baking time.

Gluten Formation

Figure 21-5. *Gluten is the protein in flour that, when combined with water, gives a baked product its structure.*

Pastry Flour. Pastry flour contains the ideal amount of gluten for pie doughs. If pastry flour is not available, a blend of 60% cake flour and 40% bread flour can be used. Pastry flour has a tendency to pack and form lumps, so it should always be sifted before use. Some bakers and pastry chefs chill pastry flour in order to keep pie dough below 70°F during the mixing period.

Wheat Flour. Whole-wheat flour is unbleached flour that still contains the bran, endosperm, and germ of the wheat kernel. It has a nuttier flavor than other flours and is used to make breads. It has more nutritional value than other types of wheat flours.

Rye Flour. Rye flour is a dark-colored flour milled from rye seeds. Rye flour lacks protein, so it is almost always combined with wheat flour when making bread. Variations of rye flour are made from different parts of rye seeds and include white, cream, dark, and pumpernickel flour. Pumpernickel flour is a meal form of rye made by grinding the whole grain.

Specialty Flours. Flours can also be made from corn, soybeans, rice, barley, oats, millet, potatoes, buckwheat, spelt, quinoa, and nuts. Almond and hazelnut flours are commonly used by bakeries and pastry shops. These nut flours are made by milling the nut into a fine meal. Specialty flours lack the gluten-forming properties of wheat, so they must always be combined with wheat flour when making breads.

Sugars and Natural Sweeteners

Sugars and natural sweeteners add color, texture, and flavor to baked products. Sugar stimulates the growth of yeast, supplies moisture, and helps prolong freshness. It also adds flavor, color, and tenderness to baked products through caramelization. Sugar should always be dissolved in liquid to ensure even distribution in baked products. Sugar comes from a variety of plant sources and is available in many different forms. Granulated sugar, pastry sugar, confectioners' sugar, turbinado sugar, brown sugar, molasses, honey, and agave nectar are a few sugars and natural sweeteners used in the bakeshop. *See Figure 21-6.*

Granulated Sugar. Granulated sugar is a white sugar composed of small, uniformly sized crystals.

Pastry Sugar. Pastry sugar is tiny crystals of granulated sugar. Pastry sugar dissolves quickly and is preferred for use in angel food cakes and meringues.

Confectioners' Sugar. Confectioners' sugar, also known as powdered sugar, is granulated sugar that has been ground into a very fine powder. A small amount of cornstarch is added to confectioners' sugar to prevent lumping. Confectioners' sugar is often used to make icings and meringues, as the cornstarch absorbs excess moisture and helps prevent weeping.

Turbinado Sugar. Turbinado sugar is a pale brown sugar that is purified and cleaned by steam before being dried into coarse-grain crystals.

Brown Sugar. Light brown sugar is a moist sugar product that contains approximately 3.5% molasses. Dark brown sugar is a moist sugar product that contains approximately 7% molasses. Brown sugar should be kept in an airtight container or it will dry out and turn hard.

Sugars and Natural Sweeteners

Granulated Sugar

Confectioners' Sugar

Brown Sugar

Molasses

Honey

Figure 21-6. *Granulated sugar, confectioners' sugar, brown sugar, molasses, and honey are a few of the sugars and natural sweeteners used in the bakeshop.*

Molasses. Molasses is the dark syrup that is left after sugar cane has been processed. Cane syrup is processed three to four times, yielding a more concentrated grade of molasses each time. Blackstrap molasses is the strongest and most processed form of molasses. Molasses is often added to bread, cake, and cookie recipes because of its distinct flavor.

Honey. Honey is a sweet, thick fluid made by honeybees from flower nectar and is approximately 1½ times as sweet as granulated sugar. The flavor of honey depends on the flowers from which the nectar was derived. Clover, buckwheat, and lavender each yield a different flavor of honey.

Agave Nectar. Agave nectar, sometimes called agave syrup, is the nectar from an agave plant native to Mexico. The blue agave contains a high percentage of fructose in its nectar and produces the premium form of agave nectar. Lighter and darker varieties of agave nectar are made from the same plant. The flavor of agave nectar is comparable, but not identical, to honey.

Fats

Fat is an essential ingredient in the bakeshop. Fat adds richness and tenderness and improves grain, texture, and shelf life of baked products. Butter, shortening, and oil are common types of fat used to make tender and moist baked products. *See Figure 21-7.*

Fats

Butter **Shortening** **Oil**

Figure 21-7. *Butter, shortening, and oil are types of fat used to make tender and moist baked products.*

Butters. Butter is made from cream that has a butterfat content of at least 80%. European-style butter contains 82% to 86% butterfat and is considered premium butter.

Shortenings. Some bread recipes call for hydrogenated shortening. Hydrogenation is a process that changes the molecular structure of an oil into a solid. When hydrogenation occurs, the melting point of fat is raised, making it more pliable when blended with flour. Another effect of hydrogenation is an increased shelf life. Hydrogenated shortening does not contain any water but is 10% air by weight.

Some recipes call for emulsified shortening. Emulsified shortenings assist with moisture absorption and are often used to make icings and cakes that contain more sugar than flour. Both hydrogenated and emulsified shortenings contain trans fats.

Oils. An oil is a fat that remains in a liquid state at room temperature. Oils are most often taken from plant sources and maintain moisture in baked products. Mildly flavored oils, such as corn, vegetable, or canola oil, are preferred for baking. Oils with a strong flavor, such as olive oil, can overpower the flavor of baked products.

Eggs

In addition to adding color and flavor, the protein in eggs gives structure to baked products. Eggs also function as a thickener in custards and add moisture, leavening, and nutritive value to cake batters. Beaten and whipped eggs trap pockets of air inside doughs and batters that expand when heated, helping baked products rise.

Both fresh and frozen eggs are used in bakeshops. *See Figure 21-8.* Egg yolks are natural emulsifiers, meaning that they hold or bind ingredients together that normally would not combine, such as water and oil. Egg yolks also add a rich flavor and golden color to baked products. An egg wash is a combination of egg yolk and either milk or water that is often brushed on top of piecrusts or bread dough before baking to add a glossy sheen and a golden color to the baked product.

Eggs

Fresh

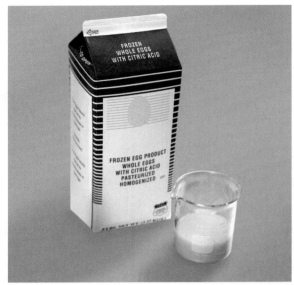

Frozen

American Egg Board

Figure 21-8. *Both fresh and frozen eggs are used in bakeshops.*

Milk

Liquid

Dried

Figure 21-9. *Milk used in the preparation of breads and sweet doughs may be liquid or dried.*

Milk

Milk and other dairy products have many different effects on baked products. Milk improves the texture and flavor, supplies moisture, and adds nutrients such as calcium, vitamin D, and protein. The proteins in milk assist in producing a finer crumb in yeast breads. When breads or pastry items containing milk are baked, the proteins and sugars in the milk absorb moisture, producing a softer crust. These same proteins and lactose (milk sugar) break down during baking and begin to caramelize, which adds a rich color to the crusts of baked products.

Milk used in the preparation of breads and sweet doughs may be liquid or dried. *See Figure 21-9.* Dry milk must be reconstituted before or during the mixing period. Milk-like products, such as soy milk, rice milk, and almond milk, can be used in place of milk. Water also can be substituted for milk in most bread recipes.

Thickening Agents

Thickening agents provide structure and stability to baked products. Common thickening agents used in the bakeshop include cornstarch, arrowroot, flour, gelatin, tapioca, and modified starches. *See Figure 21-10.* Each thickening agent has properties that make it more favorable for use in some items than others.

Thickening Agents

Figure 21-10. *Thickening agents used in the bakeshop include cornstarch, arrowroot, flour, gelatin, and tapioca.*

Cornstarch. *Cornstarch* is the white, powdery, pure starch derived from corn. Cornstarch is a popular thickening agent because it will not cloud a clear sauce and is preferred for use in pie fillings. Cornstarch should be mixed with a small amount of cold water to make a slurry before it is added to a hot mixture. Cornstarch should be thoroughly cooked or it will leave an unpleasant aftertaste.

Arrowroot. *Arrowroot* is a thickening agent that is the edible starch from the rootstock of the arrowroot plant. It is often used to make pie fillings and custards. Like cornstarch, arrowroot is made into a slurry before being added to a hot mixture.

Flours. A low-protein flour works best for use as a thickening agent. However, it requires twice as much flour as cornstarch or arrowroot to thicken a product. Any product thickened with flour should be cooked thoroughly or the flavor of raw flour will remain.

Gelatin. *Gelatin* is a flavorless thickening agent made from animal protein. It is available in powdered and sheet forms. Powdered gelatin comes in ¼ oz envelopes or in bulk form. Gelatin sheets are sold in a standard size that is equivalent to 1 tsp of powdered gelatin. Gelatin must be dissolved or soaked in cold water and then gently heated before being added to a product.

Tapioca. *Tapioca* is a thickening agent derived from the cassava plant. It is available in round, granules or as ground flour. Tapioca leaves fillings clear and glossy and has twice the thickening power of flour. When using granular tapioca, it is important to have enough liquid in a filling to ensure that the granules will dissolve.

Modified Starches. Modified starches gelatinize quickly when cooked and offset the action of fruit acids. A *modified starch* is a combination of starch and one or more vegetable gums used as a thickeners. For example, ClearJel® and similar products are blends of starch and vegetable gum that produce a finished product with a high sheen. They also maintain color, produce a smooth consistency, and do not cloud when refrigerated.

Leavening Agents

A *leavening agent* is any ingredient that causes a baked product to rise by the action of air, steam, chemicals, or yeast. The most basic way to leaven baked products is to incorporate air and steam into a dough or batter. Because the air is lighter than the surrounding dough or batter, the air forces the product to rise as it tries to escape. Steam is created when liquids in the dough or batter evaporate and force a bread to expand. Air pockets also can be created in products by whipping eggs or creaming fat and sugars.

Both natural and chemical leavening agents are used in the bakeshop. In addition to air and steam, yeast, baking powder, baking soda, cream of tartar, and ammonium bicarbonate are used to leaven breads, cakes, cookies, and doughs. *See Figure 21-11.*

Leavening Agents

Yeast

Baking Powder

Baking Soda

Cream of Tartar

Ammonium Bicarbonate

Figure 21-11. *In addition to air and steam, yeast, baking powder, baking soda, cream of tartar, and ammonium bicarbonate are used to leaven breads, cakes, cookies, and doughs.*

Yeast. *Yeast* is a microscopic, living, single-celled fungus that releases carbon dioxide and alcohol through a process called fermentation when provided with food (sugar) in a warm, moist environment. Yeast increases volume, improves flavor, and adds texture to dough. Yeast is very sensitive to temperature and, depending on the type of yeast, is activated between 70°F and 130°F. *See Figure 21-12.* If the temperature is too cold, the yeast remains dormant. If the temperature is too hot, the yeast dies. Three common varieties of yeast are compressed yeast, active dry yeast, and instant yeast.

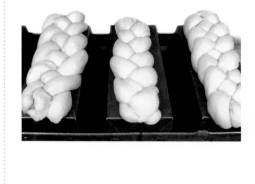

Yeast Activation

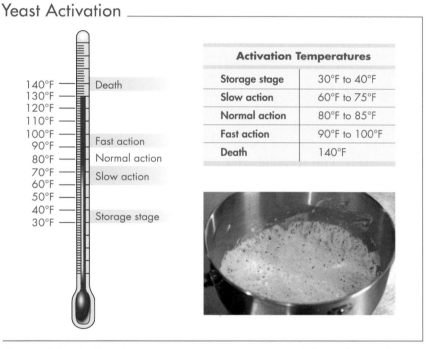

Activation Temperatures	
Storage stage	30°F to 40°F
Slow action	60°F to 75°F
Normal action	80°F to 85°F
Fast action	90°F to 100°F
Death	140°F

Figure 21-12. *Yeast is very sensitive to temperature and, depending on the type of yeast, is activated between 70°F and 130°F.*

Compressed yeast, also known as fresh yeast, is yeast that has approximately 70% moisture and is available in 1 lb cakes or blocks. Compressed yeast produces the most carbon dioxide of any yeast. Once opened, compressed yeast should be wrapped tightly in plastic wrap, refrigerated, and used within two weeks. It also can be frozen for up to four months. Its yeasty aroma will turn ammonia-like when old or spoiled. The best method for using compressed yeast is to dissolve it in double its weight of lukewarm water (between 90°F and 100°F) prior to adding it to a bread recipe.

Active dry yeast is a form of yeast that has been dehydrated and looks like small granules. It is sold in ¼ oz packages or in 1 lb vacuum-sealed packages. Because active dry yeast is lighter than compressed yeast, only half as much dry yeast (by weight) is needed. It produces the least amount of carbon dioxide. It is best activated in warm water between 105°F and 115°F. It has a shelf life of one year.

Instant yeast is a form of yeast that does not need to be hydrated and may be directly added to a warm flour mixture. It is often labeled as rapid-rise yeast and is available in both dried and vacuum-packaged form. Instant yeast is a more aggressive yeast than compressed or active dry yeast. Because it is so active, only one-fourth the amount of instant yeast is needed compared to compressed yeast and one-half the amount compared to active dry yeast. Once opened, instant yeast can be refrigerated for three to four months or frozen for up to six months.

A sourdough starter creates its own yeast spores when left at room temperature. It takes two to three weeks to grow enough yeast to leaven a loaf of bread. During this time, more flour must be added to the mixture as food for the bacteria and yeast spores. A portion of this mixture is then used to make sourdough bread. The active yeast spores are the reason for the bread's distinct tangy flavor.

Baking Powder. *Baking powder* is a chemical leavening agent that is a combination of baking soda and cream of tartar or sodium aluminum sulfate. The alkalinity of baking soda reacts with the acidity of the other ingredients to produce carbon dioxide. Double-acting baking powder reacts when it is mixed with liquids, releasing some carbon dioxide, and then reacts again when it is heated. Double-acting baking powder allows a batter to be prepared in advance without the risk that it will not rise when baked at a later time.

Baking Soda. *Baking soda,* also known as sodium bicarbonate, is an alkaline chemical leavening agent that reacts to an acidic dough or batter by releasing carbon dioxide without the addition of heat. Baking soda is often used in conjunction with acidic liquids such as buttermilk, lemon juice, or molasses. Using too much baking soda can produce a yellow color, cause brown spots, and leave a bitter aftertaste. Baking soda should not be kept longer than one year, because it absorbs moisture from the air, which reduces its strength over time.

Cream of Tartar. *Cream of tartar* is a chemical leavening agent derived from tartaric acid, a by-product of wine production. When used in recipes, it is often combined with baking soda, which is an alkaline. When the two are combined, they neutralize each other and release carbon dioxide. Cream of tartar is often added to angel food cake batter because the acid helps to stabilize the egg whites.

Ammonium Bicarbonate. *Ammonium bicarbonate* is a chemical leavening agent that reacts as soon as it is heated without the help of an acid or base. Ammonium bicarbonate is used mainly in cookie and cracker recipes. It is important to bake any product containing ammonium bicarbonate thoroughly so that all of the gases are released. Otherwise, an ammonia smell will remain in the finished product.

Flavorings

Flavorings are generally divided into two categories: natural and artificial. Natural flavorings yield a more pleasing flavor than imitation flavorings. Salts, seeds, extracts, emulsions, vanilla beans, and chocolates are a few of the key flavorings used in the bakeshop.

Salts. *Salt* is a crystalline solid composed mainly of sodium chloride and is used as a seasoning and a preservative. Salt is used to balance the sweetness in baked products while adding additional flavor. Salt should be dissolved in liquid ingredients to ensure even distribution. When used in doughs, salt improves the flavor and controls yeast growth. When used to make piecrusts, salt shortens the gluten strands in the flour, which helps to create tender crusts.

Seeds. A *seed* is the fruit or unripened ovule of a nonwoody plant. Seeds are often used to add flavor to yeast breads, quick breads, and cookies. For example, caraway seeds are commonly used to flavor rye bread. Poppy seeds are often used to garnish rolls and bagels. Ground flax seeds can be baked into yeast doughs and muffin and cookie batters.

Extracts. An *extract* is a flavorful oil that has been mixed with alcohol. Vanilla, lemon, almond, coffee, mint, and rum are a few of the extracts used in the bakeshop. Only a small amount of an extract is used since its flavor is concentrated.

Flavoring Emulsions. An *flavoring emulsion* is a flavored oil that has been mixed with water and an emulsifier. Emulsions are stronger than extracts and should be used sparingly. Lemon and orange emulsions are commonly used in the bakeshop.

Vanilla Beans. A *vanilla bean* is the dark-brown pod of a tropical orchid. *See Figure 21-13.* Vanilla is one of the most commonly used flavorings in the bakeshop and is difficult to substitute. When adding vanilla, the bean is split open and the inside is scraped into the mixture.

Chocolates. *Chocolate* is a flavoring ingredient made by roasting, skinning, crushing, and grinding cacao beans into a paste called chocolate liquor. This paste is then combined with various ingredients to create the desired flavor, from bitter to sweet. Types of chocolate commonly used in the bakeshop include cocoa powder, unsweetened chocolate, dark chocolate, milk chocolate, and white chocolate. *See Figure 21-14.*

Vanilla Beans

Figure 21-13. *A vanilla bean is the dark brown pod of a tropical orchid.*

Chocolates

| Cocoa powder | Unsweetened | Milk | White |

Figure 21-14. *Cocoa powder, unsweetened chocolate, milk chocolate, and white chocolate are a few of the types of chocolate used in the bakeshop.*

- *Cocoa powder* is a reddish-brown, unsweetened powder extracted from ground cacao beans. *Dutch-processed cocoa powder* is a dark, unsweetened cocoa powder processed with an alkali to neutralize the natural acidity. Dutch-processed cocoa powder adds a smooth flavor to baked products.

- *Unsweetened chocolate* is pure chocolate liquor that contains 50% to 58% cocoa butter. Unsweetened chocolate is melted and used in many cake and icing recipes.

Chef's Tip

Because chocolate melts at a low temperature, the heat of an open burner is not only unnecessary, but it could scorch and burn the chocolate.

- *Dark chocolate* is chocolate that contains at least 35% chocolate liquor in addition to sugar, vanilla, and cocoa butter. Two types of dark chocolate often used in baking are bittersweet and semisweet chocolate. Bittersweet and semisweet chocolate are often used interchangeably in baking.
- *Milk chocolate* is a chocolate product that contains at least 12% milk solids and 10% chocolate liquor.
- *White chocolate* is a confectionary product made from cocoa butter, sugar, milk solids, and flavorings. White chocolate does not contain any cocoa solids.

COMBINING INGREDIENTS

Terms such as blend, cream, cut-in, fold, sift, stir, and whip are used to describe different methods of combining ingredients in the bakeshop. Understanding baking terms is essential to the proper preparation of yeast breads, quick breads, cakes, cookies, pies, pastries, and specialty desserts. ***See Figure 21-15.***

Bakeshop Terminology

Term	Description
Blend	To blend is to mix two or more ingredients together until they are evenly distributed
Cream	To cream is to combine fat and sugar in order to add air
Cut-in	To cut-in is to incorporate a solid fat into a dry ingredient until pea-size lumps have formed
Fold	To fold is to gently incorporate light ingredients into heavier ones using a smooth and gentle fold-over motion
Sift	To sift is to pass dry ingredients through a sieve in order to remove lumps and incorporate air
Stir	To stir is to gently mix two or more ingredients together until they are evenly combined
Whip	To whip is to agitate ingredients in order to incorporate air

Figure 21-15. *Understanding baking terms is essential to the proper preparation of yeast breads, quick breads, cakes, cookies, pies, pastries, and specialty desserts.*

BAKESHOP EQUIPMENT

A bakeshop is equipped with special equipment, bakeware, and tools used for the preparation of yeast breads, quick breads, cakes, cookies, pies, pastries, and desserts. Some of the more frequently used pieces of equipment include bakeshop tables, bench and floor mixers, convection ovens, proofing cabinets, revolving tray ovens, and sheeters. ***See Figure 21-16.***

Common Bakeshop Equipment	
Item	**Description**
Bakeshop tables	A work table with a wood top is used to roll out doughs; a work table with a marble or quartz top is used for sugar and chocolate work
Bench and floor mixers (4 — 100 qt capacity)	Electric appliance used to mix, blend, and whip ingredients
Convection ovens	Electric appliance that circulates air evenly while a product bakes
Proofing cabinets	Temperature- and humidity-controlled box that creates the perfect environment for yeast doughs to rise
Revolving tray ovens (reel ovens)	Oven with shelves that rotate like a ferris wheel allowing a large number of items to be baked at once
Sheeters	Electric appliance used to roll out very large pieces of dough

Figure 21-16. *Some of the more frequently used pieces of equipment include bakeshop tables, bench and floor mixers, convection ovens, proofing cabinets, revolving tray ovens, and sheeters.*

Professional bakeware is typically made from stainless steel, aluminum, or silicone. Silicone bakeware is nonstick and temperature resistant. Bakeware is available in a variety of shapes and sizes, including rectangular, round, and square pans. The size and shape of the pans used depends on the items being prepared. Common bakeware includes cake pans, loaf pans, molds, muffin pans, pie pans, rings, sheet pans, springform pans, and tart pans. *See Figure 21-17.*

Common Bakeware	
Item	**Description**
Cake pans	Round, square, or specially shaped pan with short or tall sides; used to bake cakes; several pans may be required to make multiple layers of a cake
Loaf pans	Short and deep rectangular pan used to bake loaves of bread
Molds	Pan with any of several distinctive shapes that is used to hold a variety of desserts
Muffin pans	Rectangular pan with numerous round cups used to bake muffins or cupcakes; the diameter of the cups vary from miniature to jumbo
Pie pans	Round, shallow pan with sloped sides used for baking pies
Rings	Pan without a bottom that is used to give cakes and specialty desserts a round shape; available in a variety of sizes
Sheet pans	Aluminum pan with very low sides that is used to bake large amounts of biscuits, rolls, cookies, cakes, or pastries at one time; available in full, half, and quarter sizes
Springform pans	Round pan with a metal clamp on the side that allows the bottom of the pan to be separated from the sides; typically used to bake cheesecakes
Tart pans	Round, shallow baking pan with sloped sides that are smooth or fluted and may have a removable bottom; used to bake tarts with delicate crusts

Figure 21-17. *Bakeware is available in a variety of shapes and sizes, including rectangular, round, and square pans.*

Special tools used in the bakeshop make the preparation process easier. Common bakeshop tools include bench brushes, dough cutters, dough dockers, palette knives, pastry bags, pastry brushes, pastry cloths, pastry tips, pastry wheels, pie markers, rolling pins, and silicone mats. *See Figure 21-18.*

Common Bakeshop Tools	
Item	**Description**
Bench brushes	Brush with long bristles set in vulcanized rubber attached to a wood handle; used to brush excess flour from the bench (baker's table) when working with pastry or bread dough
Dough cutters (bench knives)	Flat stainless steel blade attached to a sturdy handle; used to cut dough into portions and to scrape dough off the surface of bakeshop tables
Dough dockers	Aluminum or stainless steel roller with pins that is used to perforate dough so that it will bake evenly without blistering from the oven heat
Palette knives (cake spatula)	Flat, narrow knife with a rounded, 3½–12 inch blade that varies in flexibility and is most often used to ice cakes
Pastry bags	Cone-shaped paper, canvas, or plastic bag with two open ends, the smaller of which can be fitted with a pastry tip; used to pipe icings when decorating cakes or soft foods, such as whipped potatoes, in savory applications
Pastry brushes	Small, narrow brush used to apply liquids such as egg wash or butter onto baked products; available in natural, nylon, or silicone bristles (*Note:* Nylon brushes cannot be used with hot items. Silicone bristles can withstand temperatures up to 650°F.)
Pastry cloths	Large piece of canvas used as a surface for rolling out dough; allows piecrust to be easily flipped from the cloth into pie pans
Pastry tips	Cone-shaped tip that is fitted into the narrow end of a pastry bag and used to create decorative shapes and patterns with icings or soft foods; countless tip shapes are available, including straight, star-shaped and leaf-shaped
Pastry wheels	Dough-cutting tool with a rotating disk attached to a handle that is used to cut dough into desired shapes
Pie markers	Round tool that has wire guides that leave marks indicating where to cut pies, round cakes, or pizzas into equal portions; available in various diameters and portion sizes
Rolling pins	Slim cylinder that ranges in length from 1–25 inches and is used to flatten pastry dough, bread crumbs, or other foods; available in wood, marble, ceramic, and metal and may have a handle on each end (*Note:* Many chefs prefer French rolling pins, which have tapered ends instead of handles.)
Silicone mats	Woven, nonstick mat that may be used in the refrigerator, freezer, or oven; most mats can withstand temperatures between –40°F and 580°F

Figure 21-18. *Special tools used in the bakeshop make the preparation process easier.*

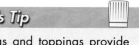
Chef's Tip

Fillings and toppings provide additional flavor and eye appeal to yeast breads. Fillings are added prior to baking and toppings are applied after baking.

YEAST BREAD PREPARATION

Yeast bread is a versatile staple that can be served at any meal. Most yeast bread doughs contain flour, liquid, sugar, salt, and yeast. Many also contain fat and eggs. Flour gives strength to dough and acts as an absorbing agent. Liquid, usually milk or water, supplies moisture and helps form gluten. Fat supplies tenderness and improves shelf life. Eggs supply structure to dough and add color. Sugar acts as a stimulant to the yeast. Salt brings out the flavor in the dough.

There are three types of yeast doughs. The three types of yeast doughs include lean doughs, rich doughs, and rolled-in doughs. *See Figure 21-19.*

Types of Yeast Doughs

National Honey Board

| Lean | Rich | Rolled-In |

Figure 21-19. *The three types of yeast doughs are lean doughs, rich doughs, and rolled-in doughs.*

- A *lean dough* is a yeast dough that is low in fat and sugar. Hard rolls, baguettes, and rye bread are made from lean doughs.

- A *rich dough* is a yeast dough that incorporates a lot of fat, sugar, and eggs into a heavy, soft structure. The finished product may be faintly yellow in color due to the large number of eggs that are used. Cinnamon rolls and doughnuts are made from rich doughs.

- A *rolled-in dough* is a yeast dough with a flaky texture that results from the incorporation of fat through a rolling and folding procedure. By alternating the layers, a very light and flaky texture is achieved in the finished product. Rolled-in doughs may be sweet, as with Danishes, or may not be sweet, as with croissants.

Yeast dough production involves scaling ingredients, mixing, kneading, fermenting, punching, scaling dough, rounding, panning, proofing, baking, cooling, and storing. These steps generally apply to all yeast doughs with slight variations depending on the product.

Scaling Ingredients

All ingredients must be scaled on either a baker's scale or a digital scale. *Scaling* is the process of weighing an ingredient using a scale. A scale must be properly balanced before use and again after adding measured ingredients. *See Figure 21-20.*

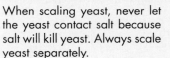

Chef's Tip

When scaling yeast, never let the yeast contact salt because salt will kill yeast. Always scale yeast separately.

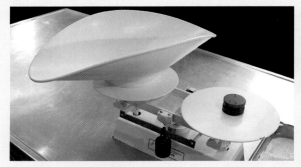 

1. Set the scale scoop on the left side of the scale.
2. Balance the scale by placing weights on the right side until the scale is balanced.
3. Adjust the scale to the desired weight of the ingredient being measured by adding the appropriate amount of weight to the right side of the scale and adjusting the ounce weight on the bar as needed.

4. Add the ingredient to the scoop until the scale is balanced.

Figure 21-20. *A scale must be properly balanced before use and again after adding measured ingredients.*

Mixing Doughs

A bench or floor mixer with one or more attachments is used to mix yeast doughs and cake batters. Common mixer attachments include the paddle, the hook, and the whip. *See Figure 21-21.* Mixing is very important for gluten development and uniform distribution of yeast throughout the dough. The three methods commonly used to mix yeast doughs are the straight-dough method, the modified straight-dough method, and the sponge method.

- The *straight-dough method* is a method of mixing yeast dough that combines yeast with warm water and then adds it to the rest of the ingredients, mixing them together all at once.

- The *modified straight-dough method* is a method of mixing yeast dough in which the ingredients are added in sequential steps.

- The *sponge method* is a method of mixing yeast dough in two steps. In the first step, 50% of the flour is added to all of the liquid and yeast to form a batter that looks like a sponge. As the sponge forms and increases in size, the flour, yeast, and liquid begin to ferment and air bubbles develop throughout the batter. In the second step, the rest of the flour and all of the fat, salt, and sugar are added. Yeast breads made using the sponge method usually have an intense flavor and a light texture.

Kneading Doughs

Kneading is the process of pushing and folding dough until it is smooth and elastic. *See Figure 21-22.* Sweet doughs are kneaded longer than other doughs because high amounts of fat and sugar inhibit gluten development. If the air is humid, a little more flour may need to be added during the kneading stage.

Mixing Equipment

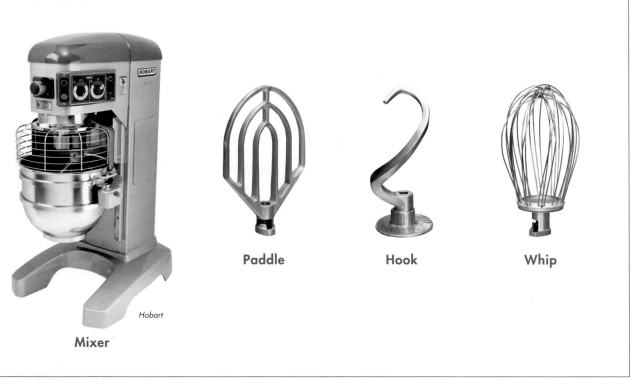

Mixer

Hobart

Paddle **Hook** **Whip**

Figure 21-21. *Common mixer attachments include the paddle, the hook, and the whip.*

Procedure for Kneading Yeast Dough

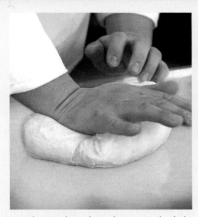 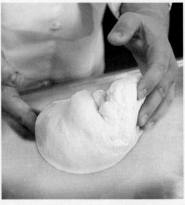 

1. Place the dough on a lightly floured surface.
2. Use the heels of the hands to push the dough down and away from the body.

3. Fold the dough in half, bringing the farthest edge up and on top of the nearest edge.

4. Repeat steps 2 and 3 until the dough is smooth.

Figure 21-22. *Kneading is the process of pushing and folding dough until it is smooth and elastic to further develop gluten.*

Fermenting Doughs

Fermentation is the process by which yeast converts sugar into carbon dioxide and alcohol. During fermentation, yeast dough should be covered and allowed to rest until it doubles in size or until it no longer springs back when pressed. *See Figure 21-23.* Gluten becomes much smoother and more elastic in the fermentation stage. Fermentation will continue until dough reaches 140°F, the temperature at which yeast dies.

Fermentation

Figure 21-23. *During fermentation, yeast dough should be covered and allowed to rest until it doubles in size or until it no longer springs back when pressed.*

Punching

Figure 21-24. *Punching is the process of folding and pressing risen yeast dough to allow carbon dioxide to escape.*

Punching Doughs

After the dough has doubled, it must be gently deflated to relax the gluten, dispel some of the carbon dioxide, and redistribute the yeast. *Punching* is the process of folding and pressing risen yeast dough to allow carbon dioxide to escape. *See Figure 21-24.* To punch dough, the edges of the dough are pulled into the center and then the dough is pressed to release the gas. Punching also helps redistribute yeast.

Scaling Doughs

Scaling dough is the process of weighing pieces of risen dough to ensure products will be consistent in size. *See Figure 21-25.* A dough cutter is used to cut dough into pieces of a desired size before they are weighed. It is important to use a dough cutter instead of tearing the dough, which destroys some of the gluten that has developed. Dough pieces can be added or taken away until the desired weight is achieved for each portion of dough.

Scaling

Edlund Co.

Figure 21-25. *Scaling dough is the process of weighing pieces of risen dough to ensure products will be consistent in size.*

Rounding Doughs

Rounding is the process of shaping scaled dough into smooth balls. Rounding enables the dough to proof evenly and have a smooth outer surface. To round dough, the dough is shaped into a ball using the fingers to form the shape, and then the dough is rotated on a flat surface while applying pressure. ***See Figure 21-26.***

When shaping rolled-out dough into a loaf, it needs to be rolled tightly. The end of the roll is referred to as a seam and should be pinched tightly to prevent it from unrolling during proofing or baking. The seam is always placed facing the bottom of the pan.

Rounding

Figure 21-26. *To round dough, the dough is shaped into a ball using the fingers to form the shape, and then the dough is rotated on a flat surface while applying pressure.*

Panning

Figure 21-27. *Panning is the process of placing rounded pieces of dough into pans.*

Panning Doughs

Panning is the process of placing rounded pieces of dough into pans. ***See Figure 21-27.*** Some yeast breads are baked in loaf pans. Rolls are typically baked on sheet pans or, in the case of cloverleaf rolls, in muffin pans. Yeast breads typically double in size during the proofing stage. It is important to leave enough space in the pan for the yeast bread to proof, or rise.

Proofing Doughs

Proofing is the process of letting yeast dough rise in a warm (85°F) and moist (80% humidity) environment until the dough doubles in size. A proofing cabinet helps reduce the time required for proofing and ensures a consistent product because the temperature and humidity are controlled. ***See Figure 21-28.*** Proofing is an extension of the fermentation process, but it is a different process. Fermentation is done at a lower temperature than proofing. During proofing, the yeast bread doubles in size again and should spring back when pressed. It is important not to overproof or underproof yeast doughs as this will result in poor texture.

After proofing, some yeast breads may require scoring or docking. *Scoring* is the process of making shallow, angled cuts across the top of unbaked bread with a sharp knife called a lame. ***See Figure 21-29.*** Scoring is done on hard-crusted breads before they are baked to allow carbon dioxide to escape during baking. If the yeast dough is not scored, the bread may bulge at the sides.

Docking is the process of making small holes in yeast dough before it is baked to allow steam to escape and to promote even baking. A dough docker or the tines of a fork can be used to dock yeast doughs.

Proofing

Cres Cor

Figure 21-28. *A proofing cabinet helps reduce the time required for proofing and ensures a consistent product because the temperature and humidity are controlled.*

Scoring

Figure 21-29. *Scoring is the process of making shallow, angled cuts across the top of unbaked bread with a sharp knife called a lame.*

Washes are often used to create color, make the outer surface shiny or dull, or make other ingredients, such as seeds, stick to the dough. A *wash* is a liquid that is brushed on the surface of a yeast dough product prior to baking. *See Figure 21-30.* The most common type of wash is an egg wash, which consists of whole eggs whisked with a little water.

Baking Yeast Breads

Yeast breads expand quickly when first placed in a hot oven. *Oven spring* is the rapid expansion of yeast dough in the oven, resulting from the expansion of gases within the dough. When the temperature of the dough reaches 140°F, the yeast dies and there is no further expansion.

Some yeast breads, such as French bread, benefit from the incorporation of steam in the oven. *Steam injection* is the process of adding water directly in the hot cavity of an oven so that steam is created. As the water sprays on the interior surface of the oven, the water vaporizes and turns into steam. Steam gives yeast breads a light, crispy crust, while the inner crumb remains moist and chewy. Bakeries have a steam injector connected to their ovens.

Cooling and Storing Yeast Breads

After baking, yeast breads must be removed from the pans to cool. Typically, breads are placed on cooling racks. If breads are left in the pans to cool, the hot bread sweats and becomes soggy. Similarly, hot bread should never be bagged until almost room temperature. If bread is bagged before it has cooled, steam will be trapped in the bag and the bread will become soggy.

Cooled yeast breads should be stored in airtight containers or plastic bags. Hard-crusted yeast breads, such as baguettes and French bread, should not be wrapped because the crust will become soft. Yeast breads should be stored at room temperature or frozen. Refrigeration causes breads to stale.

Applying a Wash

Figure 21-30. *A wash is a liquid that is brushed on the surface of a yeast dough product prior to baking.*

QUICK BREAD PREPARATION

A *quick bread* is a baked product that is made with a quick-acting leavening agent such as baking powder, baking soda, or steam. Common quick breads include biscuits, muffins, quick bread loaves, and corn bread. *See Figure 21-31.*

Biscuits

A *biscuit* is a quick bread made by mixing solid fat, baking powder or baking soda, salt, and milk with flour. Some biscuit recipes also call for sugar. Butter may be added to improve the tenderness and richness of biscuits. Flavorings may be added to create specialty products, such as buttermilk biscuits, orange biscuits, or cheese biscuits.

Flour supplies form and texture to biscuits. Solid fat is considered the most important ingredient in preparing biscuits because it creates tenderness. Baking powder and baking soda are quick-acting leavening agents that cause dough to rise when liquid is added and when heat is applied. Milk provides moisture, regulates the consistency of the dough, develops the flour, and activates the baking powder. Salt brings out the flavor of the other ingredients. When added, sugar supplies sweetness, helps retain moisture, and produces a golden brown color.

Muffins

A *muffin* is a quick bread that is made with eggs, flour, and either oil, butter, or margarine. Muffins are shaped like cupcakes and have a pebbled top and a cake-like texture. Muffins can be served with any meal.

Quick Bread Loaves

A *quick bread loaf* is a loaf of bread that is made from a batter similar to muffin batter. The shape of the loaf is determined by the type of loaf pan that is used. Quick bread loaves are often flavored with different fruits, vegetables, nuts, or spices. Banana nut bread, pumpkin bread, and zucchini bread are examples of quick bread loaves. They are similar to muffins in quality and texture. Quick bread loaves can be served warm or cold. Some quick bread loaves are topped with icing.

Corn Bread

Corn bread is a quick bread made from a batter containing cornmeal, eggs, milk, and oil. Cornmeal is dried, ground corn kernels. Available types include yellow corn or white corn and coarse grain or fine grain. Corn flour is finely ground cornmeal. Corn flour can be used in combination with wheat flour to make cakes and crêpes.

When making corn breads, the liquid must be added slowly because cornmeal does not absorb liquid quickly. If the liquid is added too rapidly, lumps will form. It is also important to avoid overmixing. The batter should only be stirred until the ingredients are mixed. A corn bread mixture should not be beaten. Corn bread batter can be poured into molds prior to baking to make corn sticks.

Quick Breads

Biscuits

Muffins

Quick Bread Loaves

Idaho Potato Commission
Corn Bread

Figure 21-31. *Common quick breads include biscuits, muffins, quick bread loaves, and corn bread.*

Mixing Quick Breads

Quick breads are easier to prepare than yeast breads because the quick-acting leavening agents allow a quick bread to be mixed and baked without waiting for the dough to rise. Quick breads are made using one of three different mixing methods. These methods include the biscuit method, the muffin method, and the creaming method.

Biscuit Mixing Method. The *biscuit mixing method* is a quick-bread mixing method used for recipes that include a cold, solid fat, such as butter. *See Figure 21-32.* Solid fat helps produce a light and flaky product.

Procedure for Using the Biscuit Mixing Method

1. Scale all ingredients.
2. Sift all dry ingredients together.

3. Cut the fat into the dry ingredients using a pastry blender or a mixer with a paddle attachment.

4. Combine liquid ingredients in a separate bowl.
5. Add liquid ingredients to the dry ingredients and mix until the mixture holds together.

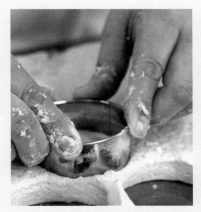

6. Place the dough on a flat, floured surface and knead lightly for 30–45 seconds. *Note:* Overkneading will toughen biscuits.

7. Roll the dough out on a floured surface.
8. Using a biscuit cutter, cut biscuits from the rolled out dough as close together as possible.

9. Place biscuits on a greased or parchment-lined sheet pan.
10. Brush the tops of the biscuits with melted butter or an egg wash for added color.

Figure 21-32. *The biscuit mixing method is used for recipes that include a cold, solid fat, such as butter.*

Muffin Mixing Method. The *muffin mixing method* is a quick-bread mixing method that uses liquid fats, such as vegetable oil or melted butter, to produce a rich and tender product. *See Figure 21-33.* To prevent the gluten from becoming tough, the dry and liquid ingredients should be placed in a mixer with a paddle attachment and mixed at a slow speed. The batter should be mixed until it is moistened, leaving it with a slightly rough appearance. Muffin batter should not be smooth. Overmixing the ingredients creates air tunnels inside the muffins.

Procedure for Using the Muffin Mixing Method

1. Scale all ingredients.
2. Sift dry ingredients together.

3. Combine liquid ingredients, including any melted butter or oil.
4. Add the liquid ingredients to the dry ingredients and mix just until moistened. *Note:* The batter should appear lumpy. Do not overmix.

5. Grease muffin pans or use paper liners.
6. Use a portion control scoop to add batter to pans for a uniform size.

Figure 21-33. *The muffin mixing method is a quick-bread mixing method that uses liquid fats to produce a rich and tender product.*

Vita-Mix® Corporation

Creaming Mixing Method. The *creaming mixing method* is a quick-bread mixing method used to produce quick bread loaves that have a fine crumb. *See Figure 21-34.* This method involves mixing room temperature fat and sugar in a mixer using the paddle attachment until the batter is light and fluffy. Eggs are then added one at a time and thoroughly incorporated into the creamed mixture. Liquid ingredients and dry ingredients are alternately added in thirds while continually mixing. The creaming method creates a finer crumb than the muffin mixing method.

1. Scale all ingredients.
2. Combine fat and sugar in a mixing bowl and mix using the paddle attachment on medium speed until light and fluffy.

3. Add eggs one at a time, mixing well after each addition.

4. Sift dry ingredients together.

5. Combine liquid ingredients in a separate bowl.

6. Alternately add dry and liquid ingredients to the creamed mixture. Do not overmix.

7. Portion batter into greased or lined pans and bake.

Figure 21-34. *The creaming mixing method is used to produce quick bread loaves that have a fine crumb.*

Baking Quick Breads

Quick breads can be baked in a variety of loaf pans to create muffins, loaves, or sticks of various sizes. When baking quick breads, pans should be filled only three-quarters full. If pans are overfilled, the batter will spill over the top as it rises during the baking process.

Cooling and Storing Quick Breads

If quick breads are not served immediately, they should be refrigerated after cooling. To extend the freshness of quick breads, they should be wrapped tightly as soon as they cool. This allows them to be sliced more easily without crumbling. Quick breads may also be frozen with excellent results.

Entourage

CAKE PREPARATION

Before mixing cake batter, all of the ingredients should be at room temperature. Each ingredient should be weighed separately on a baker's scale or a digital scale for maximum accuracy. Four common methods of mixing cake batters include the two-stage method, the creaming method, the chiffon method, and the sponge method.

Two-Stage Mixing Method

The two-stage mixing method is the easiest method of mixing cake batter. *See Figure 21-35.* After being carefully weighed, the dry ingredients are sifted together and blended with the fat and part of the liquid. In a separate bowl, the rest of the liquid and eggs are combined. The two mixtures are then mixed together until evenly distributed. At intervals throughout the mixing process, the sides of the bowl are scraped down so all ingredients are blended and a smooth batter is obtained.

Procedure for Mixing Cake Batter Using the Two-Stage Mixing Method

1. Weigh all ingredients carefully.
2. Place all dry ingredients, fat, and part of the milk in the mixing bowl. Blend on slow speed.
3. In a separate bowl, blend eggs and the remaining milk.
4. Add the egg mixture to the batter in thirds, blending well after each addition to ensure a smooth, uniform batter.

Figure 21-35. *The two-stage mixing method is the easiest method of mixing cake batter.*

Creaming Mixing Method

The creaming mixing method of mixing cakes begins by mixing room temperature fat and sugar in a mixer with a paddle attachment until the batter is light and fluffy. The eggs are then added one at a time and incorporated into the creamed fat and sugar. Next, liquid ingredients and dry ingredients are alternately added in small amounts and creamed. The creaming method of mixing cake batter produces a lighter product with a much finer crumb. *See Figure 21-36.* At intervals throughout the mixing process, the sides of the bowl are scraped down so all ingredients are blended and a smooth batter is obtained.

Chiffon Mixing Method

The chiffon mixing method involves folding whipped egg whites into a batter made from flour, egg yolks, and fat. *See Figure 21-37.* When folding in whipped products, such as egg whites, meringues, or whipped cream, it is important to add the ingredients in thirds. Each third is added using a circular folding motion, which gently incorporates air to produce a very light cake batter. It is important to make sure that each third of a whipped ingredient is fully incorporated before the next third is added and to take care not to deflate the air that has been incorporated.

Procedure for Mixing Cake Batter Using the Creaming Mixing Method

1. Scale all ingredients.
2. Combine the fat and sugar in a mixing bowl and mix with a paddle attachment on medium speed until light and fluffy.
3. Add eggs, one at a time, mixing well after each addition.
4. Sift dry ingredients together.
5. Combine liquid ingredients in a separate bowl.
6. Alternately add dry and liquid ingredients to the creamed mixture. Do not overmix.
7. Portion batter into greased or lined pans and bake.

Figure 21-36. *The creaming mixing method of mixing cake batter produces a lighter product with a much finer crumb.*

Chiffon Mixing Method

Figure 21-37. *The chiffon mixing method involves folding whipped egg whites into a batter made from flour, egg yolks, and fat.*

Sponge Mixing Method

The sponge mixing method produces a light, fluffy batter. *See Figure 21-38.* Although there are many variations of the sponge mixing method, the most common method is referred to as a genoise. In the genoise sponge mixing method, the eggs and sugar are warmed and whipped to create volume and incorporate air before any other ingredients are added. This airy whipped yolk and sugar mixture is referred to as a foam. After the foam is created, the flour and melted butter are folded in carefully to not deflate the egg foam.

Procedure for Mixing Cake Batter Using the Sponge Mixing Method

1. Warm the eggs (whole eggs, egg whites, or egg yolks as specified) and sugar to between 100°F and 105°F over a hot water bath and whisk continually.

2. Remove the mixture from the water bath and place in the mixer.

3. Using the whip attachment, whip on medium to high speed until the mixture peaks in volume and develops a thick foam. *Note:* Ensure the mixture is thick enough to form a ribbon as it runs off the whip.

4. Turn the mixer to low and slowly add any liquid and flavorings.

5. Switch to the paddle attachment and gently fold in the sifted dry ingredients to create a smooth batter. Be careful not to deflate the egg foam. *Note:* If fat is to be added, it should be done after the dry ingredients have been folded in.

Figure 21-38. *The sponge mixing method of mixing cakes produces a light, fluffy batter.*

Baking Cakes		
Cake Type	**Scaling Weight**	**Pan Size**
Layer	13–14 oz	8 inch diameter
Bar	5–6 oz	2¾ × 10 inches
Ring	10–14 oz	6½ inch diameter
Loaf	11–24 oz	3¼–7⅛ inches long
Oval loaf	8 oz	6¼ inches long
Sheet	6–7 lb	17 × 25 inches

*in °F

Figure 21-39. *The amount of cake batter required for a cake varies with the type and size of the cake.*

Baking Cakes

Most cake batters can be used to produce cakes in a variety of shapes and sizes. The amount of cake batter required for a cake varies with the type and size of the cake. *See Figure 21-39.* Many cake pans are prepared by covering the bottom of the pan with parchment paper and greasing the sides lightly. Whenever possible, cakes should be placed in the center of the oven where the heat is evenly distributed. Leaving approximately 2–3 inches around each of the pans and between the pans and oven walls allows the heat to circulate freely around each pan.

Ovens must be preheated to the required temperature. If the oven temperature is set too low, the cake will rise and then fall, producing a dense, heavy texture. If the temperature is set too high, the outside of the cake will bake too rapidly and form a crust that will burst when heat reaches the center of the cake and causes the cake to expand. Most cake recipes are developed to produce good results when baked at or near sea level. If a product is baked in a high-altitude area, adjustments in ingredient amounts are often required. Cakes bake in the four following stages:

- Stage 1—The cake is placed in the oven and starts to rise. At this stage, the lowest oven temperature called for in the baking instructions should be used to prevent overly quick browning and to keep a crust from forming.

- Stage 2—The cake continues to rise and the top surface begins to brown. The oven door should not be opened at this stage. The heat can be increased at this time if the recipe calls for it.

- Stage 3—The rising stops but the surface of the cake continues to brown. The heat can be reduced if the cake is browning too quickly.
- Stage 4—The cake starts to shrink and separates slightly from the sides of the pan. At this stage, the oven door can be opened and the cake can be tested for doneness.

Dense cakes are tested for doneness by inserting a wire tester or a toothpick in the center of the cake. The cake is done if the tester is dry when removed, with no batter adhering to it. The touch method is used for lighter cakes, such as sponge cakes. The top surface of the cake is pressed with a finger, and if the surface feels firm and the finger impression does not remain the cake is done.

Cooling and Storing Cakes

When a cake is removed from the oven, it should be placed on a wire rack or shelf so that air circulates around the pan. The cake should be allowed to cool for a minimum of 5 minutes before it is removed from the pan. The pan can then be flipped over and the cake removed from the pan. If wax or parchment paper was used on the bottom of the pan, it is also removed at this time. The cake is then placed back on the wire rack or shelf and allowed to cool thoroughly.

ICING PREPARATION

Like cake recipes, variations of icing recipes are used to obtain different flavors and textures. An *icing,* also known as frosting, is a sugar-based coating often spread on the outside of and between the layers of a baked product. Icings have three main functions when applied to a baked product such as a cake. Icings form a protective coating to seal in moisture and flavor, improve the flavor, and add eye appeal. *See Figure 21-40.* Common types of icing include flat, royal, buttercream, fondant, foam, fudge, ganache, glaze, and whipped cream.

It is important to obtain the proper consistency before applying any type of icing. The consistency needed depends on the intended use of the icing. In most cases, the consistency of icings can be controlled by adding more or less confectioners' sugar than the recipe specifies.

Flat Icings

Flat icings are the simplest icings to prepare. They are applied to sweet rolls, doughnuts, Danish pastries, and other baked products. A flat icing is made by combining water, powdered sugar, corn syrup, and flavoring and then heating the mixture to approximately 100°F. It should always be heated in a double boiler because overheating causes it to lose its gloss when it cools. Flat icing should always be covered with a damp cloth when not in use. To store, flat icing should be covered with a thin coating of water, plastic wrap, or wax paper. Previously stored flat icing must be heated to approximately 100°F over a water bath before use.

Icings

Wisconsin Milk Marketing Board, Inc.

Figure 21-40. *Icings form a protective coating to seal in moisture and flavor, improve the flavor, and add eye appeal.*

Royal Icings

A royal icing is similar to a flat icing, but the addition of egg whites produces a thicker icing that hardens to a brittle texture. Royal icing hardens when exposed to air, so it must be kept covered with a damp towel when not in use. It is often used for making decorations for cakes and sugar sculptures.

Buttercream Icings

Buttercream icing is a type of icing made by creaming together shortening or butter, confectioners' sugar, and vanilla. *See Figure 21-41.* Some types of buttercream icings incorporate eggs, giving them a richer flavor. Buttercreams are light because air cells are trapped in the icing when mixed using the creaming method. They should be stored in a cool place and covered with plastic wrap or wax paper to avoid crusting. For best results, they should not be stored in the refrigerator, because refrigeration causes them to harden.

Buttercream Icings

Carlisle FoodService Products

Figure 21-41. *Buttercream icing is made by creaming together shortening or butter, confectioners' sugar, and vanilla.*

Fondant Icings

Fondant is a rich, white, cooked icing that hardens when exposed to the air. It is prepared by cooking sugar, glucose, and water to a temperature of 240°F, letting it cool to 150°F, and then mixing it until it is creamy and smooth. Fondant is the most difficult and time-consuming icing to prepare. A powdered product known as Drifond® can be purchased in a 40 lb tin. Only water and a small amount of glucose need to be added to Drifond® to produce fondant.

When needed for use, fondant is heated to about 100°F in a double boiler while stirring constantly. This thins the icing so that it flows freely over the item to be covered. To successfully cover an item with fondant, the consistency of the icing must be correct. If the fondant is too heavy after it is heated, it can be thinned down by using a glaze consisting of 1 part glucose to 2 parts water, or a regular simple syrup may be used. Fondant can be colored and flavored as desired and used as a base for other icings.

Fondant must be kept covered with a very thin coat of water, plastic wrap, or wax paper and stored in a cool place. It can be refrigerated, but it may lose some of its gloss when reheated. Rolled fondants have a consistency similar to dough and can be rolled out and used as a thick covering for cakes and to make cake decorations. *See Figure 21-42.*

Using Fondant Icings

Rolling Fondant

Draping Fondant

Fondant Decorations

Fondant-Iced Cake

Figure 21-42. *Rolled fondants have a consistency similar to dough and can be rolled out and used as a thick covering for cakes and to make cake decorations.*

Foam Icings

Foam icing, also known as boiled icing, is prepared by combining sugar, glucose, and water, boiling the mixture to approximately 240°F, and then adding the resulting syrup to an egg white meringue. If a thin syrup is added, the result is a thin icing. If a heavy syrup is added to the meringue, a heavy icing is produced. Foam icing must be applied the same day it is prepared. This icing is applied in generous amounts and often worked into peaks.

Fudge Icings

Fudge icing is a rich, heavy-bodied icing. It is usually prepared by adding a hot liquid or syrup to the other ingredients required while whipping to obtain smoothness. Fudge icing should be used warm. If it cools, it should be reheated in a double boiler before use. Fudge icing is generally used to ice layer cakes, sheet cakes, and cupcakes. It dries rapidly when stored. To store, fudge icing should be covered with plastic wrap and placed in the refrigerator. It can be reheated over a water bath before use.

Ganache Icings

Ganache icing is a rich chocolate icing made by heating chocolate and whipping cream until the chocolate melts. The ratio (in weight) of chocolate to cream can be 1:1 for a light ganache or 2:1 or more for a firm ganache. In its liquid state, a light ganache can be used to coat cakes and pastries. In its solid state, ganache is the easiest way to make chocolate truffles. *See Figure 21-43.* Ganache can be made with dark, milk, or white chocolate and may also include butter, glucose, liquor, or a fruit purée.

Ganaches

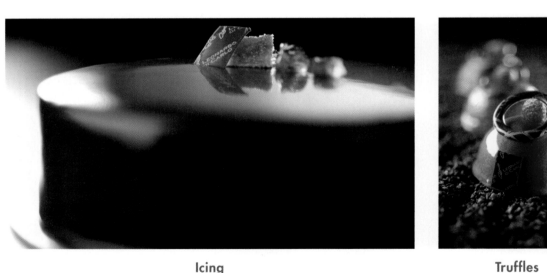

Icing   Truffles

Irinox USA

Figure 21-43. *In its liquid state, a light ganache can be used to coat cakes and pastries. In its solid state, ganache is the easiest way to make chocolate truffles.*

Glaze Icings

A glaze icing is a thin, transparent coating of icing that is poured over a baked product. Most glaze icings are made by heating and thinning fruit purées, fruit juices, chocolate, or coffee. Apricot glaze is the most common fruit glaze used in the professional bakeshop. It has a mild flavor and a nice sheen. Glazes extend shelf life by sealing in moisture.

Whipped Cream Icings

Whipped cream icings are made by beating heavy cream until soft peaks form. It is critical to chill the bowl, the beaters, and the cream before making whipped cream. Whipped creams may be sweetened by adding confectioners' sugar. Typically, 1 cup of heavy cream yields 2 cups of whipped cream.

Filling Pastry Bags

A pastry bag is a cone-shaped bag made of parchment paper, silicone, canvas, or plastic. Paper cones are simple to make and easy to handle. *See Figure 21-44.* Regardless of the materials, a separate cone should be used for each color of icing.

Procedure for Forming a Paper Pastry Bag

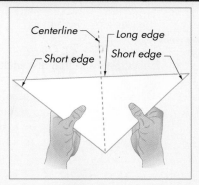

1. Cut a square of parchment paper into a large triangle.

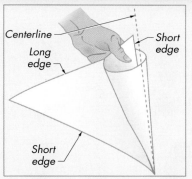

2. With the long edge on top, start to roll the paper by turning one short edge towards the centerline of the triangle.

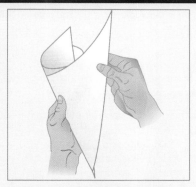

3. Continue rolling the paper into a cone shape across to the far short edge.

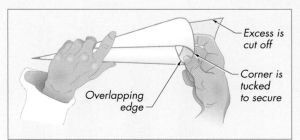

4. Tuck the top corner of the far short edge into an overlapping edge of the cone to secure. Cut off excess paper at the top of the cone.

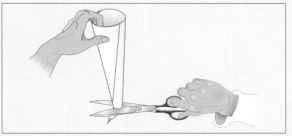

5. Cut off the bottom tip of the cone so a pastry tip can be inserted. *Note:* If too much paper is cut away, the hole will be too large to hold the pastry tip.

Figure 21-44. *Paper cones are simple to make and easy to handle.*

Filling Pastry Bags

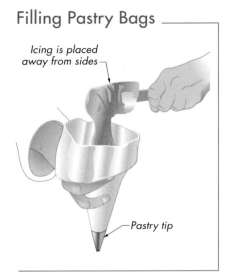

Figure 21-45. A spatula is used to insert icing into the center of a pastry bag.

To fill a pastry bag with icing, the bag is held in one hand with a light grip. The top half of the bag is folded over the hand. A spatula is then used to insert the icing into the center of the bag. *See Figure 21-45.* Each time the spatula is withdrawn, the icing is gently squeezed down into the bag. Once the pastry bag is halfway filled, it should be tightly twisted just above the filling. The twisted portion of the bag becomes the handle while piping the icing. As the bag empties, it can be twisted further to push the icing toward the tip.

Air pockets within the icing or along the sides of the pastry bag can result in bursts of air that can ruin a decorating job. To reduce the risk of air pockets, the icing should be constantly squeezed toward the tip to push out any air pockets. Air pockets should always be tested for by piping a sample of icing before applying icing onto a baked product.

Piping Icing

In order to create decorative patterns and shapes with icing, a metal or plastic pastry tip is inserted into the pastry bag before it is filled. Many different pastry tips are used to make various designs. *See Figure 21-46.* For example, a round pastry tip is used for writing.

Pastry Tips

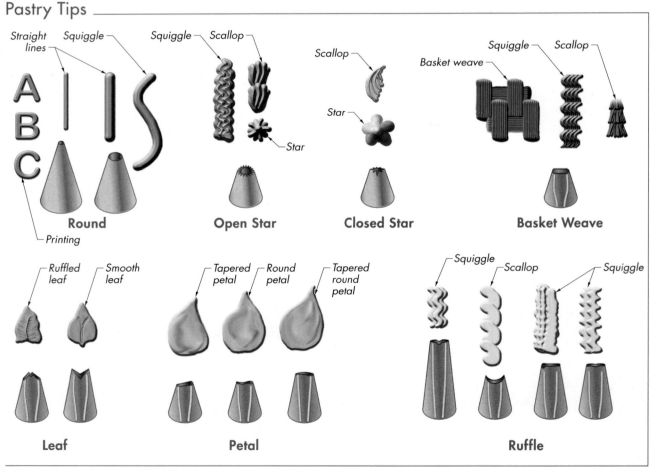

Figure 21-46. Many different pastry tips are used to make various designs.

When decorating, the top of the bag is held with one hand while the other hand is used to lightly grip the lower half. The hand at the top of the bag applies the pressure that causes the icing to flow. The hand on the lower half is the guide. In all decorating tasks, the two most important factors are holding the bag at the proper angle and applying the proper amount of pressure to obtain a smooth, even flow. *See Figure 21-47.*

COOKIE PREPARATION

Cookies are a popular dessert that can be served alone, as part of another dessert, or with fruit. The ingredients used to make cookies are similar to those used in cakes. However, cookie dough generally contains more fat and less moisture than cake batter. Cookies are usually baked at higher temperatures and for less time than cakes. Cookies are classified by texture as either soft or crisp. A *soft cookie* is a cookie prepared from dough that contains a lot of moisture. A *crisp cookie* is a cookie prepared from dough that contains a high percentage of sugar.

Types of Cookies

Cookies are often named for their preparation method. Drop, bar, rolled, refrigerator, molded, and pressed cookies are six types of cookies named for the way they are prepared. The method used is often determined by the consistency of the cookie dough or batter.

Drop Cookies. In the drop method, cookie dough is formed into consistently sized balls. *See Figure 21-48.* The cookies are prepared from a moist, soft batter. The batter should be at room temperature and then dropped in uniform balls onto parchment-covered sheet pans and baked.

Bar Cookies. In the bar method, a fairly stiff, dry cookie dough is scaled into 1 lb pieces, refrigerated until thoroughly chilled, and rolled into strips the length of a sheet pan. Three strips are placed an equal distance apart on a parchment-lined pan, flattened by hand, and then brushed with an egg wash. *See Figure 21-49.* The bars are baked and then cut after they have cooled.

Bar Method

Figure 21-49. *In the bar method, a fairly stiff cookie dough is chilled and then flattened by hand.*

Using Pastry Bags

Figure 21-47. *Holding the bag at the proper angle and applying the proper amount of pressure will result in a smooth, even flow.*

Drop Method

Figure 21-48. *In the drop method, cookie dough is formed into consistently sized balls.*

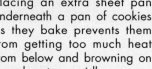

Chef's Tip

Placing an extra sheet pan underneath a pan of cookies as they bake prevents them from getting too much heat from below and browning on the edges too rapidly.

Rolled Method

Figure 21-50. *In the rolled method, chilled cookie dough is rolled out and then cut into the desired shapes.*

Refrigerator Method

Figure 21-51. *Refrigerator cookie dough is refrigerated overnight and then cut into slices before being baked.*

Molded Method

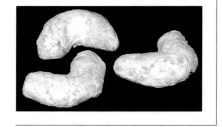

Figure 21-52. *In the molded method of cookie preparation, dough is molded into shapes by hand.*

Rolled Cookies. In the rolled method, a fairly stiff, dry cookie dough is refrigerated until chilled thoroughly and then rolled out on a floured surface to ⅛ inch thickness. The dough is then cut into desired shapes with a cookie cutter, placed on sheet pans covered with parchment paper, and baked. *See Figure 21-50.*

Refrigerator Cookies. The refrigerator method of cookie preparation, also known as the icebox method, involves scaling a fairly stiff, dry dough into 1–1¼ lb rolls approximately 16 inches long, wrapping them in parchment paper, and refrigerating them overnight. The next day the dough is cut into ¼ inch thick slices, placed on sheet pans covered with parchment paper, and then baked. *See Figure 21-51.* Thinner slices can be cut to produce crispier cookies.

Molded Cookies. In the molded method of cookie preparation, dough is molded into shapes, such as small balls or crescents, by hand. Then the cookies are placed on parchment-lined pans and baked. *See Figure 21-52.*

Pressed Cookies. Pressed cookies are made by placing soft, moist dough in a cookie press or a pastry bag to create shapes such as leaves, snowflakes, or stars. *See Figure 21-53.* The cookies are placed on sheet pans covered with parchment paper and then baked.

Cooling and Storing Cookies

Soft cookies should be placed in an airtight tin container. Crisp cookies should be stored in a tin container with a loose-fitting top and placed in a dry location. Crisp cookies can be warmed in a 225°F oven for 5 minutes just prior to serving.

Pressed Method

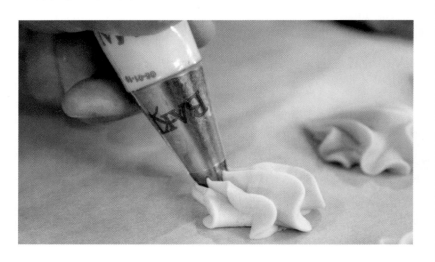

Figure 21-53. *Pressed cookies are made by placing soft, moist dough in a cookie press or a pastry bag to create shapes such as leaves, snowflakes, or stars.*

PIE PREPARATION

Pies and tarts are popular desserts. The two basic types of pies are single-crust pies and double-crust pies. *See Figure 21-54.* A single-crust pie consists of one crust on the bottom and a filling. Tarts also have a single-crust. A double-crust pie consists of two crusts, one on the bottom and one on the top, with a filling in between. The tenderness of the crust often determines how a pie is received. The filling may be of excellent quality, but a tough crust can ruin the entire preparation. Different types of pie dough vary in preparation techniques but have similar ingredients.

Most pie doughs are made primarily of flour and fat. However, the types of dough differ in how the flour and fat are mixed together and the amount of liquid added to the dough. Pie dough can be made by hand, in small batches using a pastry blender, in a food processor, or in a mixing machine. Pie dough must be mixed properly to achieve the best results. The correct amount and mixing of ingredients produces a good piecrust. The salt, sugar, and cold liquid should be blended together until the salt and sugar are thoroughly dissolved. The liquid mixture is then poured into the flour and fat mixture and mixed only until the flour absorbs the liquid.

Many pie dough recipes are written as ratios. A *ratio* is a mathematical way to represent the relationship between two or more numbers or quantities. In the professional kitchen, a ratio is expressed in terms of parts. For example, a ratio of "3 parts to 1 part" is often expressed as 3:1. A ratio is similar to a baker's formula in that the ratio does not contain actual measurements for the ingredients. Instead, a ratio describes how the ingredients are used in relation to one another. The simplest ratio used in the professional kitchen is a ratio of 1:1, also known as equal parts.

When a ratio must be used in terms of weight only or volume only, the ratio will be followed by the words "by weight" or "by volume." For example, a pie dough recipe has a ratio of 3 parts flour to 2 parts butter to 1 part water (3:2:1) by weight. If the desired yield of pie dough is 12 lb, how much of each ingredient would be required?

When using a ratio, ingredient quantities can be calculated by converting the ratio into fractions and multiplying the fractions by the desired yield. *See Figure 21-55.* The first step is to convert the ratio to fractions by dividing each number in the ratio by the total number of parts (3 + 2 + 1 = 6). The fraction for flour would be ³⁄₆ (which reduces to ½), the fraction for butter would be ²⁄₆ (which reduces to ⅓), and the fraction for water would be ⅙. By multiplying each of the fractions by the total desired amount of dough (12 lb), it can be calculated that the recipe would require 6 lb of flour (½ × 12 lb = 6 lb), 4 lb of butter (⅓ × 12 lb = 4 lb), and 2 lb of water (⅙ × 12 lb = 2 lb).

When making pie dough, most mistakes occur while the dough is being mixed. First, the flour must be sifted before the fat is cut in. Another mistake is adding extra flour or water to pie dough, causing overmixing and resulting in a tough dough.

Pie Types

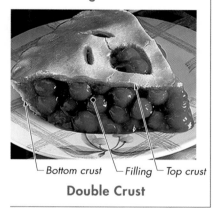

Bottom crust — *Filling*

Single Crust

Bottom crust — *Filling* — *Top crust*

Double Crust

Figure 21-54. *The two basic types of pies are single-crust pies and double-crust pies.*

Refer to DVD for
Using Ratios
Culinary Math Tutorial

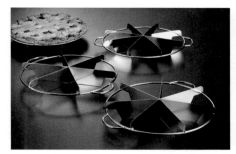

American Metalcraft, Inc.

Using Pie Dough Ratios

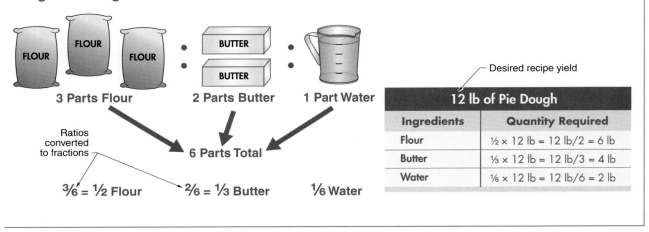

Figure 21-55. *When using a ratio, ingredient quantities can be calculated by converting the ratio into fractions and multiplying the fractions by the desired yield.*

Types of Piecrusts

Piecrusts are generally classified as mealy or flaky. *See Figure 21-56.* The basic difference between the two types of crusts is how the fat is combined with the flour. Other types of piecrusts include crumb and specialty. The type of filling to be added often determines which type of dough is the best choice.

Piecrusts		
Type	**Characteristics**	**Mixing**
Mealy	• High resistance to absorbing moisture • Used for bottom crust of pies	Cut-in fat and flour until flour is completely covered
Short-flake	• Most common type of flakey piecrust	Cut-in fat and flour until fat is pea size
Long-flake	• Absorbs the most moisture of any piecrust	Cut-in fat and flour until fat is walnut size

Figure 21-56. *Piecrusts are generally classified as mealy or flaky.*

Mealy Piecrusts. A *mealy piecrust* is a low-moisture piecrust prepared by rubbing fat into flour until the mixture resembles fine cornmeal. The flour is then unable to absorb a large amount of liquid or moisture from the filling because the flour granules are coated in fat. Mealy piecrust is used mostly for bottom-crust pies, such as fruit or custard pies, because it has high resistance to moisture.

Flaky Piecrusts. A *flaky piecrust* is a piecrust prepared by cutting fat into flour until pea-size pieces of dough form. When flaky piecrust is made with butter it is known as pâte brisée. The flour in flaky piecrust is not completely blended with the fat like it is in a mealy piecrust. There are two types of flaky crust: short-flake and long-flake.

A *short-flake crust* is a piecrust that absorbs slightly more liquid than mealy piecrust. When making short-flake crusts, the flour and fat are cut-in only until no flour spots are evident and the fat is in pea-size particles. The flour of a short-flake crust is not coated with fat to the same degree as mealy piecrust, so the flour is able to absorb slightly more liquid. A short-flake crust is the most common type of flaky piecrust and is commonly used to make lattice tops and the bottom crust of a pie filled with a cooked filling.

A *long-flake crust* is a piecrust that absorbs a large amount of liquid because the flour and fat are rubbed together very lightly, leaving the fat in chunks about the size of a walnut. Long-flake piecrust is commonly used for top crusts as well as prebaked bottom crusts. Tarts often have a long-flake crust.

Mealy crusts and short-flake crusts are handled in the same manner after mixing. Once the dough is mixed, it is wrapped with plastic wrap and refrigerated until it is firm enough to be rolled out (about 45–60 minutes). Long-flake crust must be refrigerated for a longer period, usually several hours or overnight. If long-flake dough is not refrigerated long enough, it will be too soft and difficult to roll out. When the pie dough is firm, it is removed from the refrigerator and divided into 8 oz units. Each 8 oz unit provides enough dough for one bottom or one top crust for an 8 inch or 9 inch pie.

To keep pie dough from drying out, only one unit of dough should be rolled out at a time. *See Figure 21-57.* After the dough is rolled and the bottom or top crust is formed, any remaining scraps can be pressed together and reused. Any unneeded 8 oz units should be returned to the refrigerator until ready to be rolled out.

National Honey Board

Rolling Piecrusts

Figure 21-57. *To keep pie dough from drying out, only one unit of dough should be rolled out at a time.*

Crumb Crusts. A *crumb crust* is a piecrust made from crumbled cookies or crackers that are held together with melted butter. ***See Figure 21-58.*** Crumb crusts can be made from various ingredients including graham crackers, vanilla wafers, or ginger snaps. Ground nuts may be added along with sugar. Crumb crusts are often baked for a short time to stabilize the crust. Crumb crusts are most often used for cheesecakes, unbaked pies such as cream or chiffon pies, and some tarts.

Specialty Crusts. Specialty piecrust may be prepared by adding spices, ground nuts, or other items to the pie dough. The added ingredient usually replaces up to 20% of the flour, except in the case of spices.

Crumb Crusts

Figure 21-58. *A crumb crust is made from crumbled cookies or crackers that are held together with melted butter.*

Blind Baking Crusts

Sometimes pie dough is baked for a short period before it is filled. Blind-baked crusts are often used for pies with unbaked fillings, such as chiffon or cream. *Blind baking* is a process in which a piecrust is baked for 10–15 minutes before a filling is added. Blind baking involves several important steps. ***See Figure 21-59.***

Pie Fillings

A pie filling must be thickened to the proper consistency and flavored appropriately. A thickener is typically added to a filling while it is being cooked. Approximately 3–5 oz of starch is required for each quart of liquid. More thickener is required for high-acidity fruits and juices. A pregelatinized starch can thicken a liquid without additional cooking because it does not require heat to absorb liquid and gelatinize.

1. Roll the dough out to the desired size and thickness and place it in the bottom of a pie pan.

2. Use a dough docker or a fork to prick small holes in the pie dough to allow steam to escape and prevent the dough from rising in the oven.

3. Cover the bottom, sides, and edges of the dough with parchment paper.

4. Place pie weights or dried beans on top of the paper to prevent the crust from rising as it bakes.

5. Bake the crust at 350°F for 10–15 minutes or until light golden brown.

Figure 21-59. *Blind baking is a process in which a piecrust is baked for 10–15 minutes before a filling is added.*

The proper amount of filling required for each pie is determined by weight. The prepared piecrust is placed on a baker's scale and the scale is balanced. The baker's scale is then set for the required amount of filling. Filling is then added to the piecrust until the scale balances a second time. Four common types of pie fillings are fruit, cream, soft, and chiffon fillings.

Fruit Fillings. Fruit fillings are the most popular type of pie filling. *See Figure 21-60.* Fruit used as pie filling can be fresh, frozen, canned, or dried. Each form of fruit requires different preparation for use in a pie filling.

Fruit Pie Fillings

Wisconsin Milk Marketing Board, Inc.

Figure 21-60. *Fruit fillings are the most popular type of pie filling.*

The amount of water and sugar to use in a fresh fruit filling is based on the amount and type of fresh fruit used. Typically, 65% to 70% water in comparison to the weight of the fruit is sufficient. For example, if 10 lb of fruit is used, 6½–7 lb of water is needed.

Fresh fruit fillings can be prepared using the cooked-fruit method or the cooked-juice method. Most fresh fruit fillings (except those made with fresh berries) are prepared using the cooked-fruit method, which only requires a few simple steps. *See Figure 21-61.* Dried fruits, such as raisins, currants, or figs, used as a filling are also prepared using the cooked-fruit method. These fruits need to be softened by cooking before being baked in a piecrust.

Procedure for Preparing Pie Fillings Using the Cooked Fruit Method

1. Place fruit, sugar, and a small amount of juice in saucepan with the desired spices. Bring to a boil.

2. Dissolve starch in cold water and pour slowly into the boiling fruit and juice mixture while stirring constantly.

3. Bring the mixture back to a boil and cook until clear.

4. Add salt and color (if desired) and stir until thoroughly blended.

5. Cool slightly and pour the filling into unbaked piecrusts.

Figure 21-61. *Most fresh-fruit fillings are prepared using the cooked-fruit method, which only requires a few simple steps.*

Frozen fruit is commonly used in commercial pie fillings. Frozen fruit has the same flavor characteristics and appearance as fresh fruit, except that it is available year-round. Fruit is frozen either in raw form or parboiled (partly cooked) as soon as possible after picking. The fruit is mixed with natural juices and sugar and in some cases additional color. The fruit is then packed in plastic tubs, buckets, or vacuum-sealed plastic and frozen. Frozen fruit is commonly sold in 30 lb buckets, although 6½ lb and 10 lb containers are also available.

Frozen fruit must be completely defrosted before it is used in a pie filling. The best method for defrosting frozen fruit is to place the unopened container in the refrigerator for one day. An opened container of fruit can be set in a hot water bath and stirred constantly to completely defrosted fruit for immediate use. After defrosting, the fruit is strained to remove the juice. The strained juice can be reserved, thickened, and added back to the fruit. Sugar can be added if desired. Frozen fruit must be completely defrosted or it will bleed, or continue to release juice, and cause the pie filling to separate. Thickening the juices and adding them back to the fruit prevents the fruit from bleeding.

Canned fruit is also used for commercial pie filling because it is available year-round. However, canned fruit fillings are inferior to fresh and frozen fruit fillings because canned fruit is overcooked during the canning process. Canned fruit can be purchased packed in water, syrup, or in a solid pack. A *solid pack* is a canned product with little or no water added. A solid pack is preferred for pie fillings because it contains more fruit and less sugar. This permits more sugar to be added after the juice is thickened, which produces better results.

Dried fruit is occasionally used for pie fillings. Dried fruit must be rehydrated in water or juice to restore natural moisture that was removed in the drying process. In some cases, the liquid and fruit may be brought to a boil to plump the fruit. The dried fruit is then soaked as it cools. After soaking, the liquid is drained from the fruit, thickened, flavored, and poured back over the fruit.

Cream Fillings. Smooth, flavorful, cream pie fillings only require a few simple steps to prepare. *See Figure 21-62.* After the cream filling is prepared, it is placed in a prebaked piecrust and topped with meringue or a cream topping. The most popular cream fillings are chocolate, vanilla, coconut, butterscotch, and banana.

Procedure for Preparing Cream Pie Fillings

1. Place milk in the top of a double boiler and heat.

2. In a separate container, beat eggs and add sugar, salt, and starch or flour.

3. Add cold milk while stirring constantly until a thin paste forms.

4. Add a small amount of warmed milk to the paste while whisking in the eggs.

5. Add the egg mixture to the remaining milk, whisking constantly until the mixture thickens and becomes smooth.

6. Cook until starch or flour is completely incorporated. Remove from heat.

7. Stir in flavoring and butter.

8. Pour into prebaked piecrusts and let cool.

Figure 21-62. *Smooth, flavorful, cream pie fillings only require a few simple steps to prepare.*

Soft Fillings. A soft filling is an uncooked pie filling that is baked in an unbaked piecrust. Soft-filling pies include pumpkin pie, custard pie, and pecan pie. Soft fillings are the most difficult pie fillings to make. The difficulty lies in baking the filling and crust to the proper temperature without underbaking or overbaking either part. Common mistakes can result in a soggy crust or a filling that separates or is watery.

Generally, soft pies are baked at 400°F for the first 10–15 minutes of baking, and then the temperature is reduced to 350°F or 325°F for the remaining baking time. The pie is removed from the oven as soon as the filling sets, which helps prevent the crust from becoming soggy. To check the filling for doneness, a knife can be inserted 1 inch from the center. If the knife comes out clean, the filling is done. Another method of checking for doneness is to gently shake the pie. The center of the soft filling may move slightly, but carryover cooking will continue to cook the filling after the pie is removed from the oven.

Chiffon Fillings. A chiffon filling is a light, fluffy pie filling prepared by folding a meringue into a puréed fruit or a cream pie filling. *See Figure 21-63.* In most cases, a small amount of plain gelatin is added to the fruit- or cream-based filling to help the chiffon filling set as it cools.

Procedure for Preparing Chiffon Pie Fillings

1. Prepare either a fruit or cream filling.
2. Soak plain gelatin for 5 minutes in cold water. Add gelatin to the hot fruit or cream filling, stirring until the gelatin is thoroughly dissolved.
3. Place the filling in a fairly shallow pan and let cool. Refrigerate until the filling begins to set. Stir the filling occasionally while cooling so it cools evenly. *Note:* Do not allow the filling to set completely because it will be difficult to fold in the egg whites uniformly.
4. Prepare a meringue by whipping egg whites and sugar together until the mixture forms stiff peaks.
5. Fold the meringue into the jellied fruit or cream mixture gently, preserving as many of the air cells as possible. *Note:* This step should be done quickly so that the gelatin does not set before the folding is finished.
6. Pour the chiffon filling into a baked piecrust and refrigerate until set.

Figure 21-63. *A chiffon filling is a light, fluffy pie filling prepared by folding a meringue into a puréed fruit or a cream pie filling.*

Meringues

Meringues are most often used to top pies. A *meringue* is a mixture of egg whites and sugar. Meringues can be soft or hard, depending on the ratio of sugar to egg whites. The greater the amount of sugar in proportion to egg whites, the harder (more stable) the resulting meringue. A soft meringue typically has a 1:1 ratio, 1 part sugar to 1 part egg whites. A hard meringue has a 2:1 ratio, 2 parts sugar to 1 part egg whites. Soft meringues are often used as a dessert topping, such as with lemon meringue pie.

If a meringue-topped pie is held prior to eating, a hard meringue with a ratio of 1½:1 (1½ parts sugar to 1 part egg white) would work best because it will not weep or exude drops of moisture. Hard meringues are used to make crisp crusts or disks that hold soft desserts.

Three common types of meringue are French meringue, Swiss meringue, and Italian meringue. *See Figure 21-64.* These three types of meringue are used in a variety of desserts.

Meringues

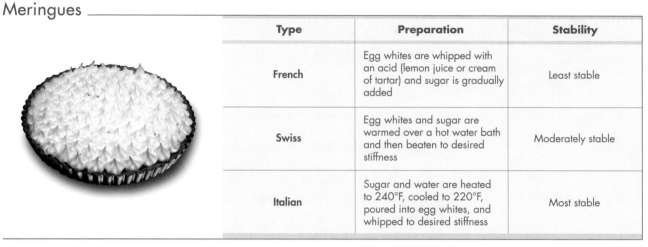

Type	Preparation	Stability
French	Egg whites are whipped with an acid (lemon juice or cream of tartar) and sugar is gradually added	Least stable
Swiss	Egg whites and sugar are warmed over a hot water bath and then beaten to desired stiffness	Moderately stable
Italian	Sugar and water are heated to 240°F, cooled to 220°F, poured into egg whites, and whipped to desired stiffness	Most stable

Figure 21-64. *Three common types of meringue are French meringue, Swiss meringue, and Italian meringue.*

French Meringue. French meringue, also known as common meringue, is made by whipping egg whites with a bit of lemon juice or cream of tartar. Then sugar is gradually added until the desired stiffness is achieved.

Swiss Meringue. Swiss meringue is a more difficult meringue to make than French meringue. Swiss meringue is made by gently warming egg whites and sugar over a hot water bath and then beating them until the desired stiffness is achieved. After the whites and sugar have been heated in a double boiler or bain-marie to approximately 110°F, the sugar dissolves thoroughly, making this meringue more stable and less prone to weeping. However, if the egg white mixture gets too hot, the egg whites will not whip properly.

Italian Meringue. Italian meringue is most commonly used in making butter-cream frosting. Italian meringue is made by heating a mixture of sugar, water, and cream of tartar to 242°F. This mixture is then cooled to 220°F and poured in a slow, steady stream into egg whites while beating continuously. The cooking of the egg whites with the very hot sugar mixture makes this meringue the most stable type of meringue. However, cooking the whites also results in a taffy-like texture that is not as light as the texture of a French or Swiss meringue.

Tips for Preparing Meringues

- Egg whites must be beaten in a clean bowl that is free of fat, especially egg yolks. Fat prohibits egg whites from increasing in volume.
- Allow egg whites to stand at room temperature for at least 30 minutes prior to whipping so the whites will whip faster.
- Add cream of tartar at the beginning of the whipping process to help stabilize and add volume to the egg whites.

Nutrition Note

Egg whites are composed of a protein known as albumen. When egg whites are whipped, the proteins unfold and bond to one another around air bubbles. As they bond, a shiny white film is produced that traps the air bubbles and creates volume. If a fat, such as egg yolk, is introduced into this process, the fat will coat the protein and prevent it from unfolding and bonding and cause a meringue to collapse.

Paderno World Cuisine

PASTRY PREPARATION

Pastries are commonly thought of as something sweet made with a light, flaky bread or crust. However, pastries can also be savory dishes depending on the ingredients used. Many different sweets, hors d'oeuvres, and savory dishes can be prepared using pastries. The word pastry comes from the word "paste," which means a mixture of flour, liquid, and fat, and typically refers to products made from pastes. Most pastries are made using laminated doughs.

Laminated Doughs

A *laminated dough* is an unleavened dough that is layered with butter and used to make pastries, such as croissants, turnovers, and strudels. To make laminated dough, an unleavened dough is wrapped around butter like a package and then rolled out and folded over. This process is called making a turn and is repeated over and over until the butter and the dough are one cohesive piece. The more turns a laminated dough is given, the more flaky the finished pastry will be. The thin layers of butter in laminated dough cause the dough to puff and rise during baking and give pastries a layered and crispy, but light appearance. *See Figure 21-65.* The butter also causes the pastry to brown.

Laminated Doughs

| Turnovers | Strudels | Phyllo Pastries |

Wisconsin Milk Marketing Board, Inc.

Figure 21-65. *The thin layers of butter in laminated dough cause the dough to puff and rise during baking and give pastries a layered and crispy, but light appearance.*

Puff Pastry. *Puff pastry dough,* also known as pâte feuilletée, is a laminated dough that does not contain sugar. Puff pastry dough rises as a result of the steam created when moisture in the layers of dough is released during baking.

Phyllo. *Phyllo* is paper-thin sheets of laminated dough used for making pastries. Frozen phyllo dough is often used in the professional kitchen. Once thawed, it must be kept moist under a damp cloth. One sheet of phyllo is brushed with butter at a time because the dough dries out very quickly. The sheets are then layered to produce pastries, such as baklava and strudels, with many thin, flaky, buttery layers.

Pâte à Choux. *Pâte à choux,* also known as éclair paste, is a pastry made by beating eggs into a paste of boiled water, butter, and flour. ***See Figure 21-66.*** The high egg content makes the dough rise.

Pâte à choux is used to make cream puffs, éclairs, and profiteroles. Cream puffs are round pastries filled with sweetened whipped cream. Éclairs are oblong pastries filled with pastry cream and topped with a chocolate glaze. Profiteroles are round pastries filled with ice cream or savory fillings.

Procedure for Preparing Pâte à Choux

1. Combine water, butter, and salt and bring to a boil. *Note:* The liquid needs to reach a rapid boil in order for the fat to disperse properly.
2. Add the flour and stir until the paste pulls away from the sides of the pan and forms a ball.
3. Allow the paste to rest and cool for a few minutes.

4. Add the eggs one at a time, vigorously beating after each addition. Stop adding eggs when the dough begins to fall away from the mixing spoon. *Note:* The eggs may not all be used, depending upon the moisture content of the flour and the size of the eggs.

5. Fill a pastry bag with the pâte à choux and pipe it into desired shapes on parchment-lined pans.
6. Bake the paste until it is completely dry.

7. Let the pâte à choux cool. Fill with desired filling and serve.

Figure 21-66. *Pâte à choux, also known as éclair paste, is made by beating eggs into a paste of boiled water, butter, and flour.*

Custards and Creams

Custards and creams come in many varieties. Custards and creams include baked custards, soufflés, crème anglaise, pastry creams, sabayons, Bavarian creams, chiffons, and mousses.

Baked Custards. Baked custards, also known as baked puddings, are most often baked in a water bath at a temperature of between 275°F and 300°F until the custard is set, but not completely cooked. Custard continues to cook after it has been removed from the oven. A water bath is used to prevent custard from drying out during cooking. Baked custard is generally a solid that retains the shape of its container and is firm enough to slice. Another method used to prepare baked custard is to cook it in a steamer. Baked custards include crème brûlée, crème caramel, cheesecake, bread pudding, and rice pudding.

- *Crème brûlée* is a baked custard that has the texture of a stirred custard beneath a hardened sugar surface. *See Figure 21-67.* A sweet custard is poured into an ovenproof, individually sized casserole dish and baked in a water bath. After it has set, it is cooled completely. For service, crème brûlée is topped with granulated sugar, and the sugar is caramelized with a torch to produce a thin, glasslike candy covering. Crème brûlée can be made savory with the addition of savory ingredients.

Crème Brûlée

National Honey Board

Figure 21-67. *Crème brûlée is a baked custard that has the texture of a stirred custard beneath a hardened sugar surface.*

- *Crème caramel,* also known as flan, is a baked custard served upside down and topped with hot, caramelized sugar. To make crème caramel, a small amount of hot caramelized sugar is poured into the bottom of a serving dish. An uncooked custard is added, and then the dish is baked in a water bath and removed when set. To serve, crème caramel is unmolded and served upside down so that the caramel sauce runs down over the baked custard. *See Figure 21-68.*

Crème Caramel

Figure 21-68. *To serve, crème caramel is unmolded and served upside down so that the caramel sauce runs down over the baked custard.*

- *Cheesecake* is a variety of baked custard that typically has a graham cracker or cookie crust. Cheesecake is made in a springform pan. There are two basic varieties of cheesecake. Italian cheesecake is light, fluffy, and creamy. New York cheesecake is more dense and rich than Italian cheesecake. *See Figure 21-69.*

Cheesecakes

Planet Hollywood International, Inc.

Italian

Wisconsin Milk Marketing Board, Inc.

New York

Figure 21-69. *The two basic varieties of cheesecake are New York cheesecake and Italian cheesecake.*

- *Bread pudding* is a baked custard that is made by pouring a custard mixture over chunks of bread and baking it in the oven. Bread pudding can be sweet if the custard is flavored with cinnamon, nutmeg, or other sweet spices and baked. It can be savory if sugar is omitted from the custard and herbs and meats are added.

- *Rice pudding* is a baked custard made from cooked rice combined with a sweet custard and often dried fruits such as raisins or currants.

Soufflés. Soufflé custards are commonly referred to as soufflés. Although soufflés are baked in the oven, they are quite different from other baked custards. A sweet soufflé is prepared by using a cooked custard base (pastry cream) combined with flavors such as chocolate, vanilla, or fruit. This base is then lightened with egg whites that have been whipped with sugar and carefully folded in. It is important that the custard base and the egg whites both be at room temperature before mixing. Egg whites whip better when not chilled, and having them at the same temperature as the base allows the base and the whites to mix more easily.

The soufflé is then put into straight-sided ramekin that has been buttered and coated with granulated sugar all the way to the rim. A properly baked soufflé rises high above the rim of the ramekin and has a golden-brown surface. *See Figure 21-70.* If any part of the interior or rim of the ramekin is not buttered and sugared, the soufflé sticks and does not rise properly.

Soufflés

Daniel NYC

Figure 21-70. *A properly baked soufflé rises high above the rim of the ramekin and has a golden brown surface.*

As a soufflé bakes, the air expansion in the egg foam causes the custard to rise as it is heated, similar to a cake. However, soufflés are not stable like cakes and easily collapse once removed from the oven. For this reason, soufflés are made to order and served immediately. *See Figure 21-71.* Soufflés can be split open at the table and served with a slightly beaten cream that is flavored with a liquor or fruit purée.

Procedure for Preparing Individual Soufflés

1. Prepare ramekins in advance by thoroughly buttering the interior and rim of each dish and then thoroughly coating the buttered surface in granulated sugar. *Note:* If any part of the ramekin is not completely buttered and sugared, it will prevent the soufflé from rising.
2. Mix the desired flavoring base into the prepared pastry cream.
3. Whip the egg whites. When soft peaks have formed, begin slowly adding sugar while continually whipping until all the sugar has been added and stiff peaks have formed.
4. Carefully fold one-third of the whipped egg whites into the soufflé base and mix well to incorporate and lighten the base. Continue to gently fold in the remaining whipped egg whites in thirds to prevent a loss of volume.
5. Pour the soufflé mixture into the prepared ramekins within ¼ inch from the top and bake.
6. Remove soufflés when they have risen well above the surface of the ramekin and turned golden brown. Serve immediately.

Figure 21-71. *Soufflés are made to order and served immediately.*

Crème Anglaise. Crème anglaise is vanilla custard sauce that is made by whisking in sugar in two stages. Half the sugar is whisked into already whisked egg yolks. The other half of the sugar is whisked into scalded milk or cream. *See Figure 21-72.* The scalded milk and sugar mixture is then whisked rapidly into the sugar and yolk mixture in three intervals in order to temper the yolks and prevent curdling. The mixture is then put back on the stove and heated until thick and smooth. Crème anglaise should be made in a nonreactive saucepan or stainless steel double boiler to prevent the eggs from scrambling or curdling.

Crème anglaise is fully cooked when it reaches 175°F and should be a nappe consistency, or able to smoothly coat the back of a spoon. Crème anglaise is well known for its use as a vanilla custard sauce served to accompany many dessert dishes. It is also used as a base for other desserts, such as ice cream and Bavarian cream. Because this stirred custard sauce is rich and smooth, it freezes well.

Pastry Creams. A pastry cream is made with egg yolks, sugar, milk, and a starch (usually cornstarch or flour). A properly prepared pastry cream will have the consistency and appearance of a smooth pudding. *See Figure 21-73.*

Unlike crème anglaise, pastry cream can be brought to a boil and simmered because the starch protects the egg yolks from curdling. Boiling allows the starch to fully gelatinize and removes any raw starch flavor from the finished product. A common use of pastry cream is to fill a tart crust and then top it with sliced fresh fruit. It is also commonly used as a filling for napoleons, cream puffs, éclairs, and many other sweet pastries. Pastry cream is often flavored with chocolate, mocha, or nut pastes such as hazelnut, praline, or pistachio.

Irinox USA

Procedure for Preparing Crème Anglaise

1. Place milk (or cream or half and half, as specified), half of the granulated sugar, and salt in a nonreactive saucepan and bring just to a scald to completely dissolve the sugar. *Note:* If vanilla bean is used, it can be added to the milk in this step. If an extract is used, it can be added in step 6.
2. In a stainless steel mixing bowl, whisk egg yolks and the remainder of granulated sugar until light and airy.
3. Temper the yolks by adding about one-third of the hot milk mixture to the yolk and sugar mixture while whisking rapidly.
4. When milk and yolks are mixed well, add the remainder of the hot milk mixture to the yolks while whisking.
5. Pour the complete mixture back into the saucepan and continue to cook while whisking until the mixture reaches 175°F or coats the back of a spoon.
6. Add vanilla extract (or another type of extract for additional flavoring).
7. Strain the sauce into a bain-marie to remove any lumps and immediately stir over an ice bath until cooled.
8. Refrigerate crème anglaise until needed.

Figure 21-72. *Crème anglaise is made by whisking in sugar in two stages. Half of the sugar is whisked into already whisked egg yolks, and the other half is whisked into scalded milk or cream.*

Procedure for Preparing Pastry Creams

1. Place three-fourths of the required milk, half of the required sugar, and the salt in a stainless steel saucepan and bring to a scald in order to completely dissolve the sugar.
2. Place yolks and half of the sugar in the mixing machine. Using the whip attachment, whip on high speed until the mixture thickens and turns pale yellow in color.
3. Whisk flour or cornstarch into yolk mixture and mix well.
4. Add remaining one-fourth of the required milk and mix well until smooth.
5. Pour one-third of the hot milk mixture into the yolk mixture while whisking continuously to temper the yolks.
6. Add the tempered yolk mixture back to the saucepan of milk while stirring rapidly. Return to the heat while stirring constantly.
7. Bring mixture to a boil while whisking vigorously. Continue to whisk for 1 minute. *Note:* As the mixture reaches the boiling point, the continuous whisking will break apart lumps and the pastry cream will appear smooth.
8. Remove the cooked pastry cream from the heat and place in a stainless steel bain-marie or mixing bowl.
9. Carefully fold in the softened butter, using caution to avoid overmixing and thinning the pastry cream.
10. Cover the entire surface of the pastry cream with plastic wrap and refrigerate until needed.

Figure 21-73. *A properly prepared pastry cream will have the consistency and appearance of a smooth pudding.*

Sabayons. A sabayon is prepared by whisking sugar and egg yolks over a double boiler until they are whipped into a silky foam. ***See Figure 21-74.*** It is made in a similar manner to stirred custards. Champagne or a high-quality sweet wine, such as Marsala, is commonly added to increase the flavor of the sauce. Heating the eggs while whisking cooks them and makes the resulting foam stable enough to retain a shape. Sabayon is commonly served as a sauce and can be served warm or cold. A classic use for sabayon is over a dish of fresh berries, because the rich, sweet sauce is balanced by the acidic quality of the berries.

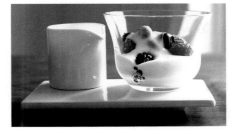

All-Clad Metalcrafters

1. Place sugar, yolks, salt, and champagne or Marsala wine in a mixer and whisk until lighter in color and thick.
2. Remove bowl from mixer and place over a pot of simmering water. Whisk continuously until the mixture is thick, is lighter in color, forms a ribbon, and reaches a temperature of 175°F.
3. Serve while warm, or return the bowl to the mixer and whip until completely cooled.
4. When cool, remove the sabayon and place it in a nonreactive bowl covered with plastic wrap pressed over the surface to prevent a skin from forming.

Figure 21-74. *Sabayon is prepared by whisking sugar and egg yolks over a double boiler until they are whipped into a silky foam.*

Bavarian Creams. Bavarian cream is a flavored custard sauce that has been stabilized with gelatin while it is still warm. After it has almost completely cooled, whipped cream is folded in. Bavarian creams can be flavored with chocolate or fruit purées. The mixture is then placed in a mold to set before being served. Once a Bavarian cream has set completely, the solid portion of the mold is dipped in warm water for a few seconds. The mold is then turned over onto a plate and sliced to order.

Chiffons. A chiffon is similar to a Bavarian cream. However, instead of folding in whipped cream, whipped egg whites are folded into the cooled custard sauce. A chiffon can be placed into a mold but is most often used as a pie filling, as in a lemon chiffon pie, or folded into cakes, such as with an Italian cream cake.

Mousses. A mousse is similar to a chiffon and a Bavarian cream in the way that it is produced. A mousse is lightened with whipped cream, whipped egg whites, or both. The main difference is that a mousse contains only a small amount of gelatin, if any at all. A mousse can be served as a dessert on its own or can be used as a cake filling.

FROZEN DESSERTS

Frozen desserts are specialty desserts that may be served alongside or on top of other desserts or eaten as a separate dish. Popular frozen desserts include ice creams, bombes, gelati, sorbets, sherbets, and granités.

Ice Creams

Ice cream can be the main attraction for a dessert course or can be an accompaniment, as with pie à la mode. *See Figure 21-75. Ice cream* is a frozen dessert made from cream, butterfat, sugar, and sometimes eggs. There are two general types of ice cream bases. The first incorporates eggs to make a crème anglaise base that produces a rich, creamy dessert. The second is a simple mixture of milk, cream, sugar, and flavorings that is heated to dissolve the sugar and then chilled prior to churning. An ice cream base must be refrigerated for at least a few hours to allow excess moisture to be absorbed and to bind to the sugar.

Wisconsin Milk Marketing Board, Inc.

Pie à la Mode

Planet Hollywood International, Inc.

Figure 21-75. *Ice cream can be the main attraction for a dessert course or can be an accompaniment, as with pie à la mode.*

Ice cream should be smooth, creamy, and rich, with enough of the main flavoring ingredient to be flavorful, yet not overpowering. The proper balance of ingredients forms the overall texture and flavor that makes ice cream so popular. For example, eggs add richness to ice cream but also act as an emulsifier. Excess moisture is bound by the lecithin in egg yolks, which helps to prevent ice crystals from forming on the surface and throughout the ice cream. Both the cream and eggs allow air to be whipped into the product during the churning process, making the end product smoother and creamier.

In addition to adding sweetness, sugar also helps to lower the freezing point of an ice cream base, which keeps it from freezing rock-hard. However, too much sugar can prevent ice cream from freezing, and too little sugar results in a product that is difficult to scoop.

Ice cream increases in volume due to overrun. *Overrun* is an increase in the volume of a frozen product as a result of the incorporation of air during churning and freezing. An overrun of 100% means that an ice cream doubled in size during churning and freezing and contains 50% air by volume. The FDA allows a maximum of 100% overrun, although the best ice creams have less. Too much overrun makes ice cream feel frothy and thin in the mouth, rather than creamy and smooth.

Bombes

A *bombe,* also known as glacée, is a French ice cream dessert with at least two varieties of ice cream that are layered inside a spherical mold, frozen, and then unmolded and decorated for service. A frozen, unmolded bombe can be dipped in tempered chocolate to create a surface shell. When three ice cream varieties are layered in a rectangular mold, the item is referred to as a Neapolitan. A Neapolitan is commonly unmolded and sliced for service.

> **Nutrition Note**
>
> Frozen yogurt contains less fat and calories than ice cream. Soy ice cream and rice milk ice cream have a similar taste and texture to ice cream, but are made without dairy products.

Gelati

A *gelato* is an Italian ice cream that has a creamier, denser texture than standard ice creams. Gelato is churned for a shorter period of time, incorporating less air and resulting in a denser product.

Sorbets

A *sorbet* is a frozen dessert made by combining puréed fresh fruit and simple syrup. *See Figure 21-76.* It is churned in an ice cream maker in the same manner as ice cream. A sorbet is naturally lower in calories and fat than ice cream because of the absence of heavy cream and egg yolks. Sorbet is traditionally served before the main course at formal dinners to cleanse the palate. Pear, watermelon, blackberry, and raspberry are popular flavors of sorbets.

Sorbets

Vita-Mix® Corporation

Figure 21-76. *A sorbet is made by combining puréed fresh fruit and simple syrup.*

Sherbets

A *sherbet* is a frozen dessert made by combining puréed fresh fruit, simple syrup, and milk. A sherbet is similar to a sorbet and originated from an ancient Persian, fruit-flavored, frozen dessert called sharbat. Along with the change in name came a slight change to the recipe. Milk is now added during the churning process, making sherbets creamier than sorbets. Sherbets are lower in fat than ice cream but, because of the presence of milk, still have a creamy texture compared to the icy texture of sorbet.

Granités

A *granité*, also known as a granita, is a frozen dessert made by frequently stirring a mixture of water, sugar, and flavorings, such as fruit juice or wine, as it is freezing. It has a grainy texture and ice crystals that are larger than those of a sorbet. A granité is often used to cleanse the palate between courses. It is also a refreshing dessert.

SUMMARY

Mastering fundamental baking and pastry methods enables foodservice professionals to prepare yeast breads, quick breads, cakes, icings, cookies, pies, pastries, and frozen desserts. Formulas are used in the bakeshop instead of recipes. Baker's percentages represent ingredient quantities as compared to the amount of the main ingredient in a formula. The bakeshop also requires the use of special equipment, bakeware, tools, and ingredients.

Bakeshop ingredients must be accurately measured to produce successful products. The type of flour, sugar, fat, egg, milk, thickening agent, leavening agent, and flavoring directly affects the structure, texture, color, nutritional value, and flavor of baked products. Special terms are used to describe how ingredients are combined.

Yeast-bread preparation involves scaling ingredients, mixing, kneading, fermentation, punching, scaling dough, rounding, panning, proofing, baking, cooling, and storing. Quick breads, such as biscuits, muffins, quick bread loaves, and corn breads, can be made using three different mixing methods. Cake batters are often blended, creamed, folded, or whipped. There are nine common types of icings used in the bakeshop. The proper use of pastry bags and tips is a useful skill to have in the professional kitchen. Cookies can be made using six different preparation methods.

Ratios are often used to make pie doughs and pastries. Ratios do not stipulate actual ingredient amounts, but provide an indication of how ingredients are used in proportion to one another. Four types of piecrusts can be used in conjunction with four different pie fillings to create a wide variety of pies and pastries. Laminated doughs are used to prepare flaky pastries such as baklava. Custards, creams, and frozen desserts make sweet finales for any meal.

Daniel NYC

Refer to DVD for
Quick Quiz® questions

Review

Refer to DVD for
Review Questions

1. Identify two types of scales used in the bakeshop.
2. Identify how ingredients are measured in the bakeshop.
3. Explain how two ingredients can have the same volume but different weights.
4. Explain how to calculate the baker's percentage of an ingredient.
5. Explain how a formula differs from a recipe.
6. Explain how to convert a recipe into a formula.
7. Identify eight types of ingredients commonly used to create baked products.
8. Identify the protein in flour that gives a baked product structure.
9. Contrast six common types of flour used in the bakeshop.
10. Explain how sugar affects baked products.
11. Identify eight types of sugar used in the bakeshop.
12. Explain how fats affect baked products.
13. Identify three fats commonly used in the bakeshop.
14. Explain the process of hydrogenation.
15. Explain how eggs affect baked products.
16. Identify two types of eggs used in the bakeshop.

17. Explain how milk and other dairy products affect baked products.
18. Identify the two primary types of milk used in the bakeshop.
19. Explain how thickening agents affect baked products.
20. Identify six common thickening agents used in a bakeshop.
21. Describe the most basic way to leaven baked products.
22. Identify three common varieties of commercial yeast.
23. Explain how flavorings affect baked products.
24. Explain the difference between an extract and an emulsion.
25. Explain how vanilla bean is added to a mixture.
26. Identify five types of chocolate used in the bakeshop.
27. Define seven terms used to describe methods of combining ingredients.
28. Identify common equipment used in the bakeshop.
29. Identify common bakeware used in the bakeshop.
30. Identify common tools used in the bakeshop.
31. Contrast lean dough, rich dough, and rolled-in yeast dough.
32. Explain how to scale ingredients.
33. Describe three methods of mixing yeast dough.
34. Explain how to knead yeast dough.
35. Define fermentation.
36. Explain why yeast dough needs to be punched.
37. Explain how to scale yeast dough.
38. Explain why yeast dough may need to be rounded.
39. Explain how to pan yeast dough.
40. Describe the purpose of proofing yeast dough.
41. Explain how to score yeast dough.
42. Explain how to dock yeast dough.
43. Explain how to wash yeast dough.
44. Explain the cause of oven spring and the temperature at which it stops.
45. Describe steam injection.
46. Explain why some yeast breads should not be wrapped for storage.
47. Identify four common types of quick breads.
48. Describe three methods of mixing quick breads.
49. List guidelines for baking, cooling, and storing quick breads.
50. Describe the two-stage method of mixing cake batters.
51. Describe the creaming method of mixing cake batters.
52. Describe the chiffon method of mixing cake batters.
53. Describe the sponge method of mixing cake batters.
54. Describe the four stages of baking cakes.
55. Describe how to test cakes for doneness.
56. Describe nine types of icings.
57. Explain the proper method of filling a pastry bag.
58. Explain the proper method of piping icing.

59. Identify six methods of preparing cookies.

60. Identify the four types of piecrusts.

61. Contrast mealy and flaky piecrust.

62. Contrast crumb crusts and specialty crusts.

63. Explain the purpose of blind baking.

64. Contrast fruit, cream, soft, and chiffon pie fillings.

65. Explain how the ratio of sugar to egg whites affects meringues.

66. Contrast three common types of meringue.

67. Explain how the number of turns used to make a laminated dough affects the flakiness of a pastry.

68. Describe eight types of custards and cream pastries.

69. Contrast six types of frozen desserts.

Applications

Refer to DVD for
Application Scoresheets

1. Use a baker's (balance) scale to accurately measure flour and sugar.

2. Use a digital scale to accurately measure baking powder and eggs.

3. Calculate the baker's percentage for each ingredient of a formula.

4. Use baker's percentages to convert a recipe to a formula.

5. Prepare a lean yeast dough. Evaluate the appearance, texture, and color.

6. Prepare a rich yeast dough. Evaluate the appearance, texture, and color.

7. Prepare a rolled-in dough. Evaluate the appearance, texture, and color.

8. Prepare a quick bread using the biscuit mixing method. Evaluate the appearance, texture, color, and flavor.

9. Prepare a quick bread using the muffin mixing method. Evaluate the appearance, texture, color, and flavor.

10. Prepare a quick bread using the creaming mixing method. Evaluate the appearance, texture, color, and flavor.

11. Use the two-stage method to mix a cake batter. Evaluate the appearance and texture.

12. Use the creaming method to mix a cake batter. Evaluate the appearance and texture.

13. Use the chiffon method to mix a cake batter. Evaluate the appearance and texture.

14. Use the sponge method to mix a cake batter. Evaluate the appearance and texture.

15. Prepare an icing. Evaluate the appearance, texture, color, and flavor.

16. Form, fill, and use a pastry bag to pipe icing. Evaluate the results.

17. Prepare cookies using six different methods. Compare the appearance, texture, color, and flavor of each type of cookie.

18. Prepare a mealy pie dough and a flaky pie dough. Compare the appearance and texture of each dough.

19. Prepare a crumb crust and a specialty crust. Evaluate the appearance, texture, and flavor of each crust.

20. Blind bake a piecrust. Evaluate the results.

21. Prepare a fruit, cream, soft, and chiffon pie filling. Evaluate each filling based on appearance, texture, color, and flavor.

22. Prepare three types of meringues. Evaluate each meringue based on appearance, stiffness, and flavor.

23. Prepare puff pastry dough and pâte à choux. Evaluate each based on appearance, texture, and color.

24. Prepare a baked custard. Evaluate the custard based on appearance, texture, color, and flavor.

25. Prepare a pastry cream. Evaluate the appearance, texture, color, and flavor.

26. Prepare a mousse. Evaluate the appearance, texture, color, and flavor.

27. Prepare a frozen dessert. Evaluate the appearance, texture, flavor, color, and presentation.

Soft Dinner Rolls

Yield: *16 servings, 2 rolls each*

Ingredients

fresh yeast	1.5 oz
lukewarm milk (85°)	1 lb 8 oz
sugar	4 oz
butter, softened	4 oz
salt	0.50 oz
bread flour	2 lb 10 oz

Egg Wash

egg, beaten slightly	1 ea
water	1 tsp

Procedures

1. Place yeast and milk together in mixing bowl.
2. Slowly add sugar. Then add butter and salt and mix until the sugar dissolves.
3. Slowly add flour to the mixture until a smooth dough forms (approximately 10 minutes).
4. Place dough in a greased bowl and cover or place in a proofer for approximately 1 hour.
5. Punch down dough and then knead.
6. Proof dough for 20 minutes.
7. Scale into 1.50 oz rolls.
8. Mix egg and water and apply egg wash to the surface of the rolls.
9. Bake rolls at 425°F for 25 minutes.

> **NUTRITION FACTS**
> **Per serving:** 386 calories, 78 calories from fat, 8.9 g total fat, 31.1 mg cholesterol, 369.8 mg sodium, 162.2 mg potassium, 64.2 g carbohydrates, 2.5 g fiber, 9.5 g sugar, 11.8 g protein

Hard Dinner Rolls

Yield: *16 servings, 1 roll each*

Ingredients

fresh yeast	1.5 oz
lukewarm milk (85°)	1 lb 8 oz
sugar	1 oz
butter	2 oz
salt	0.25 oz
egg white	1 ea
bread flour	2 lb 12 oz

Egg Wash

egg, beaten slightly	1 ea
water	1 tsp

Procedures

1. Dissolve yeast in warm milk.
2. Cream sugar and butter in a mixing bowl using the paddle attachment.
3. Add the milk and yeast mixture, salt, egg white, and flour to the creamed mixture and mix for approximately 10 minutes until smooth.
4. Place dough in a greased bowl and cover or place in a proofer until double in size (approximately 1 hour).
5. Punch down dough and proof again for 20 minutes.
6. Scale into 1.50 oz rolls.
7. Mix the egg and water and apply egg wash to the surface of the rolls.
8. Proof for 15–20 minutes.
9. Place a small pan of water in the oven to create steam during baking.
10. Bake at 400°F until golden brown (approximately 25–30 minutes).

> **NUTRITION FACTS**
> **Per serving:** 354 calories, 53 calories from fat, 6.1 g total fat, 23.5 mg cholesterol, 201.2 mg sodium, 168.1 mg potassium, 61.5 g carbohydrates, 2.6 g fiber, 4.2 g sugar, 12.4 g protein

White Bread

Yield: *64 servings (4 loaves, 16 slices per loaf)*

Ingredients

fresh yeast	1.50 oz
lukewarm milk (85°)	1 lb 8 oz
sugar	1.50 oz
salt	0.50 oz
shortening	1.50 oz
bread flour	2 lb 8 oz

Procedures

1. Dissolve yeast in one-fifth of the milk. *Note:* Milk should be heated to 85°F to ensure proper activation.
2. Place remaining milk, sugar, salt, and shortening in a mixer and mix until all ingredients are blended.
3. Alternately add the flour and yeast mixture until a smooth dough forms. Continue to mix for an additional 5–6 minutes.
4. Place dough in a greased bowl and cover or place in a proofer for approximately 1 hour.
5. Punch down dough.
6. Scale and place in well-greased loaf pans. Let rise for approximately 30 minutes.
7. Bake at 450°F until golden brown (approximately 35 minutes).

NUTRITION FACTS
Per serving: 161 calories, 23 calories from fat, 2.6 g total fat, 2.8 mg cholesterol, 182.2 mg sodium, 76.3 mg potassium, 28.6 g carbohydrates, 1.2 g fiber, 2.5 g sugar, 5.5 g protein

Whole Wheat Bread

Yield: *64 servings (4 loaves, 16 slices per loaf)*

Ingredients

fresh yeast	1.50 oz
lukewarm milk (85°)	1 lb 8 oz
sugar	1.50 oz
salt	0.50 oz
shortening	1.50 oz
whole wheat flour	1 lb 8 oz
bread flour	1 lb

Procedures

1. Dissolve yeast in one-fifth of the lukewarm milk.
2. Place remaining ingredients in mixing bowl and mix at medium speed.
3. Slowly add the yeast mixture and mix for approximately 10 minutes until a smooth dough forms.
4. Place dough in a greased bowl and cover or place in a proofer for approximately 1½ hours.
5. Punch down dough and scale to 18 oz pieces per each 1 lb loaf.
6. Place in greased loaf pans and proof for 30 minutes.
7. Bake at 400°F for 30–35 minutes.

NUTRITION FACTS
Per serving: 162 calories, 22 calories from fat, 2.4 g total fat, 2.8 mg cholesterol, 182.2 mg sodium, 77.8 mg potassium, 29.4 g carbohydrates, 1.3 g fiber, 2.5 g sugar, 5.1 g protein

German Rye Bread

Yield: *64 servings (4 loaves, 16 slices per loaf)*

Ingredients

bread flour	11.50 oz
rye flour	12.20 oz
molasses	5 oz
warm water	4 oz
active dry yeast	0.50 oz
lukewarm milk (85°F)	13 oz
butter	1 oz
sugar	1 oz
kosher salt	1 tsp

Egg Wash

egg, beaten lightly	1 ea
water	1 tsp

Procedures

1. Combine bread flour and rye flour in a bowl.
2. Combine molasses, water, and yeast and add one-third of the flour mixture. Mix for 2 minutes, cover, and set aside.
3. Let rise for approximately 1 hour until the mixture has doubled in size.
4. Add warm milk, butter, sugar, and salt to the mixture and mix.
5. Stir in the remaining flour mixture until the dough is stiff enough to knead.
6. Knead dough for 5 minutes in mixer using dough hook.
7. Place dough in a greased bowl and cover or place in a proofer for approximately 1 hour until the dough has doubled in size.
8. Punch down dough and divide to form round loaves.
9. Place on lightly greased sheet pan.
10. Mix egg and water and apply egg wash to the loaves.
11. Proof for approximately 1–1½ hours until double in size.
12. Bake at 375°F until golden brown and crispy (approximately 30–35 minutes).

NUTRITION FACTS
Per serving: 54 calories, 7 calories from fat, <1 g total fat, 4.5 mg cholesterol, 34.4 mg sodium, 69.2 mg potassium, 10.3 g carbohydrates, <1 g fiber, 2.1 g sugar, 1.6 g protein

Vienna Bread

Yield: *64 servings (4 loaves, 16 slices per loaf)*

Method: Sponge

Ingredients

lukewarm water (85°F)	1 lb 2 oz
bread flour	3 lb 10 oz
fresh yeast	2 oz
malt syrup	1 tsp
shortening	2 oz
salt	0.75 oz
sugar	0.75 oz
cornmeal	1.5 oz

Procedures

1. In a bowl, mix one-third of the water, one-third of the flour, yeast, and malt syrup to form a fermented sponge and let rest for 30 minutes.
2. Add the remainder of the water, flour, shortening, salt, and sugar to a mixer and mix for approximately 10 minutes until a smooth, stiff dough forms.
3. Dust sheet pans with cornmeal.
4. Place dough in a greased bowl and cover or place in a proofer for approximately 1 hour until dough has doubled in size.
5. Punch down, scale, and form dough into long 1 lb loaves and place on sheet pans.
6. Score tops of loaves diagonally.
7. Proof for approximately 30 minutes.
8. Bake in 400°F oven until golden brown and hollow sounding when tapped (approximately 30–35 minutes).

NUTRITION FACTS
Per serving: 143 calories, 15 calories from fat, 1.8 g total fat, <1 mg cholesterol, 173.7 mg sodium, 48.8 mg potassium, 26.6 g carbohydrates, 1.2 g fiber, <1 g sugar, 4.7 g protein

Sweet Yeast Dough Rolls

Yield: *30 servings, 1 roll each*

Method: Sponge

Ingredients

yeast	3 oz
lukewarm milk (85°)	1 lb
sugar	8 oz
bread flour	4 lb 12 oz
salt	0.5 oz
butter	8 oz
eggs	3 ea
lemon zest	0.50 oz

Egg Wash

egg, beaten lightly	1 ea
water	1 tsp

Procedures

1. Dissolve the yeast in the milk and mix with one-fourth of the sugar and one-third of the flour. Rest for 20 minutes to form sponge.
2. Add the salt, butter, eggs, and lemon zest and mix to form a smooth, stiff dough.
3. Place dough in a greased bowl and cover or place in a proofer for approximately 1 hour.
4. Punch down dough, roll out, and cut to form 4 oz rolls.
5. Mix egg and water and apply egg wash to the surface of the rolls.
6. Proof for 30 minutes.
7. Bake at 375°F until golden (approximately 35–40 minutes). *Note:* Add frosting if desired.

> **NUTRITION FACTS**
> **Per serving:** 371 calories, 76 calories from fat, 8.7 g total fat, 42.6 mg cholesterol, 202.9 mg sodium, 130.8 mg potassium, 61.7 g carbohydrates, 2.5 g fiber, 8.6 g sugar, 11.1 g protein

Cinnamon Rolls

Yield: *40 servings, 1 roll each*

Ingredients

sweet yeast dough	½ recipe
butter, softened	8 oz
brown sugar	8 oz
cinnamon	0.50 oz
pecans	16 oz

Procedures

1. Prepare sweet yeast dough recipe. After first proofing, roll out dough ½ inch thick into a rectangular shape.
2. Brush softened butter evenly on surface of rolled-out dough.
3. Mix brown sugar, cinnamon, and pecans together and sprinkle evenly onto the buttered dough.
4. Slowly roll up the dough, encasing the filling, and occasionally pinching the ends to prevent the filling from falling out.
5. Cut roll into ½ inch thick pieces and place in greased baking pan, allowing 1 inch between rolls.
6. Proof for approximately 30 minutes until double in size.
7. Bake at 375°F until golden (approximately 35–40 minutes).
8. Remove from oven and let stand for 15 minutes before removing from pan.

> **NUTRITION FACTS**
> **Per serving:** 280 calories, 137 calories from fat, 16 g total fat, 28.2 mg cholesterol, 78.3 mg sodium, 104.6 mg potassium, 30.3 g carbohydrates, 2.1 g fiber, 9.2 g sugar, 5.3 g protein

Croissants

Yield: *20 servings, 1 croissant each*

Ingredients

bread flour	1.75 lb
sugar	1 oz
salt	0.50 oz
fresh yeast	1 oz
milk (85°)	16 oz
unsalted butter, softened	1 lb

Egg Wash

egg, beaten slightly	1 ea
milk	0.50 oz

Procedures

1. In a mixer with a dough hook attachment, place the flour, sugar, and salt. Mix for 2 minutes on low speed.
2. Place yeast and milk together in a mixing bowl.
3. Add yeast mixture to flour mixture and stir until combined. Knead in mixer for approximately 10 minutes.
4. Place dough in a greased bowl and cover or place in a proofer until doubled.
5. While dough is rising, place butter between two sheets of plastic wrap and roll out the butter with a rolling pin until it has formed a rectangle, approximately 6 × 8½ inches. Place in refrigerator to chill.
6. Punch down dough and roll it into a rectangle, approximately 10 × 15 inches. Brush off excess flour. Unwrap butter and place it in the middle of the dough. Fold the corners of the dough over the butter, encasing it completely.
7. Brush off excess flour and lightly press with a rolling pin to push all the layers together. Roll the dough to form a new 10 × 15 inch rectangle.
8. Fold the dough into thirds. Wrap the dough in plastic and chill for 25 to 30 minutes.
9. Repeat rolling out the dough and folding it into thirds two more times, chilling the dough between each turn.
10. After turning a third time, chill the dough.
11. Remove dough from the refrigerator and slice into three pieces. Roll each piece into a rectangle approximately ¼ inch thick.
12. Cut the rolled-out dough into uniform triangles. Roll each triangle up starting with the large end and place on a parchment-lined sheet pan with the tip tucked under and the ends slightly curved to form a crescent.
13. Mix egg and milk to form an egg wash and wash the dough lightly.
14. Proof until the croissants double in size.
15. Bake at 375°F until golden (approximately 12–15 minutes).

NUTRITION FACTS
Per serving: 335 calories, 178 calories from fat, 20.2 g total fat, 60.6 mg cholesterol, 293.1 mg sodium, 95.3 mg potassium, 32 g carbohydrates, 1.3 g fiber, 2.8 g sugar, 6.6 g protein

Danish Pastries

Yield: *30 servings, 1 pastry each*

Ingredients

sugar	8 oz
butter	8 oz
salt	0.50 oz
nutmeg	pinch
whole eggs	16 oz
warm milk (100°F)	16 oz
instant yeast	1 oz
cake flour	17 oz
bread flour	2 lb 2 oz
butter	1.25 lb

Egg Wash
egg, beaten slightly	1 ea
water	1 tsp

Streusel
butter	10 oz
shortening	10 oz
vanilla	0.25 oz
sugar	1.25 lb
brown sugar	8 oz
bread flour	2.50 lb

Procedures

1. Cream sugar, butter, salt, and nutmeg with paddle.
2. Add eggs slowly.
3. Add warm milk and yeast.
4. Add flours. Mix 3–4 minutes at low speed.
5. Rest 15 minutes.
6. Roll dough into rectangle approximately ¼ inch thick.
7. Place butter or fat mixture on two-thirds of surface.
8. Fold the dough into thirds. Wrap the dough in plastic and chill for 15–20 minutes.
9. Roll out the dough and fold it into thirds two more times, chilling the dough between each turn.
10. Cut dough into 4 oz pieces, roll out into a thin sheet, and form into desired shapes.
11. Mix egg wash and water and apply egg wash. If desired, add a spoonful of filling to the center of each pastry.
12. Place rolls on lightly greased sheet pans, let proof and bake at 385°F until golden brown (approximately 20 minutes).
13. In a separate container, melt butter and shortening together and add vanilla.
14. Rub sugar, brown sugar, and bread flour together with butter and shortening. Reserve the mixture for streusel topping.
15. Cover pastries with streusel topping.

NUTRITION FACTS
Per serving: 814 calories, 364 calories from fat, 41.2 g total fat, 146.1 mg cholesterol, 222.5 mg sodium, 160 mg potassium, 98.5 g carbohydrates, 2.2 g fiber, 34.9 g sugar, 13 g protein

Brioche

Yield: *24 servings, 1 brioche each*

Ingredients

yeast	2 oz
lukewarm milk (85°)	8 oz
bread flour	2.50 lb
granulated sugar	4 oz
salt	0.50 oz
eggs	1.25 lb
butter	1.5 lb

Egg Wash

egg, beaten slightly	1 ea
water	1 tsp

Procedures

1. Dissolve yeast in lukewarm milk. Add one-fifth of the flour and mix 3 minutes until sponge is formed. Let sponge rest for approximately 30 minutes.
2. Add flour, sugar, salt, and eggs to the sponge and mix until smooth.
3. Add butter and mix for 5 minutes.
4. Place dough in a greased bowl and cover or place in a proofer for approximately 1 hour.
5. Punch down dough and refrigerate overnight.
6. Scale three-fourths of the dough into 3 oz pieces and round into balls. Scale remaining dough into small 1 oz balls.
7. Let rest 10 minutes.
8. Place 3 oz balls in fluted, well-greased molds.
9. Push center of each ball in and place 0.50 oz balls in hollow made by the finger.
10. When half proofed (approximately 15 minutes), mix egg and water and apply egg wash.
11. Finish proofing for approximately 15 minutes.
12. Bake at 375°F to 400°F until golden (approximately 30 minutes).

NUTRITION FACTS
Per serving: 443 calories, 235 calories from fat, 26.7 g total fat, 157.6 mg cholesterol, 275.1 mg sodium, 125.6 mg potassium, 40.7 g carbohydrates, 1.8 g fiber, 5.5 g sugar, 10.4 g protein

Baking Powder Biscuits

Yield: *20 servings, 1 biscuit each*

Ingredients

cake flour	2.50 lb
baking powder	1.50 oz
salt	1.25 oz
sugar	2 oz
milk	1.5 lb
shortening	1 lb

Egg Wash

egg, beaten slightly	1 ea
water	1 tsp

Procedures

1. Sift dry ingredients together. Cut in shortening until pea-sized clumps form.
2. Add water and mix until smooth dough forms.
3. Roll dough out on lightly floured surface to ½ inch thick.
4. Cut ½ inch thick dough into 4 oz biscuits and place onto greased pan.
5. Mix egg and water and apply egg wash the tops of the biscuits. *Note:* If desired, biscuits can be brushed with melted butter instead of egg wash.
6. Bake at 400°F for 10–15 minutes.

NUTRITION FACTS
Per serving: 446 calories, 220 calories from fat, 24.5 g total fat, 25.4 mg cholesterol, 931.5 mg sodium, 108.5 mg potassium, 49.3 g carbohydrates, <1 g fiber, 4.7 g sugar, 6 g protein

Cheddar Buttermilk Biscuits

Yield: *24 servings, 1 biscuit each*

Ingredients

cake flour	2.50 lb
sugar	2 oz
salt	1.25 oz
baking powder	1.50 oz
cheddar cheese, shredded	0.75 lb
shortening	1 lb
milk	1.5 lb

Egg Wash

egg, beaten slightly	1 ea
water	1 tsp

Procedures

1. Sift flour, salt, sugar, and baking powder and add shredded cheddar cheese.
2. Cut shortening into cubes and add the shortening to the flour mixture.
3. Cut the shortening into the flour. *Note:* The shortening should be slightly visible, and the mixture should resemble coarse cornmeal.
4. Add the milk and mix until just combined.
5. Place the dough on a lightly floured surface and roll out to ½ inch thick.
6. Cut into desired shapes, 3 oz each.
7. Mix egg and water and apply egg wash to the surface of the biscuits.
8. Bake at 400°F for 10–15 minutes.

NUTRITION FACTS
Per serving: 429 calories, 225 calories from fat, 25.1 g total fat, 36.1 mg cholesterol, 864.3 mg sodium, 104.3 mg potassium, 41.3 g carbohydrates, <1 g fiber, 4 g sugar, 8.6 g protein

Apricot and Cherry Scones

Yield: *25 servings, 1 scone each*

Ingredients

bread flour	1 lb 12 oz
baking powder	1.5 oz
salt	0.17 oz
dried cherries, diced	6 oz
dried apricots, diced	6 oz
honey	4 oz
cream	3.50 c
coarse sugar	1 oz

Procedures

1. Sift flour, baking powder, and salt together.
2. Mix fruit with flour mixture.
3. Mix in honey and 90% of the cream. Mix enough to incorporate, using caution not to overmix.
4. Let mixture rest for 5 minutes.
5. Pour out onto floured surface and roll to 1½ inches thick.
6. Cut into 3 oz triangular shapes.
7. Bake at 375°F until golden brown (approximately 10–12 minutes).
8. Remove from oven, brush with remaining cream, and press lightly into coarse sugar.

NUTRITION FACTS
Per serving: 287 calories, 114 calories from fat, 13 g total fat, 45.7 mg cholesterol, 270.1 mg sodium, 138.5 mg potassium, 39.1 g carbohydrates, 1.6 g fiber, 8.6 g sugar, 4.9 g protein

Blueberry Muffins

Yield: *30 servings, 1 muffin each*

Ingredients

sugar	12 oz
eggs	8 oz
vanilla extract	0.35 oz
milk	1 lb
shortening	9 oz
cake flour	1 lb 12 oz
salt	0.25 oz
baking powder	1 oz
blueberries	1 lb

Procedures

1. Cream sugar, eggs, vanilla extract, milk, and shortening together until smooth.
2. Sift cake flour, salt, and baking powder together. Slowly add to creamed ingredients and mix until moist.
3. Fold in blueberries carefully until just mixed.
4. Pour batter into greased and floured muffin tins.
5. Bake at 375°F for approximately 25–30 minutes.

NUTRITION FACTS
Per serving: 239 calories, 82 calories from fat, 9.2 g total fat, 33.9 mg cholesterol, 209.8 mg sodium, 70.7 mg potassium, 35.3 g carbohydrates, <1 g fiber, 13.7 g sugar, 3.7 g protein

Corn Bread

Yield: *20 servings (one 2-inch-deep half pan)*

Ingredients

sugar	8 oz
shortening	8 oz
salt	0.50 oz
eggs	4 ea
cake flour	1.5 lb
cornmeal	1.5 lb
baking powder	2 oz
milk	1 qt

Procedures

1. Lightly cream sugar, shortening, and salt together.
2. Add eggs one at a time and mix thoroughly.
3. Sift flour, baking powder, and cornmeal together.
4. Add dry ingredients alternately with the milk to the creamed mixture.
5. Place batter in greased pan.
6. Bake at 375°F until toothpick inserted comes out clean (approximately 25–30 minutes).

NUTRITION FACTS
Per serving: 428 calories, 128 calories from fat, 14.3 g total fat, 47.8 mg cholesterol, 623.1 mg sodium, 212.4 mg potassium, 67.2 g carbohydrates, 3.1 g fiber, 14.1 g sugar, 8.4 g protein

Zucchini Bread

Yield: *32 servings (2 loaves, 16 slices per loaf)*

Ingredients

butter	10 oz
sugar	1 lb 3 oz
whole eggs	6 oz
vanilla extract	0.21 oz
zucchini, grated	1 lb 5 oz
bread flour	1 lb 3 oz
salt	0.33 oz
baking powder	0.33 oz
walnut pieces	8 oz

Procedures

1. Grease and flour two loaf pans. Set aside.
2. Melt butter and then remove saucepan from heat.
3. Add sugar to butter and stir in eggs, vanilla extract, and zucchini.
4. Sift flour, salt and baking powder together.
5. Fold into zucchini mixture.
6. Fold walnut pieces into mixture.
7. Pour into greased pans to three-fourths full.
8. Bake at 350°F for 50–60 minutes.

NUTRITION FACTS
Per serving: 247 calories, 109 calories from fat, 12.7 g total fat, 38.8 mg cholesterol, 155 mg sodium, 106.8 mg potassium, 30.7 g carbohydrates, 1.1 g fiber, 17.5 g sugar, 4.1 g protein

Banana Nut Bread

Yield: *64 servings (4 loaves, 16 slices per loaf)*

Ingredients

sugar	16 oz
butter	8 oz
salt	0.50 oz
eggs, beaten	12 oz
chopped nuts	4 oz
bananas, mashed	1.25 lb
cake flour	1.50 lb
cream of tartar	0.50 oz
baking soda	0.50 oz

Procedures

1. Cream sugar, butter, salt, and eggs.
2. Add chopped nuts and bananas.
3. Sift flour, cream of tartar, and baking soda. Add to mixture.
4. Scale approximately 21 oz per loaf.
5. Bake at 375°F for 55 minutes.

NUTRITION FACTS
Per serving: 119 calories, 41 calories from fat, 4.7 g total fat, 27.4 mg cholesterol, 154.9 mg sodium, 95.6 mg potassium, 17.8 g carbohydrates, <1 g fiber, 8.3 g sugar, 1.9 g protein

Carrot Cake

Yield: *36 servings (three 8-inch round cakes, 12 slices per cake)*

Ingredients

sugar	1 lb
eggs	4 ea
vegetable oil	8 oz
all-purpose flour	1 lb 5 oz
baking powder	0.50 oz
salt	0.25 oz
cinnamon	1 tsp
nutmeg	1 tsp
shredded carrots	1 lb 8 oz
chopped walnuts	8 oz

Preparation

1. Cream the sugar and eggs in mixer until creamy and smooth.
2. Add vegetable oil and continue mixing until well combined.
3. Add flour, baking powder, salt, cinnamon, and nutmeg. Mix well.
4. Stir in shredded carrots and nuts.
5. Pour into a greased and floured baking pan. Bake in a preheated 350°F oven for 35–40 minutes, or until cake springs back to the touch.

NUTRITION FACTS
Per serving: 222 calories, 97 calories from fat, 11.2 g total fat, 20.7 mg cholesterol, 139.6 mg sodium, 114.5 mg potassium, 28.1 g carbohydrates, 1.5 g fiber, 13.7 g sugar, 3.6 g protein

Pound Cake

Yield: *32 servings (4 loaves, 8 slices per loaf)*

Ingredients

butter	1 lb
granulated sugar	1 lb
whole eggs	1 lb
salt	0.25 oz
vanilla extract	0.25 oz
lemon, squeezed	1 ea
cake flour, sifted	1 lb
baking powder	0.50 oz

Preparation

1. Cream the butter smooth.
2. Add sugar and continue to cream until light and fluffy.
3. Add eggs gradually and continue creaming.
4. Add the salt, vanilla, and lemon juice. Mix well.
5. Gradually stir in flour and baking powder.
6. Scale 1 lb 2 oz of dough each into greased loaf pans and bake at 350°F for 45–50 minutes.

NUTRITION FACTS
Per serving: 229 calories, 114 calories from fat, 13 g total fat, 83.2 mg cholesterol, 155 mg sodium, 40.5 mg potassium, 25.6 g carbohydrates, <1 g fiber, 14.3 g sugar, 3.1 g protein

Angel Food Cake

Yield: *36 servings (3 cakes, 12 slices per cake)*

Ingredients

powdered sugar	1 lb
cake flour	12 oz
salt	0.25 oz
egg whites	2 lb
cream of tartar	0.50 oz
granulated sugar	1 lb
vanilla extract	0.25 oz
almond extract	1 tsp
lemons, grated zest	2 ea

Preparation

1. Sift powdered sugar, flour, and salt together.
2. In a mixing bowl, beat egg whites until foamy. Add cream of tartar and mix at high speed with a paddle attachment.
3. Begin gradually adding granulated sugar, vanilla extract, almond extract, and lemon zest to the egg mixture.
4. Beat until eggs are glossy and stiff peaks form.
5. When stiff peaks have formed, fold the powdered sugar, flour, and salt mix into the egg mixture.
6. Place in 9 inch tube pans and bake at 375°F for 30–35 minutes.
7. Cool upside down for 30–40 minutes.

> **NUTRITION FACTS**
> **Per serving:** 147 calories, 1 calories from fat, <1 g total fat, 0 mg cholesterol, 119 mg sodium, 117.3 mg potassium, 33 g carbohydrates, <1 g fiber, 25.1 g sugar, 3.5 g protein

White Layer Cake

Yield: *12 servings (two 8-inch round layers)*

Ingredients

salt	0.50 oz
butter	5.50 oz
sugar	17 oz
cake flour	13 oz
baking powder	0.25 oz
milk	10.50 oz
vanilla or almond flavoring	1 tsp
egg whites	6 oz

Preparation

1. Cream salt, butter, and three-fourths of the sugar until fluffy and light.
2. Sift flour and baking powder.
3. Mix milk with flavoring.
4. Add flour alternately with milk to the batter at moderate speed, scraping the bottom and sides of the bowl after each addition to avoid lumps.
5. Beat egg whites stiff with remaining sugar and fold into the batter.
6. Bake at 375°F approximately 30 minutes.

> **NUTRITION FACTS**
> **Per serving:** 384 calories, 103 calories from fat, 11.7 g total fat, 30.6 mg cholesterol, 557.9 mg sodium, 94.6 mg potassium, 65.7 g carbohydrates, <1 g fiber, 41.6 g sugar, 5 g protein

Chocolate Layer Cake

Yield: *12 servings (two 8-inch round layers)*

Ingredients

nonstick pan spray	1 tsp
flour	11 oz
baking soda	0.25 oz
salt	1 tsp
cocoa powder	1.25 oz
corn oil	1.25 oz
sugar	12.50 oz
shortening	5.1 oz
eggs	2 ea
vanilla extract	1 tsp
cold water	1.25 c

Preparation

1. Spray 8 inch round layer cake pans with nonstick pan coating.
2. Combine flour, baking soda, and salt in a bowl and set aside.
3. In a small bowl, combine cocoa powder and corn oil and stir until all the cocoa powder lumps are gone. Reserve.
4. In a mixer, cream the sugar and the shortening until fluffy.
5. Gradually add the eggs and vanilla to the creamed mixture, beating well after each addition.
6. Add the cocoa and corn oil mixture to the creamed mixture and mix well.
7. Add flour mixture alternately with the water to the creamed mixture, mixing well after each addition.
8. Pour the batter into the prepared pans and bake at 350°F until cake tests done in the center with a toothpick (approximately 30 minutes).

NUTRITION FACTS
Per serving: 355 calories, 139 calories from fat, 15.5 g total fat, 37.1 mg cholesterol, 369.4 mg sodium, 85.7 mg potassium, 51.2 g carbohydrates, 1.7 g fiber, 29.7 g sugar, 4.3 g protein

Flat Icing

Yield: *32 servings, 1 oz each*

Ingredients

confectioners' sugar	2 lb 8 oz
corn syrup	4 oz
pasteurized egg whites	2 oz
water, hot	8 oz

Preparation

1. Place confectioners' sugar, corn syrup, and egg whites in mixing bowl. Mix until smooth at slow speed using a paddle attachment while adding hot water.
2. When ready to use, heat the amount of icing needed in a double boiler and use to glaze cakes or petit fours.

NUTRITION FACTS
Per serving: 149 calories, <1 calorie from fat, <1 g total fat, 0 mg cholesterol, 6.1 mg sodium, 3.7 mg potassium, 38.1 g carbohydrates, 0 g fiber, 35.6 g sugar, <1 g protein

Royal Icing

Yield: *56 servings, 1 oz each*

Ingredients

egg whites	8 oz
confectioners' sugar	3 lb
cream of tartar	0.25 oz

Preparation

1. Warm the ingredients to 96°F in a double boiler and then beat the mixture in the machine on the second speed until light and fluffy.
2. When icing is not in use, keep covered with a damp cloth to prevent a crust from forming.

NUTRITION FACTS
Per serving: 97 calories, <1 calorie from fat, <1 g total fat, 0 mg cholesterol, 7.3 mg sodium, 28 mg potassium, 24.4 g carbohydrates, 0 g fiber, 23.8 g sugar, <1 g protein

Buttercream Icing

Yield: *42 servings, 1 oz each*

Ingredients

butter, salted	12 oz
shortening	6 oz
confectioners' sugar	1 lb 8 oz
egg whites	2 oz
lemon juice	¾ tsp
vanilla extract	½ tsp

Preparation

1. Place all ingredients in mixing bowl and mix at slow speed using paddle attachment for 5 minutes.
2. Whip at medium speed for 10–15 minutes to acquire desired lightness.

NUTRITION FACTS
Per serving: 155 calories, 91 calories from fat, 10.2 g total fat, 19.5 mg cholesterol, 60.4 mg sodium, 4.6 mg potassium, 16.2 g carbohydrates, 0 g fiber, 15.9 g sugar, <1 g protein

Variation:

Chocolate Buttercream Icing: Mix 2.50 oz of cocoa and 2.50 oz of water to form a paste. Whisk cocoa paste into prepared buttercream icing.

Chocolate Ganache

Yield: *60 servings, 1 oz each*

Ingredients

sweet chocolate	2 lb
heavy cream	28 oz

Preparation

1. Chop the chocolate into small pieces.
2. Bring the cream just to a boil, stirring to prevent scorching.
3. Add the chocolate. Remove from the heat, stir, and let stand for a few minutes.
4. Stir again until the chocolate is completely melted and the mixture is smooth.
5. At this point the ganache is ready to be used as an icing or glaze.

NUTRITION FACTS
Per serving: 120 calories, 81 calories from fat, 9.6 g total fat, 19 mg cholesterol, 6.9 mg sodium, 65.6 mg potassium, 10 g carbohydrates, 0 g fiber, <1 g sugar, <1 g protein

Cream Cheese Icing

Yield: *32 servings, 1 oz each*

Ingredients

cream cheese	13 oz
powdered sugar	1 lb 2 oz
vanilla extract	1 tsp
milk	1 oz

Preparation

1. Combine all ingredients in mixer. Use a paddle attachment to mix until thoroughly combined and smooth.
2. Add additional milk if icing needs to be thinned out.

NUTRITION FACTS
Per serving: 102 calories, 35 calories from fat, 4 g total fat, 12.8 mg cholesterol, 37.7 mg sodium, 17.7 mg potassium, 16.4 g carbohydrates, 0 g fiber, 16 g sugar, <1 g protein

Chocolate Chip Cookies

Yield: *36 servings, 2 cookies each (6 dozen cookies)*

Ingredients

shortening	1 lb
sugar	1 lb 8 oz
eggs	8 oz
vanilla extract	0.25 oz
all-purpose flour	1 lb 8 oz
salt	0.50 oz
baking soda	0.25 oz
chocolate chips	1 lb 12 oz

Preparation

1. In a mixer with a paddle attachment, cream the shortening and sugar until light and fluffy.
2. Add eggs and vanilla and scrape down the sides of the mixer. Cream until fluffy.
3. In a separate bowl, mix flour, salt, and baking soda.
4. Add to creamed mixture until fully incorporated.
5. Stir in chocolate chips.
6. Using a size 30 scoop, drop cookies onto a parchment-lined sheet pan. Bake at 350°F for 12–15 minutes.
7. Remove from oven and let cool.

NUTRITION FACTS
Per serving: 370 calories, 173 calories from fat, 20 g total fat, 30.5 mg cholesterol, 218.4 mg sodium, 29.6 mg potassium, 47.3 g carbohydrates, 1.8 g fiber, 18.9 g sugar, 3.7 g protein

Oatmeal Raisin Cookies

Yield: *30 servings, 2 cookies each (5 dozen cookies)*

Ingredients

butter	1 lb
brown sugar	2 lb
eggs	8 oz
vanilla	0.67 oz
milk	2 oz
pastry flour	1.50 lb
salt	0.33 oz
baking powder	1 oz
baking soda	0.50 oz
cinnamon	0.33 oz
nutmeg	1 tsp
rolled oats	1.25 lb
raisins	1 lb

Preparation

1. In a mixer with a paddle attachment, cream the butter and sugar until light and fluffy.
2. Add eggs, vanilla, and milk and scrape down the sides of the mixer. Cream until fluffy.
3. In a separate bowl, mix flour, salt, baking powder, baking soda, cinnamon, and nutmeg.
4. Stir in oats.
5. Add to creamed mixture until fully incorporated.
6. Stir in raisins.
7. Using a size 30 scoop, drop cookies onto a parchment-lined sheet pan.
8. Bake at 375°F for 10–12 minutes.
9. Remove from oven and cool.

NUTRITION FACTS
Per serving: 440 calories, 129 calories from fat, 14.7 g total fat, 60.8 mg cholesterol, 374.7 mg sodium, 278.2 mg potassium, 72.7 g carbohydrates, 3.2 g fiber, 38.6 g sugar, 6.7 g protein

Peanut Butter Cookies

Yield: *30 servings, 2 cookies each (5 dozen cookies)*

Ingredients

butter	12 oz
brown sugar	8 oz
granulated sugar	8 oz
peanut butter	12 oz
eggs	4 oz
pastry flour	1 lb
salt	1 tsp
baking soda	1 tsp

Preparation

1. In a mixer with a paddle attachment, cream the butter, sugars, and peanut butter until light and fluffy.
2. Add eggs and scrape down the sides of the mixer. Cream until fluffy.
3. In a separate bowl, mix flour, salt, and baking soda.
4. Add to creamed mixture until fully incorporated.
5. Using a size 30 scoop, drop cookies onto a parchment-lined sheet pan.
6. Use a fork to flatten the cookies.
7. Bake at 375°F for 8–12 minutes.
8. Remove from oven and cool.

> **NUTRITION FACTS**
> **Per serving:** 266 calories, 133 calories from fat, 15.4 g total fat, 38.5 mg cholesterol, 167.4 mg sodium, 107.6 mg potassium, 29 g carbohydrates, <1 g fiber, 16 g sugar, 4.7 g protein

Sugar Cookies

Yield: *36 servings, 2 cookies each (6 dozen cookies)*

Ingredients

sugar	2 lb
shortening	1 lb 8 oz
eggs	8 oz
cake flour	2 lb 12 oz
baking powder	1.50 oz
salt	0.75 oz
skim milk	8 oz

Preparation

1. Place sugar and shortening in mixer and use paddle attachment to cream until light and fluffy.
2. Add eggs, scrape down the sides of the bowl, and mix until light and fluffy.
3. Add flour, baking powder, salt, and milk.
4. Mix until dough forms and scrape down the sides of the bowl as needed.
5. Place cookie dough in refrigerator overnight or for at least 1 hour.
6. Roll out cookie dough on floured surface and cut out desired shapes.
7. Bake on sheet pan at 375°F for 7–8 minutes.

> **NUTRITION FACTS**
> **Per serving:** 405 calories, 178 calories from fat, 19.8 g total fat, 34.2 mg cholesterol, 366.9 mg sodium, 56.5 mg potassium, 52.9 g carbohydrates, <1 g fiber, 25.6 g sugar, 3.9 g protein

Chocolate Brownies

Yield: *60 small brownies (½ sheet pan)*

Ingredients

granulated sugar	2 lb 6 oz
shortening	14 oz
vanilla extract	0.50 oz
eggs	1.25 lb
glucose	12 oz
cake flour	1 lb 5 oz
cocoa	8 oz
salt	0.75 oz

Preparation

1. Cream sugar, shortening, and vanilla.
2. Add eggs and glucose alternately, mixing well.
3. Sift together cake flour, cocoa, and salt. Add and mix until smooth.
4. Spread mixture on greased and floured sheet pan.
5. Bake at 350°F for 45–50 minutes.

NUTRITION FACTS
Per serving: 198 calories, 67 calories from fat, 7.5 g total fat, 38.5 mg cholesterol, 155.5 mg sodium, 81.9 mg potassium, 32.3 g carbohydrates, 1.4 g fiber, 19.6 g sugar, 2.7 g protein

French Butter Cookies

Yield: *30 servings, 2 cookies each (5 dozen cookies)*

Ingredients

almond paste	1.50 lb
confectioners' sugar	1.50 lb
butter	1 lb
shortening	2 lb
salt	0.25 oz
egg whites	1.50 lb
vanilla extract	0.50 oz
bread flour	3 lb
pastry flour	12 oz

Preparation

1. Mix the almond paste, confectioners' sugar, butter, shortening, and salt until a smooth paste forms.
2. Gradually add the egg whites and vanilla, and cream until light and fluffy.
3. Sift together both flours and stir into mixture.
4. Place cookie dough in pastry bag with desired tip and press out onto a parchment-lined sheet pan.
5. Bake at 375°F for 7–8 minutes.

NUTRITION FACTS
Per serving: 791 calories, 440 calories from fat, 49.7 g total fat, 49.4 mg cholesterol, 134.6 mg sodium, 170.3 mg potassium, 75.5 g carbohydrates, 2.4 g fiber, 30.8 g sugar, 11 g protein

Mealy Pie Dough

Yield: *24 servings, 1 oz each (3 crusts)*

Ingredients

pastry flour	12 oz
shortening	8 oz
water	4 oz
salt	0.25 oz
sugar	0.60 oz
dry milk	1 tsp

Preparation

1. Place flour and shortening in a large bowl. Cut shortening into flour with pastry blender until the flour is completely covered with shortening and the mixture is mealy in appearance.
2. Place the water, salt, sugar, and dry milk in a stainless steel bowl and mix well.
3. Pour the liquid mixture over the flour/shortening mixture and mix gently until the liquid is absorbed by the flour. Do not overmix.
4. Place the dough on a sheet pan. Cover and refrigerate until firm enough to roll out.
5. Remove from refrigerator, scale into 8 oz units, and refrigerate again until ready to be rolled out.

NUTRITION FACTS
Per serving: 131 calories, 78 calories from fat, 8.7 g total fat, 4.9 mg cholesterol, 115.3 mg sodium, 16.5 mg potassium, 11.8 g carbohydrates, <1 g fiber, <1 g sugar, 1.2 g protein

Flaky Pie Dough

Yield: *24 servings, 1 oz each (3 crusts)*

Ingredients

pastry flour	12 oz
shortening	8.4 oz
water	4 oz
salt	0.25 oz
sugar	0.6 oz
dry milk	1 tsp

Preparation

1. Place flour and shortening in a large bowl. Cut shortening into flour with pastry blender until the flour is completely covered with shortening and the mixture is mealy in appearance.
2. Place the water, salt, sugar, and dry milk in a stainless steel bowl and mix well.
3. Pour the liquid mixture over the flour/shortening mixture and mix gently until the liquid is absorbed by the flour. Do not overmix.
4. Place the dough on a sheet pan. Cover and refrigerate until firm enough to roll out.
5. Remove from refrigerator, scale into 8 oz units, and refrigerate again until ready to be rolled out.

NUTRITION FACTS
Per serving: 135 calories, 82 calories from fat, 9.1 g total fat, 5.1 mg cholesterol, 115.3 mg sodium, 16.5 mg potassium, 11.8 g carbohydrates, <1 g fiber, <1 g sugar, 1.2 g protein

Basic Crumb Crust

Yield: *8 servings (one 9-inch crust)*

Ingredients

cookie crumbs	8 oz
granulated sugar	3 oz
unsalted butter, melted	2 oz

Preparation

1. Combine all ingredients in a bowl.
2. Press mixture firmly in bottom and up sides of pie pan.
3. Bake at 350°F for 10–15 minutes or until golden brown and toasted.

NUTRITION FACTS
Per serving: 212 calories, 76 calories from fat, 8.6 g total fat, 15.3 mg cholesterol, 136.1 mg sodium, 40.2 mg potassium, 32.4 g carbohydrates, <1 g fiber, 19.4 g sugar, 2 g protein

Cherry Pie

Yield: *16 servings (two 8-inch pies, 8 slices per pie)*

Ingredients

mealy pie dough recipe	1 ea
cherry juice	14 oz
cornstarch	2 oz
corn syrup	3 oz
frozen cherries, thawed and drained	2 lb
granulated sugar	6 oz

Egg Wash

egg, beaten slightly	1 ea
water	1 tsp

Preparation

1. Prepare mealy pie dough recipe and blind bake the bottom crusts. Set aside.
2. Place half of the cherry juice in a saucepan and bring to a boil.
3. In a stainless steel bowl, dissolve the cornstarch in remaining cherry juice. Add the cornstarch mixture to the heated cherry juice and whisk vigorously.
4. Return mixture to a boil and cook until thickened and clear.
5. Add the corn syrup and stir until thoroughly blended. Remove from the heat.
6. Add the cherries by folding them into the thickened juice gently to avoid breaking or crushing the fruit. Let cool approximately 30 minutes.
7. Fill pie shells with filling.
8. Place a top crust or lattice crust over the filing; seal and flute the edges. If a top crust is used, slits should be cut into it to allow for the release of steam. Brush lattice crusts with egg wash and sprinkle with sugar.
9. Bake at 400°F for 50–60 minutes until golden brown.

NUTRITION FACTS
Per serving: 351 calories, 121 calories from fat, 13.5 g total fat, 19 mg cholesterol, 193.6 mg sodium, 173.5 mg potassium, 55.7 g carbohydrates, 1.8 g fiber, 24.7 g sugar, 2.9 g protein

Apple Pie

Yield: *16 servings (two 8-inch pies, 8 slices per pie)*

Ingredients

mealy pie dough recipe	1 ea
fresh tart apples, peeled, cored, and sliced thin	2 lb
granulated sugar	6 oz
salt	pinch
cinnamon	pinch
cornstarch	1 oz
lemon juice	0.33 oz
apple juice	7 oz

Preparation

1. Prepare mealy pie dough recipe and blind bake the bottom crusts. Set aside.
2. Place the apples, sugar, salt, and cinnamon in a non-reactive saucepan.
3. Dissolve the cornstarch in the lemon juice and apple juice and add the mixture to the to the apples.
4. Cover and simmer until apples have softened. Cool slightly until lukewarm.
5. Place the apple mixture into the baked pie shells and cover with a full top crust or lattice crust. Bake at 400°F until the filling is bubbly and topping is light brown, approximately 25–30 minutes.

NUTRITION FACTS
Per serving: 279 calories, 118 calories from fat, 13.1 g total fat, 7.3 mg cholesterol, 192 mg sodium, 91 mg potassium, 38.8 g carbohydrates, 1.2 g fiber, 18.9 g sugar, 2 g protein

Cream Pie

Yield: *24 servings (three 8-inch pies, 8 slices per pie)*

Ingredients

basic crumb crust	3 ea recipes
cornstarch	4 oz
whole milk	2 lbs 12 oz
whole eggs	2.5 oz
egg yolks	2.5 oz
sugar	5 oz
unsalted butter	1.5 oz
salt	pinch
vanilla extract	1 oz

Preparation

1. Prepare and pan the basic crumb crusts in 8-inch pie pans.
2. Mix the cornstarch and one-fifth of the milk to make a slurry.
3. Whisk the eggs and yolks and add to the slurry.
4. In a heavy saucepan, combine the remaining milk and sugar and bring to a boil. *Note:* Do not scorch the milk.
5. Temper the egg mixture by pouring in half of the hot milk mixture while whisking rapidly.
6. Stir the tempered egg mixture back into the remaining milk mixture and bring to a boil while stirring constantly.
7. Remove from heat and stir in the butter, salt, and vanilla extract.
8. Pour cream filling into pie shells. Refrigerate until cold.

> **NUTRITION FACTS**
> **Per serving:** 314 calories, 113 calories from fat, 12.8 g total fat, 67.3 mg cholesterol, 177 mg sodium, 118.5 mg potassium, 45.4 g carbohydrates, <1 g fiber, 28.1 g sugar, 4.5 g protein

Variations

Banana Cream Pie: Cover the bottom of a baked pie shell or crumb crust with cream pie filling. Place 1 ¾ cups of sliced bananas on the surface of each of the piecrusts. Cover the banana slices with enough pie filling to fill pie shell. Let cool completely and either top with heavy cream that has been whipped with sugar and cream of tartar or top with meringue. Bake at 450°F for 2–3 minutes until topping browns.

Coconut Cream Pie: Stir 2 lb of coconut into the cream pie filling and pour into baked pie shells. Let the filling cool and top each pie with 2 cups of meringue. Bake at 375°F for 2–3 minutes until topping browns.

Chocolate Cream Pie: Stir 12–14 oz dark or bittersweet chocolate into the cream mixture after all other ingredients have been incorporated.

Pumpkin Pie

Yield: *24 servings (three 9-inch pies, 8 slices per pie)*

Ingredients

mealy pie dough recipe	1 ea
canned pumpkin	3 lb
brown sugar	8 oz
granulated sugar	1 lb
salt	1 tsp
cake flour	2 oz
cinnamon	0.25 oz
nutmeg	1 tsp
ginger	1 tsp
eggs	11 oz
whole milk	1.50 qt

Preparation

1. Prepare a recipe of mealy pie dough and blind bake three piecrusts. Set aside.
2. Mix pumpkin, sugar, salt, flour, and spices in mixer at medium speed until thoroughly mixed.
3. Add the eggs, one at a time, alternately with the milk at low speed until all ingredients have been thoroughly blended.
4. Pour mixture into prepared pie shells and bake at 375°F for 45–50 minutes until filling is set.

NUTRITION FACTS
Per serving: 326 calories, 109 calories from fat, 12.1 g total fat, 59.3 mg cholesterol, 225.3 mg sodium, 249.6 mg potassium, 49.7 g carbohydrates, 2.1 g fiber, 33.9 g sugar, 5.6 g protein

Cheesecake

Yield: *16 servings (one 10-inch cake)*

Ingredients

cream cheese	1 lb 8 oz
sugar	6.50 oz
eggs	5 ea
sour cream	12 oz
vanilla extract	TT

Preparation

1. Cream the cream cheese at room temperature in a mixer using a paddle attachment and scrape the bowl.
2. Cream the sugar into the cream cheese, mixing thoroughly. Scrape the bowl again.
3. Slowly add eggs at room temperature one at a time and mix until smooth, scraping the sides of the bowl as needed.
4. Add sour cream and mix until smooth.
5. Add vanilla extract.
6. Place in a 10 inch springform pan lined with foil to prevent leaking.
7. Bake in a water bath at 325°F for 1 hour and 15 minutes, or until toothpick inserted in center comes out clean.

NUTRITION FACTS
Per serving: 253 calories, 178 calories from fat, 20.2 g total fat, 116 mg cholesterol, 175.8 mg sodium, 110.5 mg potassium, 14 g carbohydrates, 0 g fiber, 13.7 g sugar, 4.9 g protein

Pecan Pie

Yield: *64 servings (eight 8-inch pies, 8 slices per pie)*

Ingredients

mealy pie dough recipe	1 recipe
pastry flour	6 oz
sugar	6 oz
light corn syrup	9 lb
eggs	3 lb 4 oz
salt	0.50 oz
melted butter	8 oz
vanilla extract	1 oz
chopped pecans	1 lb 12 oz

Preparation

1. Prepare the mealy pie dough and line the 8-inch pie pans.
2. Place the flour, sugar, and syrup in a mixer with paddle attachment. Blend together at low speed.
3. Set mixer at low speed and stir in eggs, one at a time, followed by salt, melted butter, and vanilla.
4. Stir in pecans.
5. Fill pie shells with pecan mixture and bake at 325°F for 45–50 minutes.

> **NUTRITION FACTS**
> **Per serving:** 477 calories, 199 calories from fat, 22.8 g total fat, 98.2 mg cholesterol, 273.9 mg sodium, 104.1 mg potassium, 67.5 g carbohydrates, 1.5 g fiber, 21.2 g sugar, 5.5 g protein

Lemon Chiffon Pie

Yield: *24 servings (three 8-inch pies, 8 slices per pie)*

Ingredients

flaky pie dough	1 recipe
Filling Base	
water	2 lbs
sugar	1 lb
salt	0.33 oz
lemon zest	1 oz
egg yolks	0.50 lb
lemon juice	0.50 lb
cornstarch	5 oz
gelatin	0.50 oz
Meringue	
egg whites	1 lb
sugar	12 oz

Preparation

1. Prepare flaky pie dough recipe and blind bake the pie shells. Set aside to cool.
2. Place three-fours of the water and the sugar, salt, and lemon zest in a heavy saucepan and bring to a boil.
3. Whisk together the egg yolks, lemon juice, and cornstarch in a stainless steel mixing bowl.
4. Temper the egg mixture gradually by adding half of the hot mixture to the eggs while whisking rapidly. Pour back into the hot mixture while continually whipping until thickened. Remove from heat.
5. Bring the remaining water to a boil and dissolve gelatin. Gently stir into the lemon mixture.
6. Begin preparing the meringue by whipping egg whites in another bowl and slowly add sugar until stiff peaks form.
7. Gently fold meringue into the hot lemon filling.
8. Pour the filling into baked pie shells and cool in refrigerator until set.

> **NUTRITION FACTS**
> **Per serving:** 331 calories, 105 calories from fat, 11.7 g total fat, 107.7 mg cholesterol, 307.2 mg sodium, 70.5 mg potassium, 52.1 g carbohydrates, <1 g fiber, 34.8 g sugar, 4.9 g protein

Meringue

Yield: *24 servings, 2 oz each*

Ingredients

egg whites	16 oz
sugar	32 oz

Preparation

1. Using a whisk attachment on a mixer, beat egg whites at medium speed until frothy. Then, increase speed to high until the egg whites form soft peaks.
2. Gradually add sugar and whip until the mixture forms stiff peaks. Use caution to not overwhip as the whites can break.
3. Place finished whipped meringue in a pastry bag and pipe onto parchment-lined paper into desired shapes.
4. Bake at 200°F until crisp but not browned. This will take approximately 1–3 hours, depending on size. Use caution when removing from the oven as finished meringues are fragile.

NUTRITION FACTS
Per serving: 156 calories, <1 calorie from fat, <1 g total fat, 0 mg cholesterol, 31.8 mg sodium, 31.6 mg potassium, 37.9 g carbohydrates, 0 g fiber, 37.9 g sugar, 2.1 g protein

French Meringue

Yield: *24 servings, 2 oz each*

Ingredients

egg whites, room temperature	16 oz
granulated sugar	16 oz
powdered sugar	16 oz

Preparation

1. Place egg whites in bowl, start beating at moderate speed.
2. Add granulated sugar gradually as whites come up to a light froth, 3–4 oz at a time. Increase speed while adding. Stop when firm.
3. Sift powdered sugar. Incorporate by hand with spatula.

NUTRITION FACTS
Per serving: 156 calories, <1 calorie from fat, <1 g total fat, 0 mg cholesterol, 31.9 mg sodium, 31.6 mg potassium, 37.9 g carbohydrates, 0 g fiber, 37.5 g sugar, 2.1 g protein

Swiss Meringue

Yield: *24 servings, 2 oz each*

Ingredients

granulated sugar	32 oz
egg whites	16 oz

Preparation

1. Place sugar and egg whites in a round bowl. Set the bowl in a hot water bath or over a very low fire.
2. Beat mixture until sugar is dissolved and batter reaches approximately 120°F. Beat constantly to prevent egg whites from coagulating.
3. Remove from heat. Finish beating in mixer at full speed until the mixture is stiff.

NUTRITION FACTS
Per serving: 156 calories, <1 calorie from fat, <1 g total fat, 0 mg cholesterol, 31.8 mg sodium, 31.6 mg potassium, 37.9 g carbohydrates, 0 g fiber, 37.9 g sugar, 2.1 g protein

Italian Meringue

Yield: *24 servings, 2.3 oz each*

Ingredients

water	8 oz
granulated sugar	2 lb
egg whites	6 oz

Preparation

1. Place water and sugar in a saucepan along with a candy thermometer and bring to a boil.
2. Heat until all the sugar is dissolved, continually washing down sides of saucepan with a pastry brush dipped in cold water.
3. Continue to cook until the mixture reaches 240°F.
4. When the sugar mixture reaches 240°F, remove from the heat and allow to cool to 220°F.
5. Place the egg whites in the bowl of a mixer with a whip attachment and whip until soft peaks form.
6. When sugar mixture cools to 220°F, turn mixer to high speed and slowly pour the sugar mixture in a fine stream into the egg whites while continuing to mix.
7. Once all of the sugar has been added, continue to whip until the mixture has cooled to room temperature.

> **NUTRITION FACTS**
> **Per serving:** 150 calories, <1 calorie from fat, <1 g total fat, 0 mg cholesterol, 12.4 mg sodium, 12.4 mg potassium, 37.8 g carbohydrates, 0 g fiber, 37.7 g sugar, <1 g protein

Puff Pastry

Yield: *36 servings, 1 pastry each*

Ingredients

Puff Paste

bread flour	1 lb 8 oz
pastry flour	8 oz
softened butter	4 oz
salt	0.50 oz
cold water	1 lb 2 oz

Roll-in Fat

butter	2 lb
bread flour	3 oz

Preparation

1. Mix puff paste ingredients together at second speed in mixer for 5 minutes. Separate dough into two equal pieces on two separate sheet pans.
2. Cream roll-in ingredients together until fluffy.
3. Roll out each piece of dough on floured work surface until it is approximately three times as long as it is wide (approximately 12 × 15 inches). Brush off all excess flour from dough.
4. Divide the roll-in fat evenly between the two pans.
5. Roll the fat between two pieces of plastic wrap to approximately two-thirds the size of the rolled out dough (approximately 8 × 10 inches).
6. Unwrap the rolled out fat and place it towards the bottom edge of the dough leaving the top one-third of the dough with no fat.
7. Fold the top third of the dough without any fat over the dough in the center third.
8. Next fold the bottom third up so it covers the top third and middle thirds. This is referred to as a single book fold and this first folding is called the first turn.
9. Turn dough 90 degrees on work surface and roll out again to the original size.
10. Fold the dough into thirds again. Cover dough and place in refrigerator to rest for 30 minutes. This completes the second turn.
11. After 30 minutes, repeat the turning, rolling, and folding. Let the dough rest between every two turns until five turns have been completed.
12. Puff pastry is now ready to be used.

> **NUTRITION FACTS**
> **Per serving:** 303 calories, 206 calories from fat, 23.4 g total fat, 61 mg cholesterol, 156.7 mg sodium, 34.9 mg potassium, 20.4 g carbohydrates, <1 g fiber, <1 g sugar, 3.3 g protein

Éclair Paste (Pâte à Choux)

Yield: *60 servings*

Ingredients

milk	16 oz
water	16 oz
salt	0.25 oz
butter	16 oz
bread flour	16 oz
eggs	32 oz

Preparation

1. In a saucepan, bring milk, water, salt, and butter to a rolling boil.
2. Remove from heat and vigorously mix in flour with a high-heat spatula.
3. Return the saucepan to the burner and heat while mixing vigorously until mixture forms a ball and pulls away from the sides of the pan.
4. Transfer the dough to a mixing bowl with a paddle attachment and allow to cool until approximately 140°F. Begin adding the eggs, one at a time, while mixing at medium speed. The dough should begin to pull away from the sides of the bowl and the eggs should be completely absorbed into the paste. *Note:* It may not be necessary to use all the eggs, depending on the moistness of the flour mixture.
5. Fill a pastry bag with prepared paste. Pipe onto a parchment-lined sheet pan in desired shape.
6. Bake at 425°F for 10 minutes. *Note:* Do not open the oven door during baking, as this may cause pâte à choux to fall.
7. Reduce oven temperature to 375°F and bake for an additional 10–12 minutes or until pâte à choux is lightly browned and sounds hollow when tapped lightly.
8. Remove from oven and allow to cool to room temperature.
9. Fill with desired filling.

NUTRITION FACTS
Per serving: 108 calories, 70 calories from fat, 8 g total fat, 73.3 mg cholesterol, 72 mg sodium, 41.1 mg potassium, 6 g carbohydrates, <1 g fiber, <1 g sugar, 3.1 g protein

Crème Anglaise

Yield: *20 servings, 4 oz each*

Ingredients

cream	1 qt
whole milk	1 qt
vanilla beans, split in half lengthwise and scraped	2 ea
egg yolks	8 oz
sugar	8 oz
salt	½ tsp

Preparation

1. In a heavy saucepan, bring the cream, milk, and vanilla bean just to a boil.
2. Whisk the egg yolks, sugar, and salt together in a nonreactive bowl.
3. Slowly temper the egg mixture with approximately one-third of the hot cream mixture, whisking constantly.
4. Return the mixture to the saucepan, whisking the entire mixture together.
5. Return to heat and cook on medium until the mixture reaches the ribbon stage and is thick enough to coat the back of a spoon, approximately 175°F.
6. Transfer to a bowl and place over an ice bath to cool. *Note:* Cover the surface with plastic wrap before refrigerating.

NUTRITION FACTS
Per serving: 193 calories, 119 calories from fat, 13.4 g total fat, 160.7 mg cholesterol, 93.8 mg sodium, 94.9 mg potassium, 14.8 g carbohydrates, 0 g fiber, 13.9 g sugar, 3.8 g protein

Sabayon

Yield: *10 servings, 2 oz each*

Ingredients

champagne	8 oz
egg yolks	10 ea
salt	pinch
granulated sugar	8 oz

Preparation

1. Whisk together champagne, egg yolks, salt, and sugar in a nonreactive bowl.
2. Place bowl over simmering water and whisk vigorously until the mixture is thick and pale yellow in color, approximately 160°F.
3. Remove from heat and serve immediately if serving warm.
4. For chilled sabayon, place over an ice bath and whisk until completely cooled.

NUTRITION FACTS
Per serving: 177 calories, 40 calories from fat, 4.4 g total fat, 180.1 mg cholesterol, 39.4 mg sodium, 40.3 mg potassium, 26 g carbohydrates, 0 g fiber, 23 g sugar, 2.7 g protein

Tangerine Granité

Yield: *4 servings, 4 oz each*

Ingredients

honey	0.50 oz
tangerine juice	2 c
Cointreau	1.50 oz

Preparation

1. Dissolve honey in tangerine juice and combine with Cointreau.
2. Freeze the mixture in a stainless hotel pan until solid (approximately 3 hours).
3. Serve by scraping the surface with a metal spatula, breaking up large ice crystals.
4. Garnish with tangerine segments and a sprig of mint.

NUTRITION FACTS
Per serving: 105 calories, 2 calories from fat, <1 g total fat, 0 mg cholesterol, 1.4 mg sodium, 221.7 mg potassium, 19.5 g carbohydrates, <1 g fiber, 15.1 g sugar, <1 g protein

Glossary

A

abalone: A univalve contained in a brown, bowl-shaped shell with an iridescent, multicolored interior.

accommodation: The act of modifying something in response to a need or request.

achiote seed: A spice made from a red, corn-kernel-shaped seed of the annatto tree. Also known as an annatto seed.

acorn squash: A winter squash that looks like a large, dark-green acorn and can be baked, steamed, or puréed.

active dry yeast: A form of yeast that has been dehydrated and looks like small granules.

adductor muscle: A muscle that opens and closes the shell of a bivalve.

African pheasant: *See* guinea fowl.

aggregate fruit: A cluster of very tiny fruits.

aging: The period of rest that occurs for a length of time after an animal has been slaughtered.

à la carte menu: A menu that has all food and beverage items priced separately.

albumen: The clear portion of the raw egg, which makes up two-thirds of the egg and consists mostly of ovalbumin protein. Also known as the white.

allemande sauce: A sauce made by adding fresh lemon juice and a yolk-and-cream liaison to a velouté.

alligator pear: *See* avocado.

allspice: A spice made from the dried, unripe fruit of a small pimiento tree. Also known as Jamaican pepper.

almond: A teardrop-shaped fruit that grows on small almond trees.

American service: *See* plated service.

amino acid: The building block of all proteins.

ammonium bicarbonate: A chemical leavening agent that reacts as soon as it is heated without the help of an acid or base.

amuse bouche: A single, bite-sized portion of food selected by the chef and not ordered by the guest.

anadromous fish: A fish that begins life in freshwater, spends most of its life in saltwater, and returns to freshwater to spawn.

anaphylaxis: A severe allergic reaction that causes the airway to narrow and prohibits breathing.

anise: A spice made from a comma-shaped seed of the anise plant and is in the parsley family.

Anjou pear: A plump, lopsided pear that is green in color.

annatto seed: *See* achiote seed.

appetizer: Food that is larger than a single bite and is typically served as the first course of a meal.

appetizer salad: A salad that is served as a starter to a meal.

apple: A hard, round pome that can range in flavor from sweet to tart and in color from pale yellow to dark red.

apple-pear: *See* Asian pear.

apprentice: An individual enrolled in a formal training program who learns by practical experience under the supervision of a skilled professional.

apricot: A drupe that has pale orange-yellow skin with a fine, downy texture and a sweet and aromatic flesh.

aquafarming: A process of raising fish, shellfish, or related game in a controlled inland environment.

arame: A thin and wiry, shredded black sea vegetable.

Arctic char: A fatty, anadromous saltwater fish with a pale belly and dark sides that are speckled with light spots.

aromatic: An ingredient added to a food to enhance its natural flavors and aromas.

arrowroot: A thickening agent that is the edible starch from the rootstock of the arrowroot plant.

artichoke: The edible flower bud of a large, thistle-family plant that comes in many varieties.

artisan lettuces: Petite-sized lettuces that offer the full flavor, volume, and texture of mature lettuces.

Asiago: A grating cheese with a nutty, toastlike flavor.

Asian chili sauce: A red-colored Asian sauce made from puréed red chiles and garlic.

Asian pear: A round pear with the texture of an apple and yellow-colored skin. Also known as apple-pear.

asparagus: A green, white, or purple edible stem that is referred to as a spear.

aspic: A savory jelly made from a consommé or clarified meat, fish, or vegetable stock that produces a clear finish when coating foods. Also known as aspic gelée.

aspic gelée: *See* aspic.

as-purchased (AP) cost: The amount paid for a product in the form it was ordered and received.

as-purchased (AP) quantity: The original amount of a food item as it is ordered and received.

as-purchased (AP) unit cost: The unit cost of a food item based on the form in which it is ordered and received.

as-served (AS) cost: The cost of a menu item as it is served to a customer.

Atlantic oyster: A variety of oyster that has a fairly flat shell and a distinctive, salty-flavored, plump, and tender flesh. Also known as an Eastern oyster.

avant-garde cuisine: A style of cuisine, based on food chemistry, that involves the manipulation of the textures and temperatures of familiar dishes to reinvent and present them in new and creative ways. Also known as modernist cuisine.

avocado: A pear-shaped fruit with a rough green skin and a large pit surrounded by yellow-green flesh. Also known as an alligator pear.

B

baby back ribs: The meaty bones on the rib end of the pork loin.

baby T-bone: A 6–8 oz steak cut from the loin of veal.

back: A rack and a loin that are still joined.

back-of-house (BOH): The portion of a foodservice operation that is typically not open to guests.

bacon: Side pork that has been cured and usually smoked.

bacteria: Pathogens that live in soil, water, or organic matter.

bain-marie: A hot water bath used to keep foods such as sauces and soups hot.

bain-marie insert: A round stainless steel food storage container with high walls used for holding sauces or soups in a water bath or steam table.

baked egg: *See* shirred egg.

baker's cheese: A fresh cheese made from skim milk that is like cottage cheese but softer and finer grained.

baker's percentage: The weight of a particular ingredient expressed as a percentage based on the weight of the main ingredient in a formula.

baker's scale: A scale with two platforms that use a counterbalance system to measure weight. Also known as a balance scale.

baking: The dry-heat cooking method in which food is cooked uncovered in an oven.

baking powder: A chemical leavening agent that is a combination of baking soda and cream of tartar or sodium aluminum sulfate.

baking soda: An alkaline chemical leavening agent that reacts to an acidic dough or batter by releasing carbon dioxide without the addition of heat. Also known as sodium bicarbonate.

balance scale: *See* baker's scale.

ballotine: A boned poultry leg that has been stuffed with forcemeat.

balsamic vinegar: A vinegar made by aging red wine vinegar in wooden vats for many years.

bamboo shoot: A root vegetable that is the immature shoot of the bamboo plant.

banana: A yellow, elongated tropical fruit that grows in hanging bunches on a banana plant.

banquet service: A style of meal service in which servers present food to guests attending a special function.

banquette service: A style of meal service in which food is served to guests seated at a table with a bench on one side and chairs on the opposite side.

barbequing: A dry-heat cooking method in which food is slowly cooked over hot coals or burning wood.

barbeque sauce: A type of sauce used to baste a cooked protein and is often made with tomatoes, onions, mustard, garlic, brown sugar, and vinegar.

barding: The process of laying a piece of fatback across the surface of a lean cut of meat to add moisture and flavor.

bark spice: A type of spice derived from the bark portion of a plant.

barquette: A miniature, boat-shaped pastry shell that contains a savory or sweet filling.

barramundi: Large, fatty, saltwater roundfish that are silver in color.

bartender: The person responsible for serving alcoholic beverages from behind a bar.

Bartlett pear: A large, golden pear with a bell shape.

basic French vinaigrette: A temporary emulsion of oil and vinegar that may also include additional flavorings and seasonings.

basil: An herb with a pointy green leaf and is a member of the mint family.

basmati rice: A long-grain rice that only expands lengthwise when it is cooked.

bass: A lean, spiny-finned roundfish with white-colored flesh that produces sweet-tasting, delicate fillets.

basted egg: A fried egg with an unbroken yolk that is cooked like a sunny-side up egg, yet the tops are slightly cooked by tilting the pan and basting the top of the egg with hot butter.

basting: The process of using a brush or a ladle to place fat on or pour juices over an item during the cooking process to help retain moisture and enhance flavor.

batonnet cut: A stick cut that produces a stick-shaped item ¼ × ¼ × 2 inches long.

battering: The process of dipping an item in a wet mixture of flour, liquid, and fat for frying.

bay leaf: A thick, aromatic leaf that comes from the evergreen bay laurel tree.

bay scallop: A fairly small scallop harvested from shallow saltwater.

bean spice: A type of spice derived from the bean portion of a plant.

béchamel: A mother sauce that is made by thickening milk with a white roux and seasonings. Also known as a cream sauce or a white sauce.

beef: The flesh of domesticated cattle.

beef brisket: A thin section of beef that contains some of the ribs, the breastbone, and layers of lean muscle, fat, and connective tissue.

beef chuck: A shoulder primal cut of beef that contains the first five rib bones, some of the backbone, and a small amount of the arm and blade bones.

beef flank: A primal cut of beef that includes the thin, flat section of the hindquarters located beneath the loin.

beef rib: The primal cut of beef located between the chuck and short loin and contains seven rib bones, from the 6th to the 12th rib.

beef round: A primal beef cut that includes a large grouping of muscles that represent the hind hip and thigh of the carcass.

beef rump roast: A roast cut from the primal round, above the back end of the hip bone.

beef shank: A bony section of beef that is surrounded by a small amount of very tough but flavorful meat.

beef short loin: A primal cut of beef located just to the rear of the primal rib and includes the 13th rib and a small section of the backbone.

beef short plate: A primal cut of beef that includes a thin portion of the beef forequarter located just beneath the rib cut.

beef sirloin: A primal cut of beef situated just behind the short loin and contains some of the backbone and hip bone.

beef strip loin: A short loin without a tenderloin.

beef tenderloin: An eye-shaped muscle running from the primal rib cut into the primal leg.

beet: A round root vegetable with a deep reddish purple or gold color.

beet green: The green, edible leaf that grows out of the top of the beet root vegetable.

beignet: *See* fritter.

bell pepper: A fruit-vegetable with three or more lobes of crisp flesh that surround hundreds of seeds in an inner cavity.

belly: A primal cut of pork that is the lower portion of the hog between the shoulder and the leg.

bel paese: A lightly colored, semisoft dry-rind cheese with a buttery flavor that melts easily.

bench brush: A brush with long bristles set in vulcanized rubber attached to a handle.

bench knife: *See* dough cutter.

berry: A type of fruit that is small and has many tiny, edible seeds.

berry spice: A type of spice made from the berry portion of a plant.

beurre blanc: A butter-based emulsified sauce made by whisking cold, softened butter into a wine, white-wine vinegar, shallot, and peppercorn reduction. Also known as a white butter sauce.

beurre manié: A thickening agent made by kneading equal amounts, by weight, of pastry or cake flour and softened butter and can be whisked into a sauce just before service.

beverage cost percentage: A percentage that indicates how the cost of beverages relates to menu prices and beverage sales of a foodservice operation.

biological contaminant: A living microorganism that can be hazardous if it is inhaled, swallowed, or otherwise absorbed.

bird's beak knife: *See* tourné knife.

biscuit: A quick bread made by mixing solid fat, baking powder or baking soda, salt, and milk with flour.

biscuit mixing method: A quick-bread mixing method used for recipes that include a cold, solid fat, such as butter.

bisque: A form of cream soup that is typically made from shellfish.

bivalve: A mollusk with a top shell and a bottom shell connected by a central hinge that can close for protection.

blackberry: A sweet, dark-purple to black, aggregate fruit that grows on a bramble bush.

blanching: A moist-heat cooking method in which food is briefly parcooked and then shocked by placing it in ice-cold water to stop the cooking process.

blast chiller: A specialized cooling unit that rapidly reduces the temperature of foods, rendering them safe for immediate storage.

blender: A tall appliance with a slender canister that is used to chop, blend, purée, or liquefy food.

bleu cheese: *See* blue cheese.

blind baking: A process in which a piecrust is baked for 10–15 minutes before a filling is added.

blintz: A crêpe that is only cooked on one side and not flipped over to cook the other side.

blueberry: A small, dark-blue berry that grows on a shrub.

blue crab: A North American crab with blue claws and a dark blue-green, oval shell.

blue cheese: A blue-veined cheese made from cow milk and is characterized by the presence of a blue-green mold. Also known as bleu cheese.

blue potato: *See* purple potato.

blue-veined cheese: A cheese produced by inserting harmless live mold spores into the center of ripening cheese with a needle.

boiling: A moist-heat cooking method in which food is cooked by heating a liquid to its boiling point.

bok choy: An edible leaf that has tender white ribs, bright-green leaves, and a more subtle flavor than head cabbage.

bolster: A thick band of metal located where the blade joins the handle.

bombe: A French ice cream dessert with at least two varieties of ice cream that are layered inside a spherical mold, frozen, and then unmolded and decorated for service. Also known as glacée.

boning knife: A thin knife with a pointed 6–8 inch blade used to separate meat from bones with minimal waste.

booth service: A style of meal service in which food is served to guests seated at a table positioned against a wall with benches on either side.

Bosc pear: A pear with a gourd-like shape and a brown-colored to bronze-colored peel.

Boston butt: *See* pork shoulder butt.

bouchée: A puff pastry that is filled with a savory filling.

boudin: A highly seasoned Creole link sausage made of pork, pork liver, and rice.

bouillon: The liquid that is strained off after cooking vegetables, poultry, meat, or seafood in water.

bound salad: A salad made by combining a main ingredient, often a protein, with a binding agent such as mayonnaise or yogurt and other flavoring ingredients.

bouquet garni: A bundle of herbs and vegetables tied together with twine that is used to flavor stocks and soups.

bowl scraper: A curved, flexible scraping tool that is used to scrape batter or dough out of curved containers.

box grater: A stainless steel box with grids of various sizes on each side that are used to cut food into small pieces.

boysenberry: A deep-maroon, hybrid berry made by crossbreeding a raspberry, a blackberry, and a loganberry.

bracelet: A hotel rack with the breast still attached.

braising: A combination cooking method in which food is browned in fat and then cooked, tightly covered, in a small amount of liquid for a long period of time.

bran: The tough outer layer of grain that covers the endosperm.

bratwurst: A German sausage composed of veal, pork, or beef.

Brazil nut: A 2 inch long, white, richly flavored seed of very large fruit grown on a Brazilian nut tree.

braiser: *See* rondeau.

breadfruit: An exotic fruit that is native to Polynesia and has bumpy green skin and a white starchy flesh.

breading: A three-step procedure used to coat and seal an item before it is fried.

bread knife: A knife with a serrated blade 8–12 inches long and is used to cut through the crusts of breads without crushing the soft interior.

bread pudding: A baked custard that is made by pouring a custard mixture over chunks of bread and baking it in the oven.

breast: The top front portion of the flesh above the rib cage.

breast quarter: Half of a breast, a wing, and a portion of the back.

bresaola: Air-dried cured beef that is aged two to three months until it becomes hard and turns almost purple in color. Also known as brisaola.

brick cheese: A washed-rind semisoft cheese made from cow milk.

Brie: A creamy, white soft cheese with a strong odor and a sharp taste.

brigade system: A structured chain of command in which specific duties are aligned with the stations to which staff are assigned.

brine: A salt solution that usually consists of 1 cup of salt per 1 gal. of water.

brisaola: *See* bresaola.

broccoli: An edible flower that is a member of the cabbage family and has tight clusters of dark-green florets on top of a pale-green stalk with dark-green leaves.

brochette: A food that is speared onto a wooden, metal, or natural skewer and then grilled or broiled.

broiler: A large piece of cooking equipment in which the heat source is located above the food instead of below it.

broiler/fryer: A young male or female chicken under 13 weeks old that weighs 1½–3½ lb.

broiler/fryer duckling: A duck less than eight weeks old and weighs between 3–4 lb.

broiling: A dry-heat cooking method in which food is cooked directly under or over a heat source.

broken butter: A butter sauce made by heating butter until the fat, milk solids, and water separate or "break."

broth: A flavorful liquid made by simmering stock along with meat, poultry, seafood, or vegetables and seasonings.

brown rice: Rice that has had only the husk removed.

brown stock: A stock produced by simmering roasted meat, poultry, or game bones with mirepoix and an optional tomato product.

brunoise cut: A dice cut that produces a cube-shaped item with six equal sides measuring ⅛ inch each.

brussels sprout: A very small round head of tightly packed leaves that looks like a tiny cabbage.

buckwheat groat: A crushed, coarse piece of whole-grain buckwheat that can be prepared like rice.

buffalo chopper: An appliance used to process larger amounts of a product into roughly equal-size pieces.

buffet service: A style of meal service in which food is displayed on a table that guests approach for self-service.

busser: The person responsible for setting tables and removing dirty dishes from the dining area.

butcher's knife: A heavy knife with a curved tip and a blade that is 7–14 inches in length.

butcher's steel: *See* steel.

butler service: A style of meal service in which servers present prepared food on a tray to standing or seated guests for self-service.

butterflied fillet: Two single fillets from a dressed fish that are held together by the uncut back or belly of the fish.

butternut squash: A large, bottom-heavy, tan-colored winter squash.

button mushroom: A cultivated mushroom with a very smooth, rounded cap and completely closed gills atop a short stem. Also known as a white mushroom.

C

cacao bean: A bean extracted from the pods of the cacao tree.

cake pan: A round, square, or specially shaped pan with short or tall sides and is used to bake cakes.

cake spatula: *See* palette knife.

California menu: A menu that offers all food and beverage items for breakfast, lunch, and dinner throughout the entire day.

calorie: A unit of measurement that represents the amount of energy in a food.

Camembert: A soft cheese made from cow milk and has a yellow color and a waxy, creamy consistency.

Canadian bacon: A hamlike breakfast meat made from boneless, smoked, pressed pork loin.

canapé: An hors d'oeuvre that looks like a miniature open-faced sandwich.

canary melon: A fairly large, canary yellow melon with a smooth rind that is slightly waxy when ripe.

candy/deep-fat thermometer: A thermometer with a long, stainless steel stem and a large display.

canned fuel: A flammable gel that provides several hours of heat once it is lit.

canning salt: *See* pickling salt.

canola oil: A type of oil produced from rapeseeds. Also known as rapeseed oil.

can rack shelving: Shelving with rails in which cans of product can be loaded from the top.

cantaloupe: An orange-fleshed melon with a rough, deeply grooved rind.

caper: A spice from the unopened flower bud of a shrub and is only used after being pickled in strongly salted white vinegar.

capon: A surgically castrated male chicken that is less than eight months old and weighs approximately 4–7 lb.

capsaicin: A potent compound that gives chiles their hot flavor.

carambola: *See* star fruit.

caramelization: A reaction that occurs when sugars are exposed to high heat and produce browning and a change in flavor.

caraway seed: A small, crescent-shaped brown seed of the caraway plant and is used as a spice.

carbohydrate: A nutrient in the form of sugar or starch and is the human body's main source of energy.

cardamom: A spice made from the dried, immature fruit of a tropical bush in the ginger family.

career objective: A direct statement expressing the type of employment goal being sought.

carotenoid: An organic pigment found in orange or yellow vegetables.

carpaccio: Thin slices of meats or seafood that are served raw.

carrot: An elongated root vegetable that is rich in vitamin A.

carryover cooking: The rise in internal temperature of an item after it is removed from a heat source due to residual heat on the surface of the item.

cartilaginous fish: Any fish that has a skeleton composed of cartilage instead of bones.

carving: A process of slicing a large piece of cooked poultry or meat into service-sized portions.

carving knife: *See* slicer.

casaba melon: A teardrop-shaped melon with a thick, bright-yellow, ridged rind and white flesh.

cashew: A nut from a kidney-shaped kernel from the fruit of the cashew tree and has a buttery flavor.

casserole: A baked dish containing a starch (such as potatoes, grains, or pasta), other ingredients (such as meat or vegetables), and a sauce.

cassia: A spice made from the bark of a small evergreen tree. Also known as Chinese cinnamon.

cassoulet: A dish that consists of white beans stewed with duck fat, fresh sausage, and duck confit.

cast-iron skillet: A shallow-walled pan made of cast iron that can be used for pan-broiling, pan-frying, or baking.

catfish: A fatty, round freshwater fish named for the whiskers that protrude off the sides of its face.

caul fat: A meshlike fatty membrane that surrounds sheep or pig intestines.

cauliflower: An edible flower that is a member of the cabbage family and has tightly packed white florets on a short, white-green stalk with large, pale-green leaves.

caviar: The harvested roe of sturgeon.

cayenne pepper: A spice made from dried, ground berries of certain varieties of hot peppers. Also known as red pepper.

cayenne sauce: A type of pepper sauce made from cayenne peppers, vinegar, and salt. Also known as Louisiana-style hot sauce.

celeriac: A knobby, brown root vegetable cultivated from a type of celery grown for its root rather than its stalk. Also known as celery root.

celery: A green stem vegetable that has multiple stalks measuring 12–20 inches in length.

celery cabbage: *See* Napa cabbage.

celery root: *See* celeriac.

celery seed: A tiny brown seed of a wild celery plant called lovage and is used as a spice.

celiac disease: A condition in which gluten damages the small intestine's ability to absorb nutrients.

cèpe: *See* porcini mushroom.

cephalopod: Any of a variety of mollusks that do not have an external shell.

chafing dish: A hotel pan inside of a stand with a water reservoir and a portable heat source, such as canned fuel, underneath.

chalazae: Twisted cordlike strands that anchor the yolk to the center of the albumen.

champagne vinegar: A vinegar made from Champagne grapes.

channel knife: A special cutting tool with a thin metal blade within a raised channel that is used to remove a large string from the surface of a food item.

chanterelle mushroom: A trumpet-shaped mushroom that ranges in color from bright yellow to orange and has a nutty flavor and a chewy texture.

charcuterie: The art of making sausages and other preserved items such as pâtés, terrines, galantines, and ballotines.

chard: A large, dark-green, edible leaf with white or reddish stalks.

chaud froid: A French term meaning "hot-cold."

Cheddar: A hard, aged cheese that is yellow or white and ranges in taste from mild to sharp.

cheesecake: A variety of baked custard that typically has a graham cracker or cookie crust.

cheese product: A processed food made of natural cheeses that may include additional ingredients such as emulsifiers.

chef: The person responsible for all kitchen operations, including menu management, purchasing, scheduling, and food production.

chef's fork: *See* kitchen fork.

chef's knife: A large and very versatile knife with a tapering blade used for slicing, dicing, and mincing. Also known as a French knife.

chemical contaminant: Any chemical substance that can be hazardous if it is inhaled, swallowed, or otherwise absorbed into the body.

chemical sanitizing: The process of using a chemical solution to reduce harmful pathogens to a safe level.

cherry: A small, smooth-skinned drupe that grows in a cluster on a cherry tree.

chervil: An herb with dark-green, curly leaves with a flavor similar to parsley but more delicate and with a hint of licorice.

Cheshire: A hard cheese and is the oldest of the English-named cheeses.

chestnut: A nut from a rounded-off, triangular-shaped kernel found inside the burrlike fruit of the chestnut tree.

chèvre: A fresh cheese made from goat milk.

chicory: A curly, edible leaf with a slightly bitter-tasting flavor. Also known as escarole.

chiffonade cut: A slicing cut that produces thin shreds of leafy greens or herbs.

chile: A brightly colored fruit-vegetable pod with distinct mild to hot flavors.

chili flakes: *See* crushed red pepper.

chimichanga: A variety of hot sandwich that consists of a tortilla wrap filled with precooked meat and beans that is then fried.

china cap: A perforated cone-shaped metal strainer used to strain gravies, soups, stocks, sauces, and other liquids.

Chinese cinnamon: *See* cassia.

Chinese parsley: *See* cilantro.

chinois: A china cap that strains liquids through a very fine-mesh screen.

chipotle sauce: A type of pepper sauce made from dried, smoked jalapeno peppers.

chirashizushi: A type of sushi made by scattering toppings over or mixing toppings into vinegar-seasoned rice. Also known as scattered sushi.

chive: A stem herb with hollow, grass-shaped, green sprouts, and a mild onion flavor.

chlorophyll: An organic pigment found in green vegetables.

chocolate: A flavoring ingredient made by roasting, skinning, crushing, and grinding cacao beans into a paste called chocolate liquor.

cholesterol: A waxy, fat-like substance that is used to form cell membranes, vitamin D, some hormones, and bile acids.

chopping: Rough-cutting an item so that there are relatively small pieces throughout, although there is no uniformity in shape or size.

chorizo: A Mexican pork sausage flavored with paprika that can be sweet or spicy.

chowder: A hearty soup that contains visibly large chunks of the main ingredients.

cider vinegar: Honey-colored vinegar made by fermenting unpasteurized apple juice or cider until the sugars are converted into alcohol.

cilantro: An herb that comes from the stem and leaves of the coriander plant. Also known as Chinese parsley.

cinnamon: A spice made from the dried, thin, inner bark of a small evergreen tree.

citronella: *See* lemongrass.

citrus fruit: A type of fruit with a brightly colored, thick rind and pulpy, segmented flesh that grows on trees in warm climates.

clam: A bivalve found in both freshwater and saltwater.

clam knife: A small knife with a short, flat, round-tipped sharp blade that is used to open clams.

clarify: To remove impurities, sediment, cloudiness, and particles to leave a clear, pure liquid.

classical cuisine: A style of cuisine in which foods are prepared using a formalized system of cooking techniques and are presented in courses.

cleaning: The process of removing food and residue from a surface.

clearmeat: A cold, lean ground meat, fish, or poultry that is combined with an acid (such as wine, lemon juice, or a tomato product), ground mirepoix, egg whites, and an oignon brûlé.

clear plate: A rectangular slab of fat that contains a few strips of lean meat located just above the shoulder butt.

clear soup: A stock-based soup with a thin, watery consistency. Clear soups include broths and consommés.

cleaver: A heavy, rectangular-bladed knife that is used to cut through bones and thick meat.

clementine: Similar to a tangerine, but has a rougher skin.

cloud ear mushroom: *See* wood ear mushroom.

clove: A spice made from the dried, unopened bud of a tropical evergreen tree.

coagulation: The process of a protein changing from a liquid to a semisolid or a solid state when heat or friction is applied.

cockle: A 1 inch wide bivalve with a shell that has deep, straight ridges.

cocktail: A chilled appetizer served in a stemmed glass.

cocoa powder: A reddish-brown, unsweetened powder extracted from ground cacao beans.

coconut: A drupe with a white meat that is housed within a hard, fibrous brown husk.

cod: A lean, saltwater roundfish that can range from 1½–100 lb.

coffee bean: The unripe bean extracted from the fruit of a coffee tree.

colander: A bowl-shaped perforated strainer.

cold closed sandwich: A variety of cold sandwich that consists of two pieces of bread, or the top and bottom of a bun or roll, coated with a spread and one or more fillings and garnishes.

cold open-faced sandwich: A variety of cold sandwich that consists of a single slice of bread that is often toasted or grilled and then coated with a spread and topped with thin slices of poultry, seafood, meat, partially cooked or raw vegetables, or a thin layer of a bound salad and a garnish.

cold-pack cheese: A creamy cheese product made by blending natural cheeses without the addition of heat.

cold wrap sandwich: A variety of cold sandwich in which a flat bread or tortilla is coated with a spread, topped with one or more fillings and garnishes, and rolled tightly.

collagen: A soft, white, connective tissue that breaks down into gelatin when heated.

collagen casing: A casing produced from collagen.

collard: A large, dark-green, edible leaf with a thick, white vein that resembles kale. Also known as a collard green.

collard green: *See* collard.

combination cooking: Any cooking method that uses both moist and dry heat.

combination curing: The process of combining either dry curing or wet curing with artery or stitch pumping to reduce processing time.

combination oven: *See* combi oven.

combi oven: An oven that has both convection and steaming capabilities. Also known as a combination oven.

Comice pear: A fairly large pear with a rotund body and a very short, well-defined neck.

complete protein: A protein that contains all of the essential amino acids.

complex carbohydrate: A carbohydrate composed of three or more sugar units and is slowly absorbed by the body. Also known as starch.

composed salad: A salad that consists of a base, body, garnish, and dressing attractively arranged on a plate.

compound butter: A flavorful butter sauce made by mixing cold, softened butter with flavoring ingredients such as fresh herbs, garlic, vegetable purées, dried fruits, preserves, or wine reductions.

compressed yeast: Yeast that has approximately 70% moisture and is available in 1 lb cakes or blocks. Also known as fresh yeast.

concassé: A preparation method where a tomato is peeled, seeded, and then chopped or diced.

conch: A univalve that has a pinkish-orange shell with an interior that resembles a large snail.

Concord grape: A seeded grape with a deep black color.

condiment: A savory, sweet, spicy, or salty accompaniment that is added to or served with a food to impart a particular flavor that will complement the dish.

conduction: A type of heat transfer in which heat passes from one object to another through direct contact.

confit: A French term for meat that has been cooked and preserved in its own fat.

consommé: A very rich and flavorful broth that has been further clarified to remove any impurities or particles that could cloud the finished product.

contamination: A process of corrupting or infecting by direct contact or association with an intermediate carrier.

contribution margin: The amount added to the AS cost of a menu item to determine a menu price.

convection: A type of heat transfer that occurs due to the circular movement of a fluid or a gas.

convection oven: A gas or electric oven with an interior fan that circulates dry, hot air throughout the cabinet.

convection steamer: A steamer that generates steam using an internal boiler, which circulates around the food to cook it rapidly.

converting: The process of changing a measurement with one unit of measure to an equivalent measurement with a different unit of measure.

cook-and-hold oven: An oven with two separately controlled compartments within one stainless steel cabinet that can be used to cook, roast, reheat, and hold a variety of foods. Also known as a retherm oven.

cooked dressing: *See* salad dressing.

cooking: The process of heating foods in order to make them taste better, make them easier to digest, and to kill harmful microorganisms that may be present in the food.

cooking-loss yield test: A procedure used to determine the yield percentage of a food item that loses weight during the cooking process.

cooling wand: A heavy-gauge, hollow plastic paddle with a screw-on cap at the top of the handle that is filled with water and frozen prior to use.

coriander: A spice made from a light-brown, ridged seed of the coriander plant.

corn bread: A quick bread made from a batter containing cornmeal, eggs, milk, and oil.

Cornish hen: Either a female or male chicken that is five to six weeks old and weighs 1½ lb or less.

corn oil: A type of oil produced from corn.

cornstarch: The white, powdery, pure starch derived from corn.

corrective action: The point in an HACCP plan that identifies the steps that must be taken when food does not meet a critical limit.

cottage cheese: A pebble-shaped fresh cheese with a mildly sour taste.

cottage ham: The smoked, boneless meat extracted from the blade section of the shoulder butt.

coulis: A sauce typically made from either raw or cooked puréed fruits or vegetables.

count: A measurement of the actual number of items being used.

country-style forcemeat: A coarsely ground forcemeat with a strong flavor. Also known as pâté de champagne.

court bouillon: A highly flavored, aromatic vegetable broth made from simmering vegetables with herbs and a small amount of an acidic liquid (usually vinegar or wine).

couscous: A tiny, round pellet made from durum wheat that has had both the bran and germ removed.

cracked grain: A whole kernel of grain that has been cracked by being placed between rollers.

cranberry: A small, red, round berry that has a tart flavor.

crawdad: *See* crayfish.

crawfish: *See* crayfish.

crayfish: A freshwater crustacean that resembles a tiny lobster. Also known as a crawfish or crawdad.

cream cheese: A soft, fresh cheese with a rich, mild flavor and smooth consistency.

creaming mixing method: A quick-bread mixing method used to produce quick bread loaves that have a fine crumb.

cream of tartar: A chemical leavening agent derived from tartaric acid, a by-product of wine production.

cream sauce: *See* béchamel.

crème brûlée: A baked custard that has the texture of a stirred custard beneath a hardened sugar surface.

crème caramel: A baked custard served upside down and topped with hot, caramelized sugar. Also known as flan.

crenshaw melon: A large, pear-shaped melon with a yellow-green, slightly ribbed rind and an orange or salmon-colored flesh.

crêpe: A French pancake that is light and very thin.

crêpe pan: A small skillet with very short, sloped sides that is used to prepare crêpes.

crisphead lettuce: A lettuce with a large, round, tightly packed head of pale-green leaves.

critical control point (CCP): The point where a hazard can be prevented, eliminated, or reduced.

critical limit: The point in a HACCP plan where a minimum or maximum value is established for a CCP in order to prevent, eliminate, or reduce a hazard to a safe level.

crookneck squash: A yellow squash that resembles a bowling pin with a bent neck.

crudité: A group of raw vegetables arranged on a platter and served with a dipping sauce.

crumb crust: A piecrust made from crumbled cookies or crackers that are held together with melted butter.

crushed chiles: *See* crushed red pepper.

crushed red pepper: A spice made from a blend of dried, crushed berries of hot chili peppers. Also known as crushed chiles or chili flakes.

crustacean: A shellfish that has a hard, segmented shell that protects soft flesh and does not have an internal bone structure.

cucumber: A green, cylindrical fruit-vegetable that has an edible skin, edible seeds, and a moist flesh.

cumin: A spice made from a crescent-shaped amber, white, or black seed of the cumin plant and is a member of the parsley family.

curd: The thick, casein-rich part of coagulated milk.

cured sausage: A type of raw sausage that contains sodium nitrite mixed into the forcemeat.

curing: The process of using salt and sodium nitrite alone or with flavorings or sugar to preserve a food item.

curing salt: A pink mixture of table salt and sodium nitrate. Also known as TCM or curing salt.

currant: A small red, black, or golden-white berry that grows in grape-like clusters.

curry leaf: An herb with a small, shiny green leaf from the curry tree.

cuttlefish: A translucent cephalopod with two tentacles, eight sucker-equipped arms, a hard internal cuttlebone, and large eyes at the base of its head.

cycle menu: A menu written for a specific period and is repeated once that period ends.

D

dandelion green: The dark-green, edible leaf of the dandelion plant.

dark chocolate: Chocolate that contains at least 35% chocolate liquor in addition to sugar, vanilla, and cocoa butter.

date: A plump, juicy, and meaty drupe that grows on a date palm tree.

deck oven: A drawer-like oven that is commonly stacked one on top of another, providing multiple-temperature baking shelves.

deep-frying: A dry-heat cooking method in which food is completely submerged in very hot fat.

demi-glace: A sauce made by adding equal parts of espagnole and brown stock together and reducing the mixture by half.

density: The measure of how much a given volume of a substance weighs.

dessert salad: A sweet salad usually consisting of nuts, fruits, and sweeter vegetables such as carrots.

diagonal cut: A slicing cut that produces flat-sided, oval slices.

dice cuts: Precise cubes cut from uniform stick cuts.

dietary fiber: The portion of a plant that the body cannot digest.

digestion: The process the human body uses to break down food into a form that can be absorbed and used or excreted.

Dijon mustard: A light-tan mustard that has a strong, tangy flavor and is made from brown or black mustard seeds, vinegar, white wine, sugar, and salt.

dill: An herb with feathery, blue-green leaves and is a member of the parsley family. Also known as dill weed.

dill weed: *See* dill.

dining room supervisor: *See* maître d'.

dip: A sauce or condiment that is served with foods such as vegetables or chips.

disher: *See* portion control scoop.

dishwasher: The person who operates the warewashing machine and cleans all of the pots, pans, dinnerware, glassware, and flatware.

distilled vinegar: Vinegar made by fermenting diluted, distilled grain alcohol. Also known as white vinegar.

docking: The process of making small holes in yeast dough before it is baked to allow steam to escape and to promote even baking.

double boiler: A round, stainless steel pot that sits inside another slightly larger pot.

double rib lamb chop: A rib chop cut to a thickness equal to two standard rib chops.

dough cutter: A flat, stainless steel blade attached to a sturdy handle. Also known as a bench knife.

dough docker: A roller with pins that is used to perforate dough so that it will bake evenly without blistering in the oven heat.

dragon fruit: An exotic fruit of a cactus with an inedible pink skin, green scales, and white or red flesh speckled with small crunchy black seeds

drawn fish: A fish that has had only the viscera removed.

dredging: The process of lightly dusting an item in seasoned flour or fine bread crumbs.

dressed fish: A fish that has been scaled and has had the viscera, gills, and fins removed.

dried fruit: Fruit that has had most of the moisture removed either naturally or through use of a machine, such as a food dehydrator.

drummette: The innermost section of a wing located between the first wing joint and the shoulder.

drum sieve: *See* tamis.

drumstick: The lower portion of the leg located below the hip and above the knee joint.

drupe: A type of fruit that contains one hard seed or pit. Also known as a stone fruit.

dry aging: The process of aging larger cuts of meat that are hung in a well-controlled environment.

dry curing: A curing method that involves the use of salt to dehydrate the protein in food.

dry-heat cooking: Any cooking method that uses hot air, hot metal, a flame, or hot fat to conduct heat and brown food.

dry measuring cup: A metal cup with a straight handle that is used to measure dry ingredients.

dry-rind cheese: A semi-soft cheese that is allowed to ripen through exposure to air.

dry rub: A mixture of finely ground herbs and spices that is rubbed onto the surface of an uncooked food to impart flavor.

dulse: A stringy, reddish-brown sea vegetable with a fishy odor that is rich in iron, iodine, potassium, and vitamin A.

Dungeness crab: A Pacific crab with a sweet-tasting flesh.

dunnage rack shelving: Shelving consisting of reinforced platforms that keep heavy items at least 12 inches above the floor.

durian: An exotic fruit that contains several pods of sweet, yellow flesh and has a custard-like texture.

Dutch-processed cocoa powder: A dark, unsweetened cocoa powder processed with an alkali to neutralize the natural acidity.

E

early crop potato: *See* new potato.

Eastern oyster: *See* Atlantic oyster.

éclair paste: *See* pâte à choux.

Edam: A waxed-rind semisoft cheese made from cow milk and has a firm, crumbly texture.

edamame: Green soybeans housed within a fibrous, inedible pod.

edible bulb vegetable: A strongly flavored vegetable that grows underground and consists of a short stem base with one or more buds that are enclosed in overlapping membranes or leaves.

edible flowers: The flowers of nonwoody plants that are prepared as vegetables.

edible leaves: Plant leaves that are often accompanied by edible leafstalks and shoots. Also known as greens.

edible mushroom: The fleshy, spore-bearing body of an edible fungus that grows above the ground.

edible-portion (EP) quantity: The amount of a food item that remains after trimming and is ready to be served or used in a recipe.

edible-portion (EP) unit cost: The unit cost of a food or beverage item after taking into account the cost of the waste generated by trimming.

edible root vegetable: An earthy-flavored vegetable that grows underground and has leaves that extend above ground.

edible seeds: The seeds of nonwoody plants that are prepared as vegetables.

edible stem vegetable: The main trunk of a plant that develops buds and shoots instead of roots.

edible tuber: A short, fleshy, underground stem vegetable that bears buds capable of producing new plants.

edge: The sharpened part of a blade that extends from the heel to the tip.

eel: A fatty cartilaginous fish with a long, slender body that resembles a snake.

eggless egg substitute: A yellow-colored liquid composed of soy, vegetable gums, and starches derived from corn or flour.

eggplant: A deep-purple, white, or variegated fruit-vegetable with edible skin and a yellow to white, spongy flesh that contains small, brown, edible seeds.

eggshell: The thin hard covering of an egg that is composed of calcium carbonate.

egg substitute: A liquid product that is typically made from a blend of egg whites, vegetable oil, food starch, powdered milk, artificial colorings, and additives.

electronic probe thermometer: A thermocouple thermometer with a thin, stainless steel stem that is attached by wires to a battery-operated readout device.

electric braiser: *See* tilt skillet.

emergency action plan: A written plan intended to organize employees during an emergency situation.

emulsification: The process of temporarily binding two liquids that do not combine easily, such as oil and vinegar.

emulsified forcemeat: A forcemeat made from a ratio of five parts protein, four parts fat, to three parts crushed ice. Also known as the 5-4-3 style.

emulsion: A combination of two unlike liquids that have been forced to bond with each other.

endosperm: The largest component of a grain kernel and consists of carbohydrates and a small amount of protein.

English lamb chop: A 2 inch thick fabricated cut taken along the entire length of the unsplit loin.

English mustard: A type of yellow mustard that has a hot, spicy flavor and is made from ground yellow and brown or black mustard seeds, wheat flour, and turmeric.

English service: A style of meal service in which food is carved by a server and placed on a preset plate in front of the guest, yet side dishes are passed around the table for self-service.

enoki: *See* enokitake mushroom.

enokitake mushroom: A crisp, delicate mushroom that has spaghetti-like stems topped with white caps. Also known as enoki or snow puff mushroom.

en papillote: A technique in which food is steamed in a parchment paper package as it bakes in an oven.

enriched grain: A refined grain that has thiamin, riboflavin, niacin, folate, and iron added to it.

escargot: The French term for snail and refers to any variety of snail fit for human consumption.

escarole: *See* chicory.

espagnole: A mother sauce made from a full-bodied brown stock, brown roux, tomato purée, and a hearty caramelized mirepoix.

essence: A concentrated fish stock that includes a large amount of aromatic ingredients such as celery, morels, a bouquet garni, a sachet, and fennel root.

essential amino acid: An amino acid that the body cannot manufacture.

ethylene gas: An odorless gas that a fruit emits as it ripens.

European oyster: A variety of oyster with a relatively flat, cup-shaped shell and salty-sweet flavored, creamy-textured flesh.

exotic fruit: A type of fruit that comes from a hot, humid location but is not as readily available as a tropical fruit.

expediter: The person responsible for ensuring each dish is acceptable before it leaves the kitchen.

expediting: The process of speeding up the ordering, preparation, and delivery of food to guests.

external customer: An individual who uses the products or services of a business.

extract: A flavorful oil that has been mixed with alcohol.

extra-virgin olive oil: A type of olive oil produced from the first pressing of the olives without the use of heat or chemicals and has an acidic level less than 1%.

F

fabricated cut: A ready-to-cook cut that is packaged to certain size and weight specifications.

facultative bacteria: Bacteria that can grow either with or without oxygen.

family-style service: A style of meal service where all food is placed on the table for guests to pass for self-service.

fan: Taken from a muscle in the thigh and is the most tender cut of a ratite.

farm-raised game bird: A game bird that is raised for legal sale under state regulations.

fatback: The layer of fat that runs along the back of the hog.

fat cap: The fat that surrounds a muscle.

fat-soluble vitamin: A vitamin that dissolves in fat.

fennel: A celery-like stem vegetable with overlapping leaves that grow out of a large bulb at its base.

fennel seed: A spice made from an oval, light-brown, and green seed of the fennel plant.

fenugreek: A spice made from the pebble-shaped seed of the fenugreek plant in the pea family.

fermentation: The process by which yeast converts sugar into carbon dioxide and alcohol.

fermented black bean sauce: A type of Asian sauce made from fermented and salted blackened soybeans mixed with garlic and spices.

feta: A fresh cheese of Greek origin made from sheep milk or goat milk.

fiddlehead fern: The curled tip of an ostrich fern frond with a nutty and slightly bitter flavor similar to asparagus and artichokes.

fig: The small pear-shaped fruit of the fig tree.

filbert: *See* hazelnut.

filé powder: An herb that is made from the ground leaves of the sassafras tree.

finger cot: A protective sleeve placed over the finger to prevent contamination of a cut.

fine brunoise cut: A dice cut that produces a cube-shaped item with six equal sides measuring ¹⁄₁₆ inch.

fine julienne cut: A stick cut that produces a stick-shaped item ¹⁄₁₆ × ¹⁄₁₆ × 2 inches long.

fingerling potato: A small, tapered, waxy potato with butter-colored flesh and tan, red, or purple skin.

fire-suppression system: An automatic fire extinguishing system that is activated by the intense heat generated by a fire.

first-in, first-out (FIFO): The process of dating new items as they are placed into inventory and placing them behind or below older items to ensure that older items are used first.

fish: Any of a classification of animal that has fins for moving through the water, gills for breathing, and an internal skeleton made of bones or cartilage.

fish fillet: The lengthwise piece of flesh cut away from the backbone.

fish poacher: A thin, oblong pot with loop handles that is used to poach fish.

fish sauce: A type of Asian sauce made from a liquid drained from fermented, salted fish.

fish steak: A cross section of a dressed fish.

fish stock: A basic stock prepared by adding fish bones or shellfish shells, vegetables, a sachet, and cold water to a stockpot and bringing it to a simmer over medium heat.

5-4-3 style: *See* emulsified forcemeat.

fixed cost: A cost that does not change as sales increase or decrease. Rent, real estate taxes, and insurance are examples of fixed costs.

fixed menu: A menu that stays the same or rarely changes.

flaked grain: A refined grain that has been rolled to produce a flake. Also known as a rolled grain.

flaky piecrust: A piecrust prepared by cutting fat into flour until pea-size pieces of dough form.

flan: *See* crème caramel.

flashbake oven: An oven that uses both infrared radiation and light waves to cook foods quickly and evenly from above and below.

flatfish: Any thin, wide fish with both eyes located on one side of the head and a backbone that runs from head to tail through the midline of the body.

flavonoid: An organic pigment found in purple, dark-red, and white vegetables.

flavored vinegar: Any vinegar in which other items such as herbs, spices, garlic, fruits, vegetables, or flowers are added.

flavoring: An item that alters or enhances the natural flavor of food.

flavoring emulsion: A flavorful oil that has been mixed with water and an emulsifier.

flax seed: A dark-amber colored seed from the flowering Mediterranean flax plant.

flounder: A lean, saltwater flatfish with dark-brown skin on the top side.

flour: A fine powder that is created by grinding grains.

flower spice: A type of spice derived from the flower of a plant.

flow of food: The path food takes in a foodservice operation as it moves from purchasing to service.

fluffy omelet: *See* soufflé omelet.

fluted cut: A specialty cut that leaves a spiral pattern on the surface of an item by removing only a sliver with each cut.

foie gras: The fattened liver of a duck or goose.

folded omelet: An omelet that is cooked until nearly done and then folded before serving.

fond lié: *See* jus lié.

fontina: A waxed-rind semisoft cheese made from cow milk.

food allergy: A reaction by the immune system to a specific food.

food bioterrorism: The purposeful act of releasing toxins into foods with the intent to cause harm.

foodborne illness: An illness that is carried or transmitted to people through contact with or consumption of contaminated food.

food cost percentage: A percentage that indicates how the cost of food relates to the menu prices and food sales of a foodservice operation.

food intolerance: An adverse reaction to a food that does not involve the immune system.

food mill: A hand-cranked sieve with a bowl-shaped body that is used to purée soft or cooked foods.

food processor: An appliance with an S-shaped blade and a removable bowl and lid that can be used to quickly chop, purée, blend, or emulsify foods.

food spoilage indicator: A condition that signifies that food is deteriorating and is no longer safe for consumption.

food waste disposer: A food grinder mounted beneath warewashing sinks to eliminate solid food material.

forcemeat: A mixture of raw or cooked meat, poultry, fish, or vegetables, fat seasonings, and sometimes other ingredients.

forest mushroom: *See* shiitake mushroom.

formula: A recipe format in which all ingredient quantities are provided as baker's percentages.

freeze drying: The process of removing the water content from a food and replacing it with a gas.

French grill: A griddle with raised ridges that creates grill marks where the food touches the ridges.

frenching: A method of removing the meat and fat from the end of a bone and is generally applied to chops.

French knife: *See* chef's knife.

French omelet: *See* rolled omelet.

French service: A style of meal service in which food is fully prepared or finished in front of guests.

fresh cheese: A cheese that is not aged or allowed to ripen.

fresh sausage: A type of sausage that is freshly made and has not been cured, smoked, dried, or further processed in any way.

fresh yeast: *See* compressed yeast.

fried sandwich: A hot sandwich that consists of pre-cooked fillings placed within a closed or wrapped sandwich and then fried.

frittata: A traditional folded omelet served open-faced after being browned in a broiler or hot oven.

fritter: A fried donutlike item that may or may not be filled with fruit. Also known as a beignet.

front-of-house (FOH): The portion of a foodservice operation that is open to guests.

frosting: *See* icing.

fruit: The edible, ripened ovary of a flowering plant that usually contains one or more seeds.

fruit salad: A salad that is primarily made of fruits.

fruit-vegetable: A botanical fruit that is sold, prepared, and served as a vegetable.

fryer: A cooking unit used to cook foods in hot fat.

fryer/roaster turkey: A male or female turkey that is less than 16 weeks old and weighs approximately 4–9 lb.

frying: A dry-heat cooking method in which food is cooked in hot fat over moderate to high heat.

fumet: A concentrated stock made from fish bones or shellfish shells and vegetables.

funnel: A tapered bowl attached to a short tube that is used to transfer substances from one container into another container without spilling.

fusion cuisine: A style of cuisine that uses a variety of cooking techniques to combine the flavors of two or more cultural regions.

 G

galantine: Forcemeat that is wrapped in the skin of the animal the meat was taken from, shaped into a cylinder, trussed, and poached in a flavorful stock.

game bird: A wild bird that is hunted for human consumption.

gastrique: A sugar syrup made by caramelizing a small amount of granulated sugar in a saucepan and deglazing the pan with a small amount of vinegar.

garlic: A bulb vegetable made up of several small cloves that are enclosed in a thin, husklike skin.

garlic chive: A stem herb with a solid, flat, grass-shaped sprout and a mild garlic flavor.

gelatin: A flavorless thickening agent made from animal protein.

gelatinization: The process of a heated starch absorbing moisture and swelling, which thickens the liquid.

gelatin salad: A salad made from flavored gelatin.

gelato: An Italian ice cream that has a creamier, denser texture than standard ice creams.

general manager: A person who directs an operation and oversees food production, sales, and service.

geoduck: A very large Pacific clam with a meaty siphon that protrudes from its shell.

germ: The smallest part of a grain kernel and contains a small amount of natural oils as well as vitamins and minerals.

giblets: The name for the grouping of the neck, heart, gizzard, and liver of a bird.

ginger: A spice made from the bumpy root of a tropical plant grown in China, India, Jamaica, and the United States.

glace: A highly reduced stock that results in an intense flavor. For example, one gallon of a stock can be slowly reduced to one-eighth or one-tenth of its original volume to yield just one or two cups of glace.

glacée: *See* bombe.

glazing: The process of covering an item with water to form a protective coating of ice before the item is frozen.

gluten: A type of protein found in grains such as wheat, rye, barley, and some varieties of oats.

gooseberry: A smooth-skinned berry that can be green, golden, red, purple, or white and has many tiny seeds on the inside.

Gorgonzola: A blue-veined cheese that is mottled with blue-green veins.

Gouda: A waxed-rind semisoft cheese that is similar to Edam but contains more fat.

grain: The edible fruit, in the form of a kernel or seed, of a grass.

grain-fed beef: Meat from cattle that were grain-fed in confined feeding operations for 90 days to 1 year.

grande cuisine: A style of cuisine in which intricate food preparation methods and large, elaborate presentations are used. Also known as haute cuisine.

grand sauce: *See* mother sauce.

granita: *See* granité.

granité: A frozen dessert made by frequently stirring a mixture of water, sugar, and flavorings, such as fruit juice or wine, as it is freezing. Also known as a granita.

granola: A baked mixture of rolled oats, nuts, dried fruit, and honey.

grape: An oval fruit that has a smooth skin and grows on woody vines in large clusters.

grapefruit: A round citrus fruit with a thick, yellow outer rind and tart flesh.

grapeseed oil: A type of oil produced from grape seeds.

grass-fed beef: Meat from cattle that were raised on grass with little or no special feed.

gratin: Any dish prepared using the gratinée method.

gratinée: The process of topping a dish with a thick sauce, cheese, or bread crumbs and then browning it in a broiler or high-temperature oven.

gratin forcemeat: A forcemeat made from meat that is partially seared prior to grinding and a starchy binding agent.

grating cheese: A hard, crumbly, dry cheese grated or shaved onto food prior to service.

gravlax: A traditional Scandinavian dish where salmon is cured for 24–48 hours with salt, sugar, and dill.

greens: *See* edible leaves.

griddling: A dry-heat cooking method in which food is cooked on a solid metal cooking surface called a griddle.

grill: A cooking unit consisting of a large metal grate, also referred to as a grill, placed over a heat source.

grilled sandwich: A hot sandwich made by adding a precooked filling or cheese to bread that has been buttered on the exterior and then heated on a griddle, in a sauté pan, or on a panini grill after assembly.

grilling: A dry-heat cooking method in which food is cooked on open grates above a direct heat source.

grinder: A tool that consists of an auger, which pushes the forcemeat forward, and a cutting blade, which cuts the forcemeat as it is pushed through perforated stainless steel dies.

grits: A coarse type of meal made from ground corn or hominy.

griddle: A solid cooking surface made of metal on which foods are cooked.

gross pay: The total amount of an employee's pay before any deductions are made.

gross profit: The calculated difference between total revenue and the cost of goods sold.

ground poultry: Ground fabricated cuts of poultry.

grouper: A lean, white-fleshed, saltwater roundfish with a tough, strong-flavored skin that is always removed before cooking.

Gruyère: A hard cheese that is similar in texture to Emmentaler but has a sharper flavor.

guava: A small oval-shaped fruit, usually 2–3 inches in diameter, with thin edible skin that can be yellow, red, or green.

gueridon: A cart equipped with the items necessary to prepare foods tableside.

guild: An organization of craftsmen that has exclusive control of a particular craft and the production and distribution of its products.

guinea fowl: A farm-raised bird that has both light and dark flesh that is lean and tender. Also known as an African pheasant.

H

HACCP plan: A written document detailing what policies and procedures will be followed to help ensure the safety of food.

half: A full half-length of a bird split down the breast and spine.

halibut: A very large, lean, saltwater flatfish that resembles a giant flounder.

ham: *See* pork leg.

hand tool: Any of a variety of manual tools used to cut, shape, measure, strain, sift, mix, blend, turn, or lift food items.

hard cheese: A firm, somewhat pliable and supple cheese with a slightly dry texture and buttery flavor.

hard-shell clam: An Atlantic clam with a blue-grey shell that contains a chewy flesh. Also known as a quahog.

hash: Shredded and chopped meat that has been mixed and cooked with diced potatoes, onions, and seasonings.

haute cuisine: *See* grande cuisine.

Havarti: A Danish, semisoft dry-rind cheese made from cow milk and has a buttery, somewhat sharp flavor.

hazard analysis: The process of assessing potential risks in the flow of food in order to establish what must be addressed in the HACCP plan.

Hazard Analysis Critical Control Points (HACCP) system: A food safety management system that aims to identify, evaluate, and control contamination hazards throughout the flow of food.

hazardous material: A chemical present in the workplace that is capable of causing harm.

hazelnut: A grape-sized nut found in the fuzzy outer husks of the hazel tree. Also known as a filbert.

hazelnut oil: A type of oil produced from hazelnuts.

head cabbage: A tightly packed, round head of overlapping edible leaves that can be green, purple, red, or white in color.

headcheese: The spiced, pressed, and jellied meat from the head of a hog.

headed and gutted fish: *See* pan-dressed fish.

heart of palm: A slender, white, stem vegetable that is surrounded by a tough husk.

heat sanitizing: The process of using very hot water to reduce harmful pathogens to a safe level.

heel: The rear portion of a blade and is most often used to cut thick items where more force is required.

herb: A flavoring derived from the leaves or stem of a very aromatic plant.

herring: A long, thin, and somewhat fatty saltwater fish with shiny, silvery-blue skin.

hoisin sauce: A type of Asian sauce made from fermented soybean paste, garlic, vinegar, chiles, and sugar.

holding cabinet: A tall and narrow stainless steel box on wheels that accommodates standard sheet pans and contains temperature controls. Also known as a hot box.

hollandaise: A mother sauce made by thickening melted butter with egg yolks.

hominy: The hulled kernels of corn that have been stripped of their bran and germ and then dried.

honeycomb tripe: The lining of the second stomach found in cattle.

honeydew melon: A melon with a smooth outer rind that changes from a pale-green color to a creamy-yellow color as it ripens.

honing: The process of aligning a blade's edge and removing any burrs or rough spots on the blade. Also known as truing.

horned melon: *See* kiwano.

hors d'oeuvre: An elegant, bite-size portion of food that is creatively presented and served apart from a meal.

horseradish: A spice made from the large, brown-skinned root of a shrub related to the radish.

host: The person responsible for greeting and seating guests.

hot box: *See* holding cabinet.

hot closed sandwich: A sandwich made by placing one or more hot fillings between two pieces of bread or a split roll or bun.

hotel pan: A stainless steel pan that is used to cook, hold, or serve food.

hotel rack: A back that remains joined along the backbone.

hot open-faced sandwich: A sandwich consisting of one or two slices of fresh, toasted, or grilled bread, topped with one or more hot fillings, and covered with a sauce, gravy, or a melted cheese topping.

hot wrap sandwich: A sandwich made by adding a spread and precooked fillings to a flatbread and then cooking it.

hybrid: A fruit that is the result of crossbreeding two or more fruits of different species to obtain a completely new fruit.

hubbard squash: A large, oval winter squash with a bumpy, pale-green, blue-gray, or orange skin and a sweet orange flesh.

hull: *See* husk.

husk: The inedible, protective outer covering of grain. Also known as the hull.

hydrogenation: The process that chemically transforms oils into solids to improve shelf life and stabilize flavor.

I

ice cream: A frozen dessert made from cream, butterfat, sugar, and sometimes eggs.

icing: A sugar-based coating often spread on the outside of and between the layers of a baked product. Also known as frosting.

immersion blender: A narrow, handheld blender with a rotary blade that is used to purée a product in the container in which it is being prepared. Also known as a stick mixer.

immersion curing: *See* wet curing.

impinger conveyor oven: An oven that directs heat from both above and below a food item as it moves along a conveyor belt.

inarizushi: A type of sushi made with vinegar-seasoned rice and toppings stuffed into a small purse of fried tofu.

incomplete protein: A protein that does not contain all of the essential amino acids.

individually quick-frozen (IQF): A designation for products preserved using a method in which each item is glazed with a thin layer of water and frozen individually.

induction cooktop: An electromagnetic unit that uses a magnetic coil below a flat surface to heat food rapidly.

infrared oven: An oven that uses infrared radiation to evenly and efficiently bake flat foods such as pizza.

infrared thermometer: A thermometer that measures the surface temperature of an item through the use of infrared laser technology.

infused oil: Any variety of oil with added herbs, spices, or additional ingredients that increase the flavor.

insoluble fiber: Dietary fiber that will not dissolve in water.

instant-read thermometer: A stem-like thermometer attached to either a digital or mechanical display.

instant yeast: A form of yeast that does not need to be hydrated and may be directly added to a warm flour mixture.

insulated carrier: An insulated container made of heavy polyurethane that is designed to hold hotel pans of hot or cold foods during transport.

internal customer: An individual who is the recipient of other products or services within the same organization.

invoice: A document provided by a supplier that lists the items delivered to a foodservice operation and the prices of those items.

irradiation: The process of exposing food to low doses of gamma rays in order to destroy deadly organisms such as E. coli O157:H7, campylobacter, and salmonella.

Italian bacon: *See* pancetta.

J

jackfruit: An enormous, spiny, oval exotic fruit with yellow flesh that tastes like a banana and has seeds that can be boiled or roasted and then eaten.

Jamaican pepper: *See* allspice.

Japanese oyster: *See* Pacific oyster.

Jerusalem artichoke: *See* sunchoke.

jicama: A large, brown root vegetable that ranges in size from 4 oz to 6 lb and is a good source of vitamin C and potassium.

jowl: Meat from the cheek of a hog.

juice extractor: An electric machine that creates juice by liquefying raw vegetables and fruits and separating the fiber or pulp from the juice.

juicer: A device used to extract juice from fruits and vegetables.

julienne cut: A stick cut that produces a stick-shaped item ⅛ × ⅛ × 2 inches long.

juniper berry: A spice made from a small, purple berry of an evergreen bush.

jus lié: A sauce that is made by thickening a brown stock either by adding a cornstarch or arrowroot slurry or simply by a slow reduction. Also known as fond lié.

K

kabocha squash: A winter squash with a jade-green and celadon-streaked rind and a pale-orange flesh that is sweet.

kale: A large, frilly, edible leaf that varies in color from green and white to shades of purple.

kasha: Roasted buckwheat.

ketchup: A thick, tomato based product that usually includes vinegar, sugar, salt, and spices.

key lime: A variety of lime that is smaller, more acidic, and more strongly flavored than other limes.

kielbasa: A Polish pork or beef sausage flavored with garlic, pimento, and cloves.

king crab: The largest-sized variety of crab, typically weighing between 6–20 lb and measuring as much as 10 feet from the tip of one leg to the tip of the opposite leg.

kipper: A whole herring that has been split from tail to head, gutted, salted or pickled, and cold smoked.

kitchen fork: A large fork with two long prongs and is used to hold meats steady while they are being carved. Also known as a chef's fork.

kitchen shears: Heavy-gauge scissors used to trim foods during the preparation process.

kitchen spoon: A large stainless steel or silicone spoon that is used to stir or serve foods.

kiwifruit: A small, barrel-shaped tropical fruit, approximately 3 inches long and weighing between 2–4 ounces.

kiwano: An exotic fruit with jagged peaks rising from an orange and red-ringed rind that is native to Africa. Also known as a horned melon.

kneading: The process of pushing and folding dough until it is smooth and elastic.

kohlrabi: A sweet, crisp, stem vegetable that has a pale-green or purple, bulbous stem and dark-green leaves.

kombu: A long, dark-brown to purple sea vegetable that is used to flavor dashi stock.

kosher salt: A salt used to cure, season, and prepare kosher foods.

kumquat: A small, golden, oval-shaped fruit with a thin, sweet peel and tart center.

L

lactose: A sugar found in milk and dairy products.

ladle: A stainless steel, cuplike bowl attached to a long handle that is often used to serve soups, sauces, and salad dressings.

lamb: The meat from slaughtered sheep that are less than a year old.

lamb breast: A thin, flat, primal cut of lamb that contains the breastbone, the tips of the rib bones, and cartilage that is located under the shoulder and ribs.

lamb crown roast: A frenched rack with the ribs formed into a circle to resemble a crown.

lamb foresaddle: The front half of the carcass consisting of the primal shoulder, rack, and breast cuts.

lamb hindsaddle: The rear half of the carcass consisting of the loin and leg.

lamb loin: A primal cut located between the rack and leg that includes the 13th rib, the loin eye muscle, the center section of the tenderloin, the strip loin, and some flank meat.

lamb rack: Eight rib bones located between the shoulder and loin of a lamb.

lamb riblet: A rectangular strip of meat cut from the lamb breast that contains part of a rib bone.

lamb shank: A cut of lamb that contains the upper foreshank bones.

lamb shoulder: Includes the first four rib bones of each side and the arm and neck bones.

laminated dough: An unleavened dough that is layered with butter and used to make pastries, such as croissants, turnovers, and strudels.

langoustine: A small lobster that resembles a very large shrimp, except it has very long front arms with long, thin claws. Also known as a Norway lobster.

larding: The process of inserting thin strips of fatback into lean meat with a larding needle.

lavender: An herb with pale-green leaves, purple flowering tops and is a member of the mint family.

leading sauce: *See* mother sauce.

leaf herb: A type of herb derived from the leaf portion of a plant.

lean dough: A yeast dough that is low in fat and sugar.

leavening agent: Any ingredient that causes a baked product to rise by the action of air, steam, chemicals, or yeast.

leek: A long, white bulb vegetable, with long, wide, flat leaves.

leg: Poultry cut consisting of a drumstick and thigh.

leg of lamb: A primal cut of lamb that contains the last portion of the backbone, hip bone, aitchbone, round bone, hindshank, and tailbone.

leg quarter: A thigh, a drumstick, and a portion of the back.

legume: The edible seed of a nonwoody plant and grows in multiples within a pod.

lemon: A tart yellow citrus fruit with high acidity levels.

lemongrass: A stem herb with long, thin, gray-green leaves, a white scallion-like base, and a lemony flavor. Also known as citronella.

lemon zest: The grated peel of a lemon that is used as a seasoning.

lentil: A very small, dried pulse that has been split in half. There are many varieties of lentils, with the most common varieties being green, yellow, red, and brown.

lettuce: An edible leaf that is almost exclusively used in salads or as a garnish.

liaison: A thickening agent that is a mixture of egg yolks and heavy cream used to thicken sauces.

Limburger: A washed-rind semisoft cheese with a strong aroma and flavor.

lime: A small citrus fruit that can range in color from dark green to yellow-green.

line cook: The person responsible for preparing foods that are assigned to a particular station within the hot production line.

lipid: A nutrient in the form of fats, oils, and fatty acids.

liquid measuring cup: A transparent cup with a pouring lip and a loop handle and is used to measure liquid ingredients.

loaf pan: A deep rectangular pan that is used to bake loaves of bread.

lobster: A saltwater crustacean with a brown to bluish-black external shell and two large claws.

loganberry: A red-purple hybrid berry made by crossbreeding a raspberry and a blackberry.

long-flake crust: A piecrust that absorbs a large amount of liquid because the flour and fat are rubbed together very lightly, leaving the fat in chunks about the size of a walnut.

long-grain rice: A rice that is long and slender and remains light and fluffy after cooking.

long-neck clam: *See* soft-shell clam.

loss: The amount of money lost by an operation when revenue is less than expenses.

lotus root: The underwater root vegetable of an Asian water lily that looks like a solid-link chain about 3 inches in diameter and up to 4 feet in length.

Louisiana-style hot sauce: *See* cayenne sauce.

lowboy: A reach-in refrigerated unit located beneath a work surface.

lox: Salmon that has been brine cured and then cold smoked.

lychee: An exotic drupe covered with a thin, red, inedible shell and has a light-pink to white flesh that is refreshing, juicy, and sweet.

M

macadamia nut: A tan, marble-sized nut with a hard shell and is the fruit of the macadamia tree.

mace: A spice made from the lacy red-orange covering of the nutmeg kernel.

mackerel: A fatty, saltwater roundfish with a stream-lined shape and dark bluish-brown or black skin.

macromineral: A mineral that the human body requires at least 100 mg of per day.

mahi-mahi: A lean, saltwater roundfish that has colorful skin and firm, pink flesh.

Maillard reaction: A reaction that occurs when the proteins and sugars in a food are exposed to heat and merge together to form a brown exterior surface.

main course salad: A large salad containing an assortment of vegetables and a protein such as poultry, seafood, meats, eggs, or legumes and is served as an entrée.

Maine lobster: A large lobster with a dark bluish-green shell, two large, heavy claws, and four slender legs on each side of its body.

maître d': The person responsible for overseeing and coordinating all FOH activities. Also known as a dining room supervisor.

maki roll: *See* makizushi.

makizushi: A type of sushi made with vinegar-seasoned cooked rice layered with other ingredients and rolled in a dried seaweed paper called nori. Also known as a maki roll.

malt vinegar: Vinegar made from malted barley.

Manchego: A hard cheese with a tangy, slightly salty flavor and is produced from sheep milk in the La Mancha region of Spain.

mandarin: A small, intensely sweet citrus fruit that is closely related to the orange, but is more fragrant.

mandoline: A special cutting tool with adjustable steel blades used to cut food into consistently thin slices.

mango: An oval or kidney-shaped drupe with orange to orange-yellow flesh.

mangosteen: A round, sweet and juicy exotic fruit about the size of an orange with a hard, thick, dark-purple rind that is inedible.

marbling: The fat found within a muscle.

marinade: A flavorful liquid used to soak uncooked foods such as meat, poultry, and fish to impart flavor and sometimes to tenderize.

marjoram: An herb with short, oval, pale-green leaves and is a member of the mint family.

marker: A round tool that has wire guides that leave marks indicating where to cut pies, round cakes, or pizzas into equal portions.

market menu: A menu that changes frequently to coincide with changes in the availability of products.

mascarpone: A cream cheese of Italian origin that has a smooth texture, is white or pale yellow in color, and has a buttery, somewhat sweet flavor.

matignon: A uniformly cut mixture of onions or leeks, carrots, and celery and may also contain smoked bacon or ham.

mature duck: A duck more than six months of age and weighs 4–6 lb.

mature turkey: A turkey that is more than 15 months old.

mayonnaise: A thick, uncooked emulsion formed by combining oil with egg yolks and vinegar or lemon juice.

meal service style: The way in which guests are served food and beverages.

meal-specific menu: A menu that only offers a particular meal, such as breakfast, lunch, or dinner.

mealy piecrust: A low-moisture piecrust prepared by rubbing fat into flour until the mixture resembles fine cornmeal.

mealy potato: A type of potato that is higher in starch and lower in moisture than other types of potatoes.

measurement equivalent: The amount of one unit of measure that is equal to another unit of measure.

measuring spoon: A stainless steel spoon used to measure a small volume of an ingredient.

meat mallet: A hand tool used to tenderize meats prior to cooking.

medium-grain rice: A rice that contains slightly less starch than short-grain rice but is still glossy and slightly sticky when cooked.

melon: A type of fruit that has a hard outer rind (skin) and a soft inner flesh that contains many seeds.

menu: A list of items that guests may order from a foodservice operation.

menu-item food cost percentage: The AS cost of a menu item divided by the menu price, written as a percent.

menu mix: The assortment of items that may be ordered from a given menu.

meringue: A mixture of egg whites and sugar.

Mexican husk tomato: *See* tomatillo.

Meyer lemon: A cross between a lemon and an orange.

microgreens: The first sprouting leaves of an edible plant.

micromineral: A mineral that the human body requires less than 100 mg of per day.

microwave oven: A cooking unit that uses microwaves to heat the water molecules within foods.

milk chocolate: A chocolate product that contains at least 12% milk solids and 10% chocolate liquor.

milled grain: A refined grain that has been ground into a fine meal or powder.

mincing: Finely chopping an item to yield a very small cut, yet not entirely uniform, product.

mineral: An inorganic substance that is required in very small amounts to help regulate body processes.

mint: A general term used to describe a family of similar herbs, such as peppermint and spearmint.

mirepoix: A mixture of 50% onions, 25% celery, and 25% carrots, roughly cut into the appropriate size for the stock being produced.

mise en place: A French term meaning "put in place."

mixer: A versatile electric appliance with U-shaped arms that secure one of several stainless steel mixing bowls of various sizes under a rotating head that accommodates various attachments.

mixing bowl: A stainless steel or aluminum bowl used for mixing ingredients.

mixing paddle: A long-handled paddle used to stir foods in deep pots or steam kettles.

modernist cuisine: *See* avant-garde cuisine.

modified starch: A combination of starch and one or more vegetable gums used as a thickeners.

modified straight-dough method: A method of mixing yeast dough in which the ingredients are added in sequential steps.

moist-heat cooking: Any cooking method that uses liquid or steam as the cooking medium.

mollusk: A shellfish with a soft, nonsegmented body.

monkfish: A very large, lean, saltwater roundfish that can weigh up to 50 lb.

Monterey Jack: A semisoft dry-rind cheese that has a smooth texture, a creamy white color, and a mild taste.

morel mushroom: An uncultivated mushroom with a cone-shaped cap that ranges in height from 2–4 inches and in color from tan to very dark brown.

mother sauce: One of five sauces from which the small classical sauces described by Escoffier are produced. Also known as a leading or grand sauce.

mousseline-style forcemeat: A forcemeat made from less-fibrous proteins such as poultry, seafood, veal, or pork.

mozzarella: A very tender fresh cheese with a soft, elastic-like curd.

muffin: A quick bread that is made with eggs, flour, and either oil, butter, or margarine.

muffin mixing method: A quick-bread mixing method that uses liquid fats, such as vegetable oil or melted butter, to produce a rich and tender product.

muffin pan: A rectangular pan with cuplike wells and is used to bake teacakes, muffins, or cupcakes.

multidecker sandwich: Consists of three pieces of bread, a spread, and at least two layers of garnishes and fillings.

muenster: A washed-rind semisoft cheese with a flavor between that of brick cheese and Limburger.

muesli: An unbaked mixture of rolled oats, wheat flakes, oat bran, raisins, dates, sunflower seeds, hazelnuts, and wheat germ.

muskmelon: A round, orange-fleshed melon with a beige or brown, netted rind.

mussel: A freshwater or saltwater bivalve with whisker-like threads that extend outside the shell to allow the animal to attach to items for protection.

mustard: A pungent powder or paste made from the seeds of the mustard plant.

mustard green: A large, dark-green, edible leaf from the mustard plant that has a strong peppery flavor.

mustard seed: An extremely tiny seed from the mustard plant and is used as a spice.

N

nage: An aromatic court bouillon that is often used as a finishing sauce.

Napa cabbage: An elongated head of crinkly and overlapping edible leaves that are a pale yellow-green color with a white vein. Also known as celery cabbage.

nappe: The consistency of a liquid that thinly coats the back of a spoon and ensures that a sauce will cling lightly to another food.

natural casing: A casing produced from the intestines of sheep, hogs, or cattle.

nectarine: A sweet, slightly tart, orange to yellow drupe with a firm, yellow flesh and a large oval pit.

net pay: The actual amount on an employee's paycheck and is equal to the gross pay minus payroll deductions.

net profit: The calculated difference between the gross profit and operating expenses of a foodservice operation.

networking: A means of using personal connections, such as professional contacts, friends, teachers, and acquaintances, to locate possible employment opportunities.

Neufchâtel: A soft, fresh cheese made from whole or skim milk or a mixture of milk and cream.

new American cuisine: A style of cuisine that emphasizes the use of foods that are grown in America.

new potato: Refers to any variety of potato that is harvested before the sugar is converted to starch. Also known as an early crop potato.

New Zealand yam: *See* oca.

nigirizushi: A type of sushi made with a small, hand-formed mound of vinegar-seasoned cooked rice that is garnished with a topping.

noisette: A small, round, boneless medallion of meat.

nonessential amino acid: An amino acid that the body can manufacture.

nopal: The green, edible leaf of the prickly pear cactus.

nori: A thin, purple-black sea vegetable that turns green when it is toasted.

Norway lobster: *See* langoustine.

nouvelle cuisine: A style of cuisine in which foods are cooked quickly, seasoned lightly, and artistically presented in smaller portions.

nova: A brine-cured, cold-smoked salmon that is less salty than lox.

nut: A hard-shelled, dry fruit or seed that contains an inner kernel.

nutmeg: A spice made from an oval, gray-brown seed found in the yellow, nectarine-shaped fruit of a large tropical evergreen.

nutrient-dense food: A food that is high in nutrients and low in calories.

o

oat groat: An oat grain that only has the husk removed.

obesity: A medical condition characterized by an excess of body fat.

oblique cut: A slicing cut that produces wedge-shaped pieces with two angled sides. Also known as a rolled cut.

obsolescence: The removal of an item from the menu while ingredients for that particular item are still in stock.

oca: A small, knobby tuber that has a potato-like flesh and ranges in flavor from very sweet to slightly acidic. Also known as a New Zealand yam.

ocean perch: A round, saltwater fish native to the north Atlantic that has red skin and pink flesh. Also known as a redfish.

octopus: A gray cephalopod with eight sucker-equipped arms, a birdlike beak, well-developed vision, and no internal or external shell.

offal: An edible part of an animal that is not part of a primal cut.

offset spatula: A tool with a wide metal blade that bends up and back toward a handle. Also known as an offset turner.

offset turner: *See* offset spatula.

oignon brûlé: Half an onion that is charred on the cut side.

oil: A type of fat that remains in a liquid state at room temperature.

okra: A green fruit-vegetable pod that contains small, round, white seeds and a gelatinous liquid.

old-fashioned oats: *See* rolled oats.

olive: A small, green or black drupe that is grown for both the fruit and its oil.

olive oil: A type of oil produced from olives.

olive-pomace oil: A type of olive oil produced using heat, and often chemicals, to extract additional oils from the olive pulp and olive pits after the first pressing.

omelet: An egg dish made with beaten eggs and cooked into a solid form.

onion: A bulb vegetable made up of many concentric layers of fleshy leaves.

onion piquet: Half of an onion studded with cloves and a bay leaf.

orange: A round, orange-colored citrus fruit.

orange roughy: A lean, round, saltwater fish with orange to light-gray skin that has a rough appearance.

oregano: An herb with small, dark-green, slightly curled leaves and is a member of the mint family. Also known as wild marjoram.

oven: An enclosed cabinet where food is baked by being surrounded by hot air.

oven spring: The rapid expansion of yeast dough in the oven, resulting from the expansion of gases within the dough.

overall food cost percentage: The total amount of money a foodservice operation spends on food divided by the total food sales over a defined period of time, written as a percent.

over-easy egg: A fried egg with an unbroken yolk that is cooked until the egg white has gone from translucent to white on the bottom and then flipped over and cooked on the other side just until the white is no longer translucent.

over-hard egg: A fried egg with a bright, firm white that is not rubbery and a pale yellow yolk that looks almost fluffy.

overhead rack: An overhead, suspended rack with hooks that allow utensils, pots, and pans to be easily accessible.

overhead shelving: Shelving mounted on the wall or above a work surface.

overhead warmer: A heat source located above a prepared food that needs to be kept hot for service.

over-medium egg: A fried egg with a completely cooked white and a yolk that is cooked nearly all the way through.

overrun: An increase in the volume of a frozen product as a result of the incorporation of air during churning and freezing.

owner: *See* restaurateur.

oxtail: The tail from a cattle carcass.

oyster: A saltwater bivalve with a very rough shell that is coated with calcium deposits.

oyster knife: A small knife with a short, dull-edged blade with a tapered point that is used to open oysters.

oyster mushroom: A broad, fanlike or oyster-shaped mushroom that varies in color from white to gray or tan to dark brown.

oyster sauce: A type of Asian sauce made from the cooking liquid of boiled oysters, brine, and soy sauce.

P

Pacific oyster: A variety of large oyster that has fragile, curvy shells and a briny, sweet, and mild-tasting flesh. Also known as a Japanese oyster.

palette knife: A flat, narrow knife with a rounded, 3½–12 inch blade that varies in flexibility. Also known as a cake spatula.

pancetta: Unsmoked pork belly that has been cured in salt and spices, such as nutmeg and pepper, and then dried for a few months. Also known as Italian bacon.

pan-dressed fish: A dressed fish that has had its head removed. Also known as a headed and gutted fish.

pan-frying: A dry-heat cooking method in which food is cooked in a pan of hot fat.

panini grill: An Italian clamshell-style grill made specifically to cook grilled sandwiches.

panning: The process of placing rounded pieces of dough into pans.

papaya: A pear- or cylinder-shaped tropical fruit weighing 1–2 pounds with flesh that ranges in color from orange to red-yellow.

papillae: Small bumps found on the tongue that are covered with hundreds of taste buds.

paprika: A spice made from a dried, ground sweet red pepper berry that is often used as a colorful garnish.

parasite: A pathogen that relies on a host for survival in a way that benefits the organism and harms the host.

paring knife: A short knife with a stiff 2–4 inch blade used to trim and peel fruits and vegetables.

parisienne scoop: A special cutting tool that has a half-ball cup with a blade edge attached to a handle and is used to cut fruits and vegetables into uniform spheres.

Parmesan: A grating cheese that originated in Parma, Italy.

parsley: An herb with curly or flat dark-green leaves and is used as both a flavoring and a garnish.

par stock: The maximum amount of a particular product that should be kept in inventory to ensure that an adequate supply is on hand for normal production.

partial carcass: A primary division of a whole carcass.

partial tang: A shorter tail of a knife blade that has fewer rivets than a full tang.

parsnip: An off-white root vegetable, similar in shape to a carrot, that ranges from 5–10 inches in length.

passion fruit: A small, oval-shaped exotic fruit that typically weighs 2–3 ounces and has firm, inedible skin that can be either yellow or purple.

pasta: A term for rolled or extruded products made from a dough composed of flour, water, salt, oil, and sometimes eggs.

paste: *See* wet rub.

pasteurized egg: An egg that has been heated to a specific temperature for a specific period of time to kill bacteria that can cause foodborne illnesses.

pastry bag: A cone-shaped paper, canvas, or plastic bag that is fitted with a pastry tip.

pastry brush: A small, narrow brush that is used to apply liquids, such as egg wash or butter, onto baked products.

pastry chef: The person responsible for making and decorating the sweets and desserts offered by the foodservice operation.

pastry tip: A cone-shaped tip that is fitted into the narrow end of a pastry bag.

pastry wheel: A dough-cutting tool with a rotating disk attached to a handle.

pâté: A forcemeat that is typically layered with garnishes and baked in a mold.

pâte à choux: A pastry made by beating eggs into a paste of boiled water, butter, and flour. Also known as éclair paste.

pâté de campagne: *See* country-style forcemeat.

pâte feuilletée: *See* puff pastry dough.

patelquat: *See* Persian melon.

pathogen: A microorganism that can cause disease.

pattypan: A round, shallow squash with scalloped edges and is best harvested when it is no larger than 2–3 inches in diameter.

payroll expense: Any money paid to an employee who performs work for the operation.

paysanne cut: A dice cut that produces a flat square, round, or triangular cut ½ × ½ × ⅛ inch thick.

peach: A sweet, orange to yellow fruit with downy skin.

peanut: A legume that is contained in a thin, netted, tan-colored pod that grows underground.

peanut oil: A type of oil produced from peanuts.

pear: A bell-shaped pome with a thin peel and sweet flesh.

pearled grain: A refined grain with a pearl-like appearance that results from having been scrubbed and tumbled to remove the bran.

pecan: A nut with a light-brown kernel from the pecan tree.

pectin: A chemical present in all fruits that acts as a thickening agent when it is cooked in the presence of sugar and an acid.

peel: 1. A long, flat, narrow piece of wood or metal shaped like a wide, thin paddle that is used to lift items and place them into and remove them from ovens. 2. The thick outer rind of a citrus fruit.

peeler: A special cutting tool with a swiveling, double-edged blade that is attached to a handle and is used to remove the skin or peel from fruits and vegetables.

pepita: *See* pumpkin seed.

peppercorn: The dried berry of a climbing vine known as the Piper nigrum and is used whole, ground, or crushed.

pepper sauce: A type of hot and spicy sauce made from various types of chili peppers and typically includes vinegar.

perceived value pricing: The process of adjusting a target menu price based on how management thinks a customer will perceive the price.

perch: A lean, freshwater roundfish native to the Great Lakes and northern Canada.

perishable food: Food that has a short shelf life and is subject to spoilage and decay.

persimmon: A bright-orange tropical fruit that grows on trees and is similar in shape to a tomato.

Persian melon: A green muskmelon with finely textured net on the rind. Also known as a patelquat.

personal hygiene: The physical care maintained by an individual.

phyllo: Paper-thin sheets of laminated dough used for making pastries.

phyllo shell: A shell made by layering buttered sheets of phyllo dough in miniature muffin tins and baking them until golden brown.

physical contaminant: Any object that can be hazardous if it is inhaled, swallowed, or otherwise absorbed into the body.

pickling salt: A pure form of salt that contains no residual dust, iodine, or other additives. Also known as canning salt.

picnic: Fabricated from the upper part of the foreleg that includes a portion of the shoulder.

picnic shoulder: A primal cut of pork that is the lower half of the shoulder of a hog.

pie pan: A round, shallow pan with sloped sides and is used for baking pies.

pineapple: A sweet, acidic tropical fruit with a prickly, pinecone-like exterior and juicy, yellow flesh.

pine nut: An ivory-colored, torpedo-shaped nut from the pine cone of various types of pine trees.

pink salt: *See* curing salt.

pistachio: A pale-green, bean-shaped nut from the pistachio tree.

pith: The white layer just beneath the peel of a citrus fruit.

plantain: A tropical fruit that is a close relative of the banana, but is larger and has a dark brown skin when ripening.

plated service: A style of meal service in which individual portions of food are plated before being brought to guests. Also known as American service.

plum: An oval-shaped drupe that grows on trees in warm climates and comes in a variety of colors such as blue-purple, red, yellow, or green.

poaching: A moist-heat cooking method in which food is cooked in a liquid that is held between 160°F and 180°F.

poêléing: A combination cooking method that is often referred to as butter roasting.

pome: A fleshy fruit that contains a core of seeds and has an edible skin.

pomegranate: A round, bright-red tropical fruit with a hard, thick outer skin.

poppy seed: A spice made from a very small, blue-gray seed of the poppy plant.

porcini mushroom: An uncultivated, pale-brown mushroom with a smooth, meaty texture and a pungent flavor. Also known as a cèpe.

pork: The meat from slaughtered hogs that are less than a year old.

pork belly: A fatty slab of meat and skin from the side and belly of a hog.

pork leg: A primal cut of pork that is composed of the hind thigh and buttock of a hog. Also known as a ham.

pork loin: A primal cut that extends along the greater part of the backbone, from about the second rib, through the rib and loin area of a hog.

pork shoulder butt: A square, compact area of the shoulder located just above the front legs of a hog. Also known as Boston butt.

pork spareribs: The long, narrow ribs and breastbone of a hog.

pork tenderloin: A fairly long, tapered strip of lean meat taken from the underside of the loin.

porter: The person who ensures that the kitchen area is clean and in order, including the dish area, garbage area, and floors.

portfolio: A collection of items that depict the knowledge, skills, and accomplishments of an individual.

portion control: The process of ensuring that a specific amount of food or beverage is served for a given price.

portion control scoop: A stainless steel scoop of a specific size attached to a handle with a thumb-operated release lever. Also known as a disher.

portion size: The amount of a food or beverage item that is served to an individual person.

portobello mushroom: A very large and mature, brown cremini mushroom that has a flat cap measuring up to 6 inches in diameter.

Port Salut: A washed-rind semisoft cheese that has a soft, smooth, orange-colored rind and a glossy, ivory-colored interior.

POS (point-of-sale) system: A computerized network that compiles data on all sales incurred by a business.

potato: A round, oval, or elongated tuber that is the only edible part of the potato plant.

potentially hazardous food: A food that requires temperature control in order to keep it safe for consumption.

poultry: The collective term for various kinds of birds that are raised for human consumption.

poussin: A very young female or male chicken that weighs 1 lb or less.

preparation area: An area in a professional kitchen where food items are prepared and cooked.

prepared yellow mustard: A type of bright-yellow mustard that is mild in flavor and made from ground yellow mustard seeds, vinegar, and turmeric.

pressure steamer: A steamer that uses water heated within a pressure-controlled, sealed cabinet to cook foods much quicker than a convection steamer.

pricing form: A tool often used to help calculate the AS cost of a menu item and establish a menu price.

prickly pear: A pear-shaped tropical fruit with protruding prickly fibers that is a member of the cactus family.

primal cut: A large cut from a whole or a partial carcass.

prix fixe menu: A menu that offers limited choices within a collection of specific items for a multicourse meal at a set price.

processed cheese: A blend of fresh and aged natural cheeses that are heated and melted together.

processed cheese food: A cheese-based product that may contain as little as 51% cheese.

profit: The amount of money earned by an operation when revenue is greater than expenses.

profiterole: A miniature pastry made from choux paste that is filled with a sweet or savory filling.

proofer: *See* proofing cabinet.

proofing: The process of letting yeast dough rise in a warm (85°F) and moist (80% humidity) environment until the dough doubles in size.

proofing cabinet: A holding cabinet that contains both temperature and humidity controls. Also known as a proofer.

prosciutto: A type of dry-cured Italian ham that is sliced very thin and used to make hors d'oeuvres or appetizers.

protein: A nutrient that consists of one or more chains of amino acids and is essential to living cells.

provolone: A hard cheese with an elastic texture and a mild to sharp taste depending on the age.

puff pastry dough: A laminated dough that does not contain sugar. Also known as pâte feuilletée.

Pullman loaf: A long, rectangular loaf of sandwich bread with four square sides and a fine, dry texture.

pulse: A dried seed of a legume.

pumpkin: A round fruit-vegetable with a hard orange skin and a firm flesh that surrounds a cavity filled with seeds.

pumpkin seed: A green seed that comes from a pumpkin plant. Also known as a pepita.

punching: The process of folding and pressing risen yeast dough to allow carbon dioxide to escape.

purchase specification: A written form listing the specific characteristics of a product that is to be purchased from a supplier.

pure olive oil: A type of olive oil produced using heat, and often chemicals, to extract additional oils from the olive pulp after the first pressing.

purple potato: A mealy potato with smooth, thin, bluish-purple skin and purple flesh. Also known as a blue potato.

Q

quahog: *See* hard-shell clam.

quenelle: A small dumpling made from a forcemeat by using two spoons.

quiche: A baked egg dish composed of a savory custard baked in a piecrust.

quick bread: A baked product that is made with a quick-acting leavening agent such as baking powder, baking soda, or steam.

quick bread loaf: A loaf of bread that is made from a batter similar to muffin batter.

quince: A hard yellow pome that grows in warm climates.

quinoa: A small, round, gluten-free grain that is classified as a complete protein.

R

radiation: A type of heat transfer that uses electromagnetic waves to transfer energy.

radio frequency identification (RFID): A form of technology that uses electronic tags to store data that can be monitored from remote distances.

radish: A root vegetable that is small in diameter with a white flesh and a peppery taste that comes in many colors and shapes.

raft: The clearmeat that has risen to the surface.

rambutan: A fragrant and sweet exotic fruit covered on the outside with soft, hair-like spikes.

ramp: A wild leek with a flavor similar to scallions, yet with more zing.

range: A large appliance with surface burners.

rapeseed oil: *See* canola oil.

raspberry: A slightly tart, red fruit that grows in clusters.

rasp grater: A nearly flat, razor-sharp, handheld grater that shaves food into fine or very fine pieces.

ratio: A mathematical way to represent the relationship between two or more numbers or quantities.

ratite: Any of a large variety of flightless birds that have flat breastbones and small wings in relation to their body size.

rat-tail tang: A narrow rod of metal that runs the length of the knife handle but is not as wide as the handle.

raw bar: A presentation of a variety of raw and steamed seafood presented and served on a bed of ice.

raw yield test: A procedure used to determine the yield percentage of a food item that is trimmed of waste prior to being used in a recipe.

ray: *See* skate.

reach-in unit: A temperature-controlled cabinet for storing cold or frozen food items.

reamer: A manual or electric device used to extract juice from citrus fruits.

receiving area: The area of the professional kitchen where all delivered items are checked for freshness, proper amount, correct temperature, and accurate price.

redfish: *See* ocean perch.

red flame grape: A seedless grape that ranges from a light purple-red color to a dark-purple color.

red pepper: *See* cayenne pepper.

red potato: A round, waxy, red-skinned potato with white flesh.

red snapper: A lean, saltwater roundfish with pink flesh that becomes pearly white and flakes easily when cooked.

reduction: The process of gently simmering a liquid until it reduces in volume and results in a thicker liquid with a more concentrated flavor.

reel oven: *See* rotating rack oven.

refined grain: A grain that has been processed to remove the germ, bran, or both.

relish: An assortment of uncooked vegetables that are served raw, marinated, or pickled.

remouillage: A stock made from using bones that have already been used once to make a stock.

rennet: A substance that contains acid-producing enzymes or an acid.

requisition: An internally generated invoice that is used to aid in tracking inventory as it moves from storage to production.

restaurateur: The person who holds the legal title of a foodservice operation. Also known as an owner.

résumé: A document listing the education, professional experience, and interests of a job applicant.

retherm oven: *See* cook-and-hold oven.

rhubarb: A tart stem vegetable that ranges in color from pink to red and is most often prepared like a fruit.

ribbon pasta: A thin, round strand or flat, ribbonlike strand of pasta.

rib eye: A large, eye-shaped muscle within the rib that is a continuation of the sirloin muscle.

rice pudding: A baked custard made from cooked rice combined with a sweet custard and often dried fruits such as raisins or currants.

ricer: A sieve with an attached plunger that is used to purée food by pushing it through a perforated metal plate.

rice vinegar: Vinegar made from rice wine.

rich dough: A yeast dough that incorporates a lot of fat, sugar, and eggs into a heavy, soft structure.

ricotta: A creamy fresh cheese that looks similar to cottage cheese but is made from the whey of other cheeses instead of primarily from milk.

rind: An exterior layer of a food.

rind-ripened cheese: *See* soft cheese.

rivet: A metal fastener used to attach the tang of a knife to the handle.

roaster: A young male or female chicken from 3–5 months old that weighs 3½–5 lb.

roaster duckling: A duck that is less than 16 weeks old and weighs between 4–6 lb.

roasting: A dry-heat cooking method in which food is cooked uncovered at a high temperature in an oven or on a revolving spit over an open flame.

roasting pan: A rectangular pan with 4–5 inch sides.

rock lobster: *See* spiny lobster.

rolled cut: *See* oblique cut.

rolled grain: *See* flaked grain.

rolled-in dough: A yeast dough with a flaky texture that results from the incorporation of fat through a rolling and folding procedure.

rolled oats: Oats that have been steamed and flattened into small flakes. Also known as old-fashioned oats.

rolled omelet: An omelet that is cooked and then rolled onto a plate and cooked filling ingredients are added through a slit cut into the top. Also known as a French omelet.

rolling pin: A slim cylinder that is used to flatten pastry dough, bread crumbs, or other foods.

roll-in unit: An individual refrigeration unit that allows speed racks to be rolled in and out of the unit through a door opening that is just above floor height.

Romano: A grating cheese that is similar to Parmesan but softer in texture.

rondeau: A wide, shallow-walled, round pot that is used for braising, stewing, and searing meats. Also known as a braiser.

rondelle cut: A slicing cut that produces disks. Also known as a round cut.

root spice: A type of spice derived from the root portion of a plant that grows underground.

Roquefort: A blue-veined cheese made from sheep milk and is characterized by a sharp, tangy flavor.

rosemary: An herb with needlelike leaves and is a member of the evergreen family.

rotary grater: A sharp, perforated stainless steel cylinder attached to a handle that is used to shave hard cheeses, such as Parmesan, tableside.

rotating rack oven: A large oven that rotates 10–80 pans of food as it cooks. Also known as a reel oven.

rotisserie: A sideways broiler.

round cut: *See* rondelle cut.

roundfish: Any fish with a cylindrical body, an eye located on each side of the head, and a backbone that runs from head to tail in the center of the body.

rounding: The process of shaping scaled dough into smooth balls.

roux: A thickening agent made by cooking a mixture of equal amounts, by weight, of flour and fat and is used to thicken sauces and soups.

rub: A blend of ingredients that is pressed onto the surface of uncooked foods such as meat, poultry, and fish to impart flavor and sometimes to tenderize.

russet potato: A mealy potato with thin brown skin, an elongated shape, and shallow eyes.

Russian service: A style of meal service in which a server holds a tray of food and serves guests food from the tray.

rutabaga: A round root vegetable derived from a cross between a Savoy cabbage and a turnip.

S

sachet: *See* sachet d'épices.

sachet d'épices: A mixture of spices and herbs placed in a piece of cheesecloth and tied with butcher's twine. Also known as a sachet.

safety data sheet (SDS): A document that provides detailed information describing a chemical, instructions for its safe use, the potential hazards, and appropriate first-aid measures.

saffron: A spice made from the dried, bright-red stigmas of the purple crocus flower.

sage: An herb with narrow, velvety, green-gray leaves and is a member of the mint family.

salad dressing: 1. A cooked, mayonnaise-like product usually made from distilled vinegar, vegetable oil, water, sugar, mustard, salt, modified corn flour, and emulsifiers. Also known as a cooked dressing. **2.** A sauce for a salad.

salad green: An edible leaf used in raw salads or as a garnish.

salamander: A small overhead broiler that is usually attached to an open burner range.

salmon: A fatty, anadromous saltwater fish found in both the northern Atlantic Ocean and Pacific Ocean.

salsify: A white or black root vegetable, similar in shape to a carrot, and can grow up to 12 inches in length.

salt: A crystalline solid composed mainly of sodium chloride and is used as a seasoning and a preservative.

sandwich base: The edible packaging that holds the contents of a sandwich.

sandwich garnish: A complementary food item that is served on or with a sandwich.

sandwich spread: A slightly moist, flavorful substance that seals the pores of the bread and creates a thin moisture barrier.

sanitation area: A location in the professional kitchen where sanitation equipment is kept.

sanitizing: The process of destroying or reducing harmful microorganisms to a safe level.

Santa Claus melon: A large, mottled yellow and green variety of muskmelon that has a slightly waxy skin and soft stem end when ripe.

santoku knife: A knife with a razor-sharp edge and a heel that is perpendicular to the spine.

satsuma: A small, seedless variety of mandarin.

saturated fat: A lipid that is solid at room temperature.

sauce: An accompaniment that is served with a food to complete or enhance the flavor and/or moistness of a dish.

saucepan: A small, slightly shallow skillet with straight or slightly sloped sides.

saucepot: A small stockpot.

sausage: A forcemeat that is typically stuffed into a casing or shaped into a patty.

sausage stuffer: A manual or electric piece of equipment that uses a piston to pump forcemeat through a nozzle into casings.

sautéing: A dry-heat cooking method in which food is cooked quickly in a sauté pan over direct heat using a small amount of fat.

sauté pan: A round, shallow-walled pan with a long handle that is used to sauté foods. Also known as a skillet.

sauteuse: A sauté pan with sloped sides.

sautoir: A sauté pan with straight sides.

savory: An herb with smooth, slightly narrow leaves and is a member of the mint family.

savoy cabbage: A conical-shaped head of tender, crinkly, edible leaves that are blue-green on the exterior and pale green on the interior.

scaling: The process of calculating new amounts for each ingredient in a recipe when the total amount of food the recipe makes is changed.

scaling factor: The number that each ingredient amount in a recipe is multiplied by when the recipe yield is changed.

scallion: A bulb vegetable with a slightly swollen base and long, slender, green leaves that are hollow.

scallop: A bivalve with a fan-shaped shell and a cream-colored adductor muscle with a sweet, delicate flavor.

scallopini: A small, ¼ inch thick slice of veal (generally leg meat) that is 2–3 inches in diameter.

scattered sushi: *See* chirashizushi.

scimitar: A long knife with an upward curved tip that is used to cut steaks and primal cuts of meat.

scoring: The process of making shallow, angled cuts across the top of unbaked bread with a sharp knife called a lame.

searing: The process of using high heat to quickly brown the surface of a food.

sea salt: Salt produced through the evaporation of seawater.

sea scallop: A large scallop with a coarse texture that is harvested from deep saltwater.

seasoning: An ingredient used to intensify or improve the natural flavor of foods.

sea vegetables: Edible saltwater plants that contain high amounts of dietary fiber, vitamins, and minerals.

Seckel pear: A small pear that is sometimes called a honey pear or sugar pear because of its syrupy, fine-grained flesh and complex sweetness.

security cage: A lockable, wire-cage storage unit on wheels used to hold expensive items such as fine china or restricted items such as alcohol.

seed: The fruit or unripened ovule of a nonwoody plant.

seed spice: A type of spice derived from the seed portion of a plant.

semi-à la carte menu: A menu that offers entrées along with additional menu items for a set price.

semisoft cheese: A cheese that is firmer than a soft cheese but not as hard as a hard cheese.

semolina: The granular product that results from milling the endosperm of durum wheat.

sensory perception: The ability of the senses to gather information and evaluate the environment.

separate course salad: A salad served as a course of its own, typically following the main course.

server: The person responsible for taking the orders of guests and bringing food and beverages to those guests.

sesame seed: A small, flat, white or black seed found inside the pod on a sesame plant and is used as a spice.

sesame oil: A type of oil produced from sesame seeds.

shallot: A very small bulb vegetable that is similar in shape to garlic and has two or three cloves inside.

shaped pasta: A pasta that has been extruded into a complex shape such as a corkscrew, bowtie, shell, flower, or star.

shark: A lean, cartilaginous fish that is generally found in saltwater.

sheet pan: A flat pan with very low sides.

shellfish: The classification of aquatic invertebrates that may or may not have a hard, external shell.

shell membrane: A thin, skinlike material located directly under the eggshell.

sherbet: A frozen dessert made by combining puréed fresh fruit, simple syrup, and milk.

sherry vinegar: Vinegar made from sherry wine.

shirred egg: An egg that is baked on top of other ingredients in a shallow dish in the oven. Also known as a baked egg.

shiitake mushroom: An amber, tan, brown, or dark-brown mushroom with an umbrella shape and curled edges. Also known as a forest mushroom.

short-flake crust: A piecrust that absorbs slightly more liquid than mealy piecrust.

short-grain rice: Rice that is almost round in shape and has moist grains that stick together when cooked.

shrinkage: The loss of volume and weight of a piece of food as the food cooks.

shucking: The process of opening a bivalve.

side of beef: A half of a carcass split along the backbone.

side salad: A small salad served to accompany a main course.

sidework: The process of cleaning, restocking, and preparing the items needed to keep service running smoothly.

sieve: A fine-mesh sifter used to sift, aerate, and remove lumps or impurities from dry ingredients.

sifter: A cylindrical metal sieve that is hand-cranked and used to aerate and remove lumps from dry ingredients such as flour.

silicone mat: A woven, nonstick mat that may be used in the refrigerator, freezer, or oven and can withstand temperatures between −40°F and 580°F.

silverskin: A tough, rubbery, silver-white connective tissue that does not break down when heated.

simmering: A moist-heat cooking method in which food is gently cooked in a liquid that is between 185°F and 205°F.

simple carbohydrate: A carbohydrate composed of one or two sugar units and is quickly absorbed by the body. Also known as a sugar.

siphon: A tubular organ that is used to draw in or eject fluids.

skate: A lean, cartilaginous fish with two winglike sides found in saltwater. Also known as a ray.

skillet: *See* sauté pan.

skimmer: A flat, stainless steel perforated disk connected to a long handle and is used to skim impurities from soups, stocks, and sauces.

sleeper: A lobster that is dying.

slicer: **1.** A knife with a narrow blade 10–14 inches long that is used to slice roasted meats. Also known as a carving knife. **2.** An appliance that is used to uniformly slice foods such as meat and cheese.

slurry: A mixture of equal parts of cool liquid and a starch that is used to thicken other liquids.

smelt: A lean roundfish with a slender body, pointed head, and deeply forked tail.

smoked sausage: A type of cured sausage that is smoked during the curing process.

smoke point: The temperature at which an oil begins to smoke and give off an odor.

smoker oven: A gas or electric oven that generates wood smoke and is most often used to smoke or barbeque meats and poultry.

smoking: A dry-heat cooking method in which food is flavored, cooked, or preserved by exposing it to the smoke from burning or smoldering plant materials, most often wood.

snow crab: Similar to a king crab but is smaller and available in greater supply.

snow puff mushroom: *See* enokitake mushroom.

sodium bicarbonate: *See* baking soda.

soft cheese: A cheese that has been sprayed with a harmless live mold to produce a thin skin or rind. Also known as a rind-ripened cheese.

soft-shell clam: An Atlantic clam with a thin, brittle shell that breaks easily. Also known as a long-neck clam or steamer clam.

soft-shell crab: A blue crab that has been harvested within 6 hours of molting, or shedding its shell in order to grow a larger shell.

sole: A lean, saltwater flatfish with pale-brown skin on the top side.

solid pack: A canned product with little or no water added.

soluble fiber: Dietary fiber that dissolves in water.

sommelier: The person responsible for all aspects of wine service as well as wine and food pairings. Also known as a wine steward.

sorbet: A frozen dessert made by combining puréed fresh fruit and simple syrup.

sorrel: A large, green, edible leaf that ranges in color from pale green to dark green and from 2–12 inches in length.

soufflé omelet: A lighter variation of a folded omelet. Also known as a fluffy omelet.

sous chef: The person responsible for carrying out objectives, as determined by the chef, regarding all aspects of kitchen operations.

sous vide: The process of cooking vacuum-sealed food by maintaining a low temperature and warming food gradually to a set temperature.

soy nut: A legume from the soybean pod that has been soaked in water, drained, and then roasted or baked.

soy sauce: A type of Asian sauce made from mashed soybeans, wheat, salt, and water.

spaghetti squash: A dark-yellow winter squash with pale-yellow flesh that can be separated into spaghetti-like strands after it is cooked.

spatula: A scraping tool consisting of a rubber or silicone blade attached to a long handle that is used to mix foods and to scrape food from bowls, pots, and pans.

speed rack: A tall cart on wheels with rails that hold entire sheet pans of food. Also known as a tallboy.

spice: A flavoring derived from the bark, seeds, roots, flowers, berries, or beans of aromatic plants.

spider: A skimmer with an open-wire design that makes it perfect for removing hot foods from a fryer.

spinach: A dark-green, edible leaf with a slightly bitter flavor that may have flat or curly leaves, depending on the variety.

spine: The unsharpened top part of a knife blade that is opposite the edge.

spiny lobster: A lobster that has spines covering its body and five slender legs on each side. Also known as a rock lobster.

spit-roasting: The process of cooking meat by skewering it and suspending and rotating it above or next to a heat source.

sponge method: A method of mixing yeast dough in two steps.

spoodle: A solid or perforated flat-bottomed ladle.

springform pan: A round pan with a metal clamp on the side that allows the bottom of the pan to be separated from the sides.

sprout: An edible strand with an attached bud that comes from a germinated bean or seed.

squab: A young pigeon that is less than four weeks old and weighs approximately 1 lb or less.

squash blossom: The edible flower of a summer or a winter squash.

squid: A translucent, head-footed cephalopod that has two tentacles, eight sucker-equipped arms, two lateral fins, and a flat, internal cuttlebone.

standardized recipe: A list of ingredients, ingredient amounts, and procedural steps for preparing a specific quantity of a food item.

standard profit and loss statement: A form that shows the revenue, expenses, and resulting gross and net profit (or loss) over a specific period of time.

starch: *See* complex carbohydrate.

star fruit: An exotic fruit that is shaped like a star when cut perpendicular to the stem. Also known as carambola.

station chef: The person responsible for overseeing a specific production area of the kitchen.

steamer clam: *See* soft-shell clam.

steamer insert: A round stainless steel vessel with a perforate liner.

steaming: A moist-heat cooking method in which food is placed in a container that prevents steam from escaping.

steam injection: The process of adding water directly in the hot cavity of an oven so that steam is created.

steam-jacketed kettle: A large cooking kettle that has a hollow lining into which steam is pumped and a spigot at the bottom for easy draining. Also known as a steam kettle.

steam kettle: *See* steam-jacketed kettle.

steamship round roast: The beef round with the shank and rump removed.

steam table: An open-top table with heated wells that are filled with water.

steel: A steel rod approximately 18 inches long attached to a handle and is used to align the edge of knife blades. Also known as a butcher's steel.

steel-cut oats: Oat groats that have been toasted and cut into small pieces.

steer: A male calf that has been castrated prior to reaching sexual maturity.

stem herb: A type of herb that comes from the stem portion of a plant.

stewer: A female chicken that has laid eggs for one or more seasons, is usually more than 10 months old, and weighs from 3–8 lb. Also known as a stewing hen.

stewing: A combination cooking method in which bite-sized pieces of food are barely covered with a liquid and simmered for a long period of time in a tightly covered pot.

stewing hen: *See* stewer.

stick mixer: *See* immersion blender.

Stilton: A blue-veined cheese made from a cow milk with a flavor that is milder than Roquefort or Gorgonzola.

stir-frying: The process of quickly cooking items in a heated wok with a very small amount of fat while constantly stirring the items.

stock: An unthickened liquid that is flavored by simmering seasonings with vegetables, and often, the bones of meat, poultry, or fish.

stockpot: A large, round, high-walled pot that is taller than it is wide.

stone crab: An Atlantic crab with a brownish-red shell and large claws of unequal size.

stone fruit: *See* drupe.

straight-dough method: A method of mixing yeast dough that combines yeast with warm water and then adds it to the rest of the ingredients, mixing them together all at once.

straight forcemeat: A forcemeat that consists of seasoned ground meat emulsified with ground fat.

straightneck squash: A yellow squash that resembles a bowling pin.

strainer: A bowl-shaped woven mesh screen, often with a handle, that is used to strain or drain foods.

strawberry: A bright-red, heart-shaped berry covered with tiny black seeds.

stuffed pasta: A pasta that has been formed by hand or machine to hold fillings.

sturgeon: A fatty roundfish with a wedge-shaped snout and a body style similar to that of a shark.

suckling pig: A pig 4–6 weeks old that weighs 20–35 lb dressed.

sugar: *See* simple carbohydrate.

summer squash: A fruit-vegetable that grows on a vine and has edible skin, flesh, and seeds.

sunchoke: A tuber with thin, brown, knobby-looking skin. Also known as a Jerusalem artichoke.

sunflower seed: A tan seed that comes from the sunflower plant.

sunny-side up egg: A lightly fried egg with an unbroken yolk that is not flipped over to cook the other side.

supreme: The flesh from a segment of a citrus fruit that has been cut away from the membrane.

suprême sauce: A sauce made by adding cream to a chicken velouté.

surf clam: A large species of Atlantic clam that can grow to 8 inches in size.

surimi: A fish product made from a mixture of fish and/or shellfish and other ingredients.

sushi: A vinegar-seasoned rice dish garnished with raw fish, cooked seafood, eggs, or vegetables.

sweating: The process of slowly cooking food to soften its texture.

sweetbreads: The thymus glands of a calf, located in the neck.

sweet corn: A fruit-vegetable that has edible seeds called kernels that grow in rows on a spongy cob encased by thin leaves (husks), forming what is referred to as an ear of corn.

sweet potato: A tuber that grows on a vine and has a paper-thin skin and flesh that ranges in color from ivory to dark orange.

Swiss cheese: The name used to describe varieties of hard cheeses with large holes, also known as eyes.

swordfish: A very large, fatty roundfish with an upper jaw that extends to form a long and flat, sharp-edged sword.

synthetic fibrous casing: A casing produced from a plastic-like synthetic material.

Szechwan pepper: A spice made from the dried berries of an ash tree.

T

Tabasco® sauce: A type of pepper sauce made from Tabasco peppers, vinegar, and salt.

table d'hôte menu: A menu that identifies specific items that will be served for each course at a set price.

tallboy: *See* speed rack.

tamarind: A spice made from seeds found in long pods that grow on the tamarind tree.

tamis: A flat, round sieve with a wood or aluminum frame and a mesh screen bottom. Also known as a drum sieve.

tang: The unsharpened tail of a knife blade that extends into the handle.

tangerine: A small citrus fruit with a slightly red-orange peel that can be easily removed without a knife.

tangelo: A hybrid of a tangerine and a grapefruit.

tangor: A hybrid of a tangerine and a sweet orange.

tapioca: A thickening agent derived from the cassava plant.

target food cost percentage: The percentage of food sales that a foodservice operation plans to spend on purchasing food.

target price: The price that a foodservice operation needs to charge for a menu item in order to meet its target food cost percentage.

tarragon: An herb with smooth, slightly elongated leaves; best known as the flavoring in béarnaise sauce.

tartare: Freshly ground or chopped raw meat or seafood.

tartlet: A miniature, round pastry shell that contains a savory or sweet filling.

tart pan: A round, shallow baking pan with sloped sides that are smooth or fluted that may have a removable bottom.

taste bud: A cluster of cells that can detect flavor characteristics.

tatsoi: A thick, spoon-shaped, emerald-colored leaf vegetable native to Japan.

TCM: *See* curing salt.

tea: An herb with jagged leathery leaves that comes from an evergreen bush in the magnolia family.

tea sandwich: A petite and delicate cold sandwich with a trimmed crust and a soft filling.

temperature danger zone: A range of temperature, between 41°F and 135°F, in which bacteria thrive.

tempura: A very light batter that is used on vegetables, poultry, seafood, and meats served in Asian cuisine.

tender: A small strip of a breast.

tenderloin: The inner pectoral muscle that runs alongside the breastbone.

terrine: A forcemeat that is baked in a mold without a crust.

thermal immersion circulator: A device that is placed in a water bath to keep a uniform temperature for sous vide cooking.

thickening agent: A substance that adds body to a hot liquid.

thick soup: A soup having a thick texture and consistency.

thigh: The upper section of the leg located below the hip and above the knee joint.

Thompson grape: A seedless grape that is pale to light green in color.

thyme: An herb with very small gray-green leaves that is a member of the mint family.

tilapia: A lean, freshwater roundfish with firm white flesh.

tilt skillet: A versatile piece of cooking equipment with a large-capacity pan, a thermostat, a tilting mechanism, and a cover. Also known as an electric braiser.

timer: A measuring tool that indicates the amount of time that has passed or sounds an alarm when a specified time period has ended.

tip: The front quarter of a knife blade.

tomatillo: A small tomato with a thin, papery husk covering a pale-green skin that encases a pale-green flesh. Also known as a Mexican husk tomato.

tomato: A juicy fruit-vegetable that contains edible seeds.

tomato sauce: A mother sauce made by sautéing mirepoix and tomatoes, adding white stock, and thickening with a roux.

tongs: A spring-type, long metal tool used to pick up foods while retaining their shape.

tossed salad: A mixture of leafy greens, such as lettuce, spinach, chicory, or fresh herbs, and other ingredients, such as fruits, vegetables, nuts, cheese, meats, and croutons, served with a dressing.

tourné cut: A carved, football-shaped cut with seven sides and flat ends.

tourné knife: A short knife with a curved blade that is primarily used to carve vegetables into a specific shape called a tourné, which is a seven-sided football shape with flat ends. Also known as a bird's beak knife.

tree ear mushroom: *See* wood ear mushroom.

tripe: The muscular inner lining of a stomach of an animal, such as cattle or sheep.

tropical fruit: A type of fruit that comes from a hot, humid location but is readily available.

trout: A fatty, freshwater roundfish with tender flesh.

truffle: An edible fungus with a distinct taste.

truing: *See* honing.

trunnion kettle: A small steam-jacketed kettle that is tilted by pulling a lever or turning a wheel to empty the kettle.

trussing: The process of tying the legs and wings of a bird tightly to the body to keep a compact shape.

tube pasta: A pasta that has been pushed through an extruder and then fed through a cutter that cuts the tubes to the desired length.

tuna: A very large, fatty, saltwater roundfish that is a member of the mackerel family.

turban squash: A turban-shaped winter squash.

turmeric: A spice made from the root of a lily-like plant in the ginger family.

turnip: A round, fleshy root vegetable that is purple and white in color.

turnip green: A dark-green, edible leaf that grows out of the top of the turnip root vegetable.

U

ugli fruit: A large, teardrop-shaped, seedless citrus fruit. Also known as uniq fruit.

uniq fruit: *See* ugli fruit.

unit cost: The cost of a product per unit of measure.

unit of measure: A fixed quantity that is widely accepted as a standard of measurement.

univalve: A mollusk that has a single solid shell and a single foot.

unsaturated fat: A lipid that is liquid at room temperature.

unsweetened chocolate: Pure chocolate liquor that contains 50% to 58% cocoa butter.

utility knife: A multipurpose knife with a stiff 6–10 inch blade that is similar in shape to a chef's knife but much narrower at the heel.

V

vanilla bean: The dark-brown pod of a tropical orchid.

variety: A fruit that is the result of breeding two or more fruits of the same species that have different characteristics.

variable cost: A cost that increases or decreases in proportion to the volume of production.

variety spread: Any other food mixture of a spreadable consistency that can be added to a sandwich to complement or increase flavor and moisture.

veal: The flesh of calves, which are young cattle.

veal breast: A thin, flat cut of meat located under the shoulder and ribs of a calf that contains the breastbone, tips of the rib bones, and cartilage.

veal cutlet: A thin slice of veal.

veal foresaddle: The front half of a calf carcass consisting of the primal shoulder, rack, breast, and shank cuts.

veal foreshank: The upper portion of the front leg of a calf.

veal hindsaddle: The rear half of a calf carcass consisting of the loin and leg.

veal loin: A primal cut located between the primal rack and leg of a calf that includes the 12th and 13th rib, the loin eye muscle, the center section of the tenderloin, the strip loin, and flank meat.

veal rack: A primal cut of calf located between the shoulder and loin that contains seven rib bones.

veal shoulder: A primal cut that contains the first four rib bones, some of the backbone, and a small amount of the arm and blade bones.

vegetable: An edible root, bulb, tuber, stem, leaf, flower, or seed of a nonwoody plant.

vegetable salad: A type of salad that is primarily made of vegetables.

vegetable stock: A clear, light-colored stock produced by gently simmering vegetables with a white mirepoix and a sachet.

velouté: A mother sauce made from a flavorful white stock (veal, chicken, or fish stock) and a blonde roux.

ventilation system: A large exhaust system that draws heat, smoke, and fumes out of the kitchen and into the outside air.

vertical cutter/mixer (VCM): An appliance used to cut and mix foods simultaneously using high-speed blades and a mixing baffle, which is used to manually move the product into the blades.

vinegar: A sour, acidic liquid made from fermented alcohol.

virgin olive oil: A type of olive oil produced from the first pressing of the olives without the use of heat or chemicals that has an acid content of as much as 3%.

virus: A pathogen that grows inside the cells of a host.

vitamin: A nutrient composed of organic substances that is required in small amounts to help regulate body processes.

volume: A measurement of the physical space a substance occupies.

volume measure: A large, graduated aluminum container with a pouring lip and a loop handle and is used to measure larger volumes of liquid ingredients.

W

wakame: A long, tender, grayish-green sea vegetable that expands to seven times its size when soaked in water.

walk-in unit: A room-size insulated storage unit used to store bulk quantities of food.

walleye: A lean fish and a variety of perch that resembles a pike.

walnut: A nut from the fruit of the walnut tree.

walnut oil: A type of oil produced from walnuts.

warewashing: The process of cleaning and sanitizing all items used to prepare and serve food.

wasabi: A spice made from the light-green root of an Asian plant.

wash: A liquid that is brushed on the surface of a yeast dough product prior to baking.

washed-rind cheese: A semisoft cheese with an exterior rind that is washed with a brine, wine, olive oil, nut oil, or fruit juice.

water caltrop: *See* water chestnut.

water chestnut: A small tuber with brownish-black skin and white flesh. Also known as a water caltrop.

watercress: A small, crisp, dark-green, edible leaf that is a member of the mustard family.

watermelon: A sweet, extremely juicy melon that is round or oblong in shape, with pink, red, or golden flesh and green skin.

water-soluble vitamin: A vitamin that dissolves in water.

waxed-rind cheese: A semisoft cheese produced by dipping a wheel of freshly made cheese into a liquid wax and allowing the wax to harden.

waxy potato: A type of potato with a thin skin and slightly waxy flesh that is lower in starch and higher in moisture than mealy potatoes.

weight: A measurement of the heaviness of a substance.

wet aging: The process of aging meat in vacuum-sealed plastic.

wet curing: A curing method in which foods are submerged in a brine. Also known as immersion curing.

wet rub: A mixture of wet ingredients and a dry rub that is rubbed onto the surface of uncooked food to impart flavor. Also known as a paste.

whetstone: A stone used to grind the edge of a blade to the proper angle for sharpness.

whey: The watery part of milk.

whisk: A mixing tool made of stainless steel or silicone wires bent into loops and attached to a stainless steel handle.

white: *See* albumen.

white butter sauce: *See* beurre blanc.

white chocolate: A confectionary product made from cocoa butter, sugar, milk solids, and flavorings.

white mirepoix: A mirepoix made of onions, celery, and leeks or parsnips instead of carrots.

white mushroom: *See* button mushroom.

white potato: An oblong mealy potato with a thin, white or light-brown skin and tender, white flesh.

white sauce: *See* béchamel.

white stock: A light-colored stock produced by gently simmering poultry, veal, or fish bones in water with vegetables and herbs.

white vinegar: *See* distilled vinegar.

whole fish: The market form of a fish that is taken from the water and sold as is.

whole grain: A grain that only has the husk removed.

wild boar: A game animal that is similar in bone structure and muscle composition to domesticated hogs.

wild marjoram: *See* oregano.

wine steward: *See* sommelier.

wine vinegar: A vinegar made from red or white wine.

wing: A poultry wing that consists of a tip, paddle, and drummette.

wing flat: *See* wing paddle.

wing paddle: The second section of a wing located between the two wing joints. Also known as a wing flat.

wing tip: The outermost section of a wing.

winter squash: A fruit-vegetable that grows on a vine and has a thick, hard, inedible skin and firm flesh surrounding a cavity filled with seeds.

wire shelving: Shelving made of wire; primarily used to store boxed items.

wok: A round-bottom pan that is used to stir-fry, steam, braise, stew, or deep-fry foods.

wood ear mushroom: A brownish-black, ear-shaped mushroom that has a slightly crunchy texture. Also known as a cloud ear or tree ear mushroom.

wood-fired oven: An oven that is encased with masonry and heated with wood.

Worcestershire sauce: A type of sauce traditionally made with anchovies, garlic, onions, lime, molasses, tamarind, and vinegar.

work section: An area where members of the kitchen staff are all working toward the same goal at the same time.

work station: An area within a work section where specific tasks are performed by specific people.

Y

yam: A large tuber that has thick, barklike skin and flesh that varies in color from ivory to purple.

yearling turkey: A mature turkey that is less than 15 months old and weighs 10–30 lb.

yeast: A microscopic, living, single-celled fungus that releases carbon dioxide and alcohol through a process called fermentation when provided with food (sugar) in a warm, moist environment.

yellow potato: An oval, waxy potato with thin, yellowish skin and flesh and pink eyes.

yield: The total quantity of a food or beverage item that is made from a standardized recipe.

yield percentage: The edible-portion (EP) quantity of a food item divided by the as-purchased (AP) quantity and expressed as a percentage.

yolk: The yellow portion of an egg.

yogurt: A tangy, custard-like cultured dairy product produced by adding a safe bacteria and an acid to whole, low-fat, or fat-free milk.

young goose: A goose that is usually less than six months of age and weighs approximately 4–10 lb.

young guinea: A guinea fowl that has tender flesh, a flexible breastbone, and weighs between 1–1½ lb.

young turkey: A male or female turkey that is less than eight months old and is 8–22 lb.

Z

zest: The colored, outermost layer of the peel of a citrus fruit; contains a high concentration of oil.

zester: A special cutting tool with tiny blades inside of five or six sharpened holes that are attached to a handle.

zucchini: An elongated squash that resembles a cucumber and is available in green or yellow varieties.

Index

Page numbers in italic refer to figures.

headcheese, 883
health, personal, 18
hearts of palm, *524,* 524
heat sanitizing, 47
heels, knife, *66,* 66
herbal beverages, 452–453
herb leaves, *634,* 634
herbs. *See also individual herbs*
    defined, *259,* 259
    leaf herbs, *260–265,* 260–265
    and spice blends, 277, *278*
    stem herbs, *265,* 265–266
herring, 444, 776
hocks, pork, *883,* 883
Hoisin sauce, 284
hokkien noodles, *608,* 608
holding cabinets, 124
holding foods, *53,* 53, *57*
hollandaise sauces, 319–321
    Béarnaise Sauce, 350
    Choron Sauce, 350
    Grimrod Sauce, 350
    Hollandaise Sauce, 349
    Maltaise Sauce, 350
    Mousseline Sauce, 351
    Noissette Sauce, 351
    preparing, 320
    Valois Sauce, 351
home fries, *445,* 445
hominy, 586, *587*
honey, *936,* 936
honeycomb tripe, 842
honeydew melons, *480,* 480
honing knives, *77,* 77
hors d'oeuvres, *665,* 665. *See also* starters
    Ahi Tuna Nacho, 709
    Artichoke and Goat Cheese Spread
        Crostini, 719
    Bacon-Wrapped Blue Cheese Stuffed
        Dates with Balsamic Glaze, 715
    Cucumber and Smoked Salmon Canapés,
        711
    Scallop Rumaki, 715
    Seafood Canapés, 710
    Tomato and Mozzarella Canapés with
        Basil Pesto, 710
horseradish, *272,* 272
hosts, 13
hotel pans, 108–109, *109*
hotel racks, 906, 913–914
hot foods sections, 129
hot peppers, 543–544, *544*
hot sandwiches, *396,* 396–399, *397, 398, 399*
    Bistro Burgers, 414
    Philly Beef Sandwiches, 413
    Reuben Sandwiches, 414
hot storage equipment, *123,* 123–125,
    124, 125

Hubbard squash, *546,* 546
husks, 215, *583,* 583
hydrogenation, 208
hygiene, 18, 43, *44*

I

ice creams, 984–986, *985*
icings. *See individual types of icings for
    recipes*
    buttercream, *962,* 962
    Cream Cheese Icing, 1004
    defined, *961,* 961
    flat, 961
    foam, 964
    fondant, 962–963, *963*
    fudge, 964
    ganache, *964,* 964
    glaze, 965
    pastry bags, 965–966, *966*
    piping, *966,* 966–967, *967*
    royal, 962
    whipped cream, 965
illnesses, foodborne, 37, *38*
immersion blenders, *134,* 134
impinger conveyor ovens, 142, *143*
inarizushi, 679
incomplete proteins, *206,* 206
individually quick-frozen (IQF), 788
induction cooktops, 137
induction radiation, *229,* 229
infrared ovens, 141
infrared radiation, *229,* 229
infrared thermometers, *112,* 113, *116,* 116
infused oils, 294
insoluble fiber, 207
inspection tables, *118,* 118
instant coffees, 451
instant-read thermometers, 111, *112*
instant yeast, 942
institutional foodservice operations, *6,* 6
insulated carriers, *125,* 125
internal customers, 5
intestines, pork, 883
invoices, *162,* 162
irradiation, 497, 553

J

jackfruit, *495,* 496
jicamas, *516,* 516
job applications, *30,* 30
job interviews, 30–31, *31*
job searches, 29–30, *30*

jowls, pork, 883
juice extractors, 134
juicers, *134,* 134
juices, *453,* 453
julienne cuts, 82, 84
juniper berries, *275,* 275
jus lié, 317–318
Jus Lié, 337

K

kabocha squash, *546,* 546
kalbi, 862
kale, *529,* 529–530, *634,* 634
    Sesame Kale with Sautéed Shiitake
        Mushrooms, 567
kangaroo, 922
kasha, 590
ketchup, 281
key limes, 483
kielbasa, 441
kippers, 444
kitchen brigade system, 13, *14*
kitchen forks, *103,* 103
kitchen shears, 95
kitchen spoons, *100,* 100
kiwanos, *495,* 495
kiwifruit, *487,* 487
kneading dough, 948
knife cuts, 78–90
    batonnet, 82, 84
    brunoise, 84, 86
    chiffonade, 80–81
    chopping, 88
    diagonal, 79
    dice, 84–86
    fine brunoise, 84, 86
    fine julienne, 82, 84
    fluted, 89
    julienne, 82, 84
    mincing, 87
    oblique, 80
    paysanne, *86,* 86
    rondelle, 79
    slicing, 78–81
    stick, 81–83
    tourné, 89
knives. *See also* knife cuts
    gripping, 74–75
    honing, *77,* 77
    parts of, *66,* 66–69
    safety with, *73,* 73–78, *74, 76*
    sharpening, 75–76, *76*
    types of, 69–72, *70, 71*
kohlrabi, *524,* 524

pistachios, *285*, 286
pith, 482
pitting cherries, *478*, 478
pizzas, 396, *397*
  Roasted Tomato, Proscuitto, and Garlic
    Pizza, 413
plantains, *486*, 486
plant-based meals, *204*, 204
plated service, *9*, 9
platform scales, 116, *117*
plating breakfast, 454
plating sandwiches, *409*, 409–410, *410*
plums, *478*, 478
plum tomatoes, 539
poached eggs, 436–437
poaching, *245*, 246–247
  fish, 794–795
  fruit, 500
  poultry, 747
poêléing, 254, *748*, 748
polenta, 587
  Creamy Polenta, 618
  Grilled Sun-Dried Tomato Polenta, 618
pomegranates, 489, *490*
pomes, *471*, 471–475, *472*, *474*, *475*
poppy seeds, *271*, 271
porcini mushrooms, *552*, 552
pork
  Apple and Walnut Stuffed Pork Chops,
    898
  carcasses, 877
  Chipotle Braised Pork with Spicy Roja
    Sauce, 901
  cooking, *891*, 891–894
  curing, 888–890, *889*
  defined, *876*, 876
  Dijon Lemon Pepper Medallions, 900
  doneness, *891*, 891
  fabricated cuts. *See* pork: primal cuts
  fabrication, 885–887
  Five Spice Pork Tenderloin with Dark
    Cherry and Soy Demi-Glace, 897
  grades of, 885
  IMPS for, 883, *884*
  marinades for, 890
  offals, *883*, 883
  Pork Roast with Mushroom Sauce, 898
  Pork Teriyaki Lettuce Wraps, 900
  primal cuts, 877–882, *878–882*
    legs (ham), *881*, 881
    loins, 879–880, *880*
    picnic shoulders, *878*, 878
    shoulder butts, 878–879, *879*
    bellies, 881–882, *882*
  rubs, 890, *891*
  spareribs, *882*, 882
  storing, *885*, 885
  tenderloins, 879, *880*

Tex-Mex Pork Steaks, 901
Thai Red Curry Paste, 899
Thai Red Curry Pork with Basil, 899
USDA inspections, *885*, 885
USDA purchase specifications, 883, *884*
pork belly, *882*, 882
pork chops, breakfast, 444
porridge, 448
porters, 15
portfolios, 29
portion control, *181*, 181
portion control scoops, 96, *97*
portion scales, 116, *117*
portion sizes
  and nutrition, *222*, 222–223, *223*
  scaling based on, 158
  in standardized recipes, *149*, 149
portobello mushrooms, *549*, 549
Port Salut, *656*, 657
POS systems, *25*, 25
potatoes. *See also individual types of
    potatoes*
  baking, *576*, 576–578, *577*
    Rissole (Oven-Browned) Potatoes, 611
    Scalloped Potatoes, 612
    Twice-Baked Potatoes, 613
  breakfast, *445*, 445
  cooking, *574*, 575
  defined, *520*, 520, *570*, 570
  doneness, 574
  frying, *579*, 579–580
    American Fries (Home Fries), 613
    French-Fried Potatoes, 614
  grilling, *575*, 575
  market forms, 572, *573*
  mealy, *570*, 570–571
  new, *572*, 572
  receiving, *573*, 573–574
  roasting, *576*, 576
  sautéing, *579*
    Potato Pancakes, 611
    Savory Breakfast Potatoes, 465
  simmering, 580–582, *581*
    Duchess Potatoes, 612
  storing, 574
  sweet, *572*, 572
    Candied Sweet Potatoes, 614
  waxy, *571*, 571
  yams, *572*, 572
potato pancakes, 438
potentially hazardous foods, *50*, 50
pots, *105*, 105–106
poultry
  barding, 745
  cooking, 746–753
  cuts of, *734*, 737
  defined, 728
  doneness, *747*, 747

fabricated cuts, 733–735, *734*
fabricating, 736–744, *737*
  boning, 742–744
  cutting breasts, 741
  cutting into parts, 738–740
  fat, *733*, 733
  for forcemeat, 682
  giblets, *735*, 735
  grades of, *735*, 735
  marinades for, *745*, 745
  market forms of, 732–735
  for sandwiches, *392*, 392
  storing, *736*, 736
  stuffings, *745*, 745
  TTJJ method, *747*, 747
  USDA inspections, *735*, 735
  whole, 733, *734*
poussins, 728
prawns, *796*, 796
preparation areas and equipment, *111*, 128
  baking equipment, 140–142
  cooking equipment, 135–140,
    136–140
  ovens, *140*, 140–142, *141*, *142*
  preparation equipment, *131–135*,
    131–135
  work sections and stations, *128*,
    128–130, *129*, *130*, *131*
prep sections, 128, *129*
presentation, food, 255–256, *256*
pressed cookies, *968*, 968
pressure steamers, *138*, 138
price placement, menus, 201
pricing forms, *176*, 176
prickly pears, 489, *490*
prix fixe menus, *194*, 194
procedures
  beef
    beef tenderloin, trimming and cutting,
      855
    boneless strip loin, cutting into steaks,
      856
    prime rib, carving, 863
    steamship round roasts, carving, 864
  breads
    baker's scale, using, 948
    biscuit mixing method, 955
    creaming mixing method, 957
    muffin mixing method, 956
    yeast dough, kneading, 949
  breakfast
    crêpes, preparing, 440
  cakes
    cake batter, creaming, 960
    cake batter, sponge method, 960
    cake batter, two-stage mixing
      method, 958
    creaming mixing method, 959

USING THE *CULINARY ARTS PRINCIPLES AND APPLICATIONS* INTERACTIVE DVD

Before removing the Interactive DVD from the protective sleeve, please note that the book cannot be returned for refund or credit if the DVD sleeve seal is broken.

Windows System Requirements

To use this Interactive DVD on a Windows® system, your computer must meet the following minimum system requirements:

- Microsoft® Windows® 7, Windows Vista®, or Windows® XP operating system
- Intel® 1.3 GHz processor (or equivalent)
- 128 MB of available RAM (256 MB recommended)
- 335 MB of available hard disk space
- 1024 × 768 monitor resolution
- DVD drive (or equivalent optical drive)
- Sound output capability and speakers
- Microsoft® Internet Explorer® 6.0 or Firefox® 2.0 web browser
- Active Internet connection required for Internet links

Macintosh System Requirements

To use this Interactive DVD on a Macintosh® system, your computer must meet the following minimum system requirements:

- Mac OS X 10.5 (Leopard) or 10.6 (Snow Leopard)
- PowerPC® G4, G5, or Intel® processor
- 128 MB of available RAM (256 MB recommended)
- 335 MB of available hard disk space
- 1024 × 768 monitor resolution
- DVD drive (or equivalent optical drive)
- Sound output capability and speakers
- Apple® Safari® 2.0 web browser or later
- Active Internet connection required for Internet links

Opening Files

Insert the Interactive DVD into the computer DVD drive. Within a few seconds, the home screen will be displayed allowing access to all features of the Interactive DVD. Information about the usage of the Interactive DVD can be accessed by clicking on Using This Interactive DVD. The Quick Quizzes®, Illustrated Glossary, Flash Cards, Chapter Reviews, Application Scoresheets, Recipes, Culinary Math Tutorials, Media Clips, and ATPeResources.com can be accessed by clicking on the appropriate button on the home screen. Clicking on the ATP web site button (www.go2atp.com) accesses information on related educational products. Unauthorized reproduction of the material on this Interactive DVD is strictly prohibited.